Second Edition

Using *the* Engineering Literature

Second Edition

Using *the* Engineering Literature

Edited by Bonnie A. Osif

CRC Press
Taylor & Francis Group
Boca Raton   London   New York

CRC Press is an imprint of the
Taylor & Francis Group, an **informa** business

CRC Press
Taylor & Francis Group
6000 Broken Sound Parkway NW, Suite 300
Boca Raton, FL 33487-2742

Library of Congress Cataloging-in-Publication Data

Using the engineering literature / editor, Bonnie A. Osif. -- 2nd ed.
    p. cm.
Includes bibliographical references and index.
ISBN 978-1-4398-5002-2
  1. Technical literature. 2. Engineering--Computer network resources. I. Osif, Bonnie A.

T10.7.U85 2012
620.0072--dc22   2011002506

Visit the Taylor & Francis Web site at
http://www.taylorandfrancis.com

and the CRC Press Web site at
http://www.crcpress.com

This book is dedicated to present and future engineering librarians who strive to provide quality services in a critical profession that is always changing and to promote the value of evaluated information. It is also dedicated to the many faculty, librarians, and family members who have supported me in this effort. Special thanks to Thomas Osif, Thomas W. Conkling, and advisor and mentor extraordinaire, Dr. Joseph S. Schmuckler of Temple University. The faculty member I have become is largely due to his example of what a faculty member should be.

Contents

Preface..ix
Acknowledgments... xiii
Contributors xv

Chapter 1   Introduction 1
              Bonnie A. Osif

Chapter 2   General Engineering Resources...7
              John J. Meier (based on the first edition by Jean Z. Piety and John Piety)

              Minorities in Engineering...34
              Sylvia A. Nyana

Chapter 3   Aeronautical and Aerospace Engineering...............................45
              Thomas W. Conkling

Chapter 4   Agricultural and Food Engineering 61
              Kathy Fescemyer and Helen Smith

Chapter 5   Architectural Engineering...79
              Barbara Opar

Chapter 6   Bioengineering 93
              *Honora Nerz Eskridge and Linda Martinez (based on the first edition
              by Linda Martinez and Mary D. Steiner)*

Chapter 7   Chemical Engineering... 111
              Dana Roth

Chapter 8   Civil Engineering 129
              Carol Reese and Michael Chrimes

Chapter 9   Computer Engineering 203
              *Hema Ramachandran (based on the first edition
              by Hema Ramachandran and Renee McHenry)*

Chapter 10 Electrical and Electronics Engineering...241

   Larry Thompson

Chapter 11 Engineering Education..259

   Jill H. Powell and Jeremy Cusker

Chapter 12 Environmental Engineering 277

   Robert Tolliver (based on the first edition by Linda Vida and Lois Widmer)

Chapter 13 History of Engineering...317

   Nestor L. Osorio and Mary A. Osorio

Chapter 14 Industrial Engineering..341

   Nestor L. Osorio and Andrew W. Otieno

Chapter 15 Materials Science and Engineering...371

   Leena N. Lalwani and Sara Samuel (based on the first edition by Godlind Johnson)

Chapter 16 Mechanical Engineering 395

   Aleteia Greenwood and Mel DeSart

Chapter 17 Mining Engineering 427

   Jerry Kowalyk

Chapter 18 Nuclear Engineering...453

   Mary Frances Lembo

Chapter 19 Petroleum Engineering and Refining 469

   Randy Reichardt

Chapter 20 Transportation Engineering 497

   Rita Evans and Kendra K. Levine

Index..527

Preface

Since the first edition of this book was published, we have had airplanes grounded by the ash from the Icelandic volcano Eyjafjallajokul, 11 people died after an explosion on an oil rig in the Gulf of Mexico, 33 Chilean miners were rescued after 69 days underground, many other miners died in mining accidents in several countries, a collapse of a football dome, massive flooding in Pakistan, a bridge collapse in Minnesota, the crippling of the Fukushima Daiichi reactors, and the devastation in Haiti after the earthquake. All of these and many other stories in the news have a number of aspects to them—social, humanitarian, political—but, they also share engineering as one of the main features of the story. Whether it is the daring feat that allowed the successful rescue of all 33 miners in Chile, which was the result of engineers from many subdisciplines and many countries working together, or whether it was the disaster due to the lack of standards and subpar construction that resulted in the collapsed buildings in Haiti, engineering was involved.

With the almost ubiquitous encroachment of the Internet, it seems as if information is everywhere. However, there is mere information and then there is correct, appropriate, and timely information. While we might love being able to turn to Wikipedia® for encyclopedia-like information or search Google® for the thousands of links on a topic, that is not what is necessary when building new skyscrapers or developing new prosthetics for returning Iraqi veterans. Engineers need the best information, information that is evaluated, up-to-date, and complete. The first edition of *Using the Engineering Literature* used a roadmap analogy. However, we now need a three-dimensional analysis because resources have become more complex and more dynamic.

Google and its competitors dominate the searching world for many Web searchers. The more sophisticated use Google Scholar™ with its interdisciplinary collection and familiar search functions. The ability to actually have full text links included for one's institution is a great feature. However, the searching can be less exact than in a fee-based database with controlled vocabulary and indexing. And, the search results, while not as enormous as a regular Google search, can still be overwhelming, even for an experienced searcher. Quite frankly, if Google Scholar could substitute for the fee-based database, library budgets would be in a better situation. However, there is still a critical need for the carefully created databases in engineering, such as Compendex®, NTIS, Inspec, and more.

More is not always better. Engineers need efficient quality searches and missing a critical paper can be costly in time, effort, money, and loss of equipment, structure, and life. At this point, there is no definitive report on the Deepwater Horizon explosion in the Gulf of Mexico in 2010, but there are reports of some questionable decisions in construction and materials. Whether the reason was a business decision based on costs, corner cutting, or faulty engineering decisions, or none of these will be decided as the study continues. However, it is clear that the best information is always needed to provide the basis for making decisions that impact both the business bottom line as well as the people and environment affected by these decisions.

In a discussion of "how faculty members locate, obtain, read, and use scholarly articles," Carol Tenopir and co-authors find that "libraries appear to be the primary choice when articles are identified" by automated searching, citations, or mentions by others. They conclude that "much of the increase in reading that we have observed can be attributed to electronic articles available through their libraries (Tenopir, 2009, pp. 5, 17, 28). While there are differences among the subject specialties, the finding is interesting and points out the importance of libraries and skilled professionals who can help with both searches and location of resources.

Another interesting study focused on design engineers. Allard, Levine, and Tenopir (2009, p. 443) write, "In the private sector, engineers who deal with each step of the project (design, drawing, testing, etc.) coupled with tight product deadlines, are more likely to make use of information sources that are immediately available, even if those sources are not the best. This contrasts with engineers in academic or research settings, who are more likely to use documents or scholarly journals to satisfy their information needs." They found that colleagues are still very important sources of information, but the Internet is now used before consulting their colleagues, with Google their search engine of choice. They note that engineers are realizing that not everything on the Web is trustworthy, so they are relying on online communities that they know and trust to address this issue. They also found that "engineers are less likely to feel that they have as much time to spend on information-seeking tasks. This could have huge implications for the innovation process in terms of losing time on projects that might have pertinent research published, but not easily accessible through the engineer's first stop: Google" (p. 452).

"Engineers are contributors to the economic engine of a nation, and their work relies heavily on the ability to find, use, and share information" (Allard, Levine, and Tenopir, 2009, p. 453). The changes that Allard and colleagues have documented indicate that "shift toward using an Internet search engine before asking a colleague suggests that engineers are changing their values from a focus on ease of access and quality toward only considering ease of access. This suggests that engineers would benefit from tools that help them assess the quality of the information they are finding on the Web since they cannot count on it being filtered by their colleagues" (p. 454). Librarians would be great filters of that information.

In the book, *I Live in the Future and Here's How it Works*, Nick Bilton (2010) discusses a number of changes in our information world. He makes an interesting point. You can lament the changes that are happening today—tomorrow's history—convincing yourselves of the negatives and refusing to be part of a constantly changing culture. Or, you can shake off your technochondria and embrace and accept that the positive metamorphosis will continue to happen, as it has so many times before (pp. 75–76). Librarians have been at the center of many of the changes in our information world. They were early adopters and pioneers in many ways and in many technologies. If a new means of communication was created, they looked at it, evaluated it, and determined how it could be used to deliver better services to their patrons. New service models have been adopted by librarians. Some libraries have gone almost completely digital. Subject specific web portals or gateways often are created and maintained by librarians. Real-time reference through chat, text, or messaging is commonplace. Librarians are now embedded in teams, classes, and projects to provide their specialized expertise. And, librarians have left the confines of the library to provide information services in locations around campus where students and faculty can easily find them.

Yet, there are also rumblings that all of the information you need is really on the Web and using Google or Bing or whatever is the search engine flavor of the day will turn up the needed resources. However, studies show that people generally are rather imprecise searchers and do not look at many pages of search results. It goes without saying that much of the information on the Web is not reliable and should not be used in serious research. Also, much of the needed information on the Web is not free, but must come from subscriptions, again returning us to the important role libraries still play in the world of quality, accessible information in engineering.

As information floods our lives through television, radio, print, and more and more via our computers, smart phones, and whatever new technology is just on the horizon, the need to be able to cut through the chaff and get to the relevant materials is getting more difficult, not less. Technology has provided a tsunami of information and it is too easy to simply be washed away. We have the warning. Many authors have written about the glut of information, some of the poor searching skills that are used to wade through this glut, and the ways it can impact lives. The question is: Have we heeded the warnings and taken the precautions needed to find and use the best information in a timely manner?

To return to the old days of print resources or mediated searching would be ludicrous as well as impossible. The information genie is out of the bottle and I know of no one, especially librarians, who want to go back to the way it was. Information is power, and we have empowered our patrons, whether students, faculty, researchers, or administrators, to direct their own research. They can search the Web themselves and we have put databases, reference materials, journals, handbooks, reports, and many other resources literally at their fingertips. What we still need to do is be sure those resources are easily findable and usable and to be teachers and guides. Instead of being relegated to the list of obsolete professions, librarians with the tradition skills of collection, organization, and instruction are more than relevant as these skills have evolved into the electronic world. These skills are essential to engineers, though sometimes not recognized by them. Librarians and resources can make a difference.

REFERENCES

Allard, S., K. J. Levine, and C. Tenopir. 2009. Design engineers and technical professionals at work: Observing information usage in the workplace, *Journal of the American Society for Information Science and Technology* 60 (3): 443–454.

Bilton, N. 2010. *I live in the future and here's how it works.* New York: Crown Business.

Tenopir, C., D. W. King, S. Edwards, and L. Wu. 2009. Electronic journals and changes in scholarly article seeing and reading patterns, *Aslib Proceedings: New Information Perspectives* 61 (1): 5–32.

Acknowledgments

My heartfelt thanks to the people who agreed to give up countless hours to research and write their chapters. The authors were very carefully selected and represent some of the giants in the field of engineering librarianship, as well as some who are newer to the profession but are on their way to being the giants of the near future. I was honored to work with them on this book and learned so much from them. A work such as this is a long, difficult task, especially as it is done while juggling the normal workload of a librarian. They have put together a work that will help many librarians and engineers.

I would like to thank the staff of the Engineering Library at Penn State University for their support during the editing of this book. And, as always, I thank my husband, Tom, who proofed all of my writings, listened to me when I needed to vent, and encouraged me every step of the way.

Contributors

Michael Chrimes is director, Engineering Policy and Innovation at ICE, the Institution of Civil Engineers.

Thomas W. Conkling is head of the Engineering Library at the Pennsylvania State University. From 1975 to 1981, he was assistant librarian at the Princeton University Plasma Physics Laboratory Library. He has a BS in physics and math from the State University of New York at Stony Brook and an MLS from Queens College of the City University of New York. His research interests include the production and use of technical information. He is an active member of the American Society for Engineering Education (ASEE) Engineering Libraries Division. He was a co-recipient of the 1995 Engineering Information/Special Libraries Association Engineering Librarian of the Year Award, and he also received the Homer I. Bernhardt Distinguished Service Award in 2000 from the ASEE Engineering Libraries Division.

Jeremy Cusker spent his early postcollege career in Wisconsin with a group of righteous rainmakers, working for a nonprofit organization that fought health insurance companies on behalf of disabled people. After deciding he did not want to go to law school (and after winning the gold medal in his division of the 2004 National Taekwondo Championship), he enrolled at the University of Wisconsin–Madison School of Library and Information Studies. In the years that followed, Cusker worked for research foundations, public libraries, and yet more advocates for the disabled in Wisconsin before returning to his adolescent stomping grounds of upstate New York. Since 2008, he has been a librarian at the Engineering Library of Cornell University and been involved with projects related to curation of scientific data, student engineering project teams, and collaboration with engineering faculty in the classroom.

Mel DeSart is head of the Engineering Library at the University of Washington. From 1987 to 2000, he served in engineering and sci-tech librarian positions at the University of Illinois at Urbana-Champaign and at the University of Kansas. He received both his BA and MS from the University of Illinois. He is a current member of the American Library Association (ALA) and Association of College and Research Libraries (ACRL), including its Science and Technology Section and University Libraries Section, but is most active in the Engineering Libraries Division (ELD) of the American Society for Engineering Education (ASEE), having served ELD in over 15 different roles during his career. DeSart received the Homer I. Bernhardt Distinguished Service Award from ELD in 2004. His professional interests include initiating and supporting positive change in scholarly communication and in the training and mentoring of new sci-tech librarians. His nonprofessional interests include a variety of college and professional sports, in particular, baseball, and delighting in the long-standing but to date still unsuccessful hunt for the perfect microbrewed beer.

Honora Nerz Eskridge is currently director of Centennial Campus Research Services at North Carolina State University where she leads library services to the engineering community at NC State. She holds a master's degree in library and information science from The Catholic University of America and a bachelor's degree in engineering from Manhattan College.

Rita Evans is the library director at the Institute of Transportation Studies at the University of California, Berkeley, where she has worked since 2001. She previously spent 25 years in corporate libraries, including a lengthy stint at Dolby Laboratories' technical library in San Francisco. Active in the Special Libraries Association (SLA), Evans has chaired the Transportation Division and served as president of the San Francisco Bay Region chapter. Away from work, she enjoys tromping through the beautiful terrain of northern California. She received her MLS. from the University of Pittsburgh.

Kathy Fescemyer is the Life Sciences librarian in the Life Sciences Library at Penn State University. She is a graduate of the University of Illinois with a BS in zoology and an MS in library and information science. She also has a MS in entomology from Louisiana State University. She is active in the American Library Association, Council on Botanical and Horticultural Libraries, and United States Agricultural Information Network.

Aleteia Greenwood is head librarian of the Science and Engineering Library at the University of British Columbia (UBC). She is also the student, faculty, and collections development liaison to the Civil and Mechanical Engineering Departments. She received her BA in English literature (1994) and her MLIS (1999) from UBC. She devotes her off-work hours to creating innovative and colorful paintings and masks as well as print-making. Having yet to become a famous artist, she intends to remain an engineering librarian for the time being.

Jerry Kowalyk has been a public services librarian for the Cameron Science and Technology Library at the University of Alberta in Edmonton, Alberta, Canada, since 1992. Currently, he is responsible for library liaison, bibliographic instruction, and collection development in support of teaching and research in the Faculty of Engineering Departments for Civil, Mining and Petroleum, and Electrical and Computer Engineering. Prior to 1992, he was the Slavic literature and language materials cataloguer in the Bibliographic Services Division of the university library for 10 years.

Leena N. Lalwani has been the coordinator for Arts and Engineering Collection at the Art, Architecture, and Engineering Library (AAEL) at the University of Michigan since 2000. Lalwani has been with the University of Michigan library since 1995 as an engineering librarian. She is also the liaison librarian for Biomedical Engineering, Chemical Engineering, Materials Science, and Naval Architecture and Marine Engineering. In addition, she is also the patent specialist for her library. Prior to joining the University of Michigan, Lalwani has worked as librarian at Gelman Sciences and American Tobacco Company. She has an MLS degree from Catholic University of America and an MS in chemistry from the University of Mumbai, India.

Mary Frances Lembo is an information specialist for the Technical Library at the Pacific Northwest National Laboratory, where she provides in-depth reference services and classes of library products and services for laboratory staff members. She provides support for the library collection in the areas of chemistry, materials science, national security, nuclear engineering, and physics. She is active in the Special Libraries Association and co-presents the Sci/Tech 101 sessions at the annual conference. She holds a master's degree in library science from the University of Washington. The Pacific Northwest National Laboratory is operated by Battelle for the U.S. Department of Energy.

Kendra K. Levine is the Reference and Outreach librarian at the Institute of Transportation Studies Library at the University of California, Berkeley, a role that involves interactions with a wide range of transportation researchers from public agencies, universities, and the private sector. Committed to the larger transportation and library community, Levine participates in the SLA Academic and Transportation Divisions, the Transportation Research Board, and the Transportation Knowledge Networks. Outside of work, she organizes and creates finding aids for her record collection. She earned her MLIS/MSIS from Drexel University in 2008.

Linda Martinez has been library liaison to Duke University's Pratt School of Engineering and the Department of Computer Science since 1996. Prior to Duke, she was at the Massachusetts Institute of Technology as subject specialist in electrical engineering and computer science and assistant librarian for Core Information Competencies. She holds an MLS degree from Simmons College Graduate School of Library and Information Science and an EdM from Harvard University Graduate School of Education. She is also a registered yoga teacher with a focus on yoga for seniors and cancer survivors.

John J. Meier is a science librarian at the Physical and Mathematical Sciences Library in the Penn State University Libraries at University Park. His responsibilities include instruction, collection development, reference, and investigating methods of delivering library information and services. He is also the depository librarian for the U.S. Patent and Trademark Depository and liaison to the departments of mathematics and statistics. John holds an MLIS degree from the University of Pittsburgh and an MS in electrical and computer engineering from Carnegie Mellon University. He was selected as an Emerging Leader of the American Library Association for 2011 and is currently the secretary for the Science and Technology Section of the Association of College and Research Libraries (ACRL). He has received many awards including the 2009 Best Publication Award from the Engineering Libraries Division of the American Society for Engineering Education for the co-authored article "Google Scholar's Coverage of the Engineering Literature: An Empirical Study." His current research interests lie in using innovative technology to help library users and enhance the workflow of library professionals.

Sylvia A. Nyana is the Social Sciences librarian at the Pennsylvania State University. She has a BS in sociology from Cornell University and an MLS from the University at Buffalo. Her research interests include improving access and services to academic library resources, and a special research interest in the provision of information/knowledge to oral cultures. She has published in peer-reviewed articles and presented papers and conducted workshops at professional conferences.

Barbara Opar is Architecture and French librarian at Syracuse University. She holds a BA in French and English, and MA and MLS degrees, all from Syracuse. She has been employed at Syracuse University since 1975. Opar has been a member of the Art Libraries Association of North America and of Western New York (ARLIS/WNY) and of the Association of Architectural School Librarians. She has been president of both AASL and ARLIS/WNY. She prepares the booklist for the Society of Architectural Historians.

Bonnie A. Osif has been engineering reference and instruction librarian in the Engineering Library at the Penn State University since 1991. Prior to that she was a physical sciences librarian at Penn State and managed the biology library at Temple University. She holds a BS in biology from Penn State, an MS in information science from Drexel, and an EdD in science education from Temple University. She was co-recipient of the SLA Engineering Librarian of the Year Award in 1995. Other awards include the Achievement Award from the Sci-Tech Division of SLA and the Professional Achievement award from the Transportation Division of SLA, and Best Reference Work and Best Publication awards from ASEE. She was a columnist for the American Library Association's Library Administration and Management for 17 years. She is the co-author of *TMI: 25 Years Later.*

Mary A. Osorio holds a BA degree in English (minor in history) from the State University of New York (SUNY) at Buffalo and an MLS from the SUNY at Geneseo. Osorio is a reference librarian at Messenger Public Library, North Aurora, Illinois. Previous positions held include working for the Information Center of the Centro Colombo Americano and as instructor of English at Universidad del Norte in Barranquilla, Colombia.

Nestor L. Osorio received a degree in mathematics and physics from the Universidad del Atlántico, Colombia. He also holds master's degrees in physics and library and information science from the State University of New York. He joined Northern Illinois University, DeKalb, Illinois in 1982 and is currently a professor, subject specialist, and head of the Science, Engineering, and Business Department. Professor Osorio is a member of the American Society for Engineering Education, The American Society for Information Science and Technology, the Association for College and Research Libraries, and the Society for the History of Technology. He is a founder board member of the journals *Issues in Science and Technology Librarianship (ISTL)*, and *Formación Universitaria*. An author of numerous scholarly papers and bibliographies, he has presented a number of papers at national and international conferences. His areas of interest include technical information, digital libraries, history of technology, content analysis, and bibliometrics.

Andrew W. Otieno is an associate professor of Manufacturing Engineering Technology at Northern Illinois University's Department of Technology. He received his PhD in mechanical engineering from Leeds University, United Kingdom, in 1994. He has worked in various institutions in different capacities including the University of Nairobi, Kenya, and Missouri University of Science and Technology in Rolla. His research and teaching interests are in the areas of manufacturing engineering, computer applications in manufacturing, finite element modeling, structural health monitoring and damage detection, and echo-machining. He has published several research and pedagogical articles in various journals and has numerous conference presentations. He is also a member of the Society of Manufacturing Engineers, American Society of Engineering Education, American Society of Mechanical Engineers, Pi Epsilon Tau, and Phi Beta Delta. Dr. Otieno is currently involved in teaching and research in the area of manufacturing automation and environmentally friendly machining of aerospace alloys.

Jill H. Powell has a BA in German and economics from Cornell and an MLS from Syracuse University. She worked for Cornell University Press and Mayfield Publishing before coming to the Engineering Library at Cornell University in 1986. Since 2009, she has been engineering bibliographer and subject selector for the collection. She is active in the Engineering Libraries Division of the American Society for Engineering Education.

Hema Ramachandran worked for a large construction company in London, England, for six years doing reference, interlibrary loans, and ordering publications. After moving to the United States and while studying for an MLS at Florida State University, she worked at the University of Florida's NASA/STAC (an information center) conducting online searches and document delivery for fee-paying clients. After being in charge of the University of New Hampshire's Chemistry Library for three years, she joined Northwestern University's Transportation Library as reference/bibliographic instruction librarian. She left Northwestern University in 1997 and worked at San Jose State and Cal Tech. After two years as access services librarian at Pasadena City College, she took up her current position as engineering librarian at California State University–Long Beach. Ramachandran is active in the Engineering Division of the Special Libraries Association and the American Society for Engineering Education.

Carol Reese has over 20 years of experience handling civil engineering information for the American Society for Civil Engineers (ASCE). She first worked in the Publications Division where her main responsibilities included the production of indexes for all journal and book publications, the development of ASCE's Web-based Civil Engineering database and the society's full-text online publications. With these projects accomplished, she then became ASCE's cybrarian and archivist where she serves the civil engineering community by providing research assistance to those who need it. As archivist, she is responsible for organizing and maintaining the history of the society. In addition, she has been active in the Special Libraries Association and has held several positions. She holds an MS degree in library science from Florida State University and worked for over 15 years in reference at Brookdale Community College in New Jersey.

Randy Reichardt is research services librarian (engineering) at the Cameron Science and Technology Library, University of Alberta, Canada. His subject responsibilities include chemical and materials engineering, engineering management, mechanical engineering, nanotechnology, and space sciences and technology. He is active on a number of library advisory boards, including Knovel, CRC Press, Begell House, and SPIE. He published a series of columns under the title, "Digital Musings." in *Internet Reference Services Quarterly*, and is a member of Special Libraries Association and the American Society for Engineering Education. He loves movies and cable TV, as well as visiting New York City and Cambridge, Massachusetts. Reichardt is a professional guitarist, and if his favorite band, Buffalo Tom, ever needs a backup guitarist for a tour, he'll be ready and willing to answer the call.

Dana Roth has a BS in chemistry from UCLA, an MS in chemistry from Caltech, and an MLS in library service from UCLA. Employed at Caltech since 1965, he served as head of the science and engineering libraries from 1983 to 1995 and is currently the chemistry librarian. His other interests are politics and history, especially the World War II era.

Sara Samuel received her undergraduate degree from Hope College and earned an MS in information from the University of Michigan's School of Information. She is an engineering librarian at the University of Michigan in Ann Arbor.

Helen Smith is the agricultural sciences librarian at Penn State University where she provides reference, instructional, and collection development services in the Life Sciences Library with a special focus on the food sciences, animal sciences, and agricultural and biological engineering. She is active in the American Library Association, the United States Agricultural Information Network, and has published in the areas of integration of library services into course management software, and has evaluated several databases for national publications.

Larry Thompson is the sci-tech team leader and the engineering librarian at Virginia Polytechnic Institute and State University (Virginia Tech). He has been at Virginia Tech for 15 years, serving as liaison to 10 departments within the College of Engineering. Previously, he served as assistant engineering librarian at the University of Nebraska–Lincoln. He received both his BS in civil engineering and his MLS from SUNY at Buffalo. He is active in the Engineering Libraries Division (ELD) of the American Society for Engineering Education and in the past has held positions as secretary/treasurer, program chair, and division chair. He currently chairs the ELD Development Committee. In 2008, he received the ELD Homer I. Bernhardt Distinguished Service Award. One of his current professional interests is the transition of library-land from traditional resources and services to digital. His nonprofessional life centers around his family's five acre mini-farm in the beautiful Blue Ridge Mountains.

Robert Tolliver is the earth sciences librarian at the Fletcher L. Byrom Earth and Mineral Sciences Library at Penn State University in State College. Dr. Tolliver is the liaison to the Geosciences and Meteorology Departments and is also interested in Web usability and served as co-chair of the library's usability expert team during the recent redesign of the Penn State library Web site. Prior to his move to Penn State, Bob was an engineering librarian at the Art, Architecture, and Engineering Library at the University of Michigan in Ann Arbor and served as liaison to the Civil and Environmental Engineering Department. Dr. Tolliver is a past-chair of the SLA Engineering Division and has served as the division's nominating chair and is currently the discussion list owner. He also is active in the Geoscience Information Society, serving as a member of the Website

Committee, and is a member of the Atmospheric Science Librarians International. He received his MLIS from the University of South Carolina. Prior to becoming a librarian, Dr. Tolliver received geology degrees from the University of Tennessee (PhD), Western Michigan University (MS), and Michigan State University (BS), and worked as a geology instructor at the University of South Carolina at Spartanburg. His primary interests in geology are paleontology and sedimentology, with his dissertation research focusing on fossil diatoms.

CONTRIBUTORS TO THE FIRST EDITION

Godlind Johnson was head of Science and Engineering Library, Stony Brook University until retirement.

Renée McHenry has more than 20 years experience in corporate, special, and academic libraries, specializing in business and transportation-related research. She has worked at Northwestern University and is currently working as an independent contractor.

Jean Z. Piety was head of the Science and Technology Department of the Cleveland Public Library in Cleveland before retirement.

John Piety worked at the University of Wisconsin–Green Bay Library and was library director at John Carroll University before retirement.

Mary D. Steiner is assistant dean of Students at the University of Colorado, Boulder. Previously she was head of the Engineering Library and coordinator of the Physical Sciences and Engineering Libraries at the University of Pennsylvania.

Linda Vida is director of the Water Resources Center Archives located on the University of California–Berkeley campus.

Lois Widmer was manager of the Science Library at Brandeis University. She is currently associate director for E-research.

1 Introduction

Bonnie A. Osif

CONTENTS

Intended Audiences 2
The Literature of Engineering 2
The Value of Information 5
How to Use This Book 6
References 6

The first edition of this book opened with the sentence: *Pick up a newspaper or a magazine and odds are that there will be an article about the role of information in our lives.* Now, 2011, there are even newer means of getting information that are topics of the articles. It seems that almost everyone tweets and there is a Twitter tag for almost every person or topic imaginable. Libraries are using tweets to get their messages out, they are on Facebook, have posted videos on YouTube, and use all the varieties of instant messaging. Just after the publication of the first edition of *Using Engineering Literature*, Second Life® took off. There was a lot of buzz about Second Life, virtual libraries, and avatar librarians. The lure of creating a new persona and answering questions via this virtual world was intriguing and many jumped in eagerly. Interestingly, there is much less talk about it lately and a search of *Library Literature and Information Science* shows few articles on the subject. A search on the descriptor Second Life on December 13, 2010, and limited to each year showed 33 articles in 2007, 31 in 2008, 14 in 2009, and 5 in 2010. There can be several reasons for this decline, but one that needs to be considered is the role of fads in information delivery and the rapidity of rise and fall of these means of communications. What is "the hot way to deliver reference" may be frozen out by the new in-thing in a blink of an eye.

Blink of an eye is more than a figure of speech here because it seems that Internet/electronic changes are as fast as that figure of speech implies. Electronics keep getting smaller and, while hard to believe, even more ubiquitous. Everyone seems to be connected all of the time. People have to be warned not to do the obviously dangerous texting while driving, and even talking on the phone while driving has proved to be dangerous. Walking while texting is so common it is rare on a college campus to see someone who isn't multitasking this way.

The implications for the libraries are great. Speed is of the essence. All information should be right at the fingertips, right now. If it isn't, a substitution is just as good (view of the user), even if it isn't (opinion of the librarian). Google Scholar™ is a great interdisciplinary research tool, but it can fool people into thinking that what they want is available for no fee on the Web. They don't realize the connection is actually taking them to their institution's paid resources. For this reason, libraries have to worry about branding. They need their name prominently displayed to dispel this notion of the "everything is free on the Web, so why do I need a library." It is a common theme to hear librarians say that if it isn't in electronic form, it might as well not be there because the users only want e-resources. Some libraries are going almost totally digital, relegating print materials to another location.

So, in this environment of questionable evaluative skills, trends that change so fast it is easy to be out of date, and users wanting instant resource gratification, what is a librarian to do? The answer is probably the same as it always has been. Librarians are professionals who are very aware of new technologies and trends. They still utilize their professional skills of collecting, organizing, promoting access, instruction, reference, and more, but in different ways. Collections have changed as more move to the electronic format and the opportunities to share resources and means to gain access when needed have increased. While there is a wealth of resources to purchase, and budgets have grown tight, the range of collection options has allowed libraries to customize their purchases to their specific needs or to join with consortia to optimize their purchasing power. Organizing collections has changed with the granularity of object-level metadata when appropriate. Full text searching has provided the ability to find much more detailed information and to combine search terms to find very specific data, in many cases much better than that available from an index. Speed of access has changed as well, with the instant access of online publications. In addition to our traditional skills, new ones have become essential. Our professional lives are largely computer based. We search online catalogs and databases, create Web guides, create databases, digitize materials and provide detailed indexing, and do chat reference and online teaching. Flexibility is the key to providing quality service.

Quality service is the cornerstone of the profession and it is essential to providing the resources that engineers need. Engineering is at the core of so much of our lives. A glance at the chapter titles of this book indicate how engineering is involved in just about every aspect of our lives from our homes (architectural, civil, and mechanical engineering) to our meals (agricultural, chemical, and industrial engineering), trips (aeronautical, mechanical, mining, petroleum, and transportation engineering), equipment (computer, electrical and industrial engineering), a trip to the doctor (bioengineering, chemical, materials, mechanical, computer and electrical engineering), energy use (nuclear, mining, petroleum, and environmental engineering), and so many other aspects. Engineering is a basic part of our lives and the need for evaluated, appropriate, and timely information is essential for the best results.

Education is essential from the start of the engineer's training, hence, a chapter devoted specifically to this topic. Textbooks do not provide the information needed for a career in engineering; the materials will be dated far too soon. Engineers cannot depend on the textbooks and manuals they had in school. To keep up, they must be lifelong learners and this requires skills to continue to learn and the knowledge of when to go to a professional for help. Instead, engineering students need to learn the most efficient means to locate appropriate information efficiently, the ways to evaluate their information needs and the resources they locate, and to use the expertise of their librarians. Faculty need to understand the need to make the librarian an integral part of the educational team. Researchers need to know how to search, evaluate, locate, and obtain needed information and who to turn to when additional help is needed. These are the skills of the librarian, performed in a new environment, but vitally essential.

INTENDED AUDIENCES

Students of library and information science or librarians new to engineering can benefit from the expertise of the contributors in this book. They have evaluated resources in 18 areas of specialization, plus general engineering resources, to provide an overview for the beginner. Experienced engineering librarians can use it as a reference guide and as a collection development resource. Practicing engineers, especially those without a librarian in the organization, can use it as a guide to finding resources. For all three groups, it provides a detailed overview of resources in specific subdisciplines of engineering. It should be noted that the boundaries between engineering disciplines have blurred. While at one time an area of study may have been located in one specific field of engineering, now they tend to be much more interdisciplinary and knowledge of the resources needs to expand. The contributors have provided a means to navigate these interconnected subjects.

THE LITERATURE OF ENGINEERING

Engineers use a wide variety of information formats, and share many of these with other professionals, while others are more closely associated with engineering than other subject.

Periodical publications range from the scholarly journal to the news-oriented trade magazine. Journals provide relatively up-to-date information with detailed reports of the materials, methods, and results of research. The articles are normally refereed or reviewed by experts in the field who evaluate the quality of the research and determine whether it should be published. Journal articles allow the engineer to keep up with the new discoveries, trends, and issues in their discipline; determine lead researchers and institutions; and provide a record to consult when additional information is needed. Journals are published by a number of commercial companies that specialize in scientific and technical publications or by learned societies. Trade journals are not refereed and serve a different purpose. They normally cover the business trends, industry information, such as new products and methods, job opportunities, and news about the industry. Both serve a vital purpose and it is important that engineers be aware of and have access to the titles most important to their discipline and job activities. One of the major trends in journal and trade magazines is the move to electronic access. Some may be read for free on the Web, others may be subscribed to individually or through aggregators (companies that provide access to a number of titles "bundled" together). The trend toward desktop access to full text journals is extremely popular and provides fast, efficient access. Titles, length and type of access, and licensing regulations, along with cost, must be considered when evaluating this option. However, the popularity and convenience is considerable and is the trend in journal access.

Conferences are an important means of communication in engineering. Most societies sponsor some conferences and they range from general to extremely specialized topics. Conference papers are usually, but not always, published. Some conference proceedings are easily purchased, while others are only available for a short time after the conference or even made available only at the conference. Increasingly, conference proceedings are published in electronic format. Conference papers are, in some cases, the only account of a research project if the findings are not published as a journal article. This increases the importance of monitoring conference literature carefully. One added benefit of attending a conference is the ability to network, meet other researchers or practitioners, and to ask questions of the presenter to gain additional insight into the work presented. Most of the subject databases described in the subsequent chapters do index conference papers. Often, presentations will be made available on an association or participant's Web site when conference papers are not published.

While journals and conferences are very important in most academic fields, reports, while not unique to engineering, are extremely important. The report literature covers all engineering disciplines and is the official record of research that is often done with funding from the government. Reports are required of researchers under government contract, although research findings from other sources may be reported. These reports must be filed with the National Technical Information Service for inclusion in their NTIS database. Most reports are then available for purchase from NTIS, although the trend is for many reports to be available as full text on the Web. Having a Google or other search engine page open when searching NTIS is an efficient way to check for free, full text access. Copy the title of the report and paste into the phrase search box for almost instant discovery. A number of portals have made locating these reports easier and are detailed in the appropriate chapters. Increasingly, the value of reports is the fact that many of these studies will not be available in other formats, but only made public as a report. Reports tend to be very detailed, providing valuable in-depth information to the researcher, and can be published very rapidly. Even with the advent of electronic publication, there is often some delay in electronic publication of journals and conference proceedings. A review of the report literature is crucial if a thorough search is required.

Many reports are commonly referred to as gray (or grey) literature. Gray literature is defined as "information produced on all levels of government, academics, business, and industry in electronic and print formats not controlled by commercial publishing, i.e., where publishing is not the primary activity" (Grey Literature Network Services, 2010). This form of communication is an important, if somewhat elusive, form of communication. Types include reports, preprints, white papers, working papers, research in progress, presentations, datasets, and so on. Chapter 2 on general engineering resources includes a guide to gray literature. Access is improving as libraries make an effort to systematically collect gray literature that is relevant to their research needs and as more is made available on the Web. However, gray literature can be both one of the most valuable and more difficult to obtain resource type.

Monographs are extremely important in many of the liberal arts and humanities, but less so in engineering. Because so much of the relevant research is published in a more timely manner in reports, journal articles, or conference papers, or even faster as a Web paper, the time needed to write an entire book and to go through the publication process limits the usefulness of monographs. However, there are a number of books that are published in engineering and they serve many useful purposes, including introductory textbooks, reviews, analysis, historical studies, and in-depth research. While a monograph is not the primary format for much of engineering communication, it is still important and should be included in a literature review.

Handbooks, encyclopedias, and dictionaries are an integral part of the engineer's library. Engineering dictionaries define words in appropriate context and detail for the engineer and the student. Technical dictionaries are essential for understanding engineering terms because general purpose dictionaries either fail to include the many technical terms or define them in a general, relatively useless, manner. The same may be said for the benefit of using engineering encyclopedias. Encyclopedias range from the general (*AccessScience*) to the highly specific (*Dekker Encyclopedia of Nanoscience and Nanotechnology*), and they provide the same service that the encyclopedias did in our school days—they supply basic background information that is most useful when working in a new field. However, the engineering encyclopedias are often very detailed and provide an excellent introduction to the topic. Handbooks provide fast access to charts, tables, short descriptions, formulas, etc. needed to do everyday tasks. While most engineers will have a copy of the handbooks that are most relevant to their work on their desk or on their computer desktop, they will seldom have access to the wide range of handbooks to support the increasingly interdisciplinary nature of engineering. A number of companies are now bundling specialized handbooks and making these available in electronic format. *Knovel, EngNetBase*, and *AccessEngineering* are some of these and will be discussed later.

An easily overlooked resource is the dissertation or thesis. The research done by an engineering student often will be rewritten and published as an article or a conference paper, less often as a book. In some cases, it will be published in a shorter format, as a technical report because the research may have been funded by the government. However, the dissertation may be the only publication or may be available significantly before it is published elsewhere. While this in itself is a good reason to search *Dissertation Abstracts* (now called *Digital Dissertations*), an even better one is the quality of many bibliographies in a dissertation. These bibliographies will often represent the results of a thorough literature search and are an excellent short cut up to the time of the dissertation. While these bibliographies should not substitute for a well thought out and executed literature search, they can be very useful.

Standards, while important to some other disciplines, are basic to engineering. A standard "is a document, established by consensus that provides rules, guidelines, or characteristics for activities or their results" (World Intellectual Property Organization, 2010). Standards define terms, provide common methods of testing, interchangeability of parts, and work to improve quality, safety, and efficiency. Much of modern society depends on the adherence to standards so that our interconnected world can actually interconnect. General standards organizations are reviewed in Chapter 2, and more specific standards are covered in their appropriate chapters.

As defined by the U.S. Patent Office, a patent "is the grant of a property right to the inventor" and it may be for a design, a new or improved process, machine, or even for a plant. While there are very specific reasons to search patents, and to search carefully if there is interest in filing a patent, valuable information may be found in patents while doing research or trying to understand a product or process. Patent research is much easier now that there are options to search at no cost in relatively user-friendly Web sites.

The World Wide Web is ubiquitous and we ignore it at our peril. Almost all of the electronic databases we search are now on the Web. We can say almost all because there may still be locations that have workstations with a database or handbook on a CD-ROM. Full text journal articles, conference papers, technical reports, white papers, preprints, even encyclopedias, handbooks, patents, and entire monographs are hosted on the Web and accessed from our desktop. It is difficult to name an organization that does not have a Web site, usually with a wealth of information a few clicks away. Most people think everything is on the Web and a vast amount of material is.

However, one thing not recognized is that everything, or almost everything, is not available on the Web for free. While the trend is strong to make report literature available on the Web for free and there are a number of fantastic sites with chemical data or computer information free to all, most of the detailed, quality, and essential reference materials are still the product of a commercial publisher and are not free. And, these resources can be very expensive and the researchers with desktop access may not realize that their organization, usually through a library, is making this service available after contract negotiation and the payment of fees. The ease of access and the look makes it appear to be another Web product, but the difference is that this is a Web product that is not free. A number of the resources in this book are available in multiple formats; the most popular, and sometimes the most costly, is desktop access to the electronic files.

Communication patterns continue to evolve. In the first edition, RSS (real simple syndication) and blogs were discussed. Value varies widely, but in engineering a local blog by an information specialist can be an extremely valuable way to keep up with new resources and services. RSS feeds from databases and Web sites can be a real timesaver, if selected carefully. Twitter and Facebook are now options, as is setting up iGoogle pages with links to commonly visited sites and the ability to embed search widgets for databases. These pages can be used to create your own personal library homepage, saving time and reminding you of what needs to be checked. Actually, you can now set up alerts in a number of databases and they will proactively send you information based on a profile you create. These provide excellent ways to efficiently monitor the sites of interest. So, keep monitoring for new communication trends.

In addition to the format types, engineering has become increasingly interdisciplinary, not only within different disciplines of engineering, but with science and business. At different times it may be important to review the specialized databases, journals, handbooks, etc. in physics, chemistry, biology, or medicine, or to consider business models and trends. As this is a book devoted to the engineering literature, there are specialized titles that introduce these subjects and may be found by searching a catalog or asking a librarian. From this description of resource formats and increasingly blurry boundaries of disciplines, it is clear that very few libraries will have all the resources. Interlibrary loan, or cooperative borrowing among libraries, is a necessity, and librarians can assist engineers in obtaining materials not held locally or not available on the Web.

THE VALUE OF INFORMATION

"Libraries sometimes aren't used by engineers because they are unaware of important services that are provided or because they don't fully appreciate the benefits of their use" (Tenopir and King, 2004, p. 184). One of the often overlooked and under-researched areas of study is the value of information. Engineers are trained to look at value, risks, and cost/benefits as they relate to their work. It can be very useful to use these same skills to document the value of libraries and library services. While librarians have relied on anecdotes to indicate the value of their resources and services, a few documented cases can provide much stronger support. Return on investment studies have been done by several and the results can be staggering (Baldwin, 2002; Metcalf, 2003; Portugal, 2000). Baldwin's study showed a cost/benefit ratio of 12:1. A study funded by the Transportation Research Board surveyed state departments of transportation and others to obtain a number of case studies where a short literature review saved significant amounts of money in indicating technologies that could be used to perform tasks better and less expensively (Dresley and Lacombe, 1998).

In addition to the case studies with an actual dollar amount, there are a number of studies that indicate that access to information can save lives, time, equipment, and improve safety (Osif, 2004). One final way to look at the value of information is that anything that can save engineers time in doing their jobs, help avoid expensive or wasteful dead ends, save equipment, and improve safety and efficiency has real value. This translates into better work, better evaluation, and, most likely, better pay. The many ways to indicate the value of information cannot be overlooked and is a useful tool to market the value of libraries and library resources.

HOW TO USE THIS BOOK

While it is doubtful that anyone is going to sit and read this book from cover to cover, the index can lead the reader to descriptions of specific titles. In other cases, the reader will go to a specific chapter. Each of these provides some background information of the subject and then details selected resources in that subject. The book is intended to be selective and evaluative, not exhaustive. Many titles have a brief description except for those titles that are self-explanatory. It is recommended that Chapter 2, General Engineering, be reviewed as most of these resources have applications for all the other chapters and many of the titles are not repeated.

Some resources are repeated in several chapters. These are resources that are critical to the field, for example, *Compendex*. They are repeated to emphasize their importance and to guarantee the reader sees them if they don't check Chapter 2.

Another important point to make is the ever-changing URL. All URLs were checked between December 6 and 15, 2010. However, dates in the chapters represent the date the authors utilized the sites. As of these dates in December, all URLs were correct. However, they change rapidly, so if the URL is incorrect, most sites can be found by typing in the name of the site in a search engine.

As with the first edition, the expertise of the chapter authors cannot be lauded enough. While some are very experienced librarians and engineers, others are some of the upcoming leaders in the field. Care was taken in selecting these authors and the expertise they have to share is invaluable as they provide a guide to the information resources in their area of specialization.

We live in interesting information times with change constant. The resources here are some of the best, but, of course, they will change over time. However, these resources provide a good start to your search for information and the springboard to keeping abreast of new resources. And, while the delivery will change, as we have seen in the short few years between editions of this book, the professional skills will allow for engineering librarians to continue to meet the challenges of the information needs of our user groups.

REFERENCES

Baldwin, J. 2002. The crisis in special libraries. *Sci-Tech News* 56 (2): 4–11.
Dresley, S. and A. Lacombe. 1998. *Value of information and information services*. Washington, D.C.: Federal Highway Administration.
Grey Literature Network Services. http://www.greynet.org/ (accessed December 15, 2010).
Metcalf, J. 2003. The business case for translation of transportation documents. *Sci-Tech News* 57 (4): 4–9.
Osif, B. 2004. The value of information: The missing piece in the puzzle. In *Knowledge and change: Proceedings of the 12th Nordic Conference for Information and Documentation*. Copenhagen: Royal School of Library and Information Science.
Portugal, F. H. 2000. *Valuating information intangibles*. Washington, D.C.: Special Libraries Associations.
Tenopir, C. and D. W. King. 2004. *Communication patterns of engineers*. Piscataway, NJ: IEEE Press; Hoboken, NJ: Wiley-Interscience.
World Intellectual Property Organization. 2010. *Standards, intellectual property rights (IRPs) and standards-setting process*. http://www.wipo.int/sme/en/documents/ip_standards.htm (accessed December 15, 2010).

2 General Engineering Resources

John J. Meier (based on the first edition by Jean Z. Piety and John Piety)

Minorities in Engineering
Sylvia A. Nyana

CONTENTS

Introduction 8
Starting a Search 9
    Books 9
Traditional Research 9
    Bibliographies and Guides to the Literature 10
    Directories 11
    Language Dictionaries 13
    Handbooks and Manuals 13
Journals and Periodicals 14
    Periodical Titles 14
    Periodical Directories 15
    Electronic Sources for Periodicals 15
Electronic Research 16
    Books 16
    Online Tools and Indexes 17
Conferences 20
    Sources 20
Industrial Standards 21
    Books 22
    Electronic Sources for Standards 24
Patent Searching 26
    Books 26
    Electronic Sources 26
Gray Literature and Technical Reports 27
    Books 28
    Search Portals 28
    Internet Resources 28
    Blogs 29
Engineering Design, Management, and Ethics 29
    Books 30
    Design 30
    Ethics 31
    Online Sources 31

Career Information...31
Guides for Writing 32
    Books...32
Current Awareness 33
    Electronic Services..33
Minorities in Engineering 34
    Introduction..34
    Searching the Library Catalog 34
    Databases...35
    Bibliographies and Guides to the Literature 35
    Directories...35
    Encyclopedias, Dictionaries, and Handbooks...36
    Dissertations and Theses..36
    Monographs...38
    Video/TV Program 38
    Journals 38
    Journal Articles...39
    Societies, Organizations, and Associations...40
    Corporate Scholarships for Minorities in Engineering 42
References..43
Chapter Conclusion..43

INTRODUCTION

Engineering is the application of knowledge to the world around us. Whether it is designing and implementing devices, materials, structures, or systems all engineering fields share a common foundation. This chapter presents a number of basic concepts and areas that are needed to support all engineering specializations. Information resources for general engineering are presented along with sources addressing engineering studies, engineering ethics, standards, and major concepts common to all engineering subfields. It includes dictionaries, thesauri, encyclopedias, and online resources. Some of these items are classic print books, but many are large information databases.

Some sources for understanding engineering as a profession are suggested. A succinct description is found in the *Occupational Outlook Handbook*, which describes what an engineer does, what working conditions may prevail, what training and qualifications are needed, and suggests sources for more information. The handbook continues with descriptions of specific branches of engineering. *Masterworks of Technology: The Story of Creative Engineering, Architecture, and Design* is recommended for high school students thinking about careers in engineering. Examples of successful engineers are found in *Real People Working in Engineering*. This book not only gives vignettes, but also demonstrates some of the varied engineering fields. The Web sites for the National Society of Professional Engineers and the American Society for Engineering Education help in the decision-making process.

In today's global economy, a successful engineer may be employed in a company that has offices in another country. That engineer should know the industrial standards needed for implementation of the company's product, and also should understand the culture and mores of that country. Multilanguage dictionaries help with words pertinent to the professional field. Honda's *Working in Japan: An Insider's Guide for Engineers and Scientists* is just one example of the necessity of understanding what an engineer working in another country should learn about the social life and customs of that country.

Many electronic resources are expensive and not all corporations and institutions can provide total access. Thus, a number of freely available information sources are presented along with reference to earlier print editions that may be available through libraries even today. The emphasis in this chapter is on resources that help with a basic understanding of how to do research. Seeking information on one subject often leads to serendipitous ideas and encourages further research.

STARTING A SEARCH

Whether starting with a search engine, bibliographic database, or in a library catalog, it is important to know how to search effectively. Many systems support keyword searching where results reflect the frequency and location of a term, but some also offer controlled vocabulary. Library catalogs or bibliographic databases use a controlled thesaurus, a predetermined set of descriptors to describe a topic. This ensures that searching the catalog will have a more focused result than searching with uncontrolled terminology. Using the subjects in a catalog may result in a smaller, but more pertinent, quantity of material. Keyword searching is often more effective when controlled vocabulary is used perhaps directly due to the control exercised in assigning such headings. Useful tools for library catalogs include the *Library of Congress Subject Headings* and *SUPERLCCS: Gale's Library of Congress Classification Schedules*. Specific thesauri, such as *Ei Thesaurus*, also aid in developing search strategies.

With a book, the contents should be checked for an index, a table of contents, a preface, and special instructions that may be entitled "How to use this source." A site map or help section in an online resource serves a similar purpose. It should be noted that many online databases may be limited to a set group of subscribers, depending upon the license agreements, while others may be accessed for free. One element that is essential and common to most systems is that instructions should be understood before plunging into the contents, whether a book, a catalog, an index, an encyclopedia, or in searching online.

BOOKS

Ei Thesaurus, 5th ed. 2007. Hoboken, NJ: Engineering Information, Inc. This book serves as the indexing tool for *Engineering Index* and *Ei Compendex*, and can be searched online as part of *EI Engineering Village* www.engineeringvillage2.org (accessed August 5, 2010). It provides keywords that may be useful in any engineering research.

Library of Congress Subject Headings LCSH26, 21st ed. 2009. (5 vols.) Washington, D.C.: Library of Congress, Cataloging Distribution Service. Produced from records in the MARC21 subject authority format, the headings help in constructing the right terms to use in searches. Words in bold letters are real headings, while lighter ones are cross-references. Broader and narrower terms, those that are used for another word and appropriate subdivisions are also included. All are similar to the "see and see also" references formerly found in library card catalogs. The explanations found within a heading are helpful. For example, the heading: "Engineering-graphic methods" starts with: "Here are entered works on the use of scales, graph papers, ..." with other works "entered under Engineering graphics," which is a broader term. Current authorities also can be found online at http://authorities.loc.gov (accessed August 3, 2010).

Super LCCS. Class T. Technology. 2010. Detroit, MI: Gale. This series serves as companion volumes to the *Library of Congress Subject Headings* (LSCH). Individual volumes issued on each Library of Congress classification show the breakdown of the classification structure. The volume on Class T covers technology and most engineering topics. With lookup by classification code (such as TA337) as opposed to alphabetical list, it can help locate subjects in challenging research questions.

Thesaurus of Engineered Materials, 4th ed. 2003. Bethesda, MD: Cambridge Scientific Abstracts. This tool provides vocabulary in numerous materials fields, such as polymers, ceramics, and composite materials.

TRADITIONAL RESEARCH

The engineer should be aware of specific handbooks, dictionaries and encyclopedias, indexes and abstracts, standards, journals, specialized guides, and online resources. A comprehensive bibliography of the engineering literature was compiled by Charles Lord and published in 2000. Although he emphasized the literature of the 1990s, many of the resources are still useful today. His overview describes the historical roots to engineering, and the complexities and challenges to engineering today. The contents of both Lord's book and Malinowsky's *Reference Sources in Science, Engineering, Medicine, and Agriculture* serve the librarian in collection-building. While some titles were published years ago, they continue to provide useful information for current work, historical perspectives, or may still be the primary resource on the topic. Most of the information sources they detail continue to exist today, though many have gone online or experienced a name change.

Locating information online has become an even greater challenge. What the engineer will find while researching is often a lot of gray literature, resources that do not seem to fit into a standard category. Gray literature is defined as "information produced on all levels of government, academics, business and industry in electronic and print formats not controlled by commercial publishing, i.e., where publishing is not the primary activity of the producing body" (Luxembourg, 1997: expanded in New York, 2004): http://www.greynet.org/greynethome/aboutgreynet.html (accessed October 12, 2010). Auger's bibliography describes this obscure literature that offers a challenge to researchers trying to identifying sources. Conferences, standards, and reports that only exist in the digital environment also are difficult to locate via traditional research methods. A multitude of indexing and abstracting databases have replaced their print counterparts. Some, such as *Web of Science*, described in the index section, provide greater information with tools that shows where and how often an author is cited.

Several dictionaries are listed to help with definitions of the highly technical language used by engineering, as are encyclopedias that describe unfamiliar concepts. Both *Kirk–Othmer* and the McGraw-Hill encyclopedias describe concepts and processes, with bibliographic listings that lead to other sources for additional information. In the global environment, language dictionaries and fluency in foreign language become a great asset for engineers. Handbooks and manuals provide a quick look up for solving everyday problems.

As engineers work, they may be in a situation where they need to design, to create, or to develop a product, or to be a member of a professional committee that helps to develop standards. They need to know how to design, how to write, how to interpret industrial standards, and how to do a patent search. They need to understand engineering ethics. Separate sections in this chapter on general engineering provide some guides in these categories. To practice in a specific state in the United States, engineers have to be registered with that state. Most states maintain online sites for registered professional engineers, but two directories are listed as examples of sources for information.

BIBLIOGRAPHIES AND GUIDES TO THE LITERATURE

Auger, C. P. 1998. *Information Sources in Grey Literature,* 4th ed. London and New Providence, NJ: Bowker-Saur. Unique tool for showing the more difficult and obscure types of technical literature.

European Sources of Scientific and Technical Information, 11th ed. 1994. Harlow, Essex: Longman; New York: Stockton Press. The editors make an effort to include a source for every country by providing the name, address, library facility, consultant information, and types of publications for the prime technical body within each country of the world.

Lord, C. 2000. *Guide to Information Sources in Engineering.* Englewood, CO: Libraries Unlimited. This is an excellent, comprehensive bibliography of all engineering specialties. It is also available for purchase as an e-book.

Macleod, R. A. and J. Corlett. 2005. *Information Sources in Engineering,* 4th ed. New York: Bowker-Saur. Another title in the "Guides to Information Sources" series includes experts presenting and evaluating the variety of reference sources in engineering, from conferences to standards.

Malinowsky, H. R. 1994. *Reference Sources in Science, Engineering, Medicine, and Agriculture.* Phoenix, AZ: Oryx. Of particular use are the chapters listing types of sources for science and technology information and definitions for types of reference sources. Contents of most chapters list the basic books, journals, and bibliographic tools in science, engineering, medicine, and agriculture.

The Reader's Advisor, Vol. 5: The Best in Science, Technology, and Medicine, 4th ed. 1994. New Providence, NJ: Bowker. Part of a series aimed at advising the librarian and the reader about the best guides in the literature, this volume lists the classics useful for collection development in science, technology, and medicine.

Schenk, M. T. and J. K. Webster. 1984. *What Every Engineer Should Know about Engineering Information.* New York: Marcel Dekker. A good introduction to basic research information using classic print sources.

Walker, R. D. and C. D. Hurt. 1990. *Scientific and Technical Literature: An Introduction to Formal Communication.* Chicago: American Library Association. Engineers must know how to communicate their research. This book offers a philosophical approach to scientific research, but shows an engineer the vital forms, such as patents, technical reports, maps, conference proceedings, and journals.

DIRECTORIES

Encyclopedia of Associations, 49th ed. 2010. Detroit, MI: Gale. International listing of associations by name, location, or subject. Includes contact information, publications, membership, and so on. It is available through the online subscription service Associations Unlimited.

International Directory of Engineering Societies and Related Organizations, 16th *ed.* 1999/2000. Washington, D.C.: American Association of Engineering Societies. Print copy ceased with this edition. It provides a clue to engineering societies and related organizations that existed at the turn of the century.

ThomasNet, www.thomasnet.com (accessed August 3, 2010). This is an electronic version of *Thomas Register.* New York: Thomas Industrial Network. The print version was last published in 2006. It allows the searching of products and companies, CAD drawings, and more. See also Chapter 14 on industrial and manufacturing engineering.

Who's Who in Engineering, 9th *ed.* 1995. New York: American Association of Engineering Societies. Originally published by the Engineers Joint Council, this print edition seems to be the last published by the American Association of Engineering Societies. Over 15,000 engineers are included, to the exclusion of listing the numerous awards presented in the engineering fields. The Association's Engineering Workforce Commission Web site at www.ewc-online.org/ (accessed August 3, 2010) lists statistical and salary survey reports rather than directories.

Who's Who in Science and Engineering, 11th ed. 2010–2011. New Providence, NJ: Marquis Who's Who. Biographies of over 40,000 scientists, doctors, and engineers who submitted profiles to Marquis are entered in this series. Even some names from the "soft sciences" representing sociologists, economists, and psychologists have been included, plus major executives from technology and science-related businesses.

ASTM Dictionary of Engineering, Science, and Technology, 10th ed. 2005. West Conshohocken, PA: ASTM. This work defines engineering terms contained in ASTM standards, which covers many precise and technical terms. This last edition is out of print and has been replaced by the *ASTM Online Dictionary of Engineering Science and Technology* subscription service.

Brady, G. S. and H. R. Clause. 2002. *Materials Handbook,* 15th ed. New York: McGraw-Hill. This book consists of two parts; the first is an encyclopedic description of materials and substances and the second contains definitions and additional reference tables and charts.

Bucher, W. and C. Madrid. 1996. *Dictionary of Building Preservation*. New York: Preservation Press. This is an example of a dictionary that sounds specialized, but which provides useful definitions of unusual engineering terms in construction, and includes illustrations and line drawings from the Historic American Buildings Survey.

Cardarelli, F. 2008. *Materials Handbook: A Concise Desktop Reference*, 2nd ed. New York: Springer. This concise version provides quick access to physical and chemical properties of all classes of materials. Topically arranged, the usability of the chapters is reinforced with an index and a bibliography. It is also available as an e-book via Springerlink.

Eshbach, O. W., B. D. Tapley, and T. R. Poston. 1990. *Eshbach's Handbook of Engineering Fundamentals,* 5th ed. Hoboken, NJ: Wiley. From mathematics to properties of materials, this handbook covers the formulas needed in all engineering disciplines. It is also available as an e-book.

Harris, C. M. 2005. *Dictionary of Architecture and Construction,* 4th ed. New York: McGraw-Hill. Good source for terms that could apply either to architecture or engineering.

Heisler, S. I. 1998. *The Wiley Engineer's Desk Reference: A Concise Guide for the Professional Engineer,* 2nd ed. New York: Wiley. This volume includes the resources that the engineer needs for practical applications and daily problems, from mathematics to structures. A number of helpful tables range from a guide on writing for publication to the periodic table of elements.

Keller, H. and U. Erb. 2004. *Dictionary of Engineering Materials*. New York: Wiley. This comprehensive resource covers 25,000 materials and trademarks in the dictionary format, and includes an appendix listing databases, reference materials, and Web sites. Whenever possible, the trademarks list specific ingredient percentages.

Knight, E. H. 1979. *Knight's American Mechanical Dictionary: Being a Description of Tools, Instruments, Machines, Processes, and Engineering; History of Inventions; and Digest of Mechanical Appliances in Science and the Arts*. (3 vols.) St. Louis, MO: Mid-West Tool Collectors Association: Early American Industries Association. Reprint of the 1876 edition published by Hurd and Houghton, New York, this set includes one supplementary volume and is noted for its thousands of illustrations in a wide range of science and technology subjects. The original is available online through the Google Books project.

McGraw-Hill Dictionary of Scientific and Technical Terms, 7th ed. 2011. New York: McGraw-Hill. First published in 1974, the latest edition covers 5000 new terms. Credit is given to numerous sources, many of them federal government publications. Available through their electronic resources *Access Science.*

McGraw-Hill Encyclopedia of Science and Technology, 10th ed. 2007. (20 vols.) New York: McGraw-Hill. This set is a good starting point for unfamiliar terms because the extensive index leads to short, concise articles, ending with sources that provide additional information. A single-volume complement is the *McGraw-Hill Concise Encyclopedia of Science and Technology,* 6th ed. 2009. Also available through *Access Science,* see above.

Parker, S. P. 1993. *McGraw-Hill Encyclopedia of Engineering,* 2nd ed. New York: McGraw-Hill. About 700 articles were adapted from the 7th ed. of the *McGraw-Hill Encyclopedia of Science and Technology* and published in this single volume. In 2005, a smaller volume known as the *McGraw-Hill Concise Encyclopedia of Engineering* was published.

Webster, L. F. 1995. *The Contractor's Dictionary of Equipment, Tools, and Techniques*. New York: Wiley. Over 30,000 words are defined and described in the fields of civil engineering, construction, forestry, open-pit mining, and public works. It includes conversion factors.

LANGUAGE DICTIONARIES

An English–Chinese Dictionary of Engineering and Technology. 1980. Boston: Cheng & Tsui Co. An older volume for a language gaining importance for global engineering.

Callaham's Russian–English Dictionary of Science and Technology. 1996. New York: Wiley. Over 120,000 scientific and technical terms including engineering subjects.

Elsevier's Dictionary of Engineering: In English/American, German, French, Italian, Spanish and Portuguese/Brazilian. 2004. New York: Elsevier. Edited by M. Bignami, over 10,000 terms are listed in this translation source.

Ernst, R. 1989. *Dictionary of Engineering and Technology, with Extensive Treatment of the Most Modern Techniques and Processes*, 5th ed. New York: Oxford University Press. This multivolume set covers German–English and English–German and serves as a useful tool for translators.

Ernst, R. 1985. *Comprehensive Dictionary of Engineering and Technology*. (2 vols.) New York: Cambridge University Press. The set serves as a useful tool for translations. Volume 1 covers French–English and volume 2, English–French.

Thomann, A. E. 1990. *Elsevier's Dictionary of Technology: English-Spanish*. (2 vols.) New York: Elsevier. This two-volume set serves as an example for translating English terminology into Spanish. Its extensive contents are published in two parts. *Elsevier's Dictionary of Technology: Spanish–English* companion volume was published in 1993 for those translating technical terms from Spanish into English.

HANDBOOKS AND MANUALS

Berinstein, P. 1999. *The Statistical Handbook on Technology*. Phoenix, AZ: Oryx Press. Statistical information is compiled on a number of technical subjects. The book serves as a tool for watching new technologies emerge because the numbers can serve as a baseline for flourishes or declines in technology.

Bies, J. D. 1986. *Mathematics for Mechanical Technicians and Technologists: Principles, Formulas, Problem Solving*. New York: Macmillan. Audel was a publisher of practical books for the technician. This one retains the series title, "an Audel book," since it is a useful, practical guide for technical personnel. It contains the mathematical formulas for solving everyday problems, from cams and gears to economics in machining, and includes answers to the exercises.

Darton, M. and J. Clark. 1994. *The Macmillan Dictionary of Measurement*. New York: Macmillan. The dictionary format includes 4250 entries on weights and measures, values, volumes, frequencies, temperatures, and speeds. It includes a theme index.

Dorf, R. C. 2005. *The Engineering Handbook,* 2nd ed. Boca Raton, FL: CRC Press. A ready reference for practicing engineers includes key equations, mathematical formulae and tables, besides a wealth of sections covering all the engineering fields. It remains an informative, well-organized source of fundamental knowledge. Available online through *ENGnetBASE*.

ENGnetBASE CRC Press, www.crcnetbase.com/page/engineering_ebooks (accessed August 3, 2010). Electronic collection of the CRC handbooks, it provides full text searching of these valuable reference works.

Fischer, K. 1999. *Handbook of Technical Formulas*. Cincinnati, OH: Hanser Gardner. English translation of *Taschenbuch der Technischen Formeln*, originally published in 1996, includes formulas and tables, plus bibliographic references.

Ganic, E. N. and T. G. Hicks. 2002. *McGraw-Hill's Engineering Companion*. New York: McGraw-Hill. This book helps an engineer with the information needed in daily operations with handy mathematical formulas, measurement tables, engineering units, and general properties of materials. Also available as an e-book.

Gieck, K. 2006. *Engineering Formulas*, 8th ed. New York: McGraw-Hill. This translation of *Technische Formelsammlung* contains all the formulas an engineer would normally use.

Glover, T. J. 2002. *Pocket Ref,* 3rd ed. Littleton, CO: Sequoia Publishing. One example of a handy tool for the hip pocket is this little tome that includes tables ranging from those written by the American Society of Heating, Refrigeration and Air Conditioning Engineers' committees to *Water Well Handbook*. The tables go from sandpaper grit to glue types.

Grazda, E. E., M. Brenner, and W. R. Minrath. 1966. *Handbook of Applied Mathematics*, 4th ed. Princeton, NJ: Van Nostrand. In spite of its age, this handbook remains useful in its clear and practical pages on shop mathematics, business mathematics, and engineering mathematics.

Hicks, T. G., S. D. Hicks, and J. Leto. 2005. *Standard Handbook of Engineering Calculations,* 4th ed. New York: McGraw-Hill. Each chapter is devoted to one of seven specialized engineering fields (civil, architectural, mechanical, electrical, chemical and process plant, water and wastewater, and environmental).

Knovel, www.knovel.com (accessed August 3, 2010). *Knovel* is a large collection of reference works in science and engineering from a number of publishers. Full text, deep searching is possible. Value-added tools allow manipulation and analysis of the results.

Kurtz, M. 1991. *Handbook of Applied Mathematics for Engineers and Scientists*. New York: McGraw-Hill. Provides a practical approach to mathematics.

Polyanin, A. D. and A. V. Manzhirev. 2007. *Handbook of Mathematics for Engineers and Scientists*. Boca Raton, FL: Chapman and Hall/CRC. This book covers the most important formulas, definitions, methods, and theorems of mathematics in a simplified manner that benefits engineers. It also includes tables and lists of equations.

Tuma, J. J. and R. A. Walsh. 1998. *Engineering Mathematics Handbook*, 4th ed. New York: McGraw-Hill. Newer and more comprehensive than Tuma's *Technology Mathematical Handbook*, this volume also could serve as a textbook. Useful appendices cover numerical tables, glossary, units of measurement, and sample problems.

JOURNALS AND PERIODICALS

Few periodical titles are exclusively about general engineering. Even the classic *Engineering News-Record* (*ENR*) and *Engineering* (London) emphasize civil engineering, since their contents cover the construction industry more than any other field. To be knowledgeable about engineering is to know current events and trends in science and technology. Reading some of the science and technology periodicals can be helpful. *Science* and *New Scientist* are examples of newsworthy titles. Those interested in design and inventions would enjoy reading *American Heritage of Invention and Technology*. Some engineering schools publish magazines, such as MIT's *Technology Review*. Most are available by subscription or free online. Two sources to help locate titles in engineering include *The Standard Periodical Directory* and *Ulrich's Periodicals Directory*. Once a title is found, the OCLC *WorldCat* database helps to find the closest library with a copy.

PERIODICAL TITLES

American Heritage of Invention and Technology (1985–). Quarterly. New York: American Heritage (8756–7296). Subtitled "The Magazine of Innovation," this quarterly magazine encourages creativity and recognizes inventiveness. The magazine tells the story of the cultural impact of technology and innovation. Available online http://www.americanheritage.com/inventionandtechnology/ (accessed August 12, 2010).

Engineering (London) (1866–). Monthly. London: Gillard Welch Associates (0013–7782). The issues cover innovation in technology, manufacturing, and management. Subscription information is available at the Web site www.engineeringnet.co.uk (accessed August 5, 2010).

ENR (1874–). Weekly. New York: Engineering News-Record (0891–9526). Former title was *Engineering-News Record*, which is still the publisher, a subsidiary of McGraw-Hill Construction. *ENR* may be considered a construction magazine, but it also contains general business information for engineers and contractors. For more information, go to www.enr.com (accessed August 5, 2010).

New Scientist (1956–). Weekly. London: Reed Business (0262–4079). Latest news and hot topics found worldwide. Good source for keeping up-to-date in the technical world at the Web site www.newscientist.com (accessed August 5, 2010), which provides full text articles.

Science (1880–). Weekly. Washington, D.C.: American Association for the Advancement of Science (0036–8075). Science provides news of recent international developments and research in all fields of science. A subscriber must belong to the American Association for the Advancement of Science (AAAS) in order to receive *Science* in paper form, on CD-ROM, or online at www.sciencemag.org. (accessed March 23, 2011).

Science News (1921–). Weekly. Washington, D.C.: Science Service (0036–8423). Its mission is to advance the understanding and appreciation of science through publications and educational programs. Full text is available from a number of sources. Its Web site is www.sciencenews.org (accessed August 5, 2010).

Technology Review (1899–). Six issues per year. Cambridge, MA: Massachusetts Institute of Technology (1099–274X). Its strength lies in reporting on emerging technologies, their impact, and potential in the marketplace. For more details, go to http://www.techreview.com/ (accessed August 5, 2010).

PERIODICAL DIRECTORIES

The Standard Periodical Directory, 33rd ed. 2010. New York: Oxbridge (0085–6630). Published annually since 1964, this guide covers U.S. and Canadian periodicals. Although there is a section on engineering, *ENR* is listed under "construction," so some scouting helps among the categories. Title index leads to the subject categories. It is online at www.mediafinder.com (accessed August 5, 2010).

Ulrich's Periodicals Directory, 48th ed. 2010. New Providence, NJ: Bowker (0000–2010). This multivolume set has provided international periodical information since 1932. It includes a classified list of serials, annuals, and irregular publications with an alphabetical title index. Other indexes, such as ceased titles, and International Standard Serial Number (ISSN) listings are useful as well. Available online as *Ulrichsweb* www.ulrichsweb.com/ (accessed August 5, 2010).

ELECTRONIC SOURCES FOR PERIODICALS

Fulltext Sources Online, 22nd ed. (1989–). Semiannual. 2010. Medford, NJ: Information Today, Inc. (1040–8258). Each complete, cumulated directory provides sources for finding over 46,000 periodicals, newspapers, newswires, and TV or radio transcripts with full text online. Input comes from 28 major aggregator products. Online edition is updated weekly at www.fso-online.com (accessed August 5, 2010).

Worldcat http://newfirstsearch.oclc.org (accessed August 5, 2010). One of the OCLC's FirstSearch databases, *Worldcat* is the world's most comprehensive bibliography covering almost 200 million bibliographic records representing 400 languages. It covers the contents of member libraries worldwide. Some libraries include serial holdings while others list only the title, but this database remains a good search tool for obscure bibliographic information. A limited free interface is available at www.worldcat.org (accessed August 5, 2010).

ELECTRONIC RESEARCH

Traditional resources are moving increasingly toward electronic versions. Some publishers offer a choice: paper or electronic. Some services are now available exclusively on the Internet. The National Academy of Engineering still publishes the results of research in book format, but its Web site provides full documents with links in the references to online resources. Current information is important to the practicing engineer, but historical background is still significant. It is important to remember that only in the past few decades has information been produced digitally, so historic records must be imported into the digital realm. For example, *JStor* and *Project Muse* offer archives of older journal articles.

There has been simultaneously a large growth in online publishing and a decline in the number of publishers, as they are bought out by larger companies. Companies and services that existed a few years ago can disappear forever or reappear under a new name. Some information is freely available through search engines such as Google that show links to numerous sites and to books. Scientific and technical information is still a valuable commodity so many important resources are only available through subscription-based databases.

A database is a collection of citations, often with abstracts or full text, concerning a specific subject area. These citations are usually to journals in a given field and often include conference papers. Some specialized databases include *Ei Compendex*, the electronic version of *Engineering Index*, and *Inspec,* the resource published by the Institution of Engineering and Technology. Both started print versions before the twentieth century, are starting points for in-depth research, and provide thesauri to aid in searching techniques. One print and online resource, the *Gale Directory of Databases*, gives a description of all online vendors, producers, databases, and electronic search tools on the market today.

How authoritative are databases? Professional societies and recognized publishers can be easily accepted as authoritative developers of databases. With the increase of information available through the Internet, guidelines for analyzing an online source are important. Who developed and maintains the site and how current it is are some of the questions to ask. An understanding of the structure of the Web helps. Sherman and Price's book on the invisible Web offers a deeper approach to Web searching than just knowing standard universal resource locators (URLs). Other books show the development of the Internet, help with electronic terms and acronyms, provide an understanding of how to search the Web, and give some hints from those who have devised short cuts to interesting resources.

A unique resource is the *Science Citation Index* with paper subscription back to 1945 and electronic to 1964. With its multiple indexes, an author may be searched for published articles and where those articles are cited in the scientific literature. The unique features enable the Institute for Scientific Information to provide a bibliometric analysis of science journals in their database to list the most frequently used journals and those journals with the highest impact. This analysis is published as *Journal Citation Reports*.

BOOKS

Abbate, J. 1999. *Inventing the Internet.* Cambridge, MA: The MIT Press. This is a readable, extensive history of the Internet starting with packet-switching and the cold war in the 1960s, the U.S. Department of Defense Advanced Research Projects Agency (ARPANET) to today's global picture. An extensive bibliography leads to additional information on the subject.

Bidgoli, H. 2004. *The Internet Encyclopedia.* (3 vols.) Hoboken, NJ: Wiley. This three-volume work encapsulates the Internet, from active server pages (ASP) to wireless technology. In encyclopedic fashion, chapters include cross-references and references for further reading.

Calishain, T., R. Dornfest, and P. Bausch. 2006. *Google Hacks,* 3rd ed. Sebastopol, CA: O'Reilly. How to understand what Google can and cannot do is the theme of this book.

Cooke, A. 1999. *Neal–Schuman Authoritative Guide to Evaluating Information on the Internet.* New York: Neal–Schuman. This book serves as an aid to evaluating information on the Internet. In spite of its date, it helps in deciphering what is a worthwhile Internet source.

Devine, J. and F. Egger-Sider. 2009. *Going beyond Google: The Invisible Web in Learning and Teaching.* New York: Neal-Schuman. A book intended for librarians, but also a resource for discovering the large amount of information that search engines such as Google miss. The "invisible" or "deep" Web includes many specialized databases and this book provides the skill to use them effectively.

Gale Directory of Databases, 33rd ed. 2010. (2 vols.) Detroit, MI: Gale Research, Inc. This semiannual, multivolume set grew from a single volume published in 1979. Comprehensive in its coverage of the electronic services on the market, this set covers over 20,000 databases and online services. Includes contact information for vendors or distributors of database products. This product is also available as an online subscription.

Hock, R. 2004. *The Extreme Searcher's Internet Handbook: A Guide for the Serious Searcher.* Medford, NJ: CyberAge Books. With a foreword by Gary Price, this book serves as a basic tool for general information on the Web. It includes specialized directories, search engines, news resources, finding products online, and how to publish on the Internet.

Langford, D. 2000. *Internet Ethics.* New York: St. Martin's Press. Thoughts from an international team of specialists, each chapter forms an essay on a specific legal aspect of the Internet. More issues and discussion may be found in the extensive bibliography.

Moschovitis, C. J. P., ed. 1999. *History of the Internet: A Chronology, 1843 to the Present.* Santa Barbara, CA: ABC-CLIO. Capsule chronologies of telecommunications from Charles Babbage to the Internet's impact in 1998. Dated, but the early history is interesting.

Sherman, C. and G. Price. 2001. *The Invisible Web: Uncovering Information Sources Search Engines Can't See.* Medford, NJ: Information Today. Chapters give a history of the Internet and the visible Web. The contents include all kinds of searching possibilities, numerous suggestions of Web sources arranged topically, and a detailed index.

ONLINE TOOLS AND INDEXES

Applied Science and Technology Index (1958–) www.hwwilson.com (August 5, 2010). Bronx, NY: H. W. Wilson Co. One online equivalent with abstracts is *Applied Science and Technology Abstracts,* and is a basic tool for searching English-language periodical articles in numerous technical subjects. The paper edition ceased publication in 2004 as a permanent annual edition. *Applied Science and Technology Full Text* is an online combination of the abstracts and the index available via subscription. For information contact WilsonWeb at www.hwwilson.com/databases/applieds.cfm (accessed August 5, 2010).

CSA Illumina (2002–) www.csa.com (accessed August 5, 2010). Bethesda, MD: Cambridge Scientific Abstracts. Subscriptions are available in numerous combinations for major technical categories ranging from aeronautics to zoology. Contents in the categories cover a wide range of journals. CSA is continually expanding as a producer and provider of information databases. Some classic indexes, such as *Metals Index (Metadex), Abstracts in New Technologies and Engineering (ANTE),* may be accessed through this service.

CrossRef (2000–) www.crossref.org (accessed August 5, 2010). Lynnfield, MA: Publishers International Linking Association. Celebrating its 10th anniversary as a not-for-profit membership association of several international publishers, *CrossRef* is designed as the official digital object identifier (DOI) site to link together bibliographic citations, abstracts, and full text. DOI is an open standard created to facilitate online searching among the publications from a number of scientific publishing houses. Jointly, the publishers created the Publishers International Linking Association (PILA).

Dialog Corp. (1973–) www.dialog.com (accessed August 5, 2010). Cary, NC: Dialog Corporation. One of the oldest database vendors in the world, Dialog provides access to numerous business and technical files in which to search for information and numerous aids to help in searching. The corporation's newsletter, *Dialog Chronolog Summary,* was first published in 1973 when Dialog was owned by Lockheed Information Systems. As of 2010, Dialog is part of the Proquest group of information services provided by the Cambridge Information Group.

Engineering Index (1884–). Annual or Monthly. Stockton, NJ: Engineering Information, Inc. (Ei). This comprehensive interdisciplinary engineering service has an electronic version called *Ei Compendex* or simply *Compendex.* Records date back to 1884, depending on the agreement with the publisher. Both print and electronic versions cover the entire spectrum of engineering, with abstracts from over 5000 international journals, conferences, proceedings, and technical reports. The online subscription is available through several vendors. Ei, now owned by Elsevier, provides a powerful electronic platform in engineering fields in its *Engineering Village,* www.engineeringvillage2.org.

Google Scholar http://scholar.google.com (accessed August 5, 2010). Mountain View, CA: Google. This is an Internet search engine for the "scholarly" information on the Web. Google has partnered with publishers, professional societies, online repositories, and universities for the bibliographic information and sometimes full text of the World Wide Web.

High Technology Research Database, www.csa.com "… brings together in one place the most comprehensive bibliographic coverage of research, emerging technologies, applications and companies in the areas of aeronautics, astronautics, computer and information technology, electronics, communications, solid state materials and devices, and space sciences. Everything from theoretical research through practical application is covered" (from CSA's Web page).

IEEEXplore (IEL) http://ieeexplore.ieee.org (accessed August 5, 2010). Piscataway, NJ: Institute of Electrical and Electronic Engineers (IEEE). A technical information database that includes all the journals and conference proceedings produced by IEEE, indexing the literature back to 1988 and some selections back to 1893; *IEEE Xplore* offers numerous subscription packages. Browsing the table of contents and abstracts of IEEE transactions, journals, conference proceedings, and standards is free.

Inspec (1968–) www.theiet.org (accessed August 5, 2010). Stevenage, Herts., U.K. or Edison, NJ: Institution of Electrical Engineers. Founded by the Institution of Engineering and Technology (IET), the *Inspec* database was created in 1968 based on *Science Abstracts* (1898–), which split into three sections: *Physics Abstracts* (ISSN: 0036-8091), *Electrical & Electronic Abstracts* (ISSN: 0036-8105), and *Computer & Control Abstracts* (ISSN: 0036-8113). It is available online through Engineering Village or *Inspec Direct.* Paper indexes covering physics, electrical and electronics engineering, and computer and control are still published. The numerous tools available as PDF for XML that aid in refining search techniques are listed below:

Inspec Classification (2010). This publication provides the period of use of each classification entry and indicates previous codes. The codes help in refining a search. An index leads to the appropriate codes.

Inspec List of Journals. 2010. A valuable tool that lists all the serials scanned for the *Inspec* database. It is useful for locating titles in the scientific literature.

Inspec Thesaurus. 2010. Because the *Inspec* database uses controlled terms, this thesaurus aids in structuring the search strategy and shows relationships between terms.

Journal Storage Project (JSTOR) (1995–) www.jstor.org (accessed August 6, 2010), New York: ITHAKA. This site is a large digital archive of scholarly journals, providing full-text access in several scientific fields. Started as a pilot project sponsored by the Mellon Foundation to provide electronic access to the back files of journals, it has expanded to include a large range of scholarly titles, a number of which are in engineering and technology.

National Academies, www.nas.edu/ (accessed August 6, 2010). Includes the National Academy of Sciences (NAS), the National Academy of Engineering (NAE), and the Institute of Medicine (IOM). Private and nonprofit organizations, they operate to advise the federal government on science and technology issues. The NAE's "mission is to promote the technological welfare of the nation by marshaling the knowledge and insights of eminent members of the engineering profession." News links are an excellent means to monitor some engineering news. Results of their findings are published in hard copy and may be found on the Web site.

National Technical Information Service (NTIS) (1964–) www.ntis.gov (accessed August 6, 2010). Indexes government-sponsored research in all areas of science and technology. Some international reports are also included. Database available from several online providers or direct from NTIS. Reports may be ordered individually or by subscription; some reports available in full text on the Web.

Project MUSE (1995–) http://muse.jhu.edu/ (accessed August 6, 2010). Baltimore, MD: Johns Hopkins University Press. Launched in 1995 by the Johns Hopkins University Press to offer full text of scholarly journals as an archive, but now includes many journals through the most current issue. Strongest in the humanities and social sciences, it also contains science and mathematics journals.

ScienceDirect (1997–) www.sciencedirect.com (accessed August 6, 2010). New York: Elsevier. Starting as a Web database of Elsevier journals, this has become one of the largest electronic collections of science, technology, and medicine in full-text bibliographic format. Several licensing arrangements are possible.

Scirus www.scirus.com/ (accessed August 6, 2010). New York: Elsevier. The home page describes itself as "the most comprehensive scientific research tool on the Web." It searches only for specific science Web pages using academic, company, and societies' Web sites, scientists' home pages, conferences, patents, products, and e-prints/preprints. From the advanced search page, searches may be limited to specific subjects, content sources, or information types. An international advisory board, including librarians, provides guidance and aid in ongoing development.

Scopus, http://info.scopus.com (accessed August 6, 2010). A product created by Elsevier, *Scopus* is an abstract and citation database covering research and the Web, through *Scirus.* The records that are pre-1996 (about half) do not contain citation data while most of the recent articles include references. There are over 18,000 titles in the database. Subscription is required.

STN http://stnweb.cas.org/ (accessed August 6, 2010). A Web-based search produced by the CAS, a division of the American Chemical Society, which searches millions of patent and journal records from multiple databases.

UMI Dissertation Publishing, www.umi.com/en-US/products/brands/pl_umidp.shtml (accessed August 5, 2010). This is a database that has published over two million dissertations and theses from over 700 universities.

U.S. Government Agencies, http://science.gov (accessed August 6, 2010). This site is the portal for searching government science agencies. It provides links to several science Web sites that range from agriculture to science education and allows deep searching of the sites. It includes a federated search engine that looks across multiple agencies.

Web of Science, http://science.thomsonreuters.com (accessed August 6, 2010). Philadelphia: Thomson Reuters. This powerful database provides sources of where and how often an author is cited. The index covers the literature back to 1900, and may be searched by subject, author, journal, and/or author address. Coverage includes more than 10,000 major journals across 100 scientific disciplines and now includes many conference proceedings. It includes *Science Citation Index* (0036-827X), *Social Sciences Citation Index,* and *Arts and Humanities Citation Index.* Subscription options and cost may be found on the Web site. The *Web of Knowledge* package of online services also includes *Journal Citation Reports* (*JCR*), an annual analysis on the world's leading journals. The data reveals which are the largest journals, the most frequently used, and those with the highest impact.

CONFERENCES

Conferences are a main source of information in the engineering field, as a conference, a symposium, or a meeting gives an engineer the chance to express work in progress or research accomplished. They also give peers the chance to voice opinions, and they provide useful forums to listen and learn. Participants can discuss developments and present new ideas. The results often appear as published proceedings of a conference or a symposium, thus providing a time frame for work in progress or aiding future researchers from reinventing the wheel. Some are regular events while others vary in dates, sponsors, and format. Some contents of sessions never get published in their entirety. The challenge lies in trying to find the one that is needed.

Papers presented at a session may appear as a one-print document, as individual papers, in electronic format, or only as abstracts. Most of the engineering conferences fall into specific fields of interest. There are resources and guidelines that help develop the search strategy for locating any conference or paper within a conference. Two titles available through online subscriptions are the *Directory of Published Proceedings* and *Conference Proceedings Citation Index* (*ISTP*). The directory identifies the conference, the location, and its sponsor.

Some societies append a number to each paper. Attempts to index by number vary by society. SAE International provides ample resources to make retrieval possible by using the assigned number. That society produces an online digital library or a CD-ROM subscription. Print indexes cover SAE literature from 1906 to 1993 and help to locate older papers. The Institution of Electrical and Electronic Engineers (IEEE) added a file number to each of its numerous conferences. Several years ago, the library staff at General Electric published an index to the American Society of Mechanical Engineers (ASME) papers that appeared in the ASME journals. For a number of years, the American Society of Civil Engineers arranged its publications by using a coding system for each section within its proceedings. Subsequently, those sections became distinct journal titles. Searching the older years can present a challenge to those unfamiliar with the system and the titles.

SOURCES

Conference Proceedings Citation Index (1978–). Philadelphia: Thomson Reuters (0149–8088). Originally known as *Index to Scientific and Technical Proceedings* from the Institute for Scientific Information, this service provides a table of contents for published conference proceedings. Several indexes cover category, author or editor, sponsor, location, corporate, and subject of conference. It is available also in electronic formats. It is available through the Web of Science index.

Directory of Published Proceedings. Series SEMT, Science, Engineering, Medicine, and Technology (1965–2005). Harrison, NY: InterDok (0012–3293). Publication started in 1965 and is published 10 times a year, with annual and four-year accumulations. Series SEMT covering science, engineering, medicine, and technology is one of several series published by InterDok that provides citations to published conferences. Print edition ceased in 2005 and series is exclusively online. Coverage is international and contents are arranged by date of conference. Issues may be searched by name of conference, sponsors, and location. Future conferences may be searched at InterDok's Web site for its electronic database MInd: The Meetings Index, at www.interdok.com/mind/ (accessed August 6, 2010).

Index to Place of Publication of ASME Papers 1950–1977 (1981). Schenectady, NY: General Electric Company, Technology Marketing Operation. Compiled by the staff at General Electric's main library at Schenectady, the Index provides a cross reference between preprint numbers and corresponding journal volume, pages, and dates. If the number is not listed, then the paper was probably not published in ASME journals from 1950 through 1977.

INDUSTRIAL STANDARDS

Engineers depend on industrial standards for their work. Standards also simplify our daily lives. Transportation engineers ease driving by following the *Manual on Uniform Traffic Control Devices* (also included in Chapter 20, Transportation Engineering). Civil engineers surface highways following standards developed by the American Concrete Institute. Other engineers conform to building codes to erect safe structures. Electricity works safely, thanks to the National Electric Code. The operation of clocks depends on standardization. Thompson's update of Batik's *A Guide to Standards* explains the background and development of industrial standards.

Standards-developing organizations identify standards by test methods, guidelines, recommended practices, codes, specifications, and even manuals. Usually society names are reduced to acronyms in use, so books on abbreviations are useful for identification. Single volumes, such as Stahl, and multiple volumes (e.g., the set compiled by Peschke) identify initials or acronyms. Online sources provide another approach to identification. Information Handling Service developed its Standards Expert as a site that could be used to identify a specific standard or perform subject searches.

One area of confusion can be distinguishing report numbers from specification numbers. Engineers and librarians confuse letter and number combinations that are not standards, but grades of metal. Indexes and handbooks help by listing alloys to clear that confusion. *ILI Metals Infobase* identifies specific numbers, gives chemical content, and includes equivalencies. *Metals & Alloys in the Unified Numbering System* correlates many international number systems.

In the United States, one starting point to learn about standards is the National Information Standards Organization (NISO). NISO identifies and publishes information management standards, such as Z39.50. The online report, *Understanding Metadata* http://www.niso.org/publications/press/UnderstandingMetadata.pdf (accessed August 13, 2010), explains the complexity of retrieving, managing, and storing information in traditional ways (e.g., formatting catalog cards), and, in newer ways (e.g., accessing electronic resources). Another is the American National Standards Institute (ANSI). Not only is ANSI the body for approving domestic standards, but it also serves as the U.S. international representative to standards developing organizations. Over the past decade it has developed NSSN, the national resource for global standards database. Originally named the National Standards Systems Network, it grew beyond "national," but retained its acronym.

ANSI is the U.S. representative to the International Organization for Standardization (ISO), an organization that serves as the world's largest developer of standards. From testing to trading a product, officials could not operate without the network of standardization provided by ISO. International standards provide a common technological language between suppliers and customers. The World Trade Organization (WTO) recognizes the preparation, adoption, and application of standards. By understanding international standards, engineers have the framework to succeed in international projects. Engineers working in companies that want certification or have global concerns need to be familiar with the ISO 9000 and the ISO 14000 series. There are several books that explain the implementation of those standards.

Industry standards are developed by many engineering societies, acting as standards developing organizations, each covering its own specialty. Some society members are part of technical committees developing and reviewing domestic and international standards. Other societies are members of trade associations that publish in a specific area of expertise. Published standards may range from a few (e.g., Standards Engineering Society) to hundreds in specific disciplines (e.g., the *Aerospace Materials Specifications* (AMS) from SAE International).

Several federal government agencies publish safety standards. The National Institute for Occupational Safety and Health (NIOSH) addresses industrial hazards and the Occupational Safety and Health Administration (OSHA) continues with more safety standards. Other regulations come from the Environmental Protection Agency (EPA). The Federal Aviation Administration (FAA) issues safety regulations for flying. The Federal Communications Commission (FCC) regulates radio stations. Other federal branches, such as the U.S. Bureau of Mines, publish specific regulations in their area of expertise. Updates appear in the *Code of Federal Regulations*, http://www.archives.gov/federal-register/cfr/index.html (accessed August 13, 2010). The OSHA Web site cross-references the Standard Industrial Classification (SIC) four-digit codes. Since 1997, SIC has become the North American Industry Classification System (NAICS), which classifies all industry. Only a few of the federal agencies that are involved in creating standards are mentioned in this paragraph. The National Institute of Standards and Technology (NIST) provides a Web source for locating federal agencies producing standards. A number of agencies, national and international, impact upon standards, but engineers must have access to appropriate standards to accomplish their jobs.

Specific Web sites help in learning and locating standards. One example is the ISO links to other worldwide sites at the World Standards Services Network (WSSN). International Electrotechnical Commission (IEC) is also helpful in understanding about standards. Many ISOs are adopted as British Standards Institution (BSI) engineering standards. Domestic ones (e.g., American Society for Testing and Materials (ASTM), and federal government ones (e.g., U.S. Department of Defense and the National Institute for Standards and Technology (NIST)) provide starting points for learning more about standards and specifications. Because so many societies and organizations are involved in industrial standards, links to all their Web sites are helpful. Sources for linking both domestic and international sites include NSSN at www.nssn.org and the Standards Engineering Society at www.ses-standards.org (all accessed August 13, 2010).

Books

Acronyms, Initialisms & Abbreviations Dictionary, 44th ed. 2010. Detroit, MI: Gale Research Co. Not only a guide to acronyms, abbreviations, contractions, alphabetical symbols, and condensed appellations, the set also includes a reverse edition. Almost annual, this multi-volume set is updated with supplements and is available as an online subscription.

Batik, A. L. 1992. *The Engineering Standard: A Most Useful Tool.* Ashland, OH: Book Master/ EI Rancho. This book explains why standards are often overlooked and shows the broad spectrum of the standards picture, the good, the bad, and the ugly. Batik's aim is to educate executives, engineers, and government officials on the usefulness and importance of engineering standards.

Directory of Engineering Document Sources. 1997. Englewood, CO: Global Engineering Documents. A cross-reference guide helps to match the document initialism with the issuing organization, including many federal agencies. A third section offers a subject index to the standard developing organizations.

Erb, U. and H. Keller. 1994. *Dictionary of Engineering Acronyms and Abbreviations,* 2nd ed. New York: Neal-Schuman. An older volume, but useful for codes in the engineering field.

Fire Protection Handbook, 20th ed. 2008. (2 vols.) Boston, MA: National Fire Protection Association. This is a handbook for more than fire safety because, besides being a source for the fire codes, it refers to building codes, electrical codes, and chemical substances. It is an example of a quick reference tool for engineering, published by one of the largest standards developing organizations.

Frick, J. P. and N. E. Woldman. 2000. *Woldman's Engineering Alloys,* 9th ed. Materials Park, OH: ASM. A classic work originally edited by Woldman and now in its 9th edition, a good source for named engineering alloys.

Hoyle, D. 2009. *ISO 9000 Quality Systems Handbook: Using the Standards as a Framework for Business Improvement,* 6th ed. Amsterdam: Butterworh-Heinemann. Detailed, essential information on the standards.

Index and Directory of Industry Standards, 16th ed. 2001. (8 vols.) Englewood, CO: Global Engineering Documents. Last in the print series, this set provides numerical and subject indexes to numerous domestic and international standards.

Libicki, M. C. 1995. *Information Technology Standards: Quest for the Common Byte.* Boston, MA: Digital. Press. API does not stand only for American Petroleum Institute, but also for Applications Programming Interface. The author shows the complexity of information standards, why machines can communicate, and the difficulties that ensue in the process. An extensive bibliography at each chapter's end leads to more information on this complex topic. In looking for the common byte, the author projects the fate of various standards in a thought-provoking timeline.

Metals and Alloys in the Unified Numbering System. 2008, 11th ed. Warrendale, PA: Society of Automotive Engineers, or West Conshohocken, PA: American Society for Testing and Materials. Jointly developed by SAE International and ASTM International, this effort results in the Unified Numbering System (UNS) that correlates many international numbering systems, with cross-referencing that eases searching for a grade or a standard.

National Information Standards Organization. 2004. *Understanding Metadata.* Bethesda, MD: NISO Press. A revised expansion of *Metadata Made Simpler: A Guide for Libraries* published in 2001. Its title encompasses the content, explaining metadata and its implications. Available for free on their Web site, www.niso.org/standards/resources/UnderstandingMetadata.pdf (Accessed August 9, 2010).

Peschke, M. 1996. *International Encyclopedia of Abbreviations and Acronyms in Science and Technology.* (17 vols.) Munchen, Germany: K.G. Sauro. Conceived by Peter Wennrich, this is a multivolume set that includes reversed phrases and is updated with supplements, Yearbook 2005/2006 and Series A-C (most recent in 2008), all translated from the German.

Ricci, P. 1992. *Standards: A Resource and Guide for Identification, Selection, and Acquisition,* 2nd ed. Woodbury, MN: Pat Ricci Enterprises. A useful directory of domestic and international standards sources. Limited to dates in the 1990s, it provides a clue to help identify standards sources.

Ross, R. B. 1992. *Metallic Materials Specification Handbook*, 4th ed. London and New York: Chapman and Hall. A valuable source for identifying what sounds like an industrial standard, but is really a grade of metal. Useful as a source for the chemical content of a material and for locating materials that may be similar in content.

Rothery, B. 1995. *ISO 14000 and ISO 9000.* Brookfield, VT: Gower. This book explains the difference between the two international series of standards so vital to companies that want to be recognized for quality control and European Union certification. While the ISO 9000 series provides guidelines and emphasizes the implementation of quality management systems, the ISO 14000 series extends to environmental and safety issues.

Stahl, D. and K. Landen. 2001. *Abbreviations Dictionary,* 10th ed. Boca Raton, FL: CRC Press. Signs and symbols from A to Z, with appendices of special lists (i.e., airline and airport abbreviations, earthquake data, ports of the world, weather symbols, and winds of the world) that make this a source of helpful information on a variety of topics.

Standards Activities of Organizations in the United States. (NIST Special Publication 806) 1996. Washington, D.C.: U.S. Government Printing Office. U.S. public and private sector organizations that develop, publish, and revise standards are described in detail. This is a good source for locating federal agencies that issue standards. Staff at the National Institute of Standards and Technology (NIST), formerly known as the U.S. Bureau of Standards, compiled the directory.

Thompson, D.C. 2003. *A Guide to Standards*. Miami, FL: Standards Engineering Society. An update to the 1989 edition by Albert L. Batik, this is an excellent overview of the industrial standards process from initial committee work of the standards developing organizations to litigation.

Tibor, T. and I. Feldman. 1997. *Implementing ISO 14000: A Practical, Comprehensive Guide to the ISO 14000 Environmental Management Standards*. Chicago: Irwin Professional Publications. Written for companies that practice an environmental management standard, this guide also helps those who need to understand the complexities of the ISO series.

ELECTRONIC SOURCES FOR STANDARDS

American National Standards Institute (ANSI), www.ansi.org (accessed August 6, 2010). The ANSI is a private, nonprofit organization that administers and coordinates the U.S. voluntary standardization system. Although headquartered in Washington, D.C., its New York City office is the point of contact for most activities. ANSI serves as the U.S. representative to numerous international standards organizations. It oversees more than 10,000 standards. Its Web site provides a great starting point for locating domestic and global standards.

American Society for Testing and Materials (ASTM), www.astm.org/Standard/index.shtml (accessed August 6, 2010). Known today as ASTM International, this society is one of the oldest and largest standards-developing organizations. Founded in 1898, it now produces more than 12,000 standards contained in subject groups or a complete collection. The entire annual set, single volumes, or subsets are available on subscription, and individual standards may be purchased online.

Document Engineering Company (DECO) (1958–), www.doceng.com (accessed August 6, 2010). This vendor has been a source since 1958 for current and historical standards from U.S. government, industry, and international standards organizations. Standards may be ordered individually or through a subscription.

ILI, www.ili-info.com/ (accessed August 6, 2010). Started as London Information in 1949, ILI is now a producer of nine databases and a supplier of standards owned by SAI Global. Two of the databases are *Standards Infobase*, which covers hundreds of international standards organizations, and *Metals Infobase*. The *Metals Infobase* contains 70,000 world metal grades, their properties, related standards on which they are based, and the suppliers.

Information Handling Services, www.ihs.com (accessed August 6, 2010). Founded in 1959, Information Handling Services provides the world's largest collection of technical standards. The IHS Standards Expert, www.ihserc.com, is available to subscribers as a source for engineering and military standards, and specifications that may be searched with time-saving tools. Individual items may be purchased through Global Engineering Documents, www.global.ihs.com.

International Organization for Standardization (ISO), www.iso.org (accessed August 6, 2010). International standards began with the International Electrotechnical Commission (IEC) in 1906. In 1946, the International Organization for Standardization (ISO) was created to fill the need for international coordination of industrial standards. In 2004, the ISO Web site announced the joint ISO/IEC Information Centre www. standardsinfo.net (August 6, 2010) in order to facilitate world trade. ISO serves as a network of the national standards bodies of 163 countries, with a Central Secretariat in Geneva, Switzerland.

Manual on Uniform Traffic Control Devices (*MUTCD*), http://mutcd.fhwa.dot.gov/ (accessed August 6, 2010). Growing from a slim document of a single society in 1927, *MUTCD* is now an accepted standard developed by several societies and the U.S. Federal Highway Administration. Available only online with amendments through the 2009 edition, this document provides the signs and signals used for traffic control in all 50 states. It is an example of a federal government agency and several societies, the American Traffic Safety Services Association, the Institute of Transportation Engineers, and the American Association of State Highway and Transportation Officials, working together for traffic safety.

National Information Standards Organization (NISO), www.niso.org (accessed August 9, 2010). The NISO, founded in 1939, assumed its current name in 1984. As the name implies, NISO provides the information standards, such as Z39.50, which manages information retrieval, or Z39.9, which covers the International Standard Serial Number. NISO is designated by ANSI to represent U.S. interests on the ISO Technical Committee 46 on Information and Documentation.

National Institute for Standards and Technology (NIST), www.nist.gov (all accessed August 9, 2010). Formerly the National Bureau of Standards, this federal government agency provides numerous services to the industrial community. A part of the U.S. Commerce Department's Technology Administration, the NIST laboratories perform research and develop standards. NIST promotes and recognizes organizational performance excellence through the Baldrige National Quality Program, as well as partnering with business and the private sector to develop quality services. Within NIST is the National Center for Standards and Certification Information, a source for U.S. and international standards. It may be accessed at http://ts.nist.gov/standards/information/sources.cfm. To aid in locating information about the use of standards in government, NIST developed a direct portal to standards http://standards.gov.

NSSN, www.nssn.org (accessed August 9, 2010). Developed by the American National Standards Institute along with 10 of the largest standards-developing organizations as a national standards systems network, NSSN has expanded into a shopping center for standards information from many sources. It serves as a gateway connecting suppliers with seekers. The service now contains more than 300,000 records from national, regional, and international bodies.

Standards Engineering Society (SES), www.ses-standards.org (accessed August 9, 2010). Established in 1947, the SES promotes the use of standards and standardization. SES members range from information specialists to standards-developing organizations. Some societies issue hundreds of standards, but currently, SES produces and publishes two key standards entitled Recommended Practice for Designation and Organization of Standards (SES1) and Model Procedure for the Development of Standards (SES2). The Web site provides links to domestic and international standards sources.

Techstreet, www.techstreet.com (accessed August 9, 2010). Thomson Reuters owns this index to more than 500,000 international industry codes and standards. Brief descriptions are available for some records. Standards may be ordered individually or on a subscription basis.

U.S. government, www.usa.gov (all accessed August 9, 2010). This site provides the first stop for Web sites of federal government agencies. A click or two leads to the appropriate agency. Important to those needing military standards and specifications is the online site for current editions, the Department of Defense Quicksearch https://assist.daps.dla.mil/quicksearch/ (accessed March 23, 2011).

World Standards Services Network, www.wssn.net (accessed August 9, 2010). This site links to a network of standards organizations around the world through alphabetical and geographical lists. Prominently displayed are the links to ISO, IEC, and International Telecommunication Union (ITU). The objective is to simplify access to international, regional, and national standards through the Web.

PATENT SEARCHING

The United States Patent and Trademark Office (USPTO) has been automating information retrieval for decades. Since 2004, patent applications have been accessible online 18 months after being filed. The patent site www.uspto.gov (accessed August 9, 2010) contains a wealth of information for anyone searching patents and trademarks. Links at that site provide connections to additional resources for inventors and researchers. Guides help in basic understanding of domestic and international patents. With the changes in the Patent Office, print materials are rapidly dated, but some books are useful for their description of the patent process by showing the steps for searching prior literature and developing the patent application. These books also help to explain the U.S. design and plant patents, U.S. trademarks, and touch on the international process.

BOOKS

Adams, C. P. 2006. *Information Sources in Patents*, 2nd ed. München: K.G. Saur. Part of the series "Guides to Information Sources," this book covers all of the world's major patent systems. It also offers various patent searching strategies and some specialized techniques in the sciences.

Gordon, T. T. and A. S. Cookfair. 2000. *Patent Fundamentals for Scientists and Engineers,* 2nd ed. Boca Raton, FL: Lewis Publishers, Inc. This volume provides an overview of international patent systems for nonlawyers and describes patenting processes and principles.

Hitchcock, D. 2009. *Patent Searching Made Easy,* 5th ed, Berkeley, CA: Nolo. This book explains how to do patent searches on the Internet and in the library. Nolo publishes a number of other books, such as *Patent Pending in 24 Hours* and *Patent it Yourself,* which include forms and advice for inventors.

Jaffe, A. B. and M. Trajtenberg. 2002. *Patents, Citations, and Innovations: A Window on the Knowledge Economy.* Cambridge, MA: MIT Press. An interesting book on technology change; the authors measure the importance, generality, and originality of patented innovations.

Sharpe, C. C. 2000. *Patent, Trademark, and Copyright Searching on the Internet.* Jefferson, NC: McFarland. This book was researched and written in anticipation of the U.S. Patent and Trademark Office offering searchable databases for the general public. It serves as a guide for the lay person on what is a patent, what can be patented, and shows the different types of patents.

Walker, R. D. 1995. *Patents as Scientific and Technical Literature.* Metuchen, NJ: Scarecrow Press. For those unfamiliar with patents, the visuals help. The figures include descriptions of U.S. patents, U.S. design patents, and U.S. plant patents. Other interesting visuals show the 1601 Proclamation by the Queen, a British patent application, a page from the 5th edition of the *International Patent Classification*, and how the subclasses appear in the index of *Classification*.

Wherry, T. L. 2009. *Intellectual Property.* Chicago: American Library Association. Written by a librarian who taught patent searching for 30 years, this book also includes a discussion of copyright and trademarks.

ELECTRONIC SOURCES

Canadian Patents Database, http://patents1.ic.gc.ca/ (accessed August 9, 2010). Developed by the Canadian Intellectual Property Office, this site covers patent data from 1869 to the present. Patent images are available from 1920 to the present day. Online searching is limited to bibliographic data before 1978, such as titles, names and classification. Abstract and claims searching is available for the most recent patents.

Delphion, www.delphion.com/ (accessed August 9, 2010). Owned by Thomson Reuters, this subscription service provides full text searching, large file export, and many data analysis tools for international patents. It can search a number of available international patent collections and includes the *Derwent World Patents Index*, one of the oldest online patent services with access to "18.25 million inventions from 41 different international patent issuing authorities."

European Patent Office Esp@cenet, http://ep.espacenet.com (accessed August 9, 2010). Developed by the European Patent Office (EPO), this site provides access to over 45 million patents from EPO member states and over 70 countries and regions worldwide through several databases. Included are current patents and applications published by EPO or WIPO (see below), the Japan Patent Office, and the U.S. Patent Office.

Google Patents, www.google.com/patents (accessed August 9, 2010). Google has scanned and made full text searchable the entire database of U.S. patents from the USPTO (see below) from 1790 onward. There are some errors due to computer recognition of text, but full patents are available with PDF copies for download.

Intellectual Property Office of Singapore (IPOS), http://www.ipos.gov.sg (all accessed August 9, 2010). Through the intellectual property portal SurfIP www.surfip.gov.sg, access is available to patent databases from Singapore, Japan, United Kingdom, United States, Canada, WIPO, EPO, and Taipei.

Japan Patent Office, www.jpo.go.jp (accessed August 9, 2010). Besides the home page to Japan Patent Office, this site links to its Industrial Property Digital Library (IPDL) in the National Center for Industrial Property Information and Training (NCIPI) that offers public access to Japanese patents abstracts and trademarks, and information about NCIPI activities.

National Inventors Hall of Fame, www.invent.org (accessed August 9, 2010). Invent Now is a nonprofit organization that fosters the spirit and practice of invention. The National Inventors Hall of Fame in Akron, Ohio, showcases inventors' achievements and their impact on our lives. Included in the exhibits are hands-on activities, displays depicting famous inventors, and special educational exhibits that challenge creativity.

Patent Librarian's Notebook, http://patentlibrarian.blogspot.com (accessed August 9, 2010). This site is a great resource for up-to-date news on changes to patent law and other trends. It is created by famed IP librarian Michael White.

United Kingdom Patent Office, www.patent.gov.uk (accessed August 9, 2010). Information on intellectual property in the United Kingdom with a database to search patents and applications and current news in the United Kingdom.

United States Patent and Trademark Office, www.uspto.gov (all accessed August 9, 2010). Official Web site of the U.S. Patent and Trademark Office. Recently redesigned to highlight patent and trademark searching along with services for inventors. Also available online is the U.S. Manual of Classification, www.uspto.gov/web/patents/classification/, listing class schedules and linked classification definitions. This is updated regularly.

World Intellectual Property Organization (WIPO), www.wipo.int/portal/index.html.en (accessed August 9, 2010). The WIPO site includes the Patentscope search engine of international patent applications. The WIPO collects and publishes annual statistics on industrial property by country. The site includes the *PCT Gazette (Patent Cooperation Treaty)* containing bibliographic data, drawings, abstracts, and images of PCT applications issued since January 1997.

GRAY LITERATURE AND TECHNICAL REPORTS

Gray (or grey) literature has been defined as "information produced on all levels of government, academics, business, and industry in electronic and print formats not controlled by commercial publishing, i.e., where publishing is not the primary activity of the producing body" (GreyNet) www.greynet.org (accessed August 13, 2010). GreyNet hosts conferences and publishes proceedings on various aspects of grey literature. Auger's book, *Introduction to Grey Literature,* helps to explain the complexity of locating gray literature sources that may be preprints, conference papers, newsletters, white papers, working papers, reports, and other materials. The information found in this literature can be very useful and may not be published in other formats. However, these sources can be hard to identify and even harder to obtain. References to gray literature may be found in bibliographies or recommended by individuals who have used or heard of the resource. Because these resources are often difficult to obtain through routine channels, a number of portals have appeared to provide links to gray literature.

Searching for technical reports that may or may not have appeared in traditional publishing channels is particularly difficult. Report literature tends to have filing codes with letters and numbers. Many reports written under contract to a federal government agency start with distinctive letters identifying the specific agency. Others, such as the Rand Corporation, assign obvious letters for its research publications, papers, and monographs, and add numbers at the beginning of a project. *Godfrey's Dictionary of Report Series Codes* helps locate many that were results of research from World War II and later.

BOOKS

Auger, Charles P. 1994. *Information sources in grey literature.* 3rd ed. London: Bowker Saur. Good overview of gray literature.

Godfrey, L. E. and H. Redman. 1973. *Dictionary of Report Series Codes,* 2nd ed. New York: Special Libraries Association. Older but still useful, the dictionary provides a listing of report initials and numbers to research conducted by numerous agencies that contracted with the U.S. government after World War II.

SEARCH PORTALS

OSTI's Science Accelerator, www.scienceaccelerator.gov (accessed August 10, 2010). The Science Accelerator provides federated searching of full text, through Information Bridge, and bibliographic data from the U.S. Department of Energy (DOE). Research and development projects both underway and completed as far back as the Manhattan Project are searched simultaneously across multiple DOE databases.

Technical Report Archive & Image Library (TRAIL), www.technicalreports.org (accessed December 15, 2010). A resource of the Center for Research Libraries (CRL), TRAIL "is an initiative led by the University of Arizona in collaboration with CRL and other interested agencies to identify, digitize, archive, and provide access to federal technical reports issued prior to 1975," according to their Web site.

U.S. Army Corps of Engineers (USACE), www.usace.army.mil (accessed August 10, 2010). The mission, the history, the array of publications, the digitized collections, and the responsibilities of the largest public engineering, design, and construction management federal agency are covered on this site. Descriptions of the technical libraries and listings of their regulations and materials published by the Army Corp of Engineers are found here.

Virtual Technical Reports Center, www.lib.umd.edu/ENGIN/TechReports/ VirtualTechReports.html (accessed August 10, 2010). Maintained by the technical reports librarian at the University of Maryland, this site provides links to technical reports, preprints, reprints, dissertations, theses, and research reports of all kinds generated by domestic and international agencies.

INTERNET RESOURCES

In addition to Internet sites mentioned in specific categories, there are a number of sites that should be browsed periodically. These include:

American Association for the Advancement of Science, www.aaas.org (accessed August 10, 2010). Stated to provide "advance science, engineering, and innovation throughout the world for the benefit of all people" (from Web site).

American Association of Engineering Societies, www.aaes.org (accessed August 10, 2010). "A multidisciplinary organization of engineering societies dedicated to advancing the knowledge, understanding, and practice of engineering" (from Web site).

American Society for Engineering Education, http://www.asee.org (accessed August 10, 2010). This association furthers "education in engineering and engineering technology" by "providing a valuable communication link among corporations, government agencies, and educational institutions" (from Web site).

Internet Scout Project, http://scout.wisc.edu/index.php (accessed October 20, 2010). Housed at the University of Wisconsin and funded by the National Science Foundation, Mellon Foundation, and others, the Internet Scout Project "has focused on developing better tools and services for finding, filtering, and presenting online information and metadata," and produces the Scout Report, one of the Web's oldest and most respected current awareness services. Published every Friday since 1994, it is read by more than 250,000 readers every week." Scanning issues weekly or searching the archives can reveal many useful resources. (from Web page).

National Research Council Canada, www.nrc-cnrc.gc.ca (accessed August 10, 2010). Canada's national organization for research and development.

Society of Women Engineers, http://societyofwomenengineers.swe.org (accessed August 2, 2010). SWE "empowers women to succeed and advance in the field of engineering, and to be recognized for their life-changing contributions as engineers and leaders" (from Web page).

Women in Engineering Programs and Advocates Network, www.wepan.org (accessed August 10, 2010). The mission is to be a "catalyst for transforming culture in engineering education to promote the success of all women" (from Web page).

BLOGS

Englibrary, http://www.library.drexel.edu/blogs/englibrary/ (accessed October 13, 2010). Based at the Drexel University Libraries, this blog is a mix of engineering information news and research guides for engineering topics.

ReadWriteWeb, http://www.readwriteweb.com/ (accessed October 13, 2010). A blog that focuses on World Wide Web technology news and trends.

TechCrunch, http://techcrunch.com/ (accessed October 13, 2010). A business-focused technology blog that can help engineers follow entrepreneurship and innovation.

Technorati, http://technorati.com/ (accessed October 13, 2010). This site provides ranking and searching of blogs and blog posts, currently listing over one million blogs.

ENGINEERING DESIGN, MANAGEMENT, AND ETHICS

Whether engineers are employed in a large corporation that maintains a research and development department or are individual inventors, a basic understanding of design, ethics, and a commitment to avoid failures is vital. Creativity helps engineers design their ideas, while management looks at the cost factors in research and development. While familiarity with industrial standards transforms ideas into successful products, knowing ethical standards helps avoid litigation. An understanding of engineering ethics helps in making appropriate decisions.

Most of the resources on quality control and production are classified as part of the discipline of industrial engineering. Some of the newer disciplines, such as failure analysis and risk management, grew from engineering problems. Revisions to industrial standards occur after disasters. How to balance human gains of development against environmental damage becomes an ethical issue.

Environmental problems provide numerous examples in ethics and values. Vesilind uses that specialty as the subject for his textbook. Many sources on ethical issues cover specific types of research (i.e., bioengineering, genetic engineering, animal research). The U.S. National Research Council (NRC) publishes results in many of these fields. The sources listed in this chapter are offered as a starting point for the creativity found in most engineers.

Web sites provide a good starting point for finding information that will aid in an understanding of engineering ethics. Besides e-books available on the Web, succinct information is compiled on sites such as the Online Ethics Center for Engineering and Science. The National Society of Professional Engineers includes documents that help in understanding ethics and avoiding liability. The National Institute for Engineering Ethics provides direction on codes, tests, a help line, and links to other ethics sources.

BOOKS

Babcock, D. L. and L. C. Morse. 2009. *Managing Engineering and Technology: An Introduction to Management for Engineers,* 5th ed. Upper Saddle River, NJ: Prentice Hall. A textbook on basic management skills for engineers.

Bronikowski, R. J. 1985. *Managing the Engineering Design Function.* New York: Van Nostrand Reinhold. An older volume that contains useful information on basic managerial skills.

Macaulay, D. 1998. *The New Way Things Work.* Boston: Houghton Mifflin. Written in a clear, concise fashion, this work is suitable for children and understandable to everyone. The text, with numerous illustrations, explains the scientific principles and workings of hundreds of machines. This title updates his 1988 *The Way Things Work* to include new material on digital technology.

Petroski, H. 1992. *To Engineer Is Human: The Role of Failure in Successful Design.* New York: Vintage Books. As an engineer with a flair for writing, the author shares his ideas on systems failures in an understandable fashion. He is the author of several engineering books. Another similar book is his *Success through Failure*: *The Paradox of Design* published in 2006 by Princeton University Press.

Tenner, E. 2003. *Our Own Devices: The Past and Future of Body Technology.* New York: Alfred A. Knopf. "An exploration not only of inventive genius, but also of user ingenuity," this is an exploratory work for those who are curious about simple things.

DESIGN

McGowan, M. and K. Kruse. 2004. *Interior Graphic Standards.* Hoboken, NJ: Wiley. The first student edition of architectural graphic standards, it serves as a companion for all aspects of design education, from ergonomics to specifications to construction.

Pressman, A. 2007. *Ramsey/Sleeper Architectural Graphic Standards,* 11th ed. New York: Wiley. This is a classic quick reference tool that describes all types of building design and construction detail.

Watson, D. 2001. *Architectural Details: Classic Pages from Architectural Graphic Standards.* New York: Wiley. Watson compiled selections from the 1940 to 1980 editions of *Architectural Graphic Standards*, classic works by Charles Ramsey and Harold Sleeper. This volume may be used to complement a basic textbook.

Watson, D. and M. Crosbie. 2004. *Time-Saver Standards for Architectural Design Data,* 8th ed. New York: McGraw-Hill. Subtitled *Technical Data for Professional Practice*, this volume serves as one-stop shopping for architectural and building design. Contents are compiled from numerous sources, including government agencies, trade associations, manufacturers, and professionals. The good visuals provide a balance of representation and explanation of design integration.

ETHICS

Cook, R. L. 2003. *Code of Silence: Ethics of Disasters.* Jefferson City, MO: Trojan Publishing Company. The author is a professional engineer who taught engineering and technology, and in this volume asks, "How safe are we?" He covers the moral and ethical aspects of manmade disasters, from the Titanic to the World Trade Center.

Harris, C. E., Jr., M. S. Pritchard, and M. J. Rabins. 2009. *Engineering Ethics: Concepts and Cases,* 4th ed. Belmont, CA: Wadsworth Publishing. This textbook covers theory and practice by providing cases, methodology, and analysis of what is involved in practicing ethics.

Martin, M. and R. Schinzinger. 2010. *Introduction to Engineering Ethics,* 2nd ed. New York: McGraw-Hill. Key issues in engineering ethics are discussed.

Pinkus, R. L. B. 1997. *Engineering Ethics: Balancing Cost, Schedule, and Risk Lessons Learned from the Space Shuttle.* New York: Cambridge University Press. By using the space shuttle program as the framework, Pinkus and others examine the role of ethical decision making in the practice of engineering. In-depth case studies show engineers at work as they balance budgets, deadlines, and risks.

Seebauer, E. G. and R. L. Barry. 2000. *Fundamentals of Ethics for Scientists and Engineers.* New York: Oxford University Press. This is a textbook intended for ethics courses in engineering or science. It tries to emphasize ethic reasoning by including cases for graphic visualization.

Vesilind, P. A. 2010. *Introduction to Environmental Engineering,* 3rd ed. Stamford, CT: Cengage Learning. The author emphasizes materials balance and environmental ethics by incorporating ethical decision making in technical problems. The author also wrote a 60-page guide, *Doing the Right Thing: All Ethics Guide for Engineering Students*, published by Lakeshore Press (2004).

ONLINE SOURCES

National Institute for Engineering Ethics (NIEE), www.niee.org (accessed August 10, 2010). Hosted by Texas Tech University, the NIEE Web site promotes cooperation across engineering organizations toward an understanding of ethics.

National Society of Professional Engineers, www.nspe.org (accessed August 10, 2010). The site includes documents that help in professional liability and risk management. It also has career information, such as employment, licenses, and scholarships.

Online Ethics Center, www.onlineethics.org (accessed August 10, 2010). According to the mission statement, this site provides engineers, scientists, and science and engineering students with resources for understanding and addressing ethically significant problems that arise in their work. It also serves those who are promoting learning and advancing the understanding of responsible research and practice in science and engineering.

CAREER INFORMATION

While general job searching Web sites are useful for engineers, there are specific resources available both in books and on the Internet. Society Web sites can offer tailored resources to aid in the job search on their Web site, such as the American Society for Engineering Education. Also, the National Council of Examiners for Engineers and Surveying helps link engineers to certification requirements of the profession. Some of the older books below are still useful to those looking for an explanation of the engineering profession and the type of work engineers do.

Basta, N. 2003. *Opportunities in Engineering Careers*, rev. ed. Chicago: VGM Career Books. The author talks about the career opportunities in engineering professionals, including some of the specialties, working profiles, and everyday impact. Salary information and tips about colleges, licensing, and professional registration help to guide the student.

Camenson, B. 1998. *Real People Working in Engineering.* Lincolnwood, IL: VGM Career Horizons. Part of the series "On the Job," this book describes real people who choose engineering as a career.

Davis, M. 1998. *Thinking Like an Engineer.* New York: Oxford University Press. Excellent chapter on the history of engineering gives the educational background and how the engineering fields evolved. Responsibilities, code of ethics, and the social questions are woven into the decision-making process. Examples of wrongdoing are described.

Garner, G. O. 2008. *Great Jobs for Engineering Majors,* 3rd ed. New York: McGraw-Hill. Aimed at engineering majors, the book is divided into two parts, from job searching to career paths. Résumés and graduate school choices form the basis of the first section. The second section divides career paths into the industry, consulting, government, education, Internet, and nontechnical areas.

Honda, H. 2000. *Working in Japan: An Insider's Guide for Engineers and Scientist,* 2nd ed. New York: ASME Press. The theme of this book is the importance of understanding the social life and customs of the country in which the engineer works.

Lewis, E. E. 2004. *Masterworks of Technology: The Story of Creative Engineering, Architecture, and Design.* Amherst, NY: Prometheus Books. Written to inspire students to pursue a career in engineering, the author interweaves personal stories to illustrate creating in engineering. Chapter bibliographies aid in further reading to help high school students and technical program students to decide on specific careers in engineering.

National Council of Examiners for Engineers and Surveying, www.ncees.org (all accessed August 10, 2010). Site provides links to exam and license information, www.ncees.org/Exams.php. All the states and the District of Columbia require engineers and surveyors to register with the state in which they want to offer their services. This Web site provides a quick source to locate state licensing boards. The link leads to the address, telephone numbers, contacts, and Web site address.

U.S. Department of Labor, Bureau of Labor (2010–2011) *Occupational Outlook Handbook,* www.bls.gov/oco (accessed August 10, 2010). This biannual publication provides the starting point for necessary information about engineering and its specialties.

GUIDES FOR WRITING

Once research techniques are mastered, the student or engineer must learn how to communicate that information. Whether it is writing a memo, outlining a project, or composing a report, these skills are necessary. Publishers usually provide specific guidelines, but general resources can aid writing as well. The following titles will assist in writing the research paper, the report, or the memo.

BOOKS

Alexander, J. F. 2002. *Writing Better Requirements.* Boston, MA: Addison-Wesley. Emphasis is on writing software for the computer. Useful in systems engineering, it also offers guidelines that may be used in a broader context.

Alred, G. J., C. T. Brusaw, and W. E. Oliu. 2009. *Handbook of Technical Writing,* 9th ed. New York: St. Martin's Press. Arranged alphabetically with a comprehensive index, contents range from formulating abbreviations to Web design, and guidelines that may be used in a broader context than technical writing.

Beer, D. F. and D. A. McMurrey. 2009. *A Guide to Writing as an Engineer,* 3rd ed. Hoboken, NJ: John Wiley & Sons. Subject-specific guide to improving written and spoken communication.

Blicq, R. and L. Moretto. 2003. *Technically-Write!* 6th ed. Upper Saddle River, NJ: Prentice-Hall. Covers technical communication from e-mail to formal reports.

Davis, M. 2005. *Scientific Papers and Presentations,* 2nd ed. Burlington, MA: Academic Press. Covers written reports and presentations to groups.

Finkelstein, L. 2007. *Pocket Book of Technical Writing for Engineers and Scientists,* 3rd ed. New York: McGraw-Hill. Step-by-step, the author shows students the skill of technical writing. He uses practical outlines in explaining the options. This is part of McGraw-Hill's BEST (Basic Engineering Series and Tools) series of modularized textbooks for introductory courses.

Hart, H. 2009. *Introduction to Engineering Communication,* 2nd ed. Upper Saddle River, NJ: Prentice-Hall. A textbook aimed at introductory courses in engineering or computer science shows students how to communicate their ideas, whether writing documents or giving oral presentations.

Ingre, David. 2008. *Engineering Communication: A Practical Guide to Workplace Communications for Engineers.* Sydney, Australia: Thomson. Text based on the CMAPP model (context, message, audience, purpose, product).

Matthews, C. 2000. *A Guide to Presenting Technical Information: Effective Graphic Communication.* London: Professional Engineering Publications. This guide shows how to communicate technical information by providing methods and guidelines in a clear and effective way.

Microsoft Manual of Style for Technical Publications, 3rd ed. 2004. Redmond, WA: Microsoft Press. Not limited to Microsoft products, this book serves as a guide to technical writing in the computer age.

Sorby, S. A. and W. M. Bulleit. 2006. *An Engineer's Guide to Technical Communication.* Upper Saddle River, NJ: Pearson Prentice Hall. Text for both students and working engineers on effective written and oral communication.

Van Aken, D. C. 2008. *Reporting Results: A Practical Guide for Engineers and Scientists.* Cambridge, U.K.: Cambridge University Press. Focus is on laboratory reports, presentations, and research articles.

White, J. 1997. *From Research to Printout: Creating Effective Technical Documents.* New York: ASME Press. Excellent guidelines show how to write effectively, especially technical reports. References may be specific to mechanical engineers, but the principles hold true for any engineers.

CURRENT AWARENESS

Many electronic journals and databases support e-mail services in which the table of contents and/or bibliographic information is sent to users on a periodic basis or whenever a new item is available. For engineers and researches, this is a convenient way to keep up to date in specific areas. There are many free services offered by publishers or subscription database providers, though sometimes it requires an institution or an individual subscription. Some of these noted earlier were CSA Illumina, Dialog, *Science Direct*, and *Web of Science*. Some large publisher services are mentioned here, but many individual journals provide notification through RSS (Really Simple Syndication).

ELECTRONIC SERVICES

Informaworld, www.informaworld.com (accessed August 16, 2010). A free account can get alerts when a new journal issue is published or new articles meeting a previous search's criteria from all Taylor & Francis journals.

Ingenta, www.ingentaconnect.com (accessed August 16, 2010). Subscribed users can set up to five journals for table of contents and alert services from more than 13,000 academic and professional journals.

Journal TOCs, www.journaltocs.hw.ac.uk (accessed March 23, 2011). The biggest collection of Table of Contents (TOCs) available on the Internet, it organizes by subject or publisher. Registration is required for email alerts, or RSS is available on the site.

SpringerAlerts, www.springer.com/alert (accessed August 16, 2010). A free e-mail table contents service for all Springer journals as well as announcements on new books.

MINORITIES IN ENGINEERING

Sylvia A. Nyana

INTRODUCTION

With fewer than 12% of minorities graduating in engineering, representation of women and minorities in science, technology, engineering, and mathematics (STEM) is wanting, given African Americans, Native American, and Latinos constituted 32% of the population in 2010 and will be 38% by 2025 (NACME, 2007). Efforts by academic institutions, individuals, and organizations to diversify the engineering field has greatly improved the number of minorities in engineering and science fields. The National Academy of Sciences (2010) noted gains in the number of minorities in the field of engineering; however, the differences are not great enough to meet current economic demands. A study commissioned by NACME, "Confronting the 'New' American Dilemma, Underrepresented Minorities in Engineering a Data-Based Look at Diversity" showed that the rate of participation by African Americans, Native American, and Latinos in STEM fields has been minimal and, in some cases, has declined. Of the 68,000 bachelor's degrees in engineering awarded in the United States in 2006, only 8500 were awarded to minorities, and out of 6404 doctoral degree recipients in 2006, 100 were African American students, 98 were Latinos, 9 were Native Americans and Alaska Natives, and of these doctoral degree recipients, 55 were women (NACME, 2007).

The literature of minorities in engineering is scattered throughout the various fields of engineering, general and specific encyclopedias, dictionaries and other reference works, books, journals, societies, corporations, and organizations. This list is a summary of the selected resources, organizations, and Web sites with a focus on minorities in engineering fields.

SEARCHING THE LIBRARY CATALOG

Keyword searching in the library catalog (LC) can find books and other materials on women and minorities in engineering and can lead to further search terms by browsing the item record display for LC subject headings. Keyword searching is relevant for minority related fields because (1) terminology in these fields changes often, (2) subject headings do not always keep up with the changes, (3) subject headings are not specific to support minority related fields, and (4) it takes time for libraries to update subject headings to match the new terms. Library users also can find resources on women and minority engineers by searching known titles in the library catalog that deal with their topic, and then browsing the record display for Library of Congress (LOC) Subject Headings assigned to that item. For example, under the title *Women and minorities in science, technology, engineering, and mathematics: upping the numbers* yielded:

Subject: Women in technology
Subject: Minorities in science
Subject: Minorities in technology
Subject: Women in mathematics
Subject: Minorities in mathematics
Subject: Women in engineering
Subject: Minorities in engineering

DATABASES

Databases are online indexes and abstracts that provide students and scholars with access to full-text articles or abstracts of journal articles, conference papers, theses, and book reviews. There are a few indexes and abstracts available in print. The use of a multisearch function that allows searching in a number of databases at once, if available in the library catalog, is useful if you are not sure about the title of the database, especially when using "minorities in engineering" as key words. Because minorities is a subdiscipline in engineering, all of the databases in the engineering fields have some journal publications on minorities. To find recent articles on minorities in engineering, use *ERIC* (journals and nonjournals) and *Compendex*. In addition, articles on women and minorities in engineering can be found in general or interdisciplinary databases, such as *ProQuest Direct, LexisNexis Academic,* and newspaper databases.

BIBLIOGRAPHIES AND GUIDES TO THE LITERATURE

Perez, D., II. 2008. Bibliography: Underrepresented Students (Penn State University) http://www.ed.psu.edu/educ/e2020/resources/Underrepresented%20students. (accessed October 31, 2010).

Royal, K. and N. Mamaril. 2008. Women and Minorities in Engineering: A Review of the Literature. http://www.allacademic.com//meta/p_mla_apa_research_citation/2/7/3/4/3/pages273439/p273439-1.php (accessed April 10, 2010). Paper presented at the MWERA Annual Meeting on low number of women in engineering and some solutions to address the underrepresentation.

DIRECTORIES

Directories provide prospective minority engineers with relevant resources on tips for finding and applying for financial aid, scholarships, grants and loans, fellowships, and organizations and careers in engineering. Some of these sources are updated periodically, so researchers should check the date of each directory to make sure they are using the most recent edition. Many are no longer being published, but a Web search may reveal current opportunities. For specific engineering disciplines, use the library catalog, and narrow it down to a particular discipline.

Baine, C. 2004. *Is There an Engineer Inside You? A Comprehensive Guide to Career Decisions in Engineering,* 2nd ed. Belmont, CA: Professional Publications. Includes description of the profession, program selection, and career planning.

Beckham, B. 2008. *Beckham's Guide to Scholarships for Black and Minority Students*, 6th ed. Silver Springs, MD: Beckham Publishing. There are over 1000 private awards for Black and minority students.

Meiners, P. A. 1998. *National Directory of Foundation Grants for Native Americans*, Kansas City, MO: Corporate Resource Consultants. A directory of 55 private foundations funding Native American programs, American Indian communities, and those who earmark Native American studies and education programs (from CRC Publishing Company–EagleRock Books).

The Minority & Women Doctoral Directory. 2010/2011. http://www.mwdd.com/. (accessed November 16, 2010). The current edition lists approximately 4900 Blacks, Hispanics, Native Americans, Asian Americans, and women students in nearly 80 fields in the sciences, engineering, the social sciences, and the humanities.

National Research Council. 2006. *To Recruit and Advance: Women Students and Faculty in Science and Engineering,* Washington, D.C.: National Academies Press. This book describes actions taken by universities to improve the situation for women and identifies better practices for recruitment, retention, and promotion for women scientists and engineers in academia (from National Resource Council).

Schlachter, G. and R. D. Weber. *Directory of Financial Aids for Minorities.* (Annual) El Dorado Hills, CA: Reference Service Press. This directory revises and updates the previous edition, and contains information on 2182 scholarships, fellowships, loans, grants, awards, internships, state sources of educational benefits, and general financial aid directories for minorities (from Reference Services Press).

Singh, A. 2008. *The Vault Guide to Engineering Diversity Programs.* New York: Vault. The *Guide* serves as "a one stop shop" for careers in engineering, diversity related questions, and the latest statistics including minority counts (from www.vault.com).

Wilson, E. B. 1996. *Money for College: A Guide to Financial Aid for African-American Students.* New York: Plume. There are over 1000 sources of financial aid for African American students from government, colleges and universities, private foundations, and organizations. The guide lists financial aid programs by field of study and source of funds, and application process (from www.abebooks.com).

ENCYCLOPEDIAS, DICTIONARIES, AND HANDBOOKS

Students and scholars can find relevant information on minorities in engineering by browsing general and/or subject encyclopedias, dictionaries, and other engineering reference works. Below is a selection of encyclopedias, dictionaries, and handbooks that provide background information on women and minorities in engineering, sciences, and mathematics (undergraduate and graduate) education as well as careers. This is not a comprehensive listing of all available materials on this topic. For a specific encyclopedia or reference work, use the library catalog and narrow it to a particular subject/discipline.

Cantú, N. E. 2008. *Paths to Discovery: Autobiographies from Chicanas with Careers in Science, Mathematics, and Engineering.* Los Angeles: UCLA Chicano Research Studies Center Press. A group of trailblazing Chicanas trace how their interest in math and science at a young age developed into a passion fed by talent and determination. Today, they are teaching at major universities, setting public and institutional policy, and pursuing groundbreaking research (from UCLA Chicano Studies Research Center Press, http://www.chicano.ucla.edu/press/paths.asp).

Commission on Professionals in Science and Technology. 1997. *Professional Women & Minorities: A Total Human Resources Data Compendium.* Washington, D.C.: CPST. A comprehensive reference book of data on human resources presented in nearly 400 tables and charts, with breakouts by sex and race/ethnicity. Also included are data on enrollments, degrees, and the general, academic, and federal workforce by field and subfield (from The Commission on Professionals in Science and Technology (CPST) http://www.cpst.org/).

Sammons, V. O. 1989. *Blacks in Science and Medicine.* New York: Hemisphere. The book looks at contributions of 1500 blacks in various scientific disciplines.

Williams, F. M. and C. J. Emerson. 2008. *Becoming Leaders: A Practical Handbook for Women in Engineering, Science, and Technology.* Reston, VA: American Society of Civil Engineers. Information on leadership skills, career information, work–life balance for students and women.

DISSERTATIONS AND THESES

Christie, B. A. 2003. *A Qualitative Examination of the Nature and Impact of Three California Minority Engineering Programs.* PhD diss. University of Southern California. The case study examines the nature and impact of three MEPs in California and an analysis of the lack of participation by freshmen and sophomore students who qualify for these programs. The results demonstrated that a high percentage of the qualifier/nonparticipants are unaware of the programs and events on their campuses (from dissertation abstract from ProQuest).

Haden, C. 2006. *Retention of Underrepresented Students in Engineering Degree Programs: An Evaluation Study.* EdD diss. Northern Arizona University. This study examines issues related to persistence of women and minorities in engineering degree programs, and evaluates the Multicultural Engineering Program (MEP) at a public university in the western United States. The findings indicate that issues of social and academic integration and factors external to the institution influence underrepresented student persistence (from dissertation abstract from ProQuest).

Hermond, D. S. 1993. *Evaluation of Retention Strategies of Texas A&M University's Minority Engineering Program.* PhD diss. Texas A&M University. The study evaluates seven specific strategies used by Texas A&M University to retain minorities in engineering. The results found students are generally satisfied with the quality of services received from the minority engineering program (MEP). However, the MEP strategies to freshman GPAs are not statistically significant, and are not correlated with participants GPAs (from dissertation abstract from ProQuest).

Lee, J. 2006. *Getting Out the Gates: Underrepresented Minority Students' Search for Success in Introductory Chemistry Courses to Continue on the STEM Path.* PhD diss. University of Illinois. The study examines the experiences of minority students in introductory chemistry at State University, a public research institution. Findings suggest that minorities have to conform themselves to the STEM template in a way that nonminorities do not (from dissertation abstract from ProQuest).

Miracola, C. L. 2004. *A Quantitative Model to Gauge Success for Underrepresented Minorities in the Pennsylvania State University College of Engineering.* MS thesis. The Pennsylvania State University. A pilot study aimed at improving the educational development of underrepresented minorities in the College of Engineering.

Preston, S. D. 2009. *Investigating Minority Student Participation in an Authentic Science Research Experience.* PhD diss. The Pennsylvania State University. The study investigates a summer science research experience for minority students and the nature of students' participation in scientific discourse and practices within the context of the research experience. Findings revealed that students who participated in the research experience were able to successfully engage in some cultural practices of science, such as using inscriptions, constructing explanations, and collecting data (from dissertation abstract from ProQuest).

Simon, T. M. 2008. *Engineering Success: Persistence Factors of African American Doctoral Recipients in Engineering and Applied Science.* EdD diss. Columbia University. The study identifies factors that influence African Americans to pursue and complete doctoral degrees in engineering and applied science disciplines. The study found key factors impacting doctoral degree completion included peer support, faculty adviser support, support from university administrators, and family support (from dissertation abstract from ProQuest).

Weldy, E. A. 2001. *Understanding Baccalaureate Completion Rate Increases of Underrepresented Minority Students in Science and Engineering: Three Case Studies.* EdD diss. University of Illinois. The purpose of this study was to determine how individual universities successfully implemented programs and used financial resources to improve baccalaureate completion rates of underrepresented minority students in science and engineering academic disciplines at the same universities. The findings indicate federal programs and financial resources did contribute to improving minority student completion rates at the State University of New York at Albany and the University of Illinois at Urbana-Champaign, but minimally at the University of Houston (from dissertation abstract from ProQuest).

MONOGRAPHS

Burke, R. J. 2007. *Women and Minorities in Science, Technology, Engineering, and Mathematics: Upping the Numbers.* Cheltenham, U.K.: Edward Elgar. Reviews numbers of women and minorities in STEM, consequences, and solutions.

Cole, M. and P. Griffin. 1987. *Contextual Factors in Education: Improving Science and Mathematics Education for Minorities and Women.* Madison, WI: Wisconsin Center for Education Research. "The book summarizes research on the various ways that students' cultural backgrounds and innate ways of learning affect academic achievement, and offers descriptions and recommendations for improving science and mathematics education for minorities and women, based on successful programs that take these differences into account" (from Web site, full text available at http://www.eric.ed.gov/PDFS/ED288947.pdf).

Frize, M., P. R. D. Frize, and N. Faulkner. 2009. *The Bold and the Brave: A History of Women in Science and Engineering.* Ottawa: University of Ottawa Press. Both a history and a philosophical look at the contributions, advancements, and obstacles of women in the engineering and science fields.

Georges, A. 1999. *Keeping What We've Got: The Impact of Financial Aid on Minority Retention in Engineering.* "This paper evaluates the performance of engineering institutions in graduating minority freshmen ... Findings indicate that financial aid awards may be the key variables to consider at the policy level to improve the minority retention rate in engineering across all institutions." (NACME, full text available at http://www.eric.ed.gov/PDFS/ED435534.pdf).

Landis, R. B. 2005. *Retention by Design: Achieving Excellence in Minority Engineering Education.* This represents an update and republishing of *Retention by Design*, a monograph originally published in 1991 by the National Action Council for Minorities in Engineering (NACME). The "pipeline" data used to justify the need to redesign engineering college environments to meet the needs of minority students has been simplified and updated (www.csun.edu/~ecsdean/docs/retention_by_design.doc).

Long, S. 2001. *From Scarcity to Visibility: Gender Differences in the Careers of Doctoral Scientists and Engineers.* Committee on Women in Science and Engineering. Historic study documenting differences between men and women and different races and ethnic backgrounds in the 1970s and 1980s (from National Academies Press).

VIDEO/TV PROGRAM

Meeting the Challenge: Report from Penn State, a College of Engineering Key Executive Conference. 1990. University Park, PA: WPSX-TV. "Senior industry executives and university participants discussed how to enhance engineering education and the engineering profession at a two-day seminar. Out of this seminar, four topics for concern were discussed: k–12 education, undergraduate engineering education, recruitment and retention of women and minorities, and postgraduate continuing education" (from University Libraries description).

JOURNALS

Diversity/Careers in Engineering & Information Technology. (not supplied). Springfield, NJ: Renard Communications, http://www.diversitycareers.com/ (accessed March 29, 2011). This online career publication offers articles and links to news and to issues for minority engineering students and professionals.

Journal of Women and Minorities in Science and Engineering (1072–8325). Redding, CT: Begel House. http://www.begellhouse.com/journals/ (accessed March 29, 2011). The journal publishes original, peer-reviewed papers that report innovative ideas and programs for classroom teachers, scientific studies, and formulation of concepts related to the education, recruitment, and retention of under-represented groups in science and engineering.

Maes (1552–9711). 1992. Los Angeles: Society of Mexican American Engineers and Scientists. http://www.maesnationalmagazine.com/ (accessed March 29, 2011). The magazine focuses on minority engineers and scientists, ranging from students to accomplished professionals and the companies for which they work.

Minority Engineer (0884–1829). New York: Equal Opportunity Publication. http://www.eop.com/mags.ME.php (accessed March 29, 2011). First published in 1979, the magazine is provided free to qualified engineering, computer science, and information technology students and professionals who are Black, Hispanic, Native American, and Asian American, sent to their home addresses, colleges and universities, and chapters of professional engineering associations.

SWE (1070-6232). Chicago: Society of Women Engineers. http://societyofwomenengineers.swe.org/ (accessed March 29, 2011). Articles in SWE cover issues of interest to women engineers including the achievements and accomplishments of women engineers, career development, career guidance and much more.

USBE and Information Technology (not supplied). Formerly *U.S. Black Engineer and Information Technology Magazine* provides Black technology news and information about Black engineering, Black technology, Black entrepreneurs, Black engineers, Black education, Black minorities, Black Engineer of the Year Awards (BEYA), and historically Black colleges and universities (HBCU) from Black communities in the United States, the United Kingdom, the Caribbean, and Africa.

Winds of Change (0888–5612). Albuquerque, NM: America Indian Science and Engineering Society, http://aises.org/wocnews (accessed March 29, 2011). Career and educational information in the areas of science, technology, engineering, and math (STEM) for native peoples.

Woman Engineer (0887-2120). New York: Equal Opportunity Publications, http://www.eop.com/awards-WE.php (accessed March 29, 2011). Career advice for women in engineering, computer science, and information technology.

JOURNAL ARTICLES

Anderson, R. 1999. A collaborative effort to recruit and retain underrepresented engineering students. *Journal of Women and Minorities in Science and Engineering* 5 (4): 323–349. The article describes four organizations that have successfully planned and executed several partnering activities that have resulted in an increase in the enrollment and retention of students, particularly women and minority undergraduates.

Anderson, T. B., B. A. Bruschi, and W. Pearson. 1994. Minority females and precollege mathematics and science: Academic preparation and career interests. *Equity and Excellence in Education* 27 (2): 62–70. The study focuses on underrepresented minority females and their mathematics and science academic preparation and career interests. The overall data suggest that minority females are limited in their math and science skills/knowledge that begins as early as the fourth grade. The study found only one-half of Hispanic fourth-grade females and about one-third of African American females have learned third-grade math skills and prepared for fourth-grade mathematics.

Fenske, R. H., J. D. Porter, and C. P. DuBrock. 2000. Tracking financial aid and persistence of women, minority, and needy students in science, engineering, and mathematics. *Research in Higher Education* 41: 67–94. This longitudinal study followed four consecutive freshmen cohorts at a large urban public university. It found that science, engineering, and mathematics (SEM) students persisted and graduated at higher rates than non-SEM majors. Gift aid for SEM majors was more likely to be merit-based than need-based. Women, but not underrepresented minorities or needy students, graduated at higher rates than other SEM majors.

Lam, P. C. 2000. A description and evaluation of the effects of a pre-engineering program for underrepresented, low-income, and/or first generation college students at the University of Akron. *Journal of Women and Minorities in Science and Engineering* 6 (3): 221–228. The study summarizes the five-year effort of a preengineering program to improve the recruitment and retention of underrepresented college students to pursue degrees in science, mathematics, engineering, and technology at The University of Akron, and assesses the university's successful operation of the special high school Upward Bound and preengineering Academic Achievement programs.

Reichert, M. and M. Absher. 1998. Graduate engineering education of underrepresented populations. *Journal of Engineering Education* 87 (3): 257–268. This paper examined the graduate degrees awarded in engineering from the perspective of representation within the U.S. population, the U.S. engineering schools that grant significant numbers of graduate degrees to underrepresented minorities, and the measures implemented by some of these schools that are particularly effective at granting graduate degrees to underrepresented minorities.

Wentling, R. M., and C. Camacho. 2008. Women engineers: Factors and obstacles related to the pursuit of a degree in engineering. *Journal of Women and Minorities in Science and Engineering* 14 (1): 83–118. This study addressed the factors that have hindered, motivated, and assisted women who graduated with a degree in engineering. This study provided valuable insights and created a framework from which high schools, universities, researchers, and female students can directly benefit.

Societies, Organizations, and Associations

According to the National Action Council for Minorities in Engineering (NACME), privately funded financial aid from foundations, individuals, corporations, colleges and universities, and improved federal funding are keys to improving minority graduation rates in engineering (NACME, 1999). In addition to scholarships, professional organizations provide opportunities to network and learn new industry-related skills, access to jobs databases, and conferences and workshops at the local and national level.

American Indian Council of Architects and Engineers (AICAE), http://www.aicae.org/ (accessed December 15, 2010). AICAE is a professional organization of American Indian architects, engineers, and designers in the United States.

American Indian Science and Engineering Society (AISES), http://www.aises.org (accessed December 15, 2010). AISES works to increase the number of American Indians in science and engineering through education and academic support and to develop technologically informed leaders in the community.

Association for Women in Science (AWIS), http://www.awis.org/ (accessed December 15, 2010). "The AWIS is a national advocacy organization championing the interests of women in science, technology, engineering, and mathematics across all disciplines and employment sectors … By breaking down barriers and creating opportunities, AWIS strives to ensure that women in these fields can achieve their full potential" (from Web page).

Center for the Advancement of Hispanics in Science and Engineering Education (CAHSEE), http://www.cahsee.org/ (accessed December 15, 2010). The "mission of the Center is to prepare talented Hispanic and other underrepresented minority science and engineering students to achieve academic excellence and professional success through CAHSEE's pipeline of rigorous educational and leadership development programs" (from Web page).

The Hispanic Scholarship Fund (HSF), http://www.hsf.net/ (accessed December 15, 2010). The HSF is the nation's leading Hispanic scholarship organization, providing scholarships, educational and outreach support to Hispanic families and students.

Mathematics, Engineering, Science Achievement Program (MESA), http://www.csulb.edu/colleges/coe/mesa/ (accessed December 15, 2010). A partnership of the State of California and private industry, MESA Center gives academic support services to precollege students and nonremedial academic support services to college students.

Minority Science and Engineering Improvement Program (MSEIP), http://www2.ed.gov/programs/iduesmsi/index.html (accessed December 15, 2010). "This program assists predominantly minority institutions in effecting long-range improvement in science and engineering education programs and increasing the flow of underrepresented ethnic minorities, particularly minority women, into science and engineering careers" (from Web page).

National Action Council for Minorities in Engineering (NACME), http://www.nacme.org (accessed December 15, 2010). "NACME is the nation's largest privately funded source of scholarships for minorities in engineering. In addition to scholarship, NAMCE collaborates with other nonprofit organizations to provide preengineering study preparations and experience for public school and community colleges in an effort to increase the representation of African American, American Indian, and Latino women and men in engineering and technology and mathematics- and science-based careers" (from Web page).

National Association of Multicultural Engineering Program Administrators (NAMEPA), http://www.namepa.org (accessed December 15, 2010). "NAMEPA is a national network of educators and representatives from industry, government, and nonprofit organizations committed to improving the recruitment and retention of African Americans, Hispanics, and American Indians earning degrees in engineering. NAMEPA also promotes the professional development of its members and serves as an advocate for and resource to those programs and organizations that seek to recruit, educate, and employ diverse engineering talent." (from Web page).

National Coalition of Underrepresented Racial and Ethnic Advocacy Groups in Engineering and Science (NCOURAGES), http://www.ncourages.org/ (accessed December 15, 2010). "NCOURAGES' mission is to focus and align our individual efforts and activities for the purpose of dramatically increasing the racial and ethnic diversity of the nation's science and engineering workforce" (from Web page).

National Organization for the Professional Advancement of Black Chemists and Chemical Engineers (NOBCChE), http://www.nobcche.org (accessed December 15, 2010). NOBCChE works with professionals working at science-related companies and faculty at local school districts to increase the number of minorities in science, technology, and encourages college students to pursue graduate degrees in the science, technology, engineering, and mathematics (STEM) disciplines (from Web page).

National Society of Black Engineers, The (NSBE), http://national.nsbe.org (accessed December 15, 2010). NSBE is a nonprofit organization "dedicated to the academic and professional success of African-American engineering students and professionals. It offers its members leadership training, professional development, mentoring opportunities, career placement services, and more. NSBE is comprised of more than 250 collegiate, 68 professional, and 99 precollege active chapters nationwide and overseas" (from Web page).

Ronald E. McNair Postbaccalaureate Achievement Program, http://www2.ed.gov/programs/triomcnair/index.html (accessed December 15, 2010). The Ronald E. McNair Postbaccalaureate Achievement Program is funded through the U.S. Department of Education, Higher Education Programs, and is designed to increase the number of low-income, first-generation, and underrepresented minority college students who pursue and complete the doctoral degree. Its long-range mission is to help increase the diversity of college and university faculties.

Society for the Advancement of Chicanos and Native Americans in Science (SACNAS), www.sacnas.org/ (accessed December 15, 2010). "SACNAS is a national society of scientists with a 37-year dedication in fostering the success of Hispanic/Chicano and Native American scientists—from college students to professionals in attaining advanced degrees, careers, and positions of leadership" (from Web page).

Society of Hispanic Professional Engineers, The (SHPE), http://oneshpe.shpe.org (accessed December 15, 2010). Founded in 1974, the SHPE's primary function is to enhance and achieve the potential of Hispanics in engineering, math, and science by increasing educational opportunities, promoting professional and personal growth. Today, SHPE enjoys a strong but independent network of professional and student chapters throughout the nation (from Web page).

Society of Mexican American Engineers and Scientists (MAES), http://www.maes-natl.org/ (accessed December 15, 2010). Founded in 1974, the MAES mission is to promote, cultivate, and honor excellence in education and leadership among Latino engineers and scientists, and operates a number of educational and counseling programs to help young people consider and enter engineering or science careers (from Web page).

Society of Women Engineers, The (SWE), http://societyofwomenengineers.swe.org/ (accessed December 15, 2010). Founded in 1950, SWE is a not-for-profit educational and service organization that empowers women to succeed and advance in the field of engineering and to be recognized for their life-changing contributions as engineers and leaders. "SWE is the driving force that establishes engineering as a highly desirable career for women through an exciting array of training and development programs, networking opportunities, scholarships, outreach and advocacy activities, and much more" (from Web page).

Southeastern Consortium for Minorities in Engineering (SECME), http://www.secme.org (accessed December 15, 2010). "The mission of SECME is to increase the pool of historically underrepresented and underserved students who will be prepared to enter and complete postsecondary studies in science, technology, engineering, and mathematics (STEM), thus creating a diverse and globally competitive workforce" (from Web page).

United Negro College Fund (UNCF), http://www.uncf.org/ (accessed December 15, 2010). The UNCF's "mission is to build a robust and nationally-recognized pipeline of underrepresented students who, because of UNCF support, become highly qualified college graduates and to ensure that our network of member institutions is a respected model of best practice in moving students to and through college" (from Web page).

Women in Engineering ProActive Network (WEPAN), http://www.wepan.org/ (accessed December 15, 2010). It is a "national not-for-profit organization with over 600 members from engineering schools, small businesses, Fortune 500 corporations, and nonprofit organizations. WEPAN works to transform culture in engineering education to attract, retain, and graduate women. With a clear focus on research-based issues and solutions, WEPAN helps its members develop a highly prepared, diverse engineering workforce for tomorrow" (from Web page).

CORPORATE SCHOLARSHIPS FOR MINORITIES IN ENGINEERING

AT&T Labs Fellow Program (ALFP), http://www.research.att.com/internships (accessed December 15, 2010). The ALFP program offers fellowships to outstanding underrepresented minority and women students pursuing PhD studies in computing and communication.

The Gates Millennium Scholars (GMS), http://www.gmsp.org/ (accessed December 15, 2010). The GMS provides outstanding minority students with a scholarship to complete an undergraduate college education in all discipline areas, and a graduate education for those students pursuing studies in mathematics, science, engineering, education, or library science.

The General Motors Women's Retail Network Dealer Development Scholarship, http://gmsac.com/ (accessed December 15, 2010). "The GM Women's Retail Network Dealer Development Scholarship offers women who are enrolled in undergraduate, graduate and nontraditional educational institutions that offer degrees in the automotive retail field. Several scholarships will be available beginning in 2011 in award amounts up to $5000" (from Web page).

The Mellon Mays Undergraduate Fellowship Program (MMUF), http://www.mmuf.org/ (accessed December 15, 2010). Funded by the Andrew W. Mellon Foundation, "the MMUF objective is to increase the number minority students, and others with a demonstrated commitment to eradicating racial disparities, who will pursue PhDs in core fields in the arts and sciences in core fields within the arts and sciences," and to diversify the faculties at colleges and universities by providing support for qualified minority scholars (from Web page).

National Defense Science & Engineering & Graduate Fellowship Program (NDSEG), http://ndseg. asee.org/ (accessed December 15, 2010). "The National Defense Science and Engineering Graduate (NDSEG) fellowships are awarded to applicants who will pursue a doctoral degree in, or closely related to, an area of Department of Defense (DoD) interest (sciences and engineering)." Applications are particularly encouraged from women and minorities (from Web page).

REFERENCES

Confronting the 'New' American Dilemma, Underrepresented Minorities in Engineering: A Data-Based Look at Diversity. http://www.nacme.org/user/docs/NACME%2008%20ResearchReport.pdf (accessed December 15, 2010).

National Action Council for Minorities in Engineering (NACME), http://www.nacme.org/ (accessed December 15, 2010).

National Academy of Sciences. 2010. *Expanding Underrepresented minority participation: America's science and technology talent at the crossroads.* Washington, D.C.: National Academies Press.

Royal, K. and N. Mamaril 2008. Women and minorities in engineering: A review of the literature. Paper presented at the annual meeting of the MWERA Annual Meeting, Westin Great Southern Hotel, Columbus, OH, October 15. http://www.allacademic.com/meta/p273439_index.html (accessed December 15, 2010).

The college blue book: Scholarships, fellowships, grants, and loans, 35th ed. 2008. Detroit: Macmillan Reference USA.

CHAPTER CONCLUSION

This chapter offered suggestions for doing general engineering research. It also included career information to show the numerous engineering disciplines. An overview of industrial standards can guide an engineer in design and development. Ethical standards apply in all aspects of engineering. An understanding of the patent process is helpful in design engineering, but is also vital to the chemical engineer and other specialties.

Regardless of the specialty, there are basic resources and traditional methods that should be understood by any engineering student or librarian. How to find library books on the shelf and information on the Internet will help the beginner and the specialist. Roadblocks in searching for papers appearing at a general engineering conference also apply to those sponsored by specific engineering societies. Acronym dictionaries define societies' names, but the same acronym can apply to more than one society. For example, ASSE means American Society of Safety Engineers or American Society of Sanitary Engineers. Drop an "E" from IEEE (Institute of Electrical and Electronic Engineers) and it becomes IEE, the Institution of Electrical Engineers, the predecessor of IET, the Institution of Engineering and Technology. Even more confusingly, both publish resources indexed in *IEEE Xplore*. Thankfully, with more and more societies and individuals going online, the Web helps to locate society names and author locations. The invisible Web and gray literature abound in every discipline. Engineer- and librarian-developed tools to aid in document search and retrieval are vital to success. If the goal of a full-text document available online is paramount, then basic searching techniques are necessary to ensure success.

Engineering books, especially reference works, may seem trivial in the digital world, but contain information valuable as a starting point. Some books sound specific, but contain valuable information beyond a single subject area. When Paul Thrush compiled the mining dictionary for the Bureau of Mines, he complained about the limitations of subject headings to describe its content, since it covers more than mining and serves as a basic dictionary in many engineering fields. *Architectural Graphic Standards* is more than architecture. It serves as a first choice in quick reference, as it mentions sources that lead to specialties (i.e., the Americans with Disabilities Act of 1990). While there are many intricacies to searching the engineering resources, tools are available that provide excellent guidance in both print and electronic formats.

3 Aeronautical and Aerospace Engineering

Thomas W. Conkling

CONTENTS

Introduction, History, and Scope of Discipline 45
Searching the Library Catalog 46
Indexes and Abstracts 47
Databases 49
Bibliographies and Guides to the Literature 49
Directories 50
Encyclopedias and Dictionaries 51
Handbooks and Manuals 52
Monographs and Textbooks 54
Journals 54
Standards and Patents 56
Search Engines and Web Sites 57
Societies and Associations 58
Conclusion 59
References 59

INTRODUCTION, HISTORY, AND SCOPE OF DISCIPLINE

The idea of flight has always intrigued mankind, but it was only a little over 100 years ago that powered manned flight was achieved. The first successful demonstration of a flying machine is usually credited to the Montgolfier brothers' hot air balloon that flew in June of 1783. Sir George Cayley investigated the aerodynamics of fixed wing flight in the early 1800s, and designed the first man-carrying glider. Otto Lilienthal's pioneering work with manned glider flight between 1891 and 1896 advanced the field, and Samuel Langley successfully tested a number of model aircraft powered by small steam engines in 1896. The Wright brothers built and flew a number of gliders before they designed the aircraft that became the first to demonstrate powered manned flight in 1903. The theoretical foundations for the discipline of aeronautical engineering began to take form with some of the early work on fluid mechanics by Newton, Bernoulli, Euler, and d'Alembert in the seventeenth and eighteenth centuries. This work was expanded during the next century by scientists, such as Navier, Stokes, Rankine, Helmholz, Kirchhoff, and Rayleigh (Anderson, 1997), and a number of researchers compiled data files on airfoils during this period. The field received a large boost in the early 1900s when governments realized the potential military applications of this new technology and began supporting research into the principles of flight and propulsion.

The period before World War I saw many advances in aircraft design, particularly from Germany, France, and Russia. At the start of the war, aircraft were used primarily for reconnaissance and artillery spotting, but their role quickly expanded with the development of effective machine guns, better engines, and improved design. Aviation remained in the public eye after the war. Air races and other forms of competition for flying records made headlines regularly. Passenger service and airmail routes were started. Radial engines were developed, followed by liquid-cooled, inline piston engines. Metal replaced wood as the key structural element and almost all new aircraft had a single wing. A number of exceptional aircraft were designed during this period, including the Douglas DC-3. Aviation technology advanced rapidly during World War II and, before it had ended, examples of jet-powered aircraft had been developed. The sound barrier was broken in 1947 and the X-15 set many altitude and speed records in the 1960s. Propeller-driven civilian transport aircraft gave way in the 1950s to faster jet-powered craft, such as the Boeing 707. In recent years, a substantial amount of research has been done on stealth technology and pilotless aircraft for military applications.

Space exploration has been made possible by the rocket engine. The first rockets were probably built in China almost a thousand years ago, powered by black powder. By the 1500s, rockets were widely used for fireworks and in elementary weapons. Their use as weapons became more widespread by the late 1700s and into the 1800s. Rocket technology improved during World War I, but the pace of innovation and development accelerated during the 1920s and 1930s. The military became increasingly interested in rockets during this period. Progress was made in both solid and liquid propellants and in many aspects of rocket design and control. Germany had a very advanced rocket program, as did Russia. Germany developed the V-2 missile during this period, a vehicle that heavily influenced post-war launch vehicle design. The space age began in October 1957 with the Russian launch of the Sputnik satellite. The success of this mission created a sense of urgency in the United States to develop comparable missile capabilities. Military and civilian missile technology advanced rapidly as these countries attempted to build more powerful launch systems for weapons and for manned spacecraft. The Space Shuttle, first flown in 1981, is probably the most complex spacecraft and launch system developed to date.

Aeronautical engineering as an academic discipline began in the early years of the twentieth century. In the United States, the first formal course in aeronautics was taught at the University of Michigan in 1913, followed the same year by a similar course at Massachusetts Institute of Technology (MIT). The first master of science degree in aeronautical engineering and the first doctorate in the field were awarded at MIT in 1915 and 1916, respectively (McCormick, 2004). Similar courses and programs were developed in Europe during the same period. Aeronautical and aerospace engineering degree granting programs are now common worldwide.

SEARCHING THE LIBRARY CATALOG

While traditional card catalogs may have offered some unique benefits to the library user, the capabilities of Web-based catalogs are impressive. Keyword searching can help the user retrieve material in aerospace or aeronautical engineering, and also allows him/her to discover related materials in other disciplines. If a keyword search returns excessive items, searching with Library of Congress subject headings is one way to refocus a search. Following is a listing of some important subject headings for this field.

Aerodynamics
Aerofoils
Aeronautics
Airframes
Airplanes
Artificial Satellites
Astronautics
Ballistic Missiles
Boundary Layer
Drag (Aerodynamic)
Flight

Gas Turbines—Aerodynamics
Helicopters
Jet Planes
Jet Propulsion
Lift (Aerodynamic)
Reynolds Number
Rocket Engines
Rockets (Aeronautics)
Rotors (Helicopters)
Space Vehicles
Stability of Airplanes
Turbines—Aerodynamics
Turbomachines
UAVs (Unmanned aerial vehicles)
Wings

The Library of Congress classification schedule places most of the materials in the field between TL500 and TL4000. However, when searching the catalog or browsing the shelves, it will be apparent that potentially useful items lie outside of these areas. Subjects like turbomachinery, combustion, fluid flow, computational methods, electronics, computers, and structural materials are examples. Some call numbers of interest include:

TL500-589	Aeronautics
TL600-688	Aircraft
TL690-697	Electrical and communication systems
TL698	Materials
TL701-704	Aircraft engines
TL709	Jet propulsion
TL710-713	Flight
TL780	Rocket propulsion
TL787	Astronautics
TL873	Manned spaceflight
TL1050	Astrodynamics
TL3000-3285	Astrionics and electrical equipment
TL4000-4050	Ground support systems

INDEXES AND ABSTRACTS

Indexes and abstracts were originally produced to allow researchers to keep track of newly published technical information. Some indexes concentrated on articles and conference papers, while others covered only technical reports. The technical report has a special place in the communication of research results in aeronautical and aerospace engineering. Government agencies in several industrialized nations recognized the military potential of aircraft early on and began supporting research into this new technology. Research results were often issued as technical reports because this format offered rapid publication time and the ability to contain large amounts of technical data. Their distribution also could be controlled fairly easily if they contained sensitive information. Technical reports started appearing in the early 1900s, and they are still produced in quantity today, with the Web being the key distribution mechanism.

Print indexes have been superceded to a large degree by their database counterparts on the Web, but they can still play an important role in access to those older materials not yet covered online. Records are usually arranged in some type of classed subject scheme or they can be alphabetically arranged by subject heading. Print indexes almost always contain multiple indexes in each issue with annual cumulative indexes to facilitate use.

Engineering Index. 1884 –. Hoboken, NJ: Elsevier Engineering Information. The most comprehensive index to the world's engineering literature; provides very good coverage of the journal and conference literature in aeronautical and aerospace engineering. Available online as the *Compendex* database (see Chapter 2, this volume).

Government Reports Announcement and Index. 1975–1996. Springfield, VA: National Technical Information Service (NTIS). Although no longer published, this publication and its predecessors provide an in-depth guide to the technical report literature that was produced in the United States from 1946 to 1996. Some reports from other countries also are included. It provided very good coverage of the aeronautical and astronautical sciences. This index was produced by NTIS, which acts as the national clearinghouse for the technical report literature in the United States. Issues were published twice per month with annual indexes. Entries are arranged by subject category and provide bibliographic information and an abstract. This publication was preceded by the *Bibliography of Scientific and Industrial Reports* (1946–June 1949), *Bibliography of Technical Reports* (July 1949–June 1954), *U.S. Government Research Reports* (July 1954–1964), *U.S. Government Research and Development Reports* (1965–1971), and *Government Reports Announcement* (1971–1975). Technical report information can now be found in the *NTIS Database*; recently issued technical reports can be searched at the NTIS Web site.

Index of NACA Technical Publications. 1915–1958. Washington, D.C.: National Advisory Committee for Aeronautics. The definitive index to all unclassified and unlimited publications from the National Advisory Committee for Aeronautics (NACA), NASA's predecessor. Entries are arranged by subject category and contain concise information: author, title, report number, and date. Subject and author indexes are included. The period 1915 to 1949 is covered in one volume, and several volumes cover 1950 through 1958. The *NASA Technical Report Server* http://ntrs.nasa.gov (accessed September 5, 2010) now provides online access to the digitized version of many NACA reports.

International Aerospace Abstracts. 1961–. Bethesda, MD: Cambridge Scientific Abstracts. An excellent printed index to the world's published literature in aeronautics, astronautics, and the space sciences. Provides in-depth coverage of journals and conference papers. Issue contents are arranged by subject and each issue has subject, author, meeting paper, and report number indexes. Through 2000, this index was produced by the American Institute of Aeronautics and Astronautics (AIAA). AIAA now produces it in cooperation with Cambridge Scientific Abstracts. Published monthly with annual cumulative indexes. Also available as a Web-based database, *Aerospace and High Technology Database.*

Scientific and Technical Aerospace Reports (STAR). 1963–1995, print version; 1996–, Web version, Lithicum Heights, MD: NASA Center for Aerospace Information. Originally printed, but now published online in a PDF version, this index provides comprehensive coverage of the technical report literature of aeronautics, astronautics, and the space sciences. Covers technical reports from NASA and its contractors, as well as reports from other agencies, companies, and labs in the United States and elsewhere. Issue contents are arranged by subject with multiple indexes at the end of each issue. *STAR* was preceded by the *Index of NASA Technical Publications and Technical Publication Announcements* from 1958 through 1962. When used in conjunction with *International Aerospace Abstracts*, comprehensive coverage of all publications in the field is provided. The *NASA Technical Report Server* provides Web-based access to the more recent NASA-related reports in *STAR* http://www.sti.nasa.gov/Pubs/star/star.html. (accessed October 25, 2010).

DATABASES

Bibliographic databases revolutionized access to published information in the early 1970s. The tedious manual search process was replaced by rapid online searching. Today, Web-based databases deliver an enormous amount of access to the engineer's desktop. The information needs of aerospace and aeronautical engineers are well served by bibliographic databases. The databases covered in this section deal directly with publications in this field, but there are additional databases mentioned in this book that also would be useful, particularly in materials, electronics, and computer science.

> *Aerospace and High Technology Database.* 1962–. Bethesda, MD: Cambridge Scientific Abstracts. This is the best commercially availably database for aeronautical and astronautical engineering. The database is compiled by the American Institute of Aeronautics and Astronautics. Provides excellent coverage of the journal, conference, and technical report literature, and is the online equivalent of *International Aerospace Abstracts* and *STAR.* Over 3000 publications are scanned for information, and 4000 new records are added monthly; more than 8,000,000 records are in the database.
>
> *Compendex.* 1884–. Hoboken, NJ: Elsevier Engineering Information, Inc. The electronic version of *Engineering Index*, this database provides very good coverage of the journal and conference literature in aeronautics and astronautics (see Chapter 2, this volume).
>
> *DTIC Online: Public Technical Reports* http://www.dtic.mil/dtic/search/tr/index.html (accessed September 5, 2010), Fort Belvoir, VA: Defense Technical Information Center. This site provides access to the full text versions of many recent reports sponsored by the U.S. Department of Defense and gives ordering information for older ones.
>
> *NASA Technical Report Server (NTRS)* http://ntrs.nasa.gov (accessed September 5, 2010), Lithicum Heights, MD: NASA Center for Aerospace Information. *NTRS* provides free public access to NASA-generated technical reports and other publications from NASA authors. Some publications from non-NASA sources are included as well. The database indexes NASA materials back to the beginning of the agency in 1958, and also covers NACA reports back to 1917. All unclassified and unlimited NASA reports are indexed in *NTRS.* Many full text versions of NASA reports are in the database, particularly from those published in 2004 and later.
>
> *National Technical Information Service (NTIS) Database.* 1964–. Springfield, VA: National Technical Information Service. Large database (2,440,000 records) covering technical reports produced by government agencies and their contractors, including NASA, U.S. Department of Defense, and the U.S. Department of Energy. The entire database is available from commercial vendors, and a subset of it is available to the public at the NTIS Web site. Provides very good coverage of the technical report literature in aeronautics and astronautics (see Chapter 2).

BIBLIOGRAPHIES AND GUIDES TO THE LITERATURE

There are several guides to the literature that are useful in their coverage of aeronautical and aerospace engineering. Most of the guides are not exclusively written for these areas, but broader works have certain advantages in that they review resources for some of the other fields that are also of interest to aeronautical and aerospace engineers.

> Anthony, L.J., ed. 1985. *Information Sources in Engineering*, 2nd ed. London: Butterworths. Primary and secondary information sources are discussed, and these are followed with chapters on specific engineering fields. The "Aerospace Engineering" chapter has an excellent discussion of the technical information programs in a number of countries, key periodicals, abstracts, and a short list of monographs.

Auger, C.P. 1994. *Information Sources in Grey Literature*, 3rd ed. London: Bowker Saur. This book concentrates on technical reports and other grey literature. It has a very strong chapter on aerospace technical reports and the organizations who have produced them over the years.

DePetro, T.G. and T.E. Naylor, eds. 1997. *Selective Guide to Literature on Aerospace Engineering*. Washington, D.C.: American Society for Engineering Education. This brief work is specifically on the aeronautical and aerospace engineering literature. The guide covers key indexes and abstracts, databases, reference works, handbooks, directories, technical report series, and journals.

Lord, C.R. 2000. *Guide to Information Sources in Engineering*. Englewood, CO: Libraries Unlimited. This book is divided up by format (handbooks, grey literature, journals, Internet resources, etc.) and then subdivided by engineering specialty. Aeronautical and aerospace entries have their own section under most formats.

Macleod, R.A. and J. Corlett, eds. 2005. *Information Sources in Engineering,* 4th ed. New York: Bowker Saur. A collection of chapters describing primary and secondary sources in all fields of engineering, and then listing the specific literature in over 20 fields. This edition provides a chapter on aerospace engineering.

Mildren, K. W. 1976. *Use of Engineering Literature*. London: Butterworths. One of the classic guides to the engineering literature, the book discusses the important information formats for engineering (journals, technical reports, standards, etc.) and abstracts and indexes. Individual chapters are devoted to specific disciplines. The "Aeronautics and Astronautics" chapter provides excellent background information on many of the subdisciplines, such as propulsion and gas dynamics, and provides a lengthy discussion of classic texts in all of these areas.

DIRECTORIES

Aeronautical and aerospace engineers work in a large industry and their efforts result in the production of complex products. The nature of this business lends itself to the production of directories to help individuals and companies keep track of new aircraft, space, and launch vehicles, and to find vendors of related products and services. Some of the major directories in the field are presented here, but there are more including a number produced by Jane's Information Group Limited. The Web is also a rich source of similar information.

International Satellite Directory. (2 vols. annually) Sonoma, CA: Satnews Publishers. An annual directory of the satellite industry covering manufacturers, operators, service providers, and key personnel.

Isakowitz, S. J., J. P. Hopkins, Jr., and J. B. Hopkins, eds. 2004. *International Reference Guide to Space Launch Systems,* 4th ed. Reston, VA: American Institute for Aeronautics and Astronautics. International directory of launch vehicles and their related systems. Describes the vehicles (technical specifications, design, payload, and performance), flight histories, and launch pad operations. Includes appendices with abbreviations and acronyms and a bibliography of technical references for each launch vehicle covered.

Jane's All the World's Aircraft. (Annually) Alexandria, VA: Jane's Information Group. This annual is the premier directory of the world's military and civilian aircraft, available in print and Web versions. Entries are arranged alphabetically by country and then by manufacturer. Aircraft descriptions include development histories, technical and performance specifications, configurations, instrumentation, and armaments. Most entries contain photographs and diagrams. There are sections on air-launched missiles, engines, world flight records, a glossary, and a listing of first flights made during the year.

Jane's Space Systems and Industry. (Annually) Alexandria, VA: Jane's Information Group. An annual publication that provides extensive coverage of the space industry, available in print and Web versions. Thousands of entries covering the space industry, space centers, launch vehicles, propulsion systems, satellites (military and civilian), and contractors. Full of illustrations, technical specifications, launch histories, and photos.

World Aerospace Database. New York: McGraw-Hill. http://www.worldaviationdatabase. com/aw/ (accessed September 5, 2010). Comprehensive, online directory to over 22,000 companies and their products and services. Information can be retrieved by keyword searching or by browsing subject categories.

ENCYCLOPEDIAS AND DICTIONARIES

Encyclopedias fill a role in engineering by offering summaries of a subject that can be used either as an introduction to a new area for a reader or as a review of a familiar one. Dictionaries complement encyclopedias by providing very concise meanings for thousands of words, phrases, and acronyms that would be too specific or narrow to cover as encyclopedia entries. The encyclopedias and dictionaries mentioned here are representative of what is available in aeronautical and aerospace engineering. A few multilingual dictionaries are listed, but others covering additional languages have been published.

Angelo, J. A., Jr., ed. 2004. *Facts on File Dictionary of Space Technology.* New York: Facts on File. Comprehensive coverage of the basic concepts of space technology, spaceflight, and their underlying principles. Over 1500 entries, with useful photos and illustrations.

Beck, S. and S. Aslezova, eds. 2002. *Elsevier's Dictionary of Civil Aviation.* Amsterdam, The Netherlands: Elsevier. An English/Russian dictionary covering all aspects of civil aviation including air traffic control, navigation, flight, meterology, communications, and airports. The English–Russian section has 19,000 entries and the Russian–English section has 21,000.

Bristow, G. V., ed. 2003. *Encyclopedia of Technical Aviation.* New York: McGraw-Hill. An encyclopedia geared to the needs of pilots. Coverage is very wide, includes aircraft systems and engines, aerodynamics, flight, air traffic control, aviation-related operations, and selected regulations and rules. Includes many diagrams and illustrations.

Cheremisinoff, N. P., ed. 1986–1994. *Encyclopedia of Fluid Mechanics.* (13 vols.) Houston, TX: Gulf Publishing Company. This 13 volume set brings together theoretical and practical engineering information on all types of flow phenomena. Contains contributions from hundreds of engineers and scientists. Individual volumes cover particular areas (e.g., gas–liquid flow, flow phenomena, and measurement), with chapters exploring the topics in detail. All chapters contain numerous illustrations, tables, formulas, and bibliographic references.

Crocker, D., ed. 2007. *Dictionary of Aviation,* 2nd ed. London: A&C Black. This volume provides definitions of the vocabulary used by pilots, crew members, maintenance and ground personnel, and air traffic controllers.

Dickson, P. 2009. *Dictionary of the Space Age.* Baltimore, MD: Johns Hopkins University Press. This dictionary is an update of *Origins of NASA Names* (NASA SP-4402). It captures the terminology of the first 50 years of space activities including manned and unmanned missions. The volume includes a bibliography of source materials.

Gunston, B., ed. 2009. *Cambridge Aerospace Dictionary,* 2nd ed. Cambridge, U.K.: Cambridge University Press. Comprehensive dictionary of aerospace and aeronautical terms. Entries are concisely written, and those for acronyms often indicate which country or organization developed the term. Appendices cover such diverse topics as electromagnetic frequency bands, phonetic alphabets, U.S. military aircraft and missile designations, and a guide to civilian aircraft registration numbering schemes for countries around the world.

Mark, H., ed. 2003. *Encyclopedia of Space Science and Technology*. (2 vols.) Hoboken, NJ: John Wiley & Sons. A two-volume set made up of 80 articles covering major topics in the space sciences. Articles treat the subject matter in depth, and are intended for a technically literate audience. Numerous photos, illustrations, diagrams, and a bibliography accompany the articles.

Multilingual Aeronautical Dictionary. 1980. Neuilly sur Seine, France: Advisory Group for Aerospace Research and Development. A multilingual dictionary of major aeronautics and aerospace terminology. Ten languages are represented: English, French, Dutch, German, Greek, Italian, Portuguese, Turkish, Spanish, and Russian. The core is comprised of English terms and their definitions; alphabetic indexes of terms in other languages refer back to the English section.

Tomsic, J. L., and C. N. Eastlake. eds. 1998. *SAE Dictionary of Aerospace Engineering,* 2nd ed. Warrendale, PA: Society of Automotive Engineers. Includes 20,000 terms from aeronautics, aerospace, astronomy, geophysics, and computing. Intended for anyone interested in aerospace engineering as a student, educator, engineer, scientist, or technician. Some entries reference SAE-related standards.

Verger, F., I. Sourbes-Verger, and R. Ghirardi, eds. 2003. *Cambridge Encyclopedia of Space: Missions, Applications, and Exploration*. Cambridge, U.K.: Cambridge University Press. Arranged by chapters comprising in-depth articles. Covers all aspects of space science. Extensive graphics, tables, charts, illustrations, and photos. Sections include the space environment, orbits, satellites, launch vehicles, earth observations, telecommunications, and navigation.

Walker, P. M. B., ed. 1990. *Cambridge Air and Space Dictionary*. New York: Cambridge University Press. A subset of the *Cambridge Dictionary of Science and Technology*, containing 6000 definitions from aeronautics, astronomy, meteorology, and space science. Special articles are interspersed among the entries, presenting fuller treatments of topics. Entries occasionally include formulas, illustrations, graphs, and tables.

Williamson, M. 2010. *Cambridge Dictionary of Space Technology*. Cambridge, U.K.: Cambridge University Press. Covers all aspects of space science and technology. Concisely written with some formulas, illustrations, and photos. Many acronyms and abbreviations are included.

HANDBOOKS AND MANUALS

Engineering is a discipline known for an abundance of handbooks and manuals. There are several reasons for this: engineers often work across multiple fields in their assignments, and handbooks allow them to gather information that can be useful in such situations. Engineers involved in design projects often need numerical data and formulas to complete their work. Thus, handbooks in aeronautical and aerospace engineering may contain lengthy listings of materials property data, conversion factors, vehicle components, fluid mechanics, mathematical functions, formulas, and equations. Handbooks and manuals usually are excellent sources for background information on topics in the field.

Abbot, J. H. and A. E. van Doenhoff. 1958. *Theory of Wing Sections Including a Summary of Airfoil Data*. New York: Dover. Classic work on wing sections. Includes detailed theoretical and experimental data on most NACA airfoils.

Aerospace Structural Metals Handbook. 2004. (6 vols.) West Lafayette, IN: Purdue University. A six-volume set sponsored by the U.S. Department of Defense and prepared by CINDAS (Center for Information and Numerical Data Analysis and Synthesis) at Purdue University, which is also available in a Web version. The set provides comprehensive property data on alloys of interest to the aerospace industry. A wide variety of data is presented for each alloy, and references are provided to the original sources of the data.

AIAA Aerospace Design Engineers Guide, 5th ed. 2005. Reston, VA: American Institute of Aeronautics and Astronautics. Intended to help design engineers develop aerospace products. Concise collection of commonly used aeronautical, mechanical, and electrical engineering reference data. Covers mathematics, geometric section properties, conversion factors, structural elements, mechanical and electrical components, and aircraft and spacecraft design factors.

Avallone, E .A., T. Baumeister III, and A. M. Sadegh, eds. 2007. *Marks' Standard Handbook for Mechanical Engineers,* 11th ed. New York: McGraw-Hill. Classic reference tool for engineering. Useful materials for aerospace engineers and students. Chapters cover a broad range of topics: thermodynamics, strength of materials, fuels, materials properties, mechanics, instrumentation, and the design of machine elements. Sections often provide bibliographic references.

Chase, M. W., Jr. 1998. *NIST-JANAF Thermochemical Tables,* 4th ed. (2 vols.) New York: American Institute of Physics. Provides critically evaluated physical and chemical property data, primarily of interest to those involved in rocket propulsion and other areas of combustion. The volumes contain detailed thermochemical properties of hundreds of chemicals and compounds, and access is provided by chemical name and chemical formula indexes. Entries give various properties, enthalpy of formation, heat capacity and entropy, phase and decomposition data, and bibliographic references to the original sources of the data.

Damage Tolerant Design Handbook: A Compilation of Fracture and Crack Growth Data for High Strength Alloys. 1994. (5 vols.) West Lafayette, IN: Purdue University. A five-volume handbook sponsored by the U.S. Department of Defense and prepared by CINDAS at Purdue University. The set provides extensive property data on damage tolerant materials of interest to aeronautical and aerospace engineers.

Darrin, A. G. and B. L. O'Leary, eds. 2009. *Handbook of Space Engineering, Archaeology, and Heritage.* Boca Raton, FL: CRC Press. An examination of the historical, cultural, and archaeological aspects of space exploration and technology is presented. The chapters range from space basics, launch vehicles, and orbital debris to preserving the cultural heritage of space exploration.

Davies, M., ed. 2002. *The Standard Handbook for Aeronautical and Astronautical Engineers.* New York: McGraw-Hill. Contributions written by over 60 experts. Broad coverage of all areas of aeronautical and astronautical engineering including propulsion, structures, aerodynamics, stability and control, avionics and astrionics, aircraft systems, design, astrodynamics, spacecraft, space environment, aircraft safety and maintenance, human factors, and review materials on mathematics, fluid mechanics, electronics, and computers. Many diagrams, tables, graphs, charts, and illustrations, with an extensive index.

Johnson, R. W., ed. 1998. *Handbook of Fluid Dynamics.* Boca Raton, FL: CRC Press. Intended to help professionals new to the field as well as experts. Materials are arranged into six parts: basics, classic fluid dynamics, high-Reynolds number theories, numerical solutions, experimental methods, and applications. Each of these sections is made up of articles contributed by experts, providing comprehensive coverage of the field. Appendices cover mathematics, a table of dimensionless numbers, and properties of gases and vapors. An index is included.

Ley, W., K. Wittman, and W. Hallman, eds. 2009. *Handbook of Space Technology.* West Sussex, UK: Wiley. This work provides coverage of all aspects of spaceflight with each chapter written by an expert in the field. It includes many illustrations and photographs, and each chapter has a bibliography.

Matthews, C., ed. 2002. *Aeronautical Engineer's Data Book.* Oxford, U.K.: Butterworth Heinemann. Written for practicing engineers and students. Brings together a broad range of information in aeronautical engineering and aviation in a compact package. Chapters cover aerodynamics, flight dynamics, aircraft design and performance, fluid mechanics, and airport design.

Schetz, J. A. and A. E. Fuhs, eds. 1996. *Handbook of Fluid Dynamics and Fluid Machinery.* (3 vols.) New York: John Wiley & Sons. A three-volume work for the practicing engineer and researcher. The volumes are titled "Fundamentals of Fluid Dynamics," "Experimental and Computational Fluid Dynamics," and "Applications of Fluid Dynamics." Chapters provide in-depth information on specific areas within these fields.

Streeter, V. L., ed. 1961. *Handbook of Fluid Dynamics.* New York: McGraw-Hill. A classic handbook on fluid dynamics with contributions from distinguished experts. Written for engineers and scientists in the field. Deals with both fundamental concepts and applications. Covers fluid flow (one-dimensional, ideal, laminar, compressible, two phase, open channel, stratified), turbulence, boundary layers, sedimentation, turbomachinery, fluid transients, and magnetohydrodynamics. Includes many formulas, equations, tables, graphs, and illustrations. Each chapter has a bibliography and the volume has subject and author indexes.

Structural Alloys Handbook. 1996. (3 vols.) West Lafayette, IN: CINDAS/Purdue University. A companion reference work to *Aerospace Structural Metals Handbook,* these resources provide property data on hundreds of metals and alloys of importance to design and manufacturing.

Wise, J. A.,V. D. Hopkin, and D. J. Garland, eds. 2010. *Handbook of Aviation Human Factors,* 2nd ed., Boca Raton, FL: CRC Press. A comprehensive treatment of human factors in aviation. Chapters cover human performance, the design of controls and instruments, air traffic control, and other areas.

Yang, W. J. 2001. *Handbook of Flow Visualization,* 2nd ed. New York: Taylor & Francis. Covers techniques used to visualize flow in liquids and gasses. Includes both underlying theory and experimental applications. Techniques presented include Schlieren, shadowgraph, speckle, interferometry, light sheet, and plasma fluorescence. Numerous applications are given in subsequent chapters: medical, aerospace, wind tunnels, turbines, and indoor airflow.

MONOGRAPHS AND TEXTBOOKS

There is no single printed or online resource to consult for comprehensive listings of monographs and textbooks in aeronautical and aerospace engineering. However, collection development in this area is facilitated by an active group of commercial and society publishers. Important publishers for aero materials include John Wiley & Sons, McGraw-Hill, Cambridge University Press, Oxford University Press, Kluwer, Springer, CRC, Taylor & Francis, and Academic Press. The AIAA publishes many high-quality items every year. It issues texts aimed at the student audience as well as advanced monographs for practicing engineers and researchers.

A tool such as *Books in Print* (print or Web version) is very useful for discovering what is available. Amazon.com and similar sites are also good resources, as is *Worldcat*, which approximates a national online catalog for U.S. libraries. For an excellent listing and discussion of some of the older classics in aeronautical and aerospace engineering, consult the chapter devoted to these subjects in *Use of Engineering Literature* (Mildren, 1976).

JOURNALS

Scholarly journals form the cornerstone of technical information exchange in aerospace and aeronautical engineering. Journals serve as both an announcement vehicle for new advances and as an archival medium. The conference literature and the technical report literature also are important pathways for announcing research results in the field. Trade journals serve as a current awareness service for industry news, events, new products, and similar topics of interest to professionals. This section has representative scholarly and trade journals, with an emphasis on English language titles.

Acta Astronautica. 1974–. New York: Pergamon Press (0094–5765).

Advances in Energetic Materials and Chemical Propulsion. 2007–. Redding, CT: Begell House (1944–5563).

Advances in Space Research. 1981–. New York: Pergamon Press (0273–1177).

Aeronautical Journal. 1897–. London: Royal Aeronautical Society (0001–9240).

Aerospace America. 1963–. Reston, VA: American Institute of Aeronautics and Astronautics (0740–722X).

Aerospace International. 1974–. London: Royal Aeronautical Society (1467–5072).

AIAA Journal. 1963–. Reston, VA: American Institute of Aeronautics and Astronautics (0001–1452).

Atomization and Sprays. 1991–. New York: Taylor & Francis (1044–5110).

Automatica. 1963–. New York: Pergamon Press (0005–1098).

Aviation Week and Space Technology. 1916–. New York: McGraw-Hill (0005–2175).

Canadian Aeronautics and Space Journal. 1955–. Ottawa, Canada: Canadian Aeronautics and Space Institute (0008–2821).

Combustion and Flame. 1957–. New York: Elsevier (0010–2180).

Combustion Science and Technology. 1969–. London: Taylor & Francis 0010–2202).

Flight International. 1909–. Sutton, U.K.: Reed Business Information (0015–3710).

Flow, Turbulence, and Combustion. 1998–. Boston, MA: Kluwer (1386–6184).

Human Factors and Aerospace Safety. 2000–. Surrey, U.K.: Ashgate Publishing (1468–9456)

IEEE Transactions on Aerospace and Electronic Systems. 1965–. New York: Institute of Electrical and Electronic Engineers (0018–9251).

International Journal of Aerospace Innovations. 2009–. Essex, U.K.: Multi-Science Publishing (1757–2258).

International Journal of Chemical Kinetics. 1969–. New York: John Wiley & Sons (0538–8066).

International Journal of Heat and Fluid Flow. 1979–. New York: Elsevier (0142–727X).

International Journal of Heat and Mass Transfer. 1960–. New York: Pergamon Press (0017–9310).

International Journal of Robust and Nonlinear Control, 1991–. Chichester, U.K.: Wiley (1049–8923).

International Journal of Spray and Combustion Dynamics. 2009–. Essex, U.K.: Multi-Science Publishing (1756–8277).

Journal of Aerospace Computing, Information, and Communication. 2004–. Reston, VA: American Institute of Aeronautics and Astronautics (1542–9423).

Journal of Aircraft. 1964–. Reston, VA: American Institute of Aeronautics and Astronautics (0021–8669).

Journal of the British Interplanetary Society (JBIS). 1934–. London: British Interplanetary Society (0007–084X).

Journal of Chemical Physics. 1933–. New York: American Institute of Physics (0021–9606).

Journal of Computational Physics. 1966–. New York: Academic Press (0021–9991).

Journal of Fluid Mechanics. 1956–. London: Taylor & Francis (0022–1120).

Journal of Fluids Engineering. 1973–. New York: American Society of Mechanical Engineers (0098–2202).

Journal of Guidance, Control, and Dynamics. 1978–. Reston, VA: American Institute of Aeronautics and Astronautics (0731–5090).

Journal of Heat Transfer. 1959–. New York: American Society of Mechanical Engineers (0022–1481).

Journal of Intelligent Material Systems and Structures. 1990–. Lancaster, PA: Technomic Publishing (1045–389x).

Journal of Physical Chemistry. 1896–. Washington, D.C.: American Chemical Society (0022–3654).

Journal of Propulsion and Power. 1985–. Reston, VA: American Institute of Aeronautics and Astronautics (0748–4658).

Journal of Quantitative Spectroscopy and Radiative Transfer. 1961–. New York: Pergamon Press (0022–4073).

Journal of Spacecraft and Rockets. 1964–. Reston, VA: American Institute of Aeronautics and Astronautics (0022–4650).

Journal of the American Helicopter Society. 1956–. Washington, D.C.: American Helicopter Society (0002–8711).

Journal of the Astronautical Sciences. 1954–. New York: American Astronautical Society (0021–9142).

Journal of Thermophysics and Heat Transfer. 1987–. Reston, VA: American Institute of Aeronautics and Astronautics (0887–8722).

Journal of Turbomachinery. 1986–. New York: American Society of Mechanical Engineers (0889–504x).

Physics of Fluids. 1958–. New York: American Institute of Physics (1070–6631).

Proceedings of the Institution of Mechanical Engineers, Part G: Journal of Aerospace Engineering. 1989–. London: Mechanical Engineering Publications Limited (0954–4100).

Progress in Aerospace Sciences. 1961–. New York: Pergamon Press (0376–0421).

Progress in Computational Fluid Dynamics. 2001–. Geneva, Switzerland: Inderscience Publishers (1468–4349).

Propellants, Explosives, Pyrotechnics. 1982–. Weinheim, Germany: Verlag Chemie (0721–3115).

Smart Materials and Structures, 1992–. New York: Institute of Physics Publishing (0964–1726).

Spaceflight. 1956–. London: British Interplanetary Society (0038–6340).

Vertiflite. 1963–. New York: American Helicopter Society (0042–4455).

STANDARDS AND PATENTS

Designing a modern aircraft or space vehicle is an extremely complicated and expensive process. A commercial airliner may contain hundreds of thousands of components, manufactured by thousands of companies around the world. Engineering standards help ensure the quality, fit, and performance of the materials and components used in these complex products. Patents also serve as useful resources in aeronautical and aerospace engineering. Web-based resources have made it easier to locate and order these categories of materials.

Acquisitions Streamlining and Standardization Information System (ASSIST) – Quick Search, https://assist.daps.dla.mil/quicksearch/ (accessed September 5, 2010). Washington, D.C.: U.S. Department of Defense. Much of the work in the aeronautical and aerospace industries is done under U.S. Department of Defense (DOD) contracts. ASSIST was developed by the DOD to facilitate access to unrestricted DOD and government agency standards. Many of the active standards in ASSIST are available as PDF downloads.

American National Standards Institute (ANSI), http://www.ansi.org (accessed September 5, 2010). Washington, D.C.: American National Standards Institute. The American National Standards Institute serves as the main administrative and coordinating organization for the development and production of industry standards in the United States. Companies, organizations, agencies, and individuals are eligible for ANSI membership and can participate in the development of standards. These standards cover all areas of engineering practice, materials, safety, and procedures.

NSSN: A National Resource for Global Standards, http://www.nssn.org (accessed September 5, 2010). Washington, D.C.: American National Standards Institute. NSSN serves the engineering community as a one-stop location for identifying and ordering standards. The site is a cooperative effort between ANSI (American National Standards Institute), standards organizations in the United States and other countries, and government agencies. Standards can be located by keyword or document number, and the site provides ordering information.

Society of Automotive Engineers (SAE), http://www.sae.org (accessed September 5, 2010). The SAE is an active developer of standards in aerospace engineering and has many technical committees and workgroups in this area. Aerospace standards are usually issued in these series: AMS (Aerospace Material Specifications), AS (Aerospace Standards), ARP (Aerospace Recommended Practice), and AIR (Aerospace Information Report).

SEARCH ENGINES AND WEB SITES

The aerospace industry is a complex enterprise with research and production activities taking place in the commercial, military, and government sectors. There are search engines and numerous sites available for retrieving information on technical reports, parts and components, news, vendors, standards, patents, images, and companies. Several of these sites are covered in other sections of this chapter. The sites discussed here are representative of the types of additional resources that can be found on the Web.

AERADE, http://aerade.cranfield.ac.uk/ (accessed September 5, 2010). AERADE is a Web portal to information on all aspects of the worldwide aerospace industry. The site provides links to over 30,000 aerospace-related resources.

Aviation Today, http://www.aviationtoday.com (accessed July 15, 2010). This Web site covers aviation industry news and information from the commercial, business, military, and general aviation segments.

Embry-Riddle Aeronautical University, http://fusion.erau.edu/er/library/websites/rw/ (accessed July 15, 2010). The "Recommended Web sites" page on this Web site presents a wide-ranging list of sites on aeronautical and aerospace engineering topics.

The European Space Agency, http://www.esa.int/esaCP/index.html (accessed July 15, 2010). The European Space Agency, a consortium of 17 countries, provides extensive resources to space-related information and activities at their Web site.

The Google Directory: Aerospace, http://directory.google.com/Top/Science/Technology/Aerospace/ (accessed July 15, 2010). This directory offers a subject-based listing of hundreds of sites that may be useful to aerospace engineers.

JAXA: Japan Aerospace Exploration Agency, http://www.jaxa.jp (accessed July 15, 2010). JAXA has an English-language Web site that provides news and information about Japanese aeronautical and space programs.

National Aeronautics and Space Administration, http://www.nasa.gov (accessed July 15, 2010). The main NASA Web site serves as a gateway to the whole family of NASA sites for news and information on missions, technology, the NASA Centers, and space science.

Roscosmos: Russian Federal Space Agency http://www.federalspace.ru/main.php (accessed July 15, 2010). The site provides English-language coverage of news and information on Russia's space program.

SpaceRef.com, http://www.spaceref.com (accessed July 15, 2010). Worldwide news and information on space, space missions, and related technology are presented at this Web site.

Yahoo Index of Space Sciences, http://dir.yahoo.com/Science/Space/ (accessed July 15, 2010). This Yahoo page indexes hundreds of aerospace and aviation Web sites; a good place to seek information if keyword searching doesn't return appropriate sites.

SOCIETIES AND ASSOCIATIONS

Societies serve a number of important functions in engineering. They keep the membership current and informed about developments in the discipline, and they convene conferences that bring engineers together to exchange new ideas and expand networks of contacts. Their publication programs offer an outlet for research and their publications are usually considered prestigious. Committees advance the profession through the development of standards, and provide avenues for professional cooperation. Larger societies work to influence legislation and public policy at the highest levels. Trade associations in aerospace work to influence policy and promote the industry as a whole. In addition to the selected societies and associations described here, there are others of importance to the aerospace community. They include the American Society of Mechanical Engineers http://www.asme.org, Institute of Electrical and Electronics Engineers http://www.ieee.org, Acoustical Society of America http://asa.aip.org, and the Society of Automotive Engineers http://www.sae.org.

AeroSpace and Defence Industries Association of Europe (ASD), http://www.asd-europe.org/site/ (accessed August 15, 2010). A trade association dedicated to the advancement of aeronautical and space industry of Europe. The site provides news, statistical information, and publications.

Aerospace Industries Association (AIA), http://www.aia-aerospace.org/ (accessed July 20, 2010). AIA is a trade association for U.S. aerospace and defense manufacturers whose goal is to promote aviation and aerospace activities. The site has many links to statistics and other resources.

American Astronautical Society (AAS,) http://astronautical.org (accessed July 20, 2010). Professional society dedicated to advancing the space sciences, exploration, and education. AAS sponsors professional meetings, and has an active publishing program. Key objectives of the society include assessing public and private space programs, providing guidance to space planning efforts, and promoting research in the various sciences required for the exploration and utilization of space.

AHS International, http://www.vtol.org (accessed July 20, 2010). The American Helicopter Society was founded in 1943 and has dedicated itself to the advancement of the international rotorcraft industry, both military and civilian. The society conducts a variety of professional and technical meetings and publishes journals and proceedings.

American Institute of Aeronautics and Astronautics (AIAA), http://www.aiaa.org (accessed July 20, 2010). The world's largest professional society devoted to advancing engineering and science in aviation, space, and defense. AIAA has an extensive array of committees that concentrate on both technical and member-support activities. The AIAA publications program is extensive; they publish seven scholarly journals and hundreds of technical papers annually.

Association for Unmanned Vehicle Systems International (AUVSI), http://www.auvsi.org/AUVSI/AUVSI/Home/ (accessed July 20, 2010). This is a society of individuals and companies involved in the development and promotion of unmanned aerial vehicles (UAVs). The Web site has news and useful links for those in the field.

International Astronautical Federation (IAF), http://www.iafastro.com/ (accessed July 20, 2010). The IAF is a professional group whose membership is comprised of corporations, educational institutions, space agencies, research institutes, and other organizations. Its role is to advance the knowledge and development of space for the benefit of society. It regularly convenes technical and professional meetings and publishes papers presented at the various IAF congresses.

International Scientific and Technical Gliding Organisation (OSTIV), http://www.ostiv.fai.org (accessed July 20, 2010). OSTIV is a society created to encourage the "science and technology of soaring and development and use of the sailplane in pure and applied research" (from the Web page). The group holds a congress at each World Gliding Championship and publishes the papers presented at these meetings.

Royal Aeronautical Society (RAeS), http://www.raes.org.uk/ (accessed July 20, 2010). A well-known society with members in over 100 countries. Membership is open to individuals and companies. The society's goal is the advancement of the global aerospace community. The RAeS publishes a number of journals and conference proceedings.

CONCLUSION

The aeronautical and aerospace industries continue to expand in many countries around the world, driven by commercial and defense-related demand for aircraft, missiles, and satellites, and for their support services and systems. The literature of the field is broad and centered around journal articles, technical reports, and conference papers. Most of the information is publicly available, but a portion falls under various distribution controls due to its militarily sensitive or proprietary nature. The distribution mechanisms and the databases covering aerospace and aeronautical information are well established and provide engineers working in these fields with ready access to the data they require.

REFERENCES

Anderson, J. D., Jr. 1997. *A history of aerodynamics and its impact on flying machines.* Cambridge, U.K.: Cambridge University Press.

McCormick, B., C. Newberry, and E. Jumper. 2004. *Aerospace engineering education during the first century of flight.* Reston, VA: American Institute of Aeronautics and Astronautics.

Mildren, K. W. 1976. *Use of engineering literature.* London: Butterworths.

4 Agricultural and Food Engineering

Kathy Fescemyer and Helen Smith

CONTENTS

Introduction 61
Searching the Library Catalog 62
Abstracts and Indexes 63
Bibliographies and Guides to the Literature 65
Directories 66
Encyclopedias and Dictionaries 67
Handbooks and Manuals 68
Monographs and Textbooks 69
Journals 69
Standards 76
Associations, Organizations, Societies, and Conferences 76
Summary 78
References 78

INTRODUCTION

Agricultural engineering is a multidisciplinary profession that relies on expertise in both the engineering and the agricultural fields. The roots of agricultural engineering go back to the earliest civilizations with the origin of the hoe, early irrigation systems, and other early farming methods. Over time, advances in agricultural machinery resulted in more efficient crop production; irrigation and drainage developments produced more usable land resources, while conserving natural resources; advances in agricultural buildings resulted in a healthier environment for livestock and increased efficiencies in crop storage and production; and the use of electricity on the farm made possible the automation of many farm processes (Isaacs, 2003). Food engineers were also developing advances in food processing to increase the safety of foods with cost efficient technology. Research in the heating, refrigeration, drying, chemical preservation, and packaging of food has advanced the development of these areas (Farkas, 2003). The importance of alternative energy has initiated numerous studies in the area of biofuels and bioenergy. Research has also shifted toward the molecular level with interdisciplinary work between engineers and biologists, geneticists and nanotechnologies (Ochs and Patterson, 2002).

Prior to 1907, engineering for agriculture was done by mechanical, architectural, electrical, and civil engineers. As a profession, formalized agricultural engineering began with the initiation of the American Society of Agricultural Engineers in 1907. The scope of agricultural engineering has grown and evolved over time. The society is now the American Society of Agricultural and Biological Engineers (ASABE) and defines the profession as follows: "Agricultural, Food, and Biological Engineers develop efficient and environmentally sensitive methods of producing food, fiber, timber, and renewable energy sources for an ever-increasing world population" (ASABE,

2010). The ASABE has 10 specialty areas that show the depth of their interest. These areas include biological engineering; food and process engineering; information and electrical technologies; power and machinery; soil and water; structures and environment; ergonomics, safety, and health; forest engineering; aquacultural engineering; energy; and education.

Many basic engineering information resources, such as *Compendex* and *Perry's Chemical Engineers' Handbook* are essential to the study of agricultural engineering. These basic resources are described in other chapters. The chapters in this book on mechanical, electrical, civil, and architectural engineering also should be consulted for important information resources because these engineering disciplines still provide basic concepts and research that are applied to agricultural engineering. Biological engineering is a new engineering discipline that has some of its roots in agricultural engineering. Biological engineering has many variant definitions, with the common theme of integrating the biological sciences, from the molecular to the ecosystem level, with engineering principles: (Dooley, 2003–.). Because of its importance, a separate chapter on bioengineering is included in this work. This chapter encompasses the traditional areas of agricultural engineering and food engineering, and will include some broader biosystems, biotechnology, and biofuels resources that are relevant to agricultural engineering. The focus is on the information resources that will be useful to agricultural engineering librarians, upper level undergraduates, graduate students, and professionals.

SEARCHING THE LIBRARY CATALOG

The advent of keyword searching has greatly enhanced the retrieval of materials via library catalogs. However, title word searching should not be relied on to the neglect of controlled vocabulary. Controlled vocabulary searches provide a precise means of finding a specific subject in online catalogs. The following list of Library of Congress subject headings contains most of the specific agricultural and food engineering terms.

> Agricultural engineering
> Agricultural implements
> Agricultural machinery
> Agricultural mechanics
> Agricultural processing
> Agriculture safety measures
> Aquacultural engineering
> Biofuels
> Biological engineering
> Dairy engineering
> Dairy processing
> Drainage
> Electricity in agriculture
> Farm buildings
> Farm engines
> Farm equipment
> Feed processing
> Food industry and trade
> Food packaging
> Food processing industry
> Forestry engineering
> Horticultural machinery
> Irrigation engineering
> Livestock housing

Postharvest technology of crops
Seeds processing
Tractors

In addition to these terms, combining individual types or groups of plants or livestock with subheadings, such as: Housing, Equipment and supplies, Processing, or Postharvest technology e.g., "Swine—equipment and supplies" or "Corn—processing" or "Vegetables—postharvest technology" will locate relevant publications.

The following Library of Congress call number ranges cover most of the specifically agricultural engineering material, but, again, materials in other basic engineering or agricultural areas should not be ignored.

S671-S790 (agricultural engineering, equipment, machinery, buildings)
TJ1480-TJ1500 (agricultural machinery)
TK4018 (electricity in agriculture)
TP370-465 (food technology, manufacturing and processing)

ABSTRACTS AND INDEXES

The general engineering indexes included in Chapter 2 should always be consulted when searching for agricultural and food engineering information. In addition, depending on the subject of the search in question, indexes from other chapters should be consulted as well. The following resources focus specifically on agricultural and food engineering, or are basic agriculture or biological resources and should not be ignored. All these resources are available in electronic format and more information is provided at the Web sites listed.

AGRICOLA. 1970–. Beltsville, MD: National Agricultural Library, http://agricola.nal.usda. gov/ (accessed September 29, 2010). This is a core agriculture database from the National Agricultural Library that indexes journal articles as well as books, book chapters, USDA, State Experiment Station, and State Extension service publications. Agriculture is defined broadly and includes animal and plant science, entomology, agronomy, horticulture, rural sociology, agricultural economics, family living, food and nutrition, and agricultural and biosystems engineering. This database is available from many vendors as well as directly from the National Agricultural Library via a public Web access version (see above Web address).

Agricultural Engineering Abstracts/CAB Abstracts. 1976–. Wallingford, Oxon, U.K.: CAB International, http://www.cabi.org (accessed September 29, 2010). The Agricultural Engineering Abstracts database is available as a stand-alone product (online or in print) or is included in the multidisciplinary CAB Abstracts online product. It indexes all aspects of internationally published research on agricultural engineering. CAB Abstracts is a premier index to the agricultural literature, covering everything from production agriculture to nutrition and economics. Biofuels Abstracts (biofuels and bioenergy research) and Irrigation and Drainage Abstracts (all aspects of water resource management, soil water, crop irrigation, and resulting environmental aspects) are also available as stand-alone products or within CAB Abstracts. Some overlap does occur between the subsets. CAB Abstracts is available from many vendors or directly from CABI Publishing, Agricultural Engineering Abstracts, Biofuels Abstracts, and Irrigation and Drainage Abstracts are available directly from CABI Publishing.

Agricultural Engineering Index. 1950–1999. http://bae.engineering.ucdavis.edu/pages/professionals/AgIndex/aeindex.html (accessed September 29, 2010). William J. Chancellor, Compiler. This is a freely downloadable database, available courtesy of UC Davis, which includes all technical articles appearing in ASAE periodicals from 1950 to 1999 (plus some articles appearing in publications from other international societies). Searching software is included in the download. It is no longer being updated and is included here for its historical aspects.

AGRIS. 1975–. Rome: Food and Agricultural Organization of the United Nations, http://agris. fao.org/ (accessed September 29, 2010). *AGRIS* is the international information system for the agricultural sciences and technology. Participating countries input references to the literature that are produced within their boundaries. Material includes unique grey literature, such as unpublished scientific and technical reports, theses, conference papers, government publications, and more. The system identifies literature dealing with all aspects of agriculture, and covers agricultural engineering as a part of that. All searching and viewing is free. *AGRIS* today is part of the CIARD (Coherence in Information for Agricultural Research for Development) initiative, in which the CGIAR (Consultative Group on International Agricultural Research), GFAR (Global Forum on Agricultural Research), and FAO (Food and Agriculture Organization of the United Nations) collaborate to create a community for efficient knowledge sharing in agricultural research and development. All searching and viewing is free.

ASABE Technical Library. 2001–. St. Joseph, MI: American Society of Agricultural and Biological Engineers, http://asae.frymulti.com/ (accessed September 29, 2010). This is the online publication site of the American Society of Agricultural and Biological Engineers and is a core resource for agricultural and biosystems engineering information. It includes the full text of all ASABE documents (journals, conference proceedings, monographs, standards) published since 2001 with older information being added. Most of the journal content is covered from vol. 1, and the *Transactions of the ASAE* back to at least vol. 25 in 1982. The site is searchable or you can browse the table of contents of the publications. Searching and viewing the abstracts is free. Access to the full text is via ASABE membership or site license.

Biological & Agricultural Index. 1916–. New York: H. W. Wilson Co., http://www.hwwilson.com/ (accessed September 29, 2010). Although no longer being published in print (last volume was 2003–2004), it is available online as *Biological & Agricultural Index Plus*. This index cites articles from more than 380 English-language periodicals published in the United States and elsewhere with over 100 titles available as full text. Periodical coverage includes a wide range of scientific journals, from popular to professional, that pertain to biology and agriculture. Although not suitable for extensive research in any area, this database does cover the core journals in all of agriculture and can be used as a starting point. Online coverage is from 1985 to the present and it is available from H. W. Wilson Co.

BIOSIS Previews. 1969–. Philadelphia: Thompson Scientific, http://science.thomsonreuters.com/ (accessed September 29, 2010). *BIOSIS* is the premier database for the biological sciences and includes information on biochemistry, microbiology, human biology, physiology, botany, and zoology. Biosystems and biological engineering are covered as part of the biological sciences. The subset of *Biological Abstracts* includes journal article information, and *Biological Abstracts/RRM (Reports, Reviews, Meetings)* includes conference proceedings and meeting reports, book contents, and patents. With the back files, online coverage is from 1926–present and it is available via many vendors.

FSTA: Food Science & Technology Abstracts. 1969–. Reading, U.K.: International Food Information Service, http://www.foodsciencecentral.com/ (accessed September 29, 2010). This database contains comprehensive coverage of all aspects of food science and technology research, including: raw materials and ingredients; manufacturing and distribution; food safety; and product development and consumer issues. Food engineering is covered extensively from harvesting to processing and packaging technology. It is available via many vendors.

National Ag Safety Database. 1994–. Atlanta: United States Centers for Disease Control and Prevention, http://www.nasdonline.org/ (accessed September 29, 2010). This is a full-text, consumer-oriented database that seeks to provide a national resource for the dissemination of information; to educate workers and managers about occupational hazards associated with agriculture-related injuries, deaths, and illnesses; to provide prevention information; to promote the consideration of safety and health issues in agricultural operations; and to provide a convenient way for members of the agricultural safety and health community to share educational and research materials with their colleagues. Safety professionals and organizations from across the nation have contributed the information contained in NASD. Only current information is included, but there is no charge for searching or viewing.

BIBLIOGRAPHIES AND GUIDES TO THE LITERATURE

Included in this section are publications that provide information on the literature of the disciplines of agricultural engineering and food science. Many of them are useful for providing an historical overview to the literature of agricultural engineering or for verifying incomplete or inaccurate references obtained from other sources.

American Society of Agricultural Engineers. 1907/60–1986/90. *Agricultural Engineering Index.* St. Joseph, MI.: American Society of Agricultural Engineers. This five-volume set provides a complete bibliographic overview of the field of agricultural engineering. The entries are arranged alphabetically by subject.

American Society of Agricultural Engineers. 1985–1997. *Comprehensive Index of Publications.* St. Joseph, MI : American Society of Agricultural Engineers. This publication, and its predecessors *Comprehensive Index of ASAE Publications* (1979–1984) and *Comprehensive Keyword Index of ASAE Publications* (1971–1978) provides keyword and author indexes to ASAE publications and a select group of other agricultural engineering publications.

Brogdon, J. and W. C. Olsen. 1995. *The Contemporary and Historical Literature of Food Science and Human Nutrition.* Ithaca and London: Cornell University Press. Brogdon and Olsen provide quality historical background in the area of food science. In addition, there are lists of core monographs, primary journals, databases, and primary historical literature.

Cloud, G. S. 1985. *Selective Guide to Literature on Agricultural Engineering.* College Station, TX: American Society for Engineering Education, Engineering Libraries Division. This guide is a selective annotated list of agricultural engineering sources for researchers.

Green, S. 1985. *Keyguide to Information Sources in Food Science and Technology.* London and New York: Mansell Publishing Limited. This resource provides a complete overview in the area of food science and technology.

Hall, C. W. 1976. *Bibliography of Agricultural Engineering Books.* St. Joseph, MI: American Society of Agricultural Engineers. This book provides a complete listing of books focusing on agricultural engineering published before 1976.

Hall, C. W. 1976. *Bibliography of Bibliographies of Agricultural Engineering and Related Subjects.* St. Joseph, MI: American Society of Agricultural Engineers. This is a list of the bibliographies of agricultural engineering published prior to 1976.

Hall, C. W. and W. C. Olsen, eds. 1992. *The Literature of Agricultural Engineering.* Ithaca, NY: Cornell University Press. This volume surveys the traditional subjects, such as power and machinery, soil and water, structures and environment, and electric power and processing. Subject specialists provide essays describing the literature of agricultural, food, forest, and aquacultural engineering. The volume also supplies core lists of monographs, primary journals, and historical literature that made important contributions to the subject.

Hutchinson, B. S. and A. P. Greider, eds. 2002. *Using the Agricultural, Environmental, and Food Literature.* New York: Marcel Dekker. The most recently published information guide in agriculture. Chapters on agricultural engineering and food science review the most important sources in the discipline.

Morgan, B. 1985. *Keyguide to Information Sources in Agricultural Engineering.* London and New York: Mansell. Despite being 20 years old, this volume provides the complete overview of the literature of agricultural engineering except for Internet resources. Included is information on general searching of the literature, language problems, library classification, journals, directories, handbooks, monographs, and information about organizations that focus on agricultural engineering.

DIRECTORIES

Many directories are now available on the Internet and, as such, are more up to date than printed resources. Institutions or organizations can often be found via the Internet; however, some societies limit access to their membership directories to members only. The educational and equipment directories listed here may facilitate searching for addresses or contact information.

ASABE Membership Roster http://www.asabe.org (accessed September 29, 2010). American Society of Agricultural and Biological Engineers. Available only to ASABE members. From the site: "Annual listing of your colleagues by name, geographic area and employer as well as contacts for technical committees and local sections."

Directory of Universities and Approved Programs http://www.ift.org/cms/?pid=1000426 (accessed September 29, 2010). Institute of Food Technologists. The universities on this list offer undergraduate programs that "meet the IFT Undergraduate Education Standards for Degrees in Food Science" (from the Web site). Includes United States, Canadian, and Mexican institutions.

Educational Programs in Agricultural and Biological Engineering and Related Fields—United States, Canada, and Ireland http://www.asabe.org/membership/students/edprogrm.htm (accessed September 29, 2010). American Society of Agricultural and Biological Engineers. The ASABE provides this list of United States and Canadian educational programs and includes contact details for further information.

Food Master http://www.foodmaster.com/ (accessed September 29, 2010). BNP Media. This is an online database listing ingredients and equipment for food and beverage manufacturers.

Graduate Program Directory http://www.ift.org/cms/?pid=1000624 (accessed September 29, 2010). Institute of Food Technologists. The Institute of Food Technologists' listing of United States and Canadian graduate programs in Food Science.

Guide to Consultants http://www.asabe.org/resource/guideforms.html (accessed September 29, 2010). American Society of Agricultural and Biological Engineers. The ASABE provides this guide as a service to businesses and consumers interested in agricultural, food, or biological systems engineering assistance.

International Academic Programs Agricultural, Food, or Biological Engineering Departments http://www.asabe.org/membership/students/intlacademic.html (accessed September 29, 2010). American Society of Agricultural and Biological Engineers. The ASABE provides this list of links to international and U.S. educational programs and departments.

International List of Agricultural and/or Biological Engineering Societies http://www.asabe.org/membership/international.html (accessed September 29, 2010). American Society of Agricultural and Biological Engineers. The ASABE provides this directory of other Agricultural and Biological Engineering societies. It includes related professional engineering and agricultural science societies as well.

ThomasNet http://www.thomasnet.com/ (accessed September 29, 2010). Thomas Publishing Company. ThomasNet is a directory of manufacturers, distributors, and service providers in all categories. This is the online iteration of the *Thomas Register of American Manufacturers.*

Worldwide Agricultural Machinery and Equipment Directory http://www.agmachine.com/ (accessed September 29, 2010). Agmachine.com Limited. This resource is a large specialized directory of agricultural machinery and farm equipment manufacturers on the Internet. Categories are browsable, but there is no search feature.

ENCYCLOPEDIAS AND DICTIONARIES

Encyclopedias and dictionaries provide excellent basic definitions and descriptions of terms in agricultural and food engineering. The following dictionaries and encyclopedias are good sources, but none are comprehensive. One may need to consult several dictionaries to locate the definition for a specific term because all define a very different set of terms. The encyclopedias also are varied and provide different types of information. Although not listed here, older resources also are valuable in providing historical terminology and descriptions.

Arntzen, C. J. and E. M. Ritter, eds. 1994. *Encyclopedia of Agricultural Science.* (4 vols.) San Diego, CA: Academic Press. This encyclopedia provides articles in all areas of agriculture. Articles contain an outline, a glossary, cross-references to other articles in the same encyclopedia, and a bibliography.

Bains, W. 2004. *Biotechnology from A to Z,* 3rd ed. Oxford: Oxford University Press. The entries in this dictionary provide a quick description of the concept, describe related terms, and give an indication of achievements accomplished with the technology.

Bender, D. A. and A. E. Bender. 2006. *Benders' Dictionary of Nutrition and Food Technology,* 8th ed. Boca Raton, FL: CRC Press. Online: Knovel, ebrary, NetLibrary, Credo Reference. A resource of over 5,000 terms, this dictionary provides definitions for foods and food-related terms.

Caballero, B., L. Trugo, and P. Finglas. 2003. *Encyclopedia of Food Sciences and Nutrition,* 2nd ed. (10 vols.) Amsterdam: Academic Press. Online: ScienceDirect. A comprehensive 10-volume encyclopedia that covers all aspects of food and food technology, each entry is extensive, illustrated, and contains a broad list of references for further reading.

Considine, D. M. and G. D. Considine. 1982. *Foods and Food Production Encyclopedia.* (2 vols.) New York: Van Nostrand Reinhold. Though older, this encyclopedia contains information on the cultivation of food plants and livestock and the processing of the resulting food materials into refined products.

Farrall, A. W. and J. A. Basselm. 1979. *Dictionary of Agricultural and Food Engineering,* 2nd ed. Danville, IL: Interstate. An older dictionary for terms in agriculture engineering and food engineering.

Flickinger, M. C. and S. W. Drew. 2010. *Encyclopedia of Industrial Biotechnology: Bioprocess, Bioseparation, and Cell Technology.* (7 vols.) New York: John Wiley & Sons. Online: Knovel, Wiley Online Library. This new edition is based on the *Encyclopedia of Bioprocess Technology* (Flickinger and Drew, 1999) and the *Encyclopedia of Cell Technology* (Spier, 2000). The encyclopedia covers all aspects, both theoretical and practical, of industrial biological processes, techniques, and equipment.

Francis, F. J. 2000. *Encyclopedia of Food Science and Technology,* 2nd ed. (4 vols.) New York: John Wiley & Sons. Online: Knovel. This set provides articles in all areas of food science. Each entry provides a detailed description of the topic with graphs and illustrations, and a lengthy bibliography.

Hall, C. W., A. W. Farrall, and A. L. Rippen. 1986. *Encyclopedia of Food Engineering,* 2nd ed. Westport, CT: AVI Publishing. Although dated, this encyclopedia provides detailed articles with emphasis on the equipment and facilities used in food handling, manufacture, and transportation.

Heldman, D. R., ed. 2003. *Encyclopedia of Agricultural, Food and Biological Engineering*. New York: Marcel Dekker. Online: Agropedia. This encyclopedia focuses on the processes that produce raw agricultural materials and convert them in products for distribution. Each entry contains a detailed description of the topic, several illustrations, and relevant references.

International Food Information Service. 2009. *Dictionary of Food Science and Technology*, 2nd ed. Chichester, U.K.: Wiley-Blackwell; Reading, U.K.: International Food Information Service. Online: Knovel. This dictionary defines over 8,000 terms specific to food science and technology and related subjects.

Lewis, R. A. 2002. *CRC Dictionary of Agricultural Sciences*. Boca Raton FL: CRC Press. This is a specialized agricultural dictionary that defines many agricultural terms.

Philippsborn, H. E. 2002. *Elsevier's Dictionary of Nutrition and Food Processing in English, German, French and Portuguese*. Amsterdam: Elsevier. This dictionary provides translations from English to three other languages and provides indexes of German, French, and Portuguese words translated back to English.

Trimble, S. W., B. A. Stewart, and T. A. Howell. 2008. *Encyclopedia of Water Science*, 2nd ed. Boca Raton, FL: CRC Press. Online: Agropedia. This one volume encyclopedia focuses on agricultural water management. Topics include aquifers, drainage, erosion, evaporation, groundwater, irrigation, precipitation, soil water, surface water, and many others.

Tosheva, T., M. Djarova, and B. Deliiska. 2000. *Elsevier's Dictionary of Agriculture in English, German, French, Russian and Latin*. Amsterdam; New York: Elsevier. This dictionary contains 9,389 terms and over 4,000 cross references with indexes in each language.

Troeh, F. R. and R. L. Donahue. 2003. *Dictionary of Agricultural and Environmental Science*. Ames, IA: Iowa State Press. This dictionary emphasizes the terminology of the ecological aspects of agriculture.

Van der Leeden, F., F. L. Troise, and D. K. Todd. 1990. *The Water Encyclopedia*, 2nd ed. Chelsea, MI: Lewis Publishers. This book provides over 600 tables of information about the hydrologic environments. Tables include information on pollution, contamination of surface and groundwater, use of pesticides and fertilizers, waste disposal, water treatment, and other topics.

HANDBOOKS AND MANUALS

Brown, R. H. 1988. *CRC Handbook of Engineering in Agriculture*. Boca Raton, FL.: CRC Press, vol 1. Crop production engineering, vol. 2. Soil and water engineering, vol. 3. Environmental systems engineering. Although somewhat outdated, this handbook includes basic information on most agricultural engineering processes.

CIGR, the International Commission of Agricultural Engineering. *CIGR Handbook of Agricultural Engineering*. 1999. St. Joseph, MI: American Society of Agricultural Engineers, vol. 1. Land and water engineering, vol. 2. Animal production and aquacultural engineering, vol. 3. Plant production engineering, vol. 4. Agro-processing engineering, vol. 5. Energy and biomass engineering. This handbook covers all the major fields of agricultural engineering and although it is lacking in some topics, it is still quite comprehensive and valuable.

Heldman, D. R. and D. B. Lund, eds. 2007. *Handbook of Food Engineering*, 2nd ed. Boca Raton, FL: CRC Press/Taylor & Francis. This handbook has recent information on food engineering processes and systems and includes some information on the properties of food and food ingredients.

Ibarz, A. and G. V. Barbosa-Cánovas. 2003. *Unit Operations in Food Engineering*. Boca Raton, FL: CRC Press. This handbook presents the basic information required to design food processes.

Nebraska Tractor Tests http://tractortestlab.unl.edu/ (accessed September29, 2010). The University of Nebraska Tractor Test Laboratory is the officially designated tractor testing station for the United States and tests tractors according to the codes of the Organization for Economic Cooperation and Development (OECD). Specifications and performance of specific tractor models are tested. Tractor test reports from 1999 to current date are available online; others may be purchased from the Web site.

Rao, M. A., S. S. H. Rizvi, and A. K. Datta. 2005. *Engineering Properties of Foods*, 3rd ed. Boca Raton, FL: Taylor & Francis/CRC Press. This handbook is a classic in describing the physical properties of foods.

Valentas, K. J., E. Rotstein, and R. P. Singh. 1997. *Handbook of Food Engineering Practice*. Boca Raton, FL: CRC Press. An overall guide on food engineering that includes specific food processing systems (freezing, drying, etc.) as well as some general information on areas such as shelf life, packaging, and food properties.

MONOGRAPHS AND TEXTBOOKS

Agricultural engineering students will be using information from a wide spectrum of subjects. They take courses in engineering mechanics, thermodynamics, and fluid mechanics as well as basic mathematics and physics courses. These are a few recent general textbooks that focus on agricultural and food engineering.

Field, H. L. and J. B. Solie. 2007. *Introduction to Agricultural Engineering Technology: A Problem Solving Approach,* 3rd ed. New York: Springer.

Singh, R. P., and D. R. Heldman. 2009. *Introduction to Food Engineering*, 4th ed. Amsterdam, Boston: Academic Press/Elsevier.

Tollner, E. W. 2002. *Natural Resources Engineering*. Ames, IA: Iowa State Press.

Other materials are available from traditional publishers, such as Elsevier, Taylor & Francis, and CRC Press. The American Society of Agricultural and Biological Engineers also publishes excellent basic and advanced books. Two organizations that publish practical information are the Natural Resource, Agriculture, and Engineering Service at http://www.nraes.org/ (accessed September 29, 2010) and the Midwest Plan Service at http://www.mwps.org/(accessed September 29, 2010). Searching library union catalogs, library catalogs at agricultural universities, or Google Books at http://books.google.com (accessed September 29, 2010) will often retrieve these items. Use of the subject terms mentioned earlier in this chapter will assist in locating monographs in agricultural engineering.

JOURNALS

Scholarly journals are the most important form of communication in agricultural engineering and its related disciplines. The titles included on this list meet the following criteria: scholarly, peer reviewed, published in English, and actively publishing. These journals must include a significant number of articles in agricultural engineering, biofuels, biosystems engineering, biotechnology, or food engineering. Due to the interdisciplinary nature of this subject, many other basic agricultural, biological, engineering, and environmental journals also may include articles of significance to researchers and will be consulted by agricultural and biological engineers.

Agricultural Engineering International: The CIGR Journal of Scientific Research and Development. 1999–. International Commission of Agricultural and Biosystems Engineering (1682–1130). Online: http://www.cigrjournal.org/index.php/Ejounral. An open access journal that emphasizes efficiency in agricultural production, the responsible use of natural resources, sustainability, value-added processing, and other agricultural engineering subjects.

Agricultural Systems. 1976–. Amsterdam: Elsevier (0308–521X). Online: ScienceDirect. Focuses on the interactions connecting components of agricultural systems and the environment, such as the development and application of systems methodology, including system modeling, ecoregional analysis of agriculture and land use, and approaches to analyzing and improving farming systems.

Agricultural Water Management. 1976–. Amsterdam: Elsevier (0378–3774). Online: Science-Direct. Covers all aspects of water management, such as irrigation and drainage, water in relation to soil properties and vegetation cover; the role of ground and surface water in nutrient cycling, water balance problems, control of flooding, erosion, water quality and pollution, effects of land uses on water resources

Agriculture, Ecosystems & Environment. 1983–. Amsterdam: Elsevier (0167–8809). Online: ScienceDirect. Focuses on the interactions of agricultural systems and the environment, including biological and physical characteristics of agroecosystems, ecology, diversity and sustainability of agricultural systems, agroecosystems and global environmental changes, such as climate change and air pollution, changes due to soil degradation, waste application, irrigation, and mitigation options.

Agronomy Journal. 1907–. St. Joseph, MI: American Society of Agronomy (0002–1962). Online: https://www.crops.org/publications/aj (1435–0645). Subjects include soil–plant relationships; crop science; soil science; biometry; crop, soil, pasture, and range management; crop, forage, and pasture production and utilization; turfgrass; agroclimatology; agronomic modeling; statistics; production agriculture; and computer software.

AMA *Agricultural Mechanization in Asia, Africa, and Latin America.* 1971–. Tokyo: Farm Machinery Industrial Research Corp. and Shin-Norinsha Co. (0084–5841). Online: http://www.shin-norin.co.jp/ama/ama.html. Listed as one of the top agricultural engineering journals in *Journal Citation Reports*, this journal continues *Agricultural Mechanization in Southeast Asia* and is produced in Japan.

Applied and Environmental Microbiology. 1953–. Washington: American Society for Microbiology (0099–2240). Online: PubMed Central; Highwire. This journal focuses on biotechnology, microbial ecology, food microbiology, and industrial microbiology and highlight research in the development of new processes or products,

Applied Engineering in Agriculture. 1985–. St. Joseph, MI: American Society of Agricultural and Biological Engineers (0883–8542). Online: ASABE Technical Library. Covers all facets of agricultural engineering including irrigation and drainage, agricultural machinery, food processing, energy usage, storage and handling, and electronics.

Appropriate Technology. 1974–. Burnham, UK: Research Information Ltd. (0305–0920). Focusing on the developing world especially Africa, topics emphasize creating and using sustainable techniques for practical solutions to agricultural challenges.

Aquacultural Engineering. 1982–. Amsterdam: Elsevier (0144–8609). Online: ScienceDirect. Articles focus on the development, design engineering and operation of aquaculture facilities.

Biodegradation. 1990–. Dodrecht: Springer Netherlands (0923–9820) (Print). Online: SpringerLink (1572–9729). Publishes research pertaining to biotransformation, mineralization, detoxification, recycling, amelioration or treatment of chemicals or waste materials by naturally occurring microbial strains or associations or recombinant organisms.

BioEnergy Research. 2008–. New York: Springer New York LLC (1939–1234). Online: SpringerLink (1939–1242). This new journal publishes in the area of feedstock biology research related to biomass, biofuels, and bioenergy production.

Biofuels. 2010–. London: Future Science (1759–7269). Online: http://www.future-science.com/loi/bfs. New in 2010, this journal publishes articles on research on the topics of biofuels, biomass, and bioenergy production.

Biofuels, Bioproducts, and Biorefining. 2007–. Hoboken, NJ: Wiley–Blackwell (1932–104X). Online: Wiley Online Library (1932–1031). Topics include feedstock design and production, biomass treatment and conversion, separation and process technology of biomass-e derived fuels, biorefinery research, and industrial development of biomass products.

Biological Engineering. 2008–. St. Joseph, MI: American Society of Agricultural and Biological Engineers (1934–2799). ASABE Technical Library (1934–2837). Publishes research on bioprocessing, bioreactor design, tissue engineering, metabolic engineering, biosensors, bioinstrumentation, bioenvironmental systems, bioremediation, modeling of biological processes, bioconversion, and biofuels.

Biomass & Bioenergy. 1991–. Amsterdam: Elsevier (0961–9534). Online: ScienceDirect. Concentrates on biomass, biological residues, bioenergy processes, bioenergy utilization, and biomass and the environment.

Bioprocess and Biosystems Engineering. 1986–. Heidelberg: Berlin/Heidelberg: Springer (1615–7591). Online: SpringerLink (1615–7605). Focuses on multidisciplinary approaches for integrative bioprocess design, such as the rational manipulation of biosystems through metabolic engineering techniques to provide new biocatalysts and model-based design of bioprocesses, such as upstream processing, bioreactor operation, and downstream processing (from the Web page).

Bioresource Technology. 1979–. Amsterdam: Elsevier (0960–8524). Online: ScienceDirect. Topics include biofuels production, modeling and economics; bioprocesses and bioproducts biocatalysis and fermentations; biomass and feedstocks utilization; and thermal conversion of biomass: combustion, pyrolysis, gasification, catalysis.

Biosensors & Bioelectronics. 1985–. Amsterdam: Elsevier (0956–5663). Online: ScienceDirect. Studies the research, design, development, and application of biosensors and bioelectronics.

Biosystems Engineering. 1956–. Amsterdam: Elsevier (1537–5110). Online: ScienceDirect. Continues the *Journal of Agricultural Engineering Research* and publishes research in engineering and the physical sciences that represent advances in understanding or modeling of the performance of biological systems for sustainable development in land use and the environment, bioproduction processes, and the food chain.

Biotechnology and Bioengineering. 1959–. Hobeken, NJ: Wiley–Blackwell (0006–3592). Online: Wiley Online Library (1097–0290). Focuses on all aspects of applied biotechnology including cellular physiology, metabolism and energetics of cells, biocatalysis, bioseparation, biothermodynamics biofuels, biomaterials, bioprocess engineering, biosensors, plant cell biotechnology, biological aspects of biomass, and renewable resources engineering and food biotechnology.

Biotechnology for Biofuels. 2007–. London: BioMed Central Ltd. (1754–6834). Online: http://www.biotechnologyforbiofuels.com/. An open access online journal publishing research on advances in the production of biofuels from biomass, including development of plants for biofuels production, plant deconstruction, pretreatment and fractionation, enzyme production and enzymatic conversion, and fermentation and bioconversion.

Canadian Biosystems Engineering. 1959–. Saskatoon, SK: Canadian Society for Bioengineering (1492–9058) Online: Canadian Society for Engineering in Agriculture, Food, and Biological Systems http://www.engr.usask.ca/societies/csae/journal.html. Continues *Canadian Agricultural Engineering* (1959–2000) and publishes agricultural engineering research in soil and water systems, machinery systems, bioprocessing systems, biological systems, building systems, waste management, and information systems.

Compost Science & Utilization. 1993–. Emmaus, PA: JG Press, Inc. (1065–657X) Online: http://www.jgpress.com/compostscience/index.html. Focuses on management techniques to improve compost process control and product quality, and utilization of composted materials.

Computers and Electronics in Agriculture. 1985–. Amsterdam: Elsevier (0168–1699). Online: ScienceDirect. Topics include computerized decision-support aids, such as expert systems and simulation models, electronic monitoring or control of any aspect of livestock/crop production, and postharvest operations, such as drying, storage, production assessment, trimming, and dissection of plant and animal material.

Critical Reviews in Food Science and Nutrition. 1970–. Philadelphia: Taylor & Francis (1040–8398). Online: informaworld (1549–7852). Includes articles on many aspects of food science including food safety, flavor chemistry, food colors, pesticides, risk assessment, effects of processing on nutrition, food labeling, and functional/bioactive foods.

Crop Science. 1961–. Madison, WI: Crop Science Society of America (0011–183X). Online at https://www.crops.org/publications/cs (1435–0653). Subjects include crop breeding and genetics; crop physiology and metabolism; crop ecology, production, and management; seed physiology, production, and technology; turfgrass science; crop ecology, management, and quality; genomics, molecular genetics, and biotechnology; plant genetics resources; and pest management.

Current Opinion in Biotechnology. 1990. Amsterdam: Elsevier (0958–1669). Online: ScienceDirect. Subjects include analytical biotechnology, plant biotechnology, food biotechnology, environmental biotechnology, systems biology, protein technologies and commercial enzymes, biochemical engineering, tissue and cell engineering, chemical biotechnology, and pharmaceutical biotechnology.

Engineering in Life Sciences. 2001–. Weinheim, Germany: Wiley-VCH Verlag GmbH. Online: Wiley Online Library (1618–2863). Merged with *Acta Biotechnologica* in 2004, this online only journal covers all technological aspects of industrial, environmental, plant, and food biotechnology.

Food Analytical Methods. 2008–. New York: Springer (1936–9751). Online: SpringerLink (1936–976X). Covers all aspects of the development, optimization, and practical implementation of food analysis, and validation of food analytical methods for the monitoring of food safety and quality.

Food and Bioprocess Technology. 2008–. New York: Springer (1935–5130). Online: SpringerLink (1935–5149). Focuses on experimental or theoretical research to help the agrifood industry to improve process efficiency, enhance product quality, and extend shelf life of fresh and processed agrifood products.

Food and Bioproducts Processing. 1991–. Amsterdam: Elsevier (0960–3085). Online: ScienceDirect. Focuses on research for the safe processing of biological products and includes biocatalysis and biotransformations, bioprocess modeling, bioseparation, fermentation, bioreactor design, food and drink process engineering, engineering for food safety, environmental issues in food manufacture, minimal processing techniques, packaging and hygienic manufacture, and product safety.

Food Biotechnology. 1987–. Philadelphia: Taylor & Francis (0890–5436). Online: Informaworld (1532–4249). Focus is on developments and applications of modern genetics as well as enzyme, cell, tissue, and organ-based biological processes to produce and improve foods, food ingredients, and functional foods including fermentation to improve foods, food ingredients, functional foods, and food waste remediation.

Food Control. 1990–. Amsterdam: Elsevier (0956–7135). Online: ScienceDirect. Covers research in all aspects of food safety, such as microbial food safety and antimicrobial systems, mycotoxins, hazard analysis, HACCP and food safety objectives, risk assessment, including microbial risk assessment, quality assurance and control, food packaging and rapid methods of analysis and detection, including sensor technology.

Food Engineering Reviews. 2009–. New York: Springer (1866–7910). Online: SpringerLink (1866–7929). Publishes reviews on all engineering aspects of the food industry, encompassing transport phenomena in food processing, food nanoscience and nanoengineering, food equipment design, plant design, modeling food processes; microbial inactivation kinetics, preservation technologies, food packaging, shelf life, storage and distribution of foods and instrumentation, control and automation.

Food Manufacturing Efficiency. 2006–. Reading, UK: I F I S Publishing (1750–2683). Online at http://www.foodsciencecentral.com/fsc/fme. (1750–2691). Presents techniques and technologies to measure and improve the efficiency of food production.

Food Science and Technology International. 1961–. London: Sage Publications (1082–0132). Online: Sage Journals Online. Subjects include food processing engineering, composition, food safety, nutritional quality, biotechnology, quality, physical properties, microstructure, microbiology, packaging, sensory analysis, bioprocessing, and postharvest technology.

Industrial Biotechnology. 2005–. New Rochelle, NY. Mary Ann Liebert, Inc. Publishers (1550–9087). Online: Mary Ann Liebert (1931–8421). Dedicated to biobased products and processes advancing sustainable, cost- and eco-efficient production of chemicals, materials, consumer goods, and alternative energy.

Industrial Crops and Products. 1992–. Amsterdam: Elsevier (0926–6690). Online: ScienceDirect. Focuses on the development, production, harvesting, storage, and processing of nonfood crops for industrial uses, including applications of pharmaceuticals, lubricants, fuels, fibers, essential oils, biologically active materials, and uses for industrial crop by-products.

Innovative Food Science and Emerging Technologies. 2000–. Amsterdam: Elsevier (1466–8564). Online: ScienceDirect. Publishes research in food chemistry, biochemistry, microbiology, technology and nutrition, and articles dealing with engineering, scale-up, safety, sustainability, kinetics, and mechanistic aspects of food processing technologies.

International Journal of Agricultural and Biological Engineering. 2008–. Beijing: Association of Overseas Chinese Agricultural, Biological and Food Engineers (AOCABFE) and Chinese Society of Agricultural Engineering (CSAE) (1934–6352). Online: (1934–6352) http://www.ijabe.org/index.php/ijabe/index. An open access journal that publishes in six areas: power and machinery systems, land and water engineering, bio-environmental engineering, information and electrical technologies, renewable energy systems, agro-product, and food processing engineering.

International Journal of Food Engineering. 2005–. Berkeley, CA: Berkeley Electronic Press (1556–3758). Online: http://www.bepress.com/ijfe/. Focuses on research related to processing foods, such as heat, mass transfer, and fluid flow in food processing; food microstructure development and characterization; application of artificial intelligence in food engineering research and in industry; food biotechnology; and mathematical modeling and software development for food processing purposes.

International Journal of Food Science and Technology. 1966–. Hoboken, NJ: Wiley–Blackwell (0950–5423). Online: Wiley Online Library. Publishes articles from pure research associated with food to practical experiments designed to improve technical processes including from raw material composition to consumer acceptance, from physical properties to food engineering practices, and from quality assurance and safety to storage, distribution, marketing, and use.

International Journal of Forest Engineering. 1989–. Fredericton, WB: Forest Products Society (FPS) and the University of New Brunswick's Electronic Text Centre (1494–2119). Online: http://www.lib.unb.ca/Texts/JFE/. Focuses on many aspects of forest operation, such as tree harvesting, processing and transportation; stand establishment, protection, and tending; operations planning and control; machine design, management, and evaluation; forest access planning and construction; human factors engineering; and education and training.

Irrigation and Drainage. 1952–. Hoboken, NJ: Wiley-Blackwell (1531–0353). Online: Wiley Online Library (1531–0361). Provides research on irrigation, drainage, and flood control primarily in agricultural systems.

Irrigation and Drainage Systems. 1986–. Dodrecht: Springer Netherlands (0168–6291). Online: SpringerLink (1573–0654). Topics included are performance assessment of irrigation and drainage systems, the interrelationship between irrigation management and system design, design criteria of drainage systems for effective control of waterlogging and salinity, the adaptation of irrigation/drainage to avoid water-related diseases, and planning and construction methods for canals and related structures.

Irrigation Science. 1978–. Heidelberg: Springer (0342–7188). Online: SpringerLink (1432–1319). Publishes articles on all aspects of irrigation including research from the plant, soil, and atmospheric sciences, and irrigation water management modeling.

Journal of Agricultural Safety and Health. 1995–. St. Joseph, MI: American Society of Agricultural and Biological Engineers (1074–7583). Online: ASABE Technical Library. Topics include treatment and prevention of trauma and illness, engineering design and application, safety and health intervention strategies, health standards, legislation and regulation, and the development of agricultural safety.

Journal of Biobased Materials and Bioenergy. 2007–. Valencia, CA: American Scientific Publishers (1556-6560). Online at http://www.aspbs.com/jbmbe/ (1556–6579). Publishes research on biobased polymers and blends, biobased composites and nanocomposites, biobased materials processing technologies, life-cycle analysis, social and environmental impacts, and biofuels.

Journal of Bioscience and Bioengineering. 1923–. Osaka, Japan: The Society for Biotechnology, Japan (1389–1723). Online: ScienceDirect. Formerly known as *Journal of Fermentation and Bioengineering*, this journal provides research concerning fermentation technology, biochemical engineering, food technology, and microbiology.

Journal of Chemical Technology and Biotechnology. 1986–. Hoboken, NJ: Wiley-Blackwell (0268–2575). Online: Wiley Online Library (1097–4660). Publishes chemical and biological technology research that aims toward economically sustainable industrial protection including water and air pollution reduction, waste treatment/management, alternative fuels/energy sources, biomass, biofeedstocks, biofuels, and fuel cells, re-use technologies, reactor and equipment design, fermentation and industrial biotechnology, biorefining, and industrial catalysis and biocatalysis.

Journal of Environmental Quality. 1972–. Madison, WI: American Society of Agronomy, Crop Science Society of America, and Soil Science Society of America (0047–2425). Online: https://www.crops.org/publications/jeq (1537–2537). Subjects included are atmospheric pollutants, biodegredation and bioremediation, ecological risk assessment, ground water quality, landscape and watershed processes, organic compounds in the environment, remote sensing and environmental degradation, surface water quality, urban pollutants, and waste management.

Journal of Food Engineering. 1982–. Amsterdam: Elsevier (0260–8774). Online: ScienceDirect. Publishes research on engineering properties of foods, food physics and physical chemistry processing, measurement, control, packaging, storage, and distribution; engineering aspects of the design and production of novel foods; design and operation of food processes, plants, and equipment; and economics of food engineering.

Journal of Food Process Engineering. 1977–. Hoboken, NJ: Wiley-Blackwell (0145–8876). Online: Wiley Online Library (1745–4530). Publishes applications of engineering principles and concepts to food and food processes with emphasis on process simulation.

Journal of Food Processing and Preservation. 1977–. Hoboken, NJ: Wiley-Blackwell (0145–8892). Online: Wiley Online Library (1745–4549). Focuses on advances in food processing and preservation, encompassing chemical, physical, quality, and engineering properties of food materials.

Journal of Food Science. 1936–. Hoboken, NJ: Wiley-Blackwell (0022–1147). Online: Wiley Online Library (1750–3841). Coverage includes food chemistry, food engineering, and physical properties, food microbiology and safety, sensory and food quality, nanoscale food science, engineering, and technology, toxicology, and chemical food safety.

Journal of Food Science and Technology. 1964–. New Delhi, India Springer (India) (0022–1155). Online: SpringerLink (0975–8402). Publishes research on all aspects of science, technology, packaging, and engineering of foods and food products with emphasis on fundamental and applied research, which have potential for enhancing product quality, extending shelf life of fresh and processed food products, and improving process efficiency.

Journal of Irrigation and Drainage Engineering. 1956–. Reston, VA: American Society of Civil Engineers (0733–9437). Online: American Society of Civil Engineers (1943–4774). Publishes on all aspects of irrigation, drainage, engineering hydrology, and related water management subjects, such as watershed management, weather modification, water quality, groundwater, and surface water.

Journal of Soil and Water Conservation. 1946–. Ankeny, IA: Soil and Water Conservation Society (0022–4561). Online: http://www.jswconline.org/. Focuses in the conservation of soil, water, and related natural resources including erosion and sediment control, floodplain management, farmland preservation, forage management, forestry, irrigation, nonpoint source pollution, rangeland management, soil science, sustainable agriculture, watershed management, wetland restoration, and wildlife management.

Landwards. 1944–. Silsoe, UK: Institution of Agricultural Engineers (1363–8300). Continues *Agricultural Engineer* and provides articles on developments in engineering and technology for practical application in agriculture, horticulture, forestry, and environment.

Packaging Technology and Science. 1988–. Hoboken, NJ: Wiley-Blackwell (0894–3214). Wiley Online Library (1099–1522). Covers all aspects of packaging including food, agricultural chemicals, medical products, pharmaceuticals, and machinery.

Postharvest Biology and Technology. 1991–. Amsterdam: Elsevier (0925–5214). Online: ScienceDirect. Covers biological and technological postharvest research including the areas of postharvest storage, treatments and underpinning mechanisms, quality evaluation, packaging, and handling and distribution of fresh food crops, but excludes research on grains and forage.

Process Biochemistry. 1966–. Amsterdam: Elsevier (1359–5113). Online: ScienceDirect. Main topics covered are fermentation, biochemical and bioreactor engineering, biotechnology processes and their life science aspects, biocatalysts, enzyme engineering and biotransformation, downstream processing, modeling, optimization, and control techniques.

Renewable Agriculture and Food Systems. 1986–. Cambridge, UK: Cambridge University Press (1742–1705). Formerly *American Journal of Alternative Agriculture.* This journal publishes on the economic, ecological, and environmental impacts of agriculture; the effective use of renewable resources and biodiversity in agro-ecosystems; and the technological and sociological implications of sustainable food systems.

Resource: Engineering & Technology for a Sustainable World. 1994–. St. Joseph, MI: American Society of Agricultural and Biological Engineers (10763333). Online: American Society of Agricultural and Biological Engineers. Continues *Agricultural Engineering* and provides information trends in technology and progress in agricultural engineering.

Sensing and Instrumentation for Food Quality and Safety. 2007–. New York: Springer (1932–7587). Online: SpringerLink. Covers research in measurement and detection methodologies and techniques for the assessment, monitoring, and control of food quality and safety including sensing principles and mechanisms; biosensors; imaging techniques and image processing; bioimaging; signal processing; classification and pattern recognition techniques for food quality and safety applications; instrumentation systems; food quality control, and food safety and security monitoring systems.

Soil & Tillage Research. 1980–. Amsterdam: Elsevier (0167–1987). Online: ScienceDirect. Incorporating *Soil Technology* in 1988, this journal focuses on changes in the physical, chemical, and biological parameters of the soil caused by soil tillage and field traffic, including the effects of tillage, irrigation, and drainage, crops and crop rotations, fertilization, erosion, runoff environmental quality, crop establishment, root development and plant growth, and the interactions between these various effects.

Soil Science Society of America Journal. 1936–. Madison, WI: Soil Science Society of America (0361–5995). Online: https://www.crops.org/publications/sssaj (1435–0661). Subjects include soil and water management and conservation, pedology, soil fertility and plant nutrition, soil mineralogy, soil physics and nutrient management, and soil and plant analysis.

Soil Use and Management. 1985–. Hoboken, NJ: Wiley-Blackwell (0266–0032). Wiley Online Library (1475–2743). Publishes research on applying scientific principles to soil problems and how they affect crop production and environmental issues. Topics included are environmental protection, soil–crop interactions, soil erosion and conservation, pollution control, restoration and reclamation of land, evaluation of soil surveys and development of methodology.

Transactions of the ASABE. 1907–. St. Joseph, MI: American Society of Agricultural and Biological Engineers (2151–0032). Online: ASABE Technical Library. Publishes research on a broad range of topics including agricultural machinery, drainage, irrigation, electronics, biological engineering, forestry, food engineering, agricultural structures, crop production, natural resources, and soils.

Trends in Food Science and Technology. 1990–. Amsterdam: Elsevier (0924–2244). Online: ScienceDirect. Provides articles on the science and technology of food analysis, development, manufacture, storage, and marketing from the molecular and microstructural level through raw material processing to food engineering, novel processing methods, automation, quality control and assurance, microbiological safety issues, advances in preservation and packaging technologies, and sensory analysis.

World Journal of Microbiology & Biotechnology. 1985–. Dodrecht: Springer Netherlands. (0959–3993). Online: SpringerLink. Continues *MIRCEN Journal of Applied Microbiology and Biotechnology* and publishes research on all aspects of applied microbiology and biotechnology, including management of culture collections, foodstuffs, and biological control agents.

STANDARDS

Engineering standards included in other chapters should be consulted when searching for standards in agricultural or food engineering information that would be based on those areas. In addition the following resource focuses specifically on agricultural engineering.

ASAE Standards. 1985–. (Annual) St. Joseph, MI: The Society. Standards, engineering practices, and data adopted by the American Society of Agricultural Engineers. Also available online in the *ASABE Technical Library* (see information in the abstracts and indexes section of this chapter) at http://asae.frymulti.com/standards.asp (accessed September 29, 2010).

ASSOCIATIONS, ORGANIZATIONS, SOCIETIES, AND CONFERENCES

Societies provide information in many forms and often have well developed Web sites. Some types of information provided are publications, meeting schedules, history of the discipline and society, and opportunities for networking. Listed below are the major societies in agricultural and food engineering. For other societies consult the *International List of Agricultural Engineering Societies,* http://www.asabe.org/membership/international.html (accessed September 29, 2010), which provides information on agricultural engineering societies throughout the world.

American Society of Agricultural and Biological Engineering (ASABE) http://www.asabe.org/ (accessed September 29, 2010). This is the core society for the United States. The ASABE "is an educational and scientific organization dedicated to the advancement of engineering applicable to agricultural, food, and biological systems. Founded in 1907 and headquartered in St. Joseph, Michigan, ASABE comprises 9,000 members in more than 100 countries" (from the Web page). The ASABE sponsors annual meetings and specialty conferences.

Canadian Agricultural Safety Association (CASA) http://www.casa-acsa.ca/ (accessed September 29, 2010). CASA was established in 1993 in response to an identified need for a national farm safety networking and coordinating agency to address problems of illness, injuries, and accidental death in farmers, their families and agricultural workers. Since then, its mission has been to improve the health and safety conditions of those that live, and/or work on Canadian farms.

Canadian Institute of Food Science and Technology (CIFST) http://www.cifst.ca/ (accessed September 29 2010). "Founded in 1951, CIFST is the national association for food industry professionals. Its membership of more than 1,500 is comprised of scientists and technologists in industry, government, and academia who are committed to advancing food science and technology" (from the Web page).

Canadian Society for Bioengineering (CSBE) http://www.bioeng.ca/ (accessed September 29, 2010). The Web site provides publications, meetings, and news for students and members.

Canadian Society for Engineering in Agricultural, Food and Biological Systems (CSAE) http://www.csae-scgr.ca/ (accessed September 21, 2004). The Web site provides publications, meetings, and news for students and members.

Council on Forest Engineering (COFE) http://www.cofe.org/ (accessed September 29, 2010). This is an international professional organization interested in matters relating to the field of forest engineering.

Institute of Biological Engineering (IBE) http://www.ibe.org/ (accessed September 29, 2010). This institute was established to encourage inquiry and interest in biological engineering in the broadest and most liberal manner and promote the professional development of its members.

Institute of Food Science and Technology http://www.ifst.org/ (accessed September 29, 2010). The Institute of Food Science and Technology is the independent incorporated professional qualifying body for food scientists and technologists.

Institute of Food Technologists http://www.ift.org/ (accessed September 29, 2010). The Institute of Food Technologists is an international not-for-profit scientific society that was founded in 1939. Its Web site provides information on publications, continuing education and professional development, awards, and government relations and policy activities.

Institution of Agricultural Engineers http://www.iagre.org/ (accessed September 29, 2010). "The Institution of Agricultural Engineers (IAgrE) is the professional body for engineers, scientists, technologists, and managers in agricultural and allied land-based industries, including forestry, food engineering and technology, amenity, renewable energy, horticulture, and the environment" (from the Web page).

International Commission of Agricultural Engineering (CIGR, Commission Internationale du Génie Rural) http://www.cigr.org/ (accessed September 29, 2010). The commission was set up during the first International Congress of Agricultural Engineering, in Liege, Belgium, in 1930. It is an international, nongovernmental, nonprofit organization that is primarily a networking system. Formerly every five years and, from 1994, every four years CIGR convenes a *World Congress of Agricultural Engineering.*

International Union of Food Science & Technology http://www.iufost.org/ (accessed September 29, 2010). The International Union of Food Science and Technology (IUFoST), a country-membership organization, is the sole global food science and technology organization. It is a voluntary, nonprofit association of national food science organizations linking the world's best food scientists and technologists. This association supports the *World Congress of Food Science and Technology* every year, and the *International Congress on Engineering and Food* every four years (from the Web page).

National Institute for Farm Safety http://nifsagsafety.org/ (accessed September 29, 2010). "NIFS is an organization dedicated to the professional development of agricultural safety and health professionals, providing national and international leadership in preventing agricultural injuries" (from the Web page).

SUMMARY

Due to the multidisciplinary nature of agricultural and food engineering, knowledge of the basic engineering information resources mentioned in other chapters of this work is extremely important. Within the strict bounds of agricultural and food engineering, the traditional journal literature is of high importance as well as the ASAE publications and conferences. There are trends toward more specializations within the traditional realms of agricultural engineering and increasing focus on the microtechnology and cellular engineering aspects. Astute students of the subject would do well to keep abreast of such changes.

REFERENCES

ASABE (The American Society of Agricultural and Biological Engineers). http://www.asabe.org/ (accessed September 22, 2010).

Dooley, J. H. 2003. Biological engineering definition. In *Encyclopedia of agricultural food and biological engineering,* ed. D. R. Heldman, 60–63. New York: Marcel Dekker.

Farkas, D. F. 2003. Food engineering history. In *Encyclopedia of agricultural food and biological engineering,* ed. D. R. Heldman, 346–349. New York: Marcel Dekker.

Flickinger, M.C. and S.W. Drew. 1999. *Encyclopedia of Bioprocess Technology: Fermentation, Biocatalysis and Bioseparation.* 5 vols. New York: Wiley.

Isaacs, G. W. 2003. Agricultural engineering history. In *Encyclopedia of agricultural food and biological engineering,* ed. D. R. Heldman, 14–17. New York: Marcel Dekker.

Ochs, M. A. and M. E. Patterson. 2002. Agricultural and biosystems engineering. In *Using the agricultural, environmental and food literature,* eds. B. S. Hutchinson and A. P. Greider, 53–74. New York: Marcel Dekker.

Spier, R.E. 2000. *Encyclopedia of Cell Technology.* 2 vols. New York: Wiley.

5 Architectural Engineering

Barbara Opar

CONTENTS

Overview...79
The Literature of Architectural Engineering..80
Searching the Library Catalog 81
Abstracts and Indexes 82
Bibliographies and Guides to the Literature 83
Directories...83
Encyclopedias and Dictionaries..84
Handbooks and Manuals...85
Monographs and Textbooks 87
Journals 88
Patents and Standards 89
Search Engines and Important Web Sites 90
Societies 91
Conclusion 92
Acknowledgments...92

OVERVIEW

"Architectural engineering is the discipline concerned with the planning, design, construction, and operation of engineered systems for commercial, industrial, and institutional facilities. Engineered systems include electric power, communications and control; lighting; heating, ventilation, and air conditioning; and structural systems. An Architectural engineer works closely with those in all areas of the building process to design and possibly to construct the engineered systems that make buildings come to life for their inhabitants" (as quoted on the November 2004 version of the Architectural Engineering Institute Web page).

To the layman, architectural engineering is less clearly defined as a field than, say, electrical engineering, where the parameters of the profession are more firmly set. Architectural engineering is a highly complex field involving advanced knowledge of both architecture and engineering. Although licensed as engineers rather than as architects, architectural engineers collaborate with architects, and, therefore, must fully understand design concepts and the architect's vision. They are responsible for building system integration and for ensuring that the mechanical and electrical systems operate properly. Their understanding of building system integration is what differentiates them from civil engineers—although some architectural engineers do specialize in certain fields, such as structures.

Architectural engineering may be seen as an outgrowth of civil engineering, but very little has been written on the history of the profession. As architectural projects have become more sophisticated and have required more technological expertise, architectural engineers, by bridging this gap with their understanding of the design process alongside engineering issues, have emerged as part of a distinct profession.

While the American Society of Civil Engineers is over 150 years old, the Architectural Engineering Institute, the major professional organization aimed at architectural engineers, was only created on October 1, 1998, as the result of a merger between the National Society of Architectural Engineers (NSAE) and the American Society of Civil Engineers Architectural Engineering Division (AED). Seventeen universities currently offer accredited programs in architectural engineering (as opposed to 123 for architecture and 221 for civil engineering). These include:

California Polytechnic State University–San Luis Obispo
Drexel University
Illinois Institute of Technology
Kansas State University
Milwaukee School of Engineering
Missouri University of Science and Technology
North Carolina A&T
Oklahoma State University
Penn State
Tennessee State University
University of Colorado at Boulder
University of Kansas
University of Miami
University of Texas at Austin
University of Nebraska–Lincoln
University of Oklahoma
University of Wyoming

More graduate programs are emerging as well. Architectural engineering is certainly likely to grow in importance as new building systems evolve, as technology becomes more complicated, and as we search for ways to conserve our built environment. The principles of sustainable design are leading architectural engineers to find new ways to become more environmentally responsible through resource efficiency while providing better occupant comfort. High performance buildings integrate such decisions into the built work, a practice often culminating in Leadership in Engineering and Environmental Design (LEED) certification.

One of the ways in which the profession of architectural engineering will grow in importance is through its involvement with LEED. As green building becomes the norm, architectural engineers will be at the forefront of the work being done with respect to stewardship of the environment. This includes refining architects' designs with respect to increasing energy savings and water efficiency, reducing emissions, and improving indoor air quality.

Developed by the Green Building Council and its consensus based committees, LEED is an internationally recognized certification system for ranking green building. The LEED rating system applies to new as well as existing construction and is designed to work throughout the life cycle of the building. Residential, retail, school, and healthcare are among the categories addressed by LEED. State and local governments are adopting LEED for public-funded projects. Additionally, U.S. federal agencies and countries including Canada and Mexico are recognizing the importance of LEED in terms of building performance.

THE LITERATURE OF ARCHITECTURAL ENGINEERING

The monographic literature devoted to architectural engineering is somewhat limited and rather dated. Few book titles appear annually, and those that do are often cached under headings other than architectural engineering, like building systems or structural design, or under elements of a particular topic, such as sustainable building materials. Also, it can prove difficult to obtain a general overview of new developments in the field from the engineering or architecture journals because most articles focus on highly technical and specific subjects, for example, the design of concrete cladding or issues related to indoor air quality. The engineering periodical literature provides important studies on various technical aspects of architectural engineering, but the student or practitioner also should scan new developments in the architecture literature by browsing the major architectural design magazines, including *Architect*, *Architectural Record* and *Architectural Review*. Additionally, titles like *Architecture Today* often bridge the gap between building technology and design practice.

SEARCHING THE LIBRARY CATALOG

Architectural engineering is not a Library of Congress (LC) subject heading. Users are instructed to see the following LC headings:

Building
Building, iron, and steel
Strains and stresses
Strength of materials
Structural analysis (engineering)

Academic library catalogs, such as those at Columbia University and Cornell University, employ the same terminology; moreover, they also allow natural language or keyword searching, thereby permitting the use of the term *architectural engineering*. Because most of the current information on architectural engineering comes from periodicals, researchers should consult specific periodical indexes for in-depth information (see list of indexes below under Abstracts and Indexes). The *Avery Index*, the foremost architectura l periodical index, uses architectural engineering in its subject word search. The *Civil Engineering Database* includes "Architectural engineering" as a subject heading, but other appropriate search terms include building design and construction industry. The *Applied Science Indexes* denote structural engineering as a way of searching for this type of information. The researcher should consider the focus of their topic prior to selecting a heading in a catalog or database, and may wish to consider searching in more than one way. Other broad headings include architecture and technology, structural engineering, and structures. In addition to the above list, a few of the narrower headings include building materials and/or the specific material (e.g., steel, concrete); specifications; building envelope; exterior walls; air conditioning; heating, lighting, and ventilation; and seismic design. Library of Congress call numbers commonly associated with architectural engineering (other call number areas may apply) include:

NA2540-2545 Architecture in relation to special subjects
NA2640-2645 Specifications
NA2750 Architectural design
NA2800 Architectural acoustics
NA2835-3300 Architectural details
NA4050-9000 Special classes of buildings
TA151 Engineering handbooks and manuals
TA160 Engineering Research
TA170 Environmental engineering
TA178 Specifications
TA401-492 Materials of engineering and construction
TA625-695 Structural engineering
TH (entire class) Building construction
TH7005-7699 Heating and ventilation: Air conditioning

ABSTRACTS AND INDEXES

Largely because the monographic literature of architectural engineering is somewhat dated, students and professionals are advised to consult the following indexes to supplement their readings and to maintain currency in the field.

Applied Science and Technology Full Text, 1958–. Electronic resource (1983–). http://www. hwwilson.com (accessed October 15, 2010). New York: H. W. Wilson. Versions from other vendors are available. The *Applied Science & Technology Abstracts* (*ASTA*) indexes over 400 major scientific and technical publications, including those issued by trade associations and professional and scholarly organizations. With a focus on English-language publications, *ASTA* is interdisciplinary in nature, and includes specialized materials, such as conference proceedings and buyers' guides. Abstracts normally consist of one to three paragraphs describing the scope and content of the article. Indexing is comprehensive and includes any interviews, obituaries, and product surveys published in the issues (from the Web page).

Architectural Index. 1950–. Boulder, CO. Electronic database http://www.archindex. com/free/copyright.htm. Free area for years 1982–1988. Also in paper format. Jerry T. Moore, AIA, editor and publisher; Ervin J. Bell, AIA, editor and publisher, emeritus. Indexes the following major architectural journals: *Architecture, Architectural Record, Architectural Review, Builder, Building Design & Construction, Interior Design, Journal of Architectural Education, Landscape Architecture, Residential Architect,* and *The Construction Specifier.* Emphasis is on aesthetics rather than engineering. Last updated December 2007.

ASCE Civil Engineering Database. 1970–. Reston, VA. Electronic database http://www.asce. org/Content.aspx?id=2147487334 (accessed October 15, 2010). The electronic information retrieval service for all the American Society of Civil Engineers (ASCE) publications. "The *Civil Engineering Database* (*CEDB*) is designed to provide easy bibliographic access to all ASCE publications. The database covers all the journals, conference proceedings, books, standards, manuals, magazines, and newspapers. Journal papers with abstracts go back to 1970. Nonabstract journal records go back to 1958. The book records are complete dating back to the early 1900s. The database is constantly expanded and enriched. Subject areas include aerospace engineering, architectural engineering, bridges, cold regions, computer practices, construction, earthquake engineering, education, engineering mechanics, environmental engineering, forensic engineering, geotechnical engineering, geomechanics, highways, hydrology, hydraulics, irrigation and drainage, management, materials engineering, structural engineering, transportation, urban planning, water resources, waterway, port, coastal and ocean engineering" (from the Web page).

Avery Index to Architectural Periodicals. (Updated weekly). New York: Avery Architectural and Fine Arts Library, Columbia University. Electronic database. (1930s–). Selective coverage dating back to the 1860s. http://www.columbia.edu/cu/lweb/indiv/avery/avery_index. html (accessed October 15, 2010). Versions from other vendors available. Indexes more than 2,000 periodicals published worldwide on archaeology, city planning, interior design, and historic preservation as well as architecture. "*Avery* indexes not only the international scholarly and popular periodical literature, but also the publications of professional associations, U.S. state and regional periodicals, and the major serial publications on architecture and design of Europe, Asia, Latin America, and Australia. Expanded coverage includes obituary citations providing an excellent source of biographical data—often the only information available for less-published architects" (from the Web page).

BuildingGreen.com http://www.buildinggreen.com (accessed October 19, 2010). Brattleboro, VT: BuildingGreen, LLC. Electronic database that includes access to *Environmental Building News.* Independently owned and operated, "the corporate mission of

BuildingGreen, LLC is to facilitate transformation of the North American building industry into a source for local, regional, and global environmental protection; for preservation and restoration of the natural environment; and for creation of healthy indoor environments ..." (from the Web page).

Engineering Village 2 http://www.ei.org/engineering-village (accessed October 15, 2010). *Compendex* (1884–) and *Inspec.* (1968–). Electronic databases. "*Compendex®* is the most comprehensive interdisciplinary engineering database in the world. *Compendex* contains over 8 million records and references over 5,000 international engineering sources including journal, conference, and trade publications. Coverage is from 1969 to present and the database is updated weekly. Combined with the *Engineering Index Backfile*, the database includes over 9.7 million records and spans back to 1884" (from the Web page). "*Inspec®* is the world's leading bibliographic database covering the fields of physics, electronics, computing, control engineering, and information technology. *Inspec* includes over 8 million records taken from 3,400 technical and scientific journals and 2,000 conference proceedings. Coverage is from 1969 to the present," updated weekly (from the Web page).

BIBLIOGRAPHIES AND GUIDES TO THE LITERATURE

Bibliographies and guides to literature can provide shortcuts and, most often, include books and periodical articles that have been selected by subject specialists. Publications in the area of architectural engineering are limited, especially ones of a broad or generalized scope. Researchers should consult bibliographies from appropriate monographs on specific topics, e.g., exterior wall design.

Gafford, W. R. 1951. Source materials for architectural engineering. MS thesis, University of Texas at Austin.

Godel, J. B. 1977. *Sources of Construction Information*, Vol. 1: Books. Metuchen, NJ: Scarecrow Press. An annotated guide to reports, books, periodicals, standards, and codes. This reference work is divided into 11 sections, including the following topics: mechanical and electrical design, building materials, and the construction industry.

USGBC Resource Catalog http://www.usgbc.org/DisplayPage.aspx?CMSPageID=2175 (accessed December 15, 2010). Online resource launched by the U.S. Green Building` Council in 2009. The *USGBC Resource Catalog* is a collection of over 400 free digital resources that relate to green building, with topics like energy efficiency, materials, affordable housing, best practices, business cases, and more. Only educational, nonpromotional resources are included.

DIRECTORIES

Directories of relevance to the topic of architectural engineering are becoming more readily available via the World Wide Web. Again, the literature is fragmented. For example, there are no directories of architectural engineers. However, directories do exist for architects and engineers that often include lists of consultants where specialties are noted, e.g., curtain wall design. A number of directories exist for product information including paper and electronic versions.

Almanac of Architecture & Design, 10th ed. 2010. Washington, D.C.: Greenway Communications. Includes information on awards, achievements, design education, and appropriate organizations.

ARCAT Specs http://www.arcat.com/ (accessed October 15, 2010). "*ARCAT Specs* are complete, accurate, and in the CSI three-part format. They are available free for viewing and downloading in popular word processing formats. These long-form proprietary specifications are complete with technical data, ASTM standards, performance features and product attributes" (from the Web page).

Ballast, D. K. 1998. *The Encyclopedia of Association and Information Sources for Architects, Designers, and Engineers.* Armonk, NY: Sharpe Professional. In addition to listings of professional associations and organizations, this source includes information about journals, online databases, CD-ROMs, and federal government publications in the fields of architecture, design, and engineering.

ENR Directory of Construction Information Resources. 1993–. New York: ENR Special Projects. A useful source for locating contact information for construction related organizations.

GreenSpec: Product Directory with Guideline Specifications, 7th ed. 1999–. Brattleboro, VT: BuildingGreen. A comprehensive source for sustainable products with over 2,100 green product listings.

Meisel, A., *LEED Materials: A Resource Guide to Green Building.* 2010. New York: Princeton Architectural Press. This well-illustrated resource is based on the LEED Rating System Version 3.

Sweet's Catalog File. 2001–. (Various eds.: *Architects, Engineers, & Contractors Edition; Residential Edition; Facilities & Owners Edition; Structural & Civil Products Source Book.*) New York: Sweets. The most comprehensive source for standard product information. Uses CSI format.

Ultimate Civil Engineering Directory. http://www.tenlinks.com/engineering/civil/index.htm (accessed October 15, 2010). Online site containing links to general information, databases, international codes and standards, organizations, images, mailing lists and discussion groups, and more.

Who's Who in Science and Engineering, 11th ed. 2011. New York: Marquis Who's Who. A source for locating leaders in the field.

ENCYCLOPEDIAS AND DICTIONARIES

ASTM Dictionary of Engineering Science & Technology, 10th ed. 2005. West Conshohocken, PA: PA ASTM. An important dictionary of engineering terms.

Bianchina, P. 1993. *Illustrated Dictionary of Building Materials and Techniques.* New York: John Wiley & Sons. Excellent line drawings are included to illustrate specific terms.

Brooks, H. 1976. *Illustrated Encyclopedic Dictionary of Building and Construction Terms.* Englewood Cliffs, NJ: Prentice-Hall. Arranged by building function, this work includes line drawings or photographs for selected terms.

Cowan, H. J. and P. R. Smith. 2004. *Dictionary of Architectural and Building Technology,* 4th ed. London/New York: Spon Press. Substantially updated with 1750 new or revised entries, this source continues to be an important reference for the building industry.

Cowan, H. J. 1988. *Encyclopedia of Building Technology.* Englewood Cliffs, NJ: Prentice-Hall. Arranged alphabetically by topic. This source provides a good overview of various technical topics (e.g., heat engines and heat pipes, pneumatic structures, and sound absorption), often including definitions, histories, and applications.

Elsevier's Dictionary of Engineering: In English, German, French, Italian, Spanish, and Portuguese/Brazilian. 2004. Amsterdam/Oxford: Elsevier, (compiled by M. Bignami). A translation source for standard terms.

Guedes, P., ed. 1979. *Encyclopedia of Architectural Technology.* New York: McGraw-Hill Book Company. Divided into six sections, this source, while somewhat dated, does provide a solid introduction into the history of topics, such as building types, building services (mechanical and electrical systems), and building materials.

Keller, H., and U. Erb. 2004. *Dictionary of Engineering Materials.* Hoboken, NJ: John Wiley & Sons. An up-to-date source that defines key materials of the trade.

Pankratz, T. M. 2001. *Environmental Engineering Dictionary and Directory.* Boca Raton, FL: CRC Press.

Putnam, R. 1984. *Builder's Comprehensive Dictionary.* Reston, VA: Reston Publishing Company. Easy-to-read with numerous photographs and illustrations.

Schwartz, M., ed. 2002. *Encyclopedia of Smart Materials.* New York: John Wiley & Sons. Schwartz's encyclopedia outlines responsive, or smart, materials. Smart materials are those products that can change their properties, structure, or function in response to environmental stimulae.

Scott, J. S. 1984. *Dictionary of Building*, 3rd ed. New York: Halsted Press. A straightforward approach with clear definitions.

Stein, J. S. 1993. *Construction Glossary: An Encyclopedic Reference and Manual.* New York: John Wiley & Sons. Organized by building component following CSI (e.g., masonry). Very detailed and in-depth.

Webster, L., ed. 2000. *Dictionary of Environmental & Civil Engineering.* New York: Parthenon. Covers the basics of these fields.

Wilkes, J. A., ed. 1988. *Encyclopedia of Architecture: Design, Engineering & Construction.* New York: John Wiley & Sons. In five volumes, it includes historical background information on architects, building types, processes, and standard materials. Useful as an introduction.

Yeang, K. and L. Woo. 2010. *Dictionary of Ecodesign.* New York: Routledge. This is the first dictionary of the terminology of sustainable design.

HANDBOOKS AND MANUALS

ACI Manual of Concrete Practice (annually in parts). Detroit: American Concrete Institute.

Allen, E. 1985. *The Professional Handbook of Building Construction.* New York: John Wiley & Sons. Well-conceived drawings help elucidate the construction process.

American Institute of Steel Construction. 2001. *Load & Resistance Factor Design: Manual of Steel Construction*, 3rd ed. Chicago: American Institute of Steel Construction. The standard reference for steel construction.

American Institute of Timber Construction. 2005. *Timber Construction Manual.* Hoboken, NJ: John Wiley & Sons. Internet version also available. The standard reference for timber construction.

ASHRAE (American Society of Heating, Refrigerating, and Air Conditioning Engineers). 2006. *Green Guide; The Design, Construction and Operation of Sustainable Buildings.* Atlanta: ASHRAE. Clearly presented.

ASM Handbook, 10th ed. 1990. Prepared under the direction of the ASM International Handbook Committee. Materials Park, OH: ASM International. Emphasis on materials; technical in approach.

Birkhauser Construction Manuals. 2000–. Basel, Switzerland: Birkhauser. Separate editions are devoted to materials including glass, masonry, steel, timber, and building elements, such as roof construction.

Buettner, D. R. 2004. *PCI Design Handbook: Precast and Prestressed Concrete*, 6th ed. Chicago: Precast/Prestressed Concrete Institute. The source for concrete design.

Butler, R. B. 2002. *Architectural Engineering Design* (includes CD-ROM). New York: McGraw–Hill. An important resource for architectural engineering students.

Chen, W. F., and J. Y. R. Liew, eds. 2003. *Civil Engineering Handbook.* Boca Raton, FL: CRC Press. Handbook provides a solid overview of civil engineering topics.

Cowan, H. J., ed. 1991. *Handbook of Architectural Technology.* New York: Van Nostrand. Well-written and easy to read. Chapters cover materials (e.g., metal, ceramics, timber, and plastics) as well as building service design topics, such as durability, loads, and wind effects. Includes references and suggestions for further reading at the end of each chapter.

Daniels, K. 2003. *Advanced Building Systems: A Technical Guide for Architects and Engineers*. Basel, Switzerland/Boston: Birkhauser. A significant resource for designers of building systems.

Grimm, N. R. and R. C. Rosaler. 1990. *Handbook of HVAC Design*. New York: McGraw-Hill. Chapters range from conceptual and preliminary design to specific HVAC considerations, like cooling towers.

Haines, R. W. 2003. *HVAC Systems Design Handbook*, 4th ed. New York: McGraw-Hill. An important and well-conceived resource for mechanical design.

Hart, R. D. 1994. *Quality Handbook for the Architectural, Engineering, and Construction Community*. Milwaukee: ASQC Quality Press. Covers quality control aspects.

Hegger, M., M. Fuchs, T. Stark, and M. Zeumer. 2008. *Energy Manual; Sustainable Architecture*. Basel, Switzerland: Birkhauser. "This manual approaches design and construction from apparently invisible qualities: the sustainability and energy-efficiency of building" (from Preface, page 6).

Hornbostel, C. 1991. *Construction Materials: Types, Uses and Applications*, 2nd ed. New York: John Wiley & Sons. A clear, well-illustrated source that discusses the history, manufacture, types, and uses of specific materials.

LEED Reference Manuals. 2009–. Washington, D.C.: U.S. Green Building Council. These guides provide the essential reference material. Since 2009, these guides are being issued for broad areas including New Construction and Major Renovations, Core and Shell Development, Schools, Retail New Construction, Commercial Interiors, Retail Interiors, Existing Buildings: Operations and Maintenance, Neighborhood Development, and LEED for Homes.

Meier, H. W. 1989. *Construction Specifications Handbook*, 4th ed. Englewood Cliffs, NJ: Prentice-Hall. Discusses the writing of specifications, and includes draft specifications.

Pennsylvania State University Department of Architectural Engineering. 1966. *Emerging Techniques of Architectural Practice*. eds. C. H. Wheeler, Jr. et al. Washington, D.C.: American Institute of Architects. Somewhat dated, but still of historic interest.

Perry, R., and J. H. Perry. 1967. *Engineering Manual: A Practical Reference of Data and Methods in Architectural, Chemical, Civil, Electrical, Mechanical, and Nuclear Engineering*, 2nd ed. New York: McGraw-Hill. A handbook referencing technical data and calculations.

Rea, M. S., ed. 2011. *The IESNA Lighting Handbook: Reference and Application*, 10th ed. New York: Illuminating Engineering Society of North America. The standard lighting engineering manual.

Ricketts, J. T., M. K. Loftin, and F. S. Merritt, eds. 2004. *Standard Handbook for Civil Engineers*, 5th ed. New York: McGraw-Hill. A major handbook in the field.

Rush, R. D., contrib. 1986. *The Building Systems Integration Handbook*. New York: John Wiley & Sons, Includes bibliographies and index. An important source for students in the field.

Simmons, H. L. 2001. *Construction: Principles, Materials, and Methods*, 7th ed. New York: John Wiley & Sons. A solid introduction to materials and technology.

Spengler, J. D., J. F. McCarthy, and J. M. Samet, eds. 2001. *Indoor Air Quality Handbook*. New York: McGraw-Hill. A comprehensive source on an important issue.

Spiegel, R. and D. Meadows. 2006. *Green Building Materials; A Guide to Product Selection and Specification*, 2nd ed. Hoboken, NJ: John Wiley & Sons. Appendix A provides sources for further information.

Standard Handbook of Architectural Engineering. 1999. New York: McGraw-Hill. This book/disk package is designed to offer solutions to a wide range of architectural engineering problems. Includes a collection of structural, mechanical, electrical, lighting and acoustical data, universal design scenarios, design tools, and interactive formulas for sizing architectural components.

Stein, B. and J. S. Reynolds. 2005. *Mechanical and Electrical Equipment for Buildings*, 10th ed. New York: John Wiley & Sons. The standard HVAC textbook.

Time-Saver Standards. 1935–. New York: McGraw-Hill Companies. Separate editions for *Building Types, Interior Design and Space Planning, Landscape Architecture, Site Construction Details Manual, Housing and Residential Development, Building Materials and Systems*, and *Adding on and Remodeling*. Some available in electronic format. Essential tool of the trade.

Transmaterial: A Catalog of Materials That Define Our Physical Environment. 2006. New York: Princeton Architectural Press. Three catalogs have been issued.

Yeang, K. 2006. *Ecodesign: A Manual of Ecological Design*. London: Wiley-Academy. Another title by a leader in the field.

MONOGRAPHS AND TEXTBOOKS

Allen, E. 2005. *How Buildings Work: The Natural Order of Architecture*. New York: Oxford University Press. A clear and engaging explanation of building systems design.

Bachman, L. R. 2003. *Integrated Buildings: The Systems Basis of Architecture*. New York: John Wiley & Sons. Stresses integration issues between architecture and technology.

Baird, G. 2001. *The Architectural Expression of Environmental Control Systems*. London/ New York: Spon Press. Connects architectural design to the need for well-conceived mechanical systems.

Bovill, C. 1991. *Architectural Design: Integration of Structural and Environmental Systems*. New York: Van Nostrand Reinhold. Clearly explains the importance of structures and HVAC in the design process.

Building Arts Forum. 1990. *Bridging the Gap: Rethinking the Relationship of Architect and Engineer: The Proceedings of the Building Arts Forum*. Symposium held in April 1989 at the Guggenheim Museum, New York: Van Nostrand Reinhold. An interesting overview of practices and collaborative possibilities.

Fischer, R. E. 1980. *Engineering for Architecture*. New York: McGraw-Hill. An *Architectural Record* publication, this work illustrates the relationship between architecture and engineering.

Freitag, J. K. 1985. *Architectural Engineering: With Special Reference to High Building Construction, Including Many Examples of Chicago Office Buildings*. New York: John Wiley & Sons. This early title presents a solid picture of architectural engineering at the beginning of the twentieth century.

Holgate, A. 1986. *The Art in Structural Design: An Introduction and Sourcebook*. Oxford, U.K.: Clarendon Press. Shows the design potential of architectural engineering.

Kim, D.-H. 1995. *Composite Structures for Civil and Architectural Engineering*. London/New York: Spon Publishers. Discusses the integration of different materials and technology.

Komendant, A. E. 1987. *Practical Structural Analysis for Architectural Engineering*. Englewood Cliffs, NJ: Prentice-Hall. Komendant worked with Louis I. Kahn on many projects.

Kwok, A. G. and W. T. Grondzik. 2007. *The Green Studio Handbook: Environmental Strategies for Schematic Design*. Oxford, U.K.: Architectural Press. A useful tool for students and design professionals.

Larsen, O. P. 2003. *Conceptual Structural Design: Bridging the Gap between Architects and Engineers*. London: Thomas Telford. Discusses the necessary connection between architecture and engineering.

Lerum, V. 2008. *High-Performance Building*. Hoboken, NJ: John Wiley & Sons. Performance by design is addressed in this work.

Levy, M. 2002. *Why Buildings Fall Down: How Structures Fail*. New York: W. W. Norton. An important resource for students, this book discusses sources of building failure.

Liu, M., and K. Parfitt, eds. 2003. Architectural Engineering Conference. Building Integration Solutions. *Proceedings of the Architectural Engineering 2003 Conference,* September 17–20, 2003, Austin, TX. This conference covered important contemporary issues related to architectural engineering. Sponsored by Architectural Engineering Institute of American Society of Civil Engineers; Reston, VA: American Society of Civil Engineers.

Lyall, S. 2002. *Remarkable Structures: Engineering Today's Innovative Buildings.* New York: Princeton Architectural Press. A well-illustrated volume documenting key examples of innovative structures.

Mainstone, R. J. 1998. *Developments in Structural Form.* Oxford, U.K./Boston: Architectural Press. An important work on structural design.

Ritter, A. 2007. *Smart Materials in Architecture, Interior Architecture and Design.* Basel, Switzerland: Birkhauser. Includes discussion of the organic and inorganic qualities of materials.

Salvadori, M. G. 1982. *Why Buildings Stand Up: The Strength of Architecture.* New York: McGraw-Hill. Aimed at students, this book presents a clear overview of the topic.

Salvadori, M. G. and M. Levy. 1981. *Structural Design in Architecture,* 2nd ed. Englewood Cliffs, NJ: Prentice-Hall. Includes examples and problem solutions. Recommended for students.

Szokolay, S. V. 2004. *Introduction to Architectural Science: The Basis of Sustainable Design.* Amsterdam/Boston: Elsevier, Architectural Press. Discusses technology solutions toward sustainability.

West, H. H. 1989. *Analysis of Structures: An Integration of Classical and Modern Methods.* New York: John Wiley & Sons. A solid study useful for students of the field.

JOURNALS

Architectural & Engineering News. 1958–1970. New York: Hagan Publishing Corp. Indexed and of use for mid-twentieth-century developments.

Architectural Design. Engineering & architecture (whole issue), 1987. 57: 11/12. An interesting issue of an important architecture periodical.

Architectural Science Review. 1958–. Sydney: Academic Press (0003-8628). Articles on specific technical issues.

Architecture Today. 1989–. London: Architecture Today (0958-6407). Each issue includes the latest product developments in a particular area.

Builder (also called *Nation's Building News*). 1978–. Washington, D.C.: Hanley-Wood, Inc., for the National Association of Home Builders, (8750-6580). Trade association journal.

Building Design & Construction. 1960–2010. Chicago: Cahners Publishing Co. A trade publication that presents useful studies on architectural technology. Ceased publication.

Construction Review. 1997–. Washington, D.C.: U.S. Department of Commerce, Domestic and International Business Administration, Bureau of Domestic Commerce (0010–6917). For sale by Supt. of Docs., U.S. Gov. Print. Off. Begun in 1955, frequency varies. Government publication.

Construction Specifier, The. 1950–. Washington D.C.: The Construction Specifications Institute (0010-6925). A professional organization journal which presents solid information on materials specifications.

Detail: Zeitschrift fur Architektur & Baudetail & Einrichtung. 1961–. Munchen: Verlag Architektur + BaudetailGmbH, (0011–9571). An important architectural design magazine that includes fine line drawings. Now also issued in English.

ENR. Continues *Engineering News-Record.* 1917–. New York: McGraw-Hill (0891–9526). A long-running journal that covers current (largely civil) engineering projects and topics.

GreenSource: The Magazine of Sustainable Design. 2006–. New York: McGraw-Hill (1930–
9848). A graphically well-presented source for sustainable design.

Journal of Architectural Engineering. 1995–. New York: American Society of Civil Engineers
(1076–0431). Electronic version free with print subscription. "*The Journal of Architectural
Engineering* will provide a multidisciplinary forum for dissemination of practice-based
information on the engineering and technical issues concerning all aspects of building
design. Peer-reviewed papers and case studies will address issues and topics related to
buildings, such as planning and financing, analysis and design, construction and maintenance, codes applications and interpretations, conversion and renovation, and preservation" (from ASCE Web page).

Journal of Green Building. 2006–. Glen Allen, VA: College Pub. (1552–6100). Print and
online versions. "The purpose of the *Journal of Green Building* is to present the very
best of practical applications ... together with the best of peer-reviewed, contemporary
research ... in green building design and construction" (from journal vol. 4, no. 4, Fall
2009).

PATENTS AND STANDARDS

Patent and standard information can, in many instances, be searched online. A patent search may
provide insight into new technology and processes. Standards may be voluntary or required, and
they provide information on materials, testing, specifications, and practices.

American National Standards Institute http://www.ansi.org/ (accessed October 15, 2010).
"The American National Standards Institute (ANSI) has served in its capacity as administrator and coordinator of the United States private sector voluntary standardization system for more than 80 years. Founded in 1918 by five engineering societies and three
government agencies, the Institute remains a private, nonprofit membership organization
supported by a diverse constituency of private and public sector organizations" (from
ANSI Web page).

American Society for Testing and Materials http://www.astm.org/ (accessed October 15,
2010). "ASTM International is one of the largest voluntary standards development organizations in the world—a trusted source for technical standards for materials, products, systems, and services. Known for their high technical quality and market relevancy, ASTM
International standards have an important role in the information infrastructure that guides
design, manufacturing, and trade in the global economy" (from Web site).

Architectural Graphic Standards, 11ᵗʰ ed. 2007. Hoboken, NJ: John Wiley & Sons. A core
reference work, this title remains the premier source for building design and construction
information.

ASHRAE Handbook. 1970–. New York: The Society. Standards handbook for heating,
ventilation, and air conditioning. Issued in inch/pound and metric editions. Each standard (applications, fundamentals, refrigeration, systems, and equipment) is updated on
a five-year cycle, replacing the previous edition. Also available online as a subscription
service.

International Code Council International Building Code. 2003–. Falls Church, VA: The
International Building Code Council. Comprehensive in scope, and the result of cooperation within regulatory organizations. This code establishes minimum standards for building systems performance.

International Organization for Standardization http://www.iso.org/ (accessed October 15,
2010). A network of the national standards institutes of 163 countries, with a coordinating Central Secretariat in Geneva, Switzerland. "ISO is a nongovernmental organization
that forms a bridge between the public and private sectors." While many of its member

institutes are part of the governmental structure of their countries, other members are not, coming instead from the private sector as industry representatives (from Web page).

Life Safety Code. 2009. Quincy, MA: National Fire Protection Association. Includes standards for egress for specific types of buildings. NFPA 101.

National Electrical Code. 2011. Quincy, MA: National Fire Protection Association. NFPA 70.

National Information Standards Organization http://www.niso.org/ (accessed October 15, 2010. "NISO, the National Information Standards Organization, a nonprofit association accredited by the American National Standards Institute (ANSI), identifies, develops, maintains, and publishes technical standards to manage information in our changing and ever-more digital environment. NISO standards apply both traditional and new technologies to the full range of information-related needs, including retrieval, repurposing, storage, metadata, and preservation" (from Web page).

United States Patent and Trademark Office http://www.uspto.gov/index.html (accessed October 15, 2010). Since its inception over 200 years ago, the basic role of the United States Patent and Trademark Office (USPTO) has remained the same, that of promoting the progress of science and related fields by securing for inventors the exclusive right to their respective discoveries for a limited period of time under Article 1, Section 8 of the United States Constitution. "Under this system of protection, American industry has flourished. New products have been invented, new uses for old ones discovered, and employment opportunities created for millions of Americans. The strength and vitality of the U.S. economy depends directly on effective mechanisms that protect new ideas and investments in innovation and creativity. The continued demand for patents and trademarks underscores the ingenuity of American inventors and entrepreneurs. The USPTO is at the cutting edge of the Nation's technological progress and achievement" (from Web site).

SEARCH ENGINES AND IMPORTANT WEB SITES

Internet search engines grow in size and importance with each passing day. Google® has initiated a number of innovations, including Google Scholar, aimed at searching the literature of academia and research (http://scholar.google.com). While much professional literature continues to require paid subscriptions, most organizations do have free Web sites that provide introductions to their aims, goals, and services. The metasites listed below serve as portals to locating technical information.

Construction WebLinks http://www.constructionweblinks.com/ (accessed October 18, 2010). Provides links to over 10,000 Web sites, including organizations, resources, reference guides, professional topics, such as bidding and financing as well as news, RSS feeds, and blogs.

iCivilEngineer http://www.icivilengineer.com/ (accessed October 18, 2010). iCivilEngineer. com is an information portal especially designed for civil engineering professionals and students. Its goals are to collect and catalog relevant civil engineering Web-based resources enabling users to find information quickly. iCivilengineer.com also aims to explore how best to take advantage of Internet technology to serve the civil engineering community. News, career tools, and resource centers are part of the site.

Ultimate Civil Engineering Directory http://www.tenlinks.com/engineering/civil/index.htm (accessed October 18, 2010). Online site containing links to general information, databases, international codes and standards, organizations, images, mailing lists and discussion groups, and more. This site is also listed above under Directories.

The Virtual Library: Engineering http://vlib.org/Engineering.html (accessed October 18, 2010). Provides links to architectural engineering (and many other types of engineering) internet resources.

SOCIETIES

The following list includes professional organizations useful to the student or practitioner of architectural engineering.

The American Institute of Architects http://www.aia.org/ (accessed October 18, 2010). "Through a culture of innovation, The American Institute of Architects empowers its members and inspires creation of a better built environment. ... The American Institute of Architects provides guidance, service, and standards to architects around the world. The AIA continues to strive for quality, consistency, and safety in the built environment and to serve as the voice of the architecture" (from Web page, November 2004 version).

American Society of Heating, Refrigerating, and Air Conditioning Engineers, Inc. http://www.ashrae.org (accessed October 18, 2010). ASHRAE, according to its vision statement, seeks to be the global leader in the arts and sciences of heating, ventilating, air conditioning, and refrigeration. ASHRAE aims to "serve humanity and promote a sustainable world" (from Web page).

Architectural Engineering Institute http://www.aeinstitute.org/ (accessed October 18, 2010). "AEI is the home for all professionals in the building industry." AEI provides a "multidisciplinary national forum for members of, but not limited to, the architectural engineering, structural, mechanical, electrical, and architectural communities." The Architectural Engineering Institute was created through a merger of the National Society of Architectural Engineers (NSAE) and the American Society of Civil Engineers Architectural Engineering Division (AED) on October 1, 1998. (from Web page).

The Illuminating Engineering Society of North America http://www.ies.org/ (accessed October 18, 2010). "The IESNA is the recognized technical authority on illumination. For over 90 years its objective has been to communicate information on all aspects of good lighting practice to its members, to the lighting community, and to consumers through a variety of programs, publications, and services. The strength of the IESNA is its diversified membership: engineers, architects, designers, educators, students, contractors, distributors, utility personnel, manufacturers, and scientists, all contributing to the mission of the society—to advance knowledge and disseminate information for the improvement of the lighted environment to the benefit of society" (from Web page, November 2004 version).

The National Society of Professional Engineers http://www.nspe.org (accessed October 18, 2010). The National Society of Professional Engineers (NSPE) is the only engineering society that represents individual engineering professionals and licensed engineers (PEs) across all disciplines. Founded in 1934, NSPE strengthens the engineering profession by promoting engineering licensure and ethics, enhancing the engineer image, advocating and protecting PEs' legal rights at the national and state levels, publishing news of the profession, providing continuing education opportunities, and much more. NSPE serves some 60,000 members and the public through 53 state and territorial societies and more than 500 chapters" (from Web page, November 2004 version).

The Structural Engineering Institute http://www.seinstitute.org/ (accessed October 18, 2010). "SEI is a vibrant, 20,000 plus community of structural engineers within the American Society of Civil Engineers. SEI started on October 1, 1996, in order to serve the unique needs of the structural engineering community more effectively while also being their voice on broader issues that shape the entire civil engineering community. ... SEI advances our members' careers, stimulates technological advancement, and improves professional practice" (from Web page).

CONCLUSION

Research on architectural engineering will likely continue to require a multifaceted approach. Students can gain a basic understanding through general works and textbooks on architectural engineering and integrated building systems; they also should be encouraged to consult more specific works, such as those on mechanical and electrical systems. For the professional, the literature comes largely from periodicals and is normally related to more specific and specialized topics, such as seismic design or green building products. It is also important for professionals to remain current with architectural trends and to review contemporary design periodicals.

ACKNOWLEDGMENTS

Architectural engineering is a highly technical, broad-based profession, and architecture students are becoming increasingly cognizant of the many ways an architectural engineer can contribute to a project. Architecture schools now offer courses on advanced building systems in which students analyze building components, and study the integration of its various systems. At Syracuse University, such a capstone course was developed and taught by Joel Bostick, associate professor of Architecture (1947–2004). Joel helped numerous students at Syracuse understand architectural technology and the necessity for well-designed, functional building systems. This chapter is dedicated to the memory of Joel, who helped me better understand the importance of building technology.

6 Bioengineering

Honora Nerz Eskridge and Linda Martinez (based on the first edition by Linda Martinez and Mary D. Steiner)

CONTENTS

Introduction 93
Scope 94
History 94
The Literature of Bioengineering 95
Library of Congress Subject Headings (LCSH) 95
Medical Subject Headings (MeSH) 95
Keywords 96
Library of Congress Call Numbers 96
Bibliographies and Guides to the Literature 96
Abstracts and Indexes 97
Directories 98
Dictionaries 99
Encyclopedias 99
Handbooks, Data, and Manuals 101
Monographs and Textbooks 103
Journals 103
News and Trade Magazines 104
Electronic Journal Aggregators/Repositories 105
Internet Search Engines, Portals, and Discussion Forums 105
    Search Engines 105
    Portals 106
    Discussion Forums 106
Patents 107
Standards 107
Associations, Organizations, and Societies 107
Conclusion 110
Acknowledgments 110
References 110

INTRODUCTION

In 2000, *Mechanical Engineering* readers voted on the top 10 engineering achievements of the twentieth century. Bioengineering was ranked no. 9 (Rastegar, 2000, 75–76) stated:

> The 20th century was known as the Century of Physics. The 21st century is believed to be the Century of Biology. The definition of bioengineering is broad, but the key characteristics are the application of engineering principles to the study of biological sciences and biological systems. In addition, it includes various traditional fields of science: physics, chemistry, mathematics, and

computing. Bioengineering is a field that connects physical sciences to biological sciences, making it a bridge between the two centuries and, as such, one of the major fields of engineering of the 21st century.

SCOPE

As bioengineering is an extremely interdisciplinary field, it is challenging to find a single definition encompassing its entire scope. The Institute of Biological Engineering (IBE) defines biological engineering as "the interface of biological sciences, engineering sciences, mathematics, and computational sciences. It applies biological systems to enhance the quality and diversity of life." A biology-based engineering discipline that integrates life sciences with engineering in the advancement and application of fundamental concepts of biological systems from molecular to ecosystem levels, http://www.ibe.org (accessed July 12, 2010).

The terms bioengineering, biological engineering, and biomedical engineering are used interchangeably to describe the same fields of study. It is easiest to define bioengineering by describing what bioengineers do. Bioengineers are involved in designing prosthetic limbs or orthopedic implants, developing nano devices for drug delivery, testing materials for use in vascular stents, improving biomedical imaging systems, performing genetic engineering and analysis, investigating methods of tissue repair, and interventions for rehabilitation.

Universities offering bioengineering programs vary in their use of the terms bioengineering, biological engineering, and biomedical engineering in describing their programs. Generally, bioengineering describes a broader field than biomedical engineering. Historically, though not exclusively so, bioengineering programs grew out of institutions with strong agricultural schools, while biomedical engineering programs originated in universities with strong engineering and medical schools.

HISTORY

In the broadest context, bioengineering was developed through the practices of medical and engineering techniques as applied to specific problems. Bioengineering may be traced to the ancient peoples of Europe and Asia. For example, a 3,000-year-old mummy from Thebes was unearthed with a wooden prosthesis attached to its foot (http://bmes.seas.wustl.edu/WhitakerArchives/glance/history.html), (accessed July 14, 2010).

Prior to World War I, research in the areas of medicine, physics, biology, and engineering occurred in Europe and the United States, but there was very little formal communication between these research groups. In 1913, the Rockefeller Foundation supported work in the areas of physics and engineering in biological and medical research. In Germany, Friedrich Dessauer founded the Oswalt Institute for Physics in 1921 to investigate the biological effects of ionizing radiation. During the 1920s, the Johnson Foundation for Medical Physics at the University of Pennsylvania and the Biophysics Department of the Cleveland Clinic were established. After World War II, advances in technology, electronics, materials, and medicine fostered the growth of professional societies in engineering, medicine, and biology. In 1948, the first U.S. conference of engineering in medicine and biology, sponsored by the Institute of Radio Engineers, the American Institute for Electrical Engineering, and the Instrument Society of America, was held. Ten years later, more than 300 researchers attended a conference on computers in medicine and biology (Nebeker, 2002).

In the 1960s, the National Institutes of Health began supporting biomedical engineering by creating a program project committee under the General Medical Sciences Institute to evaluate program project applications in biophysics and biomedical engineering areas. It set up a biomedical engineering training study section to evaluate training grant applications, and it established two biophysics study sections. Each decade has yielded significant scientific breakthroughs that have impacted upon the fields of bioengineering. The 1950s saw the development of cardiac pacemakers, heart lung machines, x-ray imaging, and the scanning electron microscope. In the 1960s, computers began to gain importance in engineering and medical research areas. In the 1970s, new techniques of medical imaging emerged—computerized tomography and nuclear magnetic resonance. The 1960s and 1970s marked the beginning of the rise of academic programs in bioengineering and biomedical engineering. The 1980s saw the development of endoscopy, lasers, ultrasound imaging, and the increasing importance of computers in medical applications. In the 1990s, the human genome project, robotics, implantable devices, and new biomaterials yielded new research areas. Future research areas may involve cellular and tissue engineering, bionanotechnology, computational bioengineering, and genetic engineering (Nebeker, 2002). The field has evolved from engineering principles and technologies applied, adapted, or devised for biological systems to modern bioengineering, which involves the integration of engineering principles and technologies with cellular and molecular biology (Rastegar, 2000).

THE LITERATURE OF BIOENGINEERING

Bioengineering literature is dispersed widely, particularly when considering all of biomedical engineering, biomolecular engineering, biological engineering, bioinformatics, and biotechnology. Bioengineering materials, shelved in Library of Congress (LC) call numbers, would be found throughout the areas of science (Q), medicine (R), and technology (T). It includes aspects of computer science, physics, biology, chemistry, and more. Bioengineers rely heavily on journals, conference literature, and technical reports as information resources.

Keyword searching in library catalogs and indexes/abstracts is a means of discovering books, journal articles, and conference literature. Using standard subject headings will refine searches. Bioengineers should be aware of two distinct categories of subject headings: Library of Congress Subject Headings (LCSH) and Medical Subject Headings (MeSH). The former system is used in the library catalogs of the Library of Congress and most large academic institutions; the latter, developed by the National Library of Medicine, is used by many biomedical institutions and hospital libraries.

Engineers have long browsed through a library's collection by call number to foster the serendipitous discovery of materials. It is also fruitful to "virtually browse" through call number ranges in a library's online catalog, as hot topics and newer books may be checked out or increasingly shelved in different physical locations and not onsite for browsing. Most large academic and research institutions use the Library of Congress call number classification system.

LIBRARY OF CONGRESS SUBJECT HEADINGS (LCSH)

Although not exhaustive, these terms may prove helpful in searching for books and other literature in an LCSH library catalog: Biochips, Biocompatibility, Bioengineering, Bioethics, Bioinformatics, Biological transport, Biomechanics, Biomedical engineering, Biomedical materials, Biosensors, Biotechnology, Colloids in medicine, DNA microarrays, Human mechanics, Image analysis, Image processing, Imaging systems in medicine, Medical informatics, Medical instruments and apparatus, Nanotechnology, Polymeric drugs, Polymers in medicine, Prosthesis, Recombinant molecules, Tissue engineering.

MEDICAL SUBJECT HEADINGS (MESH)

These terms may prove helpful in searching a MeSH library catalog: Biocompatible materials, Biomechanics, Bioethics, Biomedical and dental materials, Biomedical engineering, Biotechnology-instrumentation, Implants/artificial, Image processing, Computer-assisted methods, Medical informatics, Nanotechnology, Polymers.

KEYWORDS

Potential keywords, in addition to the above terms, that could be used in indexes/abstracts as well as library catalogs might include: Biomaterials, Biomedicine, Clinical engineering, Drug delivery, Drug delivery system(s), Nanobiotechnology, Synthetic biology. New terminology will continue to emerge in this field.

LIBRARY OF CONGRESS CALL NUMBERS

Relevant Library of Congress call numbers are presented here in alphanumeric order, i.e., sequentially as they would appear on library shelves (or in an online catalog).

Q300-342	Cybernetics
QA76.87	Neural computers: neural networks (in computing)
QH505	Biophysics
QH507	Information theory in biology
QH508	Biological control systems
QH509	Bioelectronics
QH509.5	Biomechanics
QH513	Fluid dynamics (in biology)
QH513.5	Bioethics
QH332	Bioinformatics
QH324.2	Bioinformatics (genomics)
QP303	Mechanics; Kinesiology
QP363.3	Neural circuitry; Neural networks (in physiology)
R724-726.2	Medical ethics
R856-857	Biomedical engineering; Electronics; Instrumentation (includes biosensors, biomaterials, imaging systems, nanotechnology)
R858-859	Computer applications to medicine; Medical informatics; Medical information technology
RD130	Prosthesis; Artificial organs
RD132	Artificial implants and implant materials
RS199.5-RS201	Pharmaceutical technology; Drug delivery systems
RS210	Drug delivery devices
TA164	Bioengineering
TP248.13-TP248.65	Biotechnology (including biochemical engineering)

BIBLIOGRAPHIES AND GUIDES TO THE LITERATURE

In contrast to long-standing engineering disciplines, such as civil engineering or mechanical engineering, relatively few literature guides or bibliographies exclusive to bioengineering/biomedical engineering exist. However, the following are worth noting:

Crafts-Lighty, A. 1986. *Information Sources in Biotechnology*, 2nd ed. New York: Stockton Press. Although over 20 years old, it is very comprehensive; coverage of biotechnology science is extensive and detailed.

Hightower, C. 2002. *Guide to Selected Bioinformatics Internet Resources* http://www.library.ucsb.edu/istl/02-winter/internet.html (accessed July 16, 2010). This Web guide, which was published in the Winter 2002 issue of *Issues in Science and Technology Librarianship*, provides a thorough listing for that time of online bioinformatics resources, and includes basic definitions of terms and recommended readings, as well as bibliographic and nonbibliographic databases.

Macleod, R. A. and J. Corlett. 2005. *Information Sources in Engineering*, 4th ed. London/New Providence, NJ: Bowker-Saur. A unique work as it is organized by primary sources, then secondary sources, and finally information sources for specialized subject fields. Many of the specialized subject areas are broad (e.g., chemical engineering) and some narrow (e.g., marine technology). The biomedical engineering chapter is 20 pages long, written by an academic in biomedical engineering. In particular, the history of the field, the various societies, and the journals section continues to be relevant.

Schmidt, D., E. B. Davis, and P.F. Jacobs. 2002. *Using the Biological Literature: A Practical Guide,* 3rd ed. (revised and expanded). New York: Marcel Dekker. Sections on biochemistry, biophysics, molecular and cellular biology, genetics, and biotechnology complement other volumes in the publishers Books in Library and Information Series.

ABSTRACTS AND INDEXES

To identify recent, cutting-edge research information, engineers, scientists, and researchers consult journal articles, conference proceedings, and technical reports. Abstracts and indexes are tools to locate these types of documents. Given the sheer amount of information in the engineering and scientific disciplines, abstracts and indexes specialize in their coverage and scope.

Abstracts and indexes provide citation information: title of an article or conference paper, authors, title of the periodical or conference, volume issue, date, and pages in which the article or conference paper appears. Abstracts provide a brief description of the article or conference paper's content. Increasingly, digital abstracts and indexes provide access to the full text of the article or conference paper.

Annual Reviews. 1932–. Palo Alto, CA: Annual Reviews. Provide review articles to 40 focused disciplines within the biomedical, life, physical, and social sciences. Review articles survey primary research, including books, articles, and other sources.

Applied Science and Technology Abstracts. 1983–. Bronx, NY: H.W. Wilson. International coverage of scientific/technical journals, trade/industrial journals, and conference proceedings. Topics include: engineering, biomedical, the environment, and biotechnology.

BioCommerce Data's Biotechnology Directory. 1981–. Berkshire, U.K.: BioCommerce Data Ltd. International coverage of the commercial aspects of biotechnology. Topics include research and development projects, product news, industry news from reports, and newsletters. Includes abstracts from journals and organizational profiles; ideal for competitor monitoring, business planning, investment analysis, identifying product licensing opportunities, science and policy studies.

BioEngineering Abstracts. 1993–. Bethesda, MD: Cambridge Scientific Abstracts. International coverage on all aspects of biochemical and microbial technology as applied to bioengineering. Information drawn from journals, conference proceedings, and directories. Topics include: biomedical engineering, biotechnology, biomaterials, biomechanics and human engineering, genetic engineering, and rehabilitation engineering.

BioOne. 2000–. Washington, D.C.: BioOne. Full-text articles from 160 peer-reviewed journals. Subjects include biochemistry, biology, biophysics, environmental sciences, zoology, microbiology, and botany.

Biosis Previews. 1926–. Philadelphia, PA: Thomson Scientific. International coverage of journals, conferences, technical reports covering the life sciences, biochemistry, bioengineering, biology, biophysics and biotechnology.

Biotechnology and Bioengineering Abstracts. 1982–. Bethesda, MD: Cambridge Scientific Abstracts. International coverage of ground-breaking research, applications, regulatory developments and new patents across all areas of biotechnology and bioengineering, biomaterials including medical, pharmaceutical, and human rehabilitation engineering.

Chemical Abstracts Service (CAS). 1907–. Columbus, OH: Chemical Abstracts Service. CAS databases provide international coverage of the journals, technical reports, conference proceedings, and dissertations relevant to all aspects of chemistry, biochemistry, biotechnology, and biomolecular engineering. *SciFinder* is the software search interface to CAS databases.

Compendex. 1884–. Hoboken, NJ: Elsevier Engineering Information, Inc. International coverage of core journals, technical reports, conference proceedings, and dissertations on more than 175 engineering disciplines, including biology, biological materials, biomechanics, biomedical engineering, and biotechnology. Engineering Village is often the search interface to *Compendex*.

Derwent Biotechnology Resource. 1982–. London: Thomson Derwent. International coverage of journals, conference proceedings, and patents focused on all aspects of biotechnology and associated sciences. Topics include bioinformatics, genomics, biotechnological processes, and microbiology.

EMBASE. 1947–. New York: Elsevier Science. International coverage of journals and conferences focusing on human medicine and areas of biological sciences related to human medicine. Topics include: bioengineering, biochemistry, genetics, biophysics, and medical instrumentation.

Inspec. 1896–. London: Institution of Electrical Engineers. International coverage of journals, conference proceedings in the fields of physics, electrical and electronic engineering, computer science, nanotechnology, biomedical technology, and biophysics.

Medline. 1960–. PubMed. 1950–. Bethesda, MD: U.S. National Library of Medicine. International coverage of 4,600 biomedical journals and genetics databases focusing on clinical medicine, medical research, and healthcare; covers allied health, biological and life sciences, neuroscience, neurobiology, and information science.

Web of Science/Web of Knowledge. 1900–. New York: Thomson Reuters. Enables users to search current and retrospective multidisciplinary information from some of the most prestigious, high-impact research journals in the world. *Web of Science* also provides cited reference searching. Covers the core conferences and journals in the fields of science and engineering.

DIRECTORIES

Directories are particularly useful for finding information about people, institutions, equipment/companies, grants, and some statistics. For example, some societies maintain print and/or online directories of their membership, and several commercial publishers provide subject directories to university programs and government-funded research. The listing given below is not comprehensive and society memberships should be investigated at each society's Web site, but it does provide a sampling of sources of potential interest to the bioengineer.

Directory of Biomedical and Health Care Grants. Phoenix, AZ: Oryx Press. This annual directory provides grant titles, descriptions, requirements, amounts, application deadlines, contact information, Internet access, sponsor names and addresses, and samples of awarded grants (when available).

Health Devices Sourcebook. Plymouth Meeting, PA: ECRI. Annual print directory of medical product manufacturers in the United States and Canada. Has sections for product listings, trade names, manufacturer names, and equipment services. Also available as *Health Devices International Sourcebase*, an online database http://www.ecri.org.

Medical Device Register On-Line http://mdrweb.com (accessed June 17, 2010). Canon Communications. Online directory for sourcing medical products. Available via online subscription. Print edition is published annually and is available from Greyhouse Publishing, http://www.greyhouse.com/medical_device.htm (accessed June 17, 2010).

Research Portfolio Online Reporting Tools (RePORT) http://projectreporter.nih.gov/reporter. cfm (accessed June 17, 2010). This resource, formally known as CRISP (Computer Retrieval of Information on Scientific Projects) is a searchable database of federally funded biomedical research projects conducted at universities, hospitals, and other research institutions. The site covers awards from 1972 to the present.

DICTIONARIES

Dictionaries are alphabetical lists providing definitions of terms, concepts, and principles within a specific discipline or field. Coverage may vary from a few sentences to lengthy, in-depth explanations of a term or concept.

Arora, H. 2009. *Illustrated Dictionary of Biotechnology.* Boca Raton, FL: CRC Press. A dictionary designed for those without background education in the life sciences.

Gosling, P. J. 2002. *Dictionary of Biomedical Science.* Boca Raton, FL: CRC Press. An 8,000-term dictionary broadly covering the biomedical sciences.

Hancock, J. M. and M. J. Zvelebil. 2004. *Dictionary of Bioinformatics and Computational Biology.* Hoboken, NJ: Wiley-Liss. A dictionary of terms relating to bioinformatics or the statistical methods applied to biological data derived through experimentation. Many entries contain lengthy definitions, sometimes including images and other graphics. This title is also available in e-book format.

Juo, P. S. 2001. *Concise Dictionary of Biomedicine and Molecular Biology.* Boca Raton, FL: CRC Press. Definitions of more than 30,000 terms used in biotechnology, molecular biology, and biomedicine, understandable to researchers and students.

Lackie, J. 2010. *Oxford Dictionary of Biomedicine.* New York: Oxford University Press. This dictionary contains over 9,000 entries covering a wide range of topics in medicine and bioscience. Includes links to recommended Web sites for many of the listings.

Nill, K. R. 2005. *Glossary of Biotechnology Terms,* 4th ed. http://biotechterms.org (accessed June 17, 2010). Boca Raton, FL: CRC Press. Detailed definitions of biotechnical, biological, and chemical terms appropriate for the nonexpert.

Parker, S. P., ed. 2002. *Dictionary of Bioscience.* New York: McGraw-Hill Professional Publishing. Definitions of more than 22,000 terms from the fields of embryology, genetics, physiology, and molecular biology. Available as an e-book from http://books.mcgraw-hill.com.

Stedman, T. L. 2005. *Stedman's Medical Dictionary,* 28th ed. http://www.stedmans.com (accessed June 17, 2010). Philadelphia, PA: Lippincott Williams & Wilkins. The authority on medical terminology, this edition includes more than 5,000 new entries, totaling over 107,000 terms.

Steinberg, M. L. and S. D. Cosloy. 2001. *Facts on File Dictionary of Biotechnology and Genetic Engineering.* New York: Checkmark Books. Covers the basic and more technical terminology of modern biotechnology and genetic engineering.

Szycher, M. 1995. *Szycher's Dictionary of Medical Devices.* Lancaster, PA: Technomic Publishing Company, Inc. Definitions of terms from the disciplines of polymer chemistry, biochemistry, metallurgy, and organic chemistry as they relate to biomaterials used in medical devices.

Williams, D. F. ed. 1999. *Williams Dictionary of Biomaterials.* Liverpool, U.K.: Liverpool University Press. Defines more than 6,000 words and phrases associated with biomaterials and related disciplines. Written for clinical scientists involved in engineering matters.

ENCYCLOPEDIAS

Encyclopedias are excellent resources for obtaining an overview of a particular topic that can be understood by practitioners as well as nonexperts. The encyclopedia's authoritative articles may provide both historic and the most up-to-date information on a particular topic and usually include extensive bibliographies for further research.

Akay, M., ed. 2006. *Wiley Encyclopedia of Biomedical Engineering.* (6 vols.) New York: John Wiley & Sons. This comprehensive reference contains peer reviewed articles covering the breadth of this multidisciplinary subject. It includes articles on bioinformatics, all types of medical devices, testing methods, surgical tools and methods, and much more.

Bronzino, J. D., ed. 2000. *Biomedical Engineering Handbook,* 2nd ed. (2 vols.) Boca Raton, FL: CRC Press. Serves as a reference review of fundamental physiology and to the latest developments in various aspects of biomedical engineering. Topics include: transport phenomena, tissue engineering, prostheses and artificial organs, human performance engineering, physiological modeling, clinical engineering, medical informatics, artificial intelligence, and ethical issues.

Encyclopedia of Life Sciences. 2010. (32 vols.) New York: John Wiley & Sons. This monumental work is the backbone to any life science collection, covering the entire field in over 4,000 entries. Volumes 1–20 were published in 2002, Volumes 21–26 in 2007, and 27–32 in 2010. Articles are written for various audiences: "Introductory" for undergraduates and laypeople, "Advanced" for graduate level audiences, and "Keynote" articles "provide a platform for debate where controversial issues and 'hot topics' can be discussed."

Finn, W. E. and P. G. LoPresti. 2003. *Handbook of Neuroprosthetic Methods.* Boca Raton, FL: CRC Press. Comprehensive resource for the techniques and methodologies to design and undertake experiments within the fields of neuroprosthetics. Topics include: microelectronics, biomolecular electronics, hearing, vision, and motor prostheses.

Flickinger, M. C. 2010. *Encyclopedia of Industrial Biotechnology, Bioprocess, Bioseparation, and Cell Technology.* (7 vols.) New York: John Wiley & Sons. Covers the applications and theories of biotechnology, with a focus on industrial applications of biological processes and all of the associated procedures, equipment, and regulation. This edition expands its coverage to include pharmaceutical manufacturing.

Kirk, R. and D. F. Othmer. 2004. *Encyclopedia of Chemical Technology,* 5th ed. New York: John Wiley & Sons. This extremely well respected 27-volume reference work covers the properties, manufacturing, and uses of chemicals and materials, processes, and engineering principles.

Nalwa, H. S., ed. 2004. *Encyclopedia of Nanoscience and Nanotechnology.* (10 vols.) Stevenson Ranch, CA: American Scientific Publishers. Provides an introduction and overview of emerging fields of nanoscience and nanotechnology. Topics include nanostructured materials, nanobiotechnology, nanobiology, nanomedicines, drug delivery, and biomedical applications.

O'Neil, M. J., ed. 2006. *Merck Index: An Encyclopedia of Chemicals, Drugs, and Biologicals,* 14th ed. New York: John Wiley & Sons. More than 10,000 entries on chemicals, drugs, and natural substances; entries provide concise descriptions of a single substance or a small group of closely related compounds.

Redei, G. P. 2008. *Encyclopedia of Genetics, Genomics, Proteomics, and Informatics,* 3rd ed. (2 vols.) New York: Springer. This edition contains a significant amount of new content over the previous edition and includes articles on human and animal genetics and genomics, genetic engineering, cell biology, bioinformatics, proteomics, and much more. Also available as an e-book from Springer.

Webster, J. G., ed. 2006. *Encyclopedia of Medical Devices and Instrumentation.* (6 vols.) New York: John Wiley & Sons. Over 300 entries covering all aspects of medical devices and their use in treating illness. Medical devices are a major part of biomedical engineering, making this an important work to include in any biomedical engineering collection.

Wnek, G. E. and G. L. Bowlin. 2009. *Encyclopedia of Biomaterials and Biomedical Engineering*. (4 vols.) London: Informa Healthcare. Comprehensive coverage of approximately 300 topics; geared to clinicians, engineers, scientists, and researchers.

HANDBOOKS, DATA, AND MANUALS

There are many different types of handbooks for bioengineering; in fact, some works with the word "handbook" in the title are actually more like an encyclopedia or dictionary. Typically, handbooks provide quick look-up information, often tabular or graphical data, in a conveniently indexed form. They also can be a condensed treatment of a subject arranged in a systematic way. Material property and selection information is an important component of many handbooks. Datasets are described as organized collections of related information, a resource that is a collection of pieces of data, a homogenous collection of data records managed as a single entity, or a collection of data that originated from the same source. Datasets are of particular importance to scientific, medical, and engineering disciplines. Manuals offer guidelines on how to carry out procedures, particularly in a laboratory setting. Some of the descriptions that follow are taken from the work itself or from its Web site.

Biodynamics Database (accessed June 17, 2010). The Biodynamics Data Bank (BDB) was established in 1984 by a team of researchers at the U.S. Air Force Research Laboratory (AFRL). The primary objective was to provide a national repository of biodynamics test data accessible to the entire research community. The contents of the BOB include data from approximately 7,000 acceleration impact tests conducted at the Biodynamics and Acceleration Branch of AFRL as well as information on the associated test facilities, test programs, and test subjects. The test data include measured forces, accelerations, and motions of both human and manikin subjects. Users must register to use the site.

BioMolecular Engineering Research Center (BMERC) http://bmerc-www.bu.edu (accessed June 17, 2010). From Boston University, BMERC provides research support in the areas of biology and bioinformatics, molecular evolution, and gene regulation. Links to several statistical genome datasets are provided.

Black, J. and G. Hastings. 1998. *Handbook of Biomaterial Properties*. London/New York: Chapman & Hall. This work is organized in three parts: (1) composition and properties of biological tissues, (2) different biomaterials in use, and (3) applications issues such as biocompatibility, systems interactions, and responses.

Bronzino, J. D., ed. 2000. *The Biomedical Engineering Handbook*. See reference in this chapter under Encyclopedias.

Davis, J. R., ed. 2003. *Handbook of Materials for Medical Devices*. Materials Park, OH: ASM International. This work provides a review of the properties, processing, and selection of materials used in the environment of the human body. Among the application areas described are orthopedics (hips, knees, and spinal and fracture fixation), cardiology (stents, heart valves, pacemakers), surgical instruments, and restorative dentistry. Materials discussed include metals and alloys, ceramics, glasses, and glass-ceramics, polymeric materials, composites, coatings, and adhesives and cements.

Dyro, J. 2004. *The Clinical Engineering Handbook*. London/Burlington, MA: Elsevier Academic Press. As the title suggests, this volume concentrates on clinical engineering where the focus is healthcare and medical device technologies. Chapters cover the discipline in scientific, managerial, and societal aspects; clinical engineering in a number of countries worldwide; safety and regulatory issues; and more.

Kucklick, T. R. 2006. *The Medical Device R&D Handbook*. Boca Raton, FL: CRC Press. This one-volume handbook is designed for engineers and is focused on product development of mechanical medical devices, from design and prototyping to managing legal and regulatory issues.

Kutz, M., ed. 2009. *Biomedical Engineering and Design Handbook*. New York: McGraw-Hill. This one-volume work focuses on human biomechanics, biomaterials, and bioelectricity, then segues into design principles for medical devices, imaging systems, and prostheses.

Molecules to Go http://molbio.info.nih.gov/cgi-bin/pdb (also) http://www.ncbi.nlm.nih.gov (accessed June 17, 2010). This National Institutes of Health site provides access to the molecular data provided by the Brookhaven Protein Data Bank.

Moore, J. and G. Zouridakis, eds. 2004. *Biomedical Technology and Devices Handbook*. Boca Raton, FL: CRC Press. A good reference for anyone wanting details about medical devices and technologies, processes for measurement, imaging techniques, biological assays, tissue engineering, interventional and rehabilitation treatments.

Nalwa, H. S., ed. 2005. *Handbook of Nanostructured Biomaterials and Their Applications in Nanobiotechnology*. (2 vols.) Stevenson Ranch, CA: American Scientific. This set covers all aspects of the nanostructured biomaterials (vol. 1) and applications of biomaterials (vol. 2), including cardiovascular, orthopedic, and dental applications. Broadly speaking, all important aspects dealing with the chemistry, physics, biology, and engineering of nanostructured biomaterials and their applications in nanobiotechnology are covered.

National Center for Biotechnology Information http://www.ncbi.nlm.nih.gov (accessed June 17, 2010). From the National Library of Medicine, this site is a clearinghouse for resources in molecular biology. Links to molecular databases, genomic biology resources, and data-mining tools.

National Center for Health Statistics http://www.cdc.gov/nchs/ (accessed June 24, 2010). This site, sponsored by the Centers for Disease Control and Prevention, contains a wide variety of statistics on health, injuries, lifestyles, births, deaths, and more. The Health Data Interactive section provides interactive tables with health data that can be sorted and customized.

Protein Data Bank http://www.rcsb.org/pdb (accessed May 17, 2005). The Research Collaboratory for Structural Bioinformatics' (RCSB) international repository for the processing and distribution of 3-D structure data of large molecules of proteins and nucleic acids.

Ratner, B. D., et al., eds. 2009. *Biomedical Engineering Desk Reference*. Amsterdam/Boston: Elsevier Academic Press. This reference provides a condensed treatment of biomedical engineering topics, divided into eight section headings: modeling biosystems, biomaterials science, clinical engineering, medical devices and instrumentation, medical imaging technology, tissue engineering, and ethics. Also available in e-book format from Elsevier.

Visible Human Project http://www.nlm.nih.gov/research/visible/visible_gallery.html (accessed June 17, 2010). The National Library of Medicine's database of digital images of "complete, anatomically detailed, three dimensional representations of the normal male and female human bodies." Of interest to practitioners of clinical medicine and biomedical researchers.

von Recum, A. F., ed. 1999. *Handbook of Biomaterials Evaluation: Scientific, Technical, and Clinical Testing of Implant Materials*, 2nd ed. Philadelphia: Taylor & Francis. This volume addresses the needs of those who are involved in inventing, developing, and testing implants and are concerned about the interactions between biomaterial and body tissue. Topics include bulk and surface characterization of various materials, biocompatibility, implants, soft and hard tissue histology, regulations, and clinical trials.

Walker, J. M. and R. Rapley, eds. 2008. *Molecular Biomethods Handbook*, 2nd ed. Totowa, NJ: Humana Press. This handbook covers nanotechnology, stem cells, single-cell analysis, new techniques for mapping, and RNA expression. Written by research scientists who use the methods regularly, for students, research scientists, and technicians, the text describes theory, practical procedures, and applications; detailed laboratory protocols are referenced.

Whole Brain Atlas http://www.med.harvard.edu/AANLIB/home.html (accessed June 17, 2010). From Harvard Medical School, a collection of clinical data and magnetic resonance images of the brain. In addition to the "normal" brain, various disease states of the brain are illustrated.

Wiley Database of Polymer Properties. According to its Web site, the *Wiley Database of Polymer Properties* is the single most comprehensive source of physical property data for polymers commercially available, with experimentally determined and selected data for over 2,500 polymers. The initial content is derived from the *Polymer Handbook*, an essential print reference for polymer physical data.

MONOGRAPHS AND TEXTBOOKS

The dissemination of scientific information begins with research results/data published in conference proceedings, technical reports, and peer-reviewed journals. This process may take from one to three years to occur. Annual reviews and monographs generally deal with a specific topic in-depth in a systematic manner. These materials may appear years after the initial research.

Monographs are frequently titled "Principles of …," "Treatise on …," or "Series on …" For example, the CRC Press Biomedical Engineering Series includes titles such as *Handbook of Neuroprosthetic Methods, Medical Image Registration,* and *Noninvasive Instrumentation and Measurement in Medical Diagnosis*; or Springer's *Biological and Medical Physics, Biomedical Engineering Series* with titles such as *Molecular and Cellular Signaling, Medical Applications of Nuclear Physics,* and *Biological Imaging and Sensing.* Textbooks are designed for instruction and the focus is to provide an overview of basic information moving into more advanced topic treatment. Textbooks may appear as much as 10 to 13 years after the initial research. Textbooks frequently have titles such as "Introduction to …" or "Fundamentals of …" For example, the following textbooks were assigned to bioengineering and biomedical engineering courses from various universities: *Introduction to Bioengineering, Essential Physics of Medical Imaging, Introduction to Biomedical Engineering, An Introduction to Materials in Medicine,* and *An Introductory Text to Bioengineering.*

JOURNALS

The most important scientific communication tool for most engineers is the journal. Scientific journal publishing is undergoing much scrutiny regarding how it is produced and the source of the funding, with biomedical literature at the forefront of this dialogue. Researchers should be aware of "open access" and "self-archiving" discussions, initiatives, and mandates. Worldwide, governments and funding agencies (including the NIH) are examining the processes by which scientific literature can be most effectively disseminated, with some recommending or even mandating that the peer-reviewed content in some form or another be made quickly and freely accessible on the Internet, through institutional, discipline-specific, or personal Web sites.

Thomson Reuters' *Journal Citation Reports* (*JCR*) offers a systematic means to critically evaluate the world's leading journals, with quantifiable, statistical information based on citation data. By compiling articles-cited references, *JCR* measures research influence and impact at the journal and category levels, and displays the relationship between citing and cited journals. Each journal is assigned an "impact factor." The impact factor measures the frequency with which the "average article" in a journal has been cited in a given period of time. Impact factors indicate a journal's ranking among its peer journals in specific categories. Broad categories such as "Engineering, Biomedical," "Materials Science Biomaterials," "Biophysics," "Medical Informatics," among others, may be of interest.

In bioengineering, relevant journals cross the fields of biology, engineering, chemistry, physics, materials science, and medicine. These titles are ranked in the top 20 within the 2009 *JCR* engineering, biomedical category.

> *Acta Biomaterialia* (1742–7061)
> *Annals of Biomedical Engineering* (0090–6964)
> *Annual Review of Biomedical Engineering* (1523–9829)
> *Biomaterials* (0142–9612)
> *Biomechanics and Modeling in Mechanobiology* (1617–7959)
> *Biomedical Microdevices* (1387–2176)
> *Clinical Oral Implants Research* (0905–7161)
> *IEEE Transactions on Medical Imaging* (0278–0062)
> *IEEE Transactions on Neural Systems and Rehabilitation* (1534–4320)
> *Journal of Biomaterials Science—Polymer Edition* (0920–5063)
> *Journal of Biomechanics* (0021–9290)
> *Journal of Biomedical Materials Research, Part A* (1549–3296)
> *Journal of Neural Engineering* (1741–2560)
> *Journal of Tissue Engineering and Regenerative Medicine* (1932–6254)
> *Journal of the Mechanical Behavior of Biomedical Materials* (1751–6161)
> *Lasers in Medical Science* (0268–8921)
> *Medical Image Analysis* (1361–8415)
> *Physics in Medicine and Biology* (0031–9155)
> *Regenerative Medicine* (1746–0751)
> *Tissue Engineering and Regenerative Medicine* (1738–2696)

The Public Library of Science (PLoS) is a nonprofit organization of scientists and physicians dedicated to making scientific and medical information freely available to the public. Their journals are peer reviewed and open access.

> *PLoS Biology* (1544–9173).
> *PLoS Computational Biology* (1553–734X).
> *PLoS Genetics* (1553–7390).

NEWS AND TRADE MAGAZINES

> *Biomaterials Forum* (1527–6031) Society for Biomaterials.
> *Biomechanics: The Magazine of Body Movement and Medicine* (1075–9662) CMP Media
> *BMES Bulletin* (not supplied) Biomedical Engineering Society.
> *IEEE Engineering in Medicine and Biology Magazine* (0739–5175) IEEE Engineering in Medicine and Biology Society.
> *IFMBE News* (1741–0800) International Federation for Medical and Biological Engineering.
> *NTIS Alerts: Biomedical Technology and Human Factors Engineering* (not supplied) U.S. Department of Commerce, National Technical Information Service.
> *SCOPE* (not supplied) Institute of Physics and Engineering in Medicine.

ELECTRONIC JOURNAL AGGREGATORS/REPOSITORIES

BioMed Central http://www.biomedcentral.com/ (accessed June 24, 2010). BioMed Central is an independent publishing house committed to providing immediate free access to peer-reviewed biomedical research, offering over 100 open access journal titles covering all areas of biology and medicine, including *BioMedical Engineering OnLine*.

Directory of Open Access Journals http://www.doaj.org/ (accessed June 24, 2010). Maintained by Lund University Libraries, the site includes a growing number of freely available, full-text, quality controlled scientific and scholarly journals. Titles in Biology and Life Sciences, Health Sciences, and Technology and Engineering categories may be of interest.

PubMed Central (PMC) http://www.pubmedcentral.nih.gov/ (accessed June 24, 2010). PubMed Central is a freely available digital archive of biomedical and life sciences journal literature, developed and managed by the National Center for Biotechnology Information (NCBI), a division of the National Library of Medicine (NLM) at the U.S. National Institutes of Health (NIH).

Synthesis: The Digital Library of Engineering and Computer Science http://www. morgan-claypool.com/ (accessed June 24, 2010). Neither a journal nor a monograph, rather these are lectures. According to the publisher, Morgan & Claypool, the basic component of the library is a 50–100-page self-contained electronic document that synthesizes an important research or development topic authored by an expert contributor to the field. These lectures offer value to the reader by providing more synthesis, analysis, and depth than the typical research journal article. Synthesis *Lectures in Biomedical Engineering* began in 2006.

INTERNET SEARCH ENGINES, PORTALS, AND DISCUSSION FORUMS

The Internet has changed the landscape of scholarly communication, and this section highlights Web-based information resources and communication tools for bioengineers. Many of the descriptions are taken from the respective Web sites.

SEARCH ENGINES

National Library of Medicine (NLM) Gateway http://gateway.nlm.nih.gov/ (accessed June 24, 2010). The NLM Gateway allows users to search in multiple retrieval systems at the U.S. National Library of Medicine (NLM). The current Gateway searches *MEDLINE/PubMed*, the NLM catalog, Bookshelf full text biomedical books, TOXLINE, Developmental and Reproductive Toxicology (DART), ClinicalTrials.gov, Directory of Health Organizations, Genetics Home Reference, Household Products Database, Images from the History of Medicine, HSRProj (health services research projects), Online Mendelian Inheritance in Man, Hazardous Substances Data Bank, Integrated Risk Information System (IRIS), International Toxicity Estimates for Risk (ITER), GENE-TOX, Chemical Carcinogenesis Research Information System (CCRIS), and Profiles in Science.

National Science Digital Library (NSDL) http://nsdl.org/ (accessed June 24, 2010). NSDL is the National Science Foundation's online library of resources for science, technology, engineering, and mathematics education. Searches for "biomedical engineering" and "bio-technology" are straightforward, but for some reason index terms "bio engineering" and "bio sensors" are used (i.e., with spaces).

Science.gov: First.gov for Science http://www.science.gov/ (accessed June 24, 2010). The "Applied Science and Technologies" and "Health and Medicine" areas may be of particular interest to the bioengineer.

PORTALS

Biomaterials Network http://www.biomat.net (accessed June 24, 2010). Primarily sponsored by the Instituto de Engenharia Biomedica (INEB), of the University of Porto, Portugal, the site acts as a resource center to disclose resources, organizations, research activity, educational initiatives, scientific events, journals, books, articles, funding opportunities, industrial developments, market analyzes, jobs, and every other initiative related to biomaterials science and associated fields.

Biomedical Engineering Network (BMEnet) http://www.bmenet.org/BMEnet (accessed June 24, 2010). Developed and maintained by Purdue University's College of Engineering, the Biomedical Engineering Network may be browsed or searched for links to news, jobs, societies, upcoming conferences, funding opportunities, journals, and other information portals.

National Center for Biotechnology Information (NCBI) http://www.ncbi.nih.gov (accessed June 24, 2010). Established in 1988 as a national resource for molecular biology information, NCBI creates public databases, conducts research in computational biology, develops software tools for analyzing genome data, and disseminates biomedical information all for the better understanding of molecular processes affecting human health and disease. This is a very data-rich site, with links to literature indexes, such as *PubMed*, genetic and molecular data banks, software tools to manipulate and analyze data, educational resources, and more.

National Institute of Biomedical Imaging and Bioengineering (NIBIB), http://www.nibib.nih.gov.

National Science Foundation. Various funding opportunities, jobs, activities, news, and so on, of particular interest to the bioengineer may be found at the Directorate for Biological Sciences, http://www.nsf.gov/bio/ (accessed June 24, 2010) and the Division of Chemical, Bioengineering, Environmental and Transport (CBET) Systems, http://www.nsf.gov/div/index.jsp?org=CBET (accessed June 24, 2010).

DISCUSSION FORUMS

BIOMAT-L Electronic Biomaterials Forum http://www.lsoft.com/scripts/wl.exe?SL1=BIOMAT- L&H=NIC.SURFNET.NL (accessed June 24, 2010). This electronic discussion list is intended for members of the Australian, Canadian, European, Japanese, United States and other Societies for Biomaterials, and for anyone with an interest in biomaterials. The scope of the list includes the application of all types of materials in medicine and biology. Discussion of all aspects of biomaterials (from materials production and testing to issues of biocompatibility and tissue interactions and specific clinical applications) is encouraged. Also welcome are announcements of meetings and conferences, and discussions of particular technical problems associated with biomaterials research.

BIOMCH-L Biomechanics Electronic Discussion List http://www.biomch-l.org/ (accessed June 24, 2010). The Biomch-L electronic discussion list was started in 1988 by Herman Woltring and focuses on biomechanics and human/animal movement science.

Clinical Gait Analysis http://www.univie.ac.at/cga/ (accessed June 24, 2010). This is the companion Web site to the e-mail list cga@lists.curtin.edu.au. providing a forum for the discussion of clinical gait analysis: technical aspects, clinical cases, and new developments. Links to other biomechanics and gait analysis Web sites, including a data/software/videos section.

Nanodot http://nanodot.org/ (accessed June 24, 2010). News and discussion of coming technologies (i.e., nanotechnologies). This Web log comes from the Foresight Institute (described below in the Associations, Organizations, and Societies section of this chapter).

PATENTS

Patents are important sources for unique, detailed technical information for scientists and engineers. "According to several (albeit rather old) studies … 70–94% of information published in patent documents is not published in any other source" (Newton, 1998, p. 249). For more information, see the patents section in Chapter 2.

> Grubb, P. 2005. *Patents for Chemicals, Pharmaceuticals and Biotechnology: Fundamentals of Global Law, Practice and Strategy*, 4th ed. Cary, NC: Oxford University Press. Covers international patent processes for patent lawyers as well as scientists, researchers, and engineers.
>
> Ma, M. 2009. *Fundamentals of Patenting and Licensing for Scientists and Engineers*. Hackensack, NJ: World Scientific Publishing Company. Covers the basics of patents from the legal, technology and business perspectives.

STANDARDS

"A standard is something that is accepted as an authority or an acknowledged measure of comparison for quantitative or qualitative values. Standards become accepted through the reputation of those who create the standard" (Malinowsky, 1994, 22). Similar to patents, standards are detailed documents that can provide the engineer with a wealth of unique technical information, which is critical to design engineers.

> Association for the Advancement of Medical Instrumentation (AAMI) http://www.aami.org (accessed May 17, 2005). Composed of more than 100 technical committees responsible for standards, recommended practices, and technical reports for medical devices.
>
> Institute of Electrical and Electronics Engineers (IEEE) http://standards.ieee.org (accessed May 17, 2005). Global standards in technical areas, including biomedical areas, healthcare, information technology, and nanotechnology. International Electrotechnical Commission (IEC) http://www.iec.ch (accessed May 17, 2005). An organization of 60 nations providing standards for all electrical, electronic, and related technologies (medical instrumentation).
>
> International Organization for Standardization (ISO) http://www.iso.org (accessed May 17, 2005). An organization of 100 nations whose goal it is to develop standards for the international exchange of goods and services, including medical devices.

ASSOCIATIONS, ORGANIZATIONS, AND SOCIETIES

Associations and related groups can provide a wealth of professional information, networking opportunities, job employment information, and more technical information for scientists and engineers. Descriptions are paraphrased from the societies' Web sites.

> American Institute for Medical and Biological Engineering (AIMBE) http://www.aimbe.org (accessed July 14, 2010). Established in 1993, its mission is to establish a clear and comprehensive identity for the field of medical and biological engineering; to promote public awareness of medical and biological engineering; to establish liaisons with government agencies and other professional groups; to improve intersociety relations and cooperation within the field of medical and biological engineering; to serve and promote the national interest in science, engineering, and education; to recognize individual and group achievements and contributions to the field of medical and biological engineering.
>
> American Society of Biomechanics (ASB) http://asb-biomech.org (accessed July 14, 2010). Established in 1977, its mission is to provide a forum for the exchange of information and

ideas among researchers in biomechanics. The term biomechanics is used here to mean the study of the structure and function of biological systems using the methods of mechanics.

American Society for Engineering Education, Biomedical Engineering Division (ASEE BME) http://www.asee.org/member-resources/groups/divisions#Biomedical (accessed July 14, 2010). Established in 1893, ASEE is a nonprofit organization committed to furthering education in engineering and engineering technology. The BME Division provides a forum for those interested in biomedical engineering education through workshops, paper sessions, and panel discussions of current topics in the area.

American Society of Mechanical Engineers, Bioengineering Division http://divisions.asme.org/BED (accessed July 14, 2010). Established in 1973, the Bioengineering Division focuses on the application of mechanical engineering knowledge, skills, and principles from conception to the design, development, analysis, and operation of biomechanical systems.

Association for the Advancement of Medical Instrumentation (AAMI) http://www.aami.org (accessed July 14, 2010). Founded in 1967, with the goal of increasing the understanding and beneficial use of medical instrumentation.

Biomedical Engineering Society (BMES) http://www.bmes.org (accessed July 14, 20105). Established in 1968, its mission is to promote the increase of biomedical engineering knowledge and its utilization.

European Society for Biomaterials (ESB) http://www.esbiomaterials.eu (accessed July 14, 2010). Established in Italy in 1975, its objectives are to encourage, foster, promote, and develop research, progress, and information concerning the science of biomaterials, and to promote, initiate, sustain, and bring to a satisfactory conclusion research with others and programs of development and information in this particular field.

European Society of Biomechanics (ESB) http://esbiomech.org (accessed July 20, 2010). Established in Brussels in 1976, "To encourage, foster, promote and develop research, progress and information concerning the science of Biomechanics."

Foresight Institute http://www.foresight.org (accessed July 14, 2010). Founded by nanotechnology pioneer K. Eric Drexler, the institute is a nonprofit educational organization preparing society for anticipated advanced technologies. Primary focus is molecular nanotechnology: The ability to build materials and products with atomic precision, the development of which will have broad implications for the future of our civilization.

Gordon Research Conferences http://www.grc.org (accessed July 14, 2010). An international forum for the presentation and discussion of research in the biological, chemical, and physical sciences, and their related technologies. Conferences place a premium on the "off-the-record" presentation of previously unpublished scientific results and on the consequent ad hoc peer discussion.

IEEE Engineering in Medicine and Biology Society (EMBS) http://www.embs.org (accessed July 14, 2010). One of the earliest groups of the Institute of Electrical and Electronics Engineers (IEEE), the EMBS was established in 1952. EMBS advances the application of engineering sciences and technology to medicine and biology, promotes the profession, and provides global leadership for the benefit of its members and humanity by disseminating knowledge, setting standards, fostering professional development, and recognizing excellence.

Institute of Biological Engineering (IBE) http://www.ibe.org (accessed July 14, 2010). Encourages inquiry and interest in biological engineering in the broadest and most liberal manner and to promote the professional development of its members. It supports scholarship in education, research, and service; professional standards for engineering practices; professional and technical development of biological engineering; interactions among academia, industry, and government; and public understanding and responsible uses of biological engineering products.

Institute of Physics and Engineering in Medicine (IPEM) http://www.ipem.ac.uk/Pages/default.aspx (accessed July 14, 2010). A U.K.-based charity promoting the advancement of physics and engineering applied to medicine and biology, to advancing public education in the field, and to representing the needs and interests of engineering and physical sciences in the provision or advancement of healthcare.

International Federation for Medical and Biological Engineering (IFMBE) http://www.ifmbe.org/index.html (accessed July 14, 2010). Established in Paris in 1959, objectives are scientific, technological, literary, and educational. Within the field of medical, biological, and clinical engineering, IFMBE encourages research and the application of knowledge, the dissemination of information, and promoting collaboration.

International Society of Biomechanics (ISB) http://www.isbweb.org (accessed July 14, 2010). Established in 1971, it promotes study of the biomechanics of movement by encouraging international contacts among scientists in this field, promoting knowledge of biomechanics on an international level, and cooperating with related organizations. International Society for Prosthetics and Orthotics (ISPO) http://www.ispoint.org (accessed July 14, 2010). Established in Copenhagen in 1970, ISPO is a multidisciplinary organization composed of persons who have a professional interest in the clinical, educational, and research aspects of prosthetics, orthotics, rehabilitation engineering, and related topics.

International Union for Physical and Engineering Sciences in Medicine (IUPESM) http://www.iupesm.org (accessed July 14, 2010). IUPESM represents more than 40,000 medical physicists and biomedical engineers working on the physical and engineering science of medicine. Its principal objective is to contribute to the advancement of physical and engineering sciences in medicine for the benefit and well-being of humanity.

Materials Research Society (MRS) http://www.mrs.org (accessed July 14, 2010). Established 1973, MRS is a nonprofit organization that supports scientists, engineers, and research managers from industry, government, academia, and research laboratories in sharing findings in the research and development of new materials of technological importance. MRS sponsors two major annual meetings with numerous topical symposia, some of which may be of particular interest to those involved in biomaterials research.

Public Library of Science (PLoS) http://www.plos.org (accessed July 14, 2010). Founded in 2000, PLoS comprised a coalition of research scientists and physicians dedicated to making the world's scientific and medical literature a public resource. Their first action was to encourage scientific publishers to provide archival scientific research literature for distribution through free online public libraries of science. In 2001, the group launched their own PLoS journals. In 2003, PLoS Biology was launched, followed by PLoS Medicine in October 2004; both have open access journals and are relevant to the bioengineer.

Society for Biomaterials http://www.biomaterials.org (accessed July 14, 2010). The Society for Biomaterials promotes advances in all phases of materials research and development by encouragement of cooperative educational programs, clinical applications, and professional standards in the biomaterials field.

Society for Industrial Microbiology (SIM) http://www.simhq.org/ (accessed July 14, 2010). Established in 1949, SIM is dedicated to the advancement of microbiological sciences, as applied specifically to industrial materials, processes, products, and their associated problems. Members represent scientists employed in industry, government, and university laboratories.

Whitaker Foundation http://www.whitaker.org (accessed July 14, 2010). Founded in 1975, The Whitaker Foundation supported the development of biomedical engineering in the United States. In 2006, the foundation committed its remaining funds to a grant program focused on strengthening international collaborative links between young leaders in the BME field.

CONCLUSION

Bioengineering is a discipline (or agglomeration of interdisciplines) on the rise, as reflected by the growth of academic programs and funding for research. As a "twenty-first-century subject" at the locus of life sciences, physical sciences, and engineering, the prospects for its expansion and further development are vast. Although this compilation of resources represents only a snapshot in time, it is hoped that it will provide a useful entry point to those engaged in this exciting and evolving field.

ACKNOWLEDGMENTS

Many thanks to Duke University professors Morton Friedman and Roger Barr for their definitions of bioengineering.

REFERENCES

Malinowsky, H. R. 1994. *Reference sources in science, engineering, medicine and agriculture.* Phoenix, AX: Oryx Press.

Nebeker, F. 2002. Golden accomplishments in biomedical engineering. *IEEE Engineering in Medicine and Biology Magazine* 21 (3): 17–47.

Newton, D. 1998. Patents information: What's in it for the business information user. *Business Information Review* 15 (4): 248–253.

Rastegar, S. 2000. Life force: Bioengineering gets a burst of energy when the century of physics meets the century of biology. *Mechanical Engineering* 122 (3): 74–79.

7 Chemical Engineering

Dana Roth

CONTENTS

Introduction: History and Scope 111
    Additional Reading 112
Library Catalogs 112
Article Abstracts and Indexes 113
Bibliographies and Guides to the Literature 114
Directories 115
Dictionaries 115
Encyclopedias 115
Handbooks and Compendia of Property Data 117
    Handbooks 117
    Property Data 118
Textbooks and Monographs 120
    Textbooks 120
    Research Monographs and Symposia Publications 121
Chemical Safety 122
Journals 122
    Chemistry News Periodicals 123
    Chemical Engineering Reviews 123
    Chemical Engineering Research Journals 123
Patents 124
Associations, Organizations, Societies, and Conferences 125
Style Guides 126
Web Sites and Listservs 126
Conclusion 126
References 127

INTRODUCTION: HISTORY AND SCOPE

The first "applied chemists" date from the Bronze Age cultures of Egypt and Mesopotamia that developed techniques for extraction of metals from naturally occurring ores, for the brewing of beer, and for the production of pottery and glass. The beginnings of chemistry came later with the Greeks and their conceptions of elements and compounds. The melding of these ancient techniques and chemical concepts formed the basis for alchemy, which reigned until the eighteenth century. Chemistry then evolved with the ability to synthesize an increasing variety of chemical compounds. Chemical engineering, as we know it today, developed in the nineteenth century to meet the need for industrial-scale production of chemicals and materials. Since then, its main focus has undergone a series of transitions.

The development of a "unit operations" paradigm in the 1920s was based on the observation that many industrial chemical processes shared common basic operations (crystallization, distillation, evaporation, extraction) and these operations should be the focus rather than the production of specific products. The next transition, in the 1950s, was the development of expertise in "transport phenomena" (fluid flow, heat and mass transfer, microscopic level process rates). With these developments as a foundation, the chemical engineering profession is now coming full circle to again focus on development of specific products (e.g., biologicals, catalysts, optical fibers, polymers, semiconductors, etc.) as well as a growing recognition of the role chemical engineers can play in dealing with societal issues (Ritter, 2001; Cussler, 2002; Fan, 2008). Examples of the current diversity in chemical engineering research are exemplified by the topics listed on *CACHE: A Learning Community for Chemical Engineering* http://www.cache.org/ (accessed November 9, 2010) and the MIT OpenCourseWare Web site: http://ocw.mit.edu/courses/chemical-engineering/ (accessed November 9, 2010). Additional information on specific topics is widely available on individual professors' Web sites and in the open literature.

Chemical engineering education is currently undergoing a transformation as computing experiences now go far beyond numerical analysis to include simulation and molecular modeling. MATLAB® is the de facto standard for dynamic simulation and an extensive listing of molecular modeling and viewing programs is available on the Web. Equally important is the need for chemical engineering graduates to "know how to use a modern technical library to search for information located in electronic databases, and how to access electronic information services through the World Wide Web" (Edgar, 2000).

Because of the wide diversity of research areas and industrial applications, chemical engineering enrollments rise and fall with changes in national economies. Chemical engineering departments are fairly quick to recognize and incorporate new curricula, which explains the current emphasis on biology and nano-based engineering programs (Halford, 2004).

A layman's explanation of what chemical engineers do includes, "construct synthetic fibers that make our clothes more comfortable and water resistant; ... develop methods to mass produce drugs, making them more affordable; ... create safer, more efficient methods of refining petroleum products, making energy and chemical sources more productive and cost effective" (AIChE).

ADDITIONAL READING

These resources will provide a better understanding of the nature of chemical engineering both from a historical perspective and its current practice.

> *AccessScience.* McGraw-Hill http://www.accessscience.com/ (subscription required) (see Chapter 2, this volume). McGraw-Hill's *Access Science* is the continuously updated online version of both *McGraw-Hill Encyclopedia of Science and Technology* and *McGraw-Hill Dictionary of Scientific and Technical Terms*. Included, for example, are chapters on chemical engineering, biomedical chemical engineering, biochemical engineering, and the chemical process industry.
>
> Felder, R. Resources in Science and Engineering Education http://www4.ncsu.edu/unity/lockers/users/f/felder/public/RMF.html (accessed March 29, 2011). A comprehensive discussion of effective instructional techniques that have been validated for both individuals and groups.
>
> History of Chemical Engineering Web Page http://www.pafko.com/history/h_intro.html (accessed November 9, 2010). A light-hearted, albeit serious, look at the growth of chemical engineering as a profession, using the petroleum industry as a case study.

LIBRARY CATALOGS

Given recent trends toward adding book chapter titles in library catalog records, and the inappropriateness of many subject headings, keyword searching is often more effective than the use of Library of Congress Subject Headings. This is exemplified by the full-text, keyword-searching option in Amazon.com's Search Inside This Book program (Price, 2003), which also provides cited and citing references for books. LibWeb http://lists.webjunction.org/libweb/ (accessed November 9, 2010) is a growing list of over 8,000 Web pages from 146 countries worldwide, providing access to academic, public, national, and state libraries. It provides an excellent tool for searching catalogs for chemical titles. Amazon.com may also be considered as a "library" catalog because it lists both in-print and out-of-print books. Online library catalogs for top-ranked graduate chemical engineering programs are freely available for searching. All of these will list books in the various areas of chemical engineering for both research and collection development.

Since most academic libraries arrange their resources using the Library of Congress classification system, the major divisions are as follows:

Analytical Chemistry QD 71-145

Biochemistry QD415-431.7, QH345, QP501-801

Chemical Technology TP 1-1185

    Chemical Engineering TP 155-159

    Manufacturing and use of chemicals TP 200-248

    Fuels TP 316-360

    Food processing and manufacturing TP 368-456

    Fermentation industry TP 500-660

    Petroleum refining and products TP 690-692.4

    Polymers, plastics, and their manufacture TP 1080-1185

Chemistry (General) QD 1-69

Crystallography QD 911-999

Fluid Mechanics QA 901

Genetic Engineering QH442

Heat and Mass Transfer QC 320, TJ 260

Inorganic Chemistry QD 146-196

Kinetics QD 501-505, TP 149-157

Macromolecules QD 380-388

Molecular Biology QH506

Organic Chemistry QD 241-449

Organometallic Compounds QD 410-412

Physical and Theoretical Chemistry QD 450-731

ARTICLE ABSTRACTS AND INDEXES

Chemists and chemical engineers are extremely well served by the chemical abstracts (in print) and its online equivalent *SciFinder*. These complementary products offer very broad subject coverage, extending to the chemical aspects of astronomy, biology, education, engineering, economics, geology, history, mathematics, medicine, and physics. In addition, based on a review of the database in November 2010, the extensive format coverage includes journal articles (70%), patents (22%), conference papers (4%), dissertations (1.5%), in addition to American Chemical Society (ACS) National Meeting abstracts (1995+), books, CD-ROMS, edited research monograph chapters, preprints (2000+), and technical reports. In addition, *SciFinder* offers cited reference searching beginning with articles indexed in 1997.

    Chemical Abstracts Service. *SciFinder* http://www.cas.org/ Chemical Abstracts Services (CAS) provides an integrated family of databases (*CAplus, CA Patent Index, CA Reg File, CASreact, ChemCats, and ChemList*). *The CAplus* file (which includes the CA Patent Index) contains references to articles, patents, theses, and book chapters, while the *CA Reg File* contains chemical structures, names, and physical property data, and is linked to *CASreact*. These companion files are searchable by chemical structures/substructures, substance names, reactions, CAS Registry Numbers (sequentially assigned compound numbers), keywords, author names, patent numbers, company/organization names, and more.

ChemCats (Chemical Catalogs Online) from CAS is a catalog database containing information on nearly 44,000,000 commercially available chemicals and their worldwide suppliers. *ChemList* (Regulated Chemicals Listing) from CAS is a collection of more than 248,000 regulated chemical substances.

BIBLIOGRAPHIES AND GUIDES TO THE LITERATURE

The Web provides a wealth of library and literature resource guides for chemical engineering. These generally follow similar patterns with listings of library materials and URLs for additional information. A number of other universities and organizations maintain excellent online sites to the literature of chemical engineering. A few are noted here with a brief description. It is worthwhile to check other university subject guides for their links and suggestions in chemical engineering. Guides now tend to follow similar formats, but the resources linked from these pages vary by the objectives of the site creator, so it is useful to look at several.

American Chemical Society. 1954. *Literature Resources for Chemical Process Industries: A Collection of Papers, Advances in Chemistry Series 10.* Washington, D.C.: American Chemical Society. While dated, this has some timeless articles including: Sound and Unsound Short Cuts in Searching the Literature, Foreign Alphabetization Practices, Pitfalls of Transliteration in Indexing and Searching.

American Chemical Society. 1968. *Literature of Chemical Technology, Advances in Chemistry Series 78.* Washington, D.C.: American Chemical Society. While again dated, this provides brief introductions to virtually all fields dealing with chemical technology, for example, Photographic Chemistry, Cosmetics Industry, Synthetic Dyes, Leather and Adhesives, Rubber Industry, and Explosives.

Banville, D. L., ed. 2009. *Chemical Information Mining: Facilitating Literature-Based Discovery.* Boca Raton, FL: Taylor & Francis. The guide covers topics of interest in the networked research environment: the semantic Web, structure searching, conversion of chemical names, and more.

California State University, Long Beach, Chemical Engineering http://www.csulb.edu/library/subj/chemicalengineering.html (accessed November 10, 2010). Excellent site links to a number of resources on the Web and other guides of interest to chemical engineers.

Chemical Engineer's Resource Page. *ChE Links* http://www.cheresources.com/chelinks/index.shtml (accessed March 29, 2011). This site includes a number of useful links. In addition to expected topics, ethics, writing guides, and an introduction for beginners to the field add to the value of this resource.

Maizell, R. E. 2009. *How to Find Chemical Information: A Guide for Practicing Chemists, Educators, and Students,* 4th ed. New York: Wiley-Blackwell. Excellent tool for learning the resources necessary to access resources in chemistry and related fields.

Stanford's Swain Chemistry and Chemical Engineering Library http://lib.stanford.edu/swain (accessed November 10, 2010). Undergoing revision at the time of the book, the site has a number of links in chemical engineering and the various divisions of chemistry that provide information on constants, calculations, simulations, directories, history of the subject, and more.

The University of Alberta Library Resource Guide for Chemical Engineering http://guides.library.ualberta.ca/chemeng (accessed November 10, 2010). Topics include: Finding Books, Finding Journal Articles, Guides to the Literature of Engineering, Handbooks and Manuals, Chemical Properties and Data, Dictionaries and Encyclopedias, Chemical Engineering Design, Planning and Estimation, Pressure Vessel and Boiler Resources, Chemical and Biotechnology Guides, Directories and Catalogues, Chemical Industry, Prices and Production, Associations and Organizations, Standards and Patents, Selected Web Sites and Databases.

DIRECTORIES

Directories are an important resource for locating individuals, institutions, companies, or products. However, as in most fields, directories are increasingly available on the Web. In addition to those listed below, check organizational and company Web pages for their directories. In many cases, the most up-to-date information will be at the organizational site.

University of Texas at Austin, Department of Chemical Engineering, Chemical Engineering Faculty Directory, http://www.che.utexas.edu/che-faculty/ Records are comprised of brief biographical information (provided by the faculty member). Listings of chemical engineering departments can be displayed by institution name or geographic location (either U.S. or worldwide).

DICTIONARIES

Because there is extensive overlap between the terminologies of chemistry and chemical engineering, there are only a few dictionaries specifically for chemical engineering. The *Scifinder Registry File* provides extensive synonym listings, especially for commercially available chemical compounds.

Comyns, A .E. 2007. *Encyclopedic Dictionary of Named Processes in Chemical Technology,* 3rd ed. Boca Raton, FL: CRC Press. Concise descriptions of over 2,600 chemical processes known by special names (e.g., inventors, companies, institutions, places, acronyms) with an index that relates product names to processes.

Kent, J. A. 2007. *Riegel's Handbook of Industrial Chemistry and Biotechnology,* 11th ed. New York: Springer. Reviews the economic, pollution, and safety aspects of the chemical industry and specific processes used to create fertilizers, petroleum, adhesives, dyes, and so on. The 10th edition includes a chapter on industrial cell culture.

Lewis, R. J. 2007. *Hawley's Condensed Chemical Dictionary,* 15th ed. New York: Wiley Interscience. Primarily a dictionary of chemical substances providing physical properties, source of occurrence, CAS number, chemical formula, potential hazards, derivations, synonyms, and applications.

ENCYCLOPEDIAS

Chemical engineers and chemical engineering students are fortunate that there are two standard multivolume encyclopedias, *Kirk-Othmer* and *Ullmann's*, whose electronic versions promise to stay relatively current with the latest developments. Both also have regularly published print sets. In addition, there is a multivolume encyclopedia on chemical processing that describes the design of key unit operations, and a multivolume encyclopedia on separation science. The primary benefits of encyclopedia articles are their length and depth of coverage.

Kirk–Othmer Encyclopedia of Chemical Technology, 5th ed. 2004–. (27 vols.) Chichester: Wiley, and *Kirk-Othmer Encyclopedia of Chemical Technology* http://onlinelibrary.wiley. com/book/10.1002/0471238961. *The Kirk–Othmer Encyclopedia of Chemical Technology* has been a mainstay for chemists, biochemists, and engineers at academic, industrial, and government institutions since publication of the first edition in 1949. Because each edition focuses on currently relevant topics, research libraries will hold all print editions. This is especially important as lawyers and consultants, for example, use it for state-of-the-art in litigation and/or patent support. New and revised articles are posted monthly; older versions of such articles may still be accessed in an archived form. The electronic version allows browsing by either article title (pull-down menu ... ablative materials to zone refining) or article subject (pull-down menu ... analytical techniques to special topics). "Search in this title" searches "all text" or can be limited to article title, section title, author; author affiliation, keywords, DOI, tables, figures, Chemical Abstract Registry Numbers (CASRN), company, brand, and trade names. Boolean searching (and, or, not) is available in advanced product search. Complementary abstracts for all articles are also provided.

Ullmann's Encyclopedia of Industrial Chemistry, 7th ed. 2009–. Weinheim/ Cambridge: Wiley–VCH; 2005. In print and online editions, *Encyclopedia of Industrial Chemistry* is the European counterpart to *Kirk–Othmer* and, while providing similar coverage, better reflects European and Asian sources. First published in 1914 by Professor Fritz Ullmann in Berlin, the *Enzyklopadie der Technischen Chemie* (the 1st to 4th eds., 1914–1972, were published in German) quickly became the standard reference work in industrial chemistry. Beginning with the 5th edition, 1985, *Ullmann's* is published in English. The electronic version allows browsing by article title, article subject, or author. "Search in this title" searches "all text" or can be limited to article title, section title, author, DOI, tables, figures, Chemical Abstract Registry Numbers (CASRN). Boolean searching (and, or, not) is available in advanced product search. Additional limits include subject category and date range. An extensive listing of abbreviations is given on the home page.

These two encyclopedic works are complementary. *Kirk–Othmer* has extensive references to both journal articles and the worldwide patent literature, while *Ullmann's* primarily references the journal literature and its previous editions. *Ullmann's* also has a broader selection of figures while *Kirk–Othmer* provides more tables.

Encyclopedia of Chemical Processing and Design, The. 1976–2001. (69 vols.) New York: Marcel Dekker. This work provides detailed descriptions of both chemical processes and unit operations, such as: reactors and separation systems, process system peripherals, pilot plant design, and scale-up criteria. Research and production phases of the chemical industry are highlighted with entries on design principles (e.g., Tubular Reactor Design, Design of Extrusion Dies, Pressure Relief Valve Design, etc.) as well as engineering fundamentals (e.g., Adsorption, Fluid Flow, and Multiphase Reactors). Emerging areas (e.g., Nanotechnology, Microreactors and Microreactor Engineering, Plant Metabolic Engineering, etc.) are also covered. Sixty-nine volumes were published between 1976 and 2001. An online supplement entitled *Encyclopedia of Chemical Processing* (New York: Taylor & Francis) was published in 2005.

Encyclopedia of Separation Science. 2000. (10 vols.) San Diego: Academic Press. Articles in this encyclopedia fall into three categories (Levels). Level I provides expert overviews of a separation technique (e.g., flotation, distillation, crystallization, etc.). These articles provide an introduction to the subject and to the Level II articles. For example, the Level 1 article on chromatography provides the background for the Level II article that describes the theory, development, instrumentation, and practice of gas, liquid, and supercritical fluid chromatography. Level III articles then describe the various uses of the methods described in Levels I and II (e.g., sample preparation of pesticides or drugs and chromatographic techniques used for their analysis). The articles are all extensively cross-referenced and indexed, providing easy access to relevant information. Each article contains a selected bibliography of key books, review articles, and important research papers. The Appendix provides an extensive listing of essential guides and nomenclature recommendations. The Science Direct Web page provides access to the forward, preface, introduction, and all color figure inserts as well as the contents of each level, the appendix, and an author index.

HANDBOOKS AND COMPENDIA OF PROPERTY DATA

Handbooks provide fast access to information and are an important information resource in chemical engineering. The number of handbooks in the field is staggering and only a few will be listed here. The content of most is evident from the title. Many of the handbooks are available in electronic format as a growing number of publishers are developing electronic libraries for their reference materials. To find additional titles, search a library catalog or an Internet bookseller's list.

Property data are the sine qua non for both chemical engineers and students. An example would be vapor pressure, a measure of a liquid's volatility at a given temperature (e.g., the vapor pressure of a liquid is equal to atmospheric pressure at boiling-point). The design of storage tanks, for example, requires both hazard analysis and vent system technology, which in turn are based on the vapor pressure of the liquid being stored. Vapor-liquid operations, such as distillation, also require knowledge of vapor pressures as a function of temperature.

HANDBOOKS

Albright, L. F., ed. 2009. *Albright's Chemical Engineering Handbook*. Boca Raton, FL: CRC Press.

Branan, C. 2005. *Rules of Thumb for Chemical Engineers: A Manual of Quick, Accurate Solutions to Everyday Process Engineering Problems,* 4th ed. Boston: Elsevier. CHEMnetBASE http://www.chemnetbase.com/ (accessed December 15, 2010). CHEMnetBASE is a collection of handbooks and other reference materials from CRC and Chapman Hall. Searching the database is free, but to view the full text of the resources requires a subscription.

Chopey, N. P. 2004. *Handbook of Chemical Engineering Calculations,* 3rd ed. New York: McGraw-Hill.

Green, D. W., J. D. Maloney, and R. H. Perry. 2008. *Perry's Chemical Engineers Handbook,* 8th ed. New York: McGraw-Hill. Perry's has sections on both chemical and physical property data, and chemical engineering fundamentals; processes operations (heat transfer, distillation, kinetics); construction materials, process machinery, waste management, and safety. Also available in electronic format.

Griskey, R. G. 2000. *Chemical Engineers' Portable Handbook*. New York: McGraw-Hill.

Haynes, W. M and D. R. Lide, eds. 2010. *CRC Handbook of Chemistry and Physics,* 91st ed. Boca Raton, FL: CRC Press. The *CRC Handbook* is an excellent starting point for physical property data searches. Data tables listing both specific chemical compound data (e.g., bp, mp, mw) and physical properties (e.g., azeotropic data for binary mixtures, critical solution temperatures of polymer solutions) are provided along with literature references for additional data.

Knovel http://www.knovel.com/web/portal/main (accessed March 29, 2011). Knovel's Chemistry and Chemical Engineering section lists books on the following topics: analytical chemistry, dispersion and aggregation, electrochemistry, environmental chemistry, equipment, general references, industrial chemistry and chemicals, industrial safety, physical chemistry, polymer chemistry, separation, and transport processes. Although the book selection is both eclectic and heterogeneous, it both allows display of a book's contents and provides keyword searching results that display chapter title references. In addition, Knovel offers "free" access to several physical property databases (with personal registration): Knovel Critical Tables, *International Critical Tables of Numeric Data,* and

Smithsonian Physical Tables (9th rev. ed.). For example, searching Knovel http://www. knovel.com/knovel2/ default.jsp for "diethyl ether" and "vapor pressure," and limiting to chemistry and chemical engineering, quickly identifies the following titles (which will be easily found in academic libraries or those serving chemical engineers in industry): *Chemical Properties Handbook* (vapor pressure; organic compounds —Live Eqns), *Perry's Chemical Engineers Handbook* (vapor pressures of organic compounds, up to 1 atm), *Industrial Solvents Handbook* (vapor pressure of various ethers), and *Knovel Critical Tables* (basic physical properties of common solvents).

Reynolds, J. P., J. S. Jeris, and L. Theodore. 2007. *Handbook of Chemical and Environmental Engineering Calculations.* Hoboken, NJ: John Wiley & Sons.

Riegel, E. R. and J. A. Kent, eds. 2007. *Riegel's Handbook of Industrial Chemistry,* 11th ed. New York: Springer.

PROPERTY DATA

Daubert, T. E. and R. P. Danner. 1989–. *Physical and Thermodynamic Properties of Pure Chemicals: Data Compilation.* Washington, D.C.: Taylor & Francis. This title is an encyclopedic guide to pure chemical properties. Sponsored by the American Institute of Chemical Engineers' Design Institute for Physical Property Data (DIPPR), it contains physical and thermodynamic property data for 1,708 compounds. Tables for each compound include both physical (e.g., melting point, dipole moment, refractive index) and temperature-dependent properties (e.g., vapor pressure, heat capacity, viscosity). An excellent source of uniform data for either engineers or students. Fundamental SI units are used.

DIPPR Data Compilation of Pure Compound Properties http://dippr.byu.edu/ (subscription required). The *DIPPR Chemical Database* consists of both experimental data and temperature-dependent properties for over 2,000 pure chemicals. Data have been evaluated, correlated, and checked for thermodynamic consistency. Datasets, DIPPR-approved property constants, and regressed correlation coefficients for temperature-dependent properties are included. Calculation of temperature-dependent properties is possible in a variety of units (Standard, cgs, or British). Compounds are searched with either chemical names (or name fragments) or Chemical Abstracts Registry Numbers. Individual compound records also include both CAS and IUPAC names as well as trivial names.

Knovel. See entry under Handbooks.

Landolt-Börnstein: Substance/Property Index http://lb.chemie.uni-hamburg.de/ *The Landolt-Börnstein Substance/Property Index* has both a keyword Search and browsable Indices (Molecular Formula, Element System, CAS-RN, Chemical Name, and Zeolite) for compounds with data in the New Series volumes. The Indices entries link to compound descriptions (including structure diagrams, chemical names/synonyms), European Regulations and Landolt–Bornstein Citations/References. A physical property listing of the indexed volumes is also provided.

NIST Chemistry WebBook http://webbook.nist.gov/chemistry/ (accessed November 10, 2010). The *NIST Chemistry WebBook* provides Internet access to chemical and physical property data for nearly 50,000 chemical species (compounds, ions, radicals, etc.). The data are derived from collections maintained by both the NIST Standard Reference Data Program and outside contributors. The available data include: thermodynamic, gas phase, IR spectrum, condensed phase, mass spectrum, phase change, UV/Vis spectrum, reaction, vibrational and electronic energy levels, ion energetics, constants of diatomic molecules, ion cluster, and Henry's Law.

The *NIST Chemistry WebBook* supports a variety of searches for chemical species. Each search type has its own associated Web page. Search types currently available are general, physical property, formula, chemical name, CAS registry number, ion energetics, ionization energy, electron affinity, proton affinity, acidity, vibrational and electronic energy levels, molecular weight, structure and substructure, reaction, and author. Searches involving physical quantities also may be restricted by chemical formula. Searching for Electron Affinity = 0–10), Formula = C H and clicking "Allow more atoms of elements in formula than specified" will list all the hydrocarbons in the database with electron affinity data. Truncation is also allowed. C5H? will find all species with five carbon atoms and one or more hydrogen atoms. C5C*H? will search for species with five or more carbon atoms and one or more hydrogen atoms. Either SI or calorie based unit systems may be specified.

NIST/ASME Steam Properties Database: version 2.21 http://www.nist.gov/srd/nist10.cfm (accessed November 10, 2010) (purchase required). Thermophysical properties include: in the STEAM Database: temperature, Helmholtz energy, thermodynamic derivatives, pressure, Gibbs energy, density, fugacity, thermal conductivity, volume, isothermal compressibility, viscosity, dielectric constant, enthalpy, volume expansivity, dielectric derivatives, internal energy, speed of sound, Debye–Hlickel slopes, entropy, Joule–Thomson coefficient, refractive index, heat capacity, surface tension. The STEAM database generates tables and plots of property values. Vapor-liquid-solid saturation calculations with either temperature or pressure specified are available.

Parry, W. T. 2009. *ASME International Steam Tables for Industrial Use*, 2nd ed. New York: ASME (American Society of Mechanical Engineers). A standard set of thermodynamic and transport properties for water and steam. Based on the International Association for the Properties of Water and Steam's Formulation 1997. This is a complementary reference to the *NIST/ASME Steam Properties Database* http://www.nist.gov/srd/nist10.cfm (accessed November 10, 2010) Based upon the new International Association for the Properties of Water and Steam (lAPWS) 1995 formulation for general and scientific use for the thermodynamic properties of water, this updated version provides water properties from the international standards over a wide range of conditions.

SpringerMaterials http://www.springermaterials.com/navigation is a subscription database based on the Landolt–Börnstein New Series (LBNS), but also includes searchable files of thermophysical property data from the Dortmund Data Bank Software and Separation Technology (DDBST), the Linus Pauling Files (LPF), a solid state materials database with structure, diffraction, constitution, intrinsic property, and bibliographic information, and chemical safety documents. *SpringerMaterials* provides access to over 400 print LBNS volumes and property data for over 250,000 substances, 44,000 chemicals safety documents, over 300,000 data points from the DDBST, and data from over 190,000 documents from the LPF. *SpringerMaterials* is freely searchable and the contents of individual LB volumes are freely browsable under the Bookshelf tab. The Periodic Table tab allows searching for element systems (e.g., binary and ternary alloys of manganese). Use of Advanced Search is strongly recommended and for nonsubscribers, the [i] icon links to the bibliographic information for the LB print volumes.

Thermodex: An Index of Selected Thermodynamic and Physical Property Resources http://www.lib.utexas.edu/thermodex/ (accessed November 10, 2011). *ThermoDex* from the Mallet Chemistry Library, University of Texas–Austin is a freely searchable index to printed and Web-based compilations of thermochemical and thermophysical data. The *ThermoDex* home page also provides a guide to Finding Thermodynamic Information/Where to Start, which lists some of the standard handbooks and proprietary databases that are not indexed as well as those that are indexed. Searches for compounds, or preferably compound types, linked to a specific physical property or properties, display a list of handbooks that *may* contain data of interest. Actual data values are not displayed. *ThermoDex* records include the book's title, a brief abstract defining the scope and arrangement, properties and types of compounds, etc. Web datasets include URL links. While based primarily on the holdings of UT-Austin's Mallet Chemistry Library, most of these resources will be available in most university and major industrial libraries.

Yaws, C. 2001. *Matheson Gas Data Book,* 7th ed. Parsippany, NJ: Matheson TriGas; New
York: McGraw-Hill. The Matheson Gas Data Book contains individual sections with
property data on over 150 industrial gaseous elements and compounds (from acetylene to
xenon). Data include, thermodynamic properties, IR spectra, vapor pressure-temperature
curves, Henry's Law constants, explosion limits, and viscosity.

Zahlenwerte und Funktionen aus Physik, Chemie, Astronomie, Geophysik und Technik
(Landolt-Bornstein), 6th ed. 1950–1981. Berlin, Springer-Verlag. Commonly called *Landolt–
Bornstein* after the original editors, this massive work was intended to be a compilation of
all critically evaluated physical property data in the fields of physics, chemistry, astronomy,
geophysics, and technology. The 6th edition (in German) was published in four volumes
(Band), with multiple parts (Teile), between 1950 and 1981. The sections are: Atomic and
Molecular Physics, vol. 1, parts 1–5; Properties of Matter in Its Aggregated State, vol. 2,
parts 1–10; Astronomy and Geophysics, vol. 3, parts 1–4; Technology, vol. 4, parts 1–4.

Zahlenwerte und Funktionen aus Naturwissenschaften und Technik (Landolt-Boernstein),
Numerical Data and Functional Relationships in Science and Technology, New Series.
1961–. Berlin: Springer-Verlag.
Group I Nuclear and Particle Physics
Group II Atomic and Molecular Physics
Group III Crystal and Solid State Physics
Group IV Macroscopic and Technical Properties of Matter
Group V Geophysics and Space Research
Group VI Astronomy, Astrophysics, and Space Research
Group VII Biophysics
Group VIII Advanced Materials and Technologies

Because of the difficulty in adhering to the rigid overall plan of the 6th edition, a New Series of
Landolt-Bornstein was begun in 1961 and is arranged to accommodate future information without creating new editions. The preface, table of contents, and introductory chapter in each volume
is given both in English and German. There is an additional volume on Units and Fundamental
Constants in Physics and Chemistry, an index volume for both the 6th edition and the New Series,
and a freely available Web index for the New Series:

Subvolume a: Units in Physics and Chemistry, 1991.
Subvolume b: Fundamental Constants in Physics and Chemistry, 1992. Comprehensive (keyword) Index: 6th edition 1950–1980 and New Series 1961–1990.

TEXTBOOKS AND MONOGRAPHS

Textbooks

Textbooks are selected to support individual classes and are revised frequently, so it is difficult to
provide a definitive list. Many university libraries purchase the textbooks used in the classes and
they reflect the expert opinion of the professors as to some of the best resources available to support
their instruction. A very select list follows.

Bird, R. B. E., S. Warren, and E. N. Lightfoot, eds. 2007. *Transport Phenomena,* 2nd rev. ed.
New York: John Wiley & Sons. *Transport Phenomena* is a fundamental engineering concept with a wide variety of applications. This text and the first edition (1966) are generally
considered classics.

Callen, H. 2005. *Thermodynamics and an Introduction to Themostatics*, 2nd ed. New York:
John Wiley & Sons.

Cussler, E. L. 2009. *Diffusion: Mass Transfer in Fluid Systems,* 3rd ed. Cambridge, U.K.:
Cambridge University Press.

Felder, R. M. and R. W. Rousseau. 2005. *Elementary Principles of Chemical Processes,* 3rd ed. New York/Chichester, U.K.: John Wiley & Sons. A text that introduces thermodynamics, unit operations, kinetics, and process dynamics. The third edition is revised to reflect curriculum changes that include biotechnology, environmental engineering, and microelectronics.

Folger, H. S.. 2009. *Elements of Chemical Reaction Engineering,* 4th ed. Upper Saddle River, NJ: Prentice Hall.

Hinds, W. C. 1999. *Aerosol Technology: Properties, Behavior, and Measurement of Airborne Particles,* 2nd ed. New York: John Wiley & Sons. An upper-division/graduate-level text covering, for example, bioaerosols, Brownian motion and diffusion, respiratory deposition models, measurement, and sampling.

Incropera, F. P. 2007. *Introduction to Heat Transfer,* 5th ed. Hoboken, NJ: John Wiley & Sons.

Kyle, B. G. 1999. *Chemical and Process Thermodynamics,* 3rd ed. Upper Saddle River, NJ: Prentice-Hall PTR. A text describing techniques, applications, and mathematical analysis, with problems and examples. The CD-ROM includes spreadsheets and programs for both numerical analysis and graphics. This edition has a chapter on modeling thermodynamic systems.

Lauffenburger, D. A. and J. J. Linderman. 1993. *Receptors, Models for Binding, Trafficking and Signalling.* New York: Oxford University Press. A classic text describing the principles of modern systems biology for chemical engineers.

Sankaranarayanan, K. 2010. *Efficiency and Sustainability in the Energy and Chemical Industries: Scientific and Case Studies,* 2nd ed. Boca Raton, FL: CRC Press.

Shul, R. J. 2000. *Handbook of Advanced Plasma Processing Techniques.* Berlin/New York: Springer. A multiauthored volume covering the fundamental physics of plasmas, diagnostics, modeling, and microelectronic applications for development of transistors, sensors, and more.

Smith, J. M. H. C. Van Ness, and M. Abbott. 2005. *Introduction to Chemical Engineering Thermodynamics,* 7th ed. Boston: McGraw-Hill.

Amazon.com has an interesting feature called Listmania, which provides an individual's selection of favorite texts and handbooks. For example, listed under the Amazon entry for *Perry's Chemical Engineers Handbook,* Platinum ed., a reader will find several lists of books recommended by users of the Website. This feature should prove useful for a wider sampling of textbooks because chemical engineering curricula are heavily dependent on faculty research interests.

RESEARCH MONOGRAPHS AND SYMPOSIA PUBLICATIONS

Research monographs and symposia publications are generally focused on very specific topics. Several publishers have active publication programs related to chemical engineering including Taylor & Francis, CRC Press http://www.crcpress.com, AIChE http://www.aiche.org/Publications/, DECHEMA, and ICE and the American Chemical Society http://www.pubs.acs.org. Some representative examples include: ACS Symposium series (American Chemical Society), AIChE Symposium Series (American Institute of Chemical Engineers), Cambridge series in Chemical Engineering; (Cambridge University Press), DECHEMA Chemistry Data Series (Frankfurt/Main: Dechema; Port Washington, NY: Distributed by Scholium International), Green Chemistry and Chemical Engineering (CRC), McGraw-Hill Chemical Engineering Series, Topics in Chemical Engineering (Oxford University Press), Prentice Hall International Series in the Physical and Chemical Engineering series, and Series in Chemical and Mechanical Engineering (Taylor & Francis) are some of the major series to consider.

AIChE's 2008 Centennial Celebration Committee has identified 30 authors and their groundbreaking chemical engineering books at: http://www.aiche.org/uploadedFiles/About/Centennial/30Authors. pdf and this is an excellent review of important works in the field.

CHEMICAL SAFETY

Safety is an essential aspect of chemical engineering. There are several ways to address these including the need for every research facility that handles chemicals to have a safety manual. A useful example is Caltech's Chemical Safety Manual (2010 revision) http://www.cce.caltech.edu/resources/Safety.pdf (accessed November 10, 2010).

Della-Giustina, D. 2010. *Developing a Safety and Health Program*. Boca Raton, FL: CRC Press. Covers topics such as safety plans, writing MSDS documentation, emergency planning, and more.

Hazardous Substances Data Bank (HSDB). HSDB http://sis.nlm.nih.gov/enviro/hsdbchemicalslist.html is a peer-reviewed toxicology data file on the National Library of Medicine's (NLM) Toxicology Data Network (TOXNET) http://toxnet.nlm.nih.gov/TOXLINE (accessed November 10, 2010). The primary focus is on the toxicology of over 5,000 potentially hazardous chemicals. It also includes information on chemical and physical properties, emergency handling procedures, environmental fate, regulatory requirements, and related areas. Data are extensively referenced from a wide variety of books, government reports, and the journal literature.

Material safety data sheets and other chemical safety sheets are another means to address safety issues. Material safety data sheets (MSDS) provide the proper procedures for handling or working with specific chemicals. In addition, MSDS include physical data, toxicity, health effects, first aid, reactivity, storage, disposal, protective equipment, and spill/leak procedures. As a general rule, an MSDS should be available for every on-site chemical.

Where to Find Material Safety Data Sheets Online, http://www.ilpi.com/msds/ (accessed November 10, 2010). Provides links to company and government sites, plus other links of interest. Some of these link to only a few chemicals produced by the specific company, others to tens of thousands or more.

Internet Public Library http://www.ipl.org/IPL/Finding?Key=msds&collection=gen (accessed November 10, 2010). Has a page on the topic with relevant links.

Material Safety Data Sheets, Cornell University, Department of Environmental Health and Safety http://www.ehs.cornell.edu/msds/msds.cfm (accessed June 3, 2005). A database of over 250,000 MSDS. Oklahoma State University http://ehs.okstate.edu/links/msds.htm (accessed November 10, 2010). Has a good site with links to a number of sources, including a very useful glossary.

Oxford University's Chemical and Other Safety Information http://msds.chem.ox.ac.uk/ http://www.ilpi.com/msds/ (accessed November 10, 2010). Includes a number of useful links beyond MSDS.

JOURNALS

The journal article, in any science or engineering discipline, is its primary literature. Articles in referred journals (e.g., those with editorial boards) have undergone critical review by both editors and reviewers. A journal's quality may be assessed by its ISI impact factor, which is calculated by dividing the number of a given year's citations by the number of source items published in that journal during the previous two years (e.g., citations in 2003/source items in 2001 and 2002). In addition to the journals specific to chemical engineering, many journals identified with other fields are found in chemical engineering libraries. Because each institution's selection is generally very idiosyncratic and dependent on current faculty interests, the list below gives the top 10 chemical engineering research journals indexed by ISI in 2003. One way of determining the journals important to a given community is to analyze the references in articles they have recently authored or requested.

CHEMISTRY NEWS PERIODICALS

Chemical and Engineering News (ACS) (0009-2347) *Chemical Engineering* (CE) (0009-2460) *Chemical Engineering Progress* (AIChE) (0360-7275) *Chemical Market Reporter* (CMR) (1092-0110) *Chemical Week* (CW) (0009-272X) *Chemistry and Industry* (SCI) (0009-3068)

CHEMICAL ENGINEERING REVIEWS

Advances in Biochemical Engineering/Biotechnology (Springer) (0724-6145) *Advances in Polymer Technology* (Wiley) (0730-6679) *Catalysis Reviews: Science and Engineering* (Taylor & Francis) (0161-4940) *Chemistry and Physics of Carbon* (Dekker) 0069-3138 *Critical Reviews in Biotechnology* (Taylor & Francis) (0738-8551) *Progress in Energy and Combustion Science* (Elsevier) (0360-1285) *Reviews in Chemical Engineering* (Freund) (0167-8299)

CHEMICAL ENGINEERING RESEARCH JOURNALS

Adsorption (Springer) (0929-5607)
AIChE Journal (AIChE/Wiley) (0001-1541)
Applied Catalysis (Elsevier) (0166-9834)
Biochemical Engineering Journal (Elsevier) (1369-703X)
Bioprocess and Biosystems Engineering (Springer) (1615-7591)
Canadian Journal of Chemical Engineering (NRC) (0008-4034)
Catalysis Letters (Springer) (1011-37 2X)
Catalysis Today (Elsevier) (0920-5861)
Chemical Engineering and Processing (Elsevier) (0255-2701)
Chemical Engineering and Technology (Wiley-VCH) (0930-7516)
Chemical Engineering Communications (Taylor & Francis) (0098-6445)
Chemical Engineering Education (CEE) (0009-2479)
Chemical Engineering Journal (Elsevier) (1385-8947)
Chemical Engineering Research and Design (ICE) (0263-8762)
Chemical Engineering Science (Elsevier) (0009-2509)
Chemie-Ingenieur-Technik (Wiley) (0009-286X)
Combustion and Flame (Elsevier) (0010-2180)
Combustion Science and Technology (Taylor & Francis) (0010-2202)
Computers and Chemical Engineering (Elsevier) (0098-1354)
Energy and Fuels (ACS) (0887-0624)
Environmental Progress (AIChE/Wiley) (0278-4491)
Fluid Phase Equilibria (Elsevier) (0378-3812)
Fuel (Elsevier) (0016-2361)
Fuel Processing Technology (Elsevier) (0378-3820)
Heat and Mass Transfer (Springer) (0947-7411)
Heat Transfer Engineering (Taylor & Francis) (0145-7632)
Industrial and Engineering Chemistry Research (ACS) (0888-5885)
International Communications in Heat and Mass Transfer (Elsevier) (0735-1933
International Journal of Chemical Reactor Engineering (Berkeley Electronic Press) (1542-6580)
International Journal of Heat and Mass Transfer (Elsevier) (0017-9310)
Journal of Aerosol Science (0021-8502)
Journal of Applied Polymer Science (Wiley) (0021-8995)
Journal of Biotechnology (Elsevier) (0168-1656)
Journal of Catalysis (Elsevier) (0021-9517)
Journal of Chemical Engineering Data (ACS) (0021-9568)

Journal of Chemical Engineering of Japan (SCE) (0021-9592)
Journal of Chemical Technology and Biotechnology (SCI/Wiley) (0268-2575)
Journal of Loss Prevention in the Process Industries (Elsevier) (0950-4230)
Journal of Membrane Science (Elsevier) (0376-7388)
Journal of Process Control (Elsevier) (0959-1524)
Journal of Separation Science (Wiley) (1615-9306)
Journal of Supercritical Fluids (Elsevier) (0896-8446)
Macromolecular Materials and Engineering (Wiley) (1438-7492)
Microporous and Mesoporous Materials (Elsevier) (1387-1811)
Organic Process Research and Development (ACS) (1083-6160)
Polymer Engineering and Science (SPE/Wiley) (0032-3888)
Polymer Reaction Engineering (Taylor & Francis) (1054-3414)
Powder Technology (Elsevier) (0032-5910)
Process Biochemistry (Elsevier) (1359-5113)
Process Safety and Environmental Protection (ICE) (0957-5820)
Process Safety Progress (AIChE/Wiley) (1066-8527)
Separation and Purification Reviews (Taylor & Francis) (1542-2119)
Separation and Purification Technology (Elsevier) (1383-5866)
Separation Science and Technology (Taylor & Francis) (0149-6395)
Theoretical Foundations Chemical Engineering (Springer) (0040-5795)
Transport in Porous Media (Springer) (0040-5795)

PATENTS

Patents are an excellent source of information on particular inventions or chemical processes. A book that would be useful to all chemical engineers is Le-Nhung McLeland's *What Every Chemist Should Know about Patents*. 2002. Washington, D.C.: American Chemical Society. Patents may be searched by patent number, inventor, assignee, keywords, or classification numbers. *SciFinder* also allows searching of chemical patents with CAS Registry Numbers. Both *Web of Science* (1945+) and *SciFinder* (1997+) provide patent citation searching. The Chemical Abstracts Service routinely abstracts the first publicly available document. This is often a Japanese, German, European, or PCT (WO) patent application, which may be the first in what becomes a family of related patent documents. Chemical Abstracts' Chemical Abstracts defines a patent family as those patents (applications and granted patents that describe the same invention: http://www.cas.org/ASSETS/ E416C9A3D97E44C884AE0DC008F71BC2/patentfamily.pdf

> *SciFinder* includes information on both chemical patents and patent family members from 44 countries. Since 1907, Chemical Abstracts (CA) has abstracted and indexed over three million patents worldwide in all areas of chemistry and chemical engineering. Patent titles may be altered by the Chemical Abstracts Service to make them more descriptive. *SciFinder* is an excellent resource for article, paper, and patent searching, since both patent numbers and inventor/assignee names are searchable from 1907 to date; bibliographic and abstract information for patents from major industrial countries is available within two days of publication, full indexing is completed within twenty-seven days, and patent family information is available from 1957 to date.
>
> The *CAplus* file is an excellent resource for prior art searching because it includes both journal articles and patents. Bibliographic information and abstracts from a number of agencies, both United States and international, is available within two days of publication and full indexing is completed within 30 days. Patent Family data is available from 1957 and appears with the bibliographic information. Additional information on patent searching in *SciFinder* is available through excellent guides and tutorials from their Web site.

Several journals routinely provide listings of recent patents. These include *Applied Catalysis B: Environmental, Chemistry and Industry, Drug Discovery Today, Environmental International, Microporous and Mesoporous Materials*, and *Organic Process Research and Development*. Other works on patents that might be useful include: *Kirk–Othmer Encyclopedia of Chemical Technology's* section on Patents and Trade Secrets, a basic, step by-step approach to the practice and management of patents and trade secrets, and Patents Literature, a basic guide to sources, databases, searching, and more. J. M. Cogen's Technical Disclosures: Advanced Tactics (2001) in *Chemical Innovation* 31 (7): 33–38 is also useful reading: http://pubs.acs.org/subscribe/archive/ci/31/i07/html/07cogen.html (accessed November 10, 2010). It describes the use of technical disclosures to establish prior-art and preclude patentability and/or avoid costs of patenting.

ASSOCIATIONS, ORGANIZATIONS, SOCIETIES, AND CONFERENCES

American Chemical Society, Division of Biochemical Technology http://acsbiotn.a1whs.com/BIOT/ (accessed November 10, 2010). "The mission of this division is to promote the exchange of information among academic and industrial researchers regarding technology utilizing life-based systems to produce useful products and services" (from Web page).

American Institute of Chemical Engineers (AIChE) http://www.aiche.org/conferences/ (accessed November 10, 2010). AIChE is an international association of professions in all fields of chemical engineering.

American Society for Engineering Education, Chemical Engineering Division http://www.engr.uky.edu/~aseeched/ (accessed November 10, 2010) (includes link to conferences). "The Chemical Engineering Division of the ASEE is dedicated to the promotion and improvement of Chemical Engineering Education. The purpose of the Chemical Engineering Division site is to provide information about ASEE ChED events, awards, meetings, and more" (from Web page).

DECHEMA http://www.dechema.de/About_the_DECHEMA-lang-en.html (accessed May 18, 2005). DECHEMA (Gesellschaft fur Chemische Technik und Biotechnologie e.V.) was founded in 1926, and currently has over 5,500 personal and institutional members. Their focus is the promotion of research and technological advances in chemical engineering, biotechnology, and environmental protection. DECHEMA is responsible for the organization of the ACHEMA, ACHEMASIA, and ACHEMAMERICA exhibition congresses, as well as a wide variety of local and international workshops, congresses, and colloquia through its subject divisions (Zeolite, Catalysis, Membrane technology, Biotechnology, Adhesive technology, Safety technology, and Reaction engineering). DECHEMA is also a database publisher (CEABA-VtB —a chemical technology and biotechnology bibliographic database, CHEMSAFE—for explosion and fire protection safety parameters for flammable gases, liquids, and dusts, and DETHERM—a compilation of thermophysical properties of pure substances and mixtures).

Division of Fuel Chemistry http://www.anl.gov/PCS/acsfuel/ (accessed November 10, 2010). "The ACS Division of Fuel Chemistry provides a forum for documentation and communication to the international community of research and development results, in order to promote efficient and environmentally acceptable fuel production and use" (from Web page).

Division of Industrial and Engineering Chemistry http://iecdivision.sites.acs.org/objectives.htm (accessed November 10, 2010). "I&EC is a multidisciplinary division helping individuals convert science into commercially relevant products and processes" (from Web page). Division of Petroleum Chemistry http://petr.sites.acs.org/ (accessed May 18, 2005). "A vibrant professional network of scientists and engineers interested in the chemistry of petroleum exploration, production, refining, and utilization" (from Web page).

Division of Polymeric Materials: Science and Engineering http://pubs.acs.org/meeting-preprints/order.html (accessed May 18, 2005). "Providing a forum for the exchange of technical information on the chemistry of polymeric materials including plastics, paints, adhesives, composites and biomaterials" (from Web page).

STYLE GUIDES

The Chicago Manual of Style, 16th ed. 2010. Chicago: University of Chicago Press. While the CMS is a general standard in style guides, chemistry has an excellent subject specific guide.

Coghill, A. M. and L. R. Garson, eds. 2006. *The ACS Style Guide; Effective Communication of Scientific Information*, 3rd ed. A critical resource for those who plan to write and publish in the chemical engineering field.

WEB SITES AND LISTSERVS

The use of listservs has changed as technology changed. While a number of organizations still use listservs, many have moved to other means of communication, including wikis and blogs. For a list of groups in chemical engineering, a search of Yahoo! Groups for "engineering chemical" brings up a list to browse as does searching Google Groups. In many instances, the best listserv may be from a professional organization. To search blogs, go to Google and select blogs. A search of chemical engineering turns up a myriad links and you can search by a more specific topic.

International Directory of Chemical Engineering URLs http://www.rccostello.com/links/chem-eng.html (accessed November 10, 2010). A comprehensive listing of databases, newsgroups, publishers, organizations, companies, and more. Page is not dated so updating is unknown.

University of Delaware, Internet Resources for Chemical Engineering http://www2.lib.udel.edu/subj/chee/internet.htm (accessed November 10, 2010). Very good portal to information on organizations, reference data, and more.

University of Santa Barbara, InfoSurf: Information Resources for Chemical Engineering http://www.library.ucsb.edu/subjects/chemeng/chemeng.html (accessed November 10, 2010). Another university site with an excellent list of professional organizations.

Yahoo! Chemical Engineering http://dir.yahoo.com/Science/engineering/chemical_engineering (accessed November 10, 2010). Site includes employment opportunities as well as links to organizations, discussion groups, and other Internet sites.

CONCLUSION

Information searching is a talent that must be developed and maintained. Imagination, persistence, and, for librarians, an ethic of service to others is absolutely essential.

While a wide variety of resources are available electronically, there are a significant number of print resources available in both research and technical libraries. This highlights one of the ever-present dangers of overusing electronic resources, namely the inability to make serendipitous discoveries in books on the shelf near the one you are looking for or simply avoiding the better resource for your purpose that happens only to be available in print. As with all engineering disciplines, the chemical engineer needs to rely not only on the primary resources of his own profession, but to refer to the reference materials of chemistry. This chapter is a survey of the major works in chemical engineering. The research in both engineering and chemistry is important, supporting the significance of the role of the librarian in assisting the practitioners to stay current in their knowledge.

REFERENCES

American Institute of Chemical Engineers, What Do ChemEs Do? http://www.aiche.org/Students/CareersWhatDoChemEDo.aspx (accessed May 2, 2011)

CACHE Teaching Resource Center http://www.cache.org.

Cussler, E. L., et al. 2002. Refocusing chemical engineering. *Chemical Engineering Progress* 98, 1: 26s–31s.

Edgar, T. F. 2000. Chemical engineering education and the three C's: Computing, communication, and collaboration. *CACHE Newsletter* (Fall) http://www.che.utexas.edu/cache/newsletters/fall2000_chemengedu.pdf

Fan, L-S, et al. 2008. How can chemical engineers best impact society issues? *AIChE Centennial*. 1908–2008. http://www.aiche.org/uploadedFiles/About/DepartmentUploads/PDFs/Ch27.pdf (accessed December 13, 2010).

Halford, B. 2004. Chemical engineering education in flux. *Chemical Engineering News* 82 (10): 34.

MATLAB http://www.mathworks.com/products/matlab/

MIT OpenCourseWare Web site http://ocw.mit.edu/courses/chemical-engineering/

Molecular modeling, viewing and drawing http://ep.llnl.gov/msds/orgchem/ molmodl.html

Price, G. 2003. Amazon debuts new book search tool. *Publishers Weekly* October 27, 2003. http://searchenginewatch.com/3098831> (accessed May 18, 2005).

Ritter, S. K. 2001. The changing face of chemical engineering. *Chemical Engineering News* 79 (23): 63–66. http://pubs.acs.org/cen/education/7923/7923education.html

8 Civil Engineering

Carol Reese and Michael Chrimes

CONTENTS

Introduction 131
Literature of the Civil Engineering Field 131
General Civil Engineering 131
    Searching the Library Catalog 131
    Abstracts and Indexes 132
    Dictionaries 132
    Handbooks and Manuals 133
    Journals 134
    Web Sites 134
    Associations, Organizations, and Societies 134
Geotechnical Engineering 135
    Searching the Library Catalog 135
    Abstracts and Indexes 136
    Bibliography 136
    Dictionaries 137
    Handbooks and Manuals 137
    Monographs 138
    Journals 139
    Web Sites 140
    Professional Societies and Other Organizations 141
    Conferences 142
Maritime Engineering 143
    Searching the Library Catalog 143
    Abstracts and Indexes 143
    Handbooks and Manuals 144
    Monograph 145
    Journals 145
    Web Sites 145
    Professional Societies and Other Organizations 146
    Conferences 147
Construction Management 148
    Introduction 148
    Searching the Library Catalog 148
    Handbooks 149
    Journals 149
    Monographs 150
    Web Resources 153
    Institutions, Organizations, and Societies 153
Municipal Engineering 155
    Introduction 155
    Searching the Library Catalog 156

Abstracts and Indexing Services 156
Handbooks.. 156
Journals 158
Monographs.. 158
Web Sites.. 160
Institutions.. 160
Nanotechnology 162
Introduction.. 162
Searching the Library Catalog 162
Associations 163
Dictionary and Handbooks... 163
Monographs.. 164
Web Sites.. 165
Conferences.. 167
Journals 167
Structural Engineering 168
Introduction.. 168
Bibliography... 170
Searching the Library Catalog 170
Abstracts and Indexes.. 170
Databases.. 170
Bibliographies 171
Handbooks and Manuals 171
Monographs and General Works.. 172
Concrete 174
Iron and Steel 175
Timber 175
Masonry.. 176
Other Materials... 176
Journals 177
    General Titles 177
    Concrete... 178
    Steel 179
Web Sites.. 179
Associations, Organizations, and Societies... 180
Water Engineering.. 183
Introduction.. 183
Bibliography... 183
Searching the Library Catalog 183
Abstracting and Indexing Services... 184
Dictionaries 185
Handbooks.. 185
Journals 186
Monographs.. 187
Web Resources 189
Associations, Organizations, and Societies... 190
Standards... 193
Introduction.. 193
Societies, Associations, and Organizations... 193

Sustainable Development...195
    Searching the Library Catalog 196
    Dictionary and Handbooks..196
    Monographs...197
    Associations 198
    Conferences..199
    Journals 199
    Web Sites..200
References..201

INTRODUCTION

While people have been building structures since ancient times, the profession of civil engineering was not defined until 1828 when the Institution of Civil Engineers in London, England, received its charter. The charter stated that civil engineering was "the art of directing the great sources of power in nature for the use and convenience of man, as the means of production and of traffic in states, both for external and internal trade, as applied in the construction of roads, bridges, aqueducts, canals, river navigation, and docks for internal intercourse and exchange, and in the construction of ports, harbors, moles, breakwaters, and lighthouses, and in the art of navigation by artificial power for the purposes of commerce, and in the construction and application of machinery, and in the drainage of cities and towns" (Merdinger, 1953) (Ferguson and Chrimes, 2011).

In 1961, the American Society of Civil Engineers expanded on this definition by stating that it was "a profession in which a knowledge of the mathematical and physical sciences gained by study, experience, and practice is applied with judgment to develop ways to utilize, economically, the materials and forces of nature for the progressive well-being of humanity in creating, improving, and protecting the environment, in providing facilities for community living, industry, and transportation, and in providing structures for the use of humanity" (ASCE, 2004, p. 118).

Based on these definitions, this chapter will cover the areas of general civil engineering, construction management, geotechnical engineering, maritime engineering, municipal engineering, nanotechnology, structural engineering, sustainable development, and water resources engineering. Due to the size of the fields, energy, environmental, and transportation engineering have their own chapters, although there is much overlap between these fields of engineering and the fields discussed in this chapter.

LITERATURE OF THE CIVIL ENGINEERING FIELD

The literature of civil engineering is scattered throughout the various specialized subject areas of the field. While engineering is as old as human history, its technical literature is a recent development. For this reason, information is mainly found in the journal literature, handbooks and manuals, standards, and conference proceedings. Professional societies play an important part in the development of the literature both for civil engineering in general and for its specialized areas. Because its literature is still in the developmental state and is scattered throughout the specialized fields, there are no single guides to the literature or encyclopedias to the entire field.

GENERAL CIVIL ENGINEERING

SEARCHING THE LIBRARY CATALOG

Searching for general information on civil engineering is simple. There is one subject heading to look under: Civil engineering. Under this heading, one will find all the subheadings such as Civil engineering: biographies, Civil engineering: dictionaries, Civil engineering: handbooks and manuals. The Library of Congress classification code for civil engineering is the same for general engineering: TA 1-2040.

Abstracts and Indexes

American Society of Civil Engineers' Civil Engineering Database http://www.asce.org/ (accessed May 13, 2010). The Civil Engineering Database (CEDB) is an index to all ASCE publications. This includes all journals, proceedings, books, standards, manuals, and *CE Magazine*. Journal papers with abstracts are covered back to 1970, while those without abstracts are covered back to the 1920s. Book records are complete dating back to the early 1900s. The database is constantly being expanded.

CSA/ASCE Civil Engineering Abstracts http://www.csa.com/ (accessed May 13, 2010) provides citations, abstracts, and indexing of the serials literature in civil engineering and its complementary fields, including forensic engineering, management and marketing of engineering services, engineering education, theoretical mechanics and dynamics, and mathematics and computation. This database provides comprehensive international coverage as well as numerous nonserial publications. Many of the more recent records in the database include fields containing cited references, corresponding author's e-mail address, and publisher contact information. Sources covered include over 3,000 periodicals, conference proceedings, technical reports, trade journal/newsletter items, patents, books, and press releases.

Emerald's International Civil Engineering Abstracts http://www. emeraldinsight.com/ (accessed May 13, 2010). *International Civil Engineering Abstracts (ICEA)* provides online access to over 142,000 abstracts dating back to 1976 from the 150 most prominent journals in civil engineering. The *ICEA* accredited journal coverage list is compiled from the library holdings of civil engineering centers of excellence worldwide, such as the Institute of Civil Engineers and Imperial College, London. but does not cover, for example, all ICE journals.

ICONDA—the International Construction Database http://www.irb.fraunhofer.de/iconda/ (accessed May 13, 2010). *ICONDA®Bibliographic* is one of the most comprehensive systems worldwide for retrieval of planning and building-related publications. The database was first launched in 1986. Information provided in the bibliographical records includes bibliographical source information, keywords and/or abstracts in English language, links to the full text, or ordering information, if available (83,436 URLs, as per Jan. 1, 2008). Coverage includes types of indexed documents: periodicals, books, research reports, conference proceedings, business reports, theses, and nonconventional literature. Chronological indexing ranges from 1975 to the present (publication dates of individual sources may date back even earlier). It is available commercially via OVID.

Dictionaries

American Society of Civil Engineers. 1972 and 1991. *Biographical Dictionary of American Civil Engineers*. (2 vols.) New York: American Society of Civil Engineers. These volumes provide biographical sketches that emphasize the engineering accomplishments of individual engineers born before 1900.

Blockley, D. 2005. *New Penguin Dictionary of Civil Engineering*. New York: Penguin Group. This is a completely new and fully comprehensive edition aimed at anyone who has contact with the construction industry. Written in a simple, easily accessible fashion, the book contains major features on sustainability and the environment, hazard, vulnerability, and uncertainty management together with new categories on business and finance, sustainability, water, and transport. This edition offers easy and extensive cross referencing and definitions designed to help students to learn and associate ideas making this the perfect guide for anyone interested in any aspect of civil engineering.

Chrimes, M., A. Skempton, R. W. Rennison, R. C. Cox, T. Ruddock, and P. Cross-Rudkin, eds. 2002. *Biographical Dictionary of Civil Engineers in Great Britain and Ireland, Vol. 1: 1500–1830.* London: Thomas Telford Ltd. A reference work on the lives, works, and careers of individuals engaged in the practice of civil engineering. It provides the background, training, and achievements of engineers who began their careers before 1830.

Cross-Rudkin, P. S. M., M. M. Chrimes, M. R. Bailey, R. C. Cox, R. W. McWilliam, and R. W. Rennison, eds. 2008. *Biographical Dictionary of Civil Engineers, Vol. II: 1830–1890.* London: Thomas Telford Ltd. This book presents biographical details of nearly 800 leading practitioners of the Victorian era, including many that have never been written about until now. It outlines the lives of these engineers and lists the works for which they were responsible, and provides indexes of names and places.

Kurtz, J-P, ed. 2004. *Dictionary of Civil Engineering.* New York: Kluwer Academic. This English version of the French publication (*Dictionnaire du Genie Civil*, published in 1997) is a valuable reference tool for civil engineers. There are over 12,000 definitions accompanied by more than 1,300 charts, tables, and graphs. This dictionary is a comprehensive compilation of definitions, examples, and descriptions from the study of soils and the various materials and equipment used, including the most common architectural terms as they relate to civil engineering. This compendium will be an invaluable tool not only for civil engineers, but also for lawyers, contractors, architects, and so on, and all trade associations involved with this discipline.

Webster, L. 1997. *Wiley Dictionary of Civil Engineering and Construction.* New York: John Wiley & Sons. This dictionary provides a broad coverage of technical disciplines, such as architecture, engineering, building, construction, forestry, and mining. Entries include terms, concepts, names, abbreviations, techniques, and tools. All entries are thoroughly cross-referenced.

HANDBOOKS AND MANUALS

Blake, L. S., ed. 2004. *Civil Engineer's Reference Book,* 4th ed. London: Elsevier. This handbook provides a survey of the fundamentals, theory, and current practice in the different branches of civil engineering. It gives practical guidance for both student and practicing civil engineers. In addition, it incorporates chapters on construction, site practice, and contract management.

Chen, W-F. and J. Y. R. Liew. 2003. *Civil Engineering Handbook,* 2nd ed. Boca Raton, FL: CRC Press. This is a comprehensive handbook that covers nearly every aspect of civil engineering including design techniques, construction methods, instrumentation, material, properties, and calculations. The volume includes bibliographic references and an index.

Hicks, T. G., ed. 2007. *Handbook of Civil Engineering Calculations,* 2nd ed. New York: McGraw-Hill. Over 500 key calculations covering the entire field of civil engineering including, but not limited to, structural engineering, timber engineering, soil mechanics, fluid mechanics, and wastewater treatment. Examples of each calculation procedure are included. New civil engineering data on "green" buildings and their design, better qualifying them for LEED (Leadership in Energy and Environmental Design) ratings have been added.

Merritt, F. S., M. K. Loftin, and J. Ricketts, eds. 2003. *Standard Handbook for Civil Engineers,* 5th ed. New York: McGraw-Hill. This handbook is a guide to the principles and techniques for effective civil engineering practice. It covers all areas of the field, such as highway construction, design, and geotechnical engineering. It also contains information on the various codes and standards, and over 700 tables, formulas, and drawings to clarify every explanation and procedure.

JOURNALS

Professional societies are usually the publishers of the serials that provide general coverage of the field. Unlike the more technical, specialized journals, these publications provide the nonspecialist with a good understanding of the variety of training needed by civil engineers in order to produce the many different projects required of them. These publications are usually designated as official publications of the societies. Examples would be ASCE's *Civil Engineering Magazine*, the Canadian Society for Civil Engineering's *Canadian Civil Engineer*, and South African Society of Civil Engineering's *Civil Engineering*.

WEB SITES

There are several portals on civil engineering that can provide guidance to important Web resources.

ASCE's Infrastructure Report Card for America's Infrastructure http://www.infrastructurereportcard.org/ (accessed March 24, 2011). Since 1988, the American Society of Civil Engineers has been evaluating the state of infrastructure in the United States. These report cards cover such areas as bridges, dams, drinking water, levees, railways, and roads. Summaries of these report cards are available online.

iCivilEngineer.com http://www.icivilengineer.com (accessed June 4, 2010). This portal was designed for civil engineering professionals and students. It has sections on civil engineering news, career sites, and Web resources in civil engineering arranged by subject.

Intute http://www.intute.ac.uk/ (accessed June 4, 2010). Intute is a free online service that helps you find the best Web resources for your studies and research. With millions of resources available on the Internet, it can be difficult to find useful material. Our subject specialists review and evaluate thousands of resources to help you choose the key Web sites in your subject. Intute is created by a consortium of seven universities working together with a whole host of partners. The Intute consortium includes: University of Birmingham, University of Bristol, Heriot-Watt University, The University of Manchester, Manchester Metropolitan University, University of Nottingham, and Oxford University.

Techxtra http://www.techxtra.ac.uk/ (accessed June 4, 2010). TechXtra is a free service that can help you find articles, books, the best Web sites, the latest industry news, job announcements, technical reports, technical data, full text e-prints, the latest research, thesis and dissertations, teaching and learning resources, and more, in engineering, mathematics, and computing.

TenLinks.com: Ultimate Civil Engineering Directory http://www.tenlinks.com/engineering/civil/index.htm (accessed June 4, 2010). This site is aimed at helping professionals to find technical information faster. TenLinks uses technical professionals to comb through the Internet catalogs to locate the most relevant sites.

ASSOCIATIONS, ORGANIZATIONS, AND SOCIETIES

Most countries have their local civil engineering society. To determine if a particular country has such a society, check the online portals, such as Intute http://www.intute.ac.uk/ to find complete lists. Some of the major organizations are listed below.

American Society of Civil Engineers http://www.asce.org (accessed June 9, 2010). Founded in 1852, it is the oldest national engineering society in the United States. It supports conferences, continuing education courses, professional development of the field, and publishes journals, monographs, and conference proceedings.

Asian Civil Engineering Coordinating Council (ACECC) http://www.acecc.net/ (accessed June 9, 2010). The ACECC works to promote and advance the science and practice of civil engineering for sustainable development in the Asian region. To achieve this aim, ACECC sponsors the Civil Engineering Conference in the Asian Region (CECAR) every three years.

Canadian Society for Civil Engineering http://www.csce.ca (accessed June 9, 2010). Founded in 1887, the society works to develop and maintain high standards of civil engineering practice in Canada. It does so by supporting the work of academics, private institutions, and different organizations that deal with civil engineering. Working with sister organizations, the CSCE also promotes civil engineering among the general public and governmental institutions in Canada and abroad.

Engineer Research and Development Center, U.S. Army Corps of Engineers http://www.erdc.usace.army.mil/ (accessed June 22, 2010). The mission of the Engineer Research and Development Center (ERDC) is to provide science, technology, and expertise in engineering and environmental sciences in support of our armed forces and the nation to make the world safer and better.

Federation of Civil Engineering Associations of the Mexican Republic (FECIC) http://www.fecic.org.mx/ (accessed June 9, 2010). The society, working with the schools and compatible associations, works to develop civil engineering in Mexico in order to contribute to the viable development of their society.

Institution of Civil Engineers (ICE) http://www.ice.org.uk (accessed June 9, 2010). Established in 1818, ICE seeks to advance the knowledge, practice, and business of civil engineering. It aims to promote the value of the civil engineer's global contribution to sustainable economic growth through supporting research, conferences, and publications.

Japan Society of Civil Engineers http://www.jsce-int.org/ (accessed June 9, 2010). The society was established as an incorporated association in 1914 entrusted with the mission to contribute to the advancement of scientific culture by promoting the field of civil engineering and the expansion of civil engineering activities. Since its establishment, JSCE has endeavored to achieve the above mission, through extensive activities including scientific exchange among members, researchers/promotion of science and technologies relating to the field of civil engineering, social involvement, etc. Over the years, the JSCE membership has increased significantly and is currently engaged in various wide-ranged activities around the world.

GEOTECHNICAL ENGINEERING

Geotechnical engineering, originally called soil mechanics, is the study of the physical properties and utilization of soils, especially soils used in the planning of foundations for structures and subgrades for highways. Soil is a natural aggregate that has three phases: solid, liquid, and gaseous. Geotechnical engineers study the stresses put upon soils by the weight of different structures (Encyclopedia Britannica, 2011). Today, geotechnical engineering embraces subjects, such as rock mechanics, engineering geology, and tunneling.

SEARCHING THE LIBRARY CATALOG

Geotechnical engineering covers a wide area of subject areas. The following Library of Congress Subject Headings are a sample of just some of the major subject headings for the field:

Earthquake engineering
Earthwork
Embankments
Excavation Fills (earthwork)
Foundations
Geology Hydrogeology
Retaining walls

Sediments (Geology)
Soil dynamics
Soil mechanics: Soils, testing
Tunnels
Underground construction

Materials are also scattered throughout various classification codes.

GB3-5030 Physical geography, including Slopes GB448, Hydrogeology GB 1001-1199.8
QE1-996.5 Geology, including Mineralogy QE351-399.2, Dynamic and structural geology
    QE500-639.5, Volcanoes and earthquakes, QE521-545
TA 703-712 Engineering geology
TA 715-787 Foundations, earthwork
TA 800-820 Tunnels

ABSTRACTS AND INDEXES

Earthquake Engineering Abstracts (EEA). Available through CSA http://www.csa.com
    (accessed June 9, 2010). EEA provides coverage of such topics as geotechnical earthquake
    engineering, engineering seismology, and soil dynamics. It is produced by the National
    Information Service for Earthquake Engineering, University of California, Berkeley.

GeoRef http://www.agiweb.org (accessed June 9, 2010). Produced by the American
    Geological Institute since 1966, it provides access to the geoscience literature of the
    world. *GeoRef* is the most comprehensive database in the geosciences and continues to
    grow by more than 100,000 references a year. The database contains over 3.1 million
    references to geoscience journal articles, books, maps, conference papers, reports, and
    theses. You can gain access to this vast amount of information through searching on the
    Web, online, or on CDs.

GeoScienceworld http://www.geoscienceworld.org/ (accessed June 9, 2010). A comprehen-
    sive Internet resource for research and communications in the geosciences, built on a core
    database aggregation of peer-reviewed journals indexed, linked, and interoperable with
    GeoRef.

Quakeline Database. Produced by the Multidisciplinary Center for Earthquake Engineering
    Research http://mceer.buffalo.edu (accessed June 9, 2010), a bibliographic database devel-
    oped and maintained by the Information Service. It covers earthquakes, earthquake engi-
    neering, natural hazard mitigation, and related topics. Additional features include records
    for various publication types, such as books, journal articles, conference papers, technical
    reports, CDs, slides, and videos. *QUAKELINE®* was launched in 1987 and is updated on
    a monthly basis.

BIBLIOGRAPHY

Atkinson, J. H. 2008. *The Essence of Geotechnical Engineering: 60 Years of Geotechnique.*
    London: TTL. A set of specially commissioned papers outlining the development of geo-
    technical engineering.

Terzaghi, K. 1960. *From Theory to Practice in Soil Mechanics.* New York: John Wiley. Karl
    Terzaghi is considered to have started modern soil mechanics with his theories of consoli-
    dation, lateral earth pressures, bearing capacity, and stability. He pioneered a great range
    of methods and procedures for investigation, analysis, testing, and practice that defined
    much of the geotechnical engineering field (Goodman, 2002). These seminal papers and
    reports are important to anyone who wishes to understand the field.

DICTIONARIES

Barker, J. A. 1981 *Dictionary of Soil Mechanics and Foundation Engineering.* New York: Construction Press. This volume presents definitions of a wide variety of terms dealing with soil mechanics and foundations that range from a single sentence to an entire paragraph. It also provides cross references within the publication.

Somerville, S. H. and M. A. Paul. 1983. *Dictionary of Geotechnics.* London: Butterworths. Definitions in the areas of soil and rock mechanics, hydrology, ground stabilization, and excavation are included. This volume also provides a series of tables dealing with a variety of aspects of geotechnical engineering at the end of the book.

Van der Tuin, J. D., ed. 1989. *Elsevier's Dictionary of Soil Mechanics and Geotechnical Engineering.* New York: Elsevier Science. This dictionary provides definitions of terms in multiple languages covering such topics as soils, rocks, sediments, geological survey methods, geophysics, and the application of geotextiles.

HANDBOOKS AND MANUALS

Brown, R. W. 2000. *Practical Foundation Engineering Handbook,* 2nd ed. New York: McGraw-Hill Professional Publishing. From site assessment through design and construction to remediation of failed foundations, this handbook provides design alternatives for substandard soil and challenging site conditions with example problems for different types of structures. It also has illustrations, charts, tables, and case study examples.

Chen, W-F and C. Scawthorn. eds. 2003. *Earthquake Engineering Handbook.* Boca Raton, FL: CRC Press. This is a comprehensive resource that covers the spectrum of topics relevant to designing for and mitigating earthquakes. The handbook presents engineering practices, research, and developments in North America, Europe, and the Pacific Rim countries. Included are formulas, tables, and illustrations to answers to practical questions.

Day, R. W. 2001. *Soil Testing Manual: Procedures, Classification Data and Sampling Practices.* New York: McGraw-Hill Professional Publishing. For engineers, geologists, contractors, and onsite construction managers, this handbook simplifies each step of the process of soil testing from selecting the appropriate method to analyzing the results. It has handy tables, charts, diagrams, and formulas.

Day, R. W. 2002. *Geotechnical Earthquake Engineering Handbook.* New York: McGraw-Hill Publishing. This handbook features field and laboratory testing methods and procedures, current seismic codes, site improvement methods, in-depth analysis of soils, and problems with solutions to illustrate these analyses.

Fang, H-Y. 1991. *Foundation Engineering Handbook.* New York: Van Nostrand Reinhold. This handbook provides coverage throughout the field of applied geotechnics including explanations, methods, and examples of more efficient analysis, design, and construction of foundations.

Gunaratne, M. 2006. *Foundation Engineering Handbook.* Boca Raton, FL: CRC Press. It presents both classical and state-of-the-art design and analysis techniques for earthen structures, and covers basic soil mechanics and soil and groundwater modeling concepts along with the latest research results. It addresses isolated and shallow footings, retaining structures, and modern methods of pile construction monitoring, as well as stability analysis and ground improvement methods. The handbook also covers reliability-based design and LRFD (Load Resistance Factor Design).

Hunt, R. E. 2005. *Geotechnical Engineering Investigation Handbook,* 2nd ed. Boca Raton, FL: CRC Press. The *Geotechnical Engineering Investigation Handbook* provides the tools necessary for fusing geological characterization and investigation with critical analysis for obtaining engineering design criteria. The second edition updates this pioneering reference for the 21st century, including developments that have occurred in the 20 years since the first edition was published, such as: remotely sensed satellite imagery, global positioning systems (GPS), geophysical exploration, cone penetrometer testing, and earthquake studies.

Look, B. G. 2007. *Handbook of Geotechnical Investigation and Design Tables*. Boca Raton, FL: CRC Press. This practical handbook of properties for soils and rock contains, in a concise tabular format, the key issues relevant to geotechnical investigations, assessments, and designs in common practice. In addition, there are brief notes on the application of the tables. These data tables are compiled for experienced geotechnical professionals who require a reference document to access key information. There is an extensive database of correlations for different applications. The book should provide a useful bridge between soil and rock mechanics theory and its application to practical engineering solutions.

Macnab, A. 2002. *Earth Retention Systems Handbook*. New York: McGraw-Hill Professional. This publication discusses temporary excavation shoring and earth retention systems used to construct permanent structures inside them. Each chapter presents a different shoring system and describes how it is constructed and the equipment needed.

Pansu, M. and J. Gauthetrou. 2006. *Handbook of Soil Analysis: Mineralogical, Organic and Inorganic Methods*. New York: Springer. This handbook is a reference guide for selecting and carrying out numerous methods of soil analysis. It is written in accordance with analytical standards and quality control approaches. It covers a large body of technical information including protocols, tables, formulae, spectrum models, chromatograms, and additional analytical diagrams.

Paz, M., ed. 1994. *International Handbook of Earthquake Engineering: Codes, Programs, and Examples*. New York: Chapman & Hall. This handbook stresses the international approach to earthquake engineering. It presents national seismic codes from over 30 countries and demonstrates their application with quantitative examples.

Persson, P-A., R. Holmberg, and J. Lee. 1994. *Rock Blasting and Explosives Engineering*. Boca Raton, FL: CRC Press. This handbook covers the practical engineering aspects of the different kinds of rock blasting. It covers the fundamental sciences of rock mass and material strength, the thermal decomposition, burning, and detonation behavior of explosives. Based on practical industrial experience, the handbook provides both students and practitioners with a source for understanding the constructive use of explosives.

Smoltczyk, U., ed. 2003. *Soil Construction and Geotechnics*. (3 vols.) New York: John Wiley & Sons. Volume 1 of this handbook covers the basics in foundation engineering including laboratory and field tests. Volume 2 covers the geotechnical procedures used in manufacturing anchors and piles. The third volume deals with the basic designs of different types of foundations.

U.S. Bureau of Reclamation. 2004. *Earth Manual*. Washington, D.C.: Department of the Interior http://www.usbr.gov/pmts/writing/earth/index.html (accessed June 10, 2010). Available for downloading from the Bureau of Reclamation Web site, this manual covers the engineering of earthen structures. It also provides references and cross references of hundreds of terms.

MONOGRAPHS

Craig, R. F. 2004. *Soil Mechanics,* 7th ed. London: Spon Press. This book presents the fundamental principles of soil mechanics and shows how they can be applied in practical situations. Worked examples are included to reinforce the principles discussed. A solutions manual is available.

Das, B. M. 2009. *Principles of Geotechnical Engineering,* 7th ed. Mason, OH: Cengage. An overview of soil properties and mechanics along with a study of field practices and basic soil engineering procedures are presented. Numerous case studies are included.

Terzaghi, K., R. B. Peck, and G. Mesri. 1996. *Soil Mechanics in Engineering Practice,* 3rd ed. New York: John Wiley & Sons. This volume presents both theoretical and practical knowledge of soil mechanics in engineering. The third edition provides expanded coverage of vibration problems, mechanics of drainage, and consolidation.

Tomlinson, M. J. 2001. *Foundation Design and Construction,* 7th ed. Upper Saddle River, NJ: Longman Group. This book covers such topics as site investigations, principles of foundation design, design of specific types of foundations, foundation construction, and shoring. Appendixes provide properties of materials and conversion tables.

JOURNALS

Bulletin of the Seismological Society of America (0037-1106). Started in 1911, this journal of the SSA covers the general areas of seismology, seismicity, earthquake engineering, seismic hazards, and the effects of earthquakes.

Canadian Geotechnical Journal (0008-3674). Published by the National Research Council of Canada, this bimonthly journal was started in 1963. It presents papers on foundations, excavations, soil properties, dams, embankments, slopes, rock engineering, waste management, frozen soils, and offshore soils.

Earthquake Engineering and Structural Dynamics (0098-8847). Published since 1972, this Wiley publication covers all aspects of engineering related to earthquakes including soil amplification and failure, and ground motion characteristics.

Engineering Geology (0013-7952). Published since 1966, this Elsevier journal includes original studies, case histories, and comprehensive reviews in the field of engineering geology.

Geosynthetics International (1072-6349). This official journal of the International Geosynthetics Society, which was started in 1994, is now published by ICE Publishing, Thomas Telford. It is now only available online and includes supplementary data and links.

Geotechnical Engineering: Proceedings of the Institution of Civil Engineers (1353-2618). This journal, published by ICE Publishing, Thomas Telford, is part of the proceedings of the ICE and the journal of the British Geotechnical Association. Published since 1836, it contains papers that are aimed at both the civil engineer and the geotechnical specialist.

Geotechnique (0016-8505). Started in 1948 and published by ICE Publishing, Thomas Telford, this journal includes papers in the fields of soil and rock mechanics, engineering geology, and environmental geotechnics. It publishes detailed case histories and theoretical research in these fields.

Ground Improvement: Proceedings of the Institution of Civil Engineers (1755-0750). Published by ICE Publishing, Thomas Telford, this journal publishes high-quality, practical papers relevant to engineers, specialist contractors, and academics involved in the development, design, construction, monitoring, and quality control aspects of ground improvement. It covers a wide range of civil and environmental engineering applications, including analytical advances, performance evaluations, pilot and model studies, instrumented case histories, and innovative applications of existing technology.

International Journal of Rock Mechanics and Mining Sciences (1365-1609). Published by Elsevier since 1964, this journal includes papers concerned with original research, new developments, site measurement, and case studies in rock mechanics and rock engineering.

Journal of Cold Regions Engineering (0887-381X). Published by the American Society of Civil Engineering since 1987, this journal covers topics such as ice engineering, ice force, cold weather construction, environmental quality in cold regions, snow and ice control, and permafrost.

Journal of Earthquake Engineering (1363-2469). Published now by Taylor & Francis, this journal publishes papers on research and development in analytical, experimental, and field studies of earthquakes from an engineering seismology as well as a structural engineering viewpoint.

Journal of Geotechnical and Geoenvironmental Engineering (1090-0241). Started in 1875 by the American Society of Civil Engineers, this journal covers a broad area of practice including areas such as foundations, retaining structures, soil dynamics, slope stability, dams, earthquake engineering, environmental geotechnics, and geosynthetics.

Soil Dynamics and Earthquake Engineering (0267-7261). Begun in 1982 by Elsevier, this journal publishes papers of applied mathematicians, engineers, and other applied scientists involved in solving problems related to the field of earthquake and geotechnical engineering.

Soils and Foundations (0038-0806). Started in 1961 by the Japanese Geotechnical Society, this journal publishes research papers, reports of engineering experiences, state-of-the-art reports on certain themes, and discussions.

Web Sites

Compendium of On-Line Soil Survey Information http://www.itc.nl/~rossiter/research/rsrch_ss.html (accessed June 11, 2010). Compiled by D. G. Rossiter of the Department of Earth Systems Analysis, International Institute for Geoinformation Science and Earth Observation, it brings together online information on soil survey activities, institutions, datasets, research, and teaching materials worldwide.

GeoEngineer http://www.geoengineer.org/ (accessed June 11, 2010). Resources and educational information for geoengineering practice and research are available through this resource guide. In addition, the *International Journal of Geoengineering Case Histories* and the *Geoengineer Newsletter* are available.

Geoforum.com http://www.geoforum.com/ (accessed June 11, 2010). Geoforum.com is intended to be an information and communication service for the geotechnical industry worldwide. The site includes a directory containing a classified listing of individuals, companies, and organizations active in the area of geotechnology. A discussion forum, events postings, a multilingual dictionary containing over 700 geotechnical terms in four languages (translation between English, French, German, and Swedish is available), and a unit converter are available.

GeoIndex http://www.geoindex.com/ (accessed June 11, 2010). Developed for the geo-environmental professional by Datasurge Company, this search engine covers geotechnical, environmental, hydrogeological, geological, mining, and petroleum subject areas. This information is divided into four categories: companies, associations, education, and government.

GeoTechLinks http://www.geotechlinks.com/ (accessed June 11, 2010). GeotechLinks provides links to online free books, manuals, thesis, papers, software, and other resources in geotechnical engineering.

Mineral Resources Program http://minerals.er.usgs.gov/ (accessed June 11, 2010). The USGS Mineral Resources Program (MRP) provides scientific information for objective resource assessments and unbiased research results on mineral potential, production, consumption, and environmental effects. The MRP is the sole federal source for this information. The MRP funds research to address two major program functions: (1) research and assessment (provides information for land planners and decision makers about where mineral commodities are known and suspected in the Earth's crust and about the environmental consequences of the presence of those commodities. MRP supports an on-going effort to coordinate the development of national-scale geologic, geochemical, geophysical, and mineral resource databases and the migration of existing databases to standard models and formats that are available to both internal and external users); and (2) data collection, analysis, and dissemination (describes current production and consumption of about 100 mineral commodities, both domestically and internationally for approximately 180 countries. This program provides funding to communicate current information on available mineral resources).

National Information Service for Earthquake Engineering http://nisee.berkeley.edu (accessed June 11, 2010). The National Information Service for Earthquake Engineering (NISEE) is based at the University of California, Berkeley, and is sponsored by the National Science Foundation (NSF). NISEE provides access to technical research and development information in earthquake engineering and related fields of structural dynamics, geotechnical engineering, engineering seismology, and earthquake hazard mitigation policy.

Natural Resources Conservation Service Soils http://soils.usda.gov/ (accessed June 11, 2010). This site provides information on soil risks and hazards important to urban planners and construction contractors, soil data, and soil education.

PROFESSIONAL SOCIETIES AND OTHER ORGANIZATIONS

American Geological Institute (AGI) http://www.agiweb.org/ (accessed June 15, 2010). Founded in 1948, the AGI provides information services to the geoscience community. The institute also serves as a voice for shared interests in the profession, plays a major role in strengthening geosciences education, and works to increase public understanding of the role of the geosciences in society's use of natural resources.

American Rock Mechanics Association http://armarocks.org/ (accessed June 15, 2010). This association provides research, education and training, and a digital library for those professionals, companies, teachers, and students in the field of rock mechanics and rock engineering.

American Soil and Foundation Engineers (ASFE) http://www.asfe.org/ (accessed June 15, 2010). This trade association of earth engineering and related applied science services firms was founded in 1969. Its aim is to help its members prosper through professionalism. To do this, ASFE provides a variety of services, programs, and materials to help its members enhance their skills.

American Underground Construction Association http://www.auca.org (accessed June 15, 2010). This organization is involved in all aspects of underground facilities: planning, design, development, construction, and use.

Association of Environmental and Engineering Geologists http://www.aegweb.org/ (accessed June 15, 2010). This association provides leadership, advocacy, and applied research in environmental and engineering geology.

British Geotechnical Association (BGA) http://bga.city.ac.uk/cms/ (accessed June 15, 2010). As the principal association for geotechnical engineers in the United Kingdom, the BGA plays a coordinating role in the United Kingdom with the International Society of Soil Mechanics and Geotechnical Engineering. It arranges programs to enable discussion of topics important to geotechnical engineers with the Institution of Civil Engineers.

Canadian Geotechnical Society (CGS) http://www.cgs.ca/ (accessed June 15, 2010). The CGS exists to serve and promote the geotechnical and geoscience community in Canada.

Deep Foundations Institute (DFI) http://www.dfi.org/ (accessed June 15, 2010). Started in 1976, the DFI is a technical association of firms and individuals in the deep foundations and related industry. It offers its members international conferences, seminars, networking opportunities, and publications on the latest technologies.

Environmental and Engineering Geophysical Society (EEGS) http://www.eegs.org/ (accessed June 15, 2010). The EEGS was formed in 1992 to promote the science of geophysics, to foster common scientific interests of geophysicists, and to maintain a high professional standing among its members. The society holds an annual conference and produces various publications in the field.

Geo-Institute http://www.geoinstitute.org/ (accessed June 15, 2010). Established by the American Society of Civil Engineers in 1996 as a specialty organization focused on the geo-industry, its major aims are to advance the state of the art, and to provide leadership on the professional, business, public policy, and educational aspects of the field. To accomplish these aims, it sponsors conferences, seminars, and educational programs. The institute is involved in the publication of journals, newsletters, manuals, and geotechnical special publications.

International Geosynthetics Society (IGS) http://www.geosyntheticssociety.org (accessed June 15, 2010). The IGS works for the scientific and engineering development of geosynthetics and related technologies. It publishes newsletters, journals, and sponsors conferences.

International Society for Rock Mechanics (ISRM) http://www.isrm.net/ (accessed June 15, 2010). The ISRM was founded in Salzburg in 1962. Its main objectives are to encourage international collaboration, to encourage teaching and research, and to promote high professional standards. The ISRM sponsors international congresses and publishes a news journal.

International Society for Soil Mechanics and Geotechnical Engineering http://www.issmge.org/ (accessed June 15, 2010). This society promotes international cooperation among engineers and scientists for the advancement and dissemination of knowledge in the field of geotechnical engineering. In order to do this, its members may submit papers to many conferences, have access to the society's work in various fields, and have networking opportunities. The Web site has a growing body of documentation available.

International Tunneling and Underground Space Association http://www.ita-aites.org/ (accessed June 15, 2010). The association holds meetings, organizes studies, and publishes reports and proceedings in order to promote advances in planning, design, construction, maintenance, and safety of tunnels and underground space.

National Geophysical Data Center (NGDC) http://www.ngdc.noaa.gov/ (accessed June 15, 2010). The NGDC provides scientific leadership, products, and services for geophysical data describing the solid earth, marine, and solar-terrestrial environment. The center contains more than 400 digital and analog databases.

National Information Service for Earthquake Engineering (NISEE) http://nisee.berkeley.edu/ (accessed June 15, 2010). Sponsored by the National Science Foundation and the University of California, Berkeley, the NISEE provides access to technical research and development information in earthquake engineering and related fields.

Seismological Society of America http://www.seismosoc.org/ (accessed June 15, 2010). Founded in 1906 in San Francisco, its aim is to advance earthquake science, and it represents a variety of technical interests.

Soil Science Society of America http://www.soils.org/ (accessed June 15, 2010). The society's purpose is to advance the discipline and practice of soil science.

United States Geological Survey (USGS) http://www.usgs.gov/ (accessed June 15, 2010). Started in 1879, the USGS provides reliable scientific information to describe and understand the Earth and to manage water, biological, energy, and mineral resources.

World Organization of Dredging Associations (WODA) http://www.woda.org/ (accessed June 15, 2010). The WODA incorporates the Western Dredging Association, which serves the Americas; the Central Dredging Association, which serves Europe, Africa, and the Middle East; and the Eastern Dredging Association, which serves the Asian and Pacific regions.

CONFERENCES

Conference on Deep Foundations http://www.dfi.org/ (accessed June 15, 2010). Annually for over 30 years, contractors, academics, suppliers, and manufacturers from the deep foundations construction industry have met to learn from each other and discuss the state of the art related to the design and construction of deep foundations.

International Conference of the International Association for Computer Methods and Advances in Geomechanics http://www.iacmag.org/index.html (accessed June 15, 2010). Started in 1972 and sponsored by IACMAG, these conferences, held every three years, cover such topics as computer modeling and their applications to a wide range of geomechanical problems.

Quadriennial International Conference on Soil Mechanics and Geotechnical Engineering http://www.issmge.org (accessed June 15, 2010). This conference, first held in 1936, is the most important in the geotechnical field. Regional conferences are organized on a similar basis by continent.

MARITIME ENGINEERING

Maritime engineering deals with all aspects of waterway environments. These engineers are concerned with such issues as ocean exploration, design of offshore structures, wave action on coastlines and ports, and the protection of wetlands (Britannica, 2011).

SEARCHING THE LIBRARY CATALOG

When looking for materials related to ocean engineering, the following are some of the Library of Congress Subject Headings to search:

Coastal engineering
Harbors
Marine geotechnique
Ocean-atmosphere
Ocean engineering
Ocean waves
Offshore structures
Shore protection
Water masses

There are two main Library of Congress classification code areas for ocean engineering topics: GC and TC.

GC 1-1581 Oceanography
GC 96-97.8 Estuarine oceanography
GC 151-155 Density
GC 190-190.5 Ocean-atmosphere interaction
GC 229-296.8 Currents
TC 183-201 Dredging, submarine building
TC 203-380 Harbors and coast protective works
TC 530-537 River protective works
TC 601-791 Canals and inland navigation. Waterways

ABSTRACTS AND INDEXES

ASFA: Aquatic Sciences and Fisheries Abstracts http://www.csa.com/ (accessed June 16, 2010). ASFA provides coverage of the field since 1971. It covers publications on such topics as dynamical oceanography, underwater acoustics, marine meteorology, marine technology, offshore structures, and underwater vehicles. It is updated monthly by Cambridge Scientific Abstracts.

Oceanic Abstracts http://www.csa.com/ (accessed June 16, 2010).This publication covers the worldwide technical literature related to the marine and brackish water environment. It covers such topics as ecology, marine geology, marine pollution, nonliving marine resources, and navigation.

HANDBOOKS AND MANUALS

Clark, J. R. 1996. *Coastal Zone Management Handbook.* Boca Raton, FL: CRC Press. This handbook covers such topics as natural resources, economics, development, productivity, and diversity of coastal zones. The book discusses three aspects of management: strategies, methods, and information. The final section is a collection of project histories.

El-Hawary, F., ed. 2000. *Ocean Engineering Handbook.* Boca Raton, FL: CRC Press. This is a comprehensive compilation of information on the theory and practice of oceanic/coastal engineering. It covers such topics as modeling considerations, marine hydrodynamics, and applications of computational intelligence. The handbook also includes over 200 tables and figures on ocean engineering.

Gerwick, B. C., Jr. 2007. *Construction of Marine and Offshore Structures,* 4th ed. Boca Raton, FL: CRC Press. This handbook provides a comprehensive treatment of the construction aspects of offshore structures. It discusses floating structures, deep-water structures, ice-resistant structures, and bridge foundations. It also details all the particulars of building in the marine environment.

Herbich, J., ed. 2000. *Handbook of Dredging Engineering,* 2nd ed. New York: McGraw-Hill Professional. Providing an up-to-date guide to dredging theory and practice, this handbook covers such topics as fluid mechanics, dredging equipment, sediment, dredging methods, environmental effects, and project planning.

Kennish, M. J., ed. 2001. *Practical Handbook of Marine Science,* 3rd ed. Boca Raton, FL: CRC Press. This handbook contains information on physical oceanography, marine biology, marine chemistry, geology, and pollution. It contains over 800 tables and figures.

Mader, C. L. 2004. *Numerical Modeling of Water Waves,* 2nd ed. Boca Raton, FL: CRC Press. This manual covers all aspects of this topic from basic fluid dynamics and the basic models to the most complex, including the compressible Navier Stokes techniques to model waves generated in various ways.

McCartney, B., et al., eds. 2005. *Report on Ship Channel Design. Manual of Practice 107.* Reston, VA: American Society of Civil Engineers. This manual provides an overview of the design process and identifies the studies usually needed for a ship channel project.

Task Committee of the Waterways Committee. 1998. *Inland Navigation: Locks, Dams, and Channels.* Reston, VA: American Society of Civil Engineers. Based on the experience of the U.S. Army Corps of Engineers, this manual provides information on planning, designing, construction, and operation of the U.S. waterways used by barge traffic.

Task Committee of the Waterways and Navigation Engineering Committee of the Coasts, Oceans, Ports, and Rivers Institute of ASCE. 2008. *Navigation Engineering Practice and Ethical Standards.* Reston, VA: American Society of Civil Engineers. This manual presents engineering criteria and practices for the design, operation, and management of navigation projects and shows how to integrate them with engineering ethics.

Task Committee on Underwater Investigations. 2001. *Underwater Investigations: Standard Practice Manual.* Reston, VA: American Society of Civil Engineering. This manual presents guidelines in such areas as standards of practice, documentation and reporting, repair design, and the inspection of unique structures.

Thoresen, C. A. 2010. *Port Designer's Handbook.* London: Thomas Telford. Port design in all its aspects from a European perspective.

MONOGRAPH

Randall, R. E. 1997. *Elements of Ocean Engineering.* Jersey City, NJ: Society of Naval Architects and Marine Engineers. This text addresses the application of engineering principles for the analysis, design, development, and management of marine systems.

Reddy, D. V. 2011. *Essentials of Offshore Structures: Theory and Applications.* Boca Raton, FL: CRC Press.

JOURNALS

Coastal Engineering Journal (0578-5634). Published by World Scientific, it is a peer-reviewed medium for the publication of research achievements and engineering practices in the fields of coastal, harbor, and offshore engineering.

Coastal Engineering: An International Journal for Coastal, Harbor, and Offshore Engineers (0378-3839). Started in 1977, this Elsevier publication combines practical application with modern technological achievements. It presents studies and case histories on all aspects of coastal engineering, such as waves and currents, coastal morphology, and estuary hydraulics.

International Journal of Offshore and Polar Engineering (1053-5381). The principal journal of The International Society of Offshore and Polar Engineers (ISOPE), it publishes research in the fields of offshore, ocean, polar, marine, environment, mechanics, and materials engineering.

Journal of Atmospheric and Oceanic Technology (0739-0572). This journal, published by the American Meteorological Society, publishes papers on the instrumentation and methodology used in atmospheric and oceanic research, such as computational techniques and methods of data acquisition.

Journal of Waterway, Port, Coastal, and Ocean Engineering (0733-950X). Published by the American Society of Civil Engineers, this journal covers such topics as dredging, floods, ice, sediment transport, and wave action. It also publishes papers on the development and operation of ports, harbors, and offshore facilities.

Marine Engineering: Part of the Proceedings of the Institution of Civil Engineers (1741-7597). Started in 2004 by the ICE, this journal publishes papers that focus on such topics as fixed and moving port and harbor developments, estuarine and coastal protection, habitat creation, and seabed pipelines.

Marine Structures (0951-8339). This journal aims to provide a medium for presentation and discussion of the latest developments in research, design, fabrication, and in-service experience relating to marine structures, i.e., all structures of steel, concrete, light alloy or composite construction having an interface with the sea, including ships, fixed and mobile offshore platforms, submarine and submersibles, pipelines, subsea systems for shallow and deep ocean operations, and coastal structures such as piers.

Ocean Engineering: An International Journal of Research and Development (0029-8018). Published since 1968, this journal covers such areas as design and building of structures, submarine soil mechanics, coastal engineering, fabrication of materials, hydrodynamic properties of shells, ocean energy, propulsion systems, and underwater acoustics.

WEB SITES

Coastalmanagement.com http://www.coastalmanagement.com (accessed June 18, 2010). A portal site for resources related to coastal management. It provides links to news groups, research, courses, and to many different sectors, such as marinas, navigation, dredging, ocean engineering, and coastal environments.

EUCC Coastal Guide http://www.coastalguide.org/ (accessed June 18, 2010). A portal site aimed at professionals in coastal management, planning, conservation, and research in Europe. It provides access to such services as EUCC (European Union for Coastal Conservation) coastal news, publications, coastal guide country files, ecosystems, various relevant policies, and trends in the field.

Martindale's Calculators' Online Center/Engineering Center http://www.martindalecenter. com/Calculators4_E_Nav.html (accessed June 16, 2010). Marine Engineering and Ocean Engineering Center (calculators, applets, spreadsheets, and where applicable includes: courses, manuals, handbooks, simulations, animations, videos, etc.).

NOAA Coastal Services Center http://www.csc.noaa.gov/ (accessed June 18, 2010). This agency is devoted to serving the state and local coastal resource management programs. It provides training classes, fellowships, online mapping of coastal areas, and software tools.

Offshore Engineering Information Service (OEIS) http://www.techxtra.ac.uk/offshore/ (accessed June 18, 2010). OEIS gives information about publications and meetings dealing with: oil and gas exploration and production; offshore health, safety,, and environmental protection; resources of the seabed and renewable energy; and marine technology.

USGS Science Center for Coastal and Marine Geology http://woodshole.er.usgs.gov/ (accessed June 18, 2010). This center explores and studies many aspects of the underwater areas between shorelines and the ocean. The center has modeled various aspects of the coasts and ocean waters, such as tidal flushing and circulation. It concentrates on the Eastern coastal waters while other centers concentrate on the other coastlines.

PROFESSIONAL SOCIETIES AND OTHER ORGANIZATIONS

Association of Coastal Engineers http://www.coastalengineers.org/ (accessed June 22, 2010). Started in 1999, the association is dedicated to excellence in education, research, and the practice of coastal engineering. Among other activities, it holds an annual conference where such topics as the sciences of oceanography, design construct, and monitor coastal structures, beach nourishment, and harbor design are discussed.

Civil and Offshore Engineering Research http://www.eng.ox.ac.uk/research/cao.html (accessed June 16, 2010). Research in this area is mainly aimed at improving the design of structures, having regard for the way they are placed and used. Computation and experimentation are being developed in parallel to provide proper and safe rules for design based on validated models of behavior for a wide range of applications.

Coasts, Oceans, Ports, and Rivers Institute (COPRI) http://www.coprinstitute.org/ (accessed June 22, 2010). Established by the American Society of Civil Engineers in 2000 as a semi-autonomous institute, COPRI officially replaced the Waterway, Port, Coastal, and Ocean Division of ASCE to serve as the multidisciplinary and international leader in improving the knowledge, teaching, development, and practice of civil engineering and other disciplines working in waterway environments. It supports conferences, publications, continuing education seminars, and workshops in all areas of maritime engineering.

IEEE/Oceanic Engineering Society http://www.oceanicengineering.org/index.cfm (accessed June 16, 2010). The society's objectives shall be scientific, literary, and educational in character. The society shall strive for the advancement of the theory and practice of electrotechnology, allied branches of engineering, and related arts and sciences, applied to all bodies of water, and the maintenance of high professional and ethical standards among its members and affiliates.

Institute of Marine Engineering, Science and Technology http://www.imarest.org/ (accessed June 22, 2010). Recognizing the need to bring together marine engineers, scientists, and technologists, the institute's aim is to promote the scientific development of these three related areas by integrating their different interests into one organization. Its main activities are publishing specialized journals and books, and organizing conferences and seminars.

International Navigation Association (PIANC) http://www.pianc.org/ (accessed June 22, 2010). As a nonpolitical and nonprofit organization, PIANC brings together the best international experts on technical, economic, and environmental issues pertaining to waterborne transport infrastructure. Members include national governments and public authorities, corporations, and interested individuals. Providing expert guidance and technical advice PIANC provides guidance to public and private partners through high-quality technical reports. The association's international working groups develop regular technical updates on pressing global issues to benefit members on shared best practices.

International Society of Offshore and Polar Engineers (ISOPE) http://www.isope.org/ (accessed June 22, 2010). International in scope, ISOPE is interested in promoting engineering and scientific progress in the fields of offshore and polar engineering. It does this by sponsoring technical conferences, publications, scholarship programs, continuing education, and international cooperation. The Annual International Offshore and Polar Engineering Conference is the largest conference held on these topics.

Maritime Research Institute Netherlands http://www.marin.nl/web/show (accessed June 16, 2010). This organization has a dual mission: to provide industry with innovative design solutions, and to carry out advanced research for the benefit of the maritime sector as a whole. In this way, the institute strengthens the link between academic research and market needs. It is a unique interaction that benefits all parties concerned. The driving force behind this dual mission is a team of highly motivated and experienced people. MARIN is innovative, independent, and, above all, reliable.

Maritime Technology Society http://www.mtsociety.org (accessed June 22, 2010). Since the 1960s, the society has had a mission to disseminate marine science and technical knowledge, promote education, advance research in the field, and engender a broader understanding of the field in the general public.

Office of Coast Survey http://www.nauticalcharts.noaa.gov/ (accessed June 22, 2010). A part of the National Oceanic and Atmospheric Administration (NOAA), the Coast Survey is the oldest scientific organization in the United States. The Office of Coast Survey produces navigational charts and related publications, provides information on navigational obstructions and hydrographic surveys, and supports research in the field.

Offshore Engineering Society http://www.oes.org.uk/ (accessed June 22, 2010). The society provides networking opportunities for engineers in all related disciplines and works to disseminate technical information through lectures and other events.

Scripps Institution of Oceanography http://sio.ucsd.edu/ (accessed June 22, 2010). Founded in 1903 as a biological research laboratory, Scripps now encompasses physical, chemical, biological, geological, and geophysical studies of the oceans. Part of the University of California, it supports ongoing investigations into the topography and composition of the ocean bottom, waves and currents, and the interactions between seawater and the atmosphere.

Society of Naval Architects and Marine Engineers (SNAME) http://www.sname.org/ (accessed June 22, 2010). Started in 1893, the society works to advance the art, science, and practice of naval architecture, shipbuilding, and marine engineering. It sponsors applied research, offers career guidance, disseminates information, and supports educational programs. SNAME is interested in all types of marine ships from yachts to submersibles.

CONFERENCES

International Breakwaters Conferences http://www.ice-breakwaters.com/ (accessed April 1, 2011). This conference covers "design, performance, installation and maintenance of breakwaters, seawalls, jetties, piers and beach control structures" (from Web site).

International Conference on Coastal Engineering http://www.coprinstitute.org/ (accessed June 22, 2010). This international conference covers fields such as coastal processes and climate change, flood and coastal defense engineering, flood risk management, ports and harbors, and legislation.

International Conference on Ocean, Offshore Mechanics and Arctic Engineering http://www.asme.org/ (accessed July 13, 2010). Sponsored by the Ocean, Offshore and Arctic Engineering Division of ASME International, this conference is held to advance the development and exchange of information regarding ocean, offshore, and arctic engineering.

International Navigation Congress http://www.pianc-aipcn.org/ (accessed July 13, 2010). An international congress for the presentation of papers on subjects of current significance to waterways and maritime interests, it is organized into two sections: one is for inland navigation issues, while the other concentrates on maritime navigation topics.

Offshore Technology Conference http://www.otcnet.org/ (accessed June 22, 2010). This conference, which was started in 1969, is the foremost trade show related to the development of offshore resources in the fields of drilling, exploration, production, and environmental protection. In addition to company exhibits, volunteers develop a technical program in order to share information on advances in the field.

Ports http://www.copri.org/ (accessed June 22, 2010). This is a series of international port and harbor development specialty conferences held by ASCE on a triannual basis since 1977.

World Dredging Congress http://www.woda.org (accessed June 22, 2010). This congress brings researchers and practitioners together to explore the latest developments in all aspects of dredging, including the ability to ensure safe navigation while addressing environmental issues.

CONSTRUCTION MANAGEMENT

INTRODUCTION

Management information has become a major priority for civil engineers, and the published literature has proliferated. With competition from other professional groups in the construction sector, managerial qualifications and expertise are a prerequisite for any civil engineer aspiring to reach the top. General management gurus, such as John Adair, Tom Peters, and Alan Weiss, are as relevant as those writing specifically on construction. Within the United Kingdom, the impact of the Latham and Egan reports has been to encourage partnership rather than confrontation in construction contracts, whereas worldwide encouragement of private funding for public infrastructure projects has led to a reconsideration of traditional roles of client/consultant/contractor. There has been a proliferation of new forms of contract from relevant professional bodies and trade associations to reflect these changes, and the ICE's NEC (New Engineering Contract) has gained widespread acceptance as a result of these developments in the United Kingdom. The World Bank is actively considering its use. Much construction is done under the auspices of the building contracts issued by the Joint Contracts Tribunal rather than ICE and its partners.

SEARCHING THE LIBRARY CATALOG

Management involves industry-specific knowledge, and generic knowledge about law and regulations, as well as an understanding of human resources issues, such as psychology and motivation. Material is likely to be scattered within a library in law sections, general management, industrial management, and construction management.

Examples include:

BF Psychology
HD28-70 Management

HD61 Risk management
K623-968 Civil law
T55-5.3 Industrial safety
T55.4-60.8 Management engineering
T57.6-57.97 Operations research
T58.4 Managerial control systems
T58.6-58.62 Management information systems
T59.7-T59.77 Human engineering in industry
T60-60.8 Work measurement
TA 166-167 Human engineering
TA177.4-185 Engineering economy
TA194-197 Management of engineering works
TS15-194 Operations management handbooks

HANDBOOKS

Hicks, T. G. and J .F. Mueller, 1996. *Standard Handbook of Consulting Engineering Practice,* 2nd ed. New York: McGraw-Hill. Covers all aspects of getting started as a consulting engineer, marketing, and expanding your business.

O'Brien, J. L., ed. 1996. *Standard Handbook of Heavy Construction.* New York: McGraw-Hill. Concerned with all aspects of the management of major civil engineering projects including estimating, scheduling, earth-moving, and value engineering, with case studies of techniques and projects.

Ratay, R. T. 1996. *Handbook of Temporary Structures in Construction: Engineering Standards, Designs, Practices and Procedures,* 2nd ed. New York: McGraw-Hill. Written by a team of specialist authors with a U.S. background, this covers issues such as site safety, with more specific techniques, such as underpinning, formwork, and bracing. It covers issues often ignored elsewhere, such as loadings created by construction equipment.

Shenson, H. I. 1990. *The Contract and Fee-Setting Guide for Consultants and Contractors.* New York: John Wiley & Sons. Although published in 1990, most of the material is still relevant. It provides sample copies of proposals, interim reports, and final reports, and illustrates the marketing role of the proposal. The book is of practical help to the independent consultant.

Turner, R., ed. 2008. *Gower Handbook of Project Management,* 4th ed. Burlington, VT: Aldershot; Aldershot, U.K.: Gower. Produced with the support of the International Project Management Association and the Project Management Institute (USA), this covers the systems of project management, the context of projects including political, economic, social, technical, legal, and environmental issues; the management of performance including quality, time, cost, risk, and safety; the management of the project life cycle, commercial issues including appraisal and finance, contracts, and human resources management.

JOURNALS

Building Law Reports (0141-5875). Published by George Godwin. Although largely concerned with building contracts and contract case law, some cases relate directly to civil engineering contracts, and many have an indirect relevance.

Construction Industry Law Newsletter (0269-0039). Published by Informa. A newsletter updating legal developments, meetings, etc.

Construction Law Journal (0267-2359). Published by Sweet & Maxwell. Largely concerned with U.K. practice, contains articles and law reports.

Construction Law Reports (0950-3889). Published by Architectural Press. Like the *Building Law Reports* (above), this journal is largely concerned with U.K. case law in construction.

Construction Management and Economics (0144-6193). Published by Spon. An academic journal that is international in coverage and with frequent analysis of case studies.

Engineering, Construction and Architectural Management (0969-9988). Published by Blackwell. This journal has carried some key papers on developments in private finance and so on.

International Construction Law Review (0265-1416). Published by Lloyds/Informa. Very useful on the International Federation of Consulting Engineers (FIDIC) and also on international development of the New Engineering Contract (NEC).

International Journal of Project Management (0263-7863). Published by Elsevier. With excellent international coverage of all aspects of project management, this journal discusses the latest techniques and issues for project managers.

Journal of Construction Engineering and Management ASCE (0733-9364). Published by ASCE. This journal has a more practical emphasis than many of the other ASCE titles, which publish construction case studies.

Journal of Management in Engineering ASCE (0742-597X). Published by ASCE. Examines contemporary issues associated with leadership and management; the focus is the practicing consulting civil engineer.

Leadership and Management in Engineering ASCE (1532-6748). Published by ASCE. A publication of the Committee on Professional Practice, this journal examines contemporary issues and principles of leadership and management, including news, brief and concise leadership and management "nuggets," and short articles of interest to practicing professionals in a variety of roles and industry segments. The focus includes, but is not limited to, individuals and public and/or private entities, small and large projects, and organizations. Areas of interest include leadership, teamwork, communications, decision making, partnering, project management, mentoring, diversity, office management, professional practice and development, financial management, productivity management and tools, globalization, networking, change management, involvement in the political process, legislative and regulatory issues, and economic and environmental sustainability.

Management Procurement and Law—Proceedings of the ICE. (1751-4304) Publishes papers on all aspects of management, procurement, and associated legal issues.

MONOGRAPHS

Arup, Ove, and Partners. 1996. *Green Construction Handbook.* Bristol, U.K.: J T Design Build Limited. Advice for the construction industry on sustainable methods and materials.

Bartlett, A. 1992. *Emden's Construction Law.* London: Butterworths. Continuously updated compendium of contract documents.

Bennett, J. 1998. *Seven Pillars of Partnering.* London: Thomas Telford. Identifies the main issues in partnering for the U.K. construction industry.

Blockley, D. and P. Godfrey. 2000. *Doing It Differently: Systems for Rethinking Construction,* London: Thomas Telford. Discusses new ways of procurement philosophies based on the Egan report.

Broome, J. 2002. *Procurement Routes for Partnering: A Practical Guide.* London: Thomas Telford. A comprehensive treatment of partnering and procurement, covering issues such as incentives, target cost and cost reimbursable contracts, and strategic alliances, with discussion of the NEC.

Bunni, N. G. 2005. *The FIDIC Forms of Contract,* 3rd ed. Oxford: Wiley-Blackwell. This comprises both a discussion of the principles underlying the FIDIC forms and also a commentary on the clauses.

CESMM3 Price Database. 2010. Available from Thomas Telford. A database based on ICE Civil Engineering Standard Method of Measurement, with a carbon calculator (from 2010) developed by Franklin and Andrews.

Chartered Institute of Building (CIOB). 2000. *Code of Practice for Project Management,* 2nd ed. Harlow, U.K.: Longman. Includes the U.K. CIOB's code of practice, which is widely used by U.K. project managers.

Chung, H. W. 1999. *Understanding Quality Assurance in Construction.* London: Spon. Discusses the implementation of ISO quality systems in construction. Contains a chapter relating to small firms.

Clough, R. H., ed. 2000. *Construction Contracting,* 4th ed. New York: John Wiley & Sons. A well-established guide to planning and scheduling construction projects, with CPM software, and consideration of problems caused by inclement weather.

Clough, R. H., ed. 2005. *Construction Contracting,* 7th ed. New York: John Wiley & Sons. Concerned with all aspects of the ownership or management of a contracting construction company, including the latest regulations, contract documents, and U.S. law.

Construction Industry Board (CIB). 1997. *Partnering in the Team.* London: Thomas Telford. CIB encouragement for partnering in the U.K. construction industry.

Construction Industry Council (CIC). 2001. *Rethinking Construction Implementation Toolkit.* London: Local Government Task Force. A CIC response to Latham and Egan.

Construction Industry Research and Information Association (CIRIA). 2001. *Sustainable Construction: Company Indicators.* London: CIRIA. Provides measures for civil engineering companies to measure up to sustainability targets.

Corbett, E. C. 1991. *FIDIC: A Practical Legal Guide,* 4th ed. London: Sweet & Maxwell. A clause-by-clause legal commentary on the FIDIC condition.

Day, D. A. and N. B. H. Benjamin. 1991. *Construction Equipment Guide,* 2nd ed. New York: John Wiley & Sons. Aimed at the contractor, it summarizes the performance of commonly recognized equipment, and it deals with the physical concepts of the work, the surrounding conditions and equipment requirements, with an emphasis on controls governing the equipment's performance.

Department for the Environment, Transport, and the Regions (DETR). 1998. *Rethinking Construction.* London: DETR (Egan report). Commissioned by the then U.K. DETR, this builds upon the Latham report to try to establish a way forward.

Eggleston, B. 2001. *The ICE Conditions of Contract,* 2nd ed. Oxford, U.K.: Blackwell Scientific. A relatively simple guide to the ICE conditions.

Geddes, S. 1996. *Estimating for Building and Civil Engineering Works,* 9th ed. Oxford, U.K.: Butterworths. A long-established textbook on estimating.

Gladden, S. C. and A. Orlitt. 1996. *Marketing and Selling A/E and Other Engineering Services,* New York: ASCE. A practical guide to marketing and sales, describing seminars, and indirect marketing tools.

Godfrey, K. A., Jr. 1996. *Partnering in Design and Construction.* New York: McGraw Hill. A multiauthored work describing the development of partnering, with case studies including nonproject uses of partnering, discussions of alternative dispute resolution and with a look to the future.

Goodowens, J. B. 1996. *A User's Guide to Federal Architect Engineer Contracts.* New York: American Society of Civil Engineers. This explains all facets of the architect engineering contract from preselection to cost negotiation to architect engineer liability. It is intended to assist professionals in making informed decisions about which jobs to pursue, and how to better present their qualifications in obtaining government contracts. The book features inside details of government negotiations strategy, and perspectives on various implementation documents of the federal Departments of Defense, Logistics, Army, Navy, Air Force, and Mapping.

Halpin, D. W. and R. W. Woodhead. 1997. *Construction Management,* 2nd ed. New York: John Wiley & Sons. Looks at the skills required by a construction manager and all professional groups including engineers and contractors, the resources in terms of equipment and materials as well as financial and human. There are chapters on estimating and cost control.

Harris, F. and R. McCaffer. 2006. *Modern Construction Management,* 6th ed. Oxford, U.K.: Blackwell Scientific. A textbook that provides an introduction to most aspects of construction management from a U.K. perspective, covering bid preparation, plant management, and more.

Holroyd, T. M. 1999. *Site Management for Engineers.* London: Thomas Telford. Site management is an important aspect of any successful project, and requires skills and knowledge rarely obtained through academic study; this is a useful introduction for engineers going onsite for the first time to best practice.

Holroyd, T. M. 2000. *Principles of Estimating.* London: Thomas Telford. An introduction to estimating practice.

Holz, H. 1997. *Complete Guide to Consulting Contracts,* 2nd ed. New York: Dearbourn. Indispensable for those new to consulting contracts and agreements.

ICE. 2009. *Civil Engineering Procedure,* 6th ed. London: Thomas Telford. Based on U.K. practice, it explains all the stages involved in a construction project.

ICE. 1998. *RAMP: Risk Appraisal and Management of Projects.* London: Thomas Telford. This is a guide to risk appraisal, now a fundamental aspect of all project planning.

ICE. 2001. *Management Development in the Construction Industry,* 2nd ed. London: Thomas Telford. Although aimed at civil engineers, the basic principles enunciated in this guide are applicable to anybody interested in developing as a manager.

Illingworth, J. R. 1987. *Temporary Works: Their Role in Construction.* London: Thomas Telford. Helpful in describing temporary works and invaluable for young engineers going onsite for the first time with little knowledge of this area.

Irvine, D. J. and R. J. H. Smith. 1992. *Trenching Practice CIRIA Report 97,* rev. ed. London: CIRIA. Trenching is an area of construction that, if incorrectly carried out, can result in deaths and disasters; this provides guidance on good practice.

Joyce, R. 2001. *The CDM Regulations Explained,* 2nd ed. London: Thomas Telford. This provides guidance on the most significant aspect of health and safety legislation in the United Kingdom.

Latham, M. 1994. *Constructing the Team.* London: HMSO. A seminal report aimed at changing the face of the U.K. construction process away from confrontation and toward cooperation between all those involved in the construction process.

Mackay, E. B. 1986. *Proprietary Trench Support Systems Technical Note 95,* 3rd ed. London: CIRIA. For any engineer dealing with trench support, the prevalence of proprietary systems is a challenge. This is helpful in identifying characteristics and guidance on use.

McInnnis, A. 2001. *The New Engineering Contract: A Legal Commentary.* London: Thomas Telford. The first heavyweight commentary on the New Engineering Contract (NEC), discussing the principles behind the NEC rather than a line-by-line analysis of clauses. Heavily referenced throughout.

O'Reilly, M. 2009. *Civil Engineering Construction Contracts,* 2nd rev. ed. London: Thomas Telford. More narrowly focused than Uffs work, and a useful comparison of alternative contract forms.

Rogers, M. 2001. *Engineering Project Appraisal.* Oxford, U.K.: Blackwell Scientific. Details methods of project appraisal in engineering.

Scott, W. and B. Billing. 1998. *Communication for Professional Engineers,* 2nd ed. London: Thomas Telford. This provides good common-sense advice to engineers working to develop their communication skills in all their aspects: spoken, written, etc.

Snell, M. 1997. *Cost Benefit Analysis for Engineers.* London: Thomas Telford. CBA is a well-established technique and this serves as a basic introduction.

Spon's Civil Engineering and Highway Works Price Book. (Annual) New York: E & FN Spon. The main U.K. price book for civil engineers. (*Note*: Spon produces a number of specialist estimating books covering railway engineering and foreign regions, but not all are updated annually).

Uff, J. 2009. *Construction Law,* 10th ed. London: Sweet & Maxwell. Useful introduction to U.K. construction law.

Wallace, I. N. D. 1995. *Hudson's Building and Engineering Contracts,* 11th ed. (2 vols.) London: Sweet & Maxwell. Heavyweight commentary on British contracts. As with all such texts, changes in case law, and the introduction of new contract forms means that it can only provide guidance.

Woodward, J. F. 1997. *Construction Project Management.* London: Thomas Telford. An introduction to the subject.

WEB RESOURCES

Blue Book of Building and Construction http://www.thebluebook.com/ (accessed November 3, 2010). This is the U.S. industry's leading directory of information on construction products and firms, categorized by area and specialization.

Construction Industry http://www.construction.about.com/industry/construction.mbody.htm (accessed November 3, 2010). Provides links to sites related to every area of the construction industry, such as codes, industry news, software, equipment, and manufacturers.

Means Costworks. Kingston, MA: RS Means http://www.rsmeans.com/cworks (accessed November 3, 2010). Subscriber access to Means construction cost books, aimed at the U.S. market.

Work Zone Safety Information Clearinghouse http://wzsafety.tamu.edu/ (accessed November 3, 2010). This site provides information on the latest developments in U.S. and Canadian legislation, research, government agencies, and public and private organizations relevant to safety in traffic work zones. Spanish language version available.

INSTITUTIONS, ORGANIZATIONS, AND SOCIETIES

Association of Consulting Engineers (ACE) http://www.acenet.co.uk/ (accessed November 3, 2010). The British-based ACE produces conditions of engagement for consulting engineers' services, and works with ICE and the contractors to develop conditions of contract.

Association for Planning Supervisors http://www.aps.org.uk/ (accessed November 3, 2010). This organization has grown up as a result of changes in health and safety legislation in the U.K. under the CDM regulations, and the need for all construction projects to address health and safety issues by having accredited personnel.

Association for Project Management http://www.apm.org.uk/ (accessed November 3, 2010). The U.K. society for project managers with links to the international society. They publish a journal, *Project Management*, and a major guide to project planning. Their former trading arm, APM Group, is now independent: http://www.apmgroup.co.uk/ (accessed November 3, 2010).

Construction Best Practice Programme http://www.cbpp.org.uk/ (accessed November 3, 2010). The U.K.'s Construction Best Practice Programme provides support to individuals, companies, organizations, and supply chains in the construction industry seeking to improve the way they do business.

Construction Industry Council http://www.cic.org.uk/ (accessed November 3, 2010). A U.K. industry-wide body, given great impetus by the recommendations of the Latham and Egan reports. Full members are admitted into electoral colleges depending on their constitution: Chartered professional institutions holding a Royal Charter; professional institutes/bodies with independent status qualifying individuals in various disciplines throughout the construction process; business organizations, which represent professional services, materials/product supply, and contracting in private and public companies or partnerships; and research organizations that are dedicated to serving construction, the industry, and its clients. Associate membership embraces organizations within the construction industry that speak for a defined group (or groups) of members, but are not eligible for admission as a Full Member of Council.

Construction Management Association of America (CMAA) http://www. cmaanet.org/ (accessed November 3, 2010). The CMAA encourages excellence among construction managers. It can assist project owners by providing information about construction management practice and by helping identify qualified professional construction managers for projects. The Web site gives access to award-winning projects, and the construction management e-journal.

Construction Specifications Institute (CSI) http://www.csinet.org/ (accessed November 3, 2010). U.S.-based organization with multidisciplinary membership (over 17,000). Produces *Masterformat* (2004) as a standard for specifications, and the *Project Resources Manual: the CSI Code of Practice*. The Web site has a members' area, and online access to the journal *Construction Specifier*.

Design-Build Institute of America (DBIA) http://www.dbia.org/ (accessed November 3, 2010). The DBIA is a member organization that promotes the philosophy of design and build in construction. The site offers a database of design-build projects, an e-commerce section containing publications detailing the model contract forms available, a mission statement, membership information, courses, a members' area, and a summary on the design-build process.

Forum on the Construction Industry http://www.abanet.org/forums/construction/ (accessed November 3, 2010). A forum of the American Bar Association on the construction industry, the Web site contains past issues of the electronic newsletter, an index to *Construction Lawyer*, and an online bookshop referring to publications on design-build, which encompasses all U.S. states and Canada, fundamentals of construction law, and federal government construction contracts.

International Federation of Consulting Engineers (FIDIC) http://www.fidic.org/ (accessed April 1, 2011). An international body based in Switzerland responsible for a range of contract forms and procedures that are widely used in international civil engineering contracts.

Lean Construction Institute http://www.leanconstruction.org/ (accessed November 3, 2010). An organization that aims to extend to the construction industry the "lean production" revolution started in manufacturing. This approach maximizes value delivered to the customer while minimizing waste. The Web site describes lean construction and contains a link to the *Lean Construction Journal*, started in late 2004.

National Building Specification (NBS) http://www.thenbs.com/ (accessed November 3, 2010). The main purpose is to write, revise, and publish the *National Building Specification*. The NBS was first launched in 1973, and two years later an updating service was started with loose-leaf binders, followed shortly by a small jobs version. In 1988, the documents were completely revised to take account of the Common Arrangement of Work Sections, part of the CPI (Coordinated Project Information) initiative led by the Royal Institution of Chartered Surveyors. Its building industry focus has meant it is not entirely satisfactory for engineering work.

Society of Construction Law http://www.scl.org.uk. (accessed November 3, 2010). A U.K. society that has developed with the tremendous growth of construction and law and litigation.

World Bank http://publications.worldbank.org/online (accessed November 3, 2010). A major funder of projects in the developing world, the Web site provides a wealth of economic information and other advice for those involved in international projects.

MUNICIPAL ENGINEERING

INTRODUCTION

The tradition of municipal engineering goes back to the mid-nineteenth century when engineers were appointed to deal with the health and sanitary problems of the fast-growing towns. Earlier, in the eighteenth century, some communities had appointed county surveyors to look after their bridges. Some aspects of the specialization are covered elsewhere, such as under transport engineering and water engineering, and, increasingly, such engineering services are provided directly by, or in partnership with, the private sector. The activity of community engineering is rapidly changing to one of community focus, sustainability, and best value. Government funding across the developed world has encouraged the development of brownfield sites (sites previously developed for other purposes and often contaminated by past use) and urban regeneration rather than expansion of urban areas and the planning of new communities so much in vogue in the 1950s and 1960s.

In the municipal engineering context transport engineering is concerned with the planning and management of transportation systems, and the construction, maintenance, and renewal of the infrastructure. While this includes knowledge of the moving loads operating in the system, vehicle design is the province of the mechanical engineer, while the electrical engineer interfaces with the civil engineer in the design of electrical power systems and traffic control/signaling. Transport affects people in their daily lives and, when transport systems fail to operate efficiently, as has been witnessed in Britain and elsewhere in recent years, the work of the municipal engineer is taken into the world of politics. Investment in roads has characterized most Western economies since the 1930s, and at a local level has traditionally been the province of municipal engineers. There is an enormous literature on the subject that is generally readily accessible on subscription via the most successful cooperative database: *International Transport Research Documentation (ITRD)*, and the freely available U.S.-based *Transportation Research International Documentation (TRID)*. Although the U.K. Department of Transport/Highways Agency has produced standards and codes through its various changes of practice for many years, ready access has only been made possible via the transfer of sales to The Stationery Office (TSO) and their publication on the Internet. It is an area where American practice has played a significant role, and despite the proliferation of highways agency documentation, a large number of other bodies produce important guidelines on best practice and specialist areas, such as heavy-duty pavements. Although generally associated with road traffic congestion in towns, traffic engineering has its origins in railway signaling in the nineteenth century. Today, with rapid developments in IT and GIS, the cutting edge of traffic engineering interfaces with vehicle design and encompasses crude physical devices, such as road humps and the most sophisticated smart card technology. Many airports, although subject to (inter)national regulation, are municipally owned. Most countries have a research organization concerned with transport, generally government funded, and central governments issue regulations, which may govern safe operation as well as minimum design standards.

SEARCHING THE LIBRARY CATALOG

Municipal engineering forms a subset of sanitary engineering and environmental technology in the LC Classification scheme: TD 159-168. However, municipal engineers are likely to be interested in the legal aspects of their work as well the technical, and many are involved in other areas of civil engineering, such as TE: Highway engineering; TF: Railroad engineering where it relates to metros; TG: Bridge engineering, as well as other aspects of sanitary engineering such as Municipal waste: TD 783-812.5 and Street cleaning: TD813-870.

ABSTRACTS AND INDEXING SERVICES

International Transport Research Database (*ITRD*) (see Chapter 20 in this book on Transportation Engineering).

TRID (TRB) http://trisonline.bts.gov/search.cfm/http://www.nas.edu/trb (see Chapter 20 in this book on Transportation Engineering).

HANDBOOKS

American Railway Engineering and Maintenance of Way Association (AREMA). *Manual of Railway Engineering.* Washington, D.C.: American Railway Engineering Association. A compilation of U.S. railway standards covering all aspects of the civil engineering of rail transport.

Atkinson, K. 1997. *Highway Maintenance Handbook,* 2nd ed. London: Thomas Telford. Comprehensive treatment of the subject from a British perspective.

Brockenbrough, R. L. and K. J. Boedecke. 2003. *Highway Engineering Handbook: Building and Rehabilitating the Infrastructure,* 2nd ed. New York: McGraw-Hill. An American multiauthored handbook that covers highway planning, traffic engineering, pavement and highway structures design, street lighting and signage, as well as noise barriers and value engineering and life-cycle costs.

Cooper, A. R. 1998. *Properties of Hazardous Industrial Materials.* CRCnetBASE, Boca Raton, FL: Lewis Publishers. Available in digital format. Lists more than 25,000 hazardous materials. Each chemical is fully described.

Freeman, H. 1998. *Standard Handbook of Hazardous Waste Treatment and Disposal,* 2nd ed. New York: McGraw-Hill. Summarizes U.S. laws and regulations, with an overview of the hazardous waste problem and state-of-the-art alternative treatment and disposal processes.

Grava, S. 2003. *Urban Transportation Systems: Choices for Utilities.* New York: McGraw-Hill. Individual chapters discuss each mode in turn from walking to cable cars with a helpful guide to further reading. An excellent introduction for any covered subject.

Hess-Kosa, K. 2008. *Environmental Site Assessment, Phase I: A Basic Guide,* 3rd ed. Boca Raton, FL: CRC Press. Useful introduction to U.S. practice.

Institute of Transportation Engineers. 2009. *Traffic Engineering Handbook,* 6th ed. Washington, D.C.: ITE. Two new chapters added to this well-known text: Planning for Operations and Managing Traffic Demand to Address Congestion.

Institute of Transportation Engineers. 2009. *Trip Generation,* 8th ed. (3 vols.) Washington, D.C.: ITE. With an ever expanding set of land use values, this is essential for any traffic engineer and planner.

Institution of Structural Engineers/Institution of Highways and Transportation. 2002. *Design Recommendations for Multi-Storey and Underground Car Parks,* 3rd ed. London: Institution of Structural Engineers.

Jahm, R. K., ed. 2002. *Environmental Assessment,* 2nd ed. New York: McGraw-Hill. Covers U.S. legislative background and the implications for the National Environmental Policy Act if applied to overseas projects involving U.S. firms. There are chapters on public participation and other issues affecting the environment globally.

Jane's Urban Transport Systems. Annual. Coulsdon, U.K.: Jane's Information Systems. Comprehensive coverage of systems as existing and planned, with suppliers' details.

Karnofsky, B. 1997. *Hazardous Waste Management Compliance Handbook.* New York: Van Nostrand Reinhold. A handbook concerned with U.S. hazardous waste management regulations.

Keith, L. H. 1996. *Compilation of EPA's Sampling and Analysis Methods.* Boca Raton, FL: CRC Press. Digital format available. Information about related software programs is available at http://www.instantref.com/inst-ref.htm (accessed November 3, 2010).

Kindred Association. 1994. *A Practical Recycling Handbook.* London: Thomas Telford. Despite the proliferation of initiatives to encourage local authority recycling programs, there are relatively few textbooks on the subject.

King, R. B., et al. 1998. *Practical Environmental Bioremediation: The Field Guide,* 2nd ed. Boca Raton, FL: Lewis Publishers. Bioremediation commonly uses micro-organisms as a biological activity to reduce the effects of a pollutant in the environment. This work is illustrated with numerous case studies.

La Mantia, F., ed. 2002. *Handbook of Plastics Recycling.* Shrewsbury, U.K.: RAPRA. Plastics recycling is the major challenge for the plastics industry if it is to address the sustainability agenda. This represents the state of the art and is of value for all involved in the waste industry.

Lamm, R., ed. 1999. *Highway Design and Traffic Engineering Handbook.* New York: McGraw-Hill. The specific objective of this work was its traffic safety focus, rather than engineering design per se.

Lewis, R. J. 2007. *Sax's Dangerous Properties of Industrial Materials.* 11th ed. New York: Van Nostrand Reinhold. Digital format available. The leading source on hazardous substances, including health and safety data, regulatory standards, toxicity, and physical properties.

Minsk, L. D. 1998. *Snow and Ice Control Manual for Transportation Facilities.* New York: McGraw-Hill.

Nelson, P. M. 1997. *Transportation Noise Reference Book.* London: Butterworths.

Shell Bitumen Handbook. 2003. London: TTL. Its association with a single company is misleading regarding the coverage of this handbook, which is a valuable up-to-date source on the use of bitumen.

Strong, D. L. 1997. *Recycling in America: A Reference Handbook,* 2nd ed. Santa Barbara, CA: ABC-CLIO. Discusses legal issues surrounding recycling, and new recycling technology focusing on the role of recycling in solid waste management with reference to environmental issues, such as global warming, conservation, and the depletion of natural resources. It reviews materials currently being recycled and those that have been overlooked, and the responsibility of industry to create new materials and products in the future.

Transportation Research Board. 2000. *Highway Capacity Manual.* Washington, D.C.: National Research Council (*TRB Special Report 209*). Although other methods of traffic assessment are available and used in the United Kingdom, the methods employed in this manual are the most widely used internationally.

TSO. *Design Manual for Roads and Bridges.* http://www.tsoshop.co.uk/ (accessed November 3, 2010). Multivolume manual for highway design in the United Kingdom. Sections may be downloaded from http://www.standardsforhighways.co.uk/dmrb/ (accessed November 3, 2010).

JOURNALS

APWA Reporter APWA (0092-4873). Monthly magazine of the APWA, available via their
  Web site: http://www.apwa.net (accessed November 3, 2010).
Association of Asphalt Paving Technologists Journal (U.S.). The leading learned society
  journal in this area where most literature is trade driven.
International Journal of Pavement Engineering (1029-8436). Published by Taylor & Francis. An
  academic journal reflecting the body of university research now taking place in this area.
Journal of Infrastructure Systems ASCE (1076-0342). Published by ASCE. The only journal
  aimed at engineers that takes a holistic view of infrastructure.
Journal of Urban Planning and Development ASCE (0733-9488). The ASCE journal covering urban planning and municipal issues.
Municipal Engineer (0965-0903). Part of the ICE proceedings and published quarterly aimed at
  a municipal engineering audience; the majority of issues are based around a current theme.
Planner (RTPI). A weekly newsletter on U.K. planning issues.
Public Works http://www.pwmag.com/ (0033-3840). A monthly journal with an annual direc-
  tory/buyers' guide aimed at consulting engineers, and contractors involved with public
  works and infrastructure projects.
Surveyor (not supplied). Published by Hemming Group. The weekly for municipal engineers
  in the United Kingdom, available online to subscribers.
Traffic Engineering and Control http://www.tecmagazine.com/ (0041-0683). Editorially independent of all suppliers, with an international reputation, despite its U.K. base.
Transport (0965-092X). The ICE's journal covers all aspects of transport, including public
  transport, so often the concern of municipal engineers.
United Nations Annual Bulletin of Transport Statistics for Europe and North America
  (1027-3093). Published by UNECE. Useful for comparative statistics on transport provision and development.

MONOGRAPHS

Armstrong-Wright, A. 1993. *Public Transport in Third World Cities.* Crowthorne, U.K.: TRL.
  A review of public transport in the developing world.
Ashford, N. and P. H. Wright. 1992. *Airport Engineering,* 3rd ed. New York: John Wiley &
  Sons. Probably the most comprehensive text on all aspects of airport engineering.
Baier, R., ed. 2000. *Strassen und Plaetze Neu Gestaltet.* Bonn: Kirschbaum. Case studies of
  best practice in the design of streets for all users. The pictures tell the story.
Blow, G. J. 1996. *Airport Terminals,* 2nd ed. Oxford, U.K.: Butterworth-Heinemann. A world-
  wide review of developments in airport terminal design.
Bregman, J. I. 1999. *Environmental Impact Statements,* 2nd ed. Boca Raton, FL: Lewis
  Publishers. Useful for understanding the development of environmental impact statements
  in the United States.
Chrest, A. P., ed. 2001. *Parking Structures,* 3rd ed. Norwell, MA: Kluwer Academic Publishers.
  Written with reference to U.S. codes, this discusses all aspects of parking structures includ-
  ing structural design, signage, lighting, accessibility, maintenance, and repair.
Countryside Commission and ICE. 2002. *Rural Routes and Networks: Creating and
  Preserving Routes That Are Sustainable, Convenient, Tranquil, Attractive, and Safe,*
  London: Thomas Telford. This report addresses concerns that increases in vehicular traffic
  are changing the character of rural roads, and how they can be made sustainable for all
  road users without destroying their essential character.
Croney, D. and P. Croney. 1997. *Design and Performance of Road Pavements,* 3rd ed. New
  York: McGraw-Hill. Packed full of informed advice on pavement design based originally
  on work at the U.K. TRL.

DETR (Department for the Environment, Transport, and the Regions). 2001. *The Value of Urban Design*. London: Thomas Telford. Part of a series of reports aimed at raising the quality of urban design in the United Kingdom.

DETR (Department for the Environment, Transport, and the Regions). 2000. *Environmental Impact Assessment: A Guide to Procedures*. London: DETR. Promotes guidance on EIA in the United Kingdom; official publications may enlarge on this in specific areas.

DETR (Department for the Environment, Transport, and the Regions). 2001. *By Design: Better Places to Live*. London: Thomas Telford.

Edwards, J. T., ed. 1990. *Civil Engineering of Underground Rail Transport*. London: Butterworths. A comprehensive review of the civil engineering aspects of metro construction.

Eliasson, J. and M. Lundberg. 2003. *Road Pricing in Urban Areas*. Borlänge, Sweden: Vägverket, Butiken. A Swedish view on road pricing.

Horonjeff, R. and F. A. McKelvey. 2010. *Planning and Design of Airports*, 5th ed. New York: McGraw-Hill. Very good on the civil engineering aspects of airport design.

ICE. 1996. *Sustainability and Adaptability in Infrastructure Development*. London: Thomas Telford. Provides guidance on the main issues in the sustainability debate regarding infrastructure.

Jensen-Buttler, C., et al. 2008. *Road Pricing, the Economy and the Environment*. New York: Springer. A holistic view of road pricing.

Kendrick, P., ed. 2004. *Roadwork, Theory and Practice*, 5th ed. Oxford, U.K.: Elsevier Butterworth-Heinemann. A very practical approach to road work, aimed at technicians rather than concerned with the intricacies of highway design and planning.

Laidler, D. W., ed. 2002. *Brownfields: Managing the Development of Previously Developed Land*. London: CIRIA (C578). Very much aimed at the management of the development of brownfield sites rather than the detailed technical measures necessary to deal with issues, such as contamination.

McClintock, H. 2002. *Planning for Cycling*. Boca Raton, FL: CRC Press. A multiauthored work, focused on U.K. experience, but with contributions on European, U.S., and Australian policies and experience. Very much state-of-the-art.

McCluskey, J. 1987. *Parking: A Handbook of Environmental Design*. London: Spon. Comprehensive advice on all types of parking including multistorey structures from an aesthetic/environmental viewpoint. Describes best practice by function and location.

Motorway Archive Trust. 2002. *The Motorway Achievement*. (3 vols.) London: Thomas Telford. A history of the building of the U.K. motorway system, written largely by the engineers responsible, with general interest for all involved in developing a road-based infrastructure.

O'Flaherty, C. A. 2002. *Highways*, 4th ed. London: Butterworth-Heinemann. Focused on the engineering design of pavements, their substructure, and surface treatment. Based on U.K. practice, but with many references to practice elsewhere.

Oppenheim, N. 1995. *Urban Travel Demand Modeling*. New York: John Wiley & Sons. Discusses modeling demand under congested and uncongested conditions, transport system design, and an appendix on the mathematical background.

Ortuzar, J. D. and L. G. Willumsen. 2001. *Modelling Transport*. Chichester, U.K.: Wiley. Based on U.K.–Chilean cooperation, this is concerned with explaining the mathematical modeling techniques available, with relatively little discussion of the transport issues.

Salter, R. J. 1983. *Highway Traffic Analysis and Design*, 2nd ed. London: Macmillan. An introduction to traffic engineering aimed at students.

Salter, R. J. 1988. *Highway Design and Construction*, 2nd ed. London: Macmillan. An introduction to highway design aimed at students.

Sherwood, B., ed. 2002. *Wildlife and Roads*. London: ICP. Focuses on the ways in which roads can be integrated with nature rather than bulldozing it.

Sussman, J. 2000. *Introduction to Transportation Systems*. Boston: Artech. This is intended as an introduction to the subject and discusses all aspects of transportation systems including labor, customers, networks, models, and modes. Recent developments, such as Intelligent Transport Systems, are covered as well as trucking, railroads, and urban transport systems.

Tyler, N., ed. 2002. *Accessibility and the Bus Systems: From Concepts to Practice*. London: Thomas Telford. Written by international specialists who form the Accessibility Research Group, this deals holistically with all aspects of bus systems and is required reading for anybody responsible for bus transport.

Wells, A. T. 2000. *Airport Planning and Management,* 4th ed. New York: McGraw-Hill. Much more on the planning and management of airports than their engineering, but of value for municipal engineers with those responsibilities.

Whitelegg, J. and G. Haq. 2003. *World Transport Policy and Practice*. London: Earthscan. Arranged by continent specialists, the authors describe the state-of-the-art in urban, rural, and regional contexts around the world.

Whittles, M. J. 2003. *Urban Road Pricing: Public and Political Acceptability*. Aldershot, U.K.: Ashgate. Based on a PhD thesis, this discusses data on public attitudes to various road-pricing options and how they might be addressed.

WEB SITES

Contaminated Land http://www.contaminated-land.org/ (accessed November 3, 2010). Excellent Web site for anybody in the United Kingdom interested professionally or otherwise in contaminated land issues.

Environment Agency (EA) http://www.environment-agency.gov.uk/default.aspx (accessed November 3, 2010). The U.K. EA Web site provides important guidance on contaminated and derelict land with many CLR reports available to download.

Hazardous Waste Clean-Up Information (CLU-IN) http://www.clu-in.org (accessed November 3, 2010). This site, developed by the U.S. EPA, provides information about innovative hazardous waste treatment technology.

Highway Statistics http://www.fhwa.dot.gov/policy/ohpi/hss/hsspubs.cfm (accessed November 3, 2010). U.S. highway statistics.

Light Rail Now http://www.lightrailnow.org/ (accessed November 3, 2010). Grassroots group of volunteers (Austin, TX) supporting light rail transit. Offers transit links, light rail links, and contact details of volunteers presenting educational programs.

Motorway Archive Trust http://www.ukmotorwayarchive.org/ (accessed November 3, 2010). An online encyclopedia of U.K. motorway development.

U.S. Department of Transportation (DOT) National Transportation Library (NTL) http://ntl. bts.gov (see Chapter 20 in this book on Transportation Engineering).

U.S. Environmental Protection Agency (EPA) The Municipal Solid Waste Factbook http://www.epa.gov/epawaste/index.htm (accessed November 3, 2010). An electronic reference manual containing information about U.S. household waste management practices including the complete text of EPA's regulations for municipal solid waste landfills.

INSTITUTIONS

Advanced Transit Association (ATRA) http://advancedtransit.org/news.aspx (accessed November 3, 2010). ATRA promotes investigation and development of advanced transit technologies and strategies. Site includes a large library, links, list of publications and conferences, and advanced technology descriptions.

American Association of State Highway and Transportation Officials (AASHTO) http:// transportation.org/ (accessed November 3, 2010) (see Chapter 20 in this book on Transportation Engineering).

American Public Works Association (APWA) http://www.apwa.net/ (accessed November 3, 2010) (see Chapter 20 in this book on Transportation Engineering).

American Public Transit Association (APTA) http://www.apta.com (accessed November 3, 2010) (see Chapter 20 in this book on Transportation Engineering).

Asphalt Institute (US.) www.asphaltinstitute.org (accessed November 3, 2010). The Asphalt Institute produces a number of manuals aimed at the practitioner, together with useful FAQS on their Web site. They played a major role in the establishment of the International Society for Asphalt Pavements, which publishes a major series of conferences.

Association of Directors of Environment, Economy, Planning, and Transport (ADEPT), formerly the County Surveyors Society (CSS) http://www.cssnet.org.uk/ (accessed November 3, 2010). ADEPT acts as a forum and pressure group for local government professionals in the U.K., publishing guides on subjects of interest to members.

Association of Metropolitan Water Agencies (AMWA) http://www.amwa.org (accessed November 3, 2010). AMWA represents the interests of large, publicly owned drinking water systems. The Web site provides current news, interactive bulletin boards, summaries of legislation and regulations, and links to member Web sites, Congress, EPA, and others.

Association of Pedestrian and Bicycle Professionals (APBP) http://www.apbp.org (accessed November 3, 2010). APBP promotes excellence in the field of pedestrian and bicycle transportation. The site contains membership info, links, newsletter, job listings, and an events calendar.

Chartered Institution of Wastes Management (CIWM) http://www.ciwm.co.uk/ (accessed November 3, 2010). The U.K. society concerned with the fast-growing field of waste management and recycling.

CROW (Information and Technology Centre for Transport and Infrastructure) http://www. crow.nl/Home (accessed November 3, 2010). A Dutch organization whose research is in the area of traffic, transport, and infrastructure; standardization in this sector; transfer of knowledge and knowledge management. The center's manuals and guidelines are widely respected across Europe.

Institute of Transportation Engineers (ITE) http://www.ite.org (accessed November 3, 2010). The institute is one of five organizations designated by the U.S. Department of Transportation to develop standards. The site provides an online directory of transportation product and service providers. Other publications include the *Trip Generation* and *The Traffic Signing Handbook*.

Institution of Highways and Transportation (IHT) http://www.iht.org/ (accessed November 3, 2010). IHT has been broadening its brief and appeal outside the road engineering community that first established the institution. It publishes a journal, organizes conferences, and produces important guidelines.

International Civil Aviation Organisation (ICAO) http://www.icao.int/ (accessed November 3, 2010). Produces international standards and codes of practice for airport design.

International Society for Concrete Pavements http://www.concretepavements.org (accessed November 3, 2010). Organizes meetings and activities on rigid/concrete pavements.

National Association of County Engineers http://www.countyengineers.org/ (accessed November 3, 2010). Across the United States, county engineers have responsibility for enormous infrastructure assets, notably county bridges and roads. The former often have great historic value. Generally, states will also have their own statewide association. The Web site has an alerts service and training and career information.

National Joint Utilities Group, The (NJUG) http://www.njug.org.uk/ (accessed November 3, 2010). This is the U.K. trade grouping for the utilities in street works.

Portland Cement Association http://www.cement.org/ (accessed November 3, 2010). Produces guidance and more on road and airport pavements.

Transport Research Laboratory (TRL) http://www.trl.co.uk/default.htm (accessed November 3, 2010). Although largely concerned with highway/road-related research, this is the leading research body in the United Kingdom, a major publisher of research, and with comprehensive library holdings.

Transportation Research Board (TRB) http://www.nas.edu/trb (accessed November 3, 2010). The Transportation Research Board is part of the National Research Council. TRB publishes the TRID database. This site provides full-text publications and free access to the database.

U.S. Department of Transportation (DOT). Federal Highway Administration (FHWA), http://www.fhwa.dot.gov/ (accessed November 3, 2010). The FHWA sponsors research and issue regulations. The Web site links to a series of subsidiary Web sites on issues such as bridge technology, with a whole range of downloadable PDF documents, and an electronic library with access to TRID and regulations.

United States Federal Aviation Administration http://www.faa.gov/ (accessed November 3, 2010). Responsible for an enormous volume of research and documentation in the sector.

World Road Association, AIPCR/PIARC http://www.piarc.org/en/ (accessed November 3, 2010). PIARC deals with road infrastructure planning, design, construction, maintenance, and operation. Founded in 1909, it currently has 97 national or federal government members, 2,000 collective or individual members in 129 countries, and over 750 experts in 20 standing technical committees. In addition to organizing conferences, PIARC produces guidance on subjects, such as road tunnels via its technical committees.

NANOTECHNOLOGY

INTRODUCTION

The National Nanotechnology Initiative (NNI) of the United States Government defines nanotechnology as:

> Nanotechnology is the understanding and control of matter at dimensions between approximately 1 and 100 nanometers, where unique phenomena enable novel applications. Encompassing nanoscale science, engineering, and technology, nanotechnology involves imaging, measuring, modeling, and manipulating matter at this length scale.

A nanometer is one-billionth of a meter. A sheet of paper is about 100,000 nanometers thick; a single gold atom is about a third of a nanometer in diameter. Nanotechnology is different from older technologies because unusual physical, chemical, and biological properties can emerge in materials at the nanoscale. Web site: http://www.nano.gov/html/facts/faqs.html (accessed June 24, 2010).

According to the NNI, nanotechnology will help to provide inexpensive ways to purify water, develop new sources of energy, and provide improved materials that are lighter, stronger, and more durable.

SEARCHING THE LIBRARY CATALOG

When looking for materials related to nanotechnology, the following are some of the Library of Congress Subject Headings to search:

Nanocomposites (Materials)
Nanoelectromechanical systems
Nanochemistry
Nanomaterials and their applications
Nanoparticles

Nanoscience
Nanoscience and technology
Nanostructure science and technology
Nanostructured materials
Nanostructures
Nanotechnology

Most of the general materials on nanotechnology will be catalogued either in the Technical subject area (T174—Technical change) or in the Sciences subject area (OC176—Atomic physics). The more specialized publications will be catalogued according to their specialization. For instance, books on materials of engineering would be assigned to the TA418 subject area.

ASSOCIATIONS

Foresight Institute http://www.foresight.org/ (accessed June 2, 2010). Foresight Institute is a leading think tank and public interest organization focused on transformative future technologies. Founded in 1986, its mission is to discover and promote the upsides, and help avoid the dangers, of nanotechnology, AI, biotech, and similar life-changing developments. Foresight provides balanced, accurate, and timely information to help society understand nanotechnology through publications, public policy activities, roadmaps, prizes, and conferences.

Institute of Nanotechnology (IoN) http://www.nano.org.uk/ (accessed June 2, 2010). The Institute of Nanotechnology has been created to foster, develop and promote all aspects of nanoscience and nanotechnology in those domains where dimensions and tolerances in the range 0.1 nm to 100 nm play a critical role. It is fundamentally a membership organization, and has the aim of encouraging information exchange across scientific disciplines. The institute is also active in transferring technologies from the research base to industry and alerting industry to the economic implications of nanotechnology. The institute's site contains a news section, information on events and publications, a glossary of nano terms and abbreviations, and links to sites of related interest. In addition, there are members-only services, including information on research interests and projects.

Irish Nanotechnology Association http://www.nanotechireland.com/ (accessed June 2, 2010). This site covers nanotechnology. It includes a searchable archive of links to the full text of relevant patent documents in PDF format and news stories. A list of nanotechnology equipment at Irish universities is provided. Information on researchers can be searched by name or by keyword. There are descriptions of Irish companies involved in nanotech. Links to related sites are given.

Nanotechnology Institute http://nano.asme.org/ (accessed June 2, 2010). The Nanotechnology Institute of ASME International is dedicated to furthering the art, science, and practice of nanotechnology. The institute is a clearinghouse for ASME activities in nanotechnology and provides interdisciplinary programs and activities to bridge science, engineering, and applications. The site identifies a couple of review articles on nanotechnology, obtainable to nonsubscribers by pay-per-view, along with educational/training products and proceedings of a conference, which may be ordered online. There is a news section and a list of institute events.

DICTIONARY AND HANDBOOKS

Bhushan, B., ed. 2010. *Springer Handbook of Nanotechnology*. New York: Springer. Meets the need for a comprehensive, easily accessible source of application-oriented, authoritative information by integrating knowledge from key subfields This edition is expanded with a new part dedicated to biomimetrics. It is a highly detailed, single-volume treatment of a rapidly evolving field.

Borisenko, V. E. and S. Ossicini. 2008. *What Is What in the Nanoworld: A Handbook on Nanoscience and Nanotechnology,* 2nd ed. Hoboken, NJ: John Wiley & Sons. The second, completely revised and enlarged edition of *What Is What in the Nanoworld* summarizes the terms and definitions, most important phenomena, and regulations occurring in the physics, chemistry, technology, and application of nanostructures. A representative collection of fundamental terms and definitions from quantum physics and chemistry, special mathematics, organic and inorganic chemistry, solid state physics, material science, and technology accompanies recommended second sources (books, reviews, Web sites) for an extended study of any given subject.

Gogotsi, Y., ed. 2006. *Nanomaterials Handbook.* Boca Raton, FL: CRC Press. The *Nanomaterials Handbook* provides a comprehensive overview of the current state of nanomaterials. Employing terminology familiar to materials scientists and engineers, it provides an introduction that delves into the unique nature of nanomaterials. Looking at the quantum effects that come into play and other characteristics realized at the nano level, it explains how the properties displayed by nanomaterials can differ from those displayed by single crystals and conventional microstructured, monolithic, or composite materials. The introduction is followed by an in-depth investigation of carbon-based nanomaterials, which are as important to nanotechnology as silicon is to electronics. However, it goes beyond the usual discussion of nanotubes and nanofibers to consider graphite whiskers, cones and polyhedral crystals, and nanocrystalline diamonds. It also provides significant new information with regard to nanostructured semiconductors, ceramics, metals, biomaterials, and polymers, as well as nanotechnology's application in drug delivery systems, bioimplants, and field-emission displays.

Schramm, L. L. 2008. *Dictionary of Nanotechnology, Colloid and Interface Science.* Hoboken, NJ: John Wiley & Sons. This manageably sized dictionary covers theory, experiment, industrial practice, and applications for nanotechnology, colloid, and interface science, as well as much of what is now termed materials science. The comprehensive information is presented in several sections and formats: dictionary of terms, classification tables on colloid and nanomaterial types, and subterm glossaries for specific phenomena, properties, and methods.

MONOGRAPHS

Bittnar, P., J. M. Bartos, J. Nemecek, V. Smilauer, and J. Zeman, eds. 2009. *Nanotechnology in Construction: Proceedings of the NICOM3.* New York: Springer. The 3rd International Symposium on Nanotechnology in Construction (NICOM 3) follows the highly successful NICOM 1 (Paisley, U.K., 2003) and NICOM 2 (Bilbao, Spain, 2005) Symposia. The NICOM3 symposium was held in Prague, Czech Republic from May 31 to June 2, 2009, under the auspices of the Czech Technical University in Prague. It was a cross-disciplinary event, bringing together R&D experts and users from different fields all with interest in nanotechnology and construction.

Fan, M. I., C. P. Huang, and A. E. Bland. 2010. *Environanotechnology.* New York: Elsevier. The aim of this book is to report on the results recently achieved in different countries. It provides useful technological information for environmental scientists and will assist them in creating cost-effective nanotechnologies to solve critical environmental problems, including those associated with energy production. It presents research results from a number of countries with various nanotechnologies in multidisciplinary environmental engineering fields and gives a solid introduction to the basic theories needed for understanding how environanotechnologies can be developed cost-effectively, and when they should be applied in a responsible manner.

Koo, J. H. 2010. *Polymer Nanocomposites*. New York: McGraw-Hill. Based on the author's 26 years of experience in the field of nanotechnology, this reference offers researchers and materials scientists a complete reference to the physical concepts, techniques, applications, and principles underlying one of the most researched materials.

Sellers, K., C. Mackay, L. L. Bergeson, S. R. Clough, M. Hoyt, J. Chen, K. Henry, and J. Hamblen. 2008. *Nanotechnology and the Environment*. Boca Raton, FL: CRC Press. *Nanotechnology and the Environment* provides the fundamental basis needed to assess and understand the life cycle of nanomaterials. It begins with a general explanation of nanomaterials, their properties, and their uses and describes the processes used to manufacture nanoscale materials. Subsequent chapters furnish information on the analysis of nanomaterials in the environment and their fate and transport, including the effects of wastewater treatment on nanomaterials. The book discusses possible risks to human health and the environment, and describes developing regulations to manage those risks.

Smith, G. B. and C.-G. S. Granqvist. 2010. *Green Nanotechnology: Solutions for Sustainability and Energy in the Built Environment*. Boca Raton, FL: CRC Press. A focused exploration of the role nanotechnology plays in meeting the challenges inherent in minimizing environmental impacts while maximizing energy resources, this book provides an overview of our energy supply, increasing energy production while reducing cost, and offering novel energy sources. It explores the ways in which nanotechnologies can improve structural engineering of energy sources, create novel methods of cooling, and inspire new approaches to water supply and treatment. In addressing these critical issues, the book provides an authoritative resource that provides the foundation for new research and product development.

Theodore, L. 2005. *Nanotechnology: Basic Calculations for Engineers and Scientists*. Hoboken, NJ: John Wiley & Sons. This is a practical workbook that bridges the gap between theory and practice in the nanotechnology field. Because nanosized particles possess unique properties, nanotechnology is rapidly becoming a major interest in engineering and science. The author has developed nearly 300 problems that provide a clear understanding of this growing field in four distinct areas of study: chemistry fundamentals and principles, particle technology, applications, and environmental concerns. These problems have been carefully chosen to address the most important basic concepts, issues, and applications within each area, including such topics as patent evaluation, toxicology, particle dynamics, ventilation, risk assessment, and manufacturing. An introduction to quantum mechanics is also included in the appendix. These stand-alone problems follow an orderly and logical progression designed to develop the reader's technical understanding.

WEB SITES

Centre for Nanomaterials Applications in Construction (Spain) http://info.labein.es/nanoc/ (accessed June 29, 2010). NANOC was created within LABEIN-Tecnalia in 2002. It was conceived as a platform aimed at developing key technological capacities that will enable the construction industry to use nanotechnology as a competitive tool. As part of TECNALIA Construction and as a unit associated with the CSIC (Consejo Superior De Investigaciones Científicas), the NANOC team aims to generate knowledge in nanoscience and nanotechnology targeted toward the development of novel knowledge-based, high performance materials for the construction sector.

Good Nano Guide http://goodnanoguide.org/HomePage (accessed June 24, 2010). The Good Nano Guide is a collaboration platform designed to enhance the ability of experts to exchange ideas on how best to handle nanomaterials in an occupational setting. It is meant to be an interactive forum that fills the need for up-to-date information about current good workplace practices, highlighting new practices as they develop.

Integrated Nanosystems Group http://www.mitre.org/tech/nanotech/ (accessed June 29, 2010). MITRE is a not-for-profit corporation that works with U.S. government clients. It combines systems engineering and information technology to develop innovative solutions that are in the public interest. This Web site provides access to information on nanoelectronics and nanocomputing research both internally at MITRE and externally. This includes an introduction to nanosystems, recent papers describing interesting developments in nanoscience and nanotechnology, and a listing of conferences and events.

International Council on Nanotechnology (ICON) http://icon.rice.edu/ (accessed June 29, 2010). The creation of a sustainable nanotechnology industry requires meaningful and organized relationships among diverse stakeholders. The International Council on Nanotechnology (ICON) is the only global organization aimed at providing such interactions for a broad set of members. Managed by Rice University's Center for Biological and Environmental Nanotechnology, ICON activities promote effective nanotechnology stewardship through risk assessment, research, and communication. By pooling the resources of the nanotechnology industry, government and nongovernment organizations and academia, ICON can cost-effectively provide a wide range of synergistic projects that serve the interests of all stakeholders.

Nanofolio: What is Nanotechnology? http://www.nanofolio.org/win/index.php (accessed June 29, 2010). This introduction to nanotechnology covers the basic concepts, history and future of nanotechnological research and development. The many applications of nanotechnology are outlined, with illustrations. There are sections on the enabling technologies that make nanotechnology possible, and on concepts such as radical, evolutionary, and incremental nanotechnology. References to further sources of information are given.

Nanoforum: European Nanotechnology Gateway http://www.nanoforum.org/ (accessed June 29, 2010). Nanoforum is a pan-European nanotechnology network that was originally funded by the European Union under the Fifth Framework Programme (FP5) and which now operates as a European Economic Interest Grouping (EEIG). The Nanoforum Web site is designed to provide a gateway for researchers in Europe working in the field of nanotechnology. Areas of nanotechnology, such as chemistry and materials, energy, science, metrology, products, and health and safety, are covered. The site contains news, upcoming events, publications, forums, and education and careers links. Relevant companies across Europe can be viewed. Most articles require free registration.

Nanototechnology Knowledge Transfer Network, https://ktn.innovateuk.org/web/nanoktn/overview (accessed October 20, 2010) is the U.K. government platform, supported by its Technology Strategy Board, to encourage innovation and take-up of nanotechnology.

Nanotechnology Now http://www.nanotech-now.com/ (accessed June 29, 2010). Nanotechnology Now is a portal provided by 7thWave Inc. in the United States. It covers aspects of nanotechnology including nanospace, nanoscale, and future sciences, such as molecular machine systems, nanomedicine, and bioMEMS. The site includes an introduction to nanotechnology as well as a glossary of terms. There are a series of white papers and also interviews, predictions, a bibliography, and an events listing. Links to related sites are provided.

National Institute of Advanced Industrial Science and Technology http://www.aist.go.jp/index_en.html (accessed June 29, 2010). AIST conducts research that transcends the barriers between disciplines, to play an active role in developing advanced expertise for tomorrow's industries. AIST cover six research fields, i.e., Life Science and Biotechnology, Information Technology, Nanotechnology Materials and Manufacturing, The Environment and Energy, Geological Survey and Applied Geoscience, Metrology and Measurement Science.

National Nanotechnology Institute (NNI) http://www.nano.gov/index.html (accessed June 24, 2010). The National Nanotechnology Initiative (NNI) http://www.nano.gov/ (accessed June 29, 2010) is the program established in fiscal year 2001 to coordinate federal nanotechnology research and development. The NNI provides a vision of the long-term opportunities and benefits of nanotechnology. By serving as a central locus for communication, cooperation, and collaboration for all federal agencies that wish to participate, the NNI brings together the expertise needed to guide and support the advancement of this broad and complex field.

Safety of Nano-Materials Interdisciplinary Research Centre http://www.snirc.org/ (accessed June 2, 2010). SnIRC's aims are to develop a coherent integrated program of work that will (1) increase awareness of the issues relating to nanoparticles, health, and environment; (2) become the U.K. center for information and advice on the potential health, safety, and environmental impacts of generic or specifically engineered nanomaterials, especially nanoparticles and nanotubes; (3) generate a comprehensive and coherent body of scientific evidence, which would help toward developing relevant policies to promote U.K. nanotechnology growth while safeguarding workplace, public, and environmental health; assist U.K. industry in developing safe nanomaterials; (4) maintain and promote an international network of researchers and regulators actively involved in the safety of nanomaterials; (5) be the organization for integrating U.K. research with corresponding U.S. and European efforts; (6) maintain dialogue with the Research Coordination Group led by DEFRA in response to the Royal Society/Royal Academy of Engineering Report; and (7) raise support and funding for these activities.

Understanding Nanotechnology http://www.understandingnano.com/ (accessed June 29, 2010). This site includes excerpts from the book *Nanotechnology for Dummies* and includes explanations of nanotechnology concepts as well as links to articles and resources such as manufacturers' sites and nanotechnology stories in the news. Applications of nanotechnology in medicine, energy supply, and pollution clearing are described. There is a list of links to related sites.

CONFERENCES

Since nanotechnology is still in its infancy and has given rise to numerous novel applications to the solution of many current problems, a variety of scientific and engineering organizations are staging conferences on nanotechnology. Many can be found at Allconference.com http://www.allconferences.com/ (accessed June 2, 2010).

JOURNALS

International Journal of Nanotechnology (IJNT), Geneva, Switzerland: Inderscience Publishers. (ISSN (Online): 1741-8151; ISSN (Print): 1475-7435). The IJNT intends to provide a major reference source of comprehensive fundamental and applied knowledge in all areas of nanotechnology for researchers and scientists as well as an educational resource for students, teachers, and educators by means of reviews from experts in all fields.

Journal of Nanomaterials, New York: Hindawi Publishing Corporation (ISSN: 1687-4110; e-ISSN: 1687-4129). The overall aim of the *Journal of Nanomaterials* is to bring science and applications together on nanoscale and nanostructured materials with emphasis on synthesis, processing, characterization, and applications of materials containing true nanosize dimensions or nanostructures that enable novel/enhanced properties or functions. It is directed at both academic researchers and practicing engineers. *Journal of Nanomaterials* highlights the continued growth and new challenges in nanomaterials science, engineering, and nanotechnology, both for application development and for basic research. All papers should emphasize original results relating to experimental, theoretical, computational, and/or applications of nanomaterials ranging from hard (inorganic) materials, through soft (polymeric and biological) materials, to hybrid materials or nanocomposites. Review papers summarizing the state of the art for a particular research field or tutorial papers, especially those emphasizing multidisciplinary views of nanomaterials and those related to significant nanotechnologies, are also welcome.

Journal of Nanomechanics and Micromechanics. 2011–. Reston, VA: American Society of Civil Engineers. (ISSN - 2153-5434 / e-ISSN - 2153-5477). *The Journal of Nanomechanics and Micromechanics* brings science and applications together on nanoscale and nanostructured materials, with emphasis on mechanics, processing, characterization, design, modeling, and applications of materials containing true nanosize dimensions or nanostructures that describe novel or enhanced properties or functions that are based on tailored nanostructures. The journal is directed at both academic researchers and practicing engineers. It highlights the continued growth and new challenges in nanomechanics and micromechanics for application development and for basic research and applications. All papers should emphasize original results relating to experimental, theoretical, or computational results, as well as applications of nanomaterials ranging from hard inorganic materials, through soft polymeric and biological materials, to hybrid materials or nanocomposites.

NANO: The Magazine for Small Science http://www.nanomagazine.co.uk/ (accessed June 29, 2010). Scotland: IoN Publishing Ltd. This site includes full text articles from previous issues of the magazine covering nanotechnology as well as current news stories. Print subscription is also available. Web site provides news items, listing of relevant events, courses, and reports.

Nanotechnology, Bristol, U.K.: IOP Publishing (ISSN 0957-4484 (Print), ISSN 1361-6528 (Online)). The journal aims to publish papers at the forefront of nanoscale science and technology and especially those of an interdisciplinary nature. Here, nanotechnology is taken to include the ability to individually address, control, and modify structures, materials, and devices with nanometer precision, and the synthesis of such structures into systems of micro- and macroscopic dimensions, such as MEMS-based devices. It encompasses the understanding of the fundamental physics, chemistry, biology, and technology of nanometer-scale objects and how such objects can be used in the areas of computation, sensors, nanostructured materials, and nano-biotechnology.

Nanotoday (online only) http://www.nanotoday.com (accessed June 2, 2010), Elsevier Ltd. (ISSN: 1748-0132). *Nanotoday* publishes original articles on all aspects of nanoscience and nanotechnology. Manuscripts of three types are considered: Review Articles that inform readers of the latest research and advances in nanoscience and nanotechnology, Rapid Communications that feature exciting research breakthroughs in the field, and News and Opinions that comment on topical issues or express views on the developments in related fields.

Proceedings of the Institution of Mechanical Engineers, Part N: Journal of Nanoengineering and Nanosystems London: Professional Engineering Publishing (ISSN: 1740-3499 (Print) 2041-3092 (Online)). *The Journal of Nanoengineering and Nanosystems* is dedicated to the particular aspects of nanoscale engineering, science, and technology that involve the descriptions of nanoscale systems.

STRUCTURAL ENGINEERING

INTRODUCTION

Structural engineering as an identifiable subdivision of civil engineering developed over the second half of the nineteenth century as increasingly sophisticated analytical tools were developed to analyze arched and more particularly framed structures of iron, and latterly steel and reinforced concrete. In the 20 years before World War I, the introduction of structural steel framing and reinforced concrete encouraged widespread use of these materials in tall buildings and structures that had, in previous times, been largely the province of the architect, such as civic buildings, theaters, offices, and shops. Architects wishing to take full advantage of the economy and design potential of these new materials needed an understanding of the analytical tools that underpinned their safe use. A demand for specialist structural engineers grew, which extended beyond the engineers who had specialized in the design of iron bridges and railway structures of the previous century. This was echoed in Britain by the creation of the Institution of Structural Engineers, and internationally by the First Congress on Bridge and Structural Engineering in 1926 (see IABSE below) and the establishment of the International Association of Bridge and Structural Engineering (see below).

The best introduction to the historical development of the subject is that by Timoshenko (1951), although Charlton is helpful for the nineteenth century. For a more academic approach one may refer to Benvenuto. Addis is very useful as an introduction to the philosophy of design (1990) as well as design practice (2007), while the writings of David Billington convey an enthusiasm for the subject that is contagious. Structural engineering is the discipline within civil engineering that seizes the imagination of the public and inspires youngsters to join the profession. It emerged as a discipline in the early twentieth century with the development of reinforced concrete and steel-framed structures, most spectacularly with the skyscrapers of Chicago and New York. While much of the work of the structural engineer is involved with more mundane questions relating to building regulations approval for domestic housing extensions, etc., it is the interface of structural engineers with modern architecture that understandably attracts media attention. Since the nineteenth century, increasingly sophisticated methods of analysis have been developed to design structures, dominated today by computer programmers. Codes have changed to reflect these developments, and Eurocodes are now being phased in across Europe. An increasing number of British structures are now based on Eurocodes http://www.eurocodes.co.uk/ (accessed October 2010). The design of modern building structures is closely allied with the provision of building services, and the concept of "intelligent buildings." (These issues are dealt with in Chapter 5 on Architectural Engineering in this book.)

Structural engineers' use of materials is overwhelmingly concerned with the use of concrete and mild steel. With the properties of these materials specified in standards, the need for civil engineers to concern themselves with materials science on a daily basis might be considered limited, but, in fact, higher performance materials are frequently sought for specialist applications and to reduce the mass or cost of structures. Moreover, "new" materials such as plastics have gained an increasingly important role. More traditional materials, such as timber and glass, have enjoyed something of a revival, with jointing and joining methods facilitating timber construction, while a better understanding among engineers of the structural properties has facilitated the adventurous use of glass by architects. Materials such as cast iron are generally associated with older structures like mill buildings, but continue to have specialist uses, while a newer metal, such as aluminum, which enjoyed a minor vogue in the 1950s, continues to be important where lightness and strength are required. The concept of "zero-carbon" structures is causing a detailed reexamination of the carbon cost of all materials.

Engineering structures have a long design life—decades if not centuries—and so the durability of materials and their decay and corrosion mechanisms are ultimately as important as their strength. While standards and specifications for most materials are produced by the leading national standard organizations, when new developments take place engineers have to turn to industry standards and manufacturers' specifications for guidance until the "official" standards have caught up. The problem in seeking such information is that in many fields such specialist organizations proliferate; more than 60 are involved in concrete in the United Kingdom alone. The Internet and the concept of the "one-stop-shop" can help, and what follows is intended to provide guidance on the most important of such organizations and sources.

Bibliography

Addis, W. A. 1990. *Structural Engineering: The Nature of Theory and Design*. Chichester, U.K.: Ellis Horwood Ltd.

Addis, W. A. 2007. *Building: 3000 Years of Design Engineering and Construction*. London and New York: Phaidon.

Benvenuto, E. 1991. *An Introduction to the History of Structural Mechanics*. (2 vols.) New York: Springer.

Billington, D. P. 1983. *The Tower and the Bridge: The New Art of Structural Engineering*. New York: Basic Books.

Billington, D. P. 1990. *Rohert Maillart and the Art of Reinforced Concrete*. New York: Architectural History Foundation.

Billington, D. P. 1996. *The Innovators: The Engineering Pioneers Who Made America Modern*. New York: John Wiley & Sons.

Charlton, T. M. 1982. *A History of Theory of Structures in the Nineteenth Century*. Cambridge, U.K.: Cambridge University Press.

Timoshenko, S. P. 1953. *History of Strength of Materials*. New York: McGraw-Hill.

Searching the Library Catalog

Structural engineering is generally one specialization that fits neatly into a Library of Congress subclass (i.e., TA 630-695); while there needs to be an awareness of other areas, e.g., chemical technology namely TPO 751-762 clay industries, ceramics, glass, and TP875-888, and TP 1080-1185, this will be rare. There is a strong overlap in structural analysis with the classes TA329-348 engineering mathematics and analysis, and TA349-TA 359 applied mechanics, reflected in the works of world famous civil engineers, such as Stephen P. Timoshenko. In terms of keyword searching one must bear in mind that an engineer may be seeking information on shell structures, which the architects and general public might describe as domes.

Abstracts and Indexes

Building (Construction) References (1946-1990) (MPBW/PSA)). A useful source for references to older buildings.

Building Science Abstracts (1926-1976). In part continued in *ICONDA*, this abstract series created by the U.K. Building Research Station is very useful for tracing articles on building science (e.g., properties of materials).

International Construction Database (*ICONDA*) (1974-) (Fraunhofer-Gesellschaft Informationzentrum Raum und Bau (IRB). http://www.irb.fraunhofer.de/iconda/login/login.jsp (accessed October 20, 2010). Excellent on European coverage but poor on British coverage since 1990, due to lack of input. Available from OVID.

Lnstruct. 1923–. This database is accessible via the Web site of the Institution of Structural Engineers and covers a wide range of books, periodicals, and related publications (e.g., BRE Digests) held in the library. This includes all articles in *The Structural Engineer* (the Institution journal) from 1923 to date. These are digitized.

Databases

Barbour Index http://www.barbour-index.co.uk/content/aboutservices/cfm.asp (accessed November 3, 2010). *Barbour Index* has a range of information services for built environment professionals in the United Kingdom. It is more relevant to the building and architectural professions than TI (below), which provides more information of relevance to engineers. The *Index*, like TI, has developed from a microfiche-based product library.

Barbour produces a number of services: Building Product Expert, a building product database containing detailed technical and product information from manufacturers of building products; Specification Expert (incorporating the National Engineering Specification), a specification writing tool for building professionals and building services engineers; BSRIA Detail Drawings, a library of over 850 drawings and symbols illustrating common building services design and installation details. Construction Expert provides access to technical and product information on CD or via the Web with full-text documents from over 200 technical publishers, a directory of U.K. manufacturers, 25,000 trade names, and 50,000 product ranges with performance characteristics.

Construction Information Service http://www.ihs.com/en/uk/info/st/e/construction-information-service.aspx (accessed April 1, 2011). Developing from a microfilm-based library and product information service formerly known as TI, this is now arguably the most important U.K. provider of civil engineering documentation. It acts as authorized distributor for the British Standards Institution, ETSU, and TSO, and many other publishers. The TI agreement is produced jointly with NBS, part of RIBA Enterprises Ltd. Its target audience includes civil and structural engineers. It is in use at networked locations as well as on stand-alone PCs. The service may be delivered online or on CD-ROM; subscription includes regular updates, technical support, and access to helplines. Although it is aimed at U.K. civil engineers, there are certain important categories of documentation that are not available, such as ICE publications. On the other hand, agreements with Network Rail and the Highways Agency mean that a lot of key data are available from a single source. IHS EMEA retains hard copy documents.

BIBLIOGRAPHIES

BRE Bibliography of Structural Failures 1850–1970. Garston, U.K.: Building Research Establishment. While failures preceded 1850 and have continued to occur, this is a useful starting point for precedents.

ICE Bibliography on Prestressed Concrete 1920–1957. London: ICE. A useful introduction to the early literature on the subject although lacking particularly in early German work.

Jakkula, A. A. 1941. *A History of Suspension Bridges in Bibliographic Form.* Washington, D.C.: Public Roads Administration. Despite its age, this is a good starting point for anybody interested in the history of suspension bridges. It is weakest on European developments where the only literature is in non-English language sources.

HANDBOOKS AND MANUALS

Bickford, J. H. and S. Nassas. 1998. *The Handbook of Bolts and Bolted Joints.* New York: Marcel Dekker. The design of bolted connections requires specialist knowledge that is frequently lacking in more general handbooks.

Chen, W-F. 2002. *Structural Engineering Handbook,* 2nd ed. Boca Raton, FL: CRC Press. The early chapters deal with basic concepts and aspects of structural analysis, such as theory of plates and stability, structural analysis, seismic engineering, structural steel design, and composite construction. More specialist topics include aluminum and timber structures, bridges, shell and space framed structures, multistorey frames, and cooling towers transmissions. Elements are dealt with such as stub-girder floor systems. Other aspects covered include shock loading underground pipes and structural reliability. There is a chapter on passive energy dissipation and active control. Available at http://www.eng-netbase.com/ejournals/books/book_summary/toc.asp?id=417.

Chen, W-F. and L. Duan. 1999. *Bridge Engineering Handbook*. Boca Raton, FL: CRC Press. In seven main sections. The first section deals with fundamental issues, such as bridge aesthetics; the second, with types of bridge superstructure, including less familiar types, such as floating and movable bridges; the third, with foundations; the fourth, seismic design, including retrofit, the fifth, with construction and repair; and the sixth, with specialist aspects, such as ship collision. The final section deals with international design, including Russia and China. Given the overall coverage, it is surprising how little there is on footbridges.

Doran, D. K. 1992. *Construction Materials Reference Book*. Oxford, U.K.: Butterworth-Heinemann. Written by leading authorities in the field and covering most engineering materials, including some more generally associated with historic structures, such as cast and wrought iron.

Forde, M., ed. 2009. *ICE Manual of Structural Materials*. London: TTL. (2 vols.) Part of the ICE handbook series, with the opportunity to download individual chapters from the ICE Virtual Library.

Gaylord, E. H., C.N. Gaylord, and J. E. Stallmeyer. 1997. *Structural Engineering Handbook*. New York: McGraw-Hill. This is a comprehensive and well-established title that, with few exceptions, is as good a basic reference tool as one could desire. Some of the references to standards are out of date. Covers structural analysis, computer application, and earthquake-resistant design, geotechnical issues, structural elements by types and material, and specialist structures including suspended roofs, shells, industrial buildings, silos, steel tanks, and chimneys.

Halliday, S. 2009. *Sustainable Construction*. Oxford, U.K.: Butterworth Heinemann.

Levy, S. M. 2001. *Construction Building Envelope and Interior Finishes Databook*. New York: McGraw-Hill. This databook is intended to provide access to hundreds of tables, specifications, charts, diagrams, and illustrations covering materials and components most frequently used on a typical job.

Ryall, M. J., ed. 2000. *Manual of Bridge Engineering*. London: Thomas Telford. Covers all aspects of bridge design by material and form, including joints and foundations. Written largely by U.K. authors.

Tilly, G. 2002. *Conservation of Bridges*. London: Spon. Intended to encourage best practice in bridge conservation across the United Kingdom.

MONOGRAPHS AND GENERAL WORKS

Adler, D., ed. 1999. *Metric Handbook*. Oxford, U.K.: Architectural Press. Very useful for dimensions and guidelines on all types of structures.

ASCE. 1989. *Manual 52: Guide for Design of Steel Transmission Lines*. Reston, VA: ASCE.

ASCE. 1990. *Manual 72: Guide for Design of Steel Transmission Pole Structures*. Reston, VA: ASCE.

ASCE. 1991. *Manual 74: Guidelines for Electrical Transmission Line Structural Loading*. Reston, VA: ASCE. These three titles are the best source of published advice on the design of transmission line structures.

ASCE. 1999. *Structural Design for Physical Security*. Reston, VA: ASCE. Security is now an international issue and this is a helpful reference point for all engineers.

Bangash, M. Y. H. 2004. *Impact and Explosion*. Oxford, U.K.: Blackwell (new ed.), Springer. Contains many numerical examples, including modeling of aircraft impact on tall buildings and nuclear and conventional weapons effects.

Berlow, L. H. 1998. *The Reference Guide to Famous Engineering Landmarks of the World*. Chicago: Fitzroy Dearborn. Not nearly as comprehensive as the *Guinness Book* by Stephens (noted below), but useful for more modern structures.

Brown, G. and J. Neilsen. 1990. *Silos.* London: Spon. Silo design and construction from a U.K. perspective.

CIRIA. 2004. *Principles of Design for Deconstruction to Facilitate Reuse and Recycling.* London: CIRIA. Discusses all elements of a building in terms of life cycle and recycling, including fittings as well as structure, with a view to facilitating deconstruction and reuse through design (C607).

Cobb, G. 2009. *Structural Engineer's Pocket Book.* Oxford, U.K.: Butterworth Heineman. Aimed at the young practitioner, and rooted in British practice. A revision will be required in due course to take account of the use of Eurocodes.

Cook, N. J. 1986. *Designer's Guide to Wind Loading of Building Structures.* (2 parts.) London/Boston: Butterworths. Although not concerned with the latest codes, the basic principles of wind engineering are well covered here.

Cook, N. J. 1999. *Wind Loading: A Practiced Guide to BS 6399, Part 2.* London: Thomas Telford. The title is somewhat misleading as it also covers BS 8100 (lattice structures). Not as comprehensive as Cook (1986).

Curwell, S., ed. 2002. *Hazardous Building Materials: A Guide to the Selection of Environmentally Responsible Alternatives,* 2nd ed. London: Spon. The most helpful feature is the use of data sheets that form the bulk of this publication and provide a checklist both for a professional and homeowner chart.

Davis, J. and R. Lambert. 2002. *Engineering in Emergencies,* 2nd ed. London: ITDG. A useful guide to the appropriate technology for anybody dealing with emergencies.

Fernandez Troyano, L. 2003. *Bridge Engineering: A Global Perspective.* Madrid and London: TTL and CICCP. A magnificently presented compendium by one of the world's foremost bridge designers, which would inspire anybody to take up engineering.

Friedman, D. 2010. *Historical Building Construction,* 2nd. ed. New York: W. W. Norton. Comprehensive approach to historical U.S. building structures, although lacking in international perspective.

Hambly, E. C. 1991. *Bridge Deck Behaviour,* 2nd ed. London: E & FN Spon. A simple approach to a fundamental subject for bridge engineers.

Holmes, J. D. 2007. *Wind Loading of Structures,* 2nd ed. London: Spon/Taylor & Francis. A final chapter provides a comparison of existing standards. Good general textbook with extensive references on all aspects of wind-resistant design.

Illston, J. M. and P. L. J. Podmore, eds. 2001. *Construction Materials: Their Nature and Behaviour,* 4th ed. London: Spon. A useful introduction covering fundamentals, metals, including aluminum and copper, concrete, bitumen, brickwork, polymers and fiber composites, and cements and timber. Aimed primarily at students.

IStructE. 2010. *Appraisal of Existing Structures,* 3rd ed. London: IStructE. Almost a checklist approach to appraisal and as such generally applicable for all involved in appraisal (3rd ed. in preparation).

Jackson, N. and R. K. Dhir, eds. 1996. *Civil Engineering Materials,* 5th ed. Basingstoke, U.K.: Macmillan. Good basic introduction to materials used regularly by civil engineers. Covers metals, concrete, timber, bitumen, soils, polymers, and bricks and block work. Aimed at a student audience.

Jones, N. 1989. *Structural Impact.* Cambridge, U.K.: Cambridge University Press. Investigation of the crashworthiness of structures has grown over recent years. This text provides a good starting point.

Mainstone, R. 1998. *Developments in Structural Form,* 2nd ed. Oxford, U.K.: Architectural Press. This is perhaps the best introduction to the art of structural engineering through all ages, and very useful for those unfamiliar with various structural types.

Mays, G. C. and P. D. Smith. 1995. *Blast Effects on Buildings.* London: Thomas Telford. A useful starting point for those new to blast-resistant design.

Petroski, H. 1994. *Design Paradigms: Case Histories of Error and Judgment in Engineering.* Cambridge, U.K.: Cambridge University Press. This study of failures is as good a place as any to begin understanding the secret of good structural engineering design.

Plate, E. J. 1982. *Engineering Meteorology.* Amsterdam: Elsevier. No structure can be designed without an understanding of weather and this book provides a useful background on this aspect of engineering.

Podolny, W. and J. B. Scalzi. 1986. *Construction and Design of Cable Stayed Bridges,* 2nd ed. New York: John Wiley & Sons. There are a growing number of texts on cable-stayed bridges, but this remains the best for its historical insight, with lots of examples.

Ravenet, J. 1987. *Silos.* (3 vols.) Barcelona: EJA. The most comprehensive book on the subject, albeit in Spanish.

Reimbert, M. L. and A. M. Reimbert. 1987. *Silos,* 2nd ed. Paris: Lavoiser. The Reimberts have developed one of the most commonly used methods of silo design.

Ryall, M. J. 2010. *Bridge Management,* 2nd ed. Oxford, U.K.: Butterworth-Heinemann. A useful introduction to all aspects of bridge maintenance and assessment from a U.K. perspective.

Safarian, S. S. and E. C. Harris. 1985. *Design and Construction of Silos and Bunkers.* New York: Van Nostrand Reinhold. U.S.-based silo design.

Stafford-Smith, B. and A. Coull. 1991. *Tall Building Structures.* New York: John Wiley & Sons. These authors wrote a classic on shear wall design in the 1960s and have carried their work through into a very readable text for graduate engineers involved in design.

Stephens, J. H. 1976. *Guinness Book of Structures.* Enfield, U.K.: Guinness Superlatives. Despite its age, its comprehensive coverage, encompassing dams and tunnels, as well as building structures, makes this a valuable source of reference data on important structures.

Taranath, B. S. 1998. *Steel, Concrete and Composite Design of Tall Buildings,* 2nd ed. New York: McGraw-Hill. General introduction to tall building design covering the various loadings that must be embraced.

Young, W. C. and R. C. Budynas. 2002. *Roark's Formulas for Stress and Strain,* 7th ed. New York: McGraw-Hill. The main reference work for structural formulas.

Zaknic, I., ed. 1998. *100 of the World's Tallest Buildings.* Corte Madera, CA: Gingko. Given that such a work is continually in need of updating, this *is* a useful compendium on some of the best-known skyscrapers.

CONCRETE

Bangash, M. Y. H. 2001. *Manual of Numerical Methods in Concrete.* London: Thomas Telford. Part of an introduction to numerical methods, an important feature is the case studies and worked examples including U.S., British, and European codes. Some understanding of numerical methods is required to make best use of this.

Dobrowolski, J. A. 1998. *Concrete Construction Handbook.* New York: McGraw-Hill. A long-established handbook on U.S. practice.

Nawy, E. G. 1998. *Concrete Construction Engineering Handbook,* Boca Raton, FL: CRC Press. Deals with recent advances in materials including high-strength concrete and fiber reinforcement as well as the main fields of reinforced and prestressed concrete. There are chapters on seismic and fire-resistant design, and specialist areas, such as offshore structures and prestressed concrete bridges.

Neville, A. M. 1995. *Properties of Concrete,* 4th ed. London: Prentice Hall. Neville's work is essentially concerned with concrete practice and provides the best introduction to the properties of the material.

Reynolds, C. E., J. C. Steedman, and A. J. Threlfall. 2008. *Reinforced Concrete Designers Handbook,* 11th ed. London: Taylor & Francis/E & FN Spon. An update of the classic.

Despite criticism of its age, this handbook is a useful compendium of information on subjects not readily found elsewhere (e.g., weights of various materials).

Sutherland, R. J. M., ed. 2001. *Historic Concrete: Background to Appraisal.* London: Thomas Telford. Useful background for appraisal, with a U.K. bias, this work comprises a series of chapters by leading construction historians and engineers covering the use of concrete in all types of structures.

IRON AND STEEL

AISC. 2005. *Manual of Steel Construction, Metric,* 13th ed. Chicago: American Institute of Steel Construction. The main U.S. design manual for structural steel, widely used internationally, this edition combines the previously separated ASD and LFRD manuals.

Angus, H. T. 1976. *Cast Iron: Physical and Engineering Properties,* 2nd ed. London: Butterworth. The best guide to the properties of modern cast iron.

Bangash, M. Y. H. 2000. *Structural Detailing in Steel.* London: Thomas Telford. Deals with joints, welding, and components, with examples of buildings and bridges from U.K., U.S., and European practice.

Brockenburgh, R. L. and F. S. Merritt. 1994. *Structural Steel Designer's Handbook,* 3rd ed. New York: McGraw-Hill. Multiauthored guide, useful for American practice; although all traditional units, this handbook deals with properties, structural theory and analysis, and design of all types of buildings and bridges, including truss bridges that are often omitted from bridge designer books.

Galambos, T. V. 1998. *Guide to Stability Design Criteria for Metal Structures.* New York: John Wiley & Sons. The best monograph on this specialist aspect of structural steel design.

Hayward, A., A. C. Oakhill, and F. Wearne. 2002. *Steel Detailer's Manual,* 2nd ed. Oxford, U.K.: Blackwell. Useful for detailing U.K. standards with examples of buildings and bridges.

Hoffman, E. S., ed. 1996. *Structural Design Guide to the AISC (LRFD) Specification for Buildings.* New York: Chapman and Hall. Useful guide to the AISC specification.

Steel Construction Institute. 1997. *Appraisal of Existing Iron and Steel Structures.* Berkshire, U.K.: Steel Construction Institute. This work covers all aspects of structural appraisal, although the reference points are generally to British practice.

Steel Construction Institute. 2003. *Steel Designers Manual,* 6th ed. Oxford, U.K.: Blackwell Scientific. This work encompasses most aspects of structural steel design in the U.K. context, with a chapter on Eurocodes. Bridges and pile design are covered.

Yu, W.-W. and R. A. Laboube. 2010. *Cold-Formed Steel Design,* 4th ed. New York: John Wiley & Sons. Based on North American, including Mexican practice, this is the most comprehensive treatment of this specialist area of structural steel work.

TIMBER

American Institute of Timber Construction. 1994. *Timber Construction Manual,* 4th ed. Centennial, CO: AITC. Comprehensive manual for American practice.

Baird, J. A. and E. C. Ozelton. 2002. *Timber Designers Manual,* 3rd ed. Oxford, U.K.: Blackwell. This provides the manual to British Standards practice in the United Kingdom while incorporating references to Eurocode 5. Some chapters reflect developments in North America, which have now influenced U.K. practice.

Breyer, D., K. J. Fridley, and K.E. Cobeen. 2003. *Design of Wood Structures* (ASD), 5th ed. New York: McGraw-Hill. Widely used in teaching timber engineering at undergraduate level in the United States, the 5th edition has references to the latest *International Building Code* and *2001 National Design Specification for Wood Construction.* There is extensive treatment of joints and fixings, laminated and sandwich construction.

Centrum Holst. 1995. *Timber Engineering: STEP 1 and STEP 2*. Eindhoven, The Netherlands: Centrum Holst. This is an excellent introduction to timber design with Eurocodes in design.

Faherty, K. F. and T. G. Williamson. 2003. *Wood Engineering and Construction Handbook*, 3rd ed. New York: McGraw-Hill. Multiauthored work that deals with materials and connections, and chapters on all kinds of structures, including trusses, columns, domes and diaphragms, and foundations.

Larson, H. and V. Eryity. *Practical Design of Timber Structures to Eurocode 5*. London: TTL.

McKenzie, W. M. C. and B. Zhang. 2000. *Design of Structural Timber to Eurocode 5*. Basingstoke, U.K.: Macmillan. Aimed at undergraduates, this serves as an introduction to designs by Eurocodes.

Sunley, J. and B. Bedding, eds. 1985. *Timber in Construction*. London/Batsford: TRADA.

Yeomans, D. 1992. *The Trussed Roof*. Aldershot, Hants, U.K.: Ashgate. For anybody interested in the subject, this is a readable introduction and more.

Masonry

The term *masonry*, particularly in the United States, is used to describe mass concrete (without reinforcement), as well as natural stone and brick and other ceramics; for concrete generally, see above.

Amrhein, J. E. 1998. *Reinforced Masonry Engineering Handbook*, 5th ed. Los Angeles: Masonry Institute of America. The most comprehensive source available on the subject.

Brick Development Association. 1994. *Brick Development Association Guide to Successful Brickwork*. London: Edward Arnold. This guide provides practical advice on brick construction.

Curtin, W., ed. 1995 *Structural Masonry Designers Manual*, 2nd rev. ed. Oxford, U.K.: Blackwell Scientific. Comprehensive approach, covering relevant British standards.

Sowden, A. M., ed. 1990. *The Maintenance of Brick and Stone Masonry Structures*. New York: E & FN Spon. A series of papers by specialists on all types of masonry structures including brick lined tunnels.

Other Materials

The Aluminum Association. 2010. *Aluminum Design Manual*. Washington, D.C.: The Aluminum Association. The basic source for engineers in the English-speaking world.

Amstock, J. S. 1997. *Handbook of Glass in Construction*. New York: McGraw-Hill. The first comprehensive guide for American audiences.

Amstock, J. S. 2000. *Handbook of Adhesives and Sealants in Construction*. New York: McGraw-Hill. Chapters on the various sealant materials are followed by discussions of specifications, joint design, and performance and repair.

ASCE. 1984. *Structural Plastics Design Manual*, New York: ASCE. These two ASCE manuals remain the best source for the structural use of plastics, although there have been many advances in applications since they were written (e.g., with the use of carbon fiber-bonded plates for strengthening bridges), but such applications can be traced through conference and journal papers.

ASCE. 1985. *Structural Plastics Selection Manual*, New York: ASCE.

CIRIA. 1987. *TN128: Civil Engineering Sealants in Wet Conditions—Review of Performance and Material Guidance in Use*. London: CIRIA. A little dated, but the guidance remains helpful.

Dutton, H. and P. Rice. 1999. *Structural Glass*, 2nd ed. London: Spon. Peter Rice was probably the most important figure in popularizing the use of structural glass in Britain and France. This book originally appeared in French.

Dwight, J. 1999. *Aluminium Design and Construction*. London: E & FN Span. Useful for those designing with British codes; as Eurocode 9 is introduced, more texts are likely to appear.

Eekhout, M. 1989. *Architecture in Space Structures*. Rotterdam: Uitgeverij 010 Publishers. Eekhout, alongside Rice, is the great pioneer of structural glass.

Institution of Structural Engineers. 1999. *Guide to the Structural Use of Adhesives*. London: IStructE. Useful for those with little knowledge intending to use adhesives structurally for the first time.

Institution of Structural Engineers. 1999. *Structural Use of Glass in Buildings*. London: IStructE. For any engineer short of knowledge or experience of the structural use of glass, this provides excellent guidelines.

Mays, G. C. and A. R. Hutchinson. 1992. *Adhesives in Civil Engineering*. Cambridge, U.K.: Cambridge University Press. Reasonably comprehensive source of applications in civil engineering.

Schittich, C. 1999. *Glass Construction Manual*. Wiesbaden, Germany: Birkhauser. A useful introduction to European practice.

JOURNALS

In Britain and the United States, the institutions produce the leading journals. The Institution of Civil Engineers (ICE) now produces *Structures and Building*, while The Institution of Structural Engineers has published *The Structural Engineer* since 1923. The American Society of Civil Engineers (ASCE) produces the *Journal of Structural Engineering*, and a number of more specialist titles listed below.

General Titles

Annales de l'Institut Technique du Batiment et Travaux Publics. (Paris, France: Institute Technique du Batiment et Travaux Publics.) Mostly research papers on structures and the performance of materials.

Bauingenieur (Association of German Engineers–VDI) (Springer Verlag) (0005-6650). Contains papers on analysis, design, and construction.

Bautechnik (Ernst und Sohn) (Germany) (0932-8351). Similar to Bauingenieur, but complements rather than duplicates its coverage.

Bridge Engineering. Proceedings of the Institution of Civil Engineers (1478-4629). One of the latest ICE titles, with an emphasis on projects rather than research.

Bulletin of the International Association for Shell and Spatial Structures (IASS) (0304-3622). Well-presented specialist journal with case studies.

Computers and Structures (Elsevier) (0045-7949). The leading academic journal in this area, really intended for the specialist.

Construction and Building Materials (Butterworth-Heinemann) (0950-0618). Mostly research papers into building materials.

Engineering Structures (Elsevier) (0141-0296). Concerned with designing structures for live loads.

Forensic Engineering (ICE Publishing). (2043-9903) The latest addition to the ICE Proceedings.

Industria Italiana del Cemento (Italian Cement Association) (0019-7637). Its specialist title gives no clue to the inspirational quality of its drawings and photographs. A journal that would sit easily in an art bookshop. Sadly ceased publication in the latest downturn.

International Journal of Solids and Structures (Elsevier) (0020-7683). An academic journal concerned with the mechanics of engineering materials.

International Journal of Space Structures (Multi-Science Publishing) (0956-0599). This is much more academic in its approach than the *IASS Bulletin* (above).

Journal of Bridge Engineering (ASCE) (1084-0702). More research papers than construction examples.

Journal of Engineering Mechanics (ASCE) (0733-9399). One of the oldest of the ASCE specialty journals and very academic.

Journal of Performance of Constructed Facilities (ASCE) (0887-3828). This is one ASCE title that is of obvious interest to the practitioner rather than the academic.

Journal of Structural Engineering (ASCE) (0733-9445). The leading title on structural analysis and design.

Materials and Structures (RILEM) (France) (1359-5997). The leading international journal on materials testing for civil engineers.

National Hazards Review (ASCE). One of the new ASCE journal titles, with a more practical or state-of-the-art approach than some other titles in what is becoming an important area for civil engineers.

Practice Periodical on Structural Design and Construction (ASCE) (1084-0680). This journal is aimed at the practitioner rather than the academic and research audience of the journal of structural engineering.

Structural Design of Tall Buildings (Wiley) (1541-7794). This was the first academic journal to be concerned with the design of tall buildings.

Structural Engineering International (International Association for Bridge and Structural Engineering) (1016-8664). IABSE's magazine, with international contributions.

Structural Safety (Elsevier) (0167-4730). Concerned with analyzing probability and risk associated with structures.

Concrete

ACI Materials Journal (American Concrete Institute) (1987–) (0889-325X). Bimonthly, the world's leading journal on the material properties of concrete. Abstracts are searchable http://www.concrete.org/PUBS/JOURNALS/MJHOME.ASP (accessed October 2010).

ACI Structural Journal (American Concrete Institute) (1987–) (0889-3241). Bimonthly, the world's leading journal on the analysis of concrete structures. Abstracts are searchable http://www.concrete.org/PUBS/JOURNALS/SJHOME.ASP (accessed October 18, 2010).

Advances in Cement Research (TTL) (0951-7197). International journal on cement research; claims to be more selective than *Cement and Concrete Research*.

Beton Arme. (Société des Éditions Andre Guerrin) (France) A French title on reinforced concrete.

Beton-Kalendar. (Ernst) Written in German, this provides annual reviews of developments in concrete practice and abstracts (in some cases full text) of German and European Standards.

Beton und Stahlbetonbau (Ernst) (0005-9900). One of several German journals on concrete, this produces more on structural design and practice than do others.

Cement and Concrete Aggregates (ASTM) (0149-6123). ASTM's research journal on cement and concrete.

Cement and Concrete Research (Elsevier) (0008-8846). In terms of papers published this is the leading research journal on the subject in the world. Probably too academic to be readily beneficial for most practicing engineers.

Concrete (Concrete Society) (0010-5317). Published for the U.K. Concrete Society, this journal contains project descriptions and brief papers on design topics; similar to the ACI's *Concrete International*.

Concrete Construction (formerly Aberdeen's Concrete Construction) (1533-7316). The leading U.S. trade journal, which is often the first to announce new developments in concrete construction techniques.

Concrete International (ACI) (not supplied). The ACI monthly, which often has "special issues" featuring recent developments in concrete technology.

Concrete Science and Engineering (RILEM) (1295-2826). Until recently, RILEM published all its concrete papers in *Materials and Structures*; the launch of this journal reflects the volume of academic research in this area.

Indian Concrete Journal http://www.icjonline.com. Although generally concerned with India, a major area of civil engineering activity, this journal reports on developments elsewhere.

Magazine of Concrete Research (TTL) (0024-9831). The longest established U.K. research journal with an international circulation; all back issues are digitized.

PCI Journal (PCI) (0887-9672). The leading international source on the use of prestressed and posttensioned concrete.

Structural Concrete (TTL for FIB) (1464-4177). A relatively new entrant, but given kudos by the FIB (International Concrete Federation for Prestressed Concrete) brand.

Steel

Construction Metallique (CTICM) (France) (not supplied). The leading French-language journal on steel structures, blending research and practice.

Costruzione Metalliche (ACS-ACAI) (Italy). Similar approach to *Rivista*, but more restricted in its coverage.

Engineering Journal (AISC) (0013-8029). Design and research papers; back issues are now all available on one CD-ROM.

Modern Steel Construction (AISC) (0026-8445). Monthly magazine on structural steel developments in the United States.

New Steel Construction (Steel Construction Institute/BCSA) (0968-0098). The U.K.'s monthly magazine on steel construction.

Stahlbau (Ernst and Sohn) (Germany) (0038-9145). The leading German journal on structural steel analysis and practice.

WEB SITES

Advanced Buildings Technologies and Practices http://www.advancedbuildings.org (accessed June 18, 2010). Site guides building professionals to more than 90 examples of environmentally appropriate design for commercial, industrial, and multiunit residential buildings.

Bridge Building: Art and Science http://www.brantacan.co.uk/bridges.htm (accessed June 18, 2010). Covers all aspects of bridge building from arches to stress with lots of links and images.

Bridge Construction and Engineering http://bridgepros.com/ (accessed June18, 2010). Site presents construction and history of bridges. Covers past, current, and planned bridge projects worldwide. Learning Center discusses bridge types, and has models and links to lesson plans. Extensive links.

Bridge Site http://www.bridgesite.com/ (accessed June 18, 2010). Something for everybody. Site intends to provide a way to bring people of different communities together. Lots of links and discussion forums including one used chiefly by schoolchildren seeking help with their projects.

Council on Tall Buildings and Urban Habitat http://www.ctbuh.org/ (accessed June 18, 2010). Includes the online journal *CTBUH Review*, http://www.ctbuh.org/Publications/CTBUHJournal/tabid/72/language/en-US/Default.aspx, and subscription area.

Greatbuildingsonline http://www.greatbuildings.com/search.html (accessed June 18, 2010). Very much an architectural resource (one cannot search by name of engineer) with close links to *Architecture Week*, but it still provides access to brief details of some well-known structures. Structurae (below) is better from an engineering viewpoint.

High-Performance Buildings Research http://www.nrel.gov/ (accessed June 18, 2010). The National Renewable Energy's research site encompasses high-performance whole-building design to develop very low-energy, environmentally sensitive buildings.

Office of Bridge Technology http://www.fhwa.dot.gov/bridge/index.htm (accessed June 18, 2010). Department of Transportation, Federal Highway Administration Office of Bridge Technology. Contains an electronic library, training courses, and NBI information.

Spreadsheet Solutions for Structural Engineers http://www.yakpol.net/ (accessed June 18, 2005). A series of links to free software sites for structural engineers.

Spreadsheets for Structural Engineering http://www.structural-engineering.fsnet.co.uk/ (accessed June 18, 2010). Free software for all types of civil engineering problems.

Steel Bridges in the World http://www.sbi.se/default_en.asp (accessed June 18, 2010). Bridge statistics compiled by the Swedish Institute of Steel Construction; the site is generally of use to anyone interested in steel construction.

Structurae http://www.structurae.de (accessed June 18, 2010). Multilingual Web site that is growing rapidly in its coverage of engineers and engineering feats, with pictures and bibliography; all the work of informed amateurs.

West Point Bridge Designer http://bridgecontest.usma.edu/ (accessed June 18, 2010). Bridge Design Contest developed by the U.S. Military Academy (West Point) for U.S. students in grades K–12 aimed at promoting math, science, and technology education. Open to anyone, but U.S. students (K–12) may compete for prizes. Downloadable contest information.

ASSOCIATIONS, ORGANIZATIONS, AND SOCIETIES

Aluminum Association (AAI) http://www.aluminum.org (accessed June 18, 2010). The Aluminum Association is the U.S. trade association. It publishes *Aluminium Standards*, which are widely used internationally, and the *Aluminum Statistical Review*.

Aluminum Federation http://www.alfed.org.uk/ (accessed June 18, 2010). The U.K. equivalent body of the AAI, with more limited publications programmed.

American Ceramic Society http://www.ceramics.org/acers/acers.asp (accessed June 18, 2010). A 100-year-old nonprofit organization that serves the informational, educational, and professional needs of the international ceramics community. It is the leading such organization in the world, although many of its publications will have no direct relevance to civil engineering.

American Concrete Institute (ACI) http://www.concrete.org (accessed June 18, 2010). The world's leading learned society for concrete construction, publishing in English and Spanish. The *ACI Manual of Concrete Practice* (*MCP*) includes all the ACI standards, specifications, recommendations, and guides used for the construction, construction management, inspection, and design of concrete structures. Available online http://www.concrete.org/PUBS/MCP/MCP_HOME.HTM.

American Institute of Steel Construction (AISC) http://www.aisc.org (accessed October 20, 2010). Standards activities and publications include the *AISC Specifications and Codes*. The institute publishes the *Engineering Journal* and *Modern Steel Construction* (1964 to present), the former dealing with analysis and design while the latter is more concerned with practice. Useful FAQs dealing with issues such as progressive collapse.

American Institute of Timber Construction http://www.aitc-glulam.org/ (accessed November 3, 2010). Organization produces manuals and other key publications on U.S. practice, particularly on laminated timber.

American Iron and Steel Institute (AISI) http://www.steel.org (accessed November 3, 2010). AISI represents Canadian, U.S., and Mexican steelmakers. Produces standards, including a compilation of out-of-date sections.

American Welding Society (AWS) http://www.aws.org (accessed November 3, 2010). The AWS site includes a buyer's guide and research supplements to the *Welding Journal.*

American Wood Council http://www.awc.org/index.html (accessed November 3, 2010). The American Forest and Paper Council is the engineered wood products division of AFPA. The Web site contains a large number of downloadable technical advice notes; the council publishes standards and design manuals.

Association for Specialist Fire Protection (ASFP) http://www.asfp.org.uk/ (accessed April 4, 2011). The ASFP produces guidelines and organizes activities for U.K. specialists in structural fire protection.

Brick Development Association (BDA) http://www.brick.org.uk/Index.html (accessed November 3, 2010). The BDA publishes design guides, notes, and technical information papers.

British Constructional Steelwork Association Ltd (BCSA) http://www.steelconstruction.org/bcsa.html (accessed November 3, 2010). Publications include: *Handbook of Structural Steelwork* (with SCI), *Historical Structural Steelwork Handbook* (1984), "The red book"–*Handbook of Structural Steelwork*, "The green book"–*Joints in Simple Construction*, "The black book"–*National Structural Steelwork Specification for Building Construction*, "The orange book"–*The Contractual Handbook*, "The blue book"–*Erectors' Manual.* Their *Steel Buildings* (2003) is a useful introduction to all aspects of the design and procurement of steel buildings, useful for engineers and clients.

British Masonry Society, c/o CERAM Research (see below). This society publishes a journal and organizes regular conferences.

Building Research Establishment (BRE) http://www.bre.co.uk/ (accessed November 3, 2010). Publishes a range of material including *BRE Digests, BRE Reports, BRE Newsletter, Good Building Guides* (which have superseded the *Defect Action Sheets*), *Overseas Building Notes,* and *Information Papers.* Although library services are now minimal, this is the U.K. national source for building and geotechnical information.

Cement Mineral Products Association http://cement.mineralproducts.org/. The MPA is the U.K. organization for cement products, with links to the Concrete Society, and the Concrete Centre providing similar services to the former BCA: http://www.concretecentre.com/online_services.aspx (accessed November 3, 2010).

Centre for Window and Cladding Technology http://www.cwct.co.uk (accessed November 3, 2010). A leading information provider in the field of building envelopes and glazing.

CERAM Research http://www.ceram.co.uk/ (accessed November 3, 2010). CERAM has published research papers and special publications, and specifications such as the model specification for clay and calcium silicate structural brickwork (1988).

Civieltechnisch Centrum Uitvoering Research en Regelgeving (CUR) http://www.cur.nl/ (accessed November 3, 2010). A Dutch body responsible for a number of important design manuals, now available in English.

Concrete Reinforcing Steel Institute (CRSI) http://www.crsi.org (accessed November 3, 2010). CRSI publishes a *Design Handbook* and *Manual of Standard Practice* that relate to ACI Standards. Their site includes several full-text publications.

Concrete Society http://www.concrete.org.uk/ (accessed November 3, 2010). Publishes many technical reports, guides, and digests.

Corus http://www.corusconstruction.com/en/products/structural_steel/ (accessed November 3, 2010). Provides a wide range of information (largely free), such as the guide to Sections: Structural Sections, and corrosion protection guides.

Council on Tall Buildings and Urban Habitat http://www.ctbuh.org/ (accessed November 3, 2010). Based at Lehigh University, the council has organized many conferences and produced a significant body of research.

Federation Internationale de Beton (FIB) (International Concrete Federation) http://fib.epfl.ch (accessed April 5, 2011). An international organization formed by the merger of Federation e de la Precontrainte and Comite Eurointernationale du Beton for the development of structural concrete; it publishes the journal *Structural Concrete*, many recommendations, and state-of-the-art reports, on both reinforced and prestressed concrete.

Institution of Structural Engineers (IStructE) http://www.istructe.org.uk/ (accessed November 3, 2010). Founded in 1908 (as the Concrete Institute), IStructE publishes *The Structural Engineer*, technical reports, guidance, conference and symposium papers. In addition, it organizes CPD events. The library is an important specialist resource.

International Association for Bridge and Structural Engineering (IABSE) http://www. iabse.ethz.ch/ (accessed November 3, 2010). IABSE organizes regular international conferences on current issues in structural engineering as well as a journal and state-of-the-art reports.

International Association for Earthquake Engineering (IAEE) http://www.iaee.or.jp/ (accessed November 3, 2010). Although based in Japan, this is an international society with national and regional activities which organizes major international conferences, supports a journal (*Earthquake Engineering and Structural Dynamics*), and produces the definitive earthquake-resistant regulations world list.

International Association for Shell and Spatial Structures http://www.iass-structures.org/ (accessed November 3, 2010). Organizes international and regional conferences and publishes a journal.

International Council for Building Research Studies and Documentation (CIB) http://www. cibworld.nl/site/home/index.html (accessed November 3, 2010). A number of conferences and commercially published journals are produced under CIB auspices, normally related to CIB working groups. CIB covers all aspects of building and construction.

Masonry Institute of America http://www.masonryinstitute.org/ (accessed November 3, 2010). Produces manuals for engineers, standards, and guidance for builders. Online bookshop and industry links. National Fire Protection Association (NFPA) http://www.nfpa.org (see Chapter 5 in this book on Architectural Engineering).

Portland Cement Association (PCA) http://www.cement.org/ (accessed November 3, 2010). The PCA is a major publisher of reports on cements and their structural applications. The Web site has a number of specialist subsites aimed at practitioners.

Precast/Prestressed Concrete Institute http://www.pci.org/ (accessed November 3, 2010). Publishes a journal and a range of design manuals, making this the foremost source of information on prestressed concrete. The designers' knowledge bank is an online service intended to help answer designers' questions by referral to PCI publications and elsewhere.

RILEM http://www.rilem.org/ (accessed November 3, 2010). Reunion Internationale des Laboratoires d'Essais et de Recherches sur les Materiaux et Constructions (RILEM), the premier international body for research and testing materials, organizes conferences and publishes a growing number of journals, either directly or in partnership with other publishers, such as *Materials and Structures*.

Standing Committee on Structural Safety (SCOSS) http://www.scoss.org.uk/ (accessed November 3, 2010). This is an independent body established by the Institution of Civil Engineers and the Institution of Structural Engineers and others in 1976 to maintain a continuing review of building and civil engineering matters affecting the safety of structures. The Web site contains full text of all their reports.

Steel Construction Institute (SCI) http://www.steel-sci.org/ (accessed November 3, 2010). A membership-based research institute that produces design guides, commentaries on codes of practice, and manuals relating to the use of steel in bridges, buildings, and foundations. Although aimed largely at a British audience, as European Codes and Standards are introduced, its publications will reflect this focus.

The Welding Institute (TWI) http://www.twi.co.uk (accessed November 3, 2010). An excellent specialist source on welding whose database, WELDASEARCH, is marketed by CSA.

Timber Research and Development Association (TRADA) http://www.trada.co.uk/ (accessed November 3, 2010). TRADA, a U.K. trade and research body, publishes books, reports, and wood information sheets. The Web site has an ask TRADA feature.

United States Corps of Engineers http://www.usace.army.mil (accessed November 3, 2010). The ASCE is publishing a growing number of the Corps' engineering manuals, but its codes for blast-resistant construction and military specifications are used worldwide and are available for free download.

WATER ENGINEERING

INTRODUCTION

From extant remains of early civilizations in China, Mesopotamia, and the Nile Basin, it is evident that water resources have been an important consideration for humans for millennia. The design of works, such as the aqueduct feeding Nimes in southern France, shows evidence that the Romans had great practical knowledge of hydraulics, and indeed speculations on the theory of the flow of water may be found among the earlier printed books on engineering. Water engineers had access to serviceable hydraulic formulae from the eighteenth century, but it was not until the nineteenth century, when engineers began to address issues of urban water supply and drainage and irrigation on a large scale, that water resources engineering may be said to emerge. In this regard, the publication of Nathaniel Beardmore's *Manual of Hydrology* and papers by the Irish engineers William Thomas Mulvany and Robert Manning may be regarded as milestones, the one for coining the term *hydrology* for a scientific specialization, the others for developing a serviceable formula for run-off in catchments. Since then, our understanding of the hydrological style and its implications for water supply have improved through observations. Dealing with hydraulics and hydrology is only one aspect of the water engineers' work. Those involved in supplying potable drinking water need to have an understanding of chemistry and biology, and this has a strong interface with the handling and treatment of wastewater whether human sewage or industrial effluent. Structures have to be designed to improve and convey water, and also to control it to ensure that it can act as a channel for navigation and not pose a threat through flooding.

BIBLIOGRAPHY

Biswas, A. K. 1970. *History of Hydrology*. Amsterdam/London: North-Holland Publishing.

Fahlbusch, H., ed. 2001. *Historical Dams: Foundations of the Future Resting on the Achievements of the Past*. New Delhi: International Commission on Irrigation and Drainage.

Rouse, H. and S. Ince. 1957. *History of Hydraulics*. Iowa City: Iowa Institute of Hydraulic Research.

Wikander, O., ed. 2000. *Handbook of Ancient Water Technology*. Leiden, The Netherlands: Brill.

SEARCHING THE LIBRARY CATALOG

Library searching can be complicated because, although the main texts will be found in water resources, related material will be found in other sections, such as chemistry of water, meteorology, and other related Earth sciences. Within the Library of Congress Subject Headings, relevant classes include:

TC 160-181 Technical hydraulics
TC 401-506 River, lake, and water supply engineering

TC 530-537 River protective works, regulation, flood control
TC 540-558 Dams, barrages
TC 601-791 Canals and inland navigation, waterways
TC 801-978 Irrigation engineering, reclamation of wasteland, drainage
TD 201-500 Water supply for domestic and industrial purposes including:

TD 419-428 Water pollution
   TD 429.5-480.7 Water purification and treatment
   TD 481-493 Water distribution systems
   TD 511-780 Sewerage

and there is already a strong relationship with:

TD 169-171.8 Environmental protection
TD 172-193.5 Environmental pollution
TD 194-195 Environmental effects of industries and plants

as well as subjects such as water chemistry, rainfall, and other branches of the Earth sciences. Probably the civil engineers' most significant contribution to modern society is the provision of safe drinking water. Closely allied with this is the treatment and disposal of wastewater—both from industry and in the form of sewage and agricultural wastes. Water supply involves the design of dams, reservoirs, and water towers as well as pipelines and aqueducts. These involve knowledge of hydraulics and fluid mechanics along with geotechnical and structural engineering. Underlying it all is a careful husbanding of water resources. Water purification requires knowledge of chemistry and biology, a clear overlap with science.

Water engineering also may be seen in other aspects of the control of the natural environment (e.g., river engineering to control flooding and make navigation possible). Artificial navigation is seen in the construction of canals, and manmade channels also are required for irrigation schemes, also the work of the civil engineer. These are all closely allied with drainage engineering. Flood control is often associated with land reclamation and, most dramatically, in coastal protection schemes. The cost of such works is causing reconsideration of hard engineering in the form of seawalls, an acceptance of some loss of land to flooding and the sea, and an effort to use natural processes to protect the coasts. This area is dealt with in Maritime Engineering (above).

ABSTRACTING AND INDEXING SERVICES

The *Aqualine* database provides comprehensive coverage from the 1950s, concerning water resources, supply and treatment, wastewater and sewage treatment, and ecological and environmental effects of water pollution. Articles are drawn from approximately 300 journals as well as from conference proceedings, scientific reports, books, and theses. The *Aqualine Thesaurus* is available for aiding in online searching. Previously published by WRc in England, *Aqualine* is now produced in joint cooperation with WRc and Cambridge Scientific Abstracts, e-mail: sales@csa.com.

National Ground Water Association (NGWA) http://www.ngwa.org (accessed November 3, 2010). NGWA provides subscriber access to the database Ground Water On-Line Water Pollution Research Abstracts 1927-1975 (continued in Aqualine). A printed abstracting service focused on research.

Water Resources Abstracts http://www.csa.com (accessed November 23, 2010). A total of 1,094 journals indexed. Almost perfectly complements *Aqualine*, dealing with water resources in its broadest sense rather than engineering aspects of water supply and treatment.

DICTIONARIES

APHA/ASCE/AWWA/WPCF. 1981. *Water and Wastewater Control Engineering: Glossary.* Washington, D.C.: APHA. A comprehensive dictionary, going well beyond a focus on water supply and treatment.

ICID Multilingual Technical Dictionary on Irrigation and Drainage. 2002. New Delhi: International Commission on Irrigation and Drainage (ICID) national committees. (English/French/Italian/Spanish/Turkish/Japanese/German). A series of multilingual dictionaries; these are frequently the only specialist civil engineering dictionaries in the languages concerned available for translation into English.

ICOLD Technical Dictionary on Dams. Paris: International Commission on Large Dams (ICOLD), (various editions: French/English/German/Spanish/Italian/Portuguese). Multilingual with diagrams; one of the best dictionaries of this type.

Meinck, F. and H. Mohle. 1963. *Dictionary of Water and Sewage Engineering.* Amsterdam: Elsevier (English/German/French/Italian). Useful multilingual dictionary.

Smith, P. G. and J. S. Scott. 2002. *Dictionary of Water and Waste Management.* Oxford, U.K.: Butterworth-Heinemann, for IWA. This builds on the work of J. S. Scott (deceased), well known for his Penguin dictionary of civil engineering. Sparsely illustrated, but good definitions, up-to-date and comprehensive.

Van der Tuin, J. D. 1997. *Elsevier's Dictionary of Water and Hydraulic Engineering.* New York: Elsevier (available digitally). Reasonable coverage of the field.

HANDBOOKS

The American Society of Civil Engineers, Water Environment Federation, and American Water Works Association, sometimes in combination with other partners, produce a series of handbooks on design of municipal wastewater treatment plants, odor control in wastewater plants, wastewater and stormwater pumping stations, gravity sewer design, groundwater management, sewer evaluation and rehabilitation, aeration, urban run-off quality, urban stormwater management systems, sewer system overflows, manhole inspection, urban subsurface drainage, steel pipe design, and so on, and are the obvious first source of reference for U.S. practice.

Alley, E. R. 2000. *Water Quality Control Handbook.* New York: McGraw-Hill. From a U.S. regulatory perspective, this deals with water pollution, its treatment, and control.

Corbitt, R. A. 1998. *Standard Handbook of Environmental Engineering.* New York: McGraw-Hill. Although this handbook is concerned with all aspects of the environment, it has chapters on water quality, water supply, stormwater management, and wastewater treatment, with general environmental information.

Davis' Handbook of Applied Hydraulics, 4th ed. 1993. New York: McGraw-Hill. Written for the practicing engineer, this book is organized in the order usually followed in making regional plans for water use and control. Coverage includes basic hydraulics, reservoir hydraulics, and natural channels, river diversion and the construction of reservoirs and dams, pumped storage, navigation locks, irrigation and water, and wastewater systems. Twelve sections are completely new to this edition; the remaining 16 sections have been condensed and updated to reflect current practice. More tables and illustrations have been included than in previous editions. They provide examples of many actual projects.

Delleur, J. W. 2007. *The Handbook of Groundwater Engineering.* Boca Raton. FL: CRC Press. Comprehensive treatment dealing with aquifers, well-sinking, contamination and landfill design, and water quality. Available electronically.

De Zuane, J. 1997. *Handbook of Drinking Water Quality,* 2nd ed. New York: Van Nostrand Reinhold. Describes U.S. and WHO (World Health Organization) water quality guidelines, problems of water quality treatment and control, and related water supply engineering issues.

Gallagher, L. M. and L. M. Miller. 2003. *Clean Water Handbook,* 3rd ed. Rockville, MD: Government Institutes. Each chapter focuses on different aspects of the (U.S.) Clean Water Act with up-to-date coverage of the latest enactments and EPA regulations.

Golze, A. R. 1977. *Handbook of Dam Engineering.* New York: Van Nostrand Reinhold. Comprehensive, if dated. Very helpful for anybody unfamiliar with dam engineering.

Hunt, T., and N. Vaughan. 1996. *Hydraulic Handbook.* New York: Elsevier.

Institute of Hydrology. 1999. *Flood Estimation Handbook.* (5 vols.) Wallingford, U.K.: Institute of Hydrology. (CD-ROM of data.) Based on U.K. hydrology, a detailed handbook for flood calculations.

National Engineering Handbook. Washington, D.C: Natural Resources Conservation Services http://www.mi.nrcs.usda.gov/technicallengineering/neh.html (accessed November 3, 2010). Produced for the U.S. Department of Agriculture; sections deal with design, construction, hydrology, irrigation, and drainage. Very much aimed at the practitioner in the field. A general engineering handbook aimed at those involved in agricultural engineering.

WRc. 1990. *Design Guide for Marine Treatment Schemes.* (4 vols.) Swindon, U.K.: WRc. Design manual for waste-water treatment for marine disposal.

WRc. 2001. *Sewer Rehabilitation Manual.* 4th ed. Swindon, U.K.: WRc. Comprehensive U.K. guidance on sewer rehabilitation, including specifications.

JOURNALS

There are a large number of trade journals of interest to the industry, but with little intellectual content.

American Water Works Association Journal (0003-150X). Useful for AWWA and water industry news with lots of advertisements, issues are often themed around latest developments.

Aqua (IWA) (0003-7214). Although mostly concerned with research, this does contain papers on practical applications and operational aspects of water supply worldwide.

Circulation (BHS). News of British Hydrological Society meetings and articles.

Hydrological Sciences Journal (IAHS) (0262-6667). Contains research papers and news of IAHS meetings.

Irrigation and Drainage (Wiley) (0971-7412). The continuation of the *ICID Bulletin* is the longest established international journal in the field.

Irrigation and Drainage Systems (Kluwer) (0168-6291). A relative newcomer, with an academic flavor.

Journal of Contaminant Hydrology (Elsevier) (0169-7722). The leading academic journal in this specialist field.

Journal of Environmental Engineering (ASCE) (0733-9372). This ASCE title covers waste-water treatment.

Journal of Flood Risk Management (CIWEM/Wiley) (1753-318X). Newly established (2009) reflecting the growing interest in this area.

Journal of Hydraulic Engineering (ASCE) (0733-9429). A long-established ASCE specialty journal that has published many key papers on hydraulic design.

Journal of Hydraulic Research (IAHR/Taylor & Francis) (0022-1666) The longest established research journal in this field.

Journal of Hydroinformation (IWA/IAHR) (1464-7141) An international journal intended to be cross-disciplinary, dealing with the application of information technology to aquatic sciences; a preponderance of articles deal with numerical flow modeling.

Journal of Hydrologic Engineering (ASCE) (1084-0699). The ASCE journal for engineering hydrologists.

Journal of Hydrology (Elsevier) (0022-1694). The cost of this journal has been the cause of comment among hydrologists, but it remains the leader in the field.

Journal of Irrigation and Drainage Engineering (ASCE) (0733-9437). The leading U.S. journal in the field.

Journal of Water and Climate Change (IWA) (2040 2244). Commenced in 2010; water is regarded by many as the most sensitive commodity to climate change.

Journal of Water and Health (IWA/WHO) (1477-8920). A multidisciplinary approach to water, recognizing its basic relationship to public health.

Journal of Water Resources, Planning and Management (ASCE) (0733-9496). The ASCE's journal on water resources.

Water 21 (IWA) (1561-9508). The IWA's monthly magazine with news and feature articles on water issues worldwide.

Water and Environment Journal (CIWEM) (1747-4688). CIWEM's academic journal.

Water and Environment Magazine (CIWEM) (1362-9360). U.K. news, feature articles, and CIWEM activities.

Water Environment and Technology (Water Environment Federation) (1044-9493). News and feature articles on recent developments in water treatment.

Water Environment Research (Water Environment Federation) (1061-4303). Perhaps the best source, aside from the IWA, on the latest research into wastewater treatment.

Water Management (ICE) (1741-7589). The latest ICE proceedings title, bringing together all the papers on water supply, treatment, and resources management in the broadest sense.

Water Research (Elsevier) (0043-1354). Mostly academic research papers on all aspects of water quality and treatment.

Water Resources (AWRA). Perhaps the most widely referenced journal in the field in the United States.

Water Resources Research (American Geophysical Union) (0043-1397). Broad issues of climate are published alongside papers on water chemistry.

Water Science and Technology (2 parts) (IWA) (0273-1223). Effectively two conference series on water supply and wastewater treatment; frequently these are state-of the-art volumes from international contributors.

MONOGRAPHS

American Public Health Association. *Standard Methods for the Examination of Water and Wastewater.* New York: American Public Health Association. These are probably the most widely used methods for water analysis, both in the United States and overseas. Irregular serial.

Applegate, G. 2002. *The Complete Guide to Dowsing.* London: Vega. For those interested in the use of water divining, this is the most up-to-date source on the technique.

ASCE. 1980. *Manual 57: Operation and Maintenance of Irrigation and Drainage Systems.* New York: ASCE. For the developed world, this provides comprehensive advice of managing irrigation systems.

AWWA. 1999. *Water Quality and Treatment,* 5th ed. New York: McGraw-Hill. An introductory chapter deals with U.S. standards and regulations, and successive chapters deal succinctly with treatment methods and issues of corrosion and microbial contamination in distribution systems.

Barr, D. I. H. 1998. *Tables for the Hydraulic Design of Pipes, Sewers and Channels.* (2 vols.) London: Thomas Telford. A useful compendium of tables for pipe and channel design.

Bernstein, L. B. 1996. *Tidal Power Plants.* Seoul: KORO. In many ways the seminal text on this aspect of power from water.

BRE. 1999. *Engineering Guide to the Safety of Embankment Dams in the U.K.* London: CRC Ltd. An essential guide to safety assessment of earth dams.

CIRIA. 2004. *Sustainable Drainage Systems: Hydraulic, Structural and Water Quality Advice.* London: CIRIA. Engineering advice on the design of SUDS; intended to mimic as far as possible natural drainage systems and, thus, limit the environmental impact of a drainage scheme.

Escarameia, M. 1998. *River and Channel Revetments: A Design Manual.* London: Thomas Telford. Revetment design is an essential aspect of river control and this work covers the latest European practice.

Hansen, V. E., ed. 1980. *Irrigation Principles.* New York: John Wiley & Sons. Despite its age, this book remains on many course reading lists, and covers most aspects of irrigation engineering.

ICE. 1975. *Guide to the Reservoirs Act, 1975.* London: Thomas Telford. An essential guide to the legislation governing the safety assessment of dams in the United Kingdom.

ICE. 2001. *Learning to Live with Rivers.* London: Thomas Telford. Intended to encourage engineers to use rivers as an environmental asset, rather than controlling and hiding them.

Jansen, R. B., ed. 1988. *Advanced Dam Engineering: For Design, Construction and Rehabilitation.* New York: Van Nostrand Reinhold. A comprehensive overview of advances in dam design in the second half of the twentieth century.

Metcalf and Eddy, Inc. 2003. *Wastewater Engineering: Treatment Disposal and Reuse,* 4th ed. New York: McGraw-Hill. This work has been in print in some form or other for nearly a century. This edition incorporates SI units. It covers all aspects of water treatment and reuse, and is fully referenced. Data tables and worked design examples are included.

Nienhuis, P. H., ed. 1998. *New Concepts for Sustainable Management of River Basins.* Leiden, The Netherlands: Backhuys Publishers. A series of papers exploring sustainable management of river basins, illustrating international practice.

Novak, P. 2001. *Hydraulic Structures.* London: Spon. Covers most aspects of hydraulic structures (e.g., dams and weirs at a basic level), suitable for undergraduates or those seeking a basic understanding of the subject.

Parker, D. H., ed. 1987. *Urban Flood Protection Benefits.* Aldershot, U.K.: Gower. Cost benefit analysis of flood alleviation in an urban context.

Penning Rowsell, E. C. and J. B. Chatterton. 1977. *Benefits of Flood Alleviation.* Westmead, Hants, U.K.: Saxon House. An early and classic investigation of the cost benefit analysis of flood alleviation schemes.

Read, G. C., ed. 1997. *Sewers.* London: Arnold. A multiauthored work, largely based on British practice, dealing with sewer repair and reconstruction.

Read, G. C., ed. 2004. *Sewers: Replacement and New Construction.* Amsterdam: Elsevier. Largely complementing the above title with state-of-the-art contributions on the latest sewer construction techniques.

Sanks, R. L. 1998. *Pumping Station Design.* London: Butterworth. A comprehensive work on a subject where mechanical and civil engineers are regularly brought together.

Schiechtl, H. M. and R. Stern. 1997. *Water Bioengineering Techniques for Watercourse Bank and Shoreline Protection.* Oxford, U.K.: Blackwell Science. Explores natural alternatives to hard engineering solutions, such as concrete for slope protection.

Skogerboe, G. V. and G. P. Merkley. 1996. *Irrigation Maintenance and Operations: Learning Process.* Highlands Ranch, CO: Water Resources Publications. Provides guidelines and procedures for improved maintenance and operations of conveyance and distribution systems for irrigation water.

Stein, D. 2001. *Rehabilitation of Drains and Sewers.* Berlin: Ernst. A comprehensive treatment of sewer maintenance and repair based on German practice.

Twort, A. C. 2009. *Water Supply,* 6th ed. Oxford, U.K.: Butterworth Heinemann. The best U.K. textbook on the subject with coverage of the organization of the U.K. water industry as well as storage, distribution, and treatment of water and related aspects of hydraulics.

U.S. Bureau of Reclamation. 1987. *Design of Small Dams.* Washington, D.C.: U.S. Government Printing Office. Probably the most comprehensive treatment of dam design in a single book, now on the USBR Web site.

U.S. Bureau of Reclamation. 1997. *Water Measurement Manual.* Washington, D.C.: U.S. Government Printing Office. A useful handbook on methods of water flow measurement, and so on, available for downloading at http://www.usbr.gov/pmts/hydraulics_lab/pubsl wmm/ (accessed November 3, 2010).

U.S. Bureau of Reclamation. 1983. *Safety Evaluation of Existing Dams.* Washington, D.C.: U.S. Government Printing Office. U.S. procedures for the safety inspection of dams.

U.S. Environmental Protection Agency (EPA). 1997. *EPA Methods and Guidance for the Analysis of Water.* Rockville, MD: Government Institutes. U.S. Government Agency's guidelines for water analysis; available in digital format.

Zevenbergen, C., et al. 2010. *Urban Flood Management.* Boca Raton: CRC Press. Authored by a team of international experts in this area, this work addresses issues of flood risk and future resilience.

Web Resources

Dam decommissioning in France http://rivernet.org/welcome.htm (accessed November 3 2010). Describes the political processes and engineering challenges of decommissioning three dams in France, with references, links, and images.

DEFRA (Annual) Digest of Environmental Statistics http://www.defra.gov.uk/statistics/ environment (accessed April 4, 2011). Statistics on all aspects of the U.K. environment including water supply and the state of resources.

GRDC http://www.bafg.de/GRDC/EN/Home/homepage__node.html (accessed November 3, 2010). GRDC is an international cooperative venture under the auspices of the World Meteorological Organization and hosted within the German Federal Institute of Hydrology to provide an international databank on surface water hydrology. Although access to data from the bank is fee-based, the Web site provides free access to many other publications.

Ground-Water Remediation Technologies Analysis Centre (GWRTAC) http://www.gwrtac. org (accessed November 3, 2010). The GWRTAC site includes links to other environmental sites, full-text technical reports, and a list of U.S. regulations.

Hydrology Web http://hydrology.pnl.gov/ (November 3, 2010). Site, part of the Pacific North West National Laboratory, hosts a comprehensive list of hydrology and hydrology-related resources.

National Inventory of Dams http://geo.usace.army.mil/pgis/f?p=397:12:1701899881859912 (accessed November 3, 2010). Compiled by the U.S. Army Corps of Engineers and other agencies, this is a database giving data on 75,000 dams in the United States and Puerto Rico.

NWS Hydrologic Information Center http://www.weather.gov/ahps/hic/index.php/index/ (accessed November 3, 2010). Provides historical and current data on streamflow, drought, soil moisture, flood damage, etc.

United Kingdom Environment Agency http://www.environment-agency.gov.uk/ (November 3, 2010). Responsibility for dealing with flooding has traditionally been a divided responsibility in the United Kingdom. Around five million people, in two million properties, live in flood risk areas in England and Wales. The Environment Agency's Web site has a lot of helpful information: *Floodline* provides information on how to prepare for floods and cope with the inevitable clear-up operation in the aftermath; Current flood warnings in force allows you to view flood warnings in place and flood warnings issued. Updated every 15 minutes; Flood management and R&D, details of the agency's flood management roles, including research and development (joint with Defra) and the National Flood and Coastal Defence Database (NFCDD).

USGS SPARROW http://water.usgs.gov/nawqa/sparrow/ (accessed November 3, 2010). The SPARROW (Spatially Referenced Regressions on Watershed Attributes) model was developed to estimate the origin and fate of contaminants in streams, based on regional water quality monitoring data. Site gives examples of model's application.

Water Librarian's Homepage http://www.interleaves.org/~rteeter/waterlib.html (accessed November 3, 2010). This site contains a variety of useful links in the field of water resources with recommended Web sites.

Water Strategist Community (WSC) http://www.waterchat.com (accessed November 3, 2010). Water Strategist Community is a water news Web site that provides daily updates on press releases from U.S. environmental and water agencies as well as news on water investment. WSC provides e-mail alerts and a trade directory.

Wateright http://www.wateright.org (accessed November 3, 2010). An educational resource for irrigation water management for three audiences: homeowners, commercial turf growers, and agriculture. Presents a series of advisories and tutorials on multiple topics.

World Commission on Dams http://www.dams.org (accessed November 3, 2010). A UN Commission to review the development of large dams and to develop guidelines and standards on their use. The reports are downloadable.

ASSOCIATIONS, ORGANIZATIONS, AND SOCIETIES

American Institute of Hydrology (AIH) http://www.aihydrology.org (accessed November 3, 2010). The U.S. society for hydrologists.

American Society of Agronomy (ASA) http://www.agronomy.org/ (accessed November 3, 2010). The ASA has published a number of monographs on irrigation and drainage.

American Water Resources Association (AWRA) http://www.awra.org/ (accessed November 3, 2010). The AWRA's mission is to be the preeminent, multidisciplinary association for information exchange, professional development, and education about water resources and related issues. It publishes the leading U.S. journal (*Water Resources*) in the field. The site contains conference information, publications, employment news, e-learning courses, links, and membership information.

American Water Works Association (AWWA) http://www.awwa.org (accessed November 3, 2010). The AWWA produces standards and specifications, a journal, and conference proceedings, including an important series on water reuse.

Association of State Dam Safety Officials http://www.damsafety.org/ (accessed November 3, 2010). The association site provides dam safety information for regulators, with links, training programs, and an electronic bibliography.

CEH Wallingford (CEH) http://www.ceh.ac.uk/science/corpinfo.html (accessed November 3, 2010). The Centre for Ecology and Hydrology (formerly the Institute of Hydrology) carries out research into the effects of land use, climate, topography, and geology on the volume and character of surface water resources. A new Joint Centre for Hydrometeorological Research has been established in partnership with The Met Office.

Chartered Institution of Water and Environmental Management (CIWEM) http://www.ciwem. org.uk/ (accessed November 3, 2010). Formed as the result of a series of mergers of societies for engineers and scientists in the water and environmental field, CIWEM produces two journals, organizes conferences, and publishes state-of-the-art manuals. The multivolume *Manual of British Water Engineering Practice*, although no longer up to date, remains the best general source of information on topics such as water supply and river engineering.

Colorado Water Institute http://cwi.colostate.edu/links.asp (accessed November 3, 2010). Provides links to sources for specialty water data and information on the Web.

Computational Fluid Dynamics (CFD) http://www.cfd-online.com/ (accessed November 3, 2010). Nicely arranged directory of Web sites on the topic of fluid dynamics.

ESCAP (UN) http://www.unescap.org/ (accessed November 3, 2010). ESCAP supports the efforts of members and associate members in Asia and the Pacific regions to achieve their desired development goals in a sustainable manner. Contains sections on energy resources, environment, water and mineral resources, and space technology.

Food and Agriculture Organization of the United Nations http://www.fao.org (accessed November 3, 2010). Responsible for several series of publications relevant to the work of civil engineers.

HR Wallingford http://www.hrwallingford.co.uk (accessed November 3, 2010). Developed from the former Hydraulics Research Station, providing research, modeling, and consultancy services on hydraulic engineering in its broadest sense.

Hydraulic Engineering Research Unit http://www.ars.usda.gov/main/site_main.htm?modecode=62-17-10-00 (accessed November 3, 2010). Unit is part of the USDA Plant Science and Water Conservation Research Laboratory. Research is performed to develop criteria for design and analysis of structures and channels for control, conveyance, storage, and disposal of run-off water.

Hydrological Sciences Branch http://hsb.gsfc.nasa.gov/ (accessed November 3, 2010). NASA-based organization dedicated exclusively to the understanding, quantification, and analysis of the different components of the hydrological cycle, with emphasis on land surface hydrological processes and their interaction with the atmosphere.

International Association of Hydrogeologists http://www.iah.org/ (accessed November 3, 2010). An international body for scientists and engineers; the site provides information on conferences, publications, etc., with a member's area.

International Association of Hydrological Sciences http://iahs.info/ (accessed November 3, 2010). The society publishes a journal, and produces two series of publications, a series of symposia, and specialist reports. The association has national associated societies.

International Commission on Irrigation and Drainage (ICID) http://www.icid.org/ (accessed November 3, 2010). The ICID was established in 1950 as a nongovernmental international organization (NGO) with headquarters in New Delhi. The commission is dedicated to enhancing the worldwide supply of food for all people by improving water and land management and the productivity of irrigated and drained lands through appropriate management of water, environment, and application of irrigation, drainage, and flood management techniques. "Irrigation and drainage in the world" provides nation-by-nation summaries of major schemes.

International Commission on Large Dams (ICOLD) http://www.icold-cigb.org (accessed November 3, 2010). Created in 1928, ICOLD has membership in 82 countries. It holds regular conferences and produces bulletins of good practice, and a world register of large dams. Its publications are indexed on the Web site. Many national societies have produced national directories and histories of dams.

International Water Association (IWA) http://www.iwahq.org/ (accessed November 3, 2010). IWA sponsors a number of journals, organizes major conferences related to water supply and wastewater treatment, and is publishing an increasing number of monographs. Its Web site is increasingly rich in resources, with members' areas.

International Water Management Institute (IWMI) http://www.iwmi.cgiar.org/index.aspx (accessed November 3, 2010). IWMI is a nonprofit scientific research organization focusing on the sustainable use of water and land resources in agriculture, and on the water needs of developing countries. IWMI works with partners in the South to develop tools and methods to help these countries eradicate poverty through more effective management of their water and land resources. The Web site provides links, guidance notes, and news of international developments.

Irrigation Association (IA) http://www.irrigation.org/ (accessed November 3, 2010). The Irrigation Association's mission is to improve the products and practices used to manage water resources and to help shape the worldwide business environment of the irrigation industry. Site includes certification, water conservation policy, and brochures.

Irrigation Association of Australia http://www.irrigation.org.au (accessed November 3, 2010). This site provides comprehensive information about irrigation industry and practices in Australia.

Low Impact Development Center http://www.lowimpactdevelopment.org/ (accessed November 3, 2010). The center provides information to individuals and organizations dedicated to protecting the environment and water resources through proper site design techniques replicating preexisting hydrologic site conditions.

National Ground Water Association http://www.ngwa.org (accessed November 3, 2010). Searchable bibliographic database is restricted to subscribers. Free access areas include information on maintaining domestic water supplies plus ground water in water supplies and product specifications. The site contains membership information and activities.

National Water Resources Association (NWRA) http://www.nwra.org (accessed November 3, 2010). NWRA is concerned with appropriate use of water and land resources. Works closely with Congress and the Executive Branch in the United States.

New Mexico Water Resources Research Institute (NMWRRI) http://wrri.nmsu.edu/ (accessed November 3, 2010). NMWRRI (at New Mexico State University) is part of New Mexico's Rio Grande Research Corridor and participates in water resources planning. Site contains a resource data and information system with access to publications.

Office of Water Services (OFWAT) http://www.ofwat.gov.uk/ (accessed November 3, 2010). The regulator for the privatized water industry in the United Kingdom, producing a number of reports, industry reviews, and codes of practice, mostly available free, or downloadable from the Web site.

UNESCO http://www.unesco.org/new/en/unesco/ (accessed November 3, 2010). UNESCO publishes a number of series and reports relevant to water engineers including studies in hydrology, and discharge records for selected (major) rivers.

USGS—Water Resources of the U.S. http://water.usgs.gov/ (accessed November 3, 2010). Provides maps concerning water conditions for United States and local areas.

United States NWIS-W Data Retrieval http://waterdata.usgs.gov/usa/nwis/ (accessed November 3, 2010). Home page links to U.S. and international water gauging stations.

U.S. River Systems and Meteorology Group (RSMG) http://www.usbr.gov/pmts/rivers (accessed November 3, 2010). Part of the U.S. Bureau of Reclamation, RSMG provides expertise in meteorology, NEXRAD radar, estimates of precipitation, hydrology, hydraulic engineering, water management, water rights, statistical analysis, and user interface design and implementation.

Water, Engineering and Development Centre, Loughborough University http://wedc.lboro.ac.uk/ (accessed November 3, 2010). The U.K.'s leading institute involved in infrastructure development for the Third World.

Water Environment Federation (WEF) http://www.wef.org (accessed November 3, 2010). WEF compliments AWWA in its activities. The Web site provides weekly water environmental news in *WEF Reporter*, a product locator search tool, regulations, and legislation details, member and affiliated organizations, research information and events calendar. There are outside links, and forum discussions. Publications include journals, conferences, and manuals.

Water Research Commission (South Africa) http://www.wrc.org.za/ (accessed November 3, 2010). Promotes water research and application of research findings in South Africa. Site contains research information, publications, databases, calendar, software, reference information.

WRC information. WRc Group http://www.wrcplc.co.uk/ (accessed November 3, 2010). The former Water Research Center provides consultancy services related to water, wastewater, and the environment, and publishes numerous reports and standards/codes of practice for the U.K. water industry.

Water UK http://www.water.org.uk (accessed November 3, 2010). Membership comprises the water and wastewater suppliers of the United Kingdom. Funded by the members who all have representation on the council, it publishes reports on the water industry and statistics.

WateReuse Association http://www.watereuse.org (accessed November 3, 2010). National organization dedicated to increasing the beneficial use of recycled water. Site contains meetings calendar, technical, and membership information. Membership includes public agencies, water suppliers, with local, state and federal groups. The site has links to other sites and documentation.

Waterways Experiment Station (WES) http://www.erdc.usace.army.mil/ (accessed November 3, 2010). WES is responsible for a large amount of hydraulics research with the majority of its recent publications available to download.

World Health Organization (WHO) http://www.who.int/water_sanitation_health/ (accessed November 3, 2010). WHO produces drinking water guidelines and sponsors a number of initiatives linking water, sanitation, and health. Their Web site provides access to a wide range of documentation.

STANDARDS

INTRODUCTION

Like all engineers, civil engineers depend on local, national, and international official standards and specifications. They can be produced by a national standards body, such as the British Standards Institution, or by a professional body, or sometimes by a trade association. The status of the originating body is the key to the authority of the standard. Particularly with developing areas, the patentee/product developer is often the chief source of design information. Engineers evidently need to exercise engineering judgment in applying all standards and codes. For more information on the different types of standards produced, and the organizations involved in developing them, the Society for Standards Professionals (http://www.ses-standards.org/index.cfm) provides a detailed Frequently Asked Questions and links to local, national, and international standards-producing bodies.

Standards and specifications provide minimum performance requirements for all aspects of a construction project. A standard can involve an entire project, such as the American Society of Civil Engineer's *Minimum Design Loads for Buildings and Other Structures* or one as specific as the American Society of Mechanical Engineers' *International Boiler and Pressure Vessel Code*. In addition to technical standards, civil engineering projects are generally governed by local, state, or national regulations. The U.K. building regulations are available from TSO (http://www.tso.co.uk/bookshop/bookstore.asp). The following are some of the major standards-producing organizations about which civil engineers should be familiar.

SOCIETIES, ASSOCIATIONS, AND ORGANIZATIONS

ASM International http://www.asminternational.org (accessed June 22, 2010). Founded in 1913, ASM International serves the technical interests of metals and materials professionals.

ASTM International http://www.astm.org (accessed June 22, 2010). ASTM International is one of the largest voluntary standards development organizations in the world; a trusted source for technical standards for materials, products, systems, and services. Known for their high technical quality and market relevancy, ASTM International standards have an important role in the information infrastructure that guides design, manufacturing, and trade in the global economy.

American Association of State Highway and Transportation Officials (AASHTO) http://www.transportation.org/ (accessed June 22, 2010). AASHTO represents all five transportation modes. It works to develop specifications related to highways and public transportation.

American Concrete Institute (ACI) http://www.concrete.org/ (accessed June 22, 2010). Founded in 1904, the ACI develops codes and specifications related to the use of concrete in structures.

American Institute of Steel Construction (AISC) http://www.aisc.org (accessed June 22, 2010). Established in 1921, the AISC aims to serve the structural steel design community by developing technical information including codes and specifications related to steel construction.

American Iron and Steel Institute (AISI) http://www.steel.org (accessed June 22, 2010). AISI comprises North American steel producers and plays a lead role in the development and application of new steels and steel-making technology.

American Railway Engineering and Maintenance of Way Association http://www.arema.org/ (accessed June 22, 2010). AREMA develops standards on all aspects of railway engineering including permanent way and bridges.

American Society of Civil Engineers (ASCE) http://www.asce.org (accessed June 22, 2010). ASCE, founded in 1852, is America's oldest national engineering society. Through its technical institutes and committees it produces codes and standards related to all areas of civil engineering from building design to wastewater treatment.

American Society of Heating, Refrigeration, and Air Conditioning Engineers (ASHRAE) http://www.ashrae.org (accessed June 22, 2010). ASHRAE develops standards and guidelines in its field. The society addresses such areas as indoor air quality, thermal comfort, and energy conservation in buildings.

American Society of Mechanical Engineers (ASME) http://www.asme.org (accessed June 22, 2010). Founded in 1880, the ASME focuses on technical, educational, and research issues related to the field of mechanical engineering. It is responsible for the development of the *International Boiler and Pressure Vessel Code* that establishes rules of safety governing the design, fabrication, and inspection of boilers and pressure vessels.

American Water Works Association (AWWA) http://www.awwa.org (accessed June 22, 2010). The largest organization of water professionals, the AWWA develops codes and standards related to water quality and supply.

British Standards Institution (BSI) http://www.bsigroup.com/ (accessed June 22, 2010). BSI, initially established by the ICE, was the first national standards body in the world. The U.K. contact point for other international and national standards organizations, including: International Standards Organisation (ISO), Comite Europeen de Normalisation (CEN, European Committee for Standardisation), those of European Union nations including France (AFNOR) and Germany (DIN), Standards Association of Australia (SAA), and Standards Council of Canada (SCC). The monthly periodical *Update Standards* details development in U.K. standards. BSI's work is increasingly involved with quality assurance, while European and ISO standards are replacing the national standards.

Canadian Codes Centre http://codes.nrc.ca (accessed June 22, 2010). The center provides technical and administrative support to the Canadian Commission on Building and Fire Codes, which is responsible for the development of the national model construction codes of Canada. Some of these codes are the *National Building Code, National Fire Code,* and *National Plumbing Code.*

Deutsches Institut fuer Normung e.V. http://www.din.de/ (accessed June 22, 2010). The German Standards Institute's standards are probably the most widely used internationally aside from U.S. standards. The most important are available in English. Those relating to structural loads and reinforced and prestressed concrete and structural steelwork are widely used.

Eurocodes Expert http://www.eurocodes.co.uk/ (accessed June 22, 2010). The Eurocodes Expert Web site is an Institution of Civil Engineers (ICE) and Institution of Structural Engineers (IStructE) initiative to provide a free online information resource on the new British Standards structural Eurocodes. It is aimed primarily at civil and structural design engineers, but also intended to be useful for clients, manufacturers, building control officers, academics, trainers and students — both in the United Kingdom and in other European and non-European countries where English versions of the codes are likely to be used or adopted. From April 2010, Eurocodes superseded existing British Standards. ICE are publishing a series of guides under the editorial leadership of Haig Gulvanessian, and a compendium is available for students from BSI http://shop.bsigroup.com/en/ProductDetail/?pid=000000000030211489 (accessed November 2010).

German Geotechnical Society http://www.dggt.de/ (accessed June 22, 2010). Produces a number of internationally recognized design guidelines, in areas such as waterfront structures and contaminated land.

International Code Council (ICC) http://www.iccsafe.org (accessed November 3, 2010). The ICC is a membership association dedicated to building safety and fire prevention. ICC develops the codes and standards used to construct residential and commercial buildings to provide the highest quality codes, standards, products, and services for all concerned with the safety and performance of the built environment, including homes and schools.

International Union of Railways (UIC) http://www.uic.asso.fr/ (accessed June 22, 2010). Although each national railway system has its own standards and regulations, UIC standards are those that generally govern international routes.

National Fire Prevention Association (NFPA) http://www.nfpa.org/ (accessed June 22, 2010). The NFPA develops safety codes and standards, such as the Fire Prevention Code, that influence buildings, processes, services, and design.

U.S. Department of Commerce, National Institute of Standards and Technology (NIST) http://www.nist.gov (accessed June 22, 2010). NIST, established by Congress, is responsible for an enormous variety of standards-related research programs, including publishing a series on building, and a cooperative program with Japan on wind and seismic engineering.

SUSTAINABLE DEVELOPMENT

Sustainability is defined by the American Society of Civil Engineers (ASCE) as "sustainable civil infrastructure provides environmental, economic, and social well-being, now and for the future."

The ASCE believes that sustainable development is the challenge of meeting human needs for natural resources, industrial products, energy, food, transportation, shelter, and effective waste management while conserving and protecting environmental quality and the natural resource base essential for future development. Sustainable development requires strengthening and broadening the education of engineers and finding innovative ways to achieve needed development while conserving and preserving natural resources. To achieve these objectives, ASCE supports the following implementation strategies: (1) promote broad understanding of political, economic, social and technical issues and processes as related to sustainable development; (2) advance the skills, knowledge, and information to facilitate a sustainable future, including habitats, natural systems, system flows, and the effects of all phases of the life cycle of projects on the ecosystem; (3) advocate economic approaches that recognize natural resources and our environment as capital assets; (4) promote multidisciplinary, whole system, integrated, and multiobjective goals in all phases of project planning, design, construction, operations, and decommissioning; (5) promote reduction of vulnerability to natural, accidental, and willful hazards to be part of sustainable development; and (6) promote performance based standards and guidelines as bases for voluntary actions and for regulations, in sustainable development for new and existing infrastructure.

Searching the Library Catalog

When looking for materials related to sustainable development, the following are some of the Library of Congress Subject Headings to search:

Sustainability
Sustainability and the environment
Sustainability science and engineering
Sustainable buildings
Sustainable construction
Sustainable design
Sustainable development
Sustainable living
Sustainable urban development

Dictionary and Handbooks

Atkinson, G. S. D. and E. Neumayer, eds. 2009. *Handbook of Sustainable Development* (Elgar Original Reference). Northampton, MA: Edward Elgar Publishers. As a whole, it thoughtfully reflects upon and elucidates the fundamental ideas and contributions that have taken root in the 20 years since publication of the *Brundtland Report*. Contributors offer a comprehensive survey of the field as it stands today, starting from basic principles, through inter- and intragenerational equity, to questions of growth and development, concluding with international perspectives.

Gilpin, A. 1996. *Dictionary of Environmental and Sustainable Development.* Hoboken, NJ: John Wiley & Sons. This superb, concise dictionary comprises over 2,000 entries defining terms and concepts relating to environmental planning, management, conservation, and sustainable development.

International Union for Conservation of Nature (IUCN). 2009. *Strategies for National Sustainable Development: A Handbook for Their Planning and Implementation.* London: Earthscan Ltd. Part of the *IUCN Strategies for Sustainable Development Handbook* series, this handbook is one in a series being produced by IUCN and its partners to assist countries and communities in implementing Agenda 21, the action program of the United Nations Conference on Environment and Development. The series will include handbooks on national strategies for sustainable development, local strategies, assessing progress toward sustainability, biodiversity action plans, involving indigenous people, and on integrating population and resource use planning as well as regular companion volumes of case studies addressing the key issues of concern to strategy implementation. Many international agreements and action plans now call for countries to undertake national strategies. These strategies seek to involve communities in united approaches to sustainable development. Some are sectoral, such as tropical forest strategies, others are thematic, covering topics such as biodiversity, education, or climate change. Still others, such as national conservation strategies and national environment action plans, are evolving to become more comprehensive processes, drawing together economic, social, and environmental development actions. This handbook is for people involved in strategies. It draws on experiences in different regions of the world to present options and examples of the role of strategies in sustainable development.

Mulder, K., ed. 2006. *Sustainable Development for Engineers: A Handbook and Resource Guide*. Sheffield, U.K.: Greenleaf Publishing. Concise, but comprehensive, the book examines the key tools, skills, and techniques that can be used in engineering design and management to ensure that whole-life costs and impacts of engineering schemes are addressed at every stage of planning, implementation, and disposal. The book also aims to demonstrate through real-life examples the tangible benefits that have already been achieved in many engineering projects, and to highlight how real improvements can be, and are being, made. Each chapter ends with a series of questions and exercises for the student to undertake. *Sustainable Development for Engineers* will be essential reading for all engineers and scientists concerned with sustainable development. In particular, it provides key reading and learning materials for undergraduate and post-graduate students reading environmental, chemical, civil, or mechanical engineering, manufacturing and design, environmental science, green chemistry, and environmental management.

Roosa, S. A. 2010. *Sustainable Development Handbook,* 2nd ed. Boca Raton, FL: CRC Press. Updated throughout, the second edition of this popular resource includes updates on LEED® measurement and verification and a new chapter on cities and carbon reduction. Clarifying critical issues, this volume examines proven approaches as well as problems with failed initiatives. In addition to core concepts and trends, it explores specific renewable energy and environmental solutions. It examines global initiatives, local politics, and ways to effectively measure and track progress.

MONOGRAPHS

Abraham, M. A., ed. 2006. *Sustainability Science and Engineering* (vol. 1: Defining Principles). London: Elsevier. This book sets out a series of "sustainable engineering principles" that will help engineers design products and services to meet societal needs with minimal impact on the global ecosystem. Using specific examples and illustrations, the authors cleverly demonstrate opportunities for sustainable engineering, providing readers with valuable insight to applying these principles. This book is ideal for technical and non-technical readers looking to enhance their understanding of the impact of sustainability in a technical society.

American Society of Civil Engineers. Committee on Sustainability. 2004. *Sustainable Engineering Practice: An Introduction*. Reston, VA: American Society of Civil Engineers. *Sustainable Engineering Practice* provides a broad, fundamental understanding of sustainability principles and their application to engineering work. It is intended to fill a need for a primer on sustainability that can be introduced early in an engineer's career; it brings together all the basic dimensions of the history, concepts, and applications of sustainable engineering, and through a variety of examples and references, inspires and encourages engineers to pursue and integrate sustainable engineering into their work on a lifelong basis.

Chang, N.-B. 2010. *Systems Analysis for Sustainable Engineering Theory and Applications*. New York: McGraw-Hill. This work explains how to conduct holistic assessments in a practical manner and provides relevant case studies in various fields that demonstrate how to apply the sustainability concept. The book features a multidisciplinary approach by blending the essence of ecological engineering, green engineering, and systems engineering to illustrate the philosophy of sustainability. This pioneering guide provides a framework for applying systems analysis tools to account for environmental impacts, energy efficiencies, cost-effectiveness, socioeconomic implications, and ecosystem health in engineering solutions. Environmental engineers, civil engineers, industrial engineers, technicians, and environmental managers in the process industry will benefit from the detailed information in this authoritative volume.

Khatib, J., ed. 2009. *Sustainability of Construction Materials*. Boca Raton, FL: CRC Press. *Sustainability of Construction Materials* brings together a wealth of recent research on the subject. It provides a comprehensive and detailed analysis of the sustainability of these materials: aggregates, wood, bamboo, vegetable fibers, masonry, cement, concrete and cement replacement materials, metals and alloys, glass, and engineered wood products. Final chapters cover the use of recycled tire rubber in civil engineering works, the durability of sustainable materials, and nanotechnology in sustainable construction.

Sarte, S. B. 2010. *Sustainable Infrastructure: The Guide to Green Engineering and Design*. Hoboken, NJ: John Wiley & Sons. As more factors, perspectives, and metrics are incorporated into the planning and building process, the roles of engineers and designers are increasingly being fused together. This book explores this trend with an in-depth look at sustainable engineering practices in an urban design as it involves watershed master planning, green building, optimizing water reuse, reclaiming urban spaces, green streets initiatives, and sustainable master planning. This complete guide provides guidance on the role creative thinking and collaborative team building play in meeting solutions needed to affect a sustainable transformation of the built environment.

ASSOCIATIONS

American Society of Civil Engineers: Sustainability http://www.asce.org/sustainability (accessed June 30, 2010). Site has links to resources, policies, and activities related to sustainability issues.

International Institute for Environment and Development www.iied.org (accessed June 30, 2010). "As an independent international research organization, we are specialists in linking local to global. In Africa, Asia, the Caribbean, Central and South America, the Middle East, and the Pacific, we work with some of the world's most vulnerable people to ensure they have a say in the policy arenas that most closely affect them—from village councils to international conventions" (from Web site).

Sustainable Buildings Industry Council http://www.sbicouncil.org/ (accessed June 30, 2010). "The Sustainable Buildings Industry Council (SBIC) is an independent, nonprofit organization and a pioneer advocate of the whole building approach to sustainable facilities. We were founded in 1980 as the Passive Solar Industries Council by the major building trade groups, large corporations, small businesses, and individual practitioners who recognized that energy and resource efficient design and construction are imperative to a sustainable built environment" (from Web site).

World Association for Sustainable Development http://www.worldsustainable.org/ (accessed June 30, 2010). "World Association for Sustainable Development (WASD) is a unique global forum that brings together people from across the world to discuss key issues relating to science and technology management that impact the world sustainable development. The aims of WASD are to promote the exchange of knowledge, experience, information, and ideas among academicians, scholars, professionals, policy and decision makers, industry, executives, students ... etc. to improve the mutual understanding of the roles of science and technology in achieving sustainable development all over the world. To this end, WASD organizes different activities, which include high quality refereed publications, international conferences and seminars, development projects, etc." (from Web site).

World Business Council for Sustainable Development http://www.wbcsd.org/ (accessed June 3, 2010). The council's mission is to provide business leadership as a catalyst for change toward sustainable development, and to support the business license to operate, innovate, and grow in a world increasingly shaped by sustainable development issues.

CONFERENCES

International Symposium on Common Ground, Consensus Building and Continual Improvement: Standards and Sustainable Building, ASTM International www.astm.org (accessed July 9, 2010).

International Conference on Energy Sustainability, American Society of Mechanical Engineers www.asme.org (accessed May 26, 2010).

JOURNALS

Environment, Development and Sustainability: A Multidisciplinary Approach to the Theory and Practice of Sustainable Development, Dordrecht, The Netherlands: Springer. *Environment, Development and Sustainability* is an international and multidisciplinary journal covering all aspects of the environmental impacts of socio-economic development. It is also concerned with the complex interactions that occur between development and environment, and its purpose is to seek ways and means for achieving sustainability in all human activities aimed at such development.

International Journal of Sustainable Development and Planning (1743-7601). Southampton, U.K.: WIT Press. This journal, started in 2008, is an international, interdisciplinary journal covering the subjects of environmental design and planning, environmental management, spatial planning, environmental planning, environmental management, and sustainable development, in an integrated way as well as in accordance with the principles of sustainability.

International Journal of Sustainable Energy http://www.tandf.co.uk/journals/titles/14786451. asp (formerly *International Journal of Solar Energy*) (accessed July 9, 2010). London: Taylor & Francis (ISSN: 1478-646X). *The International Journal of Sustainable Energy* is an online-only journal that publishes experimental, theoretical, and applied results in both the science and engineering of sustainable energy. The multimedia capabilities offered by this electronic journal (including free color images and video movies) provide a unique multidisciplinary and international forum for the rapid dissemination of the latest high-quality research results. Topics covered include: photovoltaics, wind energy, bioenergy, geothermal power, solar heating, building applications, marine and hydropower, water treatment, power distribution, combined systems and meteorological data, as well as technologies for pollution control and energy conservation.

Proceedings of the ICE: Engineering Sustainability (1478-4629). London, U.K.: Institution of Civil Engineers. Published since 2003, this journal provides a forum for sharing the latest thinking from research and practice, and increasingly is presenting the "how to" of engineering a resilient future. It features refereed papers and shorter articles relating to the pursuit and implementation of sustainability principles through engineering planning, design, and application. The tensions between and integration of social, economic, and environmental considerations within such schemes are of particular relevance. Methodologies for assessing sustainability, policy issues, education, and corporate responsibility also will be included.

Sustainable Development (ISSN: 0968-0802). Hoboken, NJ: John Wiley & Sons. This is the leading international journal in its area emphasizing the need for interdisciplinary research into economic, social, and environmental aspects of sustainable development. All contributions are peer reviewed with the aim of providing readers with high quality, original material. Its emphasis is on publishing high quality research contributions that debate and discuss how sustainable development can be achieved within government, business, and civil society. It encourages papers with a national and international focus, but also recognizes the importance of local action for sustainable development. Contributions can range from practical applications of sustainable development and policy contributions through to philosophical discussions and debates. Regular reviews of books are included. What binds the contributions together is recognition of the central importance of a more sustainable development of economies and societies. The journal should be of interest to a broad interdisciplinary audience including academics, practitioners, business managers, and consultants.

Sustainability: Science, Practice, and Policy http://ejournal.nbii.org/ (accessed July 9, 2010). Reston, VA: National Biological Information Infrastructure (ISSN: 1548-7733). The e-journal *Sustainability: Science, Practice, and Policy* aims to establish "a forum for cross-disciplinary discussion of empirical and social sciences, practices, and policies related to sustainability." The full text of articles can be downloaded in HTML or PDF format.

WEB SITES

BFRL Strategic Goal: Measurement Science for Sustainable Infrastructure Materials http://www.nist.gov/bfrl/sustainable_infrastructure_materials_goal.cfm (accessed May 25, 2010). The Building and Fire Research Laboratory (BFRL) is the primary federal laboratory serving the building and fire safety industries. This strategic goal leverages the BFRL core competency in performance, durability, and service life prediction of building materials. BFRL's research in sustainability decision analysis tools and in high performance construction and building materials has been ongoing for several decades and is internationally recognized. Industrial customers continue to recognize BFRL's world-class expertise in advancing the measurement science of infrastructure materials. This recognition is evidenced by their willingness over the past two decades to establish and support ongoing NIST/ industry consortia, which have as their objectives creating and validating the measurement science necessary to effect reliable sustainability decisions for infrastructure materials.

Center for Sustainable Systems http://css.snre.umich.edu (accessed June 2, 2010). This Web site is managed by the Center for Sustainable Systems (CSS), University of Michigan. "CSS develops life cycle-based models and sustainability metrics to evaluate the performance and to guide the continuous improvement of industrial systems for meeting societal needs." The center also is committed to environmental sustainability education. It offers educational resources on sustainability issues for faculty members at higher education institutions covering major environmental sustainability issues and environmental law (exploring the influence on engineering design). Various statistics are provided to highlight some of the indicators of the demands placed on the Earth's resources. Also included is information on research methods and tools, research areas, project and publications, events and sponsors.

ECN: Energy Research Centre of the Netherlands http://www.ecn.nl/en/ (accessed June 3, 2010). This site gives information about the ECN's research activities in the areas of: policy studies, energy efficiency in industry, solar energy, wind energy, renewable energy in the built environment, biomass, clean fossil fuels, and fuel cell technology. Information is given in each section about current projects, products, and services. Also available are details about the technology and consultancy services the foundation offers and details of the activities of the Nuclear Research and Consultancy Group. Full text publications are available in PDF format along with reports and conference papers. There is also a summary of articles on ECN energy research that have recently been published in scientific journals. A Dutch-language newsletter is available. The ECN works with government and industry to achieve and maintain sustainable energy resources.

Engineering for Sustainable Development: Guiding Principles http://www.raeng.org.uk/ events/pdf/Engineering_for_Sustainable_Development.pdf (accessed June 2, 2010). This full text report is available in PDF format and covers principles for sustainable development. It is provided by the Royal Academy of Engineering. There are abbreviations of technical terms as well as diagrams, tables, and references.

InfraGuide: The National Guide to Sustainable Infrastructure http://sustainablecommunities. fcm.ca/infraguide/ (accessed June 2, 2010). This is the Web site for InfraGuide that "operated from 2001 to 2007 as a partnership between the Federation of Canadian Municipalities (FCM), the National Research Council (NRC), and Infrastructure Canada. InfraGuide's national network of experts produced a collection of case studies, best practice reports, and e-learning tools for sustainable municipal infrastructure, which are featured on the Web site. The site also features information on the background and objectives of the organization and its network, and an archive of news releases, notices, newsletters, and details of forthcoming events.

International Council for Science, Resources: Environment/Sustainability http://www.icsu. org/2_resourcecentre/RESOURCE_list_base.php4?rub=8 (accessed July 9, 2010). In order to strengthen international science for the benefit of society, ICSU mobilizes the knowledge and resources of the international science community to identify and address major issues of importance to science and society; facilitate interaction amongst scientists across all disciplines and from all countries; promote the participation of all scientists—regardless of race, citizenship, language, political stance, or gender—in the international scientific endeavor; and provide independent, authoritative advice to stimulate constructive dialogue between the scientific community and governments, civil society, and the private sector. The resource center maintains an up-to-date collection of reports from a variety of international programs.

Royal Academy of Engineering Visiting Professors Scheme in Engineering Design for Sustainable Development http://www7.caret.cam.ac.uk/ (accessed June 3, 2010). Case studies and lecture materials (with PowerPoint® slides) about sustainable development and related issues are available from this Web site as well as emerging principles and news. The Good Engineers guide is also available The scheme aims to develop teaching materials based on case studies, which will enhance the understanding and the practice of teaching sustainable development. This site was developed jointly by the University of Cambridge and the University of Hertfordshire in the United Kingdom.

Sustainable Development International http://www.sustdev.org/ (accessed June 3, 2010). This governmental organization site includes PDF format papers, reports, and case studies concerning sustainable development. There is industry news and also a calendar of events. Abstracts of current and previous journal articles are provided. Sustainable Development International is a publication body (newspaper-like), which is working in cooperation with international bodies that include United Nations Agencies (CSD, UNIDO, UNEP); World Energy Council (WEC); Investment and Banking Authorities and the International Council for Local Environmental Initiatives (ICLEI), and has a subscription database of decision makers at local, regional, and national level as well as representatives of development agencies, nongovernmental organization (NGO) communities, and international policy makers.

REFERENCES

American Society of Civil Engineers. 2004. *20104 Official Register.* Reston, VA: ASCE.
Chrimes, M. and A. Bhogal. 2001. Civil engineering: A brief history of the profession: The perspective of the Institution of Civil Engineers. In *International engineering history and heritage: Improving bridges to ASCE's 150th Anniversary*, eds. J. R. Rogers and A. J. Fredich. Reston, VA: ASCE.

Encyclopedia Britannica Online. Soil Mechanics. http://www.britannica.com/EBchecked/topic/552695/soil-mechanics (accessed March 24, 2011).

Encyclopedia Britannica Online. Harbours and Sea works. http://www.britannica.com/EBchecked/topic/254888/harbour (accessed March 24, 2011).

Ferguson, H. and M. Chrimes. 2011. *The civil engineers: A history of the Institution of Civil Engineers*. London: ICE Publishing.

Goodman, R. E. 2002. Karl Terzaghi's legacy in geotechnical engineering. *Geostrata* http://www.geoengineer.org/terzaghi2.html (accessed June 22, 2010).

Merdinger, C. J. 1953. *Civil engineering through the ages*, Washington, D.C.: Society of American Military Engineers.

9 Computer Engineering

*Hema Ramachandran (based on the first edition by
Hema Ramachandran and Renee McHenry)*

CONTENTS

Introduction..204
Overview of the Literature...205
   Classic Texts..206
Searching Library Catalogs...207
Software Engineering...207
History of Computing 208
   Bibliographies 208
   Monographic Series 209
   Other Significant Texts..209
   Serials...210
   Web Sites..210
   Book Reviews 211
Dictionaries..211
   Print Dictionaries 211
   Free Online Dictionaries 213
Encyclopedias 214
Handbooks and Manuals..216
Technical Reports...218
Institutional Repositories 219
Online Bibliographies, Preprints, and Repositories..220
Patents and Standards 222
Conferences..222
Journals 223
   Core Journals...223
   Magazines 223
   Directories..223
   Academic/Scholarly Journals...224
   Impact Factors...224
   Magazines for the Practitioner 225
   Consumer Magazines for the PC End-User Market...226
Databases 226
   Databases of Associations..226
   Comprehensive Bibliographic Databases...227
   Commercial Database Packages 227
   Individual Databases Specializing in Computer Engineering and Science...............228
   Databases in Related Fields 229
Current Awareness 229
Publishers...230
   Other..231

Associations, Organizations, and Societies...231
Careers and Education 233
Tools for the Practitioner...234
  Technical Books..234
  Market Research..235
Web Portals 235
Other Web Resources 237
  Algorithms..237
  Compilers and Interpreters..237
  Programming...238
Internet Resources: Subject Guides 238
  Metasites and Speciality Search Engines...238
  Networking and Collaborating..239
Acknowledgments...240
References..240

INTRODUCTION

Webster's *New World Dictionary of Computer Terms* (1988), even though dated, defines computer engineering as "the field of knowledge that includes the design of computer hardware systems." As simple as it is, this definition is a good starting point for the general reader, but it ignores its relationship to computer science, which the *Dictionary* describes as "the field of knowledge embracing all aspects of the design and use of computers."

The Joint Task Force on Computer Engineering Curricula, consisting of members from the Association of Computing Machinery (ACM) and the Institute of Electrical and Electronics Engineers (IEEE) Computer Society, provides us with a more detailed working definition as outlined in its *Curriculum Guidelines for Undergraduate Degree Programs in Computer Engineering* published in December 2004:

> Computer engineering as an academic field encompasses the broad areas of computer science and electrical engineering. Computer engineering is defined in this report as follows:
>
> Computer engineering is a discipline that embodies the science and technology of design, construction, implementation, and maintenance of software and hardware components of modern computing systems and computer-controlled equipment. Computer engineering has traditionally been viewed as a combination of both computer science (CS) and electrical engineering (EE) http://www.eng.auburn.edu/ece/CCCE/CCCE-FinalReport-2004Dec12.pdf (accessed October, 22, 2010).

Finally, the Web page of Trinity College Dublin provides us more context on the role of the computer engineer:

> A computer engineer needs excellent problem-solving skills, a good theoretical grounding in the fundamentals of engineering, and the practical skills to put theory into practice. Computer engineers may design computer hardware, write computer programs, integrate the various subsystems together or do all three. Computer engineers need good management skills as they often get quickly promoted to project manager-type positions. Furthermore, computer engineers need good people skills, as they have to sell their ideas to other engineers, other professionals, and members of the public.

Another way to consider this is that the computer engineer's primary goal is to implement theories developed by computer scientists to solve practical problems and create viable systems. This comparison is analogous to science and engineering in general—science often provides results that are significant to engineering. Both computer science and computer engineering owe their foundation to mathematics.

Computer engineering, also sometimes known as computer systems engineering or even computer science engineering, has been viewed in the past as just "designing computers." Now, more often than not, computer engineers are an integral part of professional teams that design computer-based systems throughout industries, such as aerospace, telecommunications, power production, manufacturing, and defense, as well as the computer and electronics industry. Professional specialty areas include system architecture, system integration, layout, RTL/chip design, package/board design, and the fabrication process, while the more theoretical areas are computer architecture, digital control theory, digital signal processing, semiconductors, and software engineering.

At the time of the publication of the first edition of this book, Web 2.0 tools were being launched. Many of the resources mentioned in this chapter are enhanced with Web 2.0 utilities and whenever these have been encountered they are mentioned, especially if they offer value-added information. It is interesting to note that the main information providers in this arena have remained the same.

OVERVIEW OF THE LITERATURE

Prior to 1947, the sparse computing literature was concerned with analog computers or analyzers and punched-card machines. Simple desk calculators existed, but usually calculations were made by hand. Journals in computing barely existed and pertinent literature, mostly technical, was scattered throughout a variety of disciplines: mathematics, statistics, physics, electrical engineering, and even sciences, such as astronomy. How different the situation is today. In less than 60 years, computers are ubiquitous, and literature about them in all shapes and forms is everywhere. From literature for the professional computer engineer to the average consumer, the volume of literature is almost breathtaking. The average consumer encounters chain bookstores with large sections devoted to computer books for the general public as well as the professional practitioner. In the bookstore's "how to" section are books (and book series) devoted to the computer neophyte; the familiar *Dummies* and *Idiots* series are known to everyone. The popular computer magazine is part of our everyday culture and found in supermarkets side-by-side with grocery items. The general news media frequently reports on the computing industry and those who work in it. This was particularly evident during the rise and fall of the dotcoms. The entrepreneurs in California's Silicon Valley are our modern day technological heroes and their rise has almost become part of folklore. The computing industry, however it is defined, is indeed big business and any economic shifts have global ramifications, as seen with the controversial subject of "outsourcing" to India and other countries.

The ACM's *Guide to Computing Literature*, published annually since 1960, may be considered a guide or bibliography of the computing literature. Indexed by author, subject, and keyword using the ACM Computing Classification System, the *Guide* includes references to scholarly and trade magazines, and books from major publishers in the computing field.

Of course, from time to time, librarians also will produce bibliographies on topics of interest. This is a particular challenge in computer engineering where a subject compilation can be out of date as soon as it is published, but it is certainly useful for retrospective literature searching. This is certainly evident with the older compilations listed below:

Hildebrandt, D. M. 1996. *Computing Information Directory*, 13th ed. Federal Way, WA: Pedaro.
It is a shame that this annual series, which ceased in 1996, has not been updated. It has been a very useful directory for librarians and information seekers alike due to its comprehensive coverage of the computing literature. The author, Darlene Myers Hildebrandt, head of the Science Libraries of Washington State University, provides a treasure trove of information. Even though it is outdated, hold on to your editions, because it still provides a good overview of the literature. The directory has listings and descriptions of journals, newsletters, books, dictionaries, indexes, abstracts, software resources, review sources, information on special issues in journals, directories, computer languages, standards, and publishers.

Rousseau, R., compiler. 1985. *Selective Guide to Literature on Computer Science*. Washington, D.C.: American Society for Engineering Education, Engineering Libraries Division.

It is interesting to compare the computer engineering literature with that of other engineering disciplines. Due to the pervasive nature of computing in society today, this may be the only engineering field where publishers cater to the needs of a whole spectrum of users hungry for information (from neophyte to expert): academicians, researchers, teachers, practitioners, students, hobbyists, and consumers. Given the fast-paced nature of the computing industry at large, material is churned out at a phenomenal rate; the literature has become a lucrative business for many publishers. The range of publishers who meet these needs covers the gamut.

The standard commercial publishers—Springer, John Wiley & Sons, Cambridge University Press, Oxford University Press, Elsevier, Newnes, and Morgan Kaufmann as well as the society publishers IEEE Press and ACM—carry many monographic titles catering to the research and educational communities. Many of these publishers also publish journals. For more information, see the Journals and Current Awareness sections. Then, we have the "practitioner" market served by such publishers as O'Reilly (the major player), Peachtree Press, Prentice Hall PTR, Que, and Sam's Publishing (see Tools for the Practitioner section).

With the development of the Internet, a medium that computer engineers not surprisingly embraced from the beginning, there is a vast amount of information freely available. The challenge now is how to make sense of the overabundance of literature and offer guidance to the computer engineer and information specialist. This chapter aims to be as comprehensive as possible within the confines of this publication. The primary audiences are the computer engineering professional and information specialist with a secondary audience of the educated layperson. The task of identifying basic resources for others, such as high school students, is left to others. At the very least, this chapter will give pointers for further investigation and serve as a reminder of sources that may have been overlooked.

Rapid dissemination of information is the hallmark of computer engineering, and for this reason access to conference proceedings is relatively more important than journals (the reverse is true in other science and engineering disciplines), although this is changing with the development of online journals. The concept of the "technical report" emerged in computer engineering as a way of communicating early research results. With the advent of the Internet, it became even easier to provide access to technical reports via FTP servers and, then later, the World Wide Web. The Internet has been a boon for the computer engineer with many tools freely available, and it is a challenge to decide which sites to include in a publication such as this, requiring one to critically evaluate each site for content, authority, usefulness, and stability.

CLASSIC TEXTS

Abelson, H., G. J. Sussman, and J. Sussman. 1996. *The Structure and Interpretation of Computer Programs,* 2nd ed. Cambridge, MA: MIT Press; New York: McGraw-Hill.

Brooks, F. P., Jr. 1995. *The Mythical Man-Month: Essays on Software Engineering,* anniversary ed. Reading, MA: Addison-Wesley.

Dahl, O-J., E. W. Dijkstra, and C. A. R. Hoare. 1972. *Structured Programming*. London/New York: New Academic Press.

Dijkstra, E. W. 1976. *A Discipline of Programming*. Englewood Cliffs, N]: Prentice-Hall.

Kernigham, B. W. and P. J. Plauger. 1978. *The Elements of Programming Style,* 2nd ed. New York: McGraw-Hill.

Knuth, D. E. 1998. *The Art of Computer Programming.* 3rd ed. (3 vols.) Reading, MA: Addison-Wesley (vol. 1. *Fundamental Algorithms*; vol. 2. *Seminumerical Algorithms*; vol. 3. *Sorting and Searching*). *Note*: As a continuation of this volume series (and to update previous volumes), Knuth has begun writing a series of books called *fascicles* to be published at regular intervals. The ultimate goal is for these new contributions to become part of the fourth edition.

Knuth, D. E. 2005. *The Art of Computer Programming Vol. 1, Fascicle 1. MMIX—A RISC Computer for the New Millennium*. Reading, MA: Addison-Wesley Professional.

Knuth, D. E. 2005. *The Art of Computer Programming Vol. 4, Fascicle 2. Generating All Tuples and Permutations*. Reading, MA: Addison-Wesley Professional.

Knuth, D. E. 2005. *The Art of Computer Programming Vol. 4, Fascicle 3, Generating All Combinations and Partitions*. Reading, MA: Addison-Wesley Professional.

Sammet, J. E. 1969. *Programming Languages: History and Fundamentals*. Englewood Cliffs, NJ: Prentice-Hall.

Weinberg, G. 1998. *The Psychology of Computer Programming*. Silver Anniversary ed. New York: Dorset House.

Wirth, N. 1986. *Algorithms and Data Structures*. Englewood Cliffs, NJ: Prentice Hall.

SEARCHING LIBRARY CATALOGS

For those wishing to identify sources within the field of computer engineering, the access points provided by *Library of Congress Subject Headings* (*LCSH*) are well defined. The subject heading Computer engineering is used for the field and Computer engineers is used for individuals within the field. The *LCSH* scope note states that:

> Here are entered work on the design of computer hardware and circuitry. Works on the logical structure that determines the way a computer executes programs are entered under Computer architecture. Works on the way a computer is constructed to implement its architecture, including what components are used and how they are connected, are entered under Computer organization http://classificationweb.net/ Auto/ (accessed October 19, 2010).

Library of Congress Subject Headings:

Computer Algorithms
Computer Architecture
Computer Graphics
Computer Input-Output Equipment (for computer hardware)
Computer Interfaces
Computer Logic
Computer Networks
Computer Organization
Computer Programming
Computer Programs
Computer Science
Computer Software
Computers
Programming Languages (also specific names of languages, such as C++, Java, XML)

SOFTWARE ENGINEERING

Where the computer engineering literature may be shelved in a library's collection depends on whether its focus is on computer science, electrical engineering, systems security, or Internet-related technologies.

Library of Congress Call Numbers:

QA 75.5–76.95	Electronic computers. Computer science
QA 76.75–76.765	Computer software
T58.5–T58.64	Information Technology

TA 157	Computer engineers
TK 5100	Computer networks
TK 7800–8360	Electronics
TK 7885–7895	Computer engineering. Computer hardware
Z 102.5–104	Cryptography
ZA 3201–3250	Information superhighway
ZA 4201–4251	Internet

HISTORY OF COMPUTING

The birth and development of computer engineering in its relatively short life has been both phenomenal and dramatic. The 1930s marked the beginning of the discipline—the era of Turing and Church—that gave us the fundamental mathematical concepts in computing. Dominant issues in the current arena include the Internet, and the growth of network technologies, such as wireless, security, and research on human–computer interfaces. The specialized field of computational science lends its techniques to all the other scientific disciplines as they try to manage huge amounts of data.

The discipline has resolved the identity issues it had in the 1950s. Does it belong in electrical engineering or is it a subdivision of mathematics? Will it develop into a science in its own right or will it fade away? The former question still sometimes poses a challenge for the field, but the latter question is no longer an issue, as it has become a well-established subject.

In its article on computer engineering, WordIQ.com states:

> … computer engineering degrees have been added to a number of schools' degree programs since the early 1990s. Some schools have integrated computer engineering, along with software engineering, into their electrical engineering departments, while others, such as MIT, have chosen to merge electrical engineering and computer science departments instead. Since computer engineers are mainly focused on electronics and computers, their course loads tend to involve fewer courses on natural sciences, such as statics or dynamics, than traditional engineering programs. Instead, courses on fundamentals of computer sciences are taught http://www.wordiq.com/definition/Computer_engineering (accessed October 21, 2010).

It is interesting to note that the computing field is at an interesting crossroads—young enough that some of the early pioneers of the field are still practicing, either teaching and/or working in industry, and yet the "second generation" has been around long enough to have heard stories from the early pioneers directly.

Luckily, students of computing history have several key resources at their disposal. The Institute of Electrical and Electronics Engineers (IEEE) and Association for Computing Machinery (ACM), the two major professional associations in computer science, have published a vast amount of information on the topic through sponsorship of conferences, journals, and books. A search of the *ACM's Guide to the Computing Literature, Inspec* and *Compendex* databases (see Databases section) will enable the user to do a more exhaustive search of the literature. The journal *IEEE Annals of the History of Computing* is worth a special mention because it provides a constant output of scholarly articles. It is one of the few periodicals to concentrate exclusively on the history of a scientific subject. The study of computer engineering is a popular topic not only for the history of science scholar but also for the general public. A list of some of the significant, seminal, and recently published studies on the topic is provided below. It should not be considered comprehensive, but a good starting point for further exploration.

BIBLIOGRAPHIES

Cortado, J. W., compiler. 1983. *Annotated Bibliography on the History of Data Processing.* Westport, CT: Greenwood Press.

Cortado, J. W., compiler. 1990. *Bibliographic Guide to the History of Computing, Computers and the Information Processing Industry*. Westport, CT: Greenwood Press.

Cortado, J. W., compiler. 1996. *Bibliographic Guide to the History of Computer Applications, 1950–1990*. Westport, CT: Greenwood Press.

These three titles provide a complete bibliographic history of computer engineering from 1800 to 1990 and cover contributions made by individuals and institutions in hardware, computing concepts, and software. The second title in the above list is a supplement to the first title. Each work is well organized with annotated entries and indexed by author and subject. There are about 10,000 entries in total over the three volumes.

Monographic Series

The MIT Press *History of Computing Series* provides significant book-length studies of the subject. The series began in 1984 and has 32 titles as of 2010. They may be viewed on the MIT Press home page http://mitpress.mit.edu; click on the topic "Computer Science and Intelligent Systems," then "Series" and "History of Computing." All of the titles in the series represent important in-depth studies and, for the sake of brevity, only a few of the well-known titles are listed below:

Aspray, W. 1990. *John von Neumann and the Origins of Modern Computing*. Cambridge, MA: MIT Press.

Aspray, W. and P. Ceruzzi. 2010. *The Internet and American Business*. Cambridge, MA: MIT Press.

Ceruzzi, P. E. 2003. *A History of Modern Computing, 2nd ed*. Cambridge, MA: MIT Press.

Cohen, I. B. 1999. *Howard Aiken: Portrait of a Computer Pioneer*. Cambridge, MA: MIT Press.

Cohen, I. B. and G. W. Welch. 1999. *Makin' Numbers: Howard Aiken and the Computer*. Cambridge, MA: MIT Press.

Pugh, E. W. 1984. *Memories that Shaped an Industry: Decisions Leading to IBM System 360*. Cambridge, MA: MIT Press.

Rojas, R. and U. Hashagen. 2000. *The First Computers—History and Architectures*. Cambridge, MA: MIT Press.

Stein, D. 1985. *Ada: A Life and Legacy*. Cambridge, MA: MIT Press.

Wilkins, M. V. 1985. *Memoirs of a Computer Pioneer*. Cambridge, MA: MIT Press.

Other Significant Texts

Agar, J. 2001. *Turing and the Universal Machine: the Making of the Modern Computer*. Cambridge, U.K.: Icon Books; distributed in the United States by National Book Network, Inc.

Aspray, W., ed. 1990. *Computing Before Computers*. Ames, IA: Iowa State University Press.

Burks, A. R. 1988. *The First Electronic Computer: The Atanasoff Story*. Ann Arbor, MI: University of Michigan Press.

Dubbey, J. M. 1978. *The Mathematical Work of Charles Babbage*. Cambridge, U.K.: Cambridge University Press.

Goldstine, H. H. 1993. *The Computer from Pascal to von Neumann*. Reprint ed. Princeton, NJ: Princeton University Press. Available as part of ACLS History E-Book Project if library subscribes at http://www.hti.umich.edu/cgi/b/bib/bibperm?q1=HEB01140 (accessed March 22, 2011).

Ifrah, G. 2001. *The Universal History of Computing: From the Abacus to the Quantum Computer*. (Translated from the French: *Histoire Universelle des Chiffres* (1994). Paris: Editions Robert Laffont.). New York: John Wiley & Sons.

Lukoff, H. 1979. *From Dits to Bits: A Personal History of the Electronic Computer.* Portland, OR: Robotics Press.

Mollenhoff, C. R. 1988. *Atanasoff: Forgotten Father of the Computer.* Ames, IA: Iowa State University Press.

Reilly, E. D. 2003. *Milestones in Computer Science and Information Technology.* Westport, CT: Greenwood Press.

Slater, R. 1987. *Portraits in Silicon.* Cambridge, MA: MIT Press.

Stern, N. B. 1981. *From ENIAC to UNIVAC: An Appraisal of the Eckert–Mauchly Computers.* Bedford, MA: Digital Press.

Wexelblat, R. L., ed. 1981. *History of programming languages.:* Proceedings of the History of Programming Languages Conference, Los Angeles, CA, June 1–3, 1978. New York: Academic Press. ACM monograph series.

Williams, M. R. 1997. *A History of Computing Technology,* 2nd ed. Los Alamitos, CA: IEEE Computer Society Press.

SERIALS

IEEE Computer Society. 1992–. *IEEE Annals of the History of Computing,* Los Alamitos, CA: IEEE Computer Society http://www.computer.org/annals/ (accessed June 30, 2010). Begun in 1979, the *Annals* are the primary source for documenting the history of the discipline. They have articles by leading scholars as well as computer pioneers. It is available in print, via *IEEE Xplore* and the IEEE Computer Society. The online journal features "Computing Lives," a series of podcasts covering "the analytical engine to the supercomputer, from Pascal to von Neumann, from punched cards to CD-ROMs" of selected articles from the journal. This Podcast series features scholarly accounts by leading computer scientists and historians as well as firsthand stories by computer pioneers. "Computing Now" presents access to multimedia content and free selected articles from *Annals of the History of Computing.*

WEB SITES

Alan Turing.net: The Turing Archive for the History of Computing, http://www.cs.usfca.edu/www AlanTuring.net/turing_archive/index.html (accessed July 30, 2010). "The documents that form the historical record of the development of computing are scattered throughout various archives, libraries, and museums around the world. Until now, to study these documents required a knowledge of where to look, and a fistful of air tickets. This virtual archive contains digital facsimiles of the documents. The archive places the history of computing, as told by the original documents, onto your own computer screen " (from Web page).

Charles Babbage Institute: Center for the History of Information Technology http://www.cbi.umn.edu/ (accessed July 30, 2010). "The Charles Babbage Institute (CBI) is an archives and research center dedicated to preserving the history of information technology and promoting and conducting research in the field" (from Web page). Of particular note is the C81 reprint series for the History of Computing. Computer Oral History Project.

The Lemelson Center for the Study of Invention and Innovation (National Museum of American History, Smithsonian Institution) http://invention.smithsonian.org/resources/fa_comporalhist_index.aspx (accessed July 30, 2010). "The Computer Oral History Collection (1969–1973, 1977) was a cooperative project of the American Federation of Information Processing Societies (AFIPS) and the Smithsonian Institution. This project began in 1967 with the main objective to collect, document, house, and make available for research source material surrounding the development of the computer. The project collected taped oral interviews with individuals who figured prominently in developing or advancing the computer field and supplemental written documentation: working papers, reports, draw-ings, and photographs" (from Web page). Some interviews are available full-text online. Virtual Museum of Computing, http://icom.museum/vlmp/computing.html (accessed July 30, 2010). This virtual museum, which "opened" in 1995, is a comprehensive site special-izing in the history of computing, and continues to be maintained by Professor Bowen of South Bank University, London.

BOOK REVIEWS

In addition to the standard book review journals, such as *Choice* and *Library Journal*, here are some specialized resources for computer engineering.

> *Computing Reviews* http://www.reviews.com (accessed July 30, 2010). *Computing Reviews* has been published for over 40 years by ACM and is the major review journal for the dis-cipline. Each review is authored by experts and covers the book and journal literature. In 2000, ACM entered into a partnership with Reviews.com and the journal metamorphosed into an online database. One needs to subscribe to Reviews.com separately from the *ACM Digital Library* to have online access. The database continues the tradition of excellent expert reviews with presentation of up-to-date material and some value-added features, such as personalized alerts, customized searching, and a blog (launched in 2006) creating a virtual community. The print version of *Computing Reviews* is still being published, and remains a good tool for those who cannot afford the database, but want to keep up with the literature. The print is included in the ACM journal print package (but not the *ACM Digital Library*). This database, either in the printed or online version, is an invaluable tool for book selection. Considered a "consumer reports" journal for information specialists, *The Charleston Advisor* in its October 2004 issue gave it an overall ranking of 4.25 out of 5 possible stars. The quality of its search engine was given 5 stars http://www.charlestonco. com/review.cfm?id= 200 (accessed May 18, 2005).
>
> TechBookReport http://www.techbookreport.com/ (accessed May 18, 2005). Tech Book Report aims to provide independent, interesting, and informative book reviews for devel-opers and technologists. Reviews are written by and for the practitioner.

DICTIONARIES

Anyone brave enough to compile a dictionary of computing terms is faced with many challenges. In short, by the time the dictionary is published it may already be out-of-date. However, every library collection or computing professional should have a few choice titles at their fingertips to clarify terms and usage. Online dictionaries offer obvious advantages for this reason. The bulk of the literature in computer engineering is published in English, but it may be prudent to have a couple of good dic-tionaries to translate terms from other languages. When the discipline was young, not very long ago, many dictionaries were published to explain basic and complex terms to the neophyte. However, pub-lishing in this area has slowed down somewhat. A current trend is for dictionaries to be published in niche or new areas. One example is the dictionary on embedded systems listed below. What follows is a hand-picked list of well-respected and established titles with the expectation so that, if desired, the serious researcher can search *Books in Print* and Amazon.com to find more titles.

PRINT DICTIONARIES

> Daintith, J., and E. Wright, eds. 2008. *A Dictionary of Computing,* 6th ed. Oxford, U.K./New York: Oxford University Press. Fully revised by a team of computer scientists, this edition pro-vides comprehensive, up-to-date coverage of the language of computing, including hardware

and software applications, programming languages, networks and communications, the Internet, and e-commerce. In addition to the clearly explained 10,000 terms, many of which are new for this edition, the dictionary offers words used in context and includes quotations from computing magazines and useful tables of programming codes and languages. It is an ideal resource for any general computer use (description based on the 5th edition).

Downing, D. A., M. A. Covington, M. Mauldin Covington, and C. A. Covington. 2009. *Dictionary of Computer and Internet Terms,* 10th ed. Hauppauge, NY: Barron's. "With 1,800 definitions and an abundance of Internet and online terms, this dictionary is perfect for new and intermediate computer users. The entries range from basic to advanced computer terminology and include historical terms and software concepts. The reference even defines and illustrates concepts and terminology for several graphics applications. This pocket-sized dictionary is packed with illustrations, charts, and examples and the comprehensive cross-referencing and clear language make it easy to use. Off- the-shelf software users will find this guide particularly useful. Approximately 1,700 key computer terms and their definitions are included in this edition with hundreds of words and expressions that apply specifically to the Internet. This pocket-sized reference offers help to prospective buyers who are mystified by hi-tech jargon. It also explains what computers do and how they work" (from Amazon.com) (accessed July 30, 2010).

Freedman, A. 2000. *The Computer Glossary: The Complete Illustrated Dictionary* (with CD-ROM), 9th ed. New York: AMACOM. *The Computer Glossary* was considered one of the best one-volume computer dictionaries on the market and an essential resource for many computer users over the past decade. Sadly, there will be no more print editions (see entry on author's *Computer Desktop Encyclopedia* for more details). This edition contains 6,000 definitions along with 175 illustrations. Alan Freedman is president of the Computer Language Company, an organization dedicated to computer education and training.

Ganssle, J. and M. Barr. 2003. *Embedded Systems Dictionary.* San Francisco, CA: CMP Books. This is a good example of a recent dictionary that deals with a segment of computer science and engineering. It is intended for engineers (on both the hardware and software side) who design embedded systems. It covers the history, technologies, tools, abbreviations, and definitions for the terms most-used in embedded systems. Serving both technical and nontechnical audiences, the dictionary contains 2,800 terms.

LaPlante, P. A., ed. 2000. *Dictionary of Computer Science, Engineering and Technology.* Boca Raton, FL: CRC Press. Written by an international team of over 80 contributors, this dictionary provides detailed definitions (including illustrations when appropriate) and practical information spanning various disciplines and industry sectors. Its 8,000 terms cover all aspects of computing and computer technology from multiple vantage points, including academic, applied, and professional. Unfortunately, this edition has not been updated.

McGraw-Hill. 2004. *McGraw-Hill Dictionary of Computing & Communications.* New York: McGraw-Hill. Compiled from the *McGraw-Hill Dictionary of Scientific and Technical Terms,* 6th ed. 2004, this dictionary contains 11,000 entries essential in the field of computer science, information technology and communications (relating to both digital and analog data). Its definitions indicate which field the term belongs to, together with synonyms, pronunciations, cross references, acronyms, and abbreviations. Includes an appendix containing charts and tables for measurement conversion and other useful data. Suitable for the general reader up to the professional. *Note*: Depending on their clientele and budget, libraries may not wish to duplicate content in their collection if they already own the larger 6th edition. Content from the sixth edition plus the *McGraw-Hill Encyclopedia of Science & Technology*, ninth edition, is available via annual subscription to *AccessScience*.

McGraw-Hill. 2004. *McGraw-Hill Dictionary of Electrical and Computer Engineering.* New York: McGraw-Hill. This dictionary, published in late 2004, covers the areas of electrical, computer, and electronics engineering and related areas of mathematics and communications. It features 15,000 entries based on terms from *McGraw-Hill Dictionary of Scientific and Technical Terms,* 6th ed. 2004. In addition to definitions, it includes synonyms, acronyms, and abbreviations and pronunciations for all terms as well as an extensive appendix. *Note*: Depending on their clientele and budget, libraries may not wish to duplicate content in their collection if they already own the larger 6th edition. The January 2005 *Choice* review states that "practicing engineers could benefit from the present dictionary because of its tighter focus and lengthy appendixes." Content from the larger sixth edition plus the *McGraw-Hill Encyclopedia of Science & Technology*, ninth edition, is available via annual subscription to *AccessScience*.

Microsoft. 2002. *Microsoft Computer Dictionary,* 5th ed. Redmond, WA: Microsoft Press. Now minus the accompanying CD-ROM, this fifth edition defines computing terms (including acronyms, jargon, and slang) and concepts pertaining to hardware, networks, programming, applications and databases. Unlike other dictionaries, it does not tackle information about companies, commercial products, or proprietary technologies (except from Microsoft itself). The more than 10,000 entries explain concepts and technical terms in easy-to-understand language making this a good resource to have for either the beginning or advanced computer user.

Narins, B., ed. 2001. *World of Computer Science*. Detroit: Gale Group/Thomson Learning. (2 vols.) A good introductory reference on the subject with an impressive list of contributors and advisors. Alphabetically arranged, the 800+ entries discuss pioneers, discoveries, theories, concepts, issues, and ethics. Of particular interest are the biographies of living computer personalities. Not aimed at an academic audience, *Choice* remarks that it will help new computer science students and professionals alike.

Newton, H. 2009. *Newton's Telecom Dictionary: Covering Telecommunications, Networking, Information Technology, the Internet, Fiber Optics, RFID, Wireless, and VoIP*, 25th anniversary ed. New York: Flatiron Books. Long considered the telecom "bible," Newton's newest dictionary editions continue to cover technical terms and acronyms in the ever-changing telecom, network, and IT industry for technology and business professionals. Each revised edition includes updated and expanded definitions including any new standards, technologies, and vendor-specific terms. With its use of nontechnical language to explain concepts, *Newton's Telecom Dictionary* is considered an "essential reference" (*PC Magazine*) for anyone involved with telecom and IT systems and services.

Pfaffenberger, B., ed. 2003. *Webster's New World Computer Dictionary,* 10th ed. Indianapolis, IN: Wiley Publishing. Useful for the novice or the professional, this updated dictionary contains over 4,750 definitions for computer terms including current coverage of standards and protocols for storage, memory, and peripherals (with excellent cross-referencing of terms). Bryan Pfaffenberger is the author of more than 75 books on personal computing (including the *HTML 4 Bible*) and teaches at the University of Virginia's Division of Technology, Culture, and Communication. This is a good reference work for any collection.

FREE ONLINE DICTIONARIES

BABEL: A Glossary of Computer Oriented Abbreviations and Acronyms http://www. cs.indiana.edu/~ysimmhan/personal/babel.html (accessed July 30, 2010). Available on the Internet since 1989, but has not been updated since 1998, Babel covers computer-related abbreviations and acronyms.

Dictionary of Algorithms and Data Structures http://www.itl.nist.gov/div897/sqg/dads/ (accessed July 30, 2010). "This is a dictionary of algorithms, algorithmic techniques, data structures, archetypical problems, and related definitions. ... We do not include algorithms particular to business data processing, communications, operating systems or distributed algorithms, programming languages, AI, graphics, or numerical analysis" (from Web page). Use the alphabetical arrangement or the search feature to find a definition.

FOLDOC: Free Online Dictionary of Computing http://foldoc.org/ (accessed July 30, 2010). "FOLDOC is a computing dictionary. It includes acronyms, jargon, programming languages, tools, architecture, operating systems, networking, theory, conventions, standards, mathematics, telecoms, electronics, institutions, companies, projects, products, history; in fact, anything you might expect to find in a computer dictionary" (from Web page).

Mathematical Programming Glossary© http://glossary.computing.society.informs.org/ (accessed July 30, 2010). Professor Harvey J. Greenberg from the Mathematics Department at University of Colorado at Denver began this glossary in 1996. It "contains terms specific to mathematical programming, and some terms from other disciplines, notably economics, computer science, and mathematics, that are directly related" (from Web page). Search the alphabetical index or jump to a letter to begin. No sources are cited.

Netlingo http://www.netlingo.com/dictionary/all.php (accessed July 30, 2010). "NetLingo is an online dictionary about the Internet. It contains thousands of words and definitions that describe the technology and community of the World Wide Web" (from Web page).

Webopedia™ http://www.webopedia.com/ (accessed July 30, 2010). Webopedia is a free online dictionary for words, phrases, and abbreviations related to computers and the Internet providing easy-to-understand definitions and avoiding the use of heavy jargon when possible so that the site is accessible to users with a wide range of computer knowledge. Full-time, experienced editors gather information from standards bodies, leading technology companies, universities, professional online technical publications, and professionals working in the field. The sources used are often listed in the links section below the definition if the sources can provide more information than was included in the definition. Every definition is verified among multiple sources; definitions are never based on just one source. The definitions on Webopedia evolve and change as technologies change, so the definitions are frequently updated to reflect trends in the field. New terms are added on a daily basis, and many of the new terms come from suggestions from the site's users (from Web page).

ENCYCLOPEDIAS

Dictionaries provide a quick answer to the meaning or definition of terms, but encyclopedias will give you an overview of the topic. Encyclopedias are particularly useful when one is embarking on research in a new discipline or topic. They will provide you with background information, brief historical context within the discipline, are written by experts in the field, and provide a short bibliography for further exploration. However, these short articles are a "snapshot" in time and it is up to information-seekers to use more up-to-date resources (such as books, periodical indexes, and Web resources) to augment their knowledge. Librarians, in particular, will find a good encyclopedia a godsend when faced with reference questions on a new topic. As with the section on dictionaries, we offer a few select titles for your computer engineering science collection. Encyclopedias are expensive, so replacing copies with new editions or new titles every year might not be economical or necessary, but replacing titles every five years is more realistic. Our highly selective list below includes the well-established and respected comprehensive encyclopedias in computer science, plus some newcomers to the arena who concentrate on a particular aspect of the discipline.

Bainbridge, W. S., ed. 2004. *Berkshire Encyclopedia of Human-Computer Interaction.* (2 vols.) Great Barrington, MA: Berkshire Group. This encyclopedia, edited by the deputy director of the National Science Foundation's Division of Information and Intelligent Systems, consists of 186 signed articles on human-computer interaction (HCI). It is the first major reference resource for a new and fast-changing field that draws upon many branches of social, behavioral, and information sciences as well as computer science, medicine, engineering, and design. The encyclopedia covers all aspects of HCI including applications, breakthroughs, challenges, interfaces, and methods. The articles are accompanied by a comprehensive bibliography, glossaries, and a "popular culture database." This would complement any of the standard encyclopedias listed in this section for its unique content.

Belzer, J., A. G. Holzman, and A. Kent, eds. 1975–2001. *Encyclopedia of Computer Science and Technology.* (43 vols.) New York: Marcel Dekker. Updated by supplements. This comprehensive reference provides access to articles on state-of-the-art computer technology. Approximately 700 articles written by 900 international authorities feature current developments and trends in computers, software, vendors, and applications as well as in-depth analysis of future directions. There are extensive bibliographies of leading figures in the field. Subscribers to the encyclopedia can stay up-to-date with book-length supplements that cover a variety of timely topics, keeping readers aware of the newest developments, the latest buzz-words, hardware and software changes, and individuals making noteworthy contributions. Volume 15 (Supplement 1) begins the Supplement series, then continues with Volume 17 (Supplement 2); Volume 16 is the index for the previous volumes. Libraries may be interested in selecting specific supplement titles of interest for their collection instead of purchasing the entire set of 43 available volumes, which carries a hefty price tag.

Bidgoli, H., ed. 2003. *The Internet Encyclopedia.* (3 vols.) Hoboken, NJ: John Wiley & Sons. This substantive three-volume set covers every aspect of fast-moving Internet technology for computer professionals, offering a broad perspective on the Internet as a business tool, an IT platform, and a medium for communications and commerce. Experts from institutions, such as Stanford University, Harvard University, and leading corporations, such as Microsoft and Sun Microsystems, describe leading-edge technology and recent developments in lengthy signed articles. Articles are supplemented with flow charts, programming samples, and technical diagrams as appropriate as well as bibliographies. The editor is a tenured professor at California State University Bakersfield, School of Business, Management Information System Department.

The Computer Desktop Encyclopedia: The Indispensable Reference on Computers. 2010. Licensed by Farlex, Inc. The Computer Language Company (Point Pleasant, PA), publisher of the encyclopedia, is the premier source for computer concepts and terminology. Updated quarterly, this powerful, authoritative reference about the computer industry has been heralded by *PC Magazine, Government Technology Magazine,* and others as the best online reference source of its type. The *Computer Desktop Encyclopedia* offers more than 18,000 definitions covering all aspects of computers and digital technology. In addition to the library of definitions, the resource also includes explanations of fundamental technology concepts, informative historical content, biographies, company backgrounders, and more. Phonetic entries allow users to source definitions for "scuzzy" (SCSI), "gooey" (GUI), and "voyp" (VoIP), while over 2,500 illustrations, photos, and charts provide additional insight and are now available free through this new publisher at http://www.thefreedictionary.com/

Marciniak, J. J., ed. 2001. *Encyclopedia of Software Engineering.* (2 vols.) New York: John Wiley & Sons. Considered the most comprehensive reference in software engineering by *Computing Reviews,* this is an important encyclopedia for practitioners who design, write, or test computer programs covering all the issues and principles of software design and engineering. This edition, although dated, is still useful, containing 500+ entries in 35 taxonomic areas as well as biographies for over 100 important contributors to the field. Among the issues discussed are the Software Engineering Body of Knowledge Project, software engineering ethics, licensing and certification of software engineering personnel, and education and training in software engineering. The terminology used complies with the standard set by IEEE. Also available as a Wiley

Interscience OnlineBook at http://www.interscience.wiley.com/onlinebooks (the online edition is regularly updated).

Ralston, A., E. D. Reilly, and D. Hemmendinger, eds. 2003. *Encyclopedia of Computer Science,* 4th ed. Chichester, U.K.; Hoboken, NJ: John Wiley & Sons. Since 1976, this has been the definitive and comprehensive reference work on computers, computing, and computer science. With over 2,000 pages, the newly revised edition contains over 100 new articles and more than 600 completely updated articles by internationally known experts. The articles are signed and the author affiliations included at the front of the book. Alphabetically arranged and classified into broad subject areas, the entries cover hardware, computer systems, information and data, software, the mathematics of computing, theory of computation, methodologies, and applications and the computing milieu. The encyclopedia skillfully combines historical perspective with practical reference information. This work is a must-have for all academic and public libraries and is an essential resource for computer professionals, engineers, mathematicians, students, scientists, and librarians. Short bibliographies, cross-references to other articles in the volume, plus a name and subject index further increase the usefulness of this publication. The nine appendices at the back of the book are worth a special mention. They cover abbreviation and acronyms, notations and units; computer journals and magazines, Ph.D. granting departments of computer science and engineering, presidents of major computing societies; key high-level languages, glossary of major terms in five languages, articles deleted from previous editions, and a timeline of significant computing milestones.

Reilly, E. D., ed. 2004. *Concise Encyclopedia of Computer Science.* Chichester, U.K.; Hoboken, NJ: John Wiley& Sons. This is the perfect encyclopedia for the nonspecialist and is based on the fourth edition of the *Encyclopedia of Computer Science* (Ralston et al. above), with shorter versions of 60 percent of the entries in the fourth edition.

Rojas, R., ed. 2001. *Encyclopedia of Computer and Computer History.* (2 vols.) Chicago: Fitzroy Dearborn. This encyclopedia aims to cover the complete subject of computers and their history from personal computing to main-frames to robotics and artificial intelligence as well the theoretical foundations of computer science. The 600+ entries cover facts, definitions, biographies, histories, and explanations of diverse topics. Contributors are scholars in computer science and computer history from around the world. This source will only serve as a starting point for more sophisticated users.

Sheldon, T. 2001. *McGraw-Hill Encyclopedia of Networking & Telecommunications,* 3rd ed. Berkeley, CA: Osborne. Accompanied by a CD-ROM. An encyclopedic reference of information with 1,400 entries and over 3,000 links on computer networking and telecommunications, covering subjects ranging from Bluetooth to mobile computing in print. Although dated now it is still useful and a free Web site, http://www.linktionary.com/about.html (accessed July 30, 2010) allows you to download the third edition free. The Web site is also the "home" for book addendums and recent topic updates.

Wah, B. 2009. *Wiley Encyclopedia of Computer Science and Engineering.* (5 vols.) Hoboken, NJ: John Wiley & Sons. Each of the 450 articles is written by experts and peer-reviewed. In addition to important topics of interest to computer scientists and engineers, there are also sections on standards, electronic commerce, financial engineering, and computer education. References and Web site of related interest accompany every article.

HANDBOOKS AND MANUALS

Engineering handbooks are ready-reference tools—compact, comprehensive sources of data and information frequently needed by engineers. They can help you find quick, factual information to support your ideas or reacquaint you with theories, formulas, and other data that may be scattered throughout the literature.

Chen, W.-K., ed. 2009. *The Circuits and Filters Handbook,* 3rd ed. Boca Raton, FL: CRC Press. Written for practicing electrical engineers, the third edition has been thoroughly updated to provide the most current, comprehensive information available in both the classical and emerging fields of circuits and filters, both analog and digital. This edition contains 29 new chapters, with significant additions in the areas of computer-aided design, circuit simulation, VLSI circuits, design automation, and active and digital filters. The volume begins with an overview of mathematics as it relates to the subject. The articles are written by experts in the field with a bibliography at the end of each chapter. This handbook is part of the *Electrical Engineering Handbook Series.* Contents of CRC Press engineering handbooks may be searched online at http://www.engnetbase.com/ (accessed March 30, 2011); access to CRC full-text resources is available with an annual subscription.

Freeman, R. L. 2001. *Reference Manual for Telecommunications Engineering,* 3rd ed. (2 vols.) New York: John Wiley & Sons. For over 15 years, the *Reference Manual for Telecommunications Engineering* has been regarded as an essential design tool for engineers and technicians who deal with communications technology. Now expanded to a two-volume set, this completely revised and updated third edition features over 3,500 pages of the latest information (organized into 41 subject areas) on designing, building, purchasing, using, and maintaining telecommunications systems. Gathered from industry, government, and academic sources, the manual contains all the technical material a telecom professional might need on a daily basis. It also includes a wealth of tables, figures, nomograms, formulas, statistics, standards, regulations, and explanatory text.

Lyons, R. G. 2010. *Understanding Digital Signal Processing,* 3rd ed. Upper Saddle River, NJ: Prentice Hall PIR. In this updated and expanded edition, the author demonstrates how engineers and other technical professionals can master and apply DSP techniques. This edition adds extensive new coverage of quadrature signals for digital communications, recent improvements in digital filtering, and contains more than twice as many "DSP Tips and Tricks."

Oklobdzija, V. G., ed. 2008. *The Computer Engineering Handbook,* 2nd ed. Boca Raton, FL: CRC Press. "After nearly six years as the field's leading reference, the second edition of this award-winning handbook reemerges with completely updated content and a brand new format. *The Computer Engineering Handbook*, second edition, is now offered as a set of two carefully focused books that together encompass all aspects of the field. In addition to complete updates throughout the book to reflect the latest issues in low-power design, embedded processors, and new standards, this edition includes a new section on computer memory and storage as well as several new chapters on such topics as semiconductor memory circuits, stream and wireless processors, and nonvolatile memory technologies and applications" (from http://www.crcpress.com/). Contents of CRC Press engineering handbooks may be searched online at http://www.engnetbase.com/; access to CRC full-text resources is available with an annual subscription.

Smith, S. W. 2002. *Digital Signal Processing: A Practical Guide for Engineers and Scientists,* 3rd ed. Amsterdam/Boston, MA: Newnes. Accompanied by CD-ROM. This guide explains DSP design, algorithms, and techniques as well as the operation and usage of DSP chips. Even though this reference work is aimed at engineers and scientists, it avoids abstract, theoretical, and mathematical explanations.

Tucker, A. B., ed. 2004. *Computer Science Handbook,* 2nd ed. Boca Raton, FL: Chapman & Hall/CRC Press. Aimed at computer scientists, software engineers, and IT professionals, this edition has broadened its scope, emphasizing a more practical and applied approach to computing. The 70+ new or revised chapters are written by over 150 recognized experts. It provides coverage across all 11 subject areas of the discipline as defined in ACM/IEEE 2001 Computing Curricula 2001.

Van Leeuwen, J., ed. 1994. *Handbook of Theoretical Computer Science.* (2 vols.) Cambridge, MA: MIT Press. This out-of-print (used copies may be available) handbook provides professionals and students with a comprehensive overview of the main results and developments in this rapidly evolving field. Volume A covers models of computation, complexity theory, data structures, and efficient computation in many recognized subdisciplines of theoretical computer science. Volume B takes up the theory of automata and rewriting systems, the foundations of modern programming languages, and logics for program specification and verification, and presents several studies on the theoretic modeling of advanced information processing. The two volumes comprise 37 chapters, with extensive chapter references and individual tables of contents for each chapter. The editor points out in the preface of the first edition (1990) that "whereas the volumes can be used independently, there are many interesting connections that show that the two areas really highly intertwined" (p. ii). Even though this book was published more than a decade ago, it is still useful given the nature of the subject area: the theoretical and mathematical underpinnings of computer science.

TECHNICAL REPORTS

Before the Internet, organizations, mostly academic computer science departments and companies, exchanged hard copies and later microfiche of computer science technical reports.

Technical reports, usually numbered report series, published by university departments, companies, and government agencies, report on current research in a timely fashion and are, as the name suggests, highly technical in nature, not peer-reviewed, and describe unsuccessful as well as successful research. They were intended to be for the rapid dissemination of technical information before it was presented at conferences or published in peer-reviewed journals. The technical report was often the first place that an innovation or development was reported and date-stamped in terms of intellectual property (especially important in the case of future patents).

In the early days of the Internet and the advent of FTP servers, many of the producers of technical reports installed FTP servers for free and easy access of their technical reports for the computer science community. During that time, directories of FTP servers were compiled, but a quick survey of these listings shows that they are, not surprisingly, out-of-date. Now with the ubiquitous nature of the Internet, a quick review of an organization's Web site will reveal if the technical report is available for download or leads to a contact address to help you track down a copy. Some institutions, such as the California Institute of Technology, have digitized their entire computer science technical reports collection and made them available globally. These reports may be searched via the CaltechCSTR Web site, http://caltechcstr.library.caltech.edu/ (accessed August 10, 2010) or Google.

The status of the "technical report" in computer science is uncertain if not on the verge of extinction. Many, if not most, researchers include information on current research projects on their Web sites, but not in the form of a technical report. However, from time to time, there is a need to track down computer science technical reports, retrospective as well as current reports. Below are some starting points that may be used to find them.

Carnegie Mellon University Engineering and Science Library houses one of the largest physical collections of computer science technical reports in the country. Find a technical report in the library collection by searching the library catalog CAMEO at http://search.library.cmu.edu/client/default (accessed August 10, 2010). For example: search "Technical Reports and Caltech."

NCSTRL http://www.ncstrl.org. *The Networked Computer Science Technical Reference Library* was established around 1995 with funding from DARPA and it was intended to be a model for other digital libraries. NCSTRL was formed from the merger of two prior projects: CS-TR (Computer Science Technical Reports) http://www.cnri.reston.va.us/cstr.html and WATERS (Wide Area Technical Report Service) http://doi.acm.org/10.1145/205323.205330. For more information see, Anderson, Greg, Rebecca Lasher, and Vicky Reich, "The Computer Science Technical Report (CS-TR) Project: A Pioneering

Digital Library Project Viewed from a Library Perspective," *The Public-Access Computer Systems Review* 7 no. 2 (1996). (All accessed September 7, 2010).

NCSTRL (pronounced "Ancestral") is a collection of computer science technical reports loosely organized as a federation of cooperating servers. *NCSTRL* was implemented and maintained by the digital library group at Cornell University from 1994–2001 with over 100 international participants and over 20,000 digital objects before it was transferred to Old Dominion University. The individual document repositories in *NCSTRL* are located at geographically distributed sites. Two methods were used for maintaining the metadata and providing indexing services: a geographically distributed set of index servers and a centralized index server. The former configuration is maintained as a research vehicle. The latter is the production system designed to provide stability to end users. Participating institutions (known as "publishing authorities") could be involved as "Standard" or "Lite" sites. A Standard site runs three services: user interface (UI), indexer, and repository. A Lite site maintains its technical reports at the home institution, but has its metadata held at a special site (the Central Server) that provides indexing services. The Central Server looks like a Standard site to the rest of the system. The architecture of *NCSTRL* was based largely on the Dienst software. With the launch and development of the Open Archives Initiative (OAI) in 1999, Old Dominion University developed an OAI-PMH version of *NCSTRL*. However, *NCSTRL* is no longer supported, but remains a good example of the concept to create a federated, geographically distributed network for computer science technical reports. *D-Lib Magazine* (http://www.dlib.org/), a free online journal, provides several more articles about the development of *NCSTRL*. I would like to thank Dr. Michael Nelson of Old Dominion University for bringing me up to date on *NCSTRL*. To paraphrase Dr. Nelson's correspondence with me (e-mail message, August 2010), there are several reasons why *NCSTRL* is no longer viable. It was overcome by better systems like the *ACM Digital Library*, Google Scholar, etc., the slow death of the computer science technical report as a concept, the use of arxiv.org for technical reports as well as placing e-prints directly on the Web, and the fact that some departments have institutional repositories, but those are often run at larger granularity than computer science departments.

Scirus (http://scirus.com) is an Elsevier product that provides access to scientists' homepages, institutional repositories, and preprint services in additional to journal articles.

TRAIL: The Technical Report Archive & Image Library (http://www.technicalreports.org) (accessed March 29, 2011) is an initiative led by the University of Arizona in collaboration with CRL and other interested agencies to identify, digitize, archive, and provide access to federal technical reports issued prior to 1975.

University of Maryland's Virtual Technical Reports Collection, http://www.lib.umd.edu/ENGIN/TechReports/Virtual-TechReports.html (accessed August 10, 2010) is a metasite that provides links to full-text or searchable extended abstracts of technical reports, preprints, reprints, dissertations, theses, and research reports in all disciplines. It was last revised July 2009.

INSTITUTIONAL REPOSITORIES

The definition of institutional repositories (IRs) from the *New World Encyclopedia* is a good starting point:

An institutional repository is an online locus for collecting, preserving, and disseminating, in digital form, the intellectual output of an institution, particularly a research institution. For a university, this would include materials such as research journal articles, peer reviews, and digital versions of theses and dissertations, but it also might include other digital assets generated by normal academic life, such as administrative documents, course notes, or learning objects. An institutional repository is published online and is basically open to the public. While most academic journal articles are available only to subscribers and not retrievable by general search engines, such as Google, research papers in an institutional repository are fully accessible by the public free of charge and are accessible by general search engines. Popular software, such as DSpace, EPrints, and Bepress, are also open sources. As of January 2009, there are about 1,239 institutional repositories in the world.

It is beyond the scope of this chapter to chart in detail the history, development, and mechanics of creating IRs or the challenges. However, the following Web sites will bring you up to speed on this important topic.

Branin, J. 2003. Institutional Repositories (draft paper). *Encyclopedia of Library and Information Science.* https://kb.osu.edu/dspace/bitstream/handle/1811/441/inst_repos.pdf;jsessionid=2EDC27BACA72193A94D3B15766DDAFEC?sequence=1 (accessed March 22, 2011).

Bepress. *Research on Institutional Repositories.* Sign up for a mailing list from Bepress at http://works.bepress.com/ir_research/ (accessed March 29, 2011).

Crow, R., preparer. 2002. *The Case for IRs: A SPARC Position Paper.* Washington, D.C.: SPARC http://scholarship.utm.edu/20/1/SPARC_102.pdf (accessed March 29, 2011).

OARiNZ. *Introduction to the OariNZ Project.* http://www.oarinz.ac.nz/ (accessed March 29, 2011). An authoritative history from OARiNZ—an online clearing house of information relevant to people involved in New Zealand repository projects.

Wolpert, A. 2002. Future of electronic data, *Nature* 420, 17–18 (November) http://www.nature.com/nature/journal/v420/n6911/full/420017a.html (accessed March 29, 2011).

It is important for the information seeker in computer engineering to be aware of the growing deployment of IRs worldwide for the intellectual output of an organization. Technical reports are a perfect genre to be included in an IR for all the reasons mentioned at the beginning of this chapter. In addition to the tools mentioned above, the following links also will be helpful in identifying IRs.

Cybermetics Lab. Ranking Web of World Repositories http://repositories.webometrics.info/ (accessed October 21, 2010).

DOAR http://www.opendoar.org (accessed October 21, 2010). OpenDOAR is an authoritative directory of academic open access repositories. Each OpenDOAR repository has been visited by project staff to check the information that is recorded here. This in-depth approach does not rely on automated analysis and gives a quality-controlled list of repositories. Search by subject area (computers and IT).

Registry of Open Access Repositories http://roar.eprints.org/ (accessed October 21, 2010). Hosted by the University of Southampton, U.K.

U.K. Institution Repository Search http://irs.mimas.ac.uk/demonstrator (accessed October 21, 2010).

ONLINE BIBLIOGRAPHIES, PREPRINTS, AND REPOSITORIES

In computer science, we find a proliferation of free online resources that augment the traditional means of publishing papers in journals and conference proceedings for fast dissemination and exchange of information. In many cases, these resources are experimental and provide a test bed for emulation not only in computer science, but also in other disciplines.

This section provides a list of the major resources in this category mostly limited to those covering the discipline as a whole. Technical reports have been covered in the previous section. For an excellent listing of current resources in subspecialties (such as cryptology, computer vision), please see the computer science metasite maintained by the State University of New York at Albany Library http://libguides.library.albany.edu/csci.

Please note that it is not advisable to use any of these resources as the sole source for bibliographic research, but in conjunction with the other traditional databases (listed in the section on databases). All of these resources are freely available.

CiteSeer http://citeseer.ist.psu.edu/ (accessed August 15, 2010). CiteSeer (formerly known as *Research Index*) was developed by the NEC Research Institute and is now hosted by The Pennsylvania State University. *Citeseer* is a "scientific literature digital library and search engine that focuses primarily on the literature in computer and information science. *CiteSeer* aims to improve the dissemination of scientific literature and to provide improvements in functionality, usability, availability, cost, comprehensiveness, efficiency, and timeliness in the access of scientific and scholarly knowledge. Rather than creating just another digital library, *CiteSeer* attempts to provide resources, such as algorithms, data, metadata, services, techniques, and software, that can be used to promote other digital libraries. *CiteSeer* has developed new methods and algorithms to index PostScript and PDF research articles on the Web (from Web page). The archive provides access to journal articles, conference papers, and technical reports in computer science. In addition to providing a list of documents by author or subject, the search engine also generates a form of citation indexing similar to the *Science Citation Index* (produced by the Institute for Scientific Information) called in this case "autonomous citation indexing."

CoGPrints http://cogprints.org/ (accessed August 15, 2010). Using Eprints3 software and in compliance with the Open Archives Initiative (OAI) http://www.openarchives.org/ (accessed August 15, 2010), *CoGPrints* allows self-archiving of papers in psychology, neuroscience, linguistics, philosophy, biology, and computer science (e.g., artificial intelligence, robotics, vision, learning, speech, neural networks). Anyone can deposit papers for consideration, but registration is required.

Collection of Computer Science Bibliographies, The http://liinwww.ira.uka.de/bibliography/index.html (accessed August 15, 2010). This resource, updated monthly, is a collection of references from almost 1,500 bibliographies covering most of computer science and related areas in mathematics. The database provides access to journal articles, conference papers, and technical reports. Searching can be limited by publication type and date.

Computing Research Repository (CoRR) http://arxiv.org/corr/home (accessed August 15, 2010). The goal of *CoRR*, supported by Cornell University Library, is to be the single repository for computer science preprints. It is sponsored by ACM, the arXiv archive, *NCSTRL*, and the American Association for Artificial Intelligence (AAAI). Established in 1998, *CoRR* allows researchers to search the repository, browse by year or subject class as well as download papers.

Digital Bibliography and Library Project (DBLP) http://dblp.uni-trier.de/ (accessed August 15, 2010). This was originally a bibliography for database systems and logic programming, but later widened its coverage to include more of the discipline. Hosted by several servers around the world, it lists over 500,000 references from major computer science journals and conference papers.

E-Print Network: Research Communications for Scientists and Engineers; Computer Technologies and Information Sciences http://www.osti.gov/eprints/pathways/computertech.shtml (accessed August 15, 2010). "The E-print Network is a set of powerful tools that facilitate access to and use of scientific and technical e-prints communicating the results of a wide range of research activities of interest to the Department of Energy" (from Web page). However, there is a section on Computer Technologies and Information Sciences. The site also includes links to scientific societies, and an alerting feature notifies users when new items are added in their areas of interest.

PATENTS AND STANDARDS

For information on how to locate patents, please refer to the patents section in Chapter 2 of this book. In the computing sector, software has always been a patent "problem child." The rather unique Web site listed below helps to sort out how to research software prior art.

> Software Patent Institute Database of Software Technologies http://www.spi.org/ (accessed July 30, 2010). "The Software Patent Institute (SPI) is a nonprofit corporation formed to provide prior art related to software technology with the intention of improving the patent process" (from Web site).

The following two sites provide good links to sources of computer standards (e.g., for hardware) or standards organizations. Please also refer to the standards section in Chapter 10 of this book on electrical engineering and electronics for additional information.

> CompInfo: The Computer Information Center: Computer Standards http://www.compinfo-center.com/itman/computer_standards.htm (accessed July 30, 2010).
> Yahoo! Directory. Computers and Internet, Standards http://dir.yahoo.com/Computers_and_Internet/Standards/?skw=standards+computers+internet/ (accessed July 30, 2010).

Some of the important standards organizations for computing-related standards are IEEE, International Standards Organization (ISO) http://www.iso.org/ (accessed July 30, 2010), and the Internet Engineering Task Force (IETF).

> IEEE Standards Online, Access to full-text standards is available to subscribers of *IEEE Xplore* http://ieeexplore.ieee.org/ (accessed July 30, 2010) or one may purchase individual computer engineering standards from Shop IEEE http://shop.ieee.org/ ieeestore/default. aspx (accessed July 30, 2010). For a small subscription fee, The IEEE Standards Association http://standards.ieee.org/ (accessed April 1, 2011) provides PDFs and VuSpec™ CD-ROMs to drafts of standards under development and historical standards.
> Internet Engineering Task Force (IETF) http://www.ietf.org// (accessed July 30, 2010). The IETF, organized around "working groups," is a large, open, international community of network designers, operators, vendors, and researchers concerned with the evolution of the Internet architecture and the smooth operation of the Internet including the development of new Internet standards.

CONFERENCES

This section provides guidance on locating information on forthcoming conferences. More information on locating papers published in conference proceedings is listed in the Databases section.

Face-to-face exchange of information at conferences is still as important as ever in computer science even though communicating via electronic media has become commonplace. This is evidenced by the large number of conferences sponsored annually by the major associations in computer engineering: ACM, IEEE Computer Society, IET (formerly IEE), and the British Computer Society, to name a few of the key players. A list of the major organizations in computer engineering is in the Associations, Societies, and Other Organizations section of this chapter. The Web sites usually compile listings not only of their own forthcoming sponsored conferences, but also others that would be of interest to their research community.

> ACM Events and Conferences http://www.acm.org/conferences / (accessed July 30, 2010).
> IEEE Computer Society Conferences http://www.computer.org/portal/web/conferences/home (accessed July 30, 2010).

Below are some additional sites specializing in conference announcements:

All Conferences.Com http://www.allconferences.com/ (accessed July 30, 2010). Online directory focusing on conferences, conventions, trade shows, exhibits, workshops, events, and business meetings with a category for "Computers and Internet." In addition to listing sites, the company also provides services to conference organizers.

Atlas Conferences Inc. http://atlas-conferences.com/index.html (accessed April 1, 2011). Founded in 2000, this Web site maintains a database of forthcoming academic conferences, meetings, and events. Browse conferences by subject, such as "Computer science," date, or country. Also provides conference support services.

JOURNALS

In line with other engineering disciplines, computer engineering has a plethora of journals. This section is a broad overview of the different types of journals and the publishers involved.

CORE JOURNALS

Encyclopedia of Computer Science. 2003. Dated as it is, provides an excellent staring point to explore this topic. It provides us with as comprehensive a list as is possible given that the number of titles is growing at a rapid pace. In Appendix III, the editors list the titles under these broad headings:

Society journals (namely from ACM, IEEE, British Computer Society, the Institution of Electrical Engineers-U.K., Society for Industrial and Applied Mathematics)

Industrial Journals

Journals of specific countries and regions

International journals

Journals and magazines in specialized areas (including related fields such as mathematics)

MAGAZINES

Major publishers in computing.

The Guide to Computing Literature http://portal.acm.org/portal.cfm (accessed April 1, 2011). Published by ACM (and part of the ACM Digital Library), but covering the whole field, it provides another good starting point. Go to the *Guide to the Computing Literature*, select "Journals" as type of literature, and you will be presented with a list of journals (1,115 as of July 2010). You can retrieve a list of journals in alphabetical order. "Browse" and "Basic" searching is free for all users (see their "Frequently Asked Questions"). For more information, you need a personal or institutional membership to the ACM Digital Library. Members can retrieve the full citation and link to the table of contents (TOCs). Keep in mind that ACM-published TOCs are kept up-to-date, but non-ACM titles may not be as up-to-date.

DIRECTORIES

To complement the resources mentioned above, there are also directories of computer journals on the Internet. Some of these may cover titles missed by the *ACM Guide* (such as trade journals).

Computer Science Journals http://www.informatik.uni-trier.de/~ley/db/journals/index.html (accessed August 2, 2010).This is a very useful listing from the DBLP (Digital Bibliography and Library Project), concentrating on scientific journals in computer science tracking ACM, IEEE, Open Access, and Science Direct (Elsevier) titles.

Directory of Open Access Journals, Computer Science http://www.doaj.org/ (accessed August 2, 2010). Open Access Journals are defined as journals that do not charge individuals or their institution for access. This directory aims to be the "one-stop shop" listing for these peer-reviewed scientific and scholarly journals. Go to the main page, click on "Technology and Engineering" and then "Computer Science."

Top 100 Magazines: Computer and Software WWW Magazines and Journals http://www. netvalley.com/top100mag.html (accessed August 2, 2010). This is a subjective view of the top IT-related journals from San Francisco-based Internet consulting and publishing company NetValley.

ACADEMIC/SCHOLARLY JOURNALS

Reviewing the resources above, it becomes obvious that scholarly/scientific journal publishing in computer engineering and science is dominated by a handful of leading associations, societies, and major publishers: ACM, IEEE Computer Society, Elsevier, Springer, and John Wiley & Sons. Belonging to one of these associations and monitoring their output on a regular basis will enable the researcher to keep up with major developments in the field.

Here are the links to the lists of journals for the above-mentioned publishers:

ACM Journals http://www.acm.org/publications/journals
IEEE Computer Society Publications http://www.computer.org/portal/web/publications
Elsevier http://www.sciencedirect.com/science/journals/computerscience (accessed August 2, 2010).
Springer http://www.springer.com /(accessed August 2010). This is the main page, then click on Subjects and then Computer Science. Springer also publishes the important series *Lecture Notes in Computer Science (LNCS)* http://www.springer.com/computer/lncs and many subseries under *LNCS*. These series, all available online and in print, have a well-deserved reputation in the computer science research community and provide an important venue for publication of new developments in the field.

A special subset to the scholarly journals in this subject area are journals from industrial organizations. They started off as "house organs," but because of their wide appeal are available outside the company. Three of the well-known titles in this category are:

Bell Labs Technical Journal http://www.wiley.com/WileyCDA/WileyTitle/productCd-BLTJ. html (accessed August 2, 2010). This journal has been through many name changes and is now published by John Wiley & Sons.
IBM Journal of Research and Development and *IBM Systems Journal* http://www.research.ibm. com/journal/ (accessed August 2, 2010). These journals are now available via IEEE Xplore.

IMPACT FACTORS

One way to determine the leading journals in the field is to look at those that are most frequently cited. The alphabetical list below is based on the Institute of Scientific Information's *Journal Citation Reports* (a *Web of Knowledge* database) under the category "Computer science, hardware and architecture." *Journal Citation Reports*, available through a subscription, is a unique tool that gives users the ability to compare journals using citation data from over 7,000 scholarly and technical journals. IEEE and ACM titles are removed from the list. In effect, this listing, plus the IEEE and ACM titles, forms the major scholarly journals in this category.

Advances in Computers (0065-2458)
Analog Integrated Circuits and Signal Processing (0925-1030)

Canadian Journal of Electrical and Computer Engineering Revue (0840-8688)
Computer Journal (0010-4620)
Computer Networks: The International Journal of Computer and Telecommunications Networking (1389-1286)
Computer Communications (0140-3664)
Computer Standards and Interfaces (0920-5489)
Computer Systems Science and Engineering (0267-6192)
Computers and Electrical Engineering (0045-7906)
Design Automation for Embedded Systems (0929-5585)
Displays (0141-9382)
IBM Journal of Research and Development (0018-8646)
Integration, The VLSI Journal (0167-9260)
International Journal of High Performance Computing Applications (1094-3420)
Journal of Circuits Systems and Computers (0218-1266)
Journal of Computer Science and Technology (1000-9000)
Journal of Computer and System Sciences (0022-0000)
Journal of High Speed Networks (0926-6801)
Journal of Information Storage and Processing Systems (1099-8047)
Journal of Network and Computer Applications (1084-8045)
Journal of Supercomputing (0920-8542)
Journal of Systems Architecture (1383-7621)
Microprocessors and Microsystems (0141-9331)
Mobile Networks and Applications (1383-469X)
Networks (0028-3045)
New Generation Computing (0288-3636)
Performance Evaluation (0166-5316)
VLDB Journal (1066-8888)
VLSI Design (1065-514X)

Magazines for the Practitioner

Trade magazines are published for and read by members of a particular profession or trade. They are usually aimed at design engineers, developers, technical, or IT managers. These range from the "how to" magazines based on a particular system, tool, or language to those for the broader electronics industry, such as *EE Times,* http://www.eetimes.com/ (accessed April 1, 2011), to those for the enterprise market. Trade journals aimed at IT professionals (see also Tools for the Practitioner section) usually carry a combination of technical, trade, and business information and are often sent free to qualified subscribers (usually individuals within a company who make major IT-related budget or purchase decisions). Some examples of these include:

CIO http://www.cio.com/ (accessed August 2, 2010).
C/C++ Users Journal, stopped publishing in 2006; past articles and source code archives will still be available through the *Dr Dobb's Journal* Web site.
Computerworld http://www.computerworld.com/ (accessed August 2. 2010).
Dr. Dobb's Journal was a journal of choice for many development managers, but it stopped publishing as a standalone monthly magazine in 2009 and became Dr. Dobb's Report in *Information Week*. However, via the Dr. Dobb's Web site, https://store.ddj.com/storefront. php (accessed April 1, 2011), you can purchase a DVD of all the past issues of *Dr. Dobb's Journal, C/C++ User Journal, Perl Journal, Dr. Dobb's Sourcebook,* and source code. The Dr. Dobb's Web site is a valuable resource for keeping up to date on the latest developments with articles under broad heading and offers RSS feeds.

InformationWeek http://www.informationweek.com/ (accessed October 21, 2010).

Linux Journal http://www.linuxjournal.com/ (accessed October 21, 2010).

Software Development Times http://www.sdtimes.com/ (accessed October 21, 2010).

The *Perl Journal*, stopped publishing in 2006 and the archive is searchable via the *Dr. Dobb's* Web site (see entry above) (accessed October 21, 2010).

CONSUMER MAGAZINES FOR THE PC END-USER MARKET

These magazines often begin life serving the "hobbyist" on a small scale, but become magazines with healthy circulation figures. In this category, we have titles such as *PC Magazine* http://www.pcmag.com/, *PC World* http://www.pcworld.com/, and *Smart Computing* http://www.Smartcomputing.com/, to name a few, each of which is accompanied by informative Web sites.

Another category is magazines with broad appeal to the general public. A good example of this type is the magazine *Wired* http://www.wired.com/ (accessed August 2, 2010).

DATABASES

DATABASES OF ASSOCIATIONS

ACM Digital Library http://portal.acm.org/dl.cfm (accessed August 2, 2010). Established in 1947, the Association of Computing Machinery (ACM) is the first and foremost computing association in the world. *The ACM Digital Library* is a one-stop, full-text database of the association's intellectual output providing access to its 40 journals, 9 magazines, over 32 SIG (Special Interest Group) newsletters, and all its conferences. ACM sponsors over 85 SIG-related conferences worldwide every year. Access to its journals, conferences, and newsletters is available in print and via its Portal, making it the single most important full-text resource in computer engineering and science. The Portal consists of two main databases: the *ACM Digital Library* (*DL*) and the *Guide to Computing Literature*. The former contains bibliographic information, abstracts, and full text of all the ACM publications, and the latter is a collection of bibliographic citations and abstracts published by ACM and other major publishers in computer science. *The Guide* provides access to more than 750,000 citations from 3,000+ publishers (including ACM) covering the entire literature of computer science in books, journal articles, conference proceedings, doctoral dissertations, master's theses, and technical reports. The ACM has two types of subscriptions: institutional or individual via ACM membership. Institutional and individual members receive access to the resources mentioned above. However, individual members have additional "personalized" services: table of content (TOC) e-mail alerts when a new issue of an ACM journal, magazine, newsletter, or proceedings has been posted in the DL and the ability to create "binders" where searches and queries may be saved and shared with colleagues. For those who do not have access to the ACM Portal via a library, becoming an ACM member provides a cost-effective way to access this vast storehouse of information. In August 2002, *Library Journal* called the ACM Portal "a bonanza of computing literature at a bargain price ... not just a valuable resource for academic use, but a powerful tool kit for IT professionals" and it continues to give value for money. Google has indexed the full-text of ACM journal articles along with scholarly research content from nine other participating publishers.

IEEE Computer Society Digital Library http://www.computer.org/publications/ dlib/ (accessed August 3, 2010). A subscription to the *IEEE Computer Society Digital Library* includes online access to 22 society periodicals and over 1,400 conference publications. This resource is available only to IEEE Computer Society members and library/institution customers. Plans for online, print, or combination subscriptions are available to libraries. OPAC (online public access catalog) links may be used for publication titles within the *Digital Library.* In addition, the IEEE Computer Society and the ACM have agreed to exchange bibliographic data and abstracts. One may now search across both digital library collections from the IEEE Computer Society and link directly to either publisher's content. Nonmembers will always have free access to abstracts and tables of contents in the *Digital Library* and may purchase individual documents.

COMPREHENSIVE BIBLIOGRAPHIC DATABASES

Engineering Village http://www.ei.org/engineering-village (accessed August 2, 2010) is produced by Engineering Information (EI), which is now owned by Elsevier. As their publicity states: "*Engineering Village* is the information discovery platform of choice for the engineering community. In a single interface, researchers get access to today's most important engineering content. *Engineering Village* is the first place users go to find answers to questions from the abstract to the precise, from the basic to the complex. Powerful search tools, plus an intuitive user interface, boosts research productivity. Hyperlinked records and full-text links give researchers meaningful results through a more comprehensive view of the information." Their cornerstone product, *Compendex* http://www.ei.org/compendex (accessed August 2, 2010) is one of the most comprehensive engineering literature databases available to engineers, with 11.3 million records across 190 engineering disciplines, and available online from 1969– (for more details about *Compendex,* please see Chapter 2 in this book). Now researchers can reach even farther back with the *Engineering Index Backfile* from 1884–1969. EI now also provides access to the *Inspec Archive* so that users can search both major databases with a common interface. In addition, *Engineering Village 2* offers the ability to access other free and add-on subscription content, such as *Referex,* patents from the U.S. Patent and Trademark Office, NTIS (National Technical Information Service), etc.

Inspec http://www.theiet.org/publishing/inspec (accessed August 3, 2010). *Inspec,* one of the most respected scientific databases in the world, was formed in 1967, based on the Science Abstracts service that has been provided by the U.K. Institution of Electrical Engineers (IEE), now the Institution of Engineering and Technology (IET) since 1898. *Inspec* is the online version of the three print indexes: *Physics Abstracts, Electrical and Electronics Abstracts*, and *Computer and Control Abstracts,* which together form *Science Abstracts.* It provides comprehensive access to the world's leading scientific and technical literature in computers and computing, information technology, electrical engineering, electronics, communications, control engineering, and physics. The database is international in scope, with indexes (full citations with lengthy abstracts), journal articles, conference proceedings, reports, dissertations, and books. In addition to its depth of coverage, its strength lies in consistent indexing by subject specialists. The scope of indexing has grown since its inception to include not only the usual Thesaurus Term and Classification Code indexes, but also the Uncontrolled Index Terms (Identifiers), Treatment Codes, Chemical Substance Indexing, Numerical Data Indexing, and Astronomical Object Indexing. This indexing helps guide the user to related key terms and subject areas, and to gain a better understanding of topics. *Inspec* is available directly from Inspec and from many vendors (e.g., ISI, Ovid, Dialog, EI, Ebsco etc.).

COMMERCIAL DATABASE PACKAGES

EBSCO Technical Package http://www.ebsco.com/home/whatsnew/ipca.asp (accessed August 2, 2010). EBSCO offers a collection of databases that provide a "comprehensive compilation of abstracts and indexing for the top journals in various fields of electronic information management and computer science. ... This collection is comprised of five databases including the following: *Inspec, IPCA (Internet and Personal Computing Abstracts—new 2004 addition), Information Science and Technology Abstracts™* (ISTA), *Computer Science Index™* (CSI), and *Computer Source* (according to an EBSCO sales representative, as well as a complementary full-text component" (from Web home page). EBSCO acquired the IPCA and ISTA files in July 2003 from Information Today, Inc.

INDIVIDUAL DATABASES SPECIALIZING IN COMPUTER ENGINEERING AND SCIENCE

UBM Computer Full-text (Dialog File 647) http://library.dialog.com/bluesheets/html/bl0647.html (accessed August 2, 2010). "*UBM Computer Fulltext* provides timely, relevant information about the computer, communications, and electronic industries. The file includes full text, cover-to-cover coverage of top-rated newspapers and magazines published by UBM LLC" (this was previously *CMP Computer Full-Text*); (from Dialog Bluesheets).

Computer Abstracts International Database http://www.emeraldinsight.com/products/abstracts/caid/index.htm (accessed August 2, 2010). Published by Emerald Abstracts, this database "provides online access to over 200,000 abstracts, dating back to 1987, from the 200 foremost journals in computer science. From its first publication in 1957, *CAID* has remained a pioneer in its field, keeping pace with an industry that moves at breathtaking speed. CAID's accredited journal coverage list is compiled from the library holdings of computing centers of excellence worldwide, such as Stanford University and MIT" (from Web page). For more than 40 years, *Computer Abstracts International Database* has been an important tool in the field and online access back to 1987. Major topics covered include: artificial intelligence, communications and networks, computer theory, data, database and information systems applications, hardware, human–computer interaction, mathematics of computing, programming, and systems organization.

Computer and Information Systems Abstracts http://www.csa.com/factsheets/computer-set-c.php (accessed August 2, 2010). This database provides a comprehensive monthly update on the latest theoretical research and practical applications around the world from over 3,000 periodicals, conference proceedings, technical reports, trade journal/newsletter items, patents, books, and press releases. It is one of the oldest databases in the field and covers artificial intelligence, computer applications, computer programming, computer systems organization, computing milieu, hardware, information systems, mathematics of computing, and software engineering.

Computer Science Index http://www.ebsco.com/ (accessed August 2, 2010). Click on Academic Databases and then choose "Computer Science" as subject. "*Computer Science Index* (formerly *Computer Literature Index*) offers abstracting and indexing of academic journals, professional publications, and other reference sources at the highest scholarly and technical levels of computer science. The collection covers more than 6,500 periodicals and books, with coverage going back to the mid-1960s. *Computer Science Index* focuses on subjects such as artificial intelligence, expert systems, system design, data structures, computer theory, computer systems and architecture, software engineering, human–computer interaction, new technologies, social and professional context, and much more. Enhancements to the original database include hundreds of new titles and searchable cited references for key academic journals. This database also includes editor-selected articles from magazine and journal titles in related areas of study" (from Web page).

Computer Source http://www.ebsco.com/ (accessed August 2, 2010). Click on Academic Databases and then choose "Computer Science" as subject. *Computer Source* contains nearly 300 full-text journals and magazines covering topics such as computer science, programming, artificial intelligence, cybernetics, information systems, robotics, and software, providing users with a balance of full-text technical journals and full-text consumer computer titles.

Gale Group Computer Database (Dialog File 275) http://library.dialog.com/bluesheets/html/ bl0275.html (accessed August 2, 2010). This database "provides comprehensive information about the computer, electronics, and telecommunications industries. Coverage includes detailed information about the evaluation, purchase, use, and support of computer and other electronic products. *Gale Group Computer Database* is designed to answer the questions of business and computer professionals about hardware, software, networks, peripherals, and services. Lengthy abstracts are available for most records from 1983 to present. Complete text is fully searchable for many records from 1988 to present" (from Dialog Bluesheets).

Internet and Personal Computing Abstracts http://www.ebsco.com/ (accessed August 2, 2010). Click on Academic Databases and then choose "Computer Science" as subject. Formerly *Microcomputer Abstracts*, this database "provides abstracts and indexing for literature related to personal computing products and developments in business, the Internet, the home, and all other applied areas. This resource contains content coverage that extends back to the 1980s. Over 400 of the most important trade publications, mainstream computer magazines, and professional journals are covered, including those that focus on specific topics, such as Macintosh and Windows platforms, programming, Web development and more. Special emphasis is also given to hardware and software reviews. The product includes content from such titles as Byte.com, *PC World. Macworld*, and *Linux Journal*. The product also includes editor-selected articles from hundreds of popular magazine titles" (from EBSCO's home page for the journal). *Note*: EBSCO purchased the back file from Information Today, Inc. in July 2003.

TecTrends (Dialog File 256) http://library.dialog.com/bluesheets/html/bl0256.html (accessed August 3, 2010). Use this file to find information about products and companies in the Information Technology industry. (Formerly *TechInfoSource* and *Softbase: Reviews, Companies, and Products*).

DATABASES IN RELATED FIELDS

lSI Web of Knowledge http://www.isiwebofknowledge.com (accessed May 18, 2005). See Chapter 2 in this book.

MathSciNet http://www.ams.org/mathscinet/ (accessed August 2, 2010). *MathSciNet* is the most comprehensive database for mathematics, produced by the American Mathematical Society (AMS), and covering the world's mathematical literature since 1940. It is analogous to the print publication *Mathematical Reviews* and also includes recent issues of *Current Mathematical Publications*. The database indexes journal articles, conference proceedings, and books. The database is organized by the Mathematics Subject Classification (MSC). This database is important to computer engineers who want to research fundamental theories in mathematics related to their discipline.

Scopus http://www.info.scopus.com/ (accessed August 2, 2010). Competing directly with *lSI's Web of Science*, the Elsevier product *Scopus* is an abstracts database comprising 27 million records with articles from 14,000 peer-reviewed titles (including a number of open access journals) from 4,000+ publishers around the globe. Through the *Scopus* interface, users also may search *Scirus*. Includes computer science-related content from publishers, such as ACM, IEEE Computer Society, CRC Press, MIT Press, Kluwer, and Springer.

CURRENT AWARENESS

Given the mass of research information published in engineering, especially the fast-paced discipline of computer engineering and science, keeping up to date is a major challenge. However, the good news is that now more than ever there are automated ways to keep abreast of the literature. For instance, e-mail alerts, although a somewhat dated technology, are still a simple way to keep up with articles added to a database on a topic of interest. Most databases allow you to set up single or multiple search strategies on your research topic, and every week (or whenever the database is updated) you will receive relevant hits via e-mail. In addition almost all databases now offer RSS feeds also.

To keep up with new books in the discipline, sign up with the various major publishers mentioned in the section below and you will receive timely alerts (see also Chapter 2 in this book). The serious researcher should be on the lookout for these time saving methods for keeping up to date. Below are examples from some of the major databases (URLs are available above).

> *ACM Digital Library.* At present, the TOC (Table of Contents Service) is only available to individual members (not institutional subscribers). The service enables a member to receive tables of contents of issues and proceedings.
>
> Cambridge Scientific Abstracts databases: *Computer Abstracts International*, *Computer Information Systems Abstracts*, and *Computer Technology*. E-mail notification service is available and when a saved search is rerun periodically.
>
> Dialog databases offer e-mail alerts daily, weekly, biweekly, monthly, or at the exact day and time you specify: http://www.dialog.com/products/alerts/ (accessed August 2, 2010).
>
> *Engineering Village.* With an institutional subscription and registration, you can set up weekly alerts and RSS feeds in *Compendex*.
>
> *IEEE Xplore.* IEEE offers Table of Content Alerts, saved search alerts, and RSS Feeds from *IEEE Xplore* and allows nonsubscribers to set up free IEEE accounts. Access to abstracts and full text of the articles depends on whether you have personal or institutional membership. *Inspec.* As has been noted in the Databases section, *Inspec* is available from many vendors. Depending on the vendor, current awareness alerts are provided. For instance, Ovid has "Autoalerts" and the ISI (*Web of Knowledge*) platform provides e-mail alerts with registration. To use the alerting feature, one needs to have access to an institutional subscription, usually through one's library.
>
> *Web of Science.* Although *Web of Science* is not a database specializing in engineering, let alone computer engineering, it is useful for tracking interdisciplinary topics. *Web of Science*, updated weekly, has three types of alerting services: a table of contents alert, citation alert (helps you track references to particular citations), and an alert by topic. One needs to have institutional access and registration is a prerequisite.

PUBLISHERS

Listed below are the major book publishers in computer engineering and science. Each has an e-mail alert service, making it easy for the librarian and practitioner alike to keep up with the latest publications in this subject area.

> Cambridge University Press Computer Science http://us.cambridge.org/computerscience/ (accessed August 2, 2010). Receive regular updates of special offers and new titles.
>
> CRC Press http://www.crcpress.com/ (accessed August 2, 2010). Click on computer science and engineering. Sign up for e-mail alerts.
>
> CSPress/JohnWiley http://www.wiley.com/WileyCDA/Section/id-301491.html (accessed August 2, 2010). IEEE Computer Society and John Wiley & Sons publish co-branded imprint of books. Sign up for John Wiley's e-mail alert. Also monthly e-mail bulletin of IEEE Computer Society's latest product offerings (http://www.computer.org/portal/web/store).
>
> Elsevier http://www.elsevierdirect.com/disciplinelanding.jsp?lid=100004 (accessed August 2, 2010). Browse new computer science titles and sign up for alerts. See also imprints Digital Press, Morgan Kaufmann, and Newnes.
>
> IEEE http://www.ieee.org/index.html (accessed August 2, 2010). Click on "Publications and Standards" or the "shop." To keep up to date, sign up for Whats New@IEEE http://whatsnew.ieee.org.

MIT Press: Computer Science and Intelligent Systems http://mitpress.mit.edu/catalog/browse/ default.asp?cid=5 (accessed August 2, 2010). Sign up for e-mail alerts and RSS feeds.

Morgan & Claypool Publishers http://www.morganclaypool.com/ (accessed August 2, 2010). Morgan-Claypool's main product is the "Synthesis Digital Library of Engineering and Computer Science" launched in October 2005. The basis of the product are 50- to 100-page Lectures; a self-contained electronic book that synthesizes an important research or development topic, authored by an expert in the field and offering more depth than a research article. According to the publisher, they are "more dynamic and convenient than traditional print or digital handbooks, contributed volumes, and monographs." The library and its lectures are organized by series. New series and lectures are added continuously and existing lectures will be revised as needed. Libraries purchase series and are given perpetual institutional access.

Morgan Kaufmann: Computing Books http://www.elsevierdirect.com/imprint.jsp?iid=100007 (accessed August 2, 2010). Morgan Kaufmann is part of Elsevier. From this page, sign up for a targeted e-mail list. Sign up for e-mail alerts, follow on Twitter and Facebook.

Newnes: Electronics and Computer Engineering http://www.elsevierdirect.com/imprint. jsp?iid=73 Newnes (accessed August 2, 2010). It is part of Elsevier.

Oxford Science Publications: Mathematics, Statistics, and Computer Science http:// ukcatalogue.oup.com/category/academic/computers.do (accessed August 2, 2010). Sign up for e-mail alerts.

Prentice-Hall Professional Technical Reference http://www.informit.com/imprint/imprint_ series.aspx?st=61089/ (accessed August 2, 2010). "Stay up to date on our new publications and special promotions by subscribing to our newsletters. You can choose to hear about all of our computer titles, or choose specific topics based on your areas of interest including: Engineering, Java, Security, Operating Systems, Business, and more" (from Web page).

Springer: Computer Science http://www.springer.com/computer?SGWID=0-146-12-70922-0 (accessed August 2, 2010). Sign up for Springer Alerts; they also provide special services for librarians. As a result of the merger of Kluwer Academic Publishers (KAP) and Springer-Verlag, Kluwer is now operating jointly under the Springer brand.

Wiley Computing http://www.wiley.com/WileyCDA/Section/id-350310.html (accessed August 2, 2010). Browse all forthcoming titles for computing or select a subject subcategory. Sign up for e-mail alerts and follow on Twitter, Facebook, and LinkedIn.

OTHER

National Technical Information Service (NTIS) http://www.ntis.gov/ (accessed August 2, 2010). Subscribe to RSS feeds on "Computers." An annual subscription to the printed version is available and averages about 80 summaries per issue. Also subscribe to the *NTIS Technical Reports Newsletter.*

Science.gov 5.0 Alerts http://www.science.gov/ helpalerts.html (accessed August 2, 2010). Science.gov, the Web portal for federal science information, searches over 42 databases and over 2,000 selected Web sites from 14 federal agencies, offering 200 million pages of authoritative U.S. government science information including research and development results, and offers an ALERT service that provides notifications of new Science.gov information in a specific area of interest. Relevant topics are computer hardware, computer security, computer networking, and computer software.

ASSOCIATIONS, ORGANIZATIONS, AND SOCIETIES

Professional associations and scholarly societies provide the foundation for structured peer-to-peer exchanges. Listed below are the most important of these organizations. A more exhaustive list of computer science scholarly societies around the world is compiled by the Scholarly Societies Project http://www.scholarly-societies.org/compsci_soc.html (accessed August 4, 2010). Research institutes related to computer science have been compiled by the DMOZ Open Directory Project http://www.dmoz.org/Computers/Computer_Science/ (accessed August 2, 2010).

American Association for Artificial Intelligence (AAAI) http://www.aaai.org/ (accessed August 3, 2010). "Founded in 1979, the American Association for Artificial Intelligence (AAAI) is a nonprofit scientific society devoted to advancing the scientific understanding of the mechanisms underlying thought and intelligent behavior and their embodiment in machines. AAAI also aims to increase public understanding of artificial intelligence, improve the teaching and training of AI practitioners, and provide guidance for research planners and funders concerning the importance and potential of current AI developments and future directions. Major AAAI activities include organizing and sponsoring conferences, symposia, and workshops, publishing a quarterly magazine for all members, publishing books, proceedings, and reports, and awarding grants, scholarships, and other honors" (from Web page). AAAI members have full access to most association publications in digital form.

Association for Computing Machinery (ACM) http://www.acm.org/ (accessed August 3. 2010). Founded in 1947, ACM is an international scientific and educational organization dedicated to advancing the arts, sciences, and applications of information technology. With 78,000 members around the world, ACM is a leading resource for computing professionals and students working in the various fields of information technology, and for interpreting the impact of information technology on society. ACM's Special Interest Groups (SIGs) in 34 distinct areas of information technology address varied interests: programming languages, graphics, computer–human interaction, and mobile communications, to name a few. Each SIG organizes itself around specific activities that best serve both its practitioner- and research-based constituencies. Many SIGs sponsor conferences and workshops, and offer members reduced rates for registration and proceedings. SIGs also produce newsletters and other publications, or support lively e-mail forums for information exchange. ACM and its SIGs sponsor more than 100 conferences around the world every year, attracting over 100,000 attendees in total. Each conference publishes a proceeding, and many have exhibitions. Many of ACM's conferences are considered "main events" in the IT industry. The ACM Press Books program covers a broad spectrum of interests in computer science and engineering.

Computing Research Association (CRA) http://cra.org/ (accessed August 3. 2010). "The Computing Research Association (CRA) is an association of more than 200 North American academic departments of computer science, computer engineering, and related fields; laboratories and centers in industry, government, and academia engaging in basic computing research; and affiliated professional societies. … CRA's mission is to strengthen research and advanced education in the computing fields, expand opportunities for women and minorities, and improve public and policymaker understanding of the importance of computing and computing research in our society" (from Web page).

IEEE Computer Society http://www.computer.org/ (accessed August 3, 2010). The IEEE Computer Society was founded in 1946 and is a leading organization of computer professionals. It is the largest of the 37 societies of the Institute of Electrical and Electronics Engineers (IEEE). Its vision is to be the leading provider of technical information and services to the world's computing professionals. The society is dedicated to advancing the theory, practice, and application of computer and information processing technology and sponsors many conferences, applications-related and research-oriented journals, local and student chapters, distance learning campus, technical committees, and standards working groups. In addition, the society promotes an active exchange of information among its thousands of members. It is a truly international society with over 40 percent of its members living and working outside the United States. Members receive *Computer Magazine* free and low-cost access to the Digital Library, free access to Safari (O'Reilley) books online, subscriptions to journals, and other resources.

International Federation for Information Processing (IFIP) http://www.ifip.org/ (accessed August 3, 2010). "Formally IFIP is a nongovernmental, nonprofit umbrella organization for national societies working in the field of information processing. It was established in 1960 under the auspices of UNESCO as a result of the first World Computer Congress held in Paris in 1959." IFIP represents IT Societies from 56 countries or regions, covering all 5 continents with a total membership of over half a million. Technical work, which is the heart of IFIP's activity, is managed by a series of Technical Committees (from Web page).

Society for Industrial and Applied Mathematics (SIAM) http://www.siam.org/ (accessed August 3, 2010). SIAM is dedicated to advancing the application of mathematics and computational science to engineering, industry, science, and society. It is very active in sponsoring conferences and publishing high-quality books, book series, and a suite of significant journals. All the 15 peer-reviewed journals are available electronically as well as in print. The product "Locus" introduced in 2005 contains the full text for every SIAM journal article published from the journal's inception through 1996, making this a valuable resource for retrospective research in applied mathematics (available via Scitation: http://scitation.aip.org/).

USENIX http://www.usenix.org/ (accessed August 3, 2010). "USENIX, the Advanced Computing Systems Association, fosters technical excellence and innovation, supports and disseminates research with a practical bias, provides a neutral forum for discussion of technical issues, and encourages computing outreach into the community at large. Since 1975, the USENIX Association has brought together the community of engineers, system administrators, scientists, and technicians working on the cutting edge of the computing world. The USENIX conferences have become the essential meeting grounds for the presentation and discussion of the most advanced information on the developments of all aspects of computing systems" (from Web page). The best papers from these conferences are available in their site's compendium.

CAREERS AND EDUCATION

Undergraduate and graduate programs in computer engineering, as well as two-year IT education programs, can all be identified by searching the well-known Peterson's Guide on the Web. Another popular resource, GradSchools.com allows one to search for specific graduate programs in computer science. For more information, see Chapter 11 in this book on Engineering Education.

GradSchools.com: Computers and Information Technology Graduate Program Directories, http://www.gradschools.com/computers_info.html (accessed August 3, 2010). The program directories are categorized by curriculum, type of program (certificate, master's, and doctor's) and subdivided by geography. Choose from topics such as computer science, information technology, or software engineering.

Technical societies remain the first and best point of entry for career-related resources, namely ACM and IEEE Computer Society. Keep in mind that other IT professional organizations, journals, or Web portals will likely have their own career-related resources of potential interest to computer engineers. See sections on Journals and Tools for the practitioner for more information.

ACM Career Resource Center http://campus.acm.org/crc/ (accessed August 3, 2010). Search and apply for jobs, find out information about CS/IS/MIS careers and industry trends, set up job alerts, get career advice from career coaches, find out about salaries, etc. In its Professional Development Center http://pd.acm.org/ (accessed August 3, 2010), ACM offers a wide variety of online courses (Element-K) to its professional and student members and member-discounted courses offered through Stevens Institute of Technology.

IEEE Computer Society Career Services Center, http://www.computer.org/portal/web/buildyour-career/home (accessed August 3, 2010) is the main Web page for careers. This main site, which has information about career and professional development, has a link to the posted jobs http://www.computer.org/portal/web/careers (accessed August 3, 2010) where you can post your resume and search open positions. The society offers its members 100 online training courses through its e-learning campus, http://www.computer.org/distancelearning/ (accessed August 3, 2010). Its Certified Software Development Associate and Professional Programs (CSDA and CSDP) http://computer.org/certification (accessed May 18, 2005) are certification programs for software engineers. Similarly, IEEE also maintains its own career site, http://www.ieee.org/education_careers/index.html, and job site, http://careers.ieee.org/ (accessed August 3, 2010).

World Lecture Hall http://wlh.webhost.utexas.edu/browse.cfm (accessed August 3, 2010). Browse by topics such as "Computer science" or "Electrical and computer engineering." "World Lecture Hall publishes links to pages created by faculty worldwide who are using the Web to deliver course materials in any language. Some courses are delivered entirely over the Internet. Others are designed for students in residence" (from Web page).

TOOLS FOR THE PRACTITIONER

Interpreting practitioners as "real world" engineers in the area of information systems and technologies, they can range from educators to programmers and analysts to management. In some cases, practitioners may have completed professional certification programs. For relevant journals, see the Journals or Web portals section (where many of the trade print journal content is available online).

TECHNICAL BOOKS

Books24x7: ITPro http://www.skillsoft.com/Books24x7/Product_Information/Collections/ITPro.asp (accessed August 3, 2010). Available as an individual, library, or corporate subscription, Books 24 × 7: ITPro provides broad and deep coverage of many topics relevant to computer engineers: enterprise computing, desktop and office applications, graphic design, hardware, Oracle technologies, IBM, networks, operating systems, etc., from publishers such as Wrox, McGraw-Hill, Microsoft Press, MIT Press, and many more. Access to this collection is also made available through a partnership with ACM for its members with a subscription via http://pd.acm.org/books/faq.cfm (accessed August 3, 2010).

FreeTechBooks http://www.freetechbooks.com/ (accessed August 3, 2010). This site provides links to free online computer books and documentation. There are over 100 books covering programming languages, scripting languages, operating systems, and other computer science topics, such as data structures, algorithms, object-oriented programming, logic programming, compiler design, and software development.

Numerical Recipes Books Online http://www.nr.com/ (accessed August 3, 2010). Numerical Recipes™ is a series of text and reference books on "the art of scientific computing" famous for its engaging text and lucid mathematical and algorithmic explanations. The link above provides information on the outdated (but still available) copies of the Numerical Recipes books from Cambridge University Press: *C: The Art of Scientific Computing, Numerical Recipes in Fortran 77, The Art of Scientific Computing, Numerical Recipes in Fortran 90,* and *The Art of Parallel Scientific Computing,* as well as information on ordering print and online copies of the current third edition.

O'Reilly Books http://oreilly.com (accessed August 3, 2010). O'Reilly Media is one of the major information sources for leading-edge computer technologies. You will find a handful of copies on the desks and shelves of most IT professionals (recognized by the animals on their covers). The company's books, conferences, and Web sites highlight the work of technology innovators. To keep up with O'Reilly news and application specific newsletters, sign up at the home page. Keep up with O'Reilly on Facebook, Twitter, YouTube, and LinkedIn.

Safari Books Online http://www.safaribooksonline.com/ (accessed August 3, 2010). O'Reilly and the Pearson Technology Group have joined forces to create Safari Books Online. Safari is a continuously updated online library that features the best IT titles not only from O'Reilly but also Adobe Press, Addison Wesley Professional, Cisco Press, New Riders, Peachpit Press, Prentice Hall PTR, Que, and Sam's Publishing. Individual, enterprise, and institutional subscriptions are available. Users can access the full text of hundreds of top-selling IT books. Safari lets you search specific topics, pinpoint the specific chapter or section of the book relevant to your issue, and download the chapter. You can update the titles in your "collection" every month or expand your subscription to access the whole collection. One of the major advantages of this is copying and pasting code to eliminate typographical errors. Safari Books Online is a good investment for companies and organizations that depend on teams of developers for their livelihood.

MARKET RESEARCH

In addition to association, society, trade, and commercial publications, IT practitioners may also want to seek out intelligence from the leading providers of technology market research in order to make better decisions regarding the buying and selling of technologies for a company. These reports are usually expensive (costing hundreds if not thousands of dollars), but provide invaluable information. Companies may subscribe to some of these offerings through enterprise subscriptions. Most of these companies also provide some free information on their Web sites; useful for those in academe who cannot afford the reports. Also, it is always worth contacting them and requesting the executive summary—sometimes they oblige—these are usually sufficient for most college assignments. Otherwise, complete reports (or specific sections) can usually be downloaded from commercial online services, such as Dialog (see your librarian). Some of the larger IT analyst firms are listed below:

Forrester Research http://www.forrester.com/ (accessed August 3, 2010). "Forrester Research is a technology and market research company that provides pragmatic advice to global leaders in business and technology. With hundreds of analysts and coverage areas, we are the only company that creates forward-thinking research specifically for your role in the organization" (from Web page). *Note*: It acquired Giga Information Group in February 2003.

Gartner, Inc. http://www.gartner.com/technology/home.jsp (accessed August 3, 2010) "Gartner is the IT professional's best first source for addressing virtually any IT issue" (from Web page). *Note*: It acquired Meta Group in December 2004.

IDC http://www.idc.com (accessed August 3, 2010). "International Data Corporation (IDC) is the premier global provider of market intelligence, advisory services, and events for the information technology, telecommunications, and consumer technology markets. IDC helps IT professionals, business executives, and the investment community make fact-based decisions on technology purchases and business strategy" (from Web page).

InfoTech Trends http://infotechtrends.com (accessed August 3, 2010), "formerly Computer Industry Forecasts, provides market data on computers, peripherals, software, storage, the Internet, and communications equipment" (from Web page).

WEB PORTALS

Representative examples of some of the most useful Web portals for IT practitioners include the following:

Builder.com http://www.builderau.com.au/ (accessed August 6, 2010)). This site is created by developers, for developers, with real-world perspective on topics from programming to architecture to management.

CNET http://www.cnet.com/ (accessed August 6, 2010). CNET provides information on personal technology, games, and entertainment, and business technology products including reviews. Now part of CBS Interactive (http://www.cbsinteractive.com/) Selected CNET portals are listed below, which may be of more interest to practitioners.

CompInfo: The Computer Information Center http://www.compinfo-center.com/ (accessed August 8, 2010). CompInfo uses a very broad interpretation of IT. Its extensive listings are organized by topic and subtopic with brief annotations.

DevX.com http://www.devx.com (accessed August 6, 2010). "DevX is the leading provider of technical information and services that enable corporate application development teams to efficiently conquer development challenges and keep projects moving. We are committed to leading the market by offering customers timely, vital resources and superior services designed to enhance their efforts of applying new technologies and techniques" (from Web page). The site is comprised of a tailored mix of "just in time" eLearning, expert tips, code libraries, collaboration and knowledge-exchange tools, and premium content and services. DevX is part of Internet.com.

EarthWeb http://www.earthweb.com (accessed August 6, 2010)). This Web site is organized into five sections: hardware and systems, Web developer, software development, IT management, and networking and communications. One of its best-known sites is http://web-developer.earthweb.com/ (accessed August 6, 2010), which has a wealth of resources for software development.

Internet.com http://www.internet.com/ (accessed August 6, 2010). "Internet.com provides enterprise IT and Internet industry professionals with the news, information resources and community they need to succeed in today's rapidly evolving IT and business environment" (from Jupiterweb.com home page). It is comprised of technology-specific Web sites, e-mail newsletters, announcement lists, and discussion lists.

itmWeb http://www.itmweb.com/ (accessed August 6, 2010). The itmWEB Site™ was established in 1996 and is recognized as a major source for information technology reference, methodology, and technical content focused on IT departmental management, technology support, and project leadership. Designed for CIOs, project managers, IT educators, and students, the site also publishes a highly regarded monthly *IT eZine* with thousands of worldwide subscribers (from Web page).

O'Reilly Network http://www.oreillynet.com/ (accessed August 6, 2010). In addition to their books, O'Reilley is an important portal for developers interested in open and emerging technologies, including new platforms, programming languages, and operating systems. The site provides in-depth technical information for the expert developer and is a forum for the O'Reilly developer community.

OSTG: Open Source Technology Group Network http://openmagazine.net/index.htm (accessed August 6, 2010). "OSTG (Open Source Technology Group), formerly Open Source Development Network (OSDN), has had its roots in the technology community since its early days as the ground-breaking tech network Andover.net. Founded in 1996 with the mission to provide unbiased content, community, and commerce for the Linux and Open Source communities, Andover.net grew in community relevance and popularity by adding the provocative community-centric sites Slashdot and freshmeat.net to its technology group, and ThinkGeek and AnimationFactory.com to its e-commerce division. After its acquisition by VA Software Corp. (NASDAQ: LNUX) in early 2000 and the introduction of SourceForge.net and Linux.com, the network cemented its position as the Internet's leading destination for the Linux and Open Source community" (accessed August 6, 2010) (from Web page).

SE Online: Software Engineering Online http://www.computer.org/portal site/seportal (accessed August 6, 2010). The IEEE Computer Society's resource for researchers and practitioners to learn and find noncommercial information about software engineering.

TechRepublic http://techrepublic.com (accessed August 2, 2010). "TechRepublic helps IT decision makers identify technologies and strategies to empower workers and streamline business processes. The site delivers a unique blend of original content by IT professionals, peer-to-peer advice from the largest community of IT leaders on the Web, and a vast library of professional resources from the leading vendors in the IT industry. TechRepublic features blogs, community forums, vendor white papers, software downloads, Webcasts, and research" (from Web page).

TechTarget Network http://www.techtarget.com (accessed August 6, 2010). "TechTarget (NASDAQ: TTGT) publishes integrated media that enable technology providers to reach targeted communities of technology professionals and executives in all phases of their decision making and purchase process. Through its industry-leading Web sites, conferences, and ROI-focused lead management services, TechTarget delivers measurable results that help marketers generate qualified leads, shorten sales cycles, and grow revenues" (from Web page).

TechWeb http://www.techweb.com (accessed August 6, 2010). Part of UBM, "TechWeb enables people and organizations to harness the transformative power of technology. Through its three core businesses—media solutions, marketing services, and paid content—UBM TechWeb powers both technology decision making and technology marketing" (from Web page). The UBM Tech Web Brand includes many well-known resources, such as *Dr. Dobb's* and *Information Week* http://www.ubmtechweb.com/brands/ (accessed August 6, 2010).

ZDNet http://www.zdnet.com (accessed August 6, 2010). ZDNet, formerly Ziffnet, and now part of CBS Interactive, can be described as a "Webzine" and offers a worldwide network of Web sites to IT professionals.

OTHER WEB RESOURCES

ALGORITHMS

Collected Algorithms of the ACM (CALGO) http://www.acm.org/calgo (accessed August 3, 2010). This site has information about the software associated with papers published in the *Transactions on Mathematical Software* (TOMS), which are available in print and through the ACM Digital Library. This software is refereed for originality, accuracy, robustness, completeness, portability, and lasting value (from Web page). Until December 2004, CALGO also was published in loose-leaf form, but this has now ceased. The associated free Web site will continue to be maintained and updated.

The Stony Brook Algorithm Repository http://www.cs.sunysb.edu/~algorith/ (accessed August 3, 2010). Professor Steven S. Skiena has mounted this site, based on his book with the same title, "to serve as a comprehensive collection of algorithm implementations for over 70 of the most fundamental problems in combinatorial algorithms" (from Web page).

COMPILERS AND INTERPRETERS

Catalog of Free Compilers and Interpreters http://free-compilers.sharnoff.org/ (accessed August 6, 2010). This directory, aimed at developers (rather than researchers), provides links to compilers, compiler generators, interpreters, translators, important libraries, assemblers, etc.

Compilers.net http://www.compilers.net/ (accessed August 6, 2010). This site started in 1997 as a student project. Since then, it has become an important source for information on compilers and programming languages. They are constantly adding new topics, including compiler links and tutorials, full text books and chapters, tutorials, and information on individuals and companies involved in compilers development.

Free Compilers and Interpreters for Programming Languages http://www.thefreecountry.com/compilers/index.shtml (accessed August 6, 2010). This site almost disappeared when the administrator, Christopher Heng, could no longer maintain it, but his "fans" inundated him with e-mails begging him to continue with offers of free hosting on their servers. He took up one of the offers and the site has been resurrected with a new look. This example indicates how sites develop and then disappear, but also shows the power of the medium to bring parties together. It is indeed an excellent site linking not only to free compilers and interpreters, but also to the whole gamut of programming tools; a site to bookmark, certainly. The site has free programming resources, Web masters' resources, security resources, and utilities.

PROGRAMMING

There are many sites on the Internet that provide programming tools for individual languages and groups of languages. It would be impossible to list them all in this chapter. The Computer Science Resource Guide at the State University of New York at Albany, maintained by Michael Knee of the Science Library, does a magnificent job in tracking programming sites: http://libguides.library.albany.edu/csci (accessed April 1, 2011). The site below could be considered a metasite for programming tools.

Programmer's Heaven http://www.programmersheaven.com/ (accessed August 3, 2010). Programmers Heaven contains a wealth of information for programmers. It is not just a Web site, but more of a "virtual community" for programmers. In addition to links to programming sites, it also has links to other resources, such as tutorials, magazines, sample chapters in books, and has a forum and a blog.

INTERNET RESOURCES: SUBJECT GUIDES

There are many good guides to the resources in computer science; however, the one listed below is one of the most comprehensive and has been continuously updated over the years.

Computer Science: A Guide to Web Resources by Michael Knee, Science Librarian, State University of New York at Albany http://libguides.library.albany.edu/csci (accessed April 1, 2011).

METASITES AND SPECIALITY SEARCH ENGINES

HCI Bibliography: Human-Computer Interaction Resources http://hcibib.org (accessed August 8, 2010). Comprehensive metasite on HCI with over 59,000 entries (as of July 2010). In addition to the book and journal literature, the bibliography covers history of the field, conferences, columns and news, Weblogs, and user interface developer resources.

Intute: Computer Science, http://www.intute.ac.uk/computing/ (accessed April 1, 2011). Intute replaced EEVL and has been one of the premier metasites in engineering. Unfortunately, Intute is only guaranteed funding from JISC (Joint Information Systems Committee of the United Kingdom) until July 2011 and sadly its fate is uncertain from that point.

Open Directory: Computers http://dmoz.org/Computers/ (accessed August 8, 2010). The DMOZ Open Directory Project is the largest, most comprehensive human-edited directory of the Web and is maintained by a global community of volunteers. DMOZ uses an all-encompassing and broad definition of computer-related resources. Its extensive listings currently total 109,786 and are subdivided by topics with brief descriptions. There is a separate category for "Business: Information technology."

Scirus http://www.scirus.com/srsapp/advanced/index.jsp (accessed May 18, 2005). For more information, see Chapter 2 in this book. Scirus is one of the most comprehensive scientific research tools on the Web. With over 370 million scientific items indexed, it allows researchers to search for not only journal content but also scientists' homepages, courseware, preprint server material, patents and institutional repository, and Web site information. Specialty sources relevant to computer science include full-text articles from ScienceDirect, Project Euclid, Scitation, and SIAM in addition to USPTO patents, e-prints from ArXiv.org and CogPrints, and NASA technical reports. Access to some resources requires registration or subscription. An innovation feature is the section "Subtopics," an authoritative distillation of current topics.

TechXtra: Engineering, Mathematics, and Computing http://www.techxtra.ac.uk/index.html (accessed April 1, 2011). TechXtra identifies articles, key Web sites, books, the latest industry news, job announcements, ejournals, eprints, technical reports, latest research, etc.

WWW Computer Architecture Page http://arch-www.cs.wisc.edu/home (accessed August 9, 2010). The WWW Computer Architecture page, hosted by University of Wisconsin-Madison, has information about the science and art of selecting and interconnecting hardware components to create computers that meet functional, performance, and cost goals. This Web site is a forum for the computer architecture community to disseminate information. Forum content is submitted by community members.

NETWORKING AND COLLABORATING

In the early days of the Internet, mailing lists and newsgroups (such as Usenet) were the only channels of communication for scientists and engineers.

Google has integrated the past 20 years of Usenet archives into its Google Groups site. Go to http://groups.google.com/ and then access the subheading "computer groups." One noteworthy resource for software engineers continues to be COMP.SOFTWARE-ENG http://groups-beta.google.com/group/comp.software-eng (accessed August 9, 2010).

IT professionals also seek out one another in many online communities based on interest in platform, operating system, language, product, or technology (too numerous to provide a comprehensive listing here). An example of an online community for developers is Java.net Forums http://www.java.net/ (accessed August 9, 2010).

With the advent of Web 2.0 technologies, there are more ways than ever before to keep up and contribute to your profession via associations, interest groups, and topics. One of the best ways to find these tools is to start with your association or Special Interest Groups (SIGs). ACM is one of the best examples of this. From the main page (http://www.acm.org/), access the "Special Interest Groups" or "Chapters" links to find out about listservs, blogs, and wikis.

IEEE Computer Society Digital Library may be found at http://bell.computer.org/rss/index.jsp (accessed April 1, 2011). "IEEE Computer Society now offers the availability of the latest magazines and transactions content through RSS (Really Simple Syndication) feeds using XML (or eXtensible Markup Language) to automatically deliver new abstracts to your desktop" (from Web page). In the previous edition of this book, we had a section entitled "Weblogs and Webfeeds" when Web 2.0 tools were very new. Now almost everyone in the information arena has a Blog, Facebook group, RSS feeds, Wiki, Twitter, etc. The list below is very selective and highlights these resources as examples using the latest Web 2.0 tools to disseminate information in computer engineering. Many of these companies have been mentioned in other sections in this chapter.

CNET Networks http://news.cnet.com/ (look for Connect with CNET).

Computing Research Policy Blog: Advocacy and Policy Analysis for the Computing Community http://www.cra.org/govaffairs/blog/ (accessed August 9, 2010).

InfoWorld http://weblog.infoworld.com/ (accessed August 9, 2010). Offers columnist blogs as well as RSS feeds on top news, columnists, test center reviews and topics such as "applications, application development, e-business solutions and strategies, end-user hardware, networking, operating systems, platforms, security, standards, storage, telecom, wireless, and Web services" (from Web page).

Internet.com, RSS feeds—"IT News and Developer Tools Delivered Straight to Your Desktop" http://www.jupiterweb.com/rss/internet.html (accessed August 9, 2010).

ITtoolboxBlogs http://it.toolbox.com/ (accessed August 9, 2010). Features professional blogs for IT professionals.

Java.net http://weblogs.java.net/blogfront (accessed August 9, 2010) for Java Technology Collaborators.

Network World http://www.networkworld.com/ (accessed August 9, 2008). Look for "More ways to stay informed." Network World is one of the premier providers of information, intelligence, and insight for Network and IT Executives with a focus on news, opinions, and analytical tools for key decision makers who architect, deploy, and manage business solutions. *Network World* magazine and NetworkWorld.com are part of the IDG (International Data Group) company.

Zdnet.com http://www.zdnet.com/ (accessed August 9, 2010).

ACKNOWLEDGMENTS

I rededicate this revised chapter to Anne Buck, university librarian, California Institute of Technology (Caltech) from 1995 to April 2003, when she lost her battle with cancer. I would also like to acknowledge my gratitude to Kimberly Douglas, current university librarian at Caltech and my colleagues there who continue to support me from afar in more ways than they realize. Thanks also to all the wonderful engineering librarians around the country whom I rely on for information and assistance.

Thanks to Amy Pacheco, MLS student, my unofficial research assistant and friend who assisted me in the initial stages of this project in checking all the entries for currency and accuracy. Renee McHenry's influence (my co-author for the first edition) especially in terms of organization is still evident in this edition and made the task easier this time around.

Thanks to Dean Roman Kochan and my colleagues at California State University–Long Beach for giving me the opportunity to pursue my passion for librarianship.

And last, but not least, my gratitude to my husband, Dr. R. Sethuraman, who held the fort and provided moral support while I tackled this and a myriad other projects over the past three years.

REFERENCES

Joint Task Force on Computer Engineering Curricula. 2004. *Computer engineering 2004: Curriculum guidelines for undergraduate degree programs: A report in the computing curricula series*, Washington, D.C.: IEEE Computer Society http://www.eng.auburn.edu/ece/CCCE/CCCE-FinalReport-2004Dec12.pdf (accessed February 22, 2005).

Knee, M. 2001. Computer science: A guide to selected resources on the Internet. *C&RL News* 62 (6): 609–615. http://w\vw.ala.org/ala/acrl/acrlpubs/crlnews/backissues2001/june1/computerscience.htm (accessed February 27, 2005).

Library of Congress. 2001. *Library of Congress subject headings*, 24th ed. Washington, D.C.: Library of Congress.

Lord, C. R. 2000. *Guide to information sources in engineering and technology.* Englewood, CO: Libraries Unlimited.

New world encyclopedia http://www.newworldencyclopedia.org/entry/Institutional_repository (accessed May 3, 2011).

Trinity College Dublin. What is computer engineering? http://www.tcd.ie/Engineering/about/what_is_eng/computer_eng_intro.html (accessed October 7, 2010).

Webster's new world dictionary of computer terms, 3rd ed. 1988. New York: Webster's New World.

10 Electrical and Electronics Engineering

Larry Thompson

CONTENTS

Introduction...241
Searching the Library Catalog 243
   Keywords, LC Subject Headings, LC Call Numbers...243
Article Indexes and Full-Text Resources 244
Databooks and Integrated Circuits..246
Handbooks, Encyclopedias, and Dictionaries...247
Monographs: Professional and Textbooks 250
Journals and Conference Proceedings 251
Standards and Codes..253
Web Sites..254
Associations and Societies...254
Green Electrical Engineering...255
Nanotechnology and Electrical Engineering 256
Conclusion 256
References...257

INTRODUCTION

There are various ways to describe electrical engineering. The following scheme (Irwin and Kerns, 1995, pp. 5–10) divides the discipline into seven areas:

Power engineering: Generating and transferring electrical energy from one location to another, and transforming it into forms that can do useful work. Power is most often generated by conversion of mechanical energy from a rotating shaft to electric energy in a generator. Power also can be produced by solar cells that convert solar energy into electrical energy or from chemical reactions, such as a battery. Power for residential and industrial use is distributed by high-voltage power lines and is used for purposes such as heating, illumination, and driving electric motors.

Electromagnetics: Concerning the interaction between magnetic fields, electric fields, and the flow of current.

Communications and signal processing: Transmitting information from one place to another via unconfined electromagnetic waves, or telephone wires, cables, or optical fibers. It includes modulating (encoding) information and demodulation (decoding) information.

Computers: Designing and developing computer hardware and software.

Electronics: Using materials in special configurations to make devices that control current flow. These devices, such as transistors or diodes, can be interconnected to make circuits. Electronics are used in devices such as circuit boards for computers, engine monitors in cars, radio receivers, and radar systems.

Systems: Using mathematical principles to model and describe complex systems.

Controls: Providing fast and accurate adjustments or placements upon command of mechanical systems (robotic arms, airplane autopilots).

A simpler scheme (Sarma, 2001, p. xxi) divides the discipline into only two areas:
1. Information systems (electrical means are used to transmit, store, and process information)
2. Power and energy systems (bulk energy is transmitted from one place to another and power is converted from one form to another).

It is impossible to specify an exact date when electrical engineering began. Discoveries involving electricity were made throughout the centuries, and electrical engineering gradually evolved into a distinct discipline. One of the first mentions of phenomena related to electrical engineering (Martin and Coles, 1919, p. 9) is credited to Thales, a Greek philosopher, who in about 600 BCE recorded that a piece of amber, when rubbed against clothing, would attract and repel light objects brought close to it. In later years (Martin and Coles, 1919, pp. 11–12), various individuals experimented with magnetism and the use of compasses to determine direction. In 1729, Stephen Grey pointed out the differences between conductors and nonconductors, and in 1745 the Leyden jar, an early capacitor, was discovered by Pieter van Musschenbroek and Ewald Jurgens von Kleist. Benjamin Franklin, famous for his 1742 kite and key experiment, invented the lightning rod as a result of that experiment.

The 1800s ushered in the more traditional beginnings of the field, with many well-known experimenters setting the foundations: Andre Marie Ampere (1775–1836), electromagnetism and Ampere's law; Greg Simon Ohm (1787–1854), Ohm's law; Michael Faraday (1791–1867), electromagnetic induction; Samuel F. B. Morse (1791–1872), telegraph; Alexander Graham Bell (1847–1922), telephone; and Thomas Alva Edison (1847–1931), incandescent light bulb and phonograph.

The electrical engineering accomplishments of the 1900s began with Guglielmo Marconi's (1874–1937) wireless transatlantic telegraph transmission in 1901. During the rest of the twentieth century, developments in areas such as electronics and energy have shaped the world in unprecedented ways.

The rapid development of electronics has been the driver behind much of the progress in the past century. Devices that were formerly difficult to transport and were large consumers of electric power have been reduced in size and increased in efficiency so that they can be taken almost anywhere.

The most widespread manifestation of electronics has been in personal computers and calculators. Although mechanical calculating machines have been available for hundreds of years, and ENIAC (electronic numerical integrator and computer) was built using over 19,000 vacuum tubes, it was not until the advent of low-cost electronics that computers became feasible for widespread use. Since that time, they have become ubiquitous, not only as stand-alone products, but also incorporated into products ranging from automobiles to military weapons.

Another benefactor of electronics developments has been communications. Electronics have made possible the development of cell phones, satellites, and other devices that enable instantaneous audio and video communication to and from any point on the globe. They have also made it possible to store huge quantities of data in readily transportable devices, such as flash drives.

Coupled with the development of electronics has been the development of energy sources to power the devices. Advances in rechargeable battery technology along with more efficient photovoltaic cells have enabled the users of electronic devices to take them almost anywhere.

Combining all these devices, it is now possible for an individual sailing on a solo circumnavigation of the globe to power up a laptop, establish a satellite link, and send out daily updates to millions of people throughout the world. This is quite a change from Marconi's 1901 transmission of the letter "s" from Poldhu, Cornwall, in the U.K. to St. John's, Newfoundland.

More recently, electrical engineers have been involved in such areas as nanotechnology, e-textiles, and biomedicine. It is impossible to predict exactly how research in these areas will be applied to products in the future. However, some possibilities are to use nanotechnology in the development of smaller and more portable devices, e-textiles to detect biohazardous chemicals used during warfare, and developments in biomedicine to produce more sophisticated diagnostic imaging machines.

The formation of professional societies paralleled the development of the discipline. In 1871 the Institution of Electrical Engineers (IEE) (U.K.) was established, followed in 1884 by the American Institute of Electrical Engineers (AIEE), and in 1912 by the Institute of Radio Engineers (IRE). In 1963, the AIEE and IRE merged to form the Institute of Electrical and Electronics Engineers (IEEE). In 2006, the Institution of Engineering and Technology (IET) was formed by the merger of the IEE and the Institution of Incorporated Engineers, whose lineage begins in 1884 with the Vulcanic Society.

SEARCHING THE LIBRARY CATALOG

KEYWORDS, LC SUBJECT HEADINGS, LC CALL NUMBERS

Because of the rapid advances being made in the field, it is difficult to search a library catalog using LC Subject Headings. These headings often lag behind current research and do not reflect the most current terminology. Therefore, in most cases, using the keyword function of the catalog is the best way to start a search. This will allow items to be retrieved without relying on the out-of-date subject headings.

After relevant entries have been retrieved using keywords, the Library of Congress Subject Headings can be reviewed, and if they appear to target the desired area, a search may be done using them. In many cases, though, the keyword will retrieve satisfactory results.

The primary Library of Congress call numbers for electrical engineering are found from TK1 to TK9971. The important subdivisions are as follows:

TK301–TK399 Electric meters
TK452–TK454.4 Electric circuits, electric networks
TK1001–TK1841 Production of electric energy or power
TK2000–TK2891 Dynamoelectric machinery including generators, motors, transformers
TK2896–TK2985 Production of electricity by direct energy conversion
TK3001–TK3521 Distribution or transmission of electric power
TK4001–TK4102 Applications of electric power
TK4125–TK4399 Electric lighting
TK4601–TK4661 Electric heating
TK5101–TK6720 Telecommunication including satellites, computer networks, telephones, television, radio
TK7800–TK8360 Electronics including electronics apparatus, computer engineering, computer hardware, optoelectronic devices
TK9001–TK9401 Atomic power
TK9900–TK9971 Electricity for amateurs

Also related to electrical engineering are selected areas in mathematics, computing, and physics:
Q300–Q390 Cybernetics
Q350–Q390 Information theory
QA75–QA76 Calculating machines, electronic computers, computer science, computer software
QC501–QC721 Electricity including electromagnetic theory, radio waves (theory), electric discharge, plasma physics, and ionized gases
QC750–QC766 Magnetism

In the social sciences, call numbers that cover industries related to electrical engineering may be of interest:
HD9684–HD9685 Lighting industries and electric utilities
HD9696 Electronic industries

HD9697 Electric industries

HE7601–HE8700.9 Telecommunications industry, including telegraph, wireless telegraph, radiotelegraphy, radio and television broadcasting

HE8701–HE9680.7 Telephone industry

HE9713–HE9714 Cellular telephone services industry and wireless telephone industry

HE9719–HE9721 Artificial satellite telecommunications

ARTICLE INDEXES AND FULL-TEXT RESOURCES

During most of publishing's history, there has been a division between article indexes and the articles indexed. Someone searching for an article by a particular author or on a particular subject would first consult the appropriate index, and then use the citation information in the index to retrieve the article.

The indexes that were first commercially available, such as *Engineering Index* (1884) and *Science Abstracts* (1898), were one-stop shops that allowed users to search for appropriate material across several publishers. This pattern of cross-publisher indexing has continued to the present, first in print format and now through online database searching.

However, in the late 1990s, a change occurred. Publishers began Web-based journal distribution and to enable users to find the desired online article, they developed indexes for their own journals. In addition to being publisher-centric, these indexes had the added disadvantage of covering only a few years of publication, rather than the decades of coverage offered by the established cross-publisher indexes.

Still, users were attracted to the new indexes because they provided a quick, free, and easy way to search, and link to, the full-text online articles. Indexes such as *Inspec* and *Engineering Index* did not yet have links to the online full text.

During the past 15 years, the publisher-based indexes have increased in influence. ACM, Elsevier, IEEE, and other publishers have not only expanded their online full-text holdings, they also have increased the search capabilities of their in-house indexes. Users find it convenient to go directly to a product such as the *ACM Digital Library, Elsevier Science Direct,* or *IEEE Xplore* to access online full-text resources. Undergraduate students, and others who do not need a comprehensive view of a topic, are understandably reluctant to search a multipublisher database when the publisher-specific database is much more efficient. After all, why go through *Compendex* or *Inspec* with their sometimes unreliable links to online full text, when a search of the *IEEE Xplore* or *ScienceDirect* sites will yield "enough" full-text articles for their project or research paper?

However, there is danger in searching the publisher-specific indexes. For the graduate student or faculty researcher, the publisher-based search imposes an unacceptable limitation on a single publisher. For these users it is not sufficient to get "just enough" articles to write a paper, rather, the objective is often to retrieve all resources on a topic. For this high-level research, it is necessary to search the comprehensive indexes, such as *Inspec* and *Compendex*, which not only index multiple publishers, but also multiple publication types to include conference proceedings, technical reports, and government documents.

It has always been necessary for librarians and users of the resources to be aware of the limitations inherent in a resource. However, it is even more necessary now that publisher-specific indexes are freely available and compete for the user's attention. The limitations inherent in the publisher indexes must be made clear.

As we examine the indexes and full text resources, we must recognize that for many users there is no separation between the two. The intertwining between index and full text is complete, with the online index providing a gateway to the online full text, and each publisher collection of full text containing its own index. For this reason, we will look at indexes and full text together.

ACM Digital Library http://portal.acm.org/dl.cfm (accessed July 15, 2010) (1954–) New York: Association for Computing Machinery. The home page of this resource states: "Full text of every article ever published by ACM." From Volume 1, Issue 1 (January, 1954) of the

Journal of the ACM through the latest SIG newsletter, this site offers excellent coverage, but unfortunately not every article ever published by ACM. A comparison between the hard copy and online offerings shows incomplete full text in titles, such as *Electronic Art and Animation Catalog*. Still, it is an excellent effort, and all publishers should strive to produce the same comprehensive full-text coverage as ACM. The site offers basic, advanced, and browse modes to its subscribers, while the basic and browse modes are freely available to the general public. The export of citations into bibliographic managers, such as Endnote or BibTeX is cumbersome. Single citations can be exported into some managers, but batch exports must be done via a "binder" created by the user.

Applied Science and Technology http://www.hwwilson.com/databases/applieds.cfm#Index (accessed July 15, 2010). New York: H. W. Wilson. Within the EE subject area, *AS&T* provides good indexing of the journals published by major societies, such as ACM, IET, and IEEE. In addition, a variety of trade journals and selected commercial titles are included. Therefore, this product can give undergraduates a reasonably good overview of the journal literature in the field. However, because the full text of the major societies (IEEE, IET, and ACM) is not provided within AS&T, the practical value is limited. In many cases, undergraduates, who usually have limited research needs, would be better served by going to the publisher specific sites (*IEEE Xplore, ACM Digital Library*, etc.) and directly retrieving the articles they need. The product is not suitable for graduate research because coverage is limited to journals. (See additional information in Chapter 2 of this book.)

Compendex http://www.engineeringvillage2.org (accessed July 15, 2010) (1884–). New York: Elsevier. This product has thorough coverage of all publication types within the EE area, and, when used alone, it provides excellent indexing for undergraduate studies. However, for graduate studies or when the most comprehensive search is needed, it should be used in conjunction with *Inspec*. Neither *Compendex* nor *Inspec* used alone provides a comprehensive search in EE. (See additional information in Chapter 2 of this book.)

Google http://www.google.com/ and Google Scholar http://scholar.google.com/ (accessed July 15, 2010). Google products have become an increasingly legitimate option for accessing the EE literature. Although the search interface is not as robust as *Engineering Village* or that of some of the publishers, its coverage is very good. Many publishers, including IEEE, Springer, and Elsevier, allow Google to index bibliographic information from their journals and books. Also, links from Google products to the publisher full-text sites are outstanding, often bettering those from commercial providers, such as Serials Solutions. Thus, although the search interface limits the effectiveness of a subject search in Google, the extensive coverage and outstanding links allow for quick full-text access at the publisher's Web site for known documents. *Google Scholar* provides good cited reference information and should be used in conjunction with ISI's *Web of Science* to obtain a more complete cited reference overview. (See additional information in Chapter 2 in this book.)

IEEE Xplore http://www.ieeexplore.ieee.org (accessed July 15, 2010). Piscataway, NJ: IEEE. *IEEE Xplore* is the online delivery system that provides access to IEEE and IET publications. Within *IEEE Xplore*, the public may use the free search-and-browse interface to access tables of contents and abstract records of IEEE journals, magazines, conference proceedings, and standards as well as IET journals and conference proceedings. IEEE members have the same browse-and-search access to all records in the database as well as access to selected full text activated in their IEEE Web account. Corporate, government, and university subscribers can search and browse all records in the database and access the full-text documents permitted in their subscription agreements. Various options for accessing the full text are available. These include flat rate subscriptions to the total library or to selected portions as well as individual article purchase. All materials included in the collection are available from 1988 to the present, and IEEE has an ongoing program to increase the back file length, with some titles already available to the 1950s.

Infotrac http://www.galegroup.com (accessed July 15, 2010) (1980–). Farmington Hills, MI: Thomson Gale. Similar to *AS&T* reviewed above, *Infotrac* provides good journal indexing (but not full-text access) of the IEEE and ACM in its Expanded Academic ASAP product. It does not index IET journals. Because full text is not available, most undergraduates will do better to go directly to the IEEE or ACM online indexes and perform their search there. Not very useful for graduate students as it indexes only the journal literature.

Inspec http://www.theiet.org/ (accessed July 26, 2010) (1898–). London: IET. *Inspec* is the combined electronic version of *Physics Abstracts, Computer and Control Abstracts* and *Electrical and Electronics Abstracts*. This has been the standard database in the field of EE and, with the extension of the back file to 1898, it encompasses the very earliest EE literature. No literature search in the field would be complete without consulting this resource and, in most cases, this database would be the first choice for a literature search. However, *Inspec* does not provide blanket coverage of all EE resources and, in order to conduct a comprehensive search, it is necessary to use it in conjunction with *Compendex*. Several vendors offer subscription and/or pay-as-you-go access to the database.

ISI—Web of Science (*Science Citation Index*). Within EE, the primary reason for using this resource is to access the citation indexing, which covers both journal and conference literature. For a subject search, *Inspec* or *Compendex* will usually provide superior results. (See additional information in Chapter 2 of this book.)

NTIS. This resource indexes technical reports not covered in any of the other indexes. It is valuable for those who need to do a comprehensive subject search, but is probably of little use to most undergraduates in the field. (See additional information in Chapter 2 of this book.)

Cambridge Scientific Abstracts Products within its array of indexes, CSA offers three that are of interest to the electrical engineer: *Computer and Information Systems Abstracts, Electronics and Communications Abstracts*, and *Solid State and Superconductivity Abstracts*. Until 2003, these three databases were produced in conjunction with Engineering Information, but since that time they have been produced solely by CSA. (See additional information in Chapter 2 of this book.)

ScienceDirect. Elsevier and its various imprints have many titles in EE, and most have back file online access to Volume 1, Issue 1. (See additional information in Chapter 2 of this book.)

Scopus. Published by Elsevier, this product indexes the EE literature well. The back file is not as extensive as that of other indexes. (See additional information in Chapter 2 of this book.)

DATABOOKS AND INTEGRATED CIRCUITS

In the past, libraries supporting an electrical engineering curriculum were expected to maintain collections of data books from various manufacturers. This could be accomplished either through obtaining the hard copy from the manufacturers or subscribing through a commercial vendor. Now, although some vendors still produce hard copy of their data books, most researchers access the information online. In addition to the online data books produced by specific manufacturers, there are several Web sites that have compiled the data from many resources.

Datasheet Archive http://www.datasheetarchive.com/ (accessed July 28, 2010). This site boasts 100 million datasheets from 7,500 manufacturers. The service is free and there is no registration required.

Electronic Engineers Master Catalog (EEM) http://www.eem.com (accessed July 28, 2010); IC Master, http://www.icmaster.com (accessed July 28, 2010). Both of these resources are provided by Hearst Business Communications. Registration is required, but is free, and the same registration will work for both resources. The EEM provides information on manufacturers and suppliers, and the various search options lead to data sheets, inventory listings, and the option to purchase components. IC Master offers a searchable collection of over 150 million parts. Searches may be done by criteria, such as part number, parameter, or description.

IHS 4DOnline Parts Universe–Information Handling Services (IHS) http://www.ihs.com (accessed July 28, 2010). IHS is a long-time commercial provider of component information, first with microfiche and now through an online service. This is a subscription-based service.

HANDBOOKS, ENCYCLOPEDIAS, AND DICTIONARIES

Traditionally, these three format types have been part of a library's reference collection. Handbooks present a subject overview and feature an author devised arrangement by subdisciplines within the subject. Encyclopedias cover similar material, but in an alphabetical arrangement. Dictionaries also use an alphabetical arrangement, but usually have more entries with significantly shorter explanations.

In the hard copy world, it was not unusual to have subject-specific encyclopedia sets relegated to a distinct section of the reference area, while handbooks and dictionaries might be found more close at hand around the reference desk. But, just as the availability of full-text journals online has blurred the distinction between indexes and the full-text, the availability of online full-text handbooks, encyclopedias, and dictionaries has blurred the distinction between the three.

Now, although the format distinctions remain, it is common for all three types to be combined in one online collection. For instance, Elsevier's *Referex* collection is composed primarily of handbooks, but also contains the *Dictionary of Video Television Technology*, while the *CRCnetBASE* contains the *Comprehensive Dictionary of Electrical Engineering* as well as hundreds of handbooks in its collection.

For users who search an online site for a particular topic, it matters little if the needed information is gleaned from a handbook, encyclopedia, or dictionary. It is because the resources have been combined online and because users are more interested in finding information than specifying a type of resource from which to find it, that all three resources are considered in this section.

There are hundreds of handbooks/encyclopedias/dictionaries available within EE. Some volumes give an overview of the entire field, while others concentrate on a specific area, such as lasers or semiconductors. Although all of these formats are useful additions to the collection, there are several reasons it is difficult to dogmatically recommend specific titles.

First, the content of books differs, even among those with similar subject emphases. Most librarians have had the experience of consulting a handbook expecting to find a specific piece of information, only to discover that the handbook does not provide it. However, another handbook with a similar title, but from a different publisher may provide just the piece of data needed. Is one handbook better than the other? Usually not; they are simply different.

Second, the programs or specific interests of researchers in an institution will influence the selection of specific titles. As noted previously, the EE field is broad and an institution with an emphasis in electronics will have different needs than one in which power generation is paramount.

Third, the change from print to online has changed the way faculty and students view these volumes. In the not too distant past, librarians and users developed a general knowledge of what individual books to consult for certain information. When confronted with a heretofore unknown situation, the librarian consulted the *Composite Index for CRC Handbooks* in order to determine what individual CRC handbook held the desired bit of information. Now, however, researchers tend to view a universe of information rather than seeking a specific title.

This view is enhanced by the advent of online products such as *CRCnetBASE, Knovel, McGraw Hill's AccessEngineering*, and Elsevier's *Referex Engineering*. With these products, the knowledge of individual titles is of little importance because one search made over hundreds (or thousands) of titles will retrieve hits across the collection.

With this in mind, and if financially feasible, an institution should subscribe to an online collection of books such as those mentioned above. Students and faculty will appreciate the convenience of desktop access, and the power of searching across multiple titles to retrieve the information they need. Librarians will appreciate the fact that online collections normally update to the most recent edition, that only one order (as opposed to hundreds) needs to be made, and that collections provide significant discounts over purchasing individual titles.

If an online collection is not within the budget, then selected hard copy or online editions should be purchased. Using the reviews in *Choice* or *SciTech Book News*, or conducting a search in *WorldCat* can be helpful to determine recent and popular titles in particular areas. For example, a *WorldCat* title search on "handbook or manual or dictionary or encyclopedia" and "electronics," sorted by publication date, gives a selection of titles that could be added to the collection.

For libraries that are not subscribing to online sets, a selected list of titles that could be considered is given below.

Bovik, A., ed. 2005. *Handbook of Image and Video Processing*. San Diego: Academic Press. This volume provides a technical context for the images and videos that have become ubiquitous in our daily lives.

Buss, E. W., and M. W. Earley. 2009. *Handbook of Electrical Safety in the Workplace,* 2nd ed. Quincy, MA: National Fire Protection Association. Contains the FNPA 70E standard for electrical safety.

Cadick, J., M. Capelli-Schellpfeffer, and D. K. Neitzel. 2006. *Electrical Safety Handbook*. New York: McGraw-Hill. This handbook takes a very practical approach, and is geared for the safety training and reference needs of a company. Electrical safety codes from agencies, such as NFPA, NEC, and OSHA are referenced.

Chang, K., ed. 2003. *Handbook of RF/Microwave Components and Engineering,* 2nd ed. Hoboken: John Wiley & Sons. Provides principles, methods, and design data for practicing engineers in the field of radio frequency and microwave engineering.

Chen, W., ed. 2005. *The Electrical Engineering Handbook*. San Diego: Academic Press. A comprehensive volume giving an overview of all aspects of electrical engineering.

Chen, W.-K. 2003. *The Circuits and Filters Handbook,* 2nd ed. Boca Raton, FL: CRC Press. Focuses on practical applications for the practicing engineer. Covers circuits and filters, both analog and digital.

Christiansen, D. and C. Alexander. 2005. *Standard Handbook of Electronic Engineering,* 5th ed. New York: McGraw-Hill. In its latest edition, with over 30 years in existence, this handbook no longer places its emphasis on computers. Although the handbook still contains all the basic material, the applications now reflect a change of direction in EE, and focus on communications, media, and medicine.

Diggers, R., ed. 2003. *Encyclopedia of Optical Engineering*. New York: Dekker. This three-volume set provides comprehensive coverage of the topic including digital image enhancement, holography, radiometry, and lasers in medicine.

Dorf, R. C., ed. 1997. *The Electrical Engineering Handbook,* 2nd ed. Boca Raton, FL: CRC Press. In over 2,700 pages, this handbook provides broad and comprehensive coverage of the major topics of electrical engineering. Although it cannot provide the specificity found in more narrowly defined volumes, it is a good choice.

Handbook of International Safety Practices. 2010. Hoboken, NJ: John Wiley & Sons. Includes evaluation of safety practices, hazard assessments, monitors, equipment, and more.

Harper, C. 2005. *Electronic Packaging and Interconnection Handbook,* 4th ed. New York: McGraw-Hill. When the phrase "electronic packaging" was first used, some thought that it referred to the box within which a device was shipped. Not so. Electronic packaging refers to the process by which an electronic component is placed within a device so that it is protected as well as enabled to connect with other electronic components.

Hart, G. V. 2010. *Ugly's Electrical Desk Reference*, rev. 2008 ed. Sudbury, MA: Jones and Bartlett. Includes *Ugly's Electrical Reference* plus additional features.

IEEE. 2000. *The Authoritative Dictionary of IEEE Standards Terms,* 7th ed. Piscataway, NJ: IEEE. Within the standards authored by IEEE, it is essential that the technical terms be defined. This volume gives the definitions for terms that have been defined within the IEEE standards. In addition to giving the definition of a word or phrase, an index gives the IEEE standard in which the word or phrase is defined.

Kaiser, K. L. 2004. *Electromagnetic Compatibility Handbook.* Boca Raton, FL: CRC Press. In order to function properly, it is necessary for electronic devices to operate without interfering with other devices in the area. This handbook provides guidelines for ensuring this electromagnetic compatibility in an increasingly electronic environment.

Kaplan, S.M. 2004. *Wiley Electrical and Electronics Engineering Dictionary.* Hoboken, NJ: Wiley–IEEE Press. This is one of the many monographs that are co-published by Wiley and IEEE. Containing over 35,000 terms, this is an up-to-date resource for definitions of electrical engineering terms and acronyms.

Kent, A. and J. G. Williams. 2001. *Encyclopedia of Computer Science.* New York: Dekker. This set contains 14 core volumes and an index. Since its initial publication, 30 supplemental volumes have been added for a total of 45 volumes.

Linden, D. and T. B. Reddy. 2002. *Handbook of Batteries,* 3rd ed. New York: McGraw-Hill. This comprehensive volume covers dozens of battery types. Separate sections cover batteries used in electric vehicles, as well as portable fuel cells that may be a competitive alternative to battery systems.

McGraw-Hill. 2004. *McGraw-Hill Dictionary of Electrical and Computer Engineering.* New York: McGraw-Hill. This volume is derived from the *McGraw-Hill Dictionary of Scientific and Technical Terms,* 6th ed. Therefore, if you have the larger, more comprehensive work in your collection, this volume would be superfluous.

Miller, M. A. 2004. *Internet Technologies Handbook: Optimizing the IP Network.* Hoboken, NJ: John Wiley & Sons. This book is application oriented, not just focusing on protocol theory, but primarily on the practical management issues facing network professionals.

Petersen, J. K. 2002. *The Telecommunications Illustrated Dictionary,* 2nd ed. Boca Raton, FL: CRC Press. This volume contains over 10,000 terms covering many aspects of telecommunications. Biographies of telecommunications pioneers, timelines, and charts enhance the volume.

Schmidt-Walter, H., and R. Kories. 2007. *Electrical Engineering: a Pocket Reference.* Boston: Artech House.

Short, T. A. 2003. *Electric Power Distribution Handbook.* Boca Raton, FL: CRC Press. This book focuses on the distribution of electricity, including reliability, equipment, safety, and distributed generation.

Skvarenina, T. L. 2001. *The Power Electronics Handbook.* Boca Raton, FL: CRC Press. Power electronics is a key component in building more energy-efficient devices, such as appliances and heat pumps. This book emphasizes the practical aspects of this growing technology.

Toliyat, H. A. and G. B. Kilman. 2004. *Handbook of Electric Motors.* New York: Marcel Dekker. This book gives details on motor types as well as guidelines for motor selection.

Warne, D. F., ed. 2005. *Newnes Electrical Power Engineer's Handbook,* 2nd ed. Oxford, U.K.: Newnes. As the name implies, this book covers all aspects of electrical power, including generators, transformers, motors, batteries, and fuel cells.

Webster, J. G. 2001. *Wiley Encyclopedia of Electrical and Electronics Engineering,* Hoboken, NJ: John Wiley & Sons. This is the most comprehensive encyclopedia available in the area of electrical engineering. Only one supplemental update has been printed at this point. An online version is available, which Wiley says is updated "regularly." Because the print is rapidly losing its currency, the online version may be a better choice.

Whitaker, J. 2005. *Standard Handbook of Broadcast Engineering.* New York: McGraw-Hill. With the advent of digital TV and radio, engineers have need of an up-to-date source of information. This volume provides the data and equations necessary for understanding these new technologies.

Whitaker, J. C., ed. 2005. *The Electronics Handbook,* 2nd ed. New York: CRC Press. Provides both the basic theory and the practical applications of electronics. Its 2,640 pages are divided into 23 sections covering not only electronics, but also safety and reliability.

Wilson, J. 2005. *Sensor Technology Handbook.* Amsterdam/Boston: Elsevier. The purpose of this handbook is to assist engineers and designers in the selection of sensors for their applications. It covers various sensor types, manufacturers, guidelines for selecting and specifying sensors as well as information on MEMS and nanotechnology applications.

MONOGRAPHS: PROFESSIONAL AND TEXTBOOKS

It is often difficult within EE to make a clear distinction between professional and textbook monographs. Imagine a continuum that has introductory circuit theory textbooks at one end and scholarly professional volumes at the other. The introductory circuit theory books are clearly oriented toward the student, and the focused scholarly works are written for the researcher or practicing professional. As one moves toward the middle of the continuum, the distinctions become blurred. Titles used as textbooks for upper undergraduate courses may be quite specific and useful as resource materials for the professional. Titles that were written primarily for the researcher may form the basis for a graduate-level course.

Publishers sometimes add to the confusion by mingling professional monographs and textbooks within the same series. For instance, the Power System Series from Springer contains *Control of Electrical Devices,* classified as a textbook for advanced students of engineering, but also the book *Insulation of High-Voltage Equipment* for experts in power and high-voltage engineering.

Book vendors (e.g., BNA and Yankee Book Peddler) try to establish some order by classifying monographs with descriptors, such as lower undergraduate, upper undergraduate, graduate, and professional in order to distinguish between the academic levels. Even though the vendor categories are not exact, they can be useful for libraries that are setting up monograph collection development policies.

Libraries vary in how they do this. Some routinely exclude lower-level undergraduate textbooks from their acquisitions plans and only purchase upper-level undergraduate and graduate-level textbooks, along with professional volumes. Other libraries take the opposite tack and specifically purchase copies of textbooks being used in classes, sometimes placing them in the library reserve collection.

Two factors combine to make textbook collection development in EE particularly challenging. First, the electrical/computer engineering field has one of the largest student enrollments of all engineering disciplines, trailing only mechanical engineering (Gibbons, 2009). This creates a demand for a large number of textbooks. Second, the field is also one of the most dynamic in engineering, with changes occurring at a rapid pace. This creates a demand for frequent revisions containing the latest material. Publishers recognize this situation and have accommodated academia with a wealth of textbooks geared to every academic level.

Because of the large number of EE textbooks available in the marketplace, it would be outside the scope of this chapter to try to select those that would be most appropriate for a collection. Indeed, as noted above, some libraries have decided that undergraduate texts are not appropriate for an academic collection.

Two articles in *IEEE Spectrum* (Nebeker, 2003a; 2003b) discuss classic textbooks in electrical engineering. Further articles may be found through searches in *IEEE Xplore* or *Inspec.*

With regard to the professional-level monographs, it is best to start with professional societies, such as IEEE, ACM, and SPIE (International Society for Optical Engineering). From there, expand into the commercial publishers and select according to the research needs of the graduate students and faculty. These groups will be the primary users of professional-level books. Most undergraduate EE students will have neither the time nor the expertise to read monographs of this type.

Be wary of standing order plans for a publisher series. They can expand and consume ever larger portions of the acquisitions budget. In addition, as noted above, the academic level may vary within a series, with a mixture of textbooks and professional monographs. If a standing order seems appropriate, be sure to check the circulation statistics for the series after a year or two to see if the volumes are being used. In many cases, usage will vary greatly across the series, and it may be financially advantageous to buy selected volumes rather than the entire series.

In addition to the traditional paper format, electronic textbooks and monographs are becoming increasingly available. This format enables distance or e-learners to enjoy access to the same titles as their on-campus counterparts. Most major publishers have sites for their own titles, and several third-party vendors are providing cross-publisher packages. Some of the packages offer the advantage of automatically updating to the latest edition of a title when it becomes available, which is of significant importance in this rapidly changing field.

A slightly different type of monograph falls into the category of tutorial, user's manual, programming manual, cookbook, and so on. Although these books are not "scholarly" in nature, they cover computing topics in a practical manner and are heavily used by undergraduates and graduates in EE. Unfortunately, their popularity makes them prime candidates for theft, and their frequent updates make it difficult to keep abreast of the newest edition. They are, however, perfectly suited for an online collection. A major online provider is *ProQuest Safari Tech Books*, which offers the popular O'Reilly books as well as titles from over 50 other publishers.

JOURNALS AND CONFERENCE PROCEEDINGS

As is true with all other engineering disciplines, the journal literature is the foundation of scholarly communication. In electrical engineering, the conference literature also plays an important part.

In electrical engineering, the place to begin is IEEE. In terms of impact, quality, and cost-effectiveness, it is the leader. In the 2009 Impact Factor rankings in the Institute for Scientific Information's (ISI) *Journal Citation Report* (*JCR*), the IEEE published 16 out of the top 20 journals in the category of Electrical and Electronic Engineering.

Although IEEE offers various journal package plans that cover a subset of its total journal collection, if the budget will allow, a subscription to the IEEE All Society Periodicals Package (ASPP) will provide a better foundation for a quality collection. This package includes all 100+ journal titles from IEEE with a 2005 back file. Many established engineering programs have cancelled the print format of the IEEE journals in favor of online-only. For new programs, the online-only strategy also makes the most sense because it immediately gives researchers access to the 2005 back file.

For programs with larger budgets, the next serials addition should be one of the IEEE Proceedings Order Plans (POP). This will give students and faculty access to the important IEEE conference proceedings. Again, the best choice would be the online format because it gives immediate access to the back file.

For the largest programs, the *IEEE/IET Electronic Library* (*IEL*), also known *as IEEE Xplore*, is the best choice. This collection offers all IEEE and IET journals and proceedings, as well as IEEE standards. A major advantage of the IEL is its extended back file, which increases to 1988 for all titles and decades earlier for selected titles.

Once the foundation has been laid with the appropriate IEEE titles, the collection may be expanded with titles from other publishers. The ISI's *JCR* may be used to identify other important titles in the field. In addition to the category of Electrical and Electronic Engineering mentioned above, the *JCR* also ranks journals in the categories of Automation and control systems, Remote sensing, Robotics, and Telecommunications. Any or all of these categories may be appropriate, depending on the research interests of the institution.

Another strategy to determine journal additions for the collection is to identify the journals in which faculty are publishing. Talking directly with faculty may give valuable information about what they read and where they publish. A less direct method is to search *Inspec* or *Compendex* by faculty name to determine past publication patterns, or search by author affiliation.

It is difficult to recommend specific journals, publishers, or package plans other than IEEE because the needs of institutions vary greatly. For smaller undergraduate programs, the IEEE ASPP along with a very few additional titles will probably meet the needs of the user group. For the largest institutions, the complete IEL, along with individual titles or packages from Elsevier, Wiley, Springer, and others may be necessary.

Regardless of the size of the electrical engineering program or the collection, online access statistics have made it possible to fine-tune the collection to an extent that was heretofore impossible. By analyzing the number of article downloads from each online journal, it is possible to calculate the cost per article for each journal and determine if it is more cost-effective to subscribe to the journal or to purchase articles on demand. It is no longer necessary to rely on inaccurate reshelving counts or other means to determine usage, and funds that are being spent on low-use journals may be used to subscribe to other journals that have the potential for higher usage.

As noted above, a journal collection should start with the IEEE and IET publications. Some of the major titles outside of the IEEE and IET publishers are listed below.

ACM Transactions on Sensor Networks (1550-4859) ACM. Focuses on research and applications of distributed, wireless, or wireline sensor and actuator networks.

Automatica (0005-1098) Pergamon–Elsevier. Covers all aspects of theoretical and experimental control theory.

Autonomous Robots (0929-5593) Springer Science. Theory and applications of autonomous robotic systems with preference given to papers that include data from actual robots in the real world.

Bioinspiration & Biomimetics (1748-3182) IOP Publishing. This journal reports on research that applies principles abstracted from natural systems to engineering and technological design and applications.

Expert Systems with Applications (0957-4174) Pergamon-Elsevier. The journal's focus is on expert and intelligent systems applied in industry, government, and universities.

GPS Solutions (1080-5370) Springer. The journal covers system design issues as well as current and emerging applications of global navigation satellite systems.

Industrial Robot—An International Journal (0143-991X) Emerald Group. The journal uses a themed issue approach as well as featuring both practical and research articles.

International Journal of Humanoid Robotics (0219-8436) World Scientific. The journal covers all subjects related to the mind and body of humanoid robots.

International Journal of Innovative Computing Information and Control (1349-4198) ICIC International. The journal publishes papers on the theory and applications of intelligent systems, information, and control.

International Journal of Robotics Research (0278-3649) Sage Publications. Covers many aspects of robotics including applied mathematics, computer science, and electrical and mechanical engineering.

ISPRS Journal of Photogrammetry and Remote Sensing (0924-2716) Elsevier. This journal is the official journal of the International Society for Photogrammetry and Remote Sensing (ISPRS).

Journal of Field Robotics (1556-4959) John Wiley & Sons. This journal publishes articles dealing with the fundamentals of robotics in unstructured environments.

Journal of Geodesy (0949-7714) Springer. The journal is concerned with the study of scientific problems of geodesy and related interdisciplinary sciences.

Journal of Machine Learning Research (1532-4435) Microtome Publishing. This unique journal is freely available online, and covers all aspects of machine learning.

Personal and Ubiquitous Computing (1617-4909) Springer. This offers research on handheld, wearable, and mobile information devices.

Progress in Electromagnetics Research—PIER (1559-8985) EMW Publishing. Covers all aspects of electromagnetic theory and applications.

Progress in Quantum Electronics (0079-6727) Pergamon-Elsevier. Publishes papers on quantum electronics and its applications.

Remote Sensing of Environment (0034-4257) Elsevier. The journal publishes in the areas of terrestrial, oceanic, and atmospheric sensing.

Robotics and Autonomous Systems (0921-8890) Elsevier. Experimental and theoretical aspects of robotics with an emphasis upon autonomous systems.

Robotics and Computer-Integrated Manufacturing (0736-5845) Pergamon-Elsevier. The emphasis is upon the application of research to the development of new or improved industrially relevant manufacturing technologies, equipment, and strategies.

With regard to conference proceedings, the IEEE and IET conferences have already been mentioned as the first choice. For most programs the next collection to consider would probably be SPIE: The International Society for Optical Engineering http://www.spie.org (accessed October 12, 2010). These volumes are available in both print and online formats. The online format has the advantage of offering an immediate back file of several years. The print format offers the advantage of customization and, therefore, cost savings, because it is only necessary to purchase those volumes that are actually needed by students and faculty.

Other society proceedings that may be of interest to electrical engineers are those sponsored by the Association for Computing Machinery, (ACM), www.acm.org, and the Society for Industrial and Applied Mathematics, (SIAM), www.siam.org. The *Lecture Notes in Computer Science* series by Springer also may be of interest.

STANDARDS AND CODES

Standards are an important part of the electrical engineer's world. In the past, it was common for libraries to maintain just-in-case hard copy collections of standards. Later, the standards became available on CD-ROM through commercial vendors, and are now available from vendors through immediate download from the Web.

The result is that libraries no longer need to invest large sums of money for collections and collection maintenance in order to provide just-in-case access to standards. These flat-rate, subscription-based packages will provide a cost advantage only for standards from the most heavily used organizations. For most standards collections, a just-in-time strategy may be used to deliver downloaded standards to clients on demand.

In most cases, standards requests are for a specific standard, rather than a request for a standard on a specific topic. Researchers will have found a reference to a specific standard in the literature, or be required to meet a specific standard in a proposal or project. The need is simply to retrieve the standard.

The major commercial standards suppliers have been identified in Chapter 2 of this book, along with a discussion of the online indexes. As noted therein, many societies do not sell standards directly, but contract with third-party vendors to do this.

Several organizations produce standards relevant to electrical engineering:

IEC http://www.iec.ch (accessed October 12, 2010). International Electrotechnical Commission. The IEC prepares and publishes international standards in the area of electrical engineering. In doing this, it works closely with organizations such as the IEEE and ISO.

IEEE http://www.ieee.org (accessed October 12, 2010). Standards may be purchased directly through the IEEE Web site as well as through third-party vendors. All current IEEE standards are included in the *IEL/IEEE Xplore* package described above.

ISO (see Chapter 2 in this book).

ITU http://www.itu.int (accessed October 12, 2010). International Telecommunication Union. The ITU publishes two major sets of recommendations, the ITU-T and the ITU-R. The ITU-T standards can be downloaded free of charge from the ITU Web site, while the ITU-R standards may be purchased at the ITU Web site as well as through third-party vendors.

NFPA http://www.nfpa.org (accessed October 12, 2010). National Fire Protection Association. The NFPA publishes the *National Electrical Code*, updated every three years, which gives guidelines for electrical installations within buildings.

UL http://www.ul.com (accessed October 12, 2010). Underwriters Laboratories. Available at http://www.comm-2000.com/ (accessed December 15, 2010) or from third-party vendors, the UL standards are probably more readily recognized by the public than any others.

WEB SITES

Many of the most important sites related specifically to EE have already been mentioned in two previous sections of this chapter: (1) Article Indexes and Full-Text Resources and (2) Standards and Codes.

A limited list of EE-related sites can be found at the WWW Virtual Library—Electrical and Electronics Engineering, http://www.cem.itesm.mx/vlee (accessed October 12, 2010). The links are grouped under the categories of Academic and research institutions, Blogs, Information resources, Journals and magazines, Products and services, and Standards. The site also allows keyword searches in order to find relevant entries.

In addition to EE-specific sites, there are sites that cover many engineering disciplines and have EE components. Two of the most helpful are the *Scout Report* and *Engineering Village*, both of which have been covered in Chapter 2 in this book. The *Scout Report*, which is freely available, has an archival search feature that retrieves a listing of reviewed Web sites matching the search criteria. Two examples of the valuable sites listed in the Scout Report follow:

Electric Power Research Institute (EPRI) http://www.epri.com (accessed October 12, 2010). Founded in 1973, EPRI is a nonprofit collaborative organization established for the purpose of research and development in electric generation, delivery, and use.

Energy Information Administration http://www.eia.doe.gov/fuelelectric.html (accessed October 12, 2010). This contains a wealth of statistical data including power generation capacities, pricing, and state-by-state information as well as the full text of many reports.

In addition, general Web directories often have EE sections within them. These EE sections contain information on topics such as discussion groups, professional societies, conferences, and university programs:

Google Electrical Engineering http://www.google.com/Top/Science/Technology/Electrical_Engineering/ (accessed October 12, 2010).

Yahoo Electrical Engineering http://dir.yahoo.com/Science/Engineering/Electrical_Engineering/ (accessed October 12, 2010).

ASSOCIATIONS AND SOCIETIES

There are numerous associations and societies related to the field of electrical engineering. Some have already been mentioned in previous sections of this chapter, but a listing of the more important ones includes:

Association for Computing Machinery http://www.acm.org (accessed October 12, 2010). Founded in 1947, this association is oriented heavily toward the computer literature, but is used by electrical engineers.

Audio Engineering Society http://www.aes.org (accessed October 12, 2010). Founded in 1948, this society consists of those who work with recording and reproducing equipment.

Institute of Electrical and Electronics Engineers (IEEE) http://www.ieee.org (accessed October 12, 2010). Founded in 1963 by the merger of the American Institute of Electrical Engineers (1884) and the Institute of Radio Engineers (1912), it is one of the premier organizations for publishing and conferences in electrical engineering. There are numerous societies within the IEEE that meet the needs of the specific subdisciplines of electrical engineering.

Institute of Engineering and Technology (IET) http://www.iet.org (accessed October 12, 2010). Dating back to 1871, the IET is the publisher of *Inspec*, one of the most important indexes in the field.

International Microelectronic and Packaging Society (IMAPS) http://www.imaps. org (accessed October 12, 2010). Founded in 1967, IMAPS holds an annual national conference as well as international, national, and regional seminars.

Associations Unlimited http://www.galegroup.com (accessed October 12, 2010) is described in Chapter 2 in this book, and is valuable for locating additional groups in the field of electrical engineering.

GREEN ELECTRICAL ENGINEERING

Electrical engineers are deeply involved with green technology. Whether the discussion involves hybrid vehicles (electric motors and batteries), alternative power generation (wind and solar), or smart homes (heating and lighting control), electrical engineering plays a pivotal role. The vast majority of journal literature related to these green fields is available from the same publishers that provide information in the traditional areas of electrical engineering.

A search conducted in *Inspec*, through *Engineering Village*, on the controlled vocabulary term *solar cells* retrieves 32,973 hits. In rank order of number of publications, the first 20 publisher names (and their imprints) listed for this set published 19,174 (58% of the total) articles. Of these first 20, the following publishers (and their imprints) contributed 18,271 (55% of the total) articles: American Chemical Society, American Institute of Physics, Elsevier, IEEE, Institute of Physics, Kluwer, Materials Research Society, SPIE, Springer, and Wiley.

Searches in *Inspec* on "smart power grids" and "hybrid electric vehicles" do not provide results quite as skewed as the "solar cells" search. Still, it is clear that in these relatively new areas of research the traditional publishers produce the majority of the articles. Certainly, in all of these results as one proceeds down the ranked list of publishers, there will be an increasing number of niche publishers listed. However, one also finds name variations and imprints of the major publishers, so these continue to increase their percent of titles published in the field.

Regarding the specific journals covering green engineering, in some instances, journals focus on green engineering. For example, titles such as *IET Renewable Power Generation, IEEE Transactions on Smart Grid, IEEE Transactions on Sustainable Energy, Renewable Energy Focus* (Elsevier), and *Solar Energy Materials and Solar Cells* (Elsevier) all have a "green" emphasis.

However, in other cases, articles addressing green engineering are incorporated into journals not specifically targeted toward green technology. Thus, although IEEE publishes a transactions journal emphasizing the smart grid, recent articles on the smart grid have also been published in *IEEE Power and Energy Magazine* as well as *IEEE Transactions on Power Systems*. Similarly, while many of Elsevier's articles on solar cells appear in *Solar Energy Materials and Solar Cells,* others are found in Elsevier's *Thin Solid Films* and *Journal of Non-Crystalline Solids*.

The dispersal of green engineering research across many journals, both green focused and broader based, simply demonstrates that this research does not exist as a standalone undertaking. It is often a continuation or modification of previous research and, thus, is reported in traditional publications that have been in existence for extended periods of time.

Those seeking to build a green electrical engineering journal collection should first look to their current holdings. If they have an adequate collection in the traditional areas of electrical engineering from the major publishers, then they likely have adequate holdings in the areas of green electrical engineering. This is particularly true if the university subscribes to a publisher package, such as the IEEE ASPP or the IEL. Difficulties may arise if the library subscribes only to specific journals in a focused area. For instance, a library subscribing to selected IEEE journals in the area of power generation and distribution would need to add the new IEEE title on smart grids in order to have complete coverage in the area. New and niche titles in green engineering can be found in the same manner as one finds traditional engineering journals: ask faculty and/or search databases to determine where faculty publish and what they cite.

Likewise, much of the monographic literature related to the academic side of green electrical engineering is published by traditional firms. Springer, CRC, Wiley, and others are well represented. In addition to the scholarly literature, this field has a significant body of guidebooks or do-it-yourself books. These cover everything from setting up a residential wind turbine to installing solar panels to converting a car to electric power. Some of the books are produced by independent publishers, such as Earthscan (*Solar Domestic Water Heating* and *Stand-Alone Solar Electric Systems*), while others are imprints of traditional publishers, although not always obviously so. For instance, the popular "for dummies" series (*Solar Power Your Home For Dummies, Wind Power For Dummies,* and *Photovoltaic Design & Installation For Dummies*) is actually a Wiley imprint.

It is a given that a collection in green electrical engineering should include the scholarly monographs. It is not as clear what books should be purchased in the how-to category. A search in *Worldcat* on *Solar Power Your Home for Dummies* reveals that the vast majority of copies are held in public and community college libraries. Conversely, a search on *Stand-Alone Solar Electric Systems*, which is billed as a practical guide "to plan, design, and install solar electric systems," reveals predominantly university holdings, including MIT, Purdue, Berkeley, and Penn State. As with any acquisition, the purchase must fit the need of the user. It is possible that a "dummies" book may provide the practical guidance a senior design student needs to finish a project on a residential solar installation.

NANOTECHNOLOGY AND ELECTRICAL ENGINEERING

Most of the comments made about green electrical engineering also can be applied to the incorporation of nanotechnology into electrical engineering. In the vast majority of cases, the journal literature relating to nanotechnology is distributed by the same publishers that are major players in the rest of electrical engineering. For those institutions that do not subscribe to a package plan, the specific journals will need to be selected to support the research interests of the faculty. As examples, the *IEEE Transactions on Nanotechnology* would most certainly be included in a nanotechnology collection, but a subscription to the *IEEE Transactions on NanoBioscience* might depend upon the specific research focus of the institution.

Within the monographic literature, most of the traditional scholarly publishers have nano-oriented offerings. Any comprehensive approval or firm order profile will bring these titles to the attention of the bibliographer for purchase.

CONCLUSION

Electrical engineering continues to be one of the most dynamic branches of engineering. Advances in telecommunications, biotechnology, nanotechnology, transportation, alternative methods of energy generation, and other areas of research all guarantee that the field will continue to expand. However, while the field expands, the number of publishers offering resources will continue to contract as mergers and consolidations take place. Also, it is likely that the number of distinct engineering handbooks and reference volumes may decrease as publishers rely more heavily on online packages rather than the sale of individual titles. Whatever the changes in format, though, as the research becomes more interdisciplinary, it will become increasingly important for electrical engineers to efficiently access the information they need.

REFERENCES

Gibbons, M. 2009. Databytes: A slow surge. *ASEE Prism* 18 (7): 22–23.

Irwin, J. D. and D. V. Kerns, Jr. 1995. *Introduction to electrical engineering*, Englewood Cliffs, NJ: Prentice-Hall.

Martin, T. C. and S. L. Coles, eds. 1919. *The story of electricity: Volume one: A popular and practical historical account of the establishment and wonderful development of the electrical industry.* New York: The Story of Electricity Company, M.M. Marcy.

Nebeker, F. 2003a. Treasured texts. *IEEE Spectrum* 40 (4): 44–49.

Nebeker, F. 2003b. More treasured texts. *IEEE Spectrum* 40 (7): 31–36.

Sarma, M. S. 2001. *Introduction to electrical engineering.* New York: Oxford University Press.

11 Engineering Education

Jill H. Powell and Jeremy Cusker

CONTENTS

Introduction 259
History of Engineering Education 260
Searching the Library Catalog 261
Research Databases..262
    Articles 263
Bibliographies and Guides to the Literature 264
Directories...264
Books 266
Reports 268
Journals 271
Conferences...272
Web Sites...272
Societies 273
Academic Discussion Lists and Newsgroups 274
Conclusion 274
Acknowledgments...274
References..275

INTRODUCTION

Engineering education is college- or university-level study that prepares students for a career as an engineer. A certain number of courses in mathematics, basic sciences, engineering sciences, engineering design, and other studies are required. Some schools offer five-year or more cooperative plans that combine classroom study with practical work in industry (commonly called *co-op*). (*Occupational Outlook Handbook*, http://www.bls.gov/oco/ocos027.htm accessed August 9, 2010.)

Besides the four-year bachelor's degree, there are programs of study leading to a master's degree, which may or may not require the production of original research or a Doctor of Engineering or PhD degree, which always requires original research. A doctorate is usually required to teach in schools of engineering.

A discussion of different types of engineers, their training, salaries, and job outlook can be found in the *Occupational Outlook Handbook*, which also describes the nature of their work: "Engineers apply the principles of science and mathematics to develop economical solutions to technical problems. Their work is the link between scientific discoveries and the commercial applications that meet societal and consumer needs" (*Occupational Outlook Handbook*).

In the United States, ABET (Accreditation Board for Engineering and Technology) is the official accreditation agency charged with granting universities the right to confer engineering degrees. Their Web site maintains a database of accredited programs, statistics, and resources at http://www.abet.org (accessed August 9, 2010). After receiving a degree, state licensing exams must then be passed by engineers offering their services directly to the public.

One of the best-regarded books describing the history of engineering education is Lawrence Grayson's *The Making of an Engineer: An Illustrated History of Engineering Education in the United States and Canada* (Grayson, 1993). This book was written to celebrate the centennial of the American Society for Engineering Education (ASEE) and contains some 350 photographs as well as an exhaustive history of engineering education. For current engineering educational methods, see John Heywood's book *Engineering Education: Research and Development in Curriculum and Instruction*. It is exhaustive and covers engineering education from the 1960s to the present (Heywood, 2005).

Research in engineering education is not complete without consulting the publications and extensive Web pages of the ASEE itself. ASEE's stated mission is "furthering education and engineering technology" and "promoting excellence in instruction, research, public service, and practice" http://www.asee.org/about/missionAndVision.cfm (accessed June 7, 2010).

A study of engineering education means looking at a number of different groups: students, faculty, practitioners, and alumni, and involves looking at courses, programs, and the means of assessment and evaluation. Topics relating to students include learning styles, admission and graduation requirements, student advising, opportunities for students to become involved in faculty research, retention, and the involvement of (or failure to involve) minority groups. Topics relating to faculty include publications, research methods, hiring, promotion, and tenure, and teaching methods used in the classroom. Consideration of engineering practitioners and alumni involve consideration of industry and collaborations between institutions of engineering education and industry. Assessment and evaluation refer to efforts to measure the effectiveness of overall engineering education, improvement of specific courses, plus topics like distance learning, design, and ethics courses.

Other topics include the history of engineering education, and the situation of engineering education within the broader context of education and learning theory. Taken together, they comprise the foundational subjects of engineering education.

HISTORY OF ENGINEERING EDUCATION

In 1802, West Point was established as the first engineering school in the United States (Hollister, 1966, p. 144). Engineering programs then began at Rensselaer Polytechnic Institute in 1824, Norwich University in 1825, and then began to be offered throughout the 1850s at the University of Michigan, Harvard, Yale, Union College, and Dartmouth (Reynolds, 1991, p. 20). There were 30 such schools by 1909 (Grayson, 1993, p. 77). In 2004, there are 343 accredited schools of engineering in the United States: http://www.asee.org/publications/ (accessed June 7, 2010).

Parallel both to engineering education and the profession as a whole, various engineering societies developed beginning in the mid- to late 1800s, including the American Society of Civil Engineers (ASCE) in 1852, American Society of Mechanical Engineers (ASME) in 1880, American Institute of Electrical Engineers in 1884, and later the American Institute of Chemical Engineers (AIChE) in 1908 (Reynolds, 1991, pp. 24, 355).

Engineering became established in the academic community with the passage of the Morrill Land Grant Act in 1862. This act provided public lands and/or funds to the states to establish colleges of agriculture and mechanical arts, an early term referring to engineering. During the 1870s and 1880s, inventions such as the steam engine, electric generator, internal combustion engine, incandescent electric lamp, phonograph, and telephone all served to accelerate and interconnect the country and the world through what became known as the Industrial Revolution. It became increasingly clear that a large group of systematically trained individuals would be needed to develop, maintain, and operate the myriad new inventions. The World Columbian Exposition of 1893, held in Chicago, showcased many such inventions and a parallel meeting of engineers at the exposition resulted in the formation of the ASEE, which provided a forum for engineering educators to identify and address such issues as entrance standards, curricula, course content, and requirements for graduation from engineering education programs. Previously, engineering education had developed in an unplanned and uncoordinated manner, but after the creation of ASEE, schools had a central resource for advice on what to teach, how to teach it, and ways to affect national policy, not the least being the state certification of trained engineers (Grayson, 1993, pp. 4–7, 43).

Europe at the time was perhaps even further along in the process of systematizing the education of engineers. Wilhelm von Humboldt, Prussian Minister of Education, oversaw the system of Technische Hochschule and Gymnasium that made Prussia a premier scientific and intellectual power in Europe in the nineteenth century. He introduced two features that influenced many universities in Europe and North America. One of the requirements was that both research and teaching be done by the same people; in other words, that professors would be not merely instructors regurgitating received knowledge, but also would be technical experts, actively generating new knowledge. The other development was the freedom for the professor to decide what to teach and research and for students to have the freedom to design their own program of study. The Humboldt University educational experience was neither practical nor focused on technical skills for a specific profession. It aimed to produce educated, well-rounded citizens and scholars for the long term. But, the revolution in educational thinking that Humboldt was to introduce had some of its most profound effects on teaching and research in science and technology (Doepke, 2003).

Today, the large increase in the ratio of students to faculty has weakened some of this ideology, particularly in regards to students designing their own course of study. Modern undergraduate programs in the United States and many other places have long since established degree programs consisting mainly of required sets of courses. However, Humboldt's ideas are still apparent in America's graduate schools. Comprehensive exams replace credit point systems, and the exams are not always linked to the specific content of courses. Students work with faculty to design their own research and the main criteria for receipt of a graduate degree is that the students produce some new piece of knowledge. American graduate schools thus still embody many of the principles of Humboldt, particularly in educating those with a deep interest in science and research (Doepke, 2003).

Challenges in engineering education that continue to this day include the debate about the best way to prepare graduates. Should they be trained as specialists or generalists? Should they be taught in industrial shops for immediate usefulness in the workplace, or educated in a laboratory, studying fundamental scientific principles? At various times in history, each of these ideas has become dominant.

Until around 1950, engineering education in the United States was vocationally rather than scientifically oriented. However, after World War II, the emphasis shifted to fundamental principles, particularly as physicists came to be valued for their role in weapons and energy development and as the boundaries between science and engineering became more fluid. World War II played several transformative roles in engineering education. It infused large sums of money into universities for the purposes of research and the government paid for many returning draftees to train as engineers. The competition from the Cold War with the Soviet Union also increased the United State's desire to educate more engineers and scientists. As the twentieth century wore on, companies in a competitive marketplace began to demand more practically trained engineers. The cycle continues, as schools respond by emphasizing engineering practice, manufacturing techniques, and concepts, such as reliability and quality (Grayson, 1993, p. x). In the late 1990s and continuing into the twenty-first century, information technology has driven the U.S. economy, and courses on entrepreneurial skills are included in the curriculum. Women, minorities, and large numbers of foreign-born students also began to receive degrees and join the workforce. This education landscape hopefully produces engineers with the skills to improve all aspects of engineering practice and to strengthen the economy.

SEARCHING THE LIBRARY CATALOG

While keyword searching will find some useful books, one can find more complete information with the use of subject headings. Listed below are some of the most useful Library of Congress Subject Headings regarding engineering education.

Engineering: study and teaching
Engineers: education
Electric engineering: study and teaching
Mechanical engineering: study and teaching
Civil engineering: study and teaching
Chemical engineering: study and teaching
Technical education
Science: study and teaching

The headings may be subdivided further by country, type of study, and type of format. For example, since many conference papers are published in conference proceedings, the term Conference or congresses may be present.

Engineering: study and teaching (graduate)
United States Engineering: study and teaching, periodicals
Engineers: education, United States
Engineering: study and teaching (Higher), computer-assisted instruction, congresses
Chemical engineering: study and teaching, congresses

Several specific headings having to do with experiential or internship-type, preprofessional training also exist and may be useful when combined with the keyword "Engineers" or "Engineering":

Education: cooperative, engineering
Industry and education

RESEARCH DATABASES

Research databases, also called abstracts and indexes, offer access to citations in journal articles, conference papers, and theses. They are vital instruments in the early steps of research. The best databases for literature covering recent engineering education are *Compendex, Inspec*, Google Scholar, and *Applied Science and Technology Abstracts*. For early historical information going back as far as the eighteenth century, *Making of America* and the *History of Science, Technology, and Medicine* databases can be useful tools. Depending on the specific type of engineering studied, additional indexes should be added.

ERIC. 1966–. New York: Ovid Technologies Inc. *ERIC* abstracts journal articles and reports on education. Many materials, such as curriculum guides, instructional materials, conference papers, and project reports are available in the *ERIC* collection. *ERIC* indexes materials on education, child development, classroom techniques, computer education, counseling, testing, communication skills, career education, science education, teacher education, evaluation, disabled children, gifted children, and library and information science. An online thesaurus shows relevant subject headings: engineering education, professional education, engineering technology, engineers, land grant universities, science education, and technical education.

Google Scholar http://scholar.google.com (accessed April 1, 2011). Free and far-ranging index across many disciplines and sources: articles, theses, books, abstracts, and from academic publishers, professional societies, online repositories, universities, and other Web sites. Google Scholar also indicates how many citations each article has received.

History of Science, Technology, and Medicine (HST). 1975–. Palo Alto, CA: Research Libraries Information Network. *HST* is the source for the earliest information on engineering education. It comprises four bibliographies: *Bibliografia Italiana di Storia*

Della Scieuza, Current Bibliography in the History of Technology, Technology and Culture, Current Work in the History of Medicine (Wellcome Library), and *Isis Current Bibliography of the History of Science.* It indexes journal articles, conference proceedings, books, book reviews, and dissertations on the history of science and technology. Relevant subject headings include engineering education, technical societies, technical education; education, technical.

Making of America: The Cornell University Library MOA Collection http://digital.library. cornell.edu/m/moa/ (accessed December 15, 2010). 1996–. Ithaca, NY: Cornell University Library. A digital library of primary sources in American history from the antebellum period through reconstruction. The collection is strong in education, psychology, American history, sociology, religion, and science and technology. Full-page images of journals are included.

When researching engineering education, one should go beyond the general science and education indexes given above. Specific disciplines are best covered by their own databases. Society publications are also indexed in these sources. For example, engineering education for computer science would be well covered by *Inspec, IEEE Xplore*, and *ACM Portal,* which indexes publications by the Association for Computing Machinery.

Articles

Dym, C. L., et al. 2005. Engineering design thinking, teaching, and learning. *Journal of Engineering Education* 94 (1): 103–120. Clear discussions of various teaching models, including design thinking, project-based learning, cornerstone courses, and how they have been used to strengthen engineering education.

Felder, R. M. 1988. Learning and teaching styles in engineering education. *Engineering Education* 78 (7): 674–681. Updated Web text http://www4.ncsu.edu/unity/lockers/users/f/felder/public/Papers/LS-1988.pdf (accessed October 27, 2010). Originally written in 1987, the author returned to it in 2002 with a new Web preface stating what features he would have preferred to see changed in the light of new research, but generally affirming its continuing value as an important reference in the field to include more up-to-date language and concepts. It discusses the differences that may exist among both teachers and students in engineering education classrooms and how best to recognize and adapt learning and teaching processes to them.

Felder, R. M., D. R. Woods, et al. 2000. The future of engineering education II: Teaching methods that work. *Chemical Engineering Education* 34 (1): 2–39.

Felder, R. M., Felder, G. N., et al. 2000. The effects of personality type on engineering student performance and attitudes. *Journal of Engineering Education* 91 (1): 3–17. This study examined the relationship of learning style to the identified Myers–Briggs index personality type of a group of 116 undergraduate engineering students.

Gordon, B., and M. Silevitch. 2009. Re-engineering engineering education. *New England Journal of Higher Education* 24(1): 18–19. This article examines shortcomings in American engineering education as a subset of larger problems related to educating enough science, technology, and mathematics students. It examines specific problems related to teaching engineering undergraduates to work as part of large-scale project teams once they become professionals.

Heitsch, A., et al. 2009. NANOLAB at the University of Texas at Austin: A model for interdisciplinary undergraduate science and engineering education. *Chemical Engineering Education* 43 (3): 225–231. This article presents a useful case study of an interdisciplinary teaching "station" aimed at integrating the experiences of students in both engineering and science disciplines. With a specific focus on nanoscale science and engineering, UT-Austin aimed to break down some of the insularity and barriers to understanding between different disciplines and fields of study as well as the somewhat abstract nature of undergraduate engineering and science education.

Mazur, Eric. 2004–2007. *Interactive Teaching: Promoting Better Learning Using Peer Instruction and Just-in-TimeTeaching.* Upper Saddle River, NJ: Pearson. DVD. Harvard University physicist demonstrates the use of peer instruction and Just-in-Time teaching in an interactive workshop.

Rugarcia, A., R.M. Felder, et al. 2000. The future of engineering education I: A vision for a new century. *Chemical Engineering Education* 34 (1): 16–25. These two articles (part two is listed under Felder) first question whether the prevailing thesis (that engineering education is severely deficient in some areas) could not be remedied with a few changes in teaching rather than a top-down revision of curricula. Then, the second article presents several discrete examples of improvements in teaching that could improve engineering education without any need for massive reorganization. Felder's article also refers to the Index of Learning Styles, a self-scoring instrument that assesses preferences on the Sensing/Intuiting, Visual/Verbal, Active/Reflective, and Sequential/Global dimensions http://www.ncsu.edu/felder-public/ILSpage.html (accessed April 4, 2011).

BIBLIOGRAPHIES AND GUIDES TO THE LITERATURE

There are several annotated bibliographies that describe research in the field of engineering education. Currency is always a concern, and one should follow up with database searches within the entries in the Research Databases section. However, these bibliographies can save time in reviewing historical papers.

Channell, D. F. 1989. *The History of Engineering Science: An Annotated Bibliography.* New York: Garland (*Bibliographies of the History of Science and Technology,* vol. 16). A chapter titled "Institutions" lists over 100 annotated references on the history of various engineering colleges around the world. Research institutes and professional organizations also are covered.

Cooper, J. and P. Robinson. 1997. *Small-Group Instruction: An Annotated Bibliography of Science, Mathematics, Engineering, and Technology Resources in Higher Education, Occasional Paper.* University of Wisconsin, ED472334. This source addresses research, theory, and practice in small groups and cooperative instruction in higher education.

Dyrud, M. A. 2004. 2003 Engineering technology education bibliography. *Journal of Engineering Technology* 21 (2): 26–40. Published yearly, this resource lists books and articles under a number of detailed subject headings.

Seymour, E. 2002. Tracking the processes of change in U.S. undergraduate engineering education in science, mathematics, engineering, and technology. *Science Education* 86 (1): 101–105. A review article that discusses the changes in engineering education over the past decade and cites over 70 articles on the subject. Excellent coverage of many hard-to-find articles.

DIRECTORIES

Directories help prospective students to evaluate the nature, scope, requirements, and statistics of various engineering programs. Many are published yearly, and most libraries will keep the current year only. Many directories have online equivalents and the URLs are noted where found; however, subscriptions may be required. This is not an exhaustive list. For specific disciplines, consult library catalogs using appropriate subject headings, e.g., Chemical engineering: study and teaching.

America's Best Graduate Schools. 1994–. Washington, D.C.: U.S. News and World Report http://www.usnews.com/usnews/edu/grad/rankings/rankindex_briefphp (accessed June 7, 2010). Controversial, subjective, and popular ranking of graduate schools, including engineering. Ranks by program. Subscription portion lists average GRE scores, enrollments, research expenditures, acceptance rates, recruiter scores, and more. This title is usually published during the month of March. The subscription section provides more in-depth information.

American Council on Education. 1928–. *American Universities and Colleges.* New York: Walter de Gruyter. Contains detailed descriptions of over 1,900 institutions in the United States. Narratives on structure of higher education, foreign students, undergraduate, graduate, and professional education. Descriptions include history, structure, admission requirements, statistical characteristics of freshmen, degrees offered, fees, enrollment, student life, library collections, and more. Indexes, tables, and appendices are included. One of the most useful indexes is the listing of colleges by degree program.

ASEE Directory of Engineering and Engineering Technology Colleges. 1997–. Washington, D.C.: American Society for Engineering Education http://www.asee.org/publications/ profiles/index.cfm (accessed June 10, 2010). Allows prospective engineering students to compare over 343 engineering and engineering technology colleges in the United States and Canada. Profiles include institutional information, undergraduate information, and graduate information. Statistical tables and profiles include information on enrollment, degrees granted, student expenses, faculty numbers, and research expenditures. Indexes by degree program, geographic location, and alphabetical by institution. The online version contains much more information, including research centers, student support programs, dual degrees, and department areas of expertise.

ASEE Directory of Graduate Engineering and Research Statistics. 1999–. Washington, D.C.: American Society for Engineering Education. Partially online at http://www.asee.org/colleges/ (accessed June 7, 2010). This directory provides extensive statistical information on graduate engineering departments. It compares programs in areas such as enrollments, degrees awarded by gender and ethnic groups, student appointments by department with average monthly stipend, and faculty statistics. Includes index by subject areas of graduate engineering-related research.

ASEE Directory of Undergraduate Engineering Statistics. 1997–. Washington, D.C.: American Society for Engineering Education. Partially online at http://www.asee.org/colleges/ (accessed June 7, 2010). This directory provides extensive statistical information on undergraduate engineering departments. It compares programs in areas such as enrollments, degrees awarded by gender and ethnic groups, faculty numbers, student expenses, and number of co-op participants.

Bear, J., M. Bear, L. McQueary, and T. C. Head. 2001. *Bear's Guide to the Best Computer Degrees by Distance Learning.* Berkeley, CA: Ten Speed Press. Compares 100 computer science degree programs with profiles and statistics.

Chemical Engineering Faculty Directory. 2000–. http://www.che.utexas.edu/che-faculty/ (accessed June 7, 2010). Previously available as *Chemical Engineering Faculties*, 1900– 1999. Basic faculty directory maintained by members at the University of Texas, Austin.

DGRWeb. 1997–. Washington, D.C.: American Chemical Society http://dgr.rints.com (accessed June 7, 2010). Includes ACS Directory of Graduate Research, publications, doctoral and master's theses in departments of chemistry, chemical engineering, biochemistry, polymer science, materials science, forensic science, marine science, toxicology, and environmental science at U.S. and Canadian universities.

Directory of Human Factors/Ergonomics Graduate Programs in the United States and Canada. 2002. Santa Monica, CA: Human Factors and Ergonomics Society http://www. hfes.org/web/students/grad_programs.html (accessed June 7, 2010). Helpful for prospective graduate students in assessing the nature, scope, and requirements of various programs.

Engineering and Technology Degrees. 1967/1968–. New York: Engineering Workforce Commission of American Association of Engineering Societies. This annual report identifies degrees awarded by race, gender, and type of degree covering 339 engineering institutions and 284 engineering technology institutions.

Engineering and Technology Enrollments. 1974–. New York: Engineering Workforce Commission of the American Association of Engineering Societies. This annual directory covers 351 engineering and 282 engineering technology institutions. Enrollment data are by institution, field, gender, race, and part-time or full-time status.

Fiske, E. B. 1988–. *The Fiske Guide to Colleges*. Naperville, IL: Sourcebooks, Inc. Contains lengthy profiles of more than 300 colleges, discussing strengths and weaknesses. Includes student quotes on academics and social life. A survey helps students select the best college.

GradSchools.com http://www.gradschools.com/listings/menus/mech_eng_menu.html (accessed June 7, 2010). This site provides access specifically to engineering graduate school programs that offer master's and PhD degrees. Provides access to universities by subject, by state, and to engineering programs available worldwide. Provides links directly to the engineering department of universities.

Graduate Programs in Engineering and Applied Sciences. 1986–. Princeton, NJ: Peterson's Guides. The directory (book five in the six-volume set) describes more than 4,000 graduate programs in 67 disciplines. Includes admission requirements, expenses, financial support, programs of study, and faculty research specialties. Typeface is very small.

Keane, C. M. 1992–. *Directory of Geoscience Departments*. Alexandria, VA: American Geological Institute. Brief information on 1,047 geoscience departments in the United States and the rest of the world. Includes faculty names, contact information, and enrollments. Federal agencies of interest to geoscientists also are included.

Occupational Outlook Handbook. 1949–. U.S. Department of Labor, Bureau of Labor Statistics. Indianapolis, IN: JIST Publications http://www.bls.gov/oco/home.htm (accessed June 7, 2010). From the Web site: "*The Occupational Outlook Handbook* is a nationally recognized source of career information, designed to provide valuable assistance to individuals making decisions about their future work lives. Revised every two years, the handbook describes what workers do on the job, working conditions, the training and education needed, earnings, and expected job prospects in a wide range of occupations."

Research Doctorate Programs in the United States: Continuity and Change. 1995. Committee for the Study of Research–Doctorate Programs in the United States, National Research Council. Washington, D.C.: National Academy of Sciences. Contains information on descriptive statistics of selected characteristics of PhD granting institutions and faculty views on program quality. Extensive narratives, tables, and figures compare institutions.

World of Learning. 1947–. London: Europa Publications. Available online to subscribers at http://www.worldoflearning.com (accessed June 7, 2010). Lists contact information, publications, faculty, and statistics on colleges and universities worldwide.

BOOKS

Galloway, P. D. 2007. *The 21st-Century Engineer: A Proposal for Engineering Education Reform*. Reston, VA: ASCE Press. Beginning with the premise that, "While engineers [of today] remain strong in terms of their technological skills, they are generally weak in terms of their management and communications capabilities" (p. 2–3), this book lays out a general series of proposals by which engineering education could be reformed. Chapters include Globalization, Communication, Ethics and Professionalism, and Diversity, among others, each one laying out ways by which the world of both technology and business have changed and how the engineering profession and engineering education should change to accommodate them.

Grasso, D. and M. Brown Burkins. 2010. *Holistic Engineering Education*. New York: Springer. (Available as an e-book via SpringerLink for subscribers). In this work, Grasso and Brown Burkins describe the need for future engineers to be "holistic" and less focused on individual fields of technology (i.e., electrical, chemical, mechanical) and the need rather for "engineers who manage, lead, and understand complex, interdisciplinary systems that bring the power of engineering thought to issues spanning and connecting technology, law, public policy, sustainability, the arts, government, and industry" (p. 1). The book's focus is broader than just the education of engineers in colleges in the United States and also has chapters on K–12 preparation for engineering education, parallel developments in liberal arts curricula, and engineering education in an international perspective.

Grayson, L. P. 1993. *The Making of an Engineer: An Illustrated History of Engineering Education in United States and Canada*. New York: John Wiley & Sons. This book was written to celebrate the centennial of the American Society for Engineering Education (ASEE) and has some 350 photographs as well as an exhaustive textual history of engineering education in the United States and Canada. Highly recommended.

Heywood, J. 2005. *Engineering Education: Research and Development in Curriculum and Instruction*. Piscataway, NJ: IEEE. This is a highly detailed book regarding the development of engineering education both in the United States and elsewhere since 1960. It also closely examines teaching techniques and research into the improvement of engineering education methods. Covers learning styles, outcomes, projects/problem-based learning, creativity, design, and new technologies. Highly recommended.

Hoag, K. 2001. *Skill Development for Engineers: An Innovative Model for Advanced Learning in the Workplace*. London: IEEE. Contains many practical examples of successful approaches, with diagrams. Contents include: Moving beyond the classroom, Management roles, Mechanisms for advanced learning, Communicating the information, The supervisor's role, and There is no such thing as a free lunch.

Kline, R. R. 1992. *Steinmetz: Engineer and Socialist*. Baltimore, MD: Johns Hopkins University Press. Chapter 7, entitled Reforming a Profession (pp. 165–199) covers the history of engineering education, the first colleges to offer engineering degrees, and Charles Steinmetz's role as college educator, professional society leader, and reformer.

Layton, E. T., Jr. 1986. *The Revolt of the Engineers: Social Responsibility and the American Engineering Profession*. Baltimore, MD: Johns Hopkins University Press. Examines the politics of engineering societies.

Leslie, S. W. 1993. *The Cold War and American Science: The Military-Industrial-Academic Complex and MIT and Stanford*. New York: Columbia University Press. Discusses the changes World War II and the Cold War funding of engineering had on the workings of these two universities.

Mazur, Eric. 1997. *Peer Instruction: A User's Manual*. Upper Saddle River, NJ: Prentice Hall. Contains CD. Highly cited book on improving learning through shortening lectures and having students teach each other concepts.

Mazur, Eric. 2004–2007. *Interactive teaching: Promoting Better Learning Using Peer Instruction and Just-in-Time Teaching*. Upper Saddle River, NJ: Pearson Prentice Hall. DVD. Harvard University physicist Eric Mazur demonstrates the use of peer instruction and Just-in-Time teaching in an interactive workshop.

Oldenziel, R. 1999. *Making Technology Masculine: Men, Women, and Modern Machines in America, 1870–1950*. Amsterdam: Amsterdam University Press. Excellent coverage of women engineers up to the founding of the Society of Women Engineers in the 1950s.

Patil, A. and P. Gray, eds. 2009. *Engineering Education Quality Assurance*. Berlin: Springer (e-book). This book examines means of assessing the effectiveness of engineering education and examines case studies of assessment of such education for certification purposes.

The book takes a highly international perspective, examining engineering education in Europe, Asia, and South America.

Project 2061, American Association for the Advancement of Science. 1993. *Benchmarks for Science Literacy.* New York: Oxford University Press. While focused on what students should know about science, mathematics, and technology by the end of grades 2, 5, 8, and 12, this book provides well-thought-out competencies, discussions, and perspectives. Engineering educators may use this book for K–12 collaborations and also as a tool to what students should know before they come to college.

Project 2061, American Association for the Advancement of Science. 2000. *Designs for Science Literacy.* New York: Oxford University Press. Contains curriculum designs, specifications, and suggestions for improvement.

Reynolds, T. S. 1991. *The Engineer in America: A Historical Anthology from Technology and Culture.* Chicago: University of Chicago Press. Containing essays that originally appeared in the journal *Technology and Culture,* this volume presents a thorough history of engineering and engineers from its earliest beginnings to the twentieth century.

Rosser, S. V. 1995. *Teaching the Majority: Breaking the Gender Barrier in Science, Mathematics, and Engineering.* New York: Teachers College Press. Discussions by various authors on how to alter teaching to include feminist and other diverse approaches into the science and engineering classroom. Chapters focus on various disciplines, such as chemistry, physics and engineering, mathematics, computer science, environmental sciences, and geosciences.

Scarl, D. 1998. *How to Solve Problems: For Success in Freshman Physics, Engineering and Beyond.* Glen Cove, NY: Dosoris Press. Teaches problem-solving methods that experienced scientists and engineers use to define a problem, solve it, and present their solution to others.

Wankat, P. C. and F. S. Oreovicz. 1993. *Teaching Engineering.* New York: McGraw-Hill. An excellent practical guide for engineering instructors, covering all aspects of teaching college-level engineering, from the component of good teaching to efficiency, designing lectures, teaching with technology, laboratories, testing, ethics, and evaluation.

Weichert, D., B. Rauhut, and R. Schmidt, eds. 2001. *Educating the Engineer for the 21st Century.* Dordrecht, the Netherlands: Kluwer Academic. Provides published papers from the *Proceedings of the 3rd Workshop on Global Engineering Education.* Topics include European, Asian, and American views of engineering education, developing personal skills, programs, curricula, educational concepts, successful university–industry partnerships, and design projects for the global engineer.

REPORTS

The following selected reports (many produced by committees) are recommended as good places to start familiarizing oneself with engineering education. Most were found from searching the *ERIC* database, *Worldcat* (OCLC), and National Academies Press Publications http://www.nap. edu/ (accessed May 3, 2011).

A National Action Plan for Addressing the Critical Needs of the U.S. Science, Technology, Engineering and Mathematics (STEM) Education System. 2009. Washington, D.C.: National Science Foundation, National Science Board. http://www.nsf.gov/nsb/documents/2007/ stem_action.pdf (accessed May 3, 2011). This report lays out a proposed action plan to the President and Congress of the United States to address shortcomings in STEM education. Specific proposals include the formation of a national but nonfederal council on STEM education, the creation of a new Assistant Secretary of Education specifically tasked with improving STEM education, and the involvement of the National Science Foundation

(NSF) in the writing of national goals and proposed standards for science, technology, and mathematics education.

Building a Workforce for the Information Economy. 2001. National Research Council. Committee on Workforce Needs in Information Technology, Board on Testing and Assessment, Board on Science, Technology, and Economic Policy. Office of Scientific and Engineering Personnel. Washington, D.C.: National Academy Press http://books.nap.edu/catalog/9830.html (accessed June 7, 2010). This report presents the results of a study on the supply and demand for information technology workers over the next 10 years. It discusses the nature of IT work, employee demographics, perceived worker shortages, older IT workers and age discrimination, foreign workers, and long-term recommendations for the future.

Educating the Engineer of 2020: Adapting Engineering Education to the New Century. 2005. Committee on the Engineer of 2020, Phase II. Washington, D.C.: Committee on Engineering Education, National Academy of Engineering http://www.nap.edu/catalog.php?record_id=11338 (accessed Aug 9, 2010). Report that responds to the questions posed by *The Engineer of 2020: Visions of Engineering in the New Century.* This report offers recommendations on how to improve engineering education so graduates are better prepared to work in a rapidly changing global economy.

Engineer of 2020, The: Visions of Engineering in the New Century. National Academy of Engineering of the National Academies. Washington, D.C.: National Academy Press. http://books.nap.edu/openbook.php?isbn=0309091624 (accessed Aug 9, 2010). This report, the first of two parts, was chartered by the National Academy of Engineering's Committee on Engineering Education to answer questions about the state of engineering in 2020. It develops a technological vision of engineering practice and desired attributes of engineers for 2020 based on scenario-based planning.

Engineering Education: Designing an Adaptive System. 1995. National Research Council. Board of Engineering Education, Commission on Engineering and Technical Systems, Office of Scientific and Engineering Personnel. Washington, D.C.: National Academy Press http://www.nap.edu/catalog/4907.html (accessed June 7, 2010). Board of Engineering Education's report on the strengths and weaknesses of engineering education and ways to achieve progress amid global political, societal, and economic changes. Includes outlines for a new engineering curriculum and specific recommendations for institutions, faculty, industry, and NSF. A key theme is "think globally, act locally."

Enhancing the Postdoctoral Experience for Scientists and Engineers: A Guide for Postdoctoral Scholars, Advisers, Institutions, Funding Organizations, and Disciplinary Societies. 2000. Committee on Science, Engineering, and Public Policy (U.S.). National Academy of Sciences, National Academy of Engineering, Institute of Medicine. Washington, DC: National Academy Press http://www.nap.edu/catalog.php?record_id=9831 (accessed Aug 10, 2010). This comprehensive guide addresses issues in the postdoctoral experience, including rights, opportunities, responsibilities, information regarding funding and advisers, and recommendations for improvements. Information is supplied from meetings with 39 focus groups at 11 universities, 7 national laboratories, and 5 private research institutes or industrial firms, plus a day-long workshop with 100 participating postdocs and administrators

Fox, M. A. and N. Hackerman, eds. 2003. *Evaluating and Improving Undergraduate Teaching in Science, Technology, Engineering, and Mathematics.* Committee on Recognizing, Evaluating, Rewarding, and Developing Excellence in Teaching of Undergraduate Science, Mathematics, Engineering, and Technology. National Research Council. Washington, D.C.: National Academy Press http://www.nap.edu/catalog.php?record_id=10024 (accessed Aug 10, 2010). This report examines effective teaching practices, and offers both methodologies and practical ways of evaluating teachers and academic

programs. Included are recommendations on how to achieve change and sample evaluation instruments.

*From Analysis to Action: Undergraduate Education in Science, Mathematics, Engineering,
and Technology.* 1996. Report from the National Research Council/National Science
Foundation Convocation on Undergraduate Science, Mathematics, and Engineering
Education. Washington, D.C.: National Academy Press http://www.nap.edu/catalog.
php?record_id=9128 (accessed June 7, 2010). Synopsis of a convocation held in April 1995
with students, faculty, and administrators to discuss such issues as access, literacy, competency, curriculum, teaching, reward systems, and evaluation systems.

Grant, H. 1993. The role of quality concepts in engineering education. In *Frontiers in
Education Conference. Twenty-Third Annual Conference Proceedings. Engineering
Education: Renewing America's Technology,* pp. 535–539. New York: IEEE. Following a
survey of deans of engineering and other research, a task force presented action items that
NSF could implement to improve the quality of education, a bibliography of basic research
in quality in engineering education, a workshop on quality of engineering education, and
alliance proposals with other groups including ABET, ASEE, and others.

Jackson, S. A. 2003. *Envisioning a 21st Century Science and Engineering Workforce for
the United States: Tasks for University, Industry, and Government.* Washington, D.C.:
National Academy of Sciences, National Academy of Engineering, Institute of Medicine
http://books.nap.edu/catalog/10647.html (accessed June 7, 2010). Written by the president
of Rensselaer Polytechnic Institute, this paper finds the competitiveness of the nation's
science and engineering talent to be eroding and reviews options to manage risks to U.S.
technological innovation.

Science and Engineering Indicators 2010. 2010. Division of Science Resource Statistics.
Arlington, VA: National Science Foundation (NSB 02-01) http://www.nsf.gov/statis-
tics/seind10/ (accessed June 7, 2010). This report contains comprehensive, quantitative
analyses of the scope and quality of the nation's science and engineering programs and
enterprises. The report presents material on science education from the elementary level
through graduate school; the scientific and engineering workforce; R&D performers, U.S.
competitiveness in high technology; public understanding of science and engineering; and
the significance of information technologies.

*Shaping the Future. New Expectations for Undergraduate Education in Science,
Mathematics, Engineering, and Technology.* 1996. Advisory Committee to the National
Science Foundation (NSF) Directorate for Education and Human Resources. Arlington,
VA: National Science Foundation. ERIC document ED404158, NSF 96-139 http://www.
nsf.gov/publications/pub_summ.jsp?ods_key=nsf96139 (accessed June 7, 2010). This report
is the product of more than a year of intensive work studying problems in undergraduate
education at both two- and four-year institutions. It addresses the state of education at the
time (1995), barriers to improvement, and ways to meet new expectations, with specific
recommendations for colleges, universities, businesses, state and federal governments, and
the National Science Foundation.

*Shaping the Future. Volume II: Perspectives on Undergraduate Education in Science,
Mathematics, Engineering and Technology (SMET).* 1998. Advisory Committee to the
National Science Foundation (NSF) Directorate for Education and Human Resources.
Arlington, VA: National Science Foundation http://www.nsf.gov/pubs/1998/nsf98128/
nsf98128.htm (accessed June 7, 2010). ERIC document ED433882, NSF 98-128. Vol.
II discusses activities that have taken place since the publication of Vol. I, and contains testimony by university deans, professors, and industry leaders presented at committee hearings. Remarks focus on developing and implementing strategies to improve
undergraduate SMET education. The report also includes focus group findings, exten-

sive background data on enrollment, expenditures, and faculty practices as well as an 18-page bibliography.

Transforming Undergraduate Education in Science, Mathematics Engineering, and Technology. 1999. Committee on Undergraduate Science Education. Center for Science, Mathematics, and Engineering Education. National Research Council. Washington, D.C.: National Academy Press. Presents six visions for raising scientific awareness of all students, not just science and engineering students. These include requiring courses, establishing admission standards, evaluating these courses, having higher education professionals share responsibility for K–12 teacher education, and having institutions provide the necessary infrastructure to achieve these visions. Extensive references.

JOURNALS

Below is a listing of journals that publish articles on engineering education and are indexed in the databases mentioned in the Research Databases section. Published articles may be categorized under such topics as students, faculty, practitioners, courses and programs, assessment and evaluation, and history of engineering education.

ASEE Prism. 1991–. Washington, D.C.: American Society for Engineering Education (1056-8077).

Chemical Engineering Education. 1965–. Gainesville, FL: Chemical Engineering Division, American Society for Engineering Education. (0009-2479).

Computer Applications in Engineering Education. 1992–. New York: John Wiley & Sons (1061-3773).

European Journal of Engineering Education. 1975–. Abingdon, U.K.: Carfax. English, French, and German (0304-3797).

Global Journal of Engineering Education. 1997–. Melbourne, Australia: UNESCO. International Centre for Engineering Education, English and German (1328-3154).

IEEE Potentials. 1982–. New York: Institute of Electrical and Electronics Engineers (0278-6648).

Innovations in Science and Technology Education. 1986–. Paris: UNESCO.

International Journal of Engineering Education. 1992–. Hamburg, Germany: Tempus. Publications (formerly *International Journal of Applied Engineering Education* 1985–1991. New York: Pergamon) (0949-149X).

International Journal of Science Education. 1987–. London/New York: Taylor & Francis, (formerly *European Journal of Science Education*, 1979-1986) (0950-0693).

Journal of Engineering Education. 1910–1969; 1993–. Other title: *Engineering Education*, 1970–1992. Washington, D.C.: American Society for Engineering Education (1069-4730).

Journal of Engineering Technology. 1984–. Washington, D.C.: Engineering Technology Division, American Society for Engineering Education (0747-9964).

Journal of Materials Education. 1983–. University Park, PA: Materials Research Laboratory, Pennsylvania State University (0738-7989).

Journal of Professional Issues in Engineering Education and Practice. 1991–. New York: American Society of Civil Engineers (formerly *Journal of Professional Issues in Engineering*, 1983–1990; *Issues in Engineering*, 1979–1982; *Engineering Issues*, 1958–1978) (1052-3928).

Journal of Science Education and Technology. 1992–. New York: Plenum Press (1059-0145).

Journal of the Learning Sciences. 1991–. Hillsdale, NJ: Lawrence Erlbaum Associates (1050-8406).

Technos. 1972–2002. Fort Collins, CO: American Society for Engineering Education. International Division (0363-308X).

CONFERENCES

Conference papers are indexed by certain databases, particularly *Engineering Village 2 Compendex* and *Inspec.*

ASEE Annual Conference Proceedings. 1978–. (formerly *Proceedings. Papers, Reports, Discussion*s, printed in the *Journal of Engineering Education* 1893–1966). Also called American Society of Engineering Education ASEE National Conference. Washington, D.C.: American Society for Engineering Education http://www.asee.org/conferences (accessed June 7, 2010). ASEE International Colloquium of Engineering Education. 2002–. American Society for Engineering Education. European Society for Engineering Education (SERI). Technical University Berlin (TUB). Chinese Academy of Engineering. Locations vary. http://www.asee.org/conferences/international/2010/index.cfm (accessed June 7, 2010).

ASEE Engineering Deans Institute. ASEE Engineering Deans Council. Includes programs such as "Engineering Education 2020: What a College of Engineering May Look Like in 20 Years" http://www.asee.org/about/events/conferences/edi.cfm (accessed June 7, 2010).

ASME International Mechanical Engineering Congress. 1989–. (Also called International Mechanical Engineering Congress and Exposition). New York: American Society of Mechanical Engineers http://www.asme.org (accessed June 7, 2010).

Frontiers in Education Conference. 1975–. New York: Institute of Electrical and Electronics Engineers. Earlier names: IEEE Conference on Frontiers in Education, International Conference on Frontiers in Education http://www.fie-conference.org/ (accessed June 7, 2010).

Proceeding Conference for Industry and Education Collaboration. 1997–. Washington, D.C.: American Society for Engineering Education (formerly College-Industry Education Conference, 1976–1996).

World Congress on Engineering Education. 1975–. World Federation of Engineering Organizations. Washington, D.C.: American Society for Engineering Education, International Division.

WEB SITES

There are many tutorials, lectures, and discussions online if one searches "interactive teaching" and limits the search to videos and a subject, such as physics. Eric Mazur, professor of physics at Harvard University, has many online clips about his interactive teaching methods in physics.

American Society for Engineering Education http://www.asee.org (accessed June 15, 2010). The most important society for engineering education in the United States provides:

- Engineering Resources: http://www.asee.org/resources/index.cfm (accessed Aug 10, 2010). Lists Web sites on distance education, multimedia course, research centers, and textbooks.
- eGFI: http://teachers.egfi-k12.org/ (accessed Aug 10, 2010). Contains lesson plans and experiments, data on outreach programs, career guidance materials, and links to readings related to engineering education.

MIT OpenCourseWare http/ocw.mit.edu/ (accessed March 23, 2011) an initiative of Massachusetts Institute of Technology (MIT) using OpenCourseWare technology to put all educational materials from its courses online and make them openly available since 2007. Many other universities are putting selected lectures online as well.

NASA http://www.nasa.gov (accessed June 15, 2010). Click on "For Educators." Includes teaching resources, including downloadable multimedia and curriculum materials separated by age group.

National Science Digital Library (NSDL) http://nsdl.org (accessed June 15, 2010). From the Web site: "NSDL is a digital library of exemplary resource collections and services, organized in support of science education at all levels. Starting with a partnership of NSDL-funded projects, NSDL is emerging as a center of innovation in digital libraries as applied to education, and a community center for groups focused on digital library-enabled science education."

NEEDS http://www.needs.org/needs/ (accessed June 15, 2010). "The National Engineering Education Delivery System (NEEDS) is a digital library of learning resources for engineering education. NEEDS provides Web-based access to a database of learning resources where the user (whether they be learners or instructors) can search for, locate, download, and comment on resources to aid their learning or teaching process" (from Web page). NEEDS and its sponsor, John Wiley & Sons, support the Premier Award, which recognizes high-quality, noncommercial courseware designed to enhance engineering education.

PBS TeacherSource http://www.pbs.org/teachersource/ (accessed June 15, 2010). Rich in educational resources, this site provides access to 4,500 free lesson plans and activities.

Teacher's Domain: Multimedia Resources for the Classroom and Professional Development http://www.teachersdomain.org/ (accessed June 15, 2010). A multimedia digital library for K–12 teachers and students, produced jointly by WGBH Boston (PBS station) and the National Science Foundation. You will find an extensive collection of classroom-ready resources as well as media-rich lesson plans and professional development resources. Each resource is catalogued by grade level and correlated to national and state standards. Teachers' Domain is produced by WGBH, with major funding from the National Science Foundation.

SOCIETIES

Accreditation Board for Engineering and Technology (ABET) http://www.abet.org (accessed June 15, 2010). Sets standards for degree programs providing qualifications for engineering practice.

American Society for Engineering Education http://www.asee.org (accessed June 15, 2010). Publishes several journals and directories: *Engineering College Research and Graduate Study, Directory of Graduate Programs in Engineering, Journal of Engineering Education.* The society sponsors conferences and is the most important society in engineering education in the United States.

American Technical Education Association http://www.ateaonline.org (accessed June 15, 2010).

Engineering Institute of Canada http://www.eic-ici.ca/ (accessed June 15, 2010). Has both English- and French-language site indices.

European Society for Engineering Education http://www.sefi.be/ (accessed June 15, 2010).

Europaische Gesellschaft fur Ingenieur-Ausbildung. See European Society for Engineering Education, above.

International Association for Continuing Education and Training http://www.iacet.org/ (accessed June 15, 2010).

National Academy Press Publications (many are online and full text) http://www.nap.edu/ (accessed June 15, 2010).

National Council of Examiners for Engineering and Surveying http://www.ncees.org/ (accessed June 15, 2010).

National Science Foundation http://www.nsf.gov/ (accessed June 15, 2010).

National Society of Professional Engineers http://www.nspe.org/ (accessed June 15, 2010).

Societe Europeanne pour la Formation des Ingenieurs. See European Society for Engineering Education, above.

Society for the History of Technology http://shot.press.jhu.edu/ (accessed June 15, 2010).

Society of Women Engineers http://www.swe.org/ (accessed June 15, 2010).

Special Libraries Association (SLA) Engineering Division http://units.sla.org/division/deng/index.html (accessed July 1, 2010). The SLA Engineering Division serves as a clearinghouse for information by, for, and about librarians who work with engineers and engineering students. Their primary membership comes from librarians working in engineering libraries at colleges and universities.

Tau Beta Pi Association http://www.tbp.org/pages/main.cfm (accessed June 15, 2010).

Triangle Fraternity http://www.triangle.org (accessed June 15, 2010).

In addition, it can be useful to consult the Web sites of many discipline-specific engineering societies, such as the Institute of Electrical and Electronics Engineers (IEEE), American Society of Civil Engineers (ASCE), American Society of Mechanical Engineers (ASME), Association for Computing Machinery (ACM), American Institute of Chemical Engineers (AIChE), American Society of Heating, Refrigerating and Air Conditioning Engineering (ASHRAE), and Society of Automotive Engineers International (SAE). A complete list of engineering societies may be found in the *International Directory of Engineering Societies and Related Organizations.* 1993–2000. Washington, D.C.: American Association of Engineering Societies. Formerly *Directory of Engineering Societies and Related Organizations* (1900s–1989). Besides contact information, it lists the organizations' officers, publications, objectives, budget, geographic, and student chapter data. Discussion of many of these organizations is located in the appropriate chapter of this book.

ACADEMIC DISCUSSION LISTS AND NEWSGROUPS

Academic discussion lists and newsgroups are e-mail forums where people discuss issues of common interest. Newsgroups messages are posted to a Web site. One can search newsgroups at www.groups.yahoo.com and www.groups.google.com. Groups that discuss the issues of engineering education include sci.engr.education, misc.education.science, sci.engr.civil, sci.engr.mech, sci.engr.chem, and sci.edu.

CONCLUSION

The Engineer of 2020: Visions of Engineering in the New Century describes attributes for an engineer in 2020 as:

> What attributes will the engineer of 2020 have? He or she will aspire to have the ingenuity of Lillian "Gilbreth, the problem-solving capabilities of Gordon Moore, the scientific insight of Albert Einstein, the creativity of Pablo Picasso, the determination of the Wright brothers, the leadership abilities of Bill Gates, the conscience of Eleanor Roosevelt, the vision of Martin Luther King, and the curiosity and wonder of our grandchildren" (*The Engineer of 2020,* p. 57).

Keeping in mind these attributes, this chapter provides a starting point for researching the history and current state of engineering education in the United States. Whether the future lies in new programs or in the older, established colleges, or both, there is much to be learned by studying their examples and the resources enumerated here.

ACKNOWLEDGMENTS

We thank Jeffrey Harris from the National Science Foundation, and Cornell Professors Ronald Kline, Simpson Linke, Anthony Ingraffea, and Michel Louge for reading this chapter and contributing useful entries and comments.

REFERENCES

ABET http://www.abet.org/ (accessed August 9, 2010). The Accreditation Board for Engineering and Technology is the recognized accreditor for college and university programs in applied science, computing, engineering, and technology.

American Society for Engineering Education http://www.asee.org/about/mission.cfm (accessed June 15, 2010).

American Society for Engineering Education http://www.asee.org/publications/ (accessed June 15, 2010).

Doepke, M. 2003. *Humboldt University—Now and then.* http://www.runder-tisch-usa.de/chicago/site/statements/matthiasdoepke.html (accessed June 15, 2010).

Engineer of 2020: Visions of Engineering in the New Century, The. National Academy of Engineering of the National Academies. Washington, D.C.: National Academy Press http://books.nap.edu/openbook.php?isbn=0309091624 (accessed Aug 9, 2010).

Grayson, L. P. 1993. *The making of an engineer: An illustrated history of engineering education in the United States and Canada.* New York: John Wiley & Sons.

Heywood, J. 2005. *Engineering education: Research and development in curriculum and instruction.* Hoboken, NJ: IEEE Press, Wiley-Interscience.

Hollister, S. C. 1966. Engineer. New York: Macmillan.

Occupational outlook handbook http://www.bls.gov/oco/ocos027.htm (accessed December 8, 2010).

Reynolds, T. S. 1991. *The engineer in America: A historical anthology from technology and culture.* Chicago: University of Chicago Press.

Rosser, S. V. 1995. *Teaching the majority: Breaking the gender barrier in science, mathematicals, and engineering,* New York: Teachers College Press.

12 Environmental Engineering

*Robert Tolliver (based on the first edition
by Linda Vida and Lois Widmer)*

CONTENTS

Introduction, History, and Scope of Discipline 278
Searching the Library of Congress Catalog 279
   Keywords, LC Subject Headings, and LC Call Numbers 279
      LC Call Numbers: Water 279
      Tips 280
      LC Subject Headings: Air 280
      LC Subject Headings: Water 280
   Keyword Searching 281
Article Indexes 281
Data Sources and Datasets 284
Bibliographies and Guides to the Literature 285
Directories 286
Encyclopedias 286
Dictionaries 287
Handbooks, Manuals, and Properties 288
   General 288
   Air Environment 291
   Water Environment 292
Monographs and Textbooks 294
   General 294
   Air Environment 296
   Hazardous Waste, Solid Waste, and Water Environment 297
Journals 298
Standards, Test Methods, Guidance, and Criteria Documents 303
   Air 303
      Federal and State Government Resources 303
      Associations, Societies, Organizations, and Commercial Resources 304
   Water 304
Criteria Documents 306
Regulations 306
Search Engines, Important Web sites, Portals, Discussion Lists, Weblogs 307
   Air Environment 307
   Water Environment 308
   Government Agencies 308
      U.S. Army Corps of Engineers (USACE) 308
      U.S. Department of Energy (DOE) 309
      U.S. Department of the Interior (DOI) 309
      U.S. Bureau of Reclamation (USBR) 309
      U.S. Fish and Wildlife Service (FWS) 309

U.S. Geological Survey (USGS) 309
U.S. Environmental Protection Agency (EPA)..310
U.S. National Oceanic and Atmospheric Administration (NOAA)....................................311
Associations, Organizations, Societies, and Conferences..311
Conclusion 315
Acknowledgments..315
Further Readings..315

INTRODUCTION, HISTORY, AND SCOPE OF DISCIPLINE

As the first decade of the twenty-first century comes to a close, we are confronted by many environmental issues. Environmental engineers have an important role to play in addressing these issues. The last few years have seen new environmental crises, such as the BP Deepwater Horizon oil spill in the Gulf of Mexico, as well as ongoing environmental threats, such as climate change, water pollution, and waste disposal. The demand for energy combined with new drilling technologies has made natural gas-rich shales, such as the Marcellus Shale in the northeastern United States, attractive sources of energy. Many of our new electronic devices rely on rare Earth elements like neodymium that are currently only mined extensively in China, increasing the interest in finding new sources of these mineral resources. Each of these developments brings environmental concerns that require the expertise of environmental engineers and the continued development of new ideas.

The history of environmental engineering is extensive, although the discipline was not called that until recently. Early concerns about water quality appear in human history dating back to 2000 BCE. The ancient Romans built an extensive aqueduct system to provide fresh water to their cities. Some evidence indicates that they had wastewater treatment collection systems as well.

The field of environmental engineering developed from a branch of civil engineering when the public became concerned about the spread of diseases. In the United States in the 1850s, engineers were primarily dedicated to providing infrastructure to transport water or to build drinking water distribution systems. Drinking water treatment became more widespread in the 1900s. The field of wastewater treatment engineering advanced more slowly until the 1950s when treatment systems were standardized.

Health concerns about air pollution can easily be documented as early as the seventeenth century, attributable to industrial development as well as to the fuel sources for domestic heating and cooking. It took such twentieth-century incidents as the 1952 London smog case and the 1948 Donora, Pennsylvania air pollution inversion, both of which produced increased particulate matter pollution leading to increased mortality, to stimulate regulatory action and steps to control air quality.

In the United States, a broad awareness of environmental issues began in the 1960s. In the 1970s, President Nixon authorized the creation of the U.S. Environmental Protection Agency (EPA) and the National Oceanic and Atmospheric Administration (NOAA) to establish and enforce environmental protection standards to improve the environment. These agencies were empowered to conduct research on the adverse effects of pollution and to develop new methods and equipment for controlling pollution. This was the beginning of an effort to gather information on pollution and to use this information to strengthen environmental programs and recommend policy changes.

A discipline that had begun to address drinking water and wastewater treatment issues expanded and now encompasses groundwater, stormwater, estuarine environments, aquatic ecology, desalination, hazardous wastes and solid wastes, acid deposition, smog, chlorofluorocarbons, indoor air quality and hazardous air pollutants, and global climate change.

Today, environmental engineering focuses on complex and interdisciplinary issues and is concerned with a broad range of environmental contaminants and their impacts on the environment. Environmental engineering requires the understanding of natural processes as well as an understanding of waste streams, the development of applied technologies, and knowledge of past and current engineering practices. Environmental engineers need to understand the fundamentals of contaminants in water, air, soils, and hazardous waste and strive to use their understanding to develop and apply technologies that will improve the environment.

SEARCHING THE LIBRARY OF CONGRESS CATALOG

KEYWORDS, LC SUBJECT HEADINGS, AND LC CALL NUMBERS

Broad terms for searching this topic and broad call numbers:

Environmental engineering, (LC call numbers TA170 through TA171)
Pollution (LC call numbers TD172 through TD193.5)
Air quality (LC call number TD883)
Air Pollution (LC call numbers TD881 through TD890, for the technology)

LC Call Numbers: Water

The Library of Congress classification system places environmental engineering materials primarily in Class T–Technology. Some of the most important areas are as follows:

GB621-628	Wetlands
GB1199	Aquifers
RA591-598.5	Water supply in relation to public health
TA170-171	Environmental engineering
TC401-506	Water supply engineering
TD1-1066	Environmental technology. Sanitary engineering
TD159-168	Municipal engineering
TD169-171.8	Environmental protection
TD172-193.5	Pollution
TD201-500	Water supply for domestic and industrial purposes
TD419-428	Water pollution
TD429.5-480.7	Water purification. Water treatment and conditioning. Saline water conversion
TD481-493	Water distribution systems
TD657	Urban runoff
TD511-780	Sewage collection and disposal systems
TD783-812.5	Municipal refuse, Solid wastes
TD878-894	Special types of environment including soil pollution, air pollution, and noise pollution
TD895-899	Industrial and factory sanitation
TD896-899	Industrial and factory wastes
TD920-934	Rural and farm sanitary engineering
TD940-949	Low temperature sanitary engineering
TD1020-1066	Hazardous substances and their disposal
TH6014-6081	Environmental engineering of buildings, Sanitary engineering of buildings
TJ212-225	Control engineering systems, Automatic machinery (general)
TP155-156	Environmental chemistry

As with most environmental issues, finding information on environmental engineering for air quality control requires searches of the specialized environmental literature as well as the engineering and technical literature.

Some useful Library of Congress Subject Headings (LCSH) for searching library catalogs are listed below. They are arranged hierarchically with the narrower subject headings indented under the broader headings to show the relationship among terms; however, each term may be searched independently.

Tips

All these subject headings have a range of subdivisions, such as a geographic location or publication type, that can be appended to the subject heading, e.g.,

Air pollution–Canada
Air pollution–Handbooks

For information on a specific substance or pollutant, search the substance name and qualify it by including the subdivision "Environmental aspects," e.g.,

Dioxins–Environmental aspects

LC Subject Headings: Air

Air analysis
Air quality
Air quality indexes
Air quality management
Atmospheric chemistry
Atmospheric deposition
Automobiles: motors, exhaust gas
Automobiles: motors, exhaust systems
Automobiles: pollution control devices
Environmental chemistry: industrial applications

Environmental engineering
Environmental health
Environmental indicators
Environmental protection
Environmental law
Environmental monitoring
Global warming
Greenhouse effect, atmospheric
Pollutants
Pollution

LC Subject Headings: Water

Aquifers (see also ground water basin)
Contaminated sediments
Desalination of water (use saline water conversion)
Drinking water
Environmental engineering (related term: environmental health)
Environmental monitoring
Factory and trade waste (used for industrial wastes)
Groundwater (a variant form is ground water) (the former term was water, underground)
Hazardous substances
Hazardous waste management industry hazardous waste sites
Hazardous wastes (used for hazardous waste disposal)
Hydraulic engineering
Hydrogeology
Hydrology
Industrial effluent (use sewage)

Industrial wastes (use factory and trade waste)
Injection wells
Marine pollution
Municipal engineering
Municipal water supply
Pollutants
Pollution
Pollution control industry
Pollution control equipment
Reverse osmosis
Runoff
Saline water conversion
Saline water barriers
Saline waters
Salinity
Sanitary engineering
Sanitary landfills
Sedimentation and deposition
Sewage
Sewage disposal
Sewage lagoons

Sewage sludge
Sewage sludge digestion
Sewerage (used for house drainage, sewers)
Sludge bulking
Solid wastes
Storage tanks
Storm sewers; stormwater retention basins
Underground storage
Underground waste disposal (use waste disposal in the ground)
Urban hydrology
Urban runoff
Waste disposal in the ground
Water
Water quality
Water reuse
Water supply engineering
Water treatment plants
Wells
Wetland ecology
Wetlands

KEYWORD SEARCHING

For database and Web searching, these same Library of Congress Subject Headings can be useful. However, for thorough coverage, synonyms should be used as well. Several resources in the Article Indexes sections have online thesauri that can suggest additional terms. Another resource for search terms is the *Thesaurus of Sanitary and Environmental Engineering,* 12th ed. (February 2003) from the World Health Organization and Pan American Health Organization http://www.cepis.org.pe/bvsair/i/manuales/tesaeng.pdf (accessed October 6, 2010).

Other useful online resources for additional keywords and search terms include:

ChemBioFinder http://www.chembiofinder.com/ (accessed October 6, 2010), for chemical names and CAS registry numbers.
EPA Topics and Terms of Environment (Glossary) http://www.epa.gov/epahome/topics.html (accessed October 16, 2010).

ARTICLE INDEXES

In the following list of indexes and abstracts for environmental engineering, the producers and their Web sites are listed as resources for additional information. In many cases, the producers both offer their own subscriptions for access and license the indexes and abstracts to a number of service providers, each of which then offers its own subscription options and search interfaces. The major service providers, likely to offer access to these indexes and abstracts, include:

Dialog, a Thomson business http://www.dialog.com/ (accessed December 15, 2010).
STN http://www.cas.org/products/ (accessed December 15, 2010).
OVID (including SilverPlatter) http://www.ovid.com/site/index.jsp (accessed December 15, 2010).
Cambridge Scientific Abstracts http://www.csa.com (accessed December 15, 2010).
OCLC FirstSearch http://www.oclc.org/firstsearch/default.htm (accessed December 15, 2010).

Most of these indexes are international in their coverage of the literature.

ASFA: Aquatic Sciences and Fisheries Abstracts (other names: *ASFA 1: Biological Sciences and Living Resources; ASFA 2: Ocean Technology, Policy and Non-living Resources; ASFA 3: Aquatic Pollution and Environmental Quality; ASFA Marine Biotechnology Abstracts; ASFA Aquaculture Abstracts*—all print counterparts), *Cambridge Scientific Abstracts.* 1971–. Updated monthly. Fee based. Types of publications indexed: Serial publications, books, reports, conference proceedings, translations, and limited distribution literature. A single database incorporating five subfiles, the ASFA series indexes research and policy on the contamination of oceans, seas, lakes, rivers, and estuaries. Major subjects covered include: aquaculture, aquatic organisms, aquatic pollution, brackish water environments, conservation, environmental quality, fisheries, freshwater environments, limnology, marine biotechnology, marine environments, meteorology, oceanography, policy and legislation, and wildlife management.

Chemical Abstracts (*CAplus, SciFinder*) (see Chapter 7, Chemical Engineering, in this book).

CSA Engineering Research Database contains the following subfiles, which may be searched separately:

ANTE: Abstracts in New Technologies and Engineering
CSA/ASCE Civil Engineering Abstracts
Earthquake Engineering Abstracts
Environmental Engineering Abstracts (see also separate entry under this name)
Mechanical and Transportation Engineering Abstracts

Content includes basic and applied research, design, construction, technological and engineering aspects of air and water quality, environmental safety, energy production, and developments in new technologies. Topics of interest to environmental engineers include: automotive design and engineering, bridges and tunnels, buildings, towers, and tanks, coastal and offshore structures, construction materials, design and properties of substructures, electric and hybrid vehicles, engineering for electric power generation, flood analysis, fuels and propellants, geotechnical engineering, hazardous materials, industrial waste and sewage, internal combustion engines, land development, irrigation and drainage, pollution, waste and water engineering, seismic engineering, seismic phenomena, site remediation and reclamation, storm water.

Ei Compendex (see Chapter 2 in this book).

Energy Science and Technology (other name: *Energy Research Abstracts*–print counterpart). U.S. Department of Energy, Office of Scientific and Technical Information http://www.ntis.gov/products/engsci.aspx (accessed October 15, 2010) (1976–). Updated biweekly. Types of publications indexed: journal literature, books, conference proceedings, papers, patents, dissertations, engineering drawings. Scientific and technical reports of the U.S. Atomic Energy Commission, U.S. Energy Research and Development Administration and its contractors, other agencies, universities, and industrial and research organizations. A comprehensive source of worldwide energy-related information, this file contains references to basic and applied scientific and technical research literature. Subject coverage of interest to this audience includes: energy sources, use, and conservation; environmental effects; waste processing and disposal; hazardous waste management; conservation technology; energy conversion; renewable energy sources; energy policy; synthetic fuels; engineering; and environmental science.

EnergyFiles: Environmental Sciences, Safety, and Health. U.S. Dept. of Energy Office of Scientific and Technical Information http://www.osti.gov/energyfiles/Environmental/main.html (accessed October 15, 2010). The environmental sciences, safety, and health subject area is defined as information on the effects of any energy-related activity on the environment, on methods for mitigating or eliminating adverse effects, and on technical aspects of ensuring that energy-related activities are environmentally safe and socially acceptable. This area covers all aspects of global climate change. Monitoring and transport of chemicals, radioactive materials, and thermal effluents within the atmospheric, terrestrial, and aquatic environs are covered.

EnergyFiles is an umbrella interface that permits searching of multiple government databases. Of potential interest to environmental engineers are:

Atmospheric Radiation Measurement Program (ARM). Global change research supported by the Department of Energy.

> *EPA Full-Text Reports–National Environmental Publications Internet Site* (NEPIS). EPA full-text documents online, archival, and current.
>
> *EPA Technical Reports.* Searchable access to the Environmental Protection Agency Web site that includes technical reports, publications, and other information.
>
> *Office of Biological and Environmental Research Abstracts Database* (OBER). Abstracts and other information on OBER-funded R&D including abstracts for Environmental Sciences from FY 94–96.
>
> *USGS Publications Warehouse.* The publications center of the U.S. Geological Survey, a federal source for science about the Earth, its natural and living resources, natural hazards, and the environment.
>
> *DOE Information Bridge.* Searchable and downloadable bibliographic records and full text of DOE research report literature from 1995 onward.
>
> *Energy Citations Database.* Bibliographic records for energy and energy-related STI from the DOE and its predecessor agencies, ERDA and AEC, from 1948 to the present.

Environment Abstracts (Environment Abstracts–print counterpart). Cambridge Scientific Abstracts (CSA). 1975–. Updated monthly. Types of publications indexed: reports, conferences, symposia, meetings, journal articles, newspaper articles. Subject coverage of interest to this audience includes: air pollution, environmental design and urban ecology, energy, general environmental topics, renewable and nonrenewable resources, oceans and estuaries, waste management, water pollution, and weather modification and geophysical change.

Environmental Engineering Abstracts: Cambridge Scientific Abstracts (CSA). Coverage from 1990 to present, with most records from 1997 or later, updated monthly. Types of publications indexed: more than 700 primary journals plus over 2,500 additional sources, including monographs and conference proceedings. A subfile of CSA Engineering Research Database, *Environmental Engineering Abstracts*, indexes the literature on technological and engineering aspects of air and water quality, environmental safety, and energy production.

Environmental Sciences and Pollution Management: Cambridge Scientific Abstracts (CSA) http://www.csa.com/ (accessed October 15, 2010). Coverage from 1967 to present, updated monthly. Indexes serials, conference proceedings, reports, monographs, books, and government publications. *Environmental Sciences and Pollution Management* includes the following subfiles, searchable separately:

ASFA 3: Aquatic Pollution and Environmental Quality
Bacteriology Abstracts (Microbiology B)
Biotechnology Research Abstracts
Ecology Abstracts
Environment Abstracts
Environmental Engineering Abstracts
Health and Safety Sciences Abstracts
Industrial and Applied Microbiology Abstracts (Microbiology A)
Pollution Abstracts
Risk Abstracts
Sustainability Science Abstracts
Toxicology Abstracts
Water Resources Abstracts

Subject coverage of interest to environmental engineers includes air quality, aquatic pollution, energy resources, environmental biotechnology, environmental engineering, environmental impact statements (U.S.), hazardous waste, industrial hygiene, microbiology related to industrial and environmental issues, pollution: land, air, water, noise, solid waste, radioactive, risk assessment, toxicology and toxic emissions, water pollution, waste management, and water resource issues.

Current Contents Connect (Current Contents–print counterpart) Thomson Reuters http:// thomsonreuters.com/products_services/science/science_products/a-z/current_contents_ connect?parentKey=591283 (accessed October 15, 2010). One year rolling file, updated weekly. Types of publications indexed: articles, editorials, meeting abstracts, commentaries, and all other significant items in recently published editions of over 1,120 journals and books. A subject search also will find related external Web sites, evaluated by Thomson Reuters editors. *Current Contents* is a weekly alerting service useful for keeping current with the most recent publications. Citations are subsequently added to *Web of Science*. Options include searching, browsing by journal title, viewing table of contents, and setting up automatic alerts delivered weekly by e-mail. The *Current Contents* weekly editions of most interest to environmental engineers are Engineering, Computing, and Technology editions, and Agriculture, Biology, and Environmental Sciences editions.

GeoRef (Bibliography and Index of Geology–print counterpart) American Geological Institute http://www.agiweb.org (accessed October 15, 2010). 1693 to present (North America coverage), 1933 to present (global coverage), updated biweekly. Types of publications indexed: publications of the U.S. Geological Survey, U.S. and Canadian theses and dissertations, journal articles, books, maps, conference papers, and reports. GeoRef provides access to the world's geoscience literature.

Global Mobility Database. Society of Automotive Engineers Inc. (SAE International). Of interest for environmental engineering is the technological information on mobility engineering, such as emissions, environment, fuels and lubricants, noise and vibration (see Chapter 20 on Transportation Engineering in this book).

Oceanic Abstracts. Cambridge Scientific Abstracts (CSA). 1981–. Updated monthly. Indexes serials. *Oceanic Abstracts* focuses on the worldwide technical literature on marine and brackish water environment. Subjects covered include marine biology, physical oceanography, fisheries, aquaculture, non-living resources, meteorology, and geology, as well as environmental, technological, and legislative topics.

SciFinder (see Chapter 7, Chemical Engineering, in this book).

TRID Online (see Chapter 20, Transportation Engineering, in this book).

Web of Science (see Chapter 2, General Engineering Resources, in this book).

It should be noted that many of the books mentioned in the sections below are available online through publishers, government agency Web sites, and e-book aggregators. Some may be available through multiple sources and the online availability of individual titles may change. Consequently, we have not made note of online availability for individual titles except for freely available publications, primarily from government sources. Details on the major providers of online books are discussed in Chapter 2, General Engineering Resources, in this book.

DATA SOURCES AND DATASETS

American Water Works Association. Water:\STATS [electronic resource]. Denver, CO: American Water Works Association. The comprehensive source of statistical information on water utilities in the United States, Canada, and internationally offers a wealth of general utility information on topics such as treatment practices, distribution systems, water quality, revenue, and financial data. The data are available on CD-ROM.

Darnay, A. J., ed. 1992. *Statistical Record of the Environment.* Detroit: Gale Research. The three editions, published from 1992 to 1995, provide statistics on the status of U.S. environmental conditions.

Handbook of Chemical Risk Assessment: Health Hazards to Humans, Plants, and Animals.
2000. Boca Raton, FL: Lewis Publishers. *Note*: Access full-text PDF file online via
ENVIROnetBASE (restricted to subscribing libraries).

NWIS Web data collected and published by the U.S. Geological Survey http://waterdata.usgs.
gov/nwis (accessed October 20, 2010). These pages provide access to water resources data
collected at approximately 1.5 million sites in all 50 states, the District of Columbia, and
Puerto Rico. Online access to this data is organized around surface water, groundwater,
and water quality. The USGS investigates the occurrence, quantity, quality, distribution,
and movement of surface and underground waters and disseminates the data to the public, state, and local governments, public and private utilities, and other federal agencies
involved with managing our water resources.

National Water Quality Assessment Data Warehouse (NAWQA). The U.S. Geological Survey
(USGS) began its NAWQA program in 1991, systematically collecting chemical, biological, and physical water quality data from 42 study units (basins) across the nation. This
database contains chemical concentrations in water, bed sediment, and aquatic organism
tissues for about 609 chemical constituents; daily stream flow information for fixed sampling sites; groundwater levels, and more http://infotrek.er.usgs.gov/apex/f?p=136:1:0::NO:::
(accessed October 6, 2010).

United Nations Environment Programme. 1987–1994. *Environmental Data Report.* Oxford,
U.K./New York: Blackwell. A biennial publication from 1987 through 1994 prepared by
the GEMS Monitoring and Assessment Research Centre in cooperation with the World
Resources Institute, it provides annotated data on environmental pollution, climate, natural
resources, human health and population, energy, wastes, and natural disasters for developed
and developing countries. The information was gathered from government agencies, environmental organizations, and published sources. It ceased publication in 1994. Records are
useful for comparison and historical studies.

BIBLIOGRAPHIES AND GUIDES TO THE LITERATURE

Printed bibliographies are much less commonly produced now that there is easier access to full-
text reports and journal articles on the World Wide Web, Web sites, and specialized Web portals.
However, a few are worth noting, especially for the ease of locating historical information.

EPA Publications Bibliography, sponsored by Library Systems Branch, U.S. Environmental
Protection Agency in cooperation with EPA Offices of Air and Waste Management.
Washington, D.C.: The Branch; Springfield, VA: National Technical Information Service,
distributor. Published quarterly, began in October/November 1977 and ceased in 2000.
Consolidates all reports published by the U.S. Environmental Protection Agency and cov-
ers all aspects of the environment.

Lane, C. N. 2003. *Acid Rain: Overview and Abstracts.* New York: Nova Science Publishers.
Covering both wet and dry deposition of acid rain, this work offers both a background
article and over 900 abstracts and book citations. Title, author, and subject indexes are
provided for easy access.

New Publications of the Geological Survey. Washington, D.C.: Geological Survey, U.S.
Government Printing Office http://pubs.er.usgs.gov/ (accessed October 6, 2010). As of
January 2004, this catalog is available monthly and annually only and only online. This
publication has had many variant titles since its inception in 1879. It is a comprehensive
listing of new materials published by the Geological Survey in the following series: digital
data, professional papers, bulletins, water supply papers, techniques of water resources
investigations, fact sheets, water resources investigation reports, open-file reports, thematic
maps and charts, and topographic maps.

Van der Leeden, F., ed. 1991. *Geraghty & Miller's Groundwater Bibliography,* 5th ed. Plainview, NY: Water Information Center. A compilation of more than 5,500 selected references covering groundwater environments and systems, groundwater contamination, saltwater intrusion, and groundwater models.

DIRECTORIES

Print directories are no longer the essential resource that they were a few years ago due to the easy access to most on the Web. Most professional societies, academic and research institutions, or government organizations have membership directories that are published and distributed to their members in print, are searchable online, or may be available for purchase.

American Academy of Environmental Engineers. Annual. *Who's Who in Environmental Engineering* www.aaee.net (accessed October 6, 2010). Annapolis, MD: The Academy. *Who's Who in Environmental Engineering* includes a roster of all the Board Certified Diplomates of Environmental Engineering who have demonstrated their expertise in one or more of seven environmental specialties. This directory provides an alphabetical listing of each Diplomate together with a biographic profile, and is cross-referenced geographically and by specialty.

American Academy of Environmental Engineers. Annual. *Environmental Engineering Selection & Career Guide* http://www.aaee.net/Website/SelectionGuide.htm (accessed October 6, 2010). Annapolis, MD: The Academy. A free public service of the American Academy of Environmental Engineers, this directory is a source for consulting firms employing board-certified engineers. It also lists individuals who have earned the title Diplomate Environmental Engineer.

AWWA Sourcebook http://sourcebook.awwa.org/ (accessed October 6, 2010). Annual. Denver, CO: American Water Works Association (AWWA). The *Sourcebook* is a comprehensive directory of drinking water products and services. It includes a directory of service providers arranged by company, a guide to suppliers, and a guide to consultants.

ENCYCLOPEDIAS

Bisio, A. and S. Boots, eds. 1995. *Encyclopedia of Energy Technology and the Environment.* (4 vols.) New York: John Wiley & Sons. Not to be confused with Wiley's condensed version of this work, this encyclopedia won the 1995 Association of American Publishers, Inc. Award for Excellence in Professional/Scholarly Publishing, Chemistry. In four volumes, the encyclopedia covers the impact of energy production technologies through discussions of such topics as acid rain, air pollution, aircraft fuel, building systems, coal combustion, computer applications for energy-efficient systems, risk assessment, solar heating, waste management planning, water power, and so on. It contains illustrations, photographs, tables, and a list of environmental organizations.

Cheremisinoff, P. N., ed. 1989– . *Encyclopedia of Environmental Control Technology.* (9 vols.) Houston, TX: Gulf Publishing. Provides in-depth coverage of specialized topics relating to environmental and industrial pollution control technology and state-of-the-art information as well as projections of future trends.

Lehr, J. H., J. Keeley, and J. Lehr. 2005. *Water Encyclopedia.* Hoboken, NJ: John Wiley & Sons. This five-volume set is a comprehensive encyclopedia of water with volumes covering the following topics: surface and agricultural water; domestic, municipal, and industrial water supply and waste disposal; water quality and resource development; oceanography, meteorology, physics and chemistry, water law, and water history, art, and culture; and groundwater.

Meyers, R. A., ed. 1998. *Encyclopedia of Environmental Analysis and Remediation.* (8 vols.) New York: John Wiley & Sons. The preface categorizes this eight-volume encyclopedia as a "professional level compendium" of all aspects of the environment. Emphasis is on sampling, analysis, and remediation, but the work also includes discussions of pollution sources, transport, regulations, and health effects. There are about 280 articles, many including tables, figures, equipment procedures, and standards. Information to assist in preparation of parts of environmental impact statements and air permitting documents is also included.

Pfafflin, J. R. and E. N. Ziegler, eds. 2006. *Encyclopedia of Environmental Science and Engineering,* 5th ed. Boca Raton, FL: CRC Press. This work is written for the practitioner and those with a science background. After brief definitions or introductions, each chapter quickly focuses on technical details. For example, the chapter on aerosols includes a table of physical interpretation for characteristic diameters of various particles, provides defining equations, and methods of aerosol particle size analysis.

Van der Leeden, F., F. L Troise, and D. K. Todd. 1990 *The Water Encyclopedia,* 2nd ed. Chelsea, MI: Lewis Publishers. This one-volume encyclopedia is really a handbook. It is a standard in the field with international coverage on all aspects of water. It contains tables, charts, graphs, and some maps. Subject matter includes climates, hydrology, surface and groundwater, water use, water quality, and water management. This encyclopedia contains unique statistical information that is often difficult to find.

DICTIONARIES

Lee, C. C., compiler and ed. 2004. *Environmental Engineering Dictionary,* 4th ed. Boca Raton, FL: Government Institutes. This newly updated dictionary provides a comprehensive reference of hundreds of environmental engineering terms in use throughout the field. This edition draws from government documents and legal and regulatory sources, and includes terms relating to pollution control technologies, monitoring, risk assessment, sampling and analysis, quality control and permitting, fuel cell technology, and basic environmental calculations. Users of this dictionary will find exact and official U.S. Environmental Protection Agency definitions for statute and regulation-related terms.

Pankratz, T. M. 2001. *Environmental Engineering Dictionary and Directory.* Boca Raton, FL: Lewis Publishers. This book includes more than 8,000 terms, acronyms, and abbreviations applying to wastewater, potable water, industrial water treatment, seawater desalination, air pollution, incineration, and hazardous waste treatment. Its most unique feature is the inclusion of 3,000 trademarks and brand names.

Pfafflin, J. R., E. N. Ziegler, and J. M. Lynch. 2008. *Dictionary of Environmental Science and Engineering,* 2nd ed. New York: Routledge. The dictionary explains specialist environmental terms concisely, provides a guide to acronyms, and describes acts, organizations, and requirements related to the legislation of the environment, particularly those of the United States.

Porteous, A. 2008. *Dictionary of Environmental Science and Technology,* 4th ed. Chichester, U.K.: New York: John Wiley & Sons. With a mix of short and long entries, this work provides basic definitions and data. For example, five pages with diagrams are devoted to anaerobic digestion, but other entries are a single line. Some include references. The dictionary covers the broad range of environmental terminology. The publisher describes this work as "over 4,000 in-depth entries on scientific and technical terminology associated with environmental protection and resource management."

Smith, P. G. and J. S. Scott. 2005. *Dictionary of Water and Waste Management,* 2nd ed. Oxford, U.K./Boston: Butterworth-Heinemann; London: IWA Publications. Reference to U.S., U.K., and European standards, legislation, and spelling ensures that the reader will find the book of global relevance. Illustrations throughout aid the reader's understanding of the explanations. Over 7,000 terms on water quality, engineering, and waste management make this the reference of choice for the professional. Over the past 20 years, areas such as air pollution control, solid waste management, hazardous waste management, pipeline management (leakage control, pipeline and sewer renewal), and environmental management systems have all become increasingly important. To reflect this shift, this completely revised and updated edition now covers water and waste management as well as treatment.

Sullivan, T. F. P., ed. 1993. *Environmental Regulatory Glossary,* 6th ed. Rockville, MD: Government Institutes. This dictionary covers terms pertaining to regulatory issues.

Webster, L. F., ed. 2000. *Dictionary of Environmental and Civil Engineering.* New York: Parthenon. Focusing on the mechanical aspects of environmental engineering, this dictionary is more comprehensive than most and includes some unique terms. Definitions are often brief, therefore, it may best be used in conjunction with other resources.

HANDBOOKS, MANUALS, AND PROPERTIES

GENERAL

Bitton, G. 1998. *Formula Handbook for Environmental Engineers and Scientists.* New York: John Wiley & Sons. This work provides formulas and equations from a range of disciplines and sources. It complements the *CRC Handbook of Physics and Chemistry.* Arranged in alphabetical order, some cross-references are included. Appendices contain conversion tables.

Bregman, J. I. 1999. *Environmental Impact Statements,* 2nd ed. Boca Raton, FL: Lewis Publishers. This extensively revised second edition addresses all the requirements for federal, state, and local environmental impact statements (EISs) and provides detailed "how to" information for their preparation.

Burke, G., B. R. Singh, and L. Theodore. 2000. *Handbook of Environmental Management and Technology,* 2nd ed. New York: John Wiley & Sons. This thoroughly updated edition offers a historical perspective on pollution problems and solutions along with an introduction to the scientific and technical literature in the field.

CFRs Made Easy (series). Rockville, MD: Government Institutes. These books provide an in-depth, focused look at environmental regulations. Each handbook provides an overview of compliance programs, guidelines for understanding requirements, and detailed explanations and checklists. *Air CFR's Made Easy* (2nd ed.), *Water CFR's Made Easy, RCRA CFR's Made Easy,* as well as *Environmental Compliance Made Easy* are of particular interest to this audience.

Cheremisinoff, N. P. 2001. *Handbook of Pollution Prevention Practices.* New York: Marcel Dekker. This book focuses on reducing manufacturing and environmental compliance costs by instituting improved operational schemes, recycling and byproduct recovery, waste minimization, and energy efficiency policies, and offers project cost accounting tools that assist in evaluating pollution prevention technologies.

Cheremisinoff, N. P. 2005. *Environmental Technologies Handbook.* Lanham, MD: Government Institutes. This handbook covers a variety of topics including human health risk assessment, air quality monitoring and air pollution control, methods and technologies in the treatment of waste and wastewater, and financial tools for environmental technologies.

Corbitt, R. A., ed. 1999. *Standard Handbook of Environmental Engineering,* 2nd ed. New York: McGraw-Hill. This second edition serves as a guide to environmental engineering, including information on tools, techniques, and regulations in the field. In addition to covering air, water, and waste treatment and handling methods, there are discussions of project management, environmental legislation and regulations, environmental assessment, and air and water quality standards. There are numerous tables and charts. One small drawback is that references do not appear to be as current as possible for a 1999 copyright date.

Eccleston, C. H. 2001. *Effective Environmental Assessments: How to Manage and Prepare NEPA EAs.* Boca Raton, FL: CRC Press. This is a comprehensive, step-by-step guide on the preparation of defensible Environmental Assessments (EAs).

ENVIROnetBASE. Boca Raton, FL: CRC Press http://www.crcnetbase.com/ (accessed October 16, 2010). A subscription is required for the online version. This is a collection within *CRCnetBASE* of over 670 CRC Press electronic handbooks, many of which are listed individually in this section. Restricted to subscribers, it is a useful resource for rapid access to precise information on environmental modeling, systems analysis, risk assessment, health and safety, chemistry and toxicology, environmental engineering, law and compliance, and water science. Visit http://www.crcnetbase.com/ for a complete list of titles and subscription information.

Environmental Law Handbook, 20th ed. 2009. Washington, D.C.: Government Institutes. A comprehensive and up-to-date analytical review of the major environmental, health, and safety laws affecting U.S. businesses and organizations. The authors provide easy-to-read interpretations of major environmental laws.

Ghassemi, A., ed. 2000. *Handbook of Pollution Control and Waste Minimization.* New York: Marcel Dekker. This handbook covers the broad spectrum of pollution prevention including process design, lifestyle analysis, risk, and decision making. The author's aim is to make environmental issues a major design consideration. The book presents the fundamentals of pollution prevention: life-cycle analysis, designs for the environment, and pollution prevention in process design. The various components of pollution prevention are discussed in detail, and current legislative and regulatory policies governing the management of waste in Europe and the United States are covered.

Gottlieb, D. W. 2003. *Environmental Technology Resources Handbook.* Boca Raton, FL: Lewis Publishers. Intended to guide users to the proper technology for solving environmental problems, this work focuses primarily on Internet resources for control, remediation, assessment, and prevention. Although it risks becoming dated quickly given the constantly changing nature of the Internet, it does guide the user to major resources that the author has carefully studied.

Khandan, N. 2002. *Modeling Tools for Environmental Engineers and Scientists.* Boca Raton, FL: CRC Press. The author describes some 50 computer models developed with 8 different software packages. Intended for nonprogrammers to develop computer-based mathematical models for natural and engineered environmental systems, this book includes a review of mathematical modeling and fundamental concepts such as material balance, reactor configurations, and fate and transport of environmental contaminants.

Lee, C. C., and S. D. Lin, eds. 2007. *Handbook of Environmental Engineering Calculations,* 2nd ed. New York: McGraw-Hill. The purpose of this handbook is to provide fully illustrated, step-by-step calculation procedures for solid waste management, air resources management, water quality assessment and control, surface water, lakes and reservoirs, groundwater, public water supply, wastewater treatment, and risk assessment/pollution prevention. Co-author C. C. Lee, with experience as an EPA Research Program Manager, integrates regulatory requirements into the discussions.

Liu, D. H. F. and B. G. Liptak, eds. 1997. *Environmental Engineers' Handbook,* 2nd ed. Boca Raton, FL: Lewis Publishers. The many contributors to this work are from industry, consulting firms, and academia and have a wide range of backgrounds from engineering, law, medicine, agriculture, meteorology, biology, and so on. This handbook on the prevention and management of pollution illustrates the technology and techniques through tables, schematic diagrams, and in-depth discussions. For example, some 140 pages are devoted to groundwater and surface water pollution; wastewater treatment covers sources, monitoring, sewers, treatment, biological treatment, and sludge disposal in 420 pages.

Mackay, D., W. Y. Shiu, K. C. Ma, and S. C. Lee. 2006. *Handbook of Physical-Chemical Properties and Environmental Fate for Organic Chemicals,* 2nd ed. Boca Raton, FL: CRC/ Taylor & Francis. This is a comprehensive series in four volumes that focuses on environmental fate prediction. These books tackle environmental fate calculations and QSAR (quantitative structure-activity relationship) plots. This shows where the chemicals will go, relative concentrations, persistence, and important intermediate transport processes.

Manly, B. F. J. 2009. *Statistics for Environmental Science and Management,* 2nd ed. Boca Raton, FL: Chapman & Hall/CRC Press. This work features a nonmathematical approach to statistical methods used in environmental data analysis. Techniques such as environmental monitoring, impact assessment, assessing site reclamation, censored data, and Monte Carlo risk assessment are discussed.

Means, R.S. 2005. *Environmental Remediation. Cost Data-assemblies.* 11th ed. R.S. Means Company; Englewood, CO: Talisman Partners. This book includes pricing for more than seventy standard remediation technologies and related tasks, and includes costs for every kind and size of project.

Means, R.S. 2005. *Environmental Remediation. Cost Data-unit Price.* 11th ed. R.S. Means Company; Englewood, CO: Talisman Partners. This book provides the detailed line items, component costs, forms, instructions, and guidelines needed to prepare or verify cost estimates for almost any type of environmental remediation project, ranging from simple underground storage tank removals to complex hazardous waste sites.

Nemerow, N. L., F. J. Agardy, and J. A. Salvato, eds. 2009. *Environmental Engineering,* 6th ed. Hoboken, NJ: John Wiley & Sons. Applies the principles of sanitary science and engineering to sanitation and environmental health.

Patnaik, P. 2010. *Handbook of Environmental Analysis: Chemical Pollutants in Air, Water, Soil, and Solid Wastes,* 2nd ed. Boca Raton, FL: CRC Press. The handbook provides thorough treatment of the analysis of toxic pollutants in the environment, addressing ambient air, groundwater, surface water, industrial wastewater, and soils and sediments.

Reynolds, J. P., J. S. Jeris, and L. Theodore. 2002. *Handbook of Chemical and Environmental Engineering Calculations,* New York: John Wiley & Sons. The scientific and mathematical cross-over between chemical and environmental engineering is the key to solving a host of environmental problems. Many problems included in this book demonstrate this cross-over, as well as the integration of engineering with current regulations and environmental media, such as air, soil, and water. Solutions to the problems are presented in a programmed instructional format.

Singh, V. P., S. K. Jain, and A. Tyagi. 2007. *Risk and Reliability Analysis: A Handbook for Civil and Environmental Engineers.* Reston, VA: ASCE Press. This is an excellent introduction to risk and uncertainty for environmental engineers. The book covers the fundamentals of decision making, probability, risk, and reliability analysis. This title was named an Outstanding Title in the May 2008 *CHOICE* review. (*Choice Reviews* online, American Library Association http://www.cro2.org (accessed March 23, 2011).

Spellman, F. R. and N. E. Whiting. 2005. *Environmental Engineer's Mathematics Handbook.* Boca Raton, FL: CRC Press. This handbook covers advanced mathematical principles in environmental engineering with chapters on modeling, algorithms, and pollution assessment and control calculations.

Wang, L. K., series ed. 2004–. *Handbook of Environmental Engineering.* Totowa, NJ: Humana Press (newer volumes published by Springer). This series covers various aspects of environmental engineering with currently published volumes on air, noise, water, and waste related topics. Titles include: *Air Pollution Control Engineering, Advanced Air and Noise Pollution Control, Physicochemical Treatment Process, Water Resources and Control Processes, Advanced Physicochemical Treatment Technologies, Biosolids Treatment Processes, Biosolids Engineering and Management, Biological Treatment Processes, Advanced Biological Treatment Processes, Environmental Biotechnology, Environmental Bioengineering, Flotation Technology,* and *Membrane and Desalination Technologies.*

AIR ENVIRONMENT

Alley, E. R., L. B. Stevens, and W. L. Cleland. 1998. *Air Quality Control Handbook.* New York: McGraw-Hill. Starting with an overview of air pollution control management, this work discusses the sources and effects of air pollution, and then explains the available pollution treatment and control technologies from the perspective of compliance with the 1990 Clean Air Act. Air-moving, vapor, and particulate control equipment are included as well as ambient air, continuous air, and stack monitoring and various forms of pollution testing.

Beim, H. J., J. Spero, and L. Theodore. 1998. *Rapid Guide to Hazardous Air Pollutants.* New York: Van Nostrand Reinhold. This portable, concise guide is a convenient compilation of information on the 189 substances deemed to be hazardous air pollutants (HAPs) under the 1990 Amendments to the Clean Air Act. In addition to chemical substance names and synonyms, each record provides the CAS Registry number, uses, physical and chemical properties, acute and chronic health risks, hazardous risks, and regulatory information.

Brownell, F. W. 1998. *Clean Air Handbook,* 3rd ed. Rockville, MD: Government Institutes. The third edition provides an overview of the regulatory requirements of the Clean Air Act and its amendments. Chapters cover such topics as the federal–state partnership, control technology regulation, operating and preconstruction permitting programs, acid deposition control program, and regulation of mobile sources of air pollution.

Cheremisinoff, N. P. 2002. *Handbook of Air Pollution Prevention and Control.* Boston, MA: Butterworth-Heinemann. For process engineers and environmental managers, this work focuses on the industrial sector, providing discussions of hardware, cost accounting, estimation methods for emissions, indoor air quality overview, point source, and regulations.

Compilation of Air Pollutant Emission Factors, Stationary Point and Area Sources (AP-42). Volume I, Supplements A-F, and Updates 2001–2009. 1995, with continuous updates. Washington, D.C.: U.S. Environmental Protection Agency, Office of Air Quality Planning and Standards, Office of Air and Radiation http://www.epa.gov/ttn/chief/ap42/ (accessed October 16, 2010). Prepared by the Emission Factor and Inventory Group (EFIG) of the U.S. Environmental Protection Agency's (EPA) Office of Air Quality Planning and Standards (OAQPS), the AP-42 series is the principal means of documenting stationary point and area sources emission factors. These factors are cited in numerous other EPA publications and electronic databases, but without the process details and supporting reference material provided in AP-42. As noted in the introduction to this work, emission estimates are important for developing emission control strategies, determining applicability of permitting and control programs, and ascertaining the effects of sources and appropriate mitigation strategies. Therefore, emission factors are frequently the best or only method available for estimating emissions, in spite of their limitations.

Cooper, A. R. 2004. *Air CFRs Made Easy,* 2nd ed. Rockville, MD: Government Institutes. Addressing the air regulations covered in CFR Tide 40, beginning with Part 104, and in Tide 33, beginning with Part 320, core elements of each program are explained. Subjects covered include accident prevention regulations, acid rain control, emissions offsets, enforcement provisions, hazard assessments, mobile sources, national ambient air quality standards, and state implementation plans.

Davis, W. T. ed. 2000. *Air Pollution Engineering Manual,* 2nd ed. Air & Waste Management Association; New York: John Wiley & Sons. Organized into 20 chapters by industry, this manual discusses the different processes that generate air pollution, equipment used with all types of gases and particulate matter, and emissions control for such industries as graphic arts, chemical processes, metallurgy, and water treatment plants. This updated second edition reflects recent emission factors and control measures for reducing air pollutants, providing technological and regulatory information for compliance with the air pollution standards. The work contains detailed flow charts and photographs as well as Internet resources.

Hewitt, C. N. and A. V. Jackson, eds. 2003. *Handbook of Atmospheric Science: Principles and Applications.* Malden, MA: Blackwell. Divided into two major parts with contributions from 28 authors, this work begins with a discussion of the behavior of the Earth's atmosphere. Part 2 then examines the problems of air pollution and the tools and practices for air quality management. Emissions monitoring and modeling tools are discussed for large-scale applications. Charts, maps, and color plates are used to illustrate the concepts, and extensive references are included for each chapter.

Pluschke, P., ed. 2005. *Indoor Air Pollution.* New York: Springer. *Indoor Air Pollution* is Volume 4, part F in the *Handbook of Environmental Chemistry.* It contains nine chapters by various authors and includes an international perspective with chapters on air pollution issues in developing countries and China.

Schifftner, K. C. 2002. *Air Pollution Control Equipment Selection Guide.* Boca Raton, FL: Lewis Publishers. This guide assists with appropriate selection of equipment for air pollution control. Organized by primary technology employed (i.e., quenching, cooling, particulate removal, gas absorption), each section covers basic physical forces used in the technology, common sizes, and common uses. Text includes current photographs or drawings of typical equipment within that device type.

Schnelle, K. B. and C. A. Brown. 2002. *Air Pollution Control Technology Handbook.* Boca Raton, FL: CRC Press. The handbook is a resource for commonly used air pollution control technology in stationary sources for the control of gaseous pollutants and particulate matter. Selection, evaluation, and design are covered as well as alternative air pollution control processes. Environmental regulations and costs are also addressed.

Spicer, C. W. 2002. *Hazardous Air Pollutant Handbook: Measurements, Properties, and Fate in Ambient Air.* Boca Raton, FL: Lewis Publishers. This work describes the 188 substances designated as hazardous air pollutants (HAPs) under the Clean Air Act Amendments of 1990. Selected chemical and physical properties, measurement methods in ambient air, mean and range of ambient levels reported in the literature, and reaction mechanisms are covered for each substance.

WATER ENVIRONMENT

Alley, E. R. 2007. *Water Quality Control Handbook,* 2nd ed. New York: McGraw-Hill. The second edition of this handbook includes expanded coverage of treatment systems for specific pollutants, the latest water quality regulations, and new content on wastewater treatment operations, membrane treatment processes, and cost-saving treatment design methods.

Boulding, J. R. 2004. *Practical Handbook of Soil, Vadose Zone, and Ground-Water Contamination: Assessment, Prevention, and Remediation,* 2nd ed. Boca Raton, FL: CRC Press. This handbook provides a comprehensive yet practical guide to soil, vadose zone, and groundwater contamination and remediation. It discusses the basics of soils, hydrology, and hydrogeology as well as addressing assessment and monitoring, and prevention.

Burton, G. A., Jr., and R. E. Pitt. 2002. *Stormwater Effects Handbook: A Toolbox for Watershed Managers, Scientists, and Engineers.* Boca Raton, FL: Lewis Publishers. The handbook assists in determining when storm water runoff causes adverse effects in receiving waters. It includes case studies, many photographs, and figures that allow easy visualization of methods.

Cheremisinoff, N. P. 2001. *Handbook of Water and Wastewater Treatment Technologies.* Boston, MA: Butterworth-Heinemann. A handbook aimed at process and plant engineers, water treatment operators, and environmental consultants, this reference contains practical information for the treatment of drinking water and wastewater.

Cheremisinoff, N. P. 2003. *Handbook of Solid Waste Management and Waste Minimization Technologies.* Boston, MA: Butterworth-Heinemann. An essential tool for plant managers, process engineers, environmental consultants, and site remediation specialists that focuses on practices for handling a broad range of industrial solid waste problems and presents information on waste minimization practices. Included in the text are sidebar discussions, questions for consideration and discussion, recommended resources (print and Web) for the reader, and a comprehensive glossary.

De Zuane, J. 1997. *Handbook of Drinking Water Quality,* 2nd ed. New York: Van Nostrand Reinhold. This handbook can be a quick reference for anyone dealing with water quality issues. Appendices include World Health Organization Guidelines and European Drinking Water Directives.

Gallagher, L. M. 2003. *Clean Water Handbook.* Rockville, MD: Government Institutes. This handbook was designed to provide a comprehensive road map to the requirements, legal theories, and critical issues of water pollution control.

Gallant, B. 2006. *Hazardous Waste Operations and Emergency Response Manual.* Hoboken, NJ: Wiley-Interscience. Covers regulations, agencies, site safety plans, decontamination, and more.

HDR Engineering, Inc. 2001. *Handbook of Public Water Systems.* New York: John Wiley & Sons. This book has become a standard reference in this field. The numerous contributors are experts on the subject and they have been involved in the development of several new water treatment processes that have been incorporated into water treatment plant designs.

Letterman, R. D. ed. 1999. *Water Quality and Treatment: A Handbook of Community Water Supplies,* 5th ed. New York: McGraw-Hill; American Water Works Association. This book consists of 18 illustrated chapters detailing state-of-the-art technologies and methods. It features updated discussions of all water treatment processes.

Lewis, R. J. Sr. 2008. *Hazardous Chemicals Desk Reference,* 6th ed. New York: Wiley-Interscience. This book is intended to be a quick reference and includes over 6,000 common industrial and laboratory materials. The information is taken from the more comprehensive *Sax's Dangerous Properties of Industrial Materials* by Irving Sax, but it is less technical.

Lin, S. D. and C. C. Lee. 2007. *Water and Wastewater Calculations Manual,* 2nd ed. New York: McGraw-Hill. This manual offers streamlined, step-by-step procedures for problems in water and wastewater engineering from the simple to the complex.

Mays, L. W., ed. 2001 *Stormwater Collection Systems Design Handbook.* New York: McGraw-Hill. This book is a comprehensive reference of state-of-the-art design of stormwater collection systems and their component parts. It includes problem examples, discussion of what can go wrong in the design process, references, formulae, and diagrams.

Montgomery, J. H. 2007. *Groundwater Chemicals Desk Reference,* 4th ed. Boca Raton, FL: CRC Press. This comprehensive work deals with hazardous chemicals in groundwater.

Each entry gives a complete description of the chemical, including physical and chemical properties.

Organisation for Economic Cooperation and Development. 2007. *Guidance Manual on Environmentally Sound Management of Waste.* Paris: OECD. Recommendation adopted by OECD countries on environmental standards.

RCRA Orientation Manual. 2008. Manual was developed by the U.S. Environmental Protection Agency, Office of Solid Waste, Communications, Information, and Resources Management Division. Washington, D.C.: U.S. Environmental Protection Agency, Solid Waste and Emergency Response. Available online at www.epa.gov/epawaste/inforesources/pubs/orientat/rom.pdf (accessed October 16, 2010). This updated manual provides introductory information on the solid and hazardous waste management programs under the Resource Conservation and Recovery Act (RCRA). Designed for EPA and state staff, members of the regulated community, and the general public who wish to better understand the RCRA program.

Rizzo, J. A,. ed. 1998. *Underground Storage Tank Management: A Practical Guide,* 5th ed. Rockville, MD: Government Institutes. This guide includes all updates and requirements to comply with the U.S. EPA's federal requirements including soil sampling, analytical guidelines, and the evolution of tank testing strategies.

Standard Methods for the Examination of Water and Wastewater. 2005. Prepared and published jointly by the American Public Health Association, American Water Works Association, Water Environment Federation (21st ed.). Joint Editorial Board: Lenore S. Clesceri, Arnold E. Greenberg, and Andrew D. Eaton; managing editor Mary Ann H. Franson. Washington, D.C.: American Public Health Association. This book is a standard in the field. The methods presented in this book, as in previous editions, are believed to be the best available and generally accepted procedures for the analysis of water and wastewater. The selection of methods that are included, as well as the formal procedure for approval, have been reviewed by a broad range of experts in the field.

Tchobanoglous, G., and F. Kreith. 2002. *Handbook of Solid Waste Management,* 2nd ed. New York: McGraw-Hill. This handbook offers an integrated approach to the planning, design, and management of economical and environmentally responsible solid waste disposal systems.

White, G. C. 2010. *White's Handbook of Chlorination and Alternative Disinfectants,* 5th ed. New York: John Wiley & Sons. This book discusses all aspects of water purification using chlorination.

MONOGRAPHS AND TEXTBOOKS

GENERAL

CRC Press, one of the leading publishers of engineering books, has created three valuable resource collections for environmental engineers: *ENGnetBASE, ENVIROnetBASE*, and *STATSnetBASE*. These book collections are now available online at *CRCnetBASE* http://www.crcnetbase.com/ (accessed December 15, 2010). Using a simple search screen, you enter the search terms you want either as a single word, phrase, or with Boolean operators. These three databases are only available as an annual subscription. Once you are a subscriber, any new books added during the year are yours to browse for no additional cost.

Berthouex, P. M. and L. C. Brown. 2002. *Statistics for Environmental Engineers,* 2nd ed. Boca Raton, FL: CRC Press (electronic resource and print). This edition consists of 54 short, stand-alone chapters with each chapter addressing a particular environmental problem or statistical technique.

Bhandari, A., et al., eds. 2009. *Contaminants of Emerging Concern.* Reston, VA: ASCE. Prepared by the Technical Committee on Hazardous, Toxic, and Radioactive Waste Management, this book has chapters on contaminants of increasing importance, such as pharmaceuticals, antimicrobials and antibiotics, hormones, plasticizers, and surfactants as well as other contaminants.

Davis, L. R. 1999. *Fundamentals of Environmental Discharge Modeling.* Boca Raton, FL: CRC Press. This book focuses on engineering and mathematical models for documenting and approving mechanical and environmental discharges with an emphasis on wastewater and atmospheric discharges. Various diffuser and surface discharge models are discussed as well as the fundamentals of turbulent jet mixing. Case studies are used to illustrate problems.

Gupta, R. S. 2004. *Introduction to Environmental Engineering and Science,* 2nd ed. Rockville, MD: ABS Consulting, Government Institute. An introduction to the fundamental principles common to most environmental problems is followed by major sections on water pollution, hazardous wastes and risk assessment, waste treatment, air pollution, global climate change, and hazardous substances. Includes problems to develop skills learned in the text.

Jørgensen, S. E. 2000. *Principles of Pollution Abatement: Pollution Abatement for the 21st Century.* New York/Amsterdam: Elsevier. This is a revised and expanded version of the 1988 *Principles of Environmental Science* by the same author. Contents include mass conservation, energy conservation, risks and effects, water and wastewater problems, solid waste problems, and air pollution problems. The work features new tools such as ecotechnology, cleaner technology, life-cycle analysis, and new environmental management techniques by changes in products and production methods.

Manly, B. F. J. 2009. *Statistics for Environmental Science and Management,* 2nd ed. Boca Raton, FL: Chapman & Hall/CRC Press. The use of appropriate statistical methods is essential when working with environmental data. This book is intended to introduce environmental scientists and managers to the statistical methods that will be useful in their work. The book is not meant to be a complete introduction to statistics, but Appendix A provides a quick refresher.

Masters, G. M. and W. P. Ela. 2008. *Introduction to Environmental Engineering and Science.* Upper Saddle River, NJ: Prentice Hall. A good, undergraduate-level introductory text for environmental engineering. Chapters include such topics as environmental chemistry, risk assessment, water pollution, water quality control, air pollution, and solid waste management and resource recovery.

Mulligan, C. N. 2002. *Environmental Biotreatment: Technologies for Air, Water, Soil, and Waste.* Rockville, MD: Government Institutes. The author summarizes the application of 26 bioremediation techniques to cleaning air, soil, water, and wastes. For each method, advantages, disadvantages, costs, and other considerations are discussed, and then each method is compared to more conventional technology. Case studies are presented and extensive references are included.

Nazaroff, W. W. and L. Alvarez-Cohen. 2001. *Environmental Engineering Science,* New York: John Wiley & Sons. This textbook covers the fundamentals of environmental engineering and applications in water quality, air quality, and hazardous waste management.

Salvato, J. A., N. L. Nemerow, and F.J. Agardy, eds. 2003. *Environmental Engineering,* 5th ed. Hoboken, NJ: John Wiley & Sons. This book applies principles of sanitary science and engineering to sanitation and environmental health.

Wiersma, G. B., ed. 2004. *Environmental Monitoring.* Boca Raton, FL: CRC Press. The book is a compilation that brings together the activities and complex approaches to monitoring air, water, and land.

AIR ENVIRONMENT

Baumbach, G. 1996. *Air Quality Control: Formation and Sources, Dispersion, Characteristics and Impact of Air Pollutants–Measuring Methods, Techniques for Reduction of Emissions and Regulations for Air Quality Control*. New York: Springer. A translation from the German, this English edition has been revised to include situations and regulations applicable in the United States. It received "Outstanding Title!" status and was strongly recommended in the June 1997 *CHOICE* review.

Devinny, J. S., M. A. Deshusses, and T. S. Webster. 2008. *Biofiltration for Air Pollution Control*, 2nd ed. Boca Raton, FL: CRC Press. This is a comprehensive discussion of biofiltration technology (the use of micro-organisms growing on porous media) for air pollution control. Materials, designs, monitoring methods, as well as examples of successful applications are included.

Friedlander, S. K. 2000. *Smoke, Dust, and Haze: Fundamentals of Aerosol Dynamics*, 2nd ed. New York: Oxford University Press. Updating a much earlier edition, this work summarizes current aerosol knowledge. It was awarded "Outstanding Title!" status in the December 2000 *CHOICE* review.

Godish, T. 2004. *Air Quality*, 4th ed. Boca Raton, FL: Lewis Publishers. This fourth, revised edition of a classic offers a comprehensive overview of air quality issues, including atmospheric chemistry, analysis of the control of emissions from stationary sources, the effects of pollution on public health and the environment, public policy concerns, and technology and regulatory practices. Among the new sections are toxicological principles and risk assessment. According to the preface, the work is useful as a supplement to engineering curricula for the design and operation of pollution control equipment.

Heinsohn, R. J. and J. M. Cimbala. 2003. *Indoor Air Quality Engineering: Environmental Health and Control of Indoor Pollutants*. New York: Basel. Intended as an upper-level textbook and reference for professionals, this monograph seeks to bridge engineering and industrial hygiene. In addition to the 10 chapters covering risk, contaminant concentration, respiratory systems, design criteria, ventilation, particle emission, and emission rates, the authors provide some supplementary information at http://www.mne.psu.edu/cimbala/ Heinsohn_Cimbala_book/index.htm (accessed March 23, 2011). Here they provide MS Excel and MathCAD files corresponding to examples in the book and a free program for calculating flow in ducts.

Kowalski, W. J. 2003. *Immune Building Systems Technology*. New York: McGraw-Hill. Advertised as a one-stop guide to building ventilation and air treatment systems design, this book aims for a comprehensive approach to the protection of buildings against biological pathogens. Designing, retrofitting, and building state-of-the-art ventilation and air treatment systems are covered.

Sportisse, B. 2010. *Fundamentals in Air Pollution: From Processes to Modelling*. Dordrecht, The Netherlands: Springer. Includes chapters on atmospheric radiative transfer; atmospheric boundary layer, gas-phase atmospheric chemistry; aerosols, clouds, and rain; and numerical simulations.

Stern, A. C., ed. 1976. *Air Pollution*, 3rd ed. (8 vols.) New York: Academic Press. Despite the publication date, Stern's three-volume second edition is still cited regularly and is available in numerous libraries. Written for professionals and practitioners, the 60 contributors address the cause, effect, transport, measurement, and control of air pollution.

Wang, L. K., N. C. Pereira, and Y.-T. Hung. 2004–2005. *Air Pollution Control Engineering*. (2 vols.) Totowa, NJ: Humana Press. Available either as an electronic book or as hard copy, this work surveys principles and practices underlying control processes, illustrating them with detailed design examples. Chapters include fabric filtration, cyclones, electrostatic precipitation, wet and dry scrubbing, condensation as a basis for intelligent planning of abatement systems, flare processes, thermal oxidation, catalytic oxidation, gas-phase activated carbon adsorption, and gas-phase biofiltration. Best Available Technologies (BAT) for air pollution control are detailed and cost data provided as well as engineering methods for the design, installation, and operation of air pollution process equipment.

Zhang, Y. 2005. *Indoor Air Quality Engineering*, Boca Raton, FL: CRC Press. Based on the author's own course lecture notes, this text covers properties and mechanics of airborne pollutants, measurement and sampling, and indoor air quality control technologies. Since it was written as a textbook, it provides discussion topics, problems, and references to further reading for each topic.

HAZARDOUS WASTE, SOLID WASTE, AND WATER ENVIRONMENT

Asano, T., ed. 1998. *Wastewater Reclamation and Reuse*. Lancaster, PA: Technomic Publications. Experts from around the world contributed to this useful and unique text that analyzes and reviews aspects of wastewater reclamation, recycling, and reuse in countries around the world. This is volume 10 of an 11-volume series, the *Water Quality Management Library*, which thoroughly addresses issues in wastewater treatment, sludge, nonpoint pollution, toxicity reduction, and groundwater remediation.

Bagchi, A. 2004. *Design of Landfills and Integrated Solid Waste Management,* 3rd ed. Hoboken, NJ: John Wiley & Sons. This book combines integrated solid waste management with the traditional coverage of landfills. This new edition offers the first comprehensive guide to managing the entire solid waste cycle, from collection, to recycling, to eventual disposal. Includes new material on source reduction, recycling, composting, contamination soil remediation, incineration, and medical waste management.

Baruth, E. E., tech. ed. 2005. *Water Treatment Plant Design,* 4th ed. New York: McGraw-Hill. Water treatment plant design has become increasingly complex. This book has become the standard reference for engineers involved in this area. It contains articles by over 40 international design experts and is a comprehensive reference on modernizing existing water treatment facilities and planning new ones—from initial plans and permits through design, construction, and start-up. The third edition of this book was written as a companion to AWWA's *Water Quality and Treatment: A Handbook of Community Water Supplies.* American Water Works Association, 1999.

Chin, D. A. 2006. *Water-Quality Engineering in Natural Systems*. Hoboken, NJ: Wiley-Interscience. An excellent book focusing on the fate and transport of contaminants in all natural water systems. This book was voted an "Outstanding Title" in *CHOICE* 2006.

Ghosh, S. N. and V. R. Desai. 2006. *Environmental Hydrology and Hydraulics: Eco-Technological Practices for Sustainable Development*. Enfield, NH: Science Publishers. Focuses on small-scale water projects and sustainable practices covering water uses, hydraulic principles and design, water hazards and management, and eco-technological practices for sustainable development.

Grady, C. P. L. Jr., G. T. Daigger, and H. C. Lim. 1999. *Biological Wastewater Treatment,* 2nd ed. New York: Marcel Dekker. This book integrates the principles of the biochemical processes with applications to the design, operation, and optimization of biochemical operations. It is volume 19 in the *Environmental Science and Pollution Control* series that includes 27 volumes. This comprehensive series covers a vast array of topics pertinent to environmental engineers, including pollution prevention practices and control, bioremediation of contaminated soils, combustion and control, biosolids treatment and management, groundwater contamination, and hazardous waste.

Hammer, D. A., ed. 1989. *Constructed Wetlands for Wastewater Treatment: Municipal, Industrial, and Agricultural*. Chelsea, MI: Lewis Publishers. New advancements in this area continue to be covered in the journal literature; however, this book is a standard in the field for its comprehensive coverage of this topic. It covers general principles of wetlands, ecology, hydrology, soil chemistry, vegetation, and wildlife as well as case studies and specific applications.

Jordening, H.-J. and J. Winter. 2005. *Environmental Biotechnology: Concepts and Applications.* Weinheim, Germany: Wiley-VCH. Covers four areas in which biotechnology is used for waste treatment: wastewater, soil, solid waste, and waste gases.

Kadlec, R. H. and S. Wallace. 2008. *Treatment Wetlands,* 2nd ed. Boca Raton, FL: CRC Press. The most complete guide to using wetlands for water quality management and habitat creation. This book provides scientific and operational information on treatment wetlands technologies for entry-level practitioners. Contents include an introduction to wetlands for treatment, wetland structure and function, effects of wetlands on water quality, wetland project planning and design, wetland treatment system establishment, operation and maintenance, and wetland data case histories.

Lottermoser, B. G. 2007. *Mine Wastes: Characterization, Treatment, Environmental Impacts,* 2nd ed. Berlin: Springer. A comprehensive reference on mine wastes with chapters on sulfidic mine wastes, mine water, tailings, cyanidation wastes of gold–silver ores, radioactive wastes of uranium ores, and wastes of phosphate and potash ores. This book was named an Outstanding Title in the January 2008 *CHOICE* review.

Metcalf & Eddy, Inc. (revised by Tchobanoglous, G., F. Burton, and H. D. Stensel) 2003. *Wastewater Engineering: Treatment and Reuse*, 4th ed. Boston, MA: McGraw-Hill. Gives a solid overall perspective on wastewater engineering.

Methods, H. and S. R. Durrans, eds. 2003. *Stormwater Conveyance Modeling and Design.* Waterbury, CT: Haestad Press. The academic version includes a CD-ROM, and StormCAD Stand-Alone, PondPack, CulvertMaster, and FlowMaster software. This book guides you through the design and analysis process using an approach that combines theoretical fundamentals with practical design guidance and hydraulic monitoring techniques.

Miller, J. R. and S. M. O. Miller. 2007. *Contaminated Rivers: A Geomorphological-Geochemical Approach to Site Assessment and Remediation.* Dordrecht, The Netherlands: Springer. Covers the geomorphological and geochemical processes that pertain to contamination of river systems and remediation methods, both *ex-situ* and *in-situ*.

Mitsch, W. J. and J. G. Gosselink. 2007. *Wetlands,* 4th ed. New York: John Wiley & Sons. The fourth edition, this standard textbook is more international in scope and has been expanded in other areas. It includes information about all types of wetlands, and chapters on wetland law and regulation and wetland delineation.

Pierzynski, G. M., J. T. Sims, and G. F. Vance. 2005. *Soils and Environmental Quality,* 3rd ed. Boca Raton, FL: CRC Press. This revised and updated text provides detailed discussions about soils science, hydrology, and the classification of pollutants.

Todd, D. K. and L. W. Mays. 2005. *Groundwater Hydrology,* 3rd ed., Hoboken, NJ: John Wiley & Sons. All aspects of groundwater hydrology are covered; includes chapters on the quality and pollution of groundwater and saline water intrusion.

Turovskiy, I. S. and P. K. Mathai. 2006. *Wastewater Sludge Processing.* Hoboken, NJ: John Wiley & Sons. A comprehensive text on wastewater sludge processing that describes new, emerging, and international methods, compares different processing methods, and covers both municipal and industrial treatment technologies.

JOURNALS

Unless otherwise noted, all journals are peer reviewed. For further identification, the ISSN (International Standard Serial Number) appears after each title. When a separate ISSN for electronic format is available, it is indicated as the second number.

Aerosol Science and Technology: The Journal of the American Association for Aerosol Research (0278-6826) (1521-7388). This journal covers theoretical and experimental investigations of aerosol and closely related phenomena as well as papers on fundamental and applied topics.

Ambio: A Journal of the Human Environment (0044-7447) (1654-7209). This journal has been published since 1972 by the Royal Swedish Academy of Sciences, an independent, nongovernmental organization that aims to promote research in mathematics and the natural sciences.

Annual Review of Environment and the Resources (1543-5938) (1545-2050). This review focuses on the emerging scientific and policy issues at the crux of sustainable development. The reviews assess critical scientific, policy, technological, and methodological issues related to the Earth's global life support systems, sectors of human use of environment and resources, and the human dimensions and management of resources and environmental change.

Applied Catalysis. B, Environmental (0926-3373) (1873-3883). This journal covers the catalytic chemistry of polluting substances.

Archives of Environmental Contamination and Toxicology (0090-4341) (1432-0703). This journal includes significant, full-length articles describing original experimental or theoretical research work pertaining to the scientific aspects of contaminants in the environment.

Atmospheric Environment (1352-2310) (1873-2844). Focusing on the consequences of natural and human-induced perturbations on the Earth's atmosphere, this journal covers air pollution research and its applications, air quality and its effects, dispersion and transport, deposition, biospheric-atmospheric exchange, global atmospheric chemistry, radiation, and climate.

Biodegradation (0923-9820) (1572-9729). This journal publishes papers on all aspects of science pertaining to the detoxification, recycling, amelioration, or treatment of waste materials and pollutants by naturally occurring microbial strains or recombinant organisms.

Building and Environment (0360-1323) (1873-684X). This journal publishes original papers and review articles on building research and its applications, and on the social, cultural, and technological contexts of building research and architectural science.

Bulletin of Environmental Contamination and Toxicology (0007-4861) (1432-0800). This journal provides rapid publication of significant advances and discoveries in the fields of air, soil, water, and food contamination and pollution as well as articles on methodology and other disciplines concerned with the introduction, presence, and effects of toxicants in the total environment.

Chemosphere (0045-6535) (1879-1298). This international, multidisciplinary journal disseminates original articles describing new discoveries or developments in fields related to the environment and human health and developing areas of environmental science.

Clean Water Report (0009-8620) (1545-7435). This journal provides comprehensive coverage of drinking water and sewer systems, lakes, rivers and streams, coastal protection, tributaries and bays, as well as the Safe Drinking Water Act, the Clean Water Act, and other major legislative initiatives. It also includes articles on biosolids, pathogens, arsenic, chlorine, dioxin, and other pollutants, problems such as flooding, silting, sedimentation, nutrients, and more.

Climate Research (0936-577X) (1616-1572). This journal evaluates, selects, and disseminates important new information about basic and applied research devoted to all aspects of climate (present, past, and future); effects of human societies and organisms on climate; and effects of climate on the ecosphere.

Cold Regions Science and Technology (0165-232X) (1872-7441). This journal primarily addresses problems related to the freezing of water, and especially with the many forms of ice, snow, and frozen ground. Emphasis is given to applied science, mainly in the physical sciences, with broad coverage of the physics, chemistry, and mechanics of ice, ice-water systems, and ice-bonded soils.

Critical Reviews in Environmental Science and Technology (1064-3389) (1547-6537). This journal serves as an international forum for the critical review of current knowledge on the broad range of topics in environmental science. It addresses current problems and the scientific basis for new pollution control technologies.

Desalination (0011-9164) (1873-4464). Desalination covers all desalting fields—distillation, membranes, reverse osmosis, electrodialysis, ion exchange, freezing, water purification, water reuse, and wastewater treatment—and aims to provide a forum for any innovative concept or practice.

Ecological Engineering (0925-8574) (1872-6992). This journal is meant for ecologists who are involved in designing, monitoring, or constructing ecosystems and serves as a bridge between ecologists and engineers.

Environmental Engineering Science (1092-8758) (1557-9018). This journal publishes papers on environmental science topics that include the development and application of fundamental principles toward solving problems in air, land, and water media, including environmental applications of the basic science.

Environmental Health Perspectives (EHP) (0091-6765) (1552-9924). A peer-reviewed journal of the National Institute of Environmental Health Sciences, this is an important vehicle for the dissemination of environmental health information and research findings.

Environment International (0160-4120) (1873-6750). This journal covers all disciplines engaged in the field of environmental research, and seeks to quantify the impact of contaminants in the human environment, and to address human impacts on the natural environment itself. It covers the entire spectrum of sources, pathways, sinks, and interactions between environmental pollutants, whether chemical, biological, or physical.

Environmental Management (0364-152X) (1432-1009). This journal serves to improve cross-disciplinary communication. Contributions are drawn from biology, botany, climatology, ecology, ecological economics, environmental engineering, fisheries, environmental law, management science, forest sciences, geography, geology, information science, law politics, public affairs, and zoology.

Environmental Modelling and Software: With Environment Data News (1364-8152) (1873-6726). This journal publishes contributions in the form of research articles, reviews, and short communications as well as software and data news on recent advances in environmental modeling and/or software to improve the capacity to represent, understand, predict, or manage the behavior of environmental systems at all practical scales.

Environmental Pollution (0269-7491) (1873-6424). This is an international journal that addresses issues relevant to the nature, distribution, and ecological effects of all types and forms of chemical pollutants in air, soil, and water. It includes articles based on original research, findings from reexamination and interpretation of existing data, and articles on new methods of detection and remediation of environmental pollutants.

Environmental Science and Technology (0013-936X) (1520-5851). This journal is a unique source of information for scientific and technical professionals in a wide range of environmental disciplines. Contributed materials may appear as current research papers, policy analyses, or critical reviews. Also includes a magazine section that provides authoritative news and analysis of the major developments, events, and challenges shaping the field.

Ground Water (0017-467X) (1745-6584). This technical publication is strictly for groundwater hydrogeologists. Each issue of the journal contains peer-reviewed scientific articles on pertinent groundwater subjects.

Ground Water Monitoring and Remediation (1069-3629) (1745-6592). Since its inception in 1981, this journal has been the leader in the field of groundwater monitoring and cleanup.

It contains a mixture of original columns authored by industry leaders, industry news, EPA updates, product and equipment news, and peer-reviewed papers.

Hazardous Waste Consultant (0738-0232). This journal has articles about the newest developments relating to hazardous waste assessment, treatment, storage, and disposal as well as waste minimization technologies. In addition, summaries of federal and state legal cases with a focus on regulatory interpretation and enforcement related to hazardous waste compliance are reported.

Hazardous Waste/Superfund Week (1521-2882). This journal covers federal and state regulations and their interpretations and the effect they have on organizations as well as contract opportunities and lawsuit coverage. Every other week, there is comprehensive, behind-the-scenes coverage of congressional action, EPA initiatives, Department of Defense cleanups, Superfund sites, regulatory changes, court cases, enforcement news, contract opportunities, research findings, and business developments.

Indoor and Built Environment: The Journal of the International Society of the Built Environment (1420-326X) (1423-0070). This journal publishes reports on any topic pertaining to the quality of indoor or built environment, and how this may affect the health, performance, efficiency, and comfort of persons in these environments.

Journal of Aerosol Science (0021-8502) (1879-1964). Covering all aspects of basic and applied aerosol research, the original papers in this journal describe recent theoretical and experimental research relating to the basic physical, chemical, and biological properties of systems of airborne particles of all types; their measurement, formation, transport, deposition and effects; and industrial, medical, and environmental applications.

Journal of the Air and Waste Management Association (1096-2247). One of the oldest continuously published, peer-reviewed technical environmental journals in the world, this journal serves those occupationally involved in air pollution control and waste management. It covers air pollution, hazardous waste, management, regulations, measuring, modeling, emissions, testing, monitoring, and more.

Journal of Atmospheric Chemistry (0167-7764) (1573-0662). This journal includes, in particular, studies of the composition of air and precipitation and the physiochemical processes in the Earth's atmosphere; the role of the atmosphere in biogeochemical cycles; the chemical interaction of the oceans, land surface, and biosphere with the atmosphere, as well as laboratory studies of the mechanics in transformation processes, and descriptions of major advances in instrumentation.

Journal of the American Water Works Association (0003-150X) (1551-8833). Both a professional and a scholarly publication, this journal contains information about water quality, water resources, and supply as well as the management and operation of water utilities.

Journal of Environmental Engineering (0733-9372) (1943-7870). This journal presents broad interdisciplinary information on the practice and status of research in environmental engineering science, systems engineering, and sanitation. Contributors include consultants, practicing engineers, and researchers.

Journal of Environmental Management (0301-4797) (1095-8630). This publication, formerly *Advances in Environmental Research*, contains original full-length research papers, case studies, notes, and critical reviews on major advances in protection of the quality of air, water, and land environments; improvements to existing technology; and contributions to the knowledge of transport and fate of pollutants in the environment.

Journal of Environmental Science and Health. Part A, Toxic/Hazardous Substances and Environmental Engineering (1093-4529) (1532-4117). This is a comprehensive journal that provides an international forum for the rapid publication of essential information including the latest engineering innovations, effects of pollutants on health, control systems, laws, and projections pertinent to environmental problems whether in the air, water, or soil.

Journal of Environmental Systems (0047-2433) (1541-3802). This journal includes articles that range from case studies of particular environmental/energy/waste problems or technologies, to assessments of overall environmental system (or cost, risk, energy) impacts, to broad discussions of issues of theory, methodology, and policy. The emphasis is on practical environmental problems, such as recycling and waste minimization.

Journal of Hazardous Materials (0304-3894) (1873-3336). This journal publishes full-length research papers, reviews, project reports, case studies, and short communications that improve understanding of the hazards and risks that certain materials pose to people and the environment, and with ways of controlling the hazards and associated risks. The journal is published in two parts: *Part A: Risk Assessment and Management*; and *Part B: Environmental Technologies.*

Journal of Hydrologic Engineering (1084-0699) (1943-5584). This journal disseminates information on the development of new hydrologic methods, theories, and applications to current engineering problems. It publishes papers on analytical, numerical, and experimental methods for the investigation and modeling of hydrological processes.

Journal of Hydrology (0022-1694). Papers comprise but are not limited to the physical, chemical, biogeochemical, stochastic and systems aspects of surface and groundwater hydrology, hydrometeorology, and hydrogeology. Relevant topics in related disciplines, such as climatology, water resource systems, hydraulics, geomorphology, soil science, instrumentation, remote sensing, and civil and environmental engineering, are also included.

Journal of Water Resources Planning and Management (0733-9496) (1943-5452). This journal reports on all phases of planning and management of water resources. Papers examine social, economic, environmental, and administrative concerns related to the use and conservation of water. Social and environmental objectives in fish and wildlife management, water-based recreation, and wild and scenic river use are assessed.

Noise Control Engineering Journal (0736-2501) (1021-643X). A refereed journal published by the Institute of Noise Control Engineering, it contains information about noise control solutions and manufacturers.

Pollution Engineering (0032-3640) (1937-4437). Includes feature articles devoted to practical engineering applications useful for the recognition, measurement, control, and disposal of hazardous solids, air, and liquid containments.

Remediation: The Journal of Environmental Cleanup Costs, Technologies and Techniques (1051-5658) (1520-6831). This quarterly journal focuses on the practical application of remediation techniques and technologies; how to diagnose problems at hazardous waste disposal sites; and how to select the best, most cost-effective cleanup technology.

Resources, Conservation, and Recycling (0921-3449). This journal emphasizes the processes involved in more sustainable production and consumption systems. Emphasis is upon technological, economic, institutional, and policy aspects of specific resource management practices, such as conservation, recycling, and resource substitution, and of strategies, such as restructuring of production and consumption profiles, and the transformation of industry.

Science of the Total Environment (0048-9697) (1879-1026). With an emphasis on applied environmental chemistry and environmental health, this international journal publishes research about changes in the natural level and distribution of chemical elements and compounds that may affect man and the natural world.

Tellus. Series B, Chemical and Physical Meteorology (0280-6509) (1600-0889). The series focuses on air chemistry, surface exchange processes, long-range and global transport, aerosol science, and cloud physics including related radiation transfer.

Waste Management (0956-053X) (1879-2456). This is the official journal of the International Waste Working Group (IWWG). It was started to create an intellectual forum to encourage and support economical and ecological (integrated and sustainable) waste management worldwide, and to promote scientific advancement in the field. This aim will be accomplished by learning from the past, analyzing the present for developing new ideas, and visions for the future.

Water, Air, and Soil Pollution (0049-6979) (1573-2932). This international, interdisciplinary journal covers all aspects of pollution and solutions to pollution in the biosphere. This includes chemical, physical, and biological processes affecting flora, fauna, water, air, and soil in relation to environmental pollution.

Water Environment and Technology (1044-9493) (1938-193X). This journal covers issues such as expansions and upgrades, nutrient removal, biosolids management, new technology, safety, permitting and regulations, collection systems, disinfection, pumps, odor control, watershed management, stormwater, and groundwater cleanup.

Water Environment Research: A Research Publication of the Water Environment Federation (1061-4303) (1554-7531). An environmental journal for the dissemination of fundamental and applied research in all scientific and technical areas related to water quality and pollution control. Topics of interest include hazardous wastes, groundwater and surface water, drinking water, source water protection, remediation and treatment systems, reuse, and environmental risk and health.

Water Research (0043-1354) (1879-2448). Covering all aspects of the science and technology of water quality and its management worldwide, the journal's scope includes treatment processes for water and wastewater; water quality standards and monitoring; studies on inland, tidal, or coastal waters and urban waters; limnology of lakes, impoundments, and rivers; solid and hazardous waste management; soil and groundwater remediation; analysis of the interfaces between sediments and water, and water/atmosphere interactions; modeling techniques and public health and risk assessment.

Water Resources Research (0043-1397) (1944-7973). This is an interdisciplinary journal that contains original articles about hydrology; in the physical, chemical, and biological sciences; and in the social and policy sciences, including economics, systems analysis, sociology, and law.

Water Science and Technology: A Journal of the International Association on Water Pollution Research (0273-1223) (1996-9732). This journal selects the best papers from biennial, regional, and specialized conferences sponsored by the International Water Association (IWA) encompassing important developments in all aspects of water quality management and pollution control. It is a useful resource for those unable to attend the conferences.

Wetlands Ecology and Management (0923-4861) (1572-9834). An international journal that publishes original articles in the field of wetlands ecology, the science of the structure and functioning of wetlands for their transformation, utilization, preservation, and management on a sustainable basis. The journal covers pure and applied science dealing with biological, physical, and chemical aspects of freshwater, brackish, and marine coastal wetlands.

STANDARDS, TEST METHODS, GUIDANCE, AND CRITERIA DOCUMENTS

AIR

Federal and State Government Resources

California Air Resources Board Ambient Air Quality Standards http://www.arb.ca.gov/research/aaqs/aaqs.htm (accessed October 13, 2010). Both the Air Resources Board (ARB) and the U.S. Environmental Protection Agency (USEPA) are authorized to set ambient air quality standards. This Web site includes both California and U.S. federal ambient air quality standards.

National Ambient Air Quality Standards for Criteria Pollutants (NAAQS) http://www.epa.gov/air/criteria.html (accessed October 15, 2010). The Clean Air Act, which was last amended in 1990, requires EPA to set National Ambient Air Quality Standards for pollutants considered harmful to public health and the environment. The Clean Air Act established two types of national air quality standards. Primary standards set limits to protect public health, including the health of "sensitive" populations, and Secondary standards set limits to protect public welfare, including protection against decreased visibility, damage to animals, crops, vegetation, and buildings.

U.S. Department of Labor, Occupational Health and Safety Administration (OSHA) http://www.osha.gov/SLTC/indoorairquality/index.html#contents (accessed October 13, 2010). To quote from this federal Web site: "The purpose of this Web page is to give workers and employers useful, up-to-date information to identify, correct, and prevent IAQ problems" (from Web page).

U.S. Environmental Protection Agency, Air and Radiation, National Ambient Air Quality Standards (NAAQS) http://www.epa.gov/air/criteria.html (accessed October 13, 2010). The Clean Air Act requires EPA to set National Ambient Air Quality Standards for pollutants considered harmful to public health and the environment. The EPA Office of Air Quality Planning and Standards (OAQPS) http://www.epa.gov/air/oaqps/ (accessed October 15, 2010) has set National Ambient Air Quality Standards for six principal pollutants (ground level ozone, lead, carbon monoxide, sulfur dioxide, nitrogen oxides, and respirable particulate matter), which are called "criteria" pollutants.

Associations, Societies, Organizations, and Commercial Resources

Acoustical Society of America (ASA) http://www.acosoc.org/standards/ (accessed October 13, 2010). The ASA maintains four standards committees accredited by the American National Standards Institute (ANSI): SI on Acoustics, S2 on Mechanical Vibration and Shock, S3 on Bioacoustics, and S12 on Noise. These four accredited standards committees also provide the U.S. input to several international committees.

American Society of Heating, Refrigeration and Air-conditioning Engineers, Inc. (ASHRAE) http://www.ashrae.org/technology/ (accessed October 13, 2010). The ASHRAE develops standards for both its members and others professionally concerned with refrigeration processes and the design and maintenance of indoor environments, for example, "Standard 62–1989: Ventilation for Acceptable Indoor Air Quality." This is a voluntary standard for "minimum ventilation rates and indoor air quality acceptable to human occupants and intended to avoid adverse health effects." See list of standards at http://xp20.ashrae.org/STANDARDS/standa.htm (accessed October 13, 2010).

Society of Automotive Engineers, SAE International, Technical Standards Development http://www.sae.org/standardsdev/ (accessed October 13, 2010). SAE International maintains over 8,300 technical standards and related documents, which may be purchased through their Web site. Selected standards related to fuels, emissions, and the environment are listed in the ground vehicle, and aerospace sections may be of interest to this audience.

WATER

American Water Works Association (1990-) AWWA Standards http://www.awwa.org/Resources/ (accessed October 18, 2010). AWWA standards are approved as American National Standards and are used by water supply professionals throughout North America and in other parts of the world. AWWA develops and maintains standards to improve drinking water quality and supply for the public.

EPA Test Methods Collections http://www.epa.gov/osa/fem/methcollectns.htm (accessed October 13, 2010). This resource provides a series of links to related EPA Web sites containing methods and standards for air, drinking water, and wastewater, solid and hazardous waste, including underground storage tanks, pesticides and toxic substances, and microbial methods.

Franson, M. A. H. and A. D. Eaton, eds. 2005. *Standard Methods for the Examination of Water and Wastewater,* 21st ed. Washington, D.C.: American Public Health Association, American Water Works Association, Water Environment Federation, prepared and published jointly. This standard work details comprehensive tests for all major pollutants, giving precise instructions for procedures, apparatus setup, calibrations, and current reference data. Subscription information at http://www.standardmethods.org/ (accessed March 23, 2011).

Index to EPA Test Methods http://www.epa.gov/region1/info/testmethods/pdfs/testmeth.pdf (accessed October 15, 2010). Lists over 700 air, water, and waste methods. EPA test methods are approved procedures for measuring the presence and properties of chemical substances or measuring the effects of substances under various conditions.

Keith, L. H. 1996. *Compilation of EPA's Sampling and Analysis Methods,* 2nd ed. Boca Raton, FL: CRC Press/Lewis Publishers. This is a printed edition of the EPA database. The book and the database are intended to help people select the most appropriate methods of sampling and analysis for a particular situation.

Kopp, J. F. and G. D. McKee. 1983. *Methods for Chemical Analysis of Water and Wastes,* Cincinnati, OH: Environmental Monitoring and Support Laboratory, Office of Research and Development, U.S. Environmental Protection Agency, 1983. Contains chemical analytical procedures used in EPA laboratories for the examination of ground and surface waters, domestic and industrial waste effluents, and treatment of process samples. These methods are also included on Selected Office of Water methods and guidance (electronic resource)/U.S. Environmental Protection Agency, Office of Water. Version 4. Washington, D.C.: U.S. Environmental Protection Agency, Office of Water (2002). This series is being frequently updated and includes recently developed and requested analytical methods and related guidance documents.

National Primary Drinking Water Standards and National Secondary Drinking Water Standards. Visit the EPA Web site at http://water.epa.gov/lawsregs/rulesregs/sdwa/curren-tregulations.cfm (accessed October 15, 2010).

Telliard, W. A. 2004. *Selected Office of Water Methods and Guidance* (electronic resource). v. 5, Washington, D.C.: U.S. Environmental Protection Agency, Office of Water. This CD-ROM includes more than 500 recently developed and frequently requested analytical methods and related guidance documents for the detection of trace metals, microbiological contaminants, and organics in wastewater and drinking water; chemical and biological methods for biosolids; guidance documents on the Office of Water's revised approach to method approval, whole effluent toxicity testing, and oil and grease analysis. Version 5 contains all the methods and guidance contained on previous versions. CD-ROM available from NTIS (National Technical Information Service).

Test Methods for Evaluating Solid Waste: Physical/Chemical Methods. 1986. Washington, D.C.: U.S. Environmental Protection Agency, Office of Solid Waste and Emergency Response (Supt. of Docs., U.S. G.P.O., distributor). "The EPA publication SW-846, entitled *Test Methods for Evaluating Solid Waste, Physical/Chemical Methods*, is Waste's official compendium of analytical and sampling methods that have been evaluated and approved for use in complying with the RCRA regulations. SW-846 functions primarily as a guidance document setting forth acceptable, although not required, methods for the regulated and regulatory communities to use in responding to RCRA-related sampling and analysis requirements" (from Web page). Go to the Web site for more information: http://www.epa.gov/epawaste/hazard/testmethods/sw846/index.htm (accessed October 13, 2010).

CRITERIA DOCUMENTS

Criteria documents accurately reflect the latest scientific knowledge and are based solely on data and scientific judgments on the relationship between pollutant concentrations and environmental and human health effects. Criteria provide guidance to EPA when promulgating federal regulations and developing standards.

> *National Recommended Water Quality Criteria.* 2002. Washington, D.C.: U.S. Environmental Protection Agency, Office of Water, Office of Science and Technology http://water.epa.gov/scitech/swguidance/waterquality/standards/current/index.cfm (accessed October 15, 2010). This is a compilation of recommended water quality criteria for 157 pollutants and is an update of *Quality Criteria for Water*, Washington, D.C.: U.S. Environmental Protection Agency, Office of Water Regulations and Standards. For sale by the Supt. of Docs., U.S. G.P.O., 1986, and 1995 update. This book was also known as the Gold Book. 1995 updates: water quality criteria documents for the protection of aquatic life in ambient water/United States Environmental Protection Agency, Office of Water. Washington, D.C.: U.S. Environmental Protection Agency, Office of Water; Springfield, VA: U.S. Department of Commerce, National Technical Information Service (distributor) (1996). A Clean Water Act Section 304(a) water quality criterion is a qualitative or quantitative estimate of the concentration of a contaminant or pollutant in ambient waters, which, when not exceeded, will ensure water quality is sufficient to protect a specified water use.

REGULATIONS

Laws and regulations are a major tool in protecting the environment. Congress passes the laws that govern in the United States. To put those laws to work, Congress authorizes certain government agencies to create and enforce regulations. Regulations state specific details about what is required by various communities to comply with the law and regulations set specific levels of pollutants, and so on. Once an authorized agency (such as the EPA) decides that a regulation is needed, it lists the proposed regulation in the Federal Register (FR). The public is allowed to comment on the impact of the regulation. The agency considers the comments, revises the regulation, and issues a final rule. At each stage in the process, the agency publishes a notice in the FR. These notices include the original proposal, requests for public comment, notices about meetings that are open to the public where the proposal will be discussed, and the text of the final regulation. The FR is published every work day of the federal government.

Twice a year, each agency publishes a comprehensive report that describes all the regulations it is working on or has recently finished. These are published in the *Federal Register*, usually in April or October, as the Unified Agenda of Federal and Regulatory and Deregulatory Actions.

Once a regulation is completed and has been printed in the FR as a final rule, it is codified and published in the *Code of Federal Regulations* (*CFR*). The *CFR* is the official record of all regulations created by the federal government. It is divided into 50 volumes, called titles, each with a specific focus. Almost all environmental regulations appear in *Title 40 CFR*. The *CFR* is revised yearly with a quarter of the volumes updated every three months. Title 40 is revised every July 1 (from the U.S. EPA Web site).

> Bureau of National Affairs (BNA) http://www.bna.com (accessed October 13, 2010). BNA is the foremost publisher of print and electronic news, analysis, and reference products, providing intensive coverage of legal and regulatory developments for professionals in

business and government. BNA's practical and thorough professional tools make it easier to track international, federal, and state requirements, and help you assess the impact of developments on your company or clients. Review the wealth of materials BNA has in its environment, health, and safety library at http://www.bna.com/products/ens/index.html (accessed October 13, 2010).

Code of Federal Regulations (CFR). Washington, D.C.: General Services Administration, National Archives and Records Service, Office of the Federal Register. Also online at http://www.gpoaccess.gov/cfr/index.html (accessed October 13, 2010).

Federal Register. Government Printing Office, Washington, D.C.: Office of the Federal Register, National Archives and Records Service, General Services Administration: Supt. of Docs., U.S. G.P.O., distributor. Available online at http://www.gpoaccess.gov/fr/ (accessed October 13, 2010). Published each federal work day. Includes proposed and final regulations by federal regulatory agencies.

Government Printing Office (GPO) Access http://www.gpoaccess.gov/ (accessed October 13, 2010). Service of the U.S. Government Printing Office that provides free electronic access to information products produced by the federal government. Included are congressional hearings, committee print and directory information, federal regulations, public laws, the *Congressional Record*, and many more titles.

SEARCH ENGINES, IMPORTANT WEB SITES, PORTALS, DISCUSSION LISTS, WEBLOGS

E-Print Network—Research Communications for Scientists and Engineers. Office of Scientific and Technical Information, Department of Energy http://www.osti.gov/eprints/ (accessed October 13, 2010). Provides a searchable gateway to thousands of science and technology Websites and databases worldwide and is a unique Deep Web search through scientific and technical e-prints and preprints generated by researchers. Subject areas of interest to this audience include biology and medicine; chemistry; energy storage, conversion, and utilization; engineering; environmental management and restoration technologies; environmental sciences and ecology; fossil fuels; geosciences; power transmission, distribution and plants; and renewable energy. It is possible to browse or search by these subject headings.

National Environmental Methods Index (NEMI) by the U.S. Geological Survey is a free, Web-based online clearinghouse of environmental monitoring methods http://www.nemi. gov (accessed October 13, 2010). The NEMI database contains chemical, microbiological, and radiochemical method summaries of laboratory and field protocols for regulatory and nonregulatory water quality analyses. In the future, NEMI will be expanded to meet the needs of the monitoring community. The tool also allows monitoring data to be shared among different agencies and organizations that use different methods at different times. This database was developed in conjunction with the U.S. Environmental Protection Agency (USEPA), and other partners in the federal, state, and private sectors.

Science.gov—Environment and Environmental Quality http://www.science.gov/browse/ w_123.htm (accessed October 13, 2010). A browsable list of environment and environmental quality resources at Science.gov http://www.science.gov/ (accessed October 13, 2010).

Air Environment

Global Change Research Information Office (GCRIO) http://www.gcrio.org/ (accessed October 13, 2010). The U.S. GCRIO provides access to data and information on climate change research, adaptation/mitigation strategies and technologies, and global change-related educational resources on behalf of the various U.S. federal agencies that are involved in the U.S. Global Change Research Program (USGCRP).

WATER ENVIRONMENT

StormwaterAuthority.org http://www.stormwaterauthority.org/ (accessed October 13, 2010). This Web site is the first comprehensive Internet resource for all aspects of the stormwater industry, covering news, events, state regulations, education, and more. The mission of StormwaterAuthority.org is to assist professionals in making educated and environmentally sound stormwater decisions.

WaterWiser—The Water Efficiency Clearinghouse—American Water Works Association http://www.awwa.org/Resources/Waterwiser.cfm (accessed October 13, 2010). WaterWiser is the premier water efficiency and water conservation online information resource.

GOVERNMENT AGENCIES

There are several federal agencies that regulate different aspects of the environment and are essential to the field of environmental engineering. These agencies include:

U.S. Army Corps of Engineers http://www.usace.army.mil/ (accessed October 14, 2010)
U.S. Bureau of Reclamation http://www.usbr.gov/ (accessed October 14, 2010)
U.S. Department of Energy http://www.energy.gov/ (accessed October 14, 2010)
U.S. Department of the Interior http://www.doi.gov/ (accessed October 14, 2010)
U.S. Environmental Protection Agency http://www.epa.gov (accessed October 14, 2010)
U.S. Fish and Wildlife Service http://www.fws.gov/ (accessed October 14, 2010)
U.S. Geological Survey http://www.usgs.gov (accessed October 14, 2010)
U.S. National Oceanic and Atmospheric Administration http://www.noaa.gov/ (accessed October 14, 2010)

The Web sites for these key agencies contain a wealth of information. However, it is important to point out some essential elements of the individual Web sites and specific work of the agencies.

U.S. Army Corps of Engineers (USACE)

The U.S. Army Corps of Engineers http://www.usace.army.mil/ (accessed October 14, 2010) is made up of approximately 34,00 civilians and soldiers and is headquartered in Washington, D.C. The Corps staff provides engineering services that include planning, designing, building, and operating water resources and other civil works projects (e.g., navigation, flood control, environmental protection). They also design and manage the construction of military facilities for the army and air force, and provide design and construction management support for other defense and federal agencies.

The Corps has eight regional divisions in the United States and 45 subordinate districts throughout the United States, Asia, and Europe. The Engineer Research and Development Center (ERDC) http://www.erdc.usace.army.mil/ (accessed March 23, 2011) consists of seven laboratories with each conducting specialized research. These laboratories include the Waterways Experiment Station (WES), Coastal and Hydraulics, Cold Regions Research and Engineering, Construction Engineering Research, Geotechnical and Structures, Environmental, Information Technology, and Topographical Engineering.

Another organization of interest to environmental engineers within the Corps is the Institute for Water Resources (IWR) http://www.iwr.usace.army.mil/ (accessed March 23, 2011). The IWR develops and applies new planning evaluation methods, policies, and data in anticipation of changing water resources management conditions.

For publications, search http://www.usace.army.mil/publications (all accessed October 14, 2010).

U.S. Department of Energy (DOE)

The DOE was established and effective from October 1, 1977 and consolidated the federal energy functions into a cabinet-level department. The DOE's mission is to advance the national, economic, and energy security of the United States, to promote scientific and technological innovation, and to ensure the environmental cleanup of the national nuclear weapons complex. Several programs that are part of the DOE are the Energy Information Administration, National Nuclear Security Administration, National Laboratories and Technology Centers, Power Marketing Administration, and Operations Offices and Field Organizations.

U.S. Department of the Interior (DOI)

The Department of the Interior manages public lands and mineral resources, national parks, national wildlife refuges, Western water resources, surface mined lands, and upholds federal trust responsibilities to Indian tribes. The DOI is a large agency with extensive responsibilities essential to environmental engineers. Of particular importance are the Bureau of Reclamation, U.S. Fish and Wildlife Service, and the U.S. Geological Survey.

U.S. Bureau of Reclamation (USBR)

The Reclamation Service was established in 1902 to manage, develop, and protect water and related resources in an environmentally and economically sound manner. It is best known for the dams, power plants, and canals it constructed in 17 Western states to secure a year-round water supply for irrigation. These projects led to homesteading and promoted the economic development of the West.

The Reclamation Service was initially created within the U.S. Geological Survey, but in 1907, the Reclamation Service was separated from USGS and in 1923 was renamed the Bureau of Reclamation. The Bureau is made up of five regions with project facilities that include over 600 reservoirs and dams, thousands of miles of canals and other distribution facilities, and 58 hydroelectric plants. For publications, search http://www.usbr.gov/library/ (accessed March 23, 2011).

U.S. Fish and Wildlife Service (FWS)

The FWS employs approximately 8,700 people at facilities across the United States. It is a decentralized organization with a headquarters office in Washington, D.C., eight geographic regional offices, and nearly 700 field units. The service manages the 96-million-acre National Wildlife Refuge System and thousands of small wetlands and other special management areas. It also operates 70 National Fish Hatcheries, 65 fishery resource offices, and 78 ecological services field stations.

The FWS enforces federal wildlife laws, protects and manages migratory birds, restores nationally significant fisheries, conserves and restores wildlife habitat, such as wetlands, and helps foreign governments. For publications, search http://library.fws.gov/ (accessed March 23, 2011).

U.S. Geological Survey (USGS)

The USGS was established in 1879. The USGS serves the nation by providing reliable scientific information to describe and understand the Earth; minimize loss of life and property from natural disasters; manage water, biological, energy, and mineral resources; and enhance and protect our quality of life. To attain these objectives, the USGS prepares maps, collects and interprets data on energy and mineral resources; conducts nationwide assessments of quality, quantity, and use of the nation's water resources; performs fundamental and applied research in the sciences and techniques involved; and publishes their investigations through maps and a variety of technical publications.

The USGS employs 10,000 staff and maintains an extensive network of regional offices and research centers. The Water Resources Division http://water.usgs.gov (accessed March 23, 2011) is particularly relevant to environmental engineers. The National Water Information System (NWIS) http://waterdata.usgs.gov/nwis (accessed March 23, 2011) Web page contains historical and real-time data on surface water, groundwater, and water quality. The USGS manages water information at offices located throughout the United States. Although all offices are tied together through a national network, each office collects data and conducts studies in a particular area. Local information is best found at the following sites: Cooperative Water Program, National Streamflow Information Program (NSIP), National Water Quality Assessment Program (NAWQA), Toxic Substances Hydrology (TOXICS) Program, Groundwater Resources Program, Hydrologic Research and Development (HRD) Program, State Water Resources Research Institute Program. The agency also conducts international research. For publications, search http://pubs.er.usgs.gov/ (accessed March 23, 2011).

U.S. Environmental Protection Agency (EPA)

The mission of the EPA is to protect human health and the environment. The EPA employs over 17,000 people across the country at its headquarters in Washington, D.C., 10 regional offices, and more than a dozen laboratories.

There are more than a dozen statutes and laws that form the legal basis for the programs that the EPA administers. A complete list may be found on the Web site. The following list contains the laws that are most applicable to the environmental engineering community:

Clean Air Act (CAA)
Clean Water Act (CWA)
Comprehensive Environmental Response, Compensation, and Liability Act (CERCLA or Superfund)
Emergency Planning and Community Right-to-Know Act (EPCRA)
Endangered Species Act (ESA)
Federal Insecticide, Fungicide, and Rodenticide Act (FIFRA)
National Environmental Policy Act (NEPA)
Oil Pollution Act (OPA)
Pollution Prevention Act (PPA)
Resource Conservation and Recovery Act (RCRA)
Safe Drinking Water Act (SDWA)
Solid Waste Disposal Act
Superfund Amendments and Reauthorization Act (SARA)
Toxic Substances Control Act (TSCA)

The EPA operates extensive information networks. Some of the most important are listed below:

Hotlines and clearinghouses
Technology transfer networks
Dockets
Libraries
Publications
Databases and software

Visit the Web pages of the following EPA program offices:

Air and Radiation
Compliance and Enforcement
Emergency Response
Hazardous Waste Program
Indoor Air–Radon
National Center for Environmental Assessment (NCEA)

National Center for Environmental Research (NCER)
National Estuary Program
Persistent Bioaccumulative and Toxic (PBT) Chemical Program
Research and Development
Science Advisory Board
Solid Waste and Emergency Response
Superfund Information
Underground Injection Control (UIC) Program
Underground Storage Tanks (UST)
Water: aquatic ecosystems, drinking water, groundwater, stormwater, surface water, waste-
water, water pollution, water quality monitoring
Watersheds
Wetlands Program

U.S. National Oceanic and Atmospheric Administration (NOAA)

Although NOAA began in 1970, it was formed from agencies that are among the oldest in the federal government. The agencies included the U.S. Coast and Geodetic Survey formed in 1807, the Weather Bureau formed in 1870, and the Bureau of Commercial Fisheries formed in 1871. Individually, these organizations were America's first physical science agency, America's first agency dedicated specifically to the atmospheric sciences, and America's first conservation agency. Much of America's scientific heritage resides in these agencies. They brought their cultures of scientific accuracy and precision, stewardship of resources, and protection of life and property to the newly formed agency. NOAA researches and gathers data on weather, oceans, satellites, fisheries, climate, coasts, and charting and navigation. NOAA is made up of six divisions: National Weather Service (NWS), National Environmental Satellite, Data, and Information Service (NESDIS), National Marine Fisheries Service (NMFS), National Ocean Service (NOS), Office of Oceanic and Atmospheric Research (OAR), and Office of Program Planning and Integration (PPI). For publications, search http://www.lib.noaa.gov/ (accessed March 23, 2011).

ASSOCIATIONS, ORGANIZATIONS, SOCIETIES, AND CONFERENCES

Most of the entities listed here alphabetically offer annual meetings, education opportunities, publications, buyer's guides, job listings, and membership directories. More recently, some have begun to offer online discussion groups, electronic mailing lists, and virtual seminars.

Air and Waste Management Association (A&WMA) http://www.awma.org/ (accessed October 14, 2010). An environmental, educational, and technical organization, A&WMA seeks to provide a neutral forum for the exchange of technical information on a wide variety of environmental topics. Committees and divisions include Air, Environmental management, and Waste. Formerly Smoke Prevention Association of America (1987), Air Pollution Control Association (APCA) (1989). Special events or services: Annual conference and exhibition, with the International Urban Air Quality Forum as a satellite event.

American Academy of Environmental Engineers (AAEE) http://www.aaee.net/ (accessed October 14, 2010). Environmentally oriented registered professional engineers are certified by examination as Diplomats of the Academy. Purposes include: to improve the standards of environmental engineering, to certify those with special knowledge of environmental engineering, and to furnish lists of those certified to the public. The AAEE works with other professional organizations on environmentally oriented activities. Special events or services: Certification program.

American Association for Aerosol Research (AAAR) http://www.aaar.org/ (accessed October 14, 2010). AAAR promotes and communicates technical advances in the field of aerosol research. Special events or services: Annual conference.

American Chemical Society (ACS), http://portal.acs.org/portal/acs/corg/content (accessed October 14, 2010). ACS is a self-governed individual membership organization that consists of more than 159,000 members at all degree levels and in all fields of chemistry. The Division of Environmental Chemistry sponsors an annual specialty meeting on Green Chemistry and Engineering. Special events or services: Annual national meeting and regional meetings.

American Conference of Governmental Industrial Hygienists (ACGIH) http://www.acgih.org/home.htm (accessed October 14, 2010). The best known of ACGIH's activities is the Threshold Limit Values for Chemical Substances (TLV®-CS). Today, ACGIH has 11 committees focusing on a range of topics that include air sampling instruments, bioaerosols, exposure indices, and industrial ventilation. Special events or services: Partners with Society of Toxicology, AIHA, and AIHce in their conferences.

American Industrial Hygiene Association (AIHA) http://www.aiha.org/Pages/default.aspx (accessed October 14, 2010). Serving the needs of occupational and environmental health professionals practicing industrial hygiene, AIHA's more than 30 technical committees deal with such concerns as exposure and risk assessment strategies, indoor environmental quality, workplace environmental exposure levels, noise hazards, and respiratory protection. Special events or services: AIHce annual conference, co-sponsored by AIHA and ACGIH. Web site offers an electronic discussion list.

American Membrane Technology Association (AMTA) http://www.membranes-amta.org/ (accessed October 14, 2010). Water supply agencies, manufacturers of desalting equipment, design and construction companies, advanced water sciences and technologies consultants, individuals, academicians, and librarians form the membership. The AMTA advances research and development programs in desalination, wastewater reclamation, and other water sciences; promotes programs of water supply and urban environment improvement; and sponsors training of water treatment plant operators. Formerly National Water Supply Improvement Association (1993); American Desalting Association (2003). Special events or services: Various technology conferences; Web site hosts a message board.

American Water Resources Association (AWRA) http://www.awra.org/ (accessed October 14, 2010). Membership encompasses engineers; natural, physical, and social scientists; other persons engaged in any aspect of the field of water resources; business concerns and other organizations. AWRA seeks to advance water resources research, planning, development, and management. Special events or services: Annual conference. Web site offers a virtual mentoring program.

American Water Works Association (AWWA) http://www.awwa.org/ (accessed October 14, 2010). Membership is drawn from water utility managers, superintendents, engineers, chemists, bacteriologists, and other individuals interested in public water supply; municipal- and investor-owned water departments; boards of health; manufacturers of waterworks equipment; government officials and consultants. AWWA develops standards and supports research programs in waterworks design, construction, operation, and management, conducts in-service training schools and prepares manuals for waterworks personnel. The database, Waternet, is available on Dialog as file 245 and on CD-ROM. Affiliated with the Water Environment Federation. Special events or services: Annual conference and exposition as well as numerous regional conferences and workshops. Electronic discussion forums available through Web site.

Association of Environmental and Engineering Geologists (AEG) http://www.aegweb.org/i4a/pages/index.cfm?pageid=1 (accessed October 14, 2010). Serving professionals in groundwater, environmental, and engineering geology, the mission of AEG is to provide leadership in the development and application of geologic principles and knowledge to serve engineering, environmental, and public needs. Special events or services: Annual meeting.

Association of Environmental Engineering and Science Professors (AEESP) http://www.aeesp.org/ (accessed October 14, 2010). The association works to improve education and research programs in the science and technology of environmental protection. It provides information to government agencies and the public, encourages graduate education by supporting research and training for students, and maintains a speaker's bureau. Formerly Association of Environmental Engineering Professors (1999). Special events or services: Annual conference.

Association of State and Interstate Water Pollution Control Administrators (ASIWPCA) http://www.asiwpca.org/ (accessed October 14, 2010). With a membership of administrators of state and interstate governmental agencies legally responsible for the prevention, abatement, and control of water pollution, the association promotes coordination among state agency programs and those of the Environmental Protection Agency, Congress, and other federal agencies. Special events or services: Annual and mid-year meetings.

Association of State Drinking Water Administrators (ASDWA) http://www.asdwa.org/ (accessed October 14, 2010). Comprises managers of state and territorial drinking water programs and state regulatory personnel. The ASDWA works to meet communication and coordination needs of state drinking water program managers; facilitates the exchange of information and experience among state drinking water agents; acts as a collective voice for the protection of public health through assurance of high-quality drinking water; and oversees the implementation of the Safe Drinking Water Act. Special events or services: Annual conference.

Environmental and Engineering Geophysical Society (EEGS) http://www.eegs.org/ (accessed October 14, 2010). This applied scientific organization's mission is to promote the science of geophysics, especially as it is applied to environmental and engineering problems and to foster common scientific interests of geophysicists and their colleagues in other related sciences and engineering. Special events or services: Annual meeting.

Institute of Environmental Sciences and Technology (IEST) http://www.iest.org (accessed October 14, 2010). IEST, an ANSI-accredited developer of American National Standards, focuses on education and the development of recommended practices and standards. Absorbed the American Association for Contamination Control (1973). Formed by merger of the Institute of Environmental Engineers and Society of Environmental Engineers (1994). Formerly Institute of Environmental Sciences (1999). Special events or services: ESTECH, the IEST annual technical meeting and exposition. Web site offers a conferencing and chat service.

Institute of Noise Control Engineering of the USA (INCE-USA) http://www.inceusa.org/ (accessed October 14, 2010). Promoting engineering solutions to environmental noise problems, INCE/USA is a member society of the International Institute of Noise Control Engineering, an international consortium of organizations with interests in acoustics and noise control. Special events or services: National conference, which often includes the meeting of the Acoustical Society of America.

International Environmental Modelling and Software Society (iEMSs) http://www.iemss.org/ (accessed October 14, 2010). Dealing with environmental modeling, software, and related topics, the aims of the iEMSs include the development and use of environmental modeling and software tools to advance the science and improve decision making with respect to resource and environmental issues. iEMSs was founded in 2000. Special events or services: Biennial general meeting usually held on even years.

International Institute of Noise Control Engineering (I-INCE) http://www.i-ince.org/ (accessed October 14, 2010). This is a worldwide consortium of organizations concerned with noise control, acoustics, and vibration. The primary focus of the institute is on unwanted sounds and on vibrations producing such sounds when transduced. Special events or services: Annual congress.

International Society of Exposure Science (ISES) http://www.isesweb.org/ (accessed October 14, 2010). Established in 1989 to foster and advance the science of exposure analysis related to environmental contaminants, both for human populations and ecosystems, the membership promotes communication among all disciplines involved in exposure analysis, recommends exposure analysis approaches to address substantive or methodological concerns, and works to strengthen the impact of exposure assessment on environmental policy. Special events or services: Annual conference.

International Society of Indoor Air Quality and Climate (ISIAQ) http://www.isiaq.org/ (accessed October 14, 2010). An international scientific organization, the purpose of ISIAQ is to support the creation of healthy, comfortable, and productive indoor environments by advancing the science and technology of indoor air quality and climate as it relates to indoor environment design, construction, operation and maintenance, air quality measurement, and health sciences. Special events or services: Annual conference.

International Water Association (IWA) http://www.iwahq.org/Home/ (accessed October 14, 2010). Integrating the leading edge of professional thought on research and practice across the drinking water, wastewater, and stormwater disciplines, the IWA was founded in September 1999 by the merger of the International Association of Water Quality (IAWQ) and the International Water Supply Association (IWSA). Special events or services: Biennial World Water Congress plus regional and specialty conferences. Web site offers a members-only Extranet.

National Association of Clean Air Agencies (NACAA) http://www.4cleanair.org/ (and) http://www.cleanairworld.org/ (accessed October 14, 2010). NACAA (formerly STAPPA and ALAPCO) is the national association representing air pollution control agencies in 54 states and territories and over 165 major metropolitan areas across the United States. Formed over 30 years ago, their purpose is to improve their effectiveness as managers of air quality programs through the exchange of information among air pollution control officials, communication and cooperation among federal, state, and local regulatory agencies, and management of air resources.

National Association of Environmental Professionals (NAEP) http://www.naep.org/ (accessed October 14, 2010). NAEP is a multidisciplinary, professional association dedicated to the promotion of ethical practices, technical competency, and professional standards in the environmental fields. Members have access to the most recent developments in environmental practices, research, technology, law, and policy. Special events or services: Annual conference. Professional certification program.

National Ground Water Association (NGWA) http://www.ngwa.org/ (and) http://www.wellowner.org (accessed October 14, 2010). With a membership of groundwater geologists and hydrologists, engineers, groundwater contractors, manufacturers, and suppliers of groundwater-related products and services, the association's purpose is to provide guidance to members, government representatives, and the public for sound scientific, economic, and beneficial development, protection, and management of the world's groundwater resources. (Formerly the National Water Well Association.) Special events or services: Annual exposition. Professional certification programs.

Society of Environmental Toxicology and Chemistry (SETAC) http://www.setac.org/ (accessed October 14, 2010). A nonprofit, worldwide professional society, SETAC promotes the advancement and application of scientific research related to contaminants and other stressors in the environment, education in the environmental sciences, and the use of science in environmental policy and decision making. Special events or services: SETAC North America Annual Meeting, an annual World Congress, other international meetings.

Water Environment Federation (WEF) http://www.wef.org/ (accessed October 14, 2010). The WEF seeks to advance fundamental and practical knowledge concerning the nature, collection, treatment, and disposal of domestic and industrial wastewaters, and the design, construction, operation, and management of facilities for these purposes. Formerly Federation of Sewage Works Associations (1949), Federation of Sewage and Industrial Wastes Associations (1959), Water Pollution Control Federation (1991). Special events or services: WEFTEC, the annual technical exhibition and conference.

CONCLUSION

Environmental engineering is a diverse and challenging discipline. Environmental engineers are learning more about our environment every day and they are finding solutions to many complex problems. However, as test methods and instrumentation become more sophisticated, environmental engineers are discovering more pollutants and realizing that the number of unanswered questions continues to grow.

The information resources recommended in this chapter are starting points to help penetrate the increasing volume and complexity of knowledge about environmental engineering as a means of managing human impact on the environment. With its broad range of specialties and subspecialties—air pollution control, industrial hygiene, radiation protection, hazardous waste management, toxic materials control, water supply, wastewater management, stormwater management, solid waste disposal, public health, and land management—keeping up-to-date in one's area of expertise and aware of the bigger picture is becoming increasingly difficult.

We encourage those teaching, studying, and working in environmental engineering to further the scientific knowledge in support of good stewardship of the planet.

ACKNOWLEDGMENTS

I would like to thank Linda Vida and Lois Widmer for all of the hard work they put into the first edition of this chapter. I would also like to thank Bonnie Osif for her advice and assistance in updating this chapter.

FURTHER READINGS

Balachandran, S., ed. 1993. *Encyclopedia of environmental information sources: A subject guide to about 34,000 print and other sources of information on all aspects of the environment.* Detroit: Gale Research.

Lord, C. R. 2000. *Guide to information resources in engineering.* Englewood, CO: Libraries Unlimited.

Nazaroff, W. W. and L. Alvarez-Cohen. 2001. *Environmental engineering science.* New York: John Wiley & Sons.

13 History of Engineering

Nestor L. Osorio and Mary A. Osorio

CONTENTS

Introduction, History, and Scope of Discipline 317
Searching the Library Catalog 319
   Keywords, LC Subject Headings, and LC Call Numbers 319
Indexes and Abstracts 321
Private and Institutional Databases 323
Bibliographies and Guides to the Literature 323
Encyclopedias and Dictionaries 325
Handbooks, Manuals, and Guides 327
Monographs and Textbooks 328
Journals and Monographic Series 330
Conferences and Symposia 332
Museums and Special Collections 332
   Libraries 333
   Museums 334
Internet Resources 336
Societies and Research Institutions 337
   Societies, Associations, and Organizations 338
   Research Institutions 339
Conclusion 340
References 340

INTRODUCTION, HISTORY, AND SCOPE OF DISCIPLINE

According to the *Oxford English Dictionary* (1989), one of the first uses of the word *engineering* in the English language dates back to 1681; in 1729, Bernard Fores de Belidor published one of the first books with the word *engineering* in the title: *La Science de Ingenieures.* These are two literary indications that engineering began to solidify as a distinct profession during the Industrial Revolution. There have been continuing discussions among experts about the differences and the definitions of science, technology, and engineering. The numerous sources presented in this section might help the readers interested in these types of academic analyses, but the focus of this chapter is far from presenting a critical review of their definitions and relationships. Nevertheless, it is important to raise some of the basic concepts related to the history of engineering. It is appropriate to bring up the connection between applied sciences and engineering, to make a distinction about the meaning of technology and engineering, and to present a brief chronology of engineering in modern times.

To better understand the state of engineering history today, it is necessary to look back at its origins. If one were to choose one magnificent human creation that signified the ingenuity of early engineering wonders, it would be the pyramids of Egypt. The golden age of the Egyptian pyramid-building occurred between 2700 and 2550 BCE. At this time, and without the use of the wheel, about 11 million cubic meters of stone were transported (Garrison, 1999). Early civilizations in Mesopotamia, Greece, Rome, India, and China—to mention a few—showed highly sophisticated activities that utilized resources for the production of goods. Progress continued through the years, but a new, invigorated era of technological advances and scientific knowledge began during the Renaissance. Domenico Fontana is an example of an engineer of this era. He "(1) mastered difficult technical problems, (2) designed road and hydraulic work, (3) worked at practical building construction, and (4) had knowledge of mathematics, geometry, and the deliberate use of calculation" (Garrison, 1999). Some of the remarkable innovations that occurred prior to the development of steam power include new techniques for working with raw materials, the use of new agriculture techniques, the extraction of chemical products, the beginning of the textile industry, the first stage of industrial mechanization, the mastering of clocks construction, and the improvement of land and sea transportation (Daumas, 1969–1979).

In modern times, the relationship between engineering and applied science is readily noted by observing the number of colleges that include these two disciplines in their name. Several engineering specialties have evolved from the sciences: chemical engineering from chemistry; metallurgical engineering from geology, chemistry, and physics; and electrical engineering from mathematics and physics. Scientific knowledge can attain practical applications that can be used by engineers in the development of goods and services that produce usable and economic benefits to society. Because of this strong connection between applied sciences and engineering, the study of the history of science overlaps with the study of the history of technology. In this chapter, science resources that are part of this overlap have been presented.

As in the case of applied sciences, there is no clear line of demarcation between engineering and technology. According to the following definitions, engineering deals with the utilization of resources and the production of goods and services, while technology deals with the tools and techniques utilized for achieving those goals. This is how the *McGraw-Hill Encyclopedia of Science and Technology* (2010) defines engineering: "Most simply, the art of directing the great sources of power in nature for the use and the convenience of humans. In its modern form engineering involves people, money, materials, machines, and energy. It is differentiated from science because it is primarily concerned with how to direct to useful and economical ends the natural phenomena that scientists discover and formulate into acceptable theories. Engineering, therefore, requires above all the creative imagination to innovate useful applications of natural phenomena. It is always dissatisfied with present methods and equipment. It seeks newer, cheaper, better means of using natural sources of energy and materials to improve the standard of living and to diminish toil."

The same encyclopedia defines technology as: "Systematic knowledge and action, usually of industrial processes, but applicable to any recurrent activity. Technology is closely related to science and to engineering. Science deals with humans' understanding of the real world about them—the inherent properties of space, matter, energy, and their interactions. Engineering is the application of objective knowledge to the creation of plans, designs, and means for achieving desired objectives. Technology deals with the tools and techniques for carrying out the plans." The terms *engineering* and *technology* are in many cases inclusive, interchangeable, and sometimes misused. That is the reason why a good number of the sources listed in this chapter include the term *technology*. An effort has been made to include only those sources that have an engineering content.

According to Auyang (2004), there are three overlapping phases in the history of modern engineering: the Industrial Revolution, the second Industrial Revolution, and the Information Revolution.

The Industrial Revolution, which includes the eighteenth and early nineteenth centuries, is characterized by the changes that occurred in civil, mechanical, mining, and metallurgical engineering when they went from being practical arts to becoming scientific professions. This is also the time when the first engineering schools were founded. Some of the most significant engineering advances of this era are steam engines, the improvements made to water power, the beginning of the study of the strength of materials, the development of the iron industry, and the use of cement as building material (Kirby et al., 1956).

The second Industrial Revolution occurred in the twentieth century before World War II (Auyang, 2004). It is characterized by the improvement of steel production, the manufacturing of the rapid-fire gun, the shift from steam power to the internal combustion engine, the further development of metallurgy, the generation of electric power, the filament lamp, the transmission of sound over long distances, and the beginning of radio communication, radar, and television. Later in this era, the aircraft industry, systems of mass production, and the automobile were also created (Engineering and Materials, 2010).

World War II served as a catalyst to many innovations that highlight the Information Revolution. It may be summarized by listing some of the major engineering accomplishments: space travel, atomic power, electronic computers, and the advent of the microchip. In this last era, the capacity to create, store, and exchange data in its multiple forms has propelled industry, commerce, the production of services and products, and is present today in all aspects of human endeavor.

The future of engineering as a profession is assured because it will continue to have a significant impact in the transformation of society. In writing about the future of engineering, Bell and Dooling (2000) stated, "Some of the challenges will require a purely technical approach. Others will be primarily societal, requiring a balancing of material priorities with sociological values—with the outcome expressed by funding availability and regulations. All involve engineering techniques and expertise. And never before have humans been so technically well-prepared." The history of engineering is the story of how engineers and their organizations have brought about changes throughout society through the ages.

SEARCHING THE LIBRARY CATALOG

KEYWORDS, LC SUBJECT HEADINGS, AND LC CALL NUMBERS

Searching the online catalog is an effective way of finding documents about the history of engineering. Online catalogs of libraries in the United States, in Europe, in Japan, and in most countries of the world do not just list books. Their databases include data about libraries' holdings in many other formats, such as journals, films, audiovisuals, federal and state government publications, reports, and dissertations. To find noncataloged documents held by special collections in libraries and museums, the best option is to get direct information from their curators. There are several ways of searching for documents in an online catalog, as will be shown through the following examples:

This is a list of subject terms used by the Library of Congress (LC) of the United States. The first group includes some of the most commonly used terms in finding materials in major general subject areas relating to the history of engineering:

Engineering	History
Technology	History
Science	History
Inventions	History
Technology and civilization	

This second group of terms shows examples of how to find information in more specific areas in the history of engineering:

Chemical engineering	History
Civil engineering	History
Electric engineering	History
Human engineering	History
Industrial engineering	History
Mechanical engineering	History
Petroleum engineering	History
Women in technology	History

Specific areas of engineering also can be limited to geographical locations as indicated in the examples below:

Aerospace engineering	United States, History
Electric industry workers	Great Britain, History
Electronics	United States, History
Engineering	Great Britain, History
Industrial management	United States, History
Mass production	United States, History
Technology	Social aspects, United States, History
Telecommunication	United States, History

The names of relevant people in engineering can be searched in the subject field by using their last name followed by the first name. For example:

Alexanderson, Ernst Fredrik Werner, 1878–1975
Jervis, John B. (John Bloomfield), 1795–1885
Coulomb, Charles Augustin de, 1736–1806

The name of corporations or institutions also can be searched in the subject field, for example:

H. H. Franklin Manufacturing Company	History
American Telephone and Telegraph Company	History

The following subject terms show how to find biographical documents about a person or a group of persons:

Civil engineers	United States, Biography
Electric engineers	United States, Biography
Electric engineers	Inventors, Biography, Dictionaries

Finally, readers could limit their search as well to a specific period of time as is indicated below:

Civil engineering	Early works to 1850
Engineering	Europe, History, to 1500
Statics	Early works to 1800
Technology	History, 20th century
Technology	United States, History, 19th century

Browsing the shelves is another way of finding documents. These are some of the areas in the LC classification system where readers can find materials:

T14.5 Technology	Social aspects, History
T15-T40 Technology	History. It includes general works, ancient, medieval, modern, nineteenth—twentieth centuries, special counties, United States, Canada, Latin America, South America, Europe, Asia, Africa, Australia, Pacific islands, arctic regions, lost arts, historical atlases, industrial archaeology, and biographies.
T144-145 Technology	General works, before nineteenth century, nineteenth century and later.

The historical perspective is covered under the range 15 to 145 for the following engineering areas:

TA15-T145   General engineering and civil engineering
TC15-T145   Hydraulic engineering
TD15-T145   Environmental technology, Sanitary engineering
TE15-T145   Highway engineering, Road and pavements
TF15-T145   Railroad engineering and operation
TG15-T145   Bridge engineering
TH15-T145   Building construction
TJ15-T145   Mechanical engineering and machinery
TK15-T145   Electrical engineering, Electronics, Nuclear engineering
TL15-T145   Motor vehicles, Aeronautics, Astronautics
TN15-T145   Mining engineering, Metallurgy
TP15-T145   Chemical technology
TS15-T145   Manufactures

In addition, materials on the history of specific engineering fields may be found throughout the entire T subject classification of the LC system. For example, history of automation control in TJ213, history of petroleum engineering in TN871, and history of mechanical drawing in T353.

Furthermore, because there is a close relationship between the applied sciences and engineering, other works on the history of engineering may be found in these LC sections: mathematics (QA), physics (QC), chemistry (QD), Geology (QE), as well as in psychology (BF), and Industrial management (HF). In this case, and to identify specific areas, it is better to first search the online catalog by using keywords or subject terms.

The Library of Congress classification system is the most widely used system in North America. It is advisable that readers in other countries follow the classification and subject heading guidelines of their native system.

INDEXES AND ABSTRACTS

Indexing and abstracting databases are excellent sources for finding contemporary publications on the history of engineering. They index articles published in journals and in conference proceedings that represent a major portion of the publications on this subject. Indexes and abstracts also list reviews of books, conference proceedings, dissertations, and reports. To search for older materials it is advisable to consult with a librarian.

America: History and Life. 1964–. Ipswich, MA: EBSCO. This is a basic, but comprehensive bibliographic resource for the study of U.S. and Canadian history. The print edition has two parts: *Part A, Article Abstracts and Citations* and *Part B, Index to Book Reviews and Doctoral Dissertations.* The online database included some retrospective coverage of leading history journals that go back to the late nineteenth century.

Bibliografia Italiana di Storia della Scienza. 1982–1998. Florence, Italy: L. S. Olschki. It is produced by the staff of the Istituto e Museo di Storia della Scienza (IMSS), Florence, Italy. This bibliography includes mainly books, articles in periodicals, book chapters, and book reviews primarily in Italian. Paper issues of this publication were last published in 1998. Since 1999, the bibliography is available on the Web, searchable from the Museo Galileo library catalog http://www.museogalileo.it/esplora/biblioteche/biblioteca/bibliografiaitalianastoriascienza.html (accessed November 10, 2010). It has been incorporated into the *History of Science, Technology, and Medicine* database.

British Humanities Index. 1962–. Bethesda, MD: Cambridge Scientific Abstracts. This is an abstract and index database that covers about 400 British journals and newspapers. It includes citations on the history of technology within the coverage of philosophy, economics, politics, history, and society. Major obituaries of prominent figures are noted. It is arranged according to abstract number, followed by the subject index term and title. It was originally published in London by Library Association Publishing. The online database is a ProQuest interface.

Current Bibliography in the History of Technology. 1964–2002. Baltimore, MD: The Johns Hopkins University Press. This publication is a supplement of the journal *Technology and Culture of the Society for the History of Technology.* The paper edition ended in 2002. Since then, it is published only in electronic format as part of the *History of Science, Technology and Medicine* database. It is divided into 17 subject classifications for each of the 5 defined historical time periods. It provides extensive coverage in all formats of the literature on the history of technology.

Francis. 1984–. Paris: Institut de L'Information Scientifique et Technique. Since 1994, *Francis* is published in electronic format only. It covers articles in journals, conference proceedings, books, research reports, and French dissertations from 1972 to the present. It is an index for the humanities and social sciences including the history of science and technology. Prior to 1994, it was published in paper. The sections on the history of technology are known as: *Francis Bulletin Signalétique. 522, Histoire des Sciences et des Techniques,* 1991–1994; *Bulletin Signalétique. 522, Histoire des Sciences et des Techniques,* 1969–1990; and *Bulletin Signalétique. 22, Histoire des Sciences et des Techniques,* 1961–1968. The online database is a ProQuest interface.

Historical Abstracts. 1955–. Ipswich, MA: EBSCO. This is considered to be the most comprehensive index of abstracts for articles, book reviews, essays, papers, conference proceedings, and dissertations in world history and the social sciences. The time frame of the topics in history abstracts starts from the 1450s and continues into the present. It covers 2,100 journals, book reviews, and dissertations. The online database included some retrospective coverage of leading history journals that go back to the late nineteenth century.

History of Science, Technology, and Medicine. 1976–. Charlottesville, VA: Department of Science, Technology & Society; University of Virginia. Available through FirstSearch, this database contains four bibliographies: *Isis Current Bibliography of the History of Science and its Cultural Influences* (1975–), *Current Bibliography in the History of Technology* (1987–), *Bibliografia Italiana di Storia della Scienza* (1982–), and the *Wellcome Bibliography for the History of Medicine* (1991–2004). This agglomeration makes this database the most comprehensive index to the literature on the history of engineering, technology, and related fields. It covers all formats.

Isis Current Bibliography of the History of Science and its Cultural Influences. 1975–. Chicago: University of Chicago Press. It is published as a supplement of the *History of Science Society* journal *Isis.* It indexes mainly books, articles in journals and symposia, dissertations, book chapters, and book reviews from all areas of the history of science. These data are being incorporated into the *History of Science, Technology and Medicine* database. It has a journal list and is divided into 24 subject sections.

The Making of the Modern World: The Goldsmiths'–Kress Library of Economic Literature. 2004–. Farmington Hills, MI: Thomson Gale. This is a comprehensive collection, fully indexed, containing 61,000 monographs published from 1450 to 1850 and 466 serials published before 1906. It is very strong in economics and related subjects including history, banking, transportation, and manufacturing. This collection has more than 11 million pages. Materials were collected from the Goldsmiths' Library of Economic Literature at the University of London Library, the Kress Library of Business and Economics at the Harvard University Graduate School of Business Administration, the Seligman Collection in the Butler Library at Columbia University, and from the libraries of Yale University.

PRIVATE AND INSTITUTIONAL DATABASES

The previous section has a selected list of indexing services that include mainly contemporary materials on this subject. In this section, some examples of sources have been included where older materials can be found. Also, numerous institutions, such as museums and historical archives, have their own databases. To localize materials from collections in most archives and museums, it is necessary to search their own databases. Another option is to contact or visit them.

Bibliografía Histórica de la Ciencia y la Técnica en España http://www.ihmc.uv-csic.es/buscador.php (accessed November 12, 2010). This database is maintained by the Instituto de Historia de la Medicina y de la Ciencia López Piñero, Universidad de Valencia-CSIC. Formerly know as the Bibliografía Española de Historia de la Ciencia y de la Técnica. It covers all works done about the history of science and technology in Spain since 1989.

British Library, The—Manuscripts Catalogue http://www.bl.uk/catalogues/manuscripts/ (accessed October 20, 2010). "Use this Website to search the main catalogues of the British Library's collection of Western manuscripts, covering handwritten documents of all kinds from pre-Christian, classical, medieval and modern times" (from BL Web page).

HathiTrust Digital Library http://www.hathitrust.org/home (accessed November 12, 2010). This collection includes an extensive number of full-text documents from the Middle Ages to modern times, and in all languages. It makes accessible primary documents, and is open to the public. It represents a conglomerate of metadata and virtual documents collected by a large number of academic libraries.

The History Journals Guide. Periodicals Directory http://www.history-journals.de/journals/index.html (accessed October 20, 2010). This database is maintained by Stefan Blachke of IC Media Informations systeme GmbH, Augsburg, Germany. It includes nearly 6,000 serials titles from all periods, all regions, and all fields of history.

National Cataloging Unit for the Archives of Contemporary Scientists (NCUACS) http://www.bath.ac.uk/ncuacs/ (accessed October 20, 2010). "The (NCUACS) was established in 1973 to locate, sort, index, and catalogue the manuscript papers of distinguished contemporary British scientists and engineers. They include correspondence of all kinds, professional or technical documents, such as laboratory notebooks, experimental drawings and calculations, lecture notes, engagement diaries and journals" (from NCUACS Web page).

BIBLIOGRAPHIES AND GUIDES TO THE LITERATURE

Ferguson's bibliography on the history of technology listed in this section is considered one of the foremost printed bibliographies in the field. It includes a good number of valuable sources for researchers in the history of engineering. Channell's work is one of the first bibliographies in the English language on the history of engineering, and both titles are highly recommended. Included are more recent bibliographies on contemporary topics, such as gender, human factors, and computing. Two general guides on the history of engineering are also included.

Bibliography of British and Irish History http://www.history.ac.uk/projects/bbih (accessed November 12, 2010). "The *Bibliography of British and Irish History* (BBIH) replaced the *RHS Bibliography* on 1 January 2010. It contains all of the data previously available in the *RHS Bibliography*, including its partner projects, London's Past Online and (up to the end of 2009) Irish History Online. *BBIH* is updated three times a year and now includes over 490,000 records describing books and articles about British and Irish history" (from RHS Web page).

Bibliography for History of Engineering http://www.tc.umn.edu/~tmisa/biblios/hist_engineering.html (accessed November 9, 2010). This Web bibliography is maintained by T. J. Misa of the Charles Babbage Institute, University of Minnesota, Minneapolis–St. Paul,

MN. Last updated in August of 2007, it contains a listing of books without annotations and is organized into two sections: Chronology and Topics, and Branches of Engineering.

Bindocci, C. G. 1993. *Women and Technology: An Annotated Bibliography.* New York: Garland. The purpose of this bibliography is to identify scholarly research about women and technology, an area of research that has been greatly overlooked through the years. The bibliography is meant to be the starting point for research by summarizing what others have already accomplished in this field. This work includes only secondary works in the English language published from 1979 to 1991 in the forms of articles, books, published conference proceedings, and dissertations.

Channell, D. F. 1989. *The History of Engineering Science: An Annotated Bibliography* New York: Garland. The author of this work intended it to be a guide for students, scholars, and researchers into the history of engineering science, a field that can encompass both the history of science and the history of technology. Numerous entries may be overlooked by both of them when it comes to finding scholarly work in the history of engineering science. This comprehensive bibliography fills the void. Primary and secondary materials are found.

Cortada, J. W. 1996. *Second Bibliographic Guide to the History of Computing, Computers, and the Information Processing Industry.* Westport, CT: Greenwood Press. This volume primarily emphasizes historically consequential published materials since the late 1980s. The titles are annotated and include both an author and subject index. Each citation is numbered and all references are to item numbers instead of pages. It favors documenting hardware and industry literature. This bibliography is meant to be a reputable source for the student of the history of computing, computers, and information processing, and for all those who wish to gain valuable insights into the world of computer technology.

Dekosky, R. K. 1995. Science, technology, and medicine. In *The American Historical Association's Guide to Historical Literature,* 3rd ed., eds. M. B. Norton and P. Gerardi, New York: Oxford University Press. This guide presents the highest quality and the most insightful books and articles from every field of historical scholarship. By providing a critical overview of the best historical scholarship, it helps to regain the unifying or integral vision that can be lost in a time of many specialists and specialties. Section 4 in volume 1 delves into the area of science, technology, and medicine. This work contains a list of journals, an author index, and a subject index.

Ferguson, E. S. 1968. *Bibliography of the History of Technology.* Cambridge, MA: Society for the History of Technology. The purpose of this bibliography is to gather all relevant information for a comprehensive introduction to primary and secondary sources in the history of technology. It is geared to the students, by sending them to the tools and resources of the scholar. However, all readers who have an interest in the subject profit greatly through its use. Not only is the "what" and "how" of technology probed, but the "why" of technology is also studied, since it is essential to the understanding of the culture and its implications on technology.

Fritze, R. H., B. E. Coutts, and L. Vyhnanek. 2004. *Reference Sources in History: An Introductory Guide,* 2nd ed., Santa Barbara, CA: ABC-CLIO. This is a general reference book that consists of 930 reference works listed in numbered bibliographic entries with annotations. It is an introduction to major historical works for all time periods of history and for all geographical areas. Chapter 10 is devoted to biographical sources that will aid in the search for persons in the field of engineering or technology, while chapter 13 is a resource for information from archives, manuscripts, special collections, and digital sites from around the world.

Green, R. J., H. C. Self, and T. S. Ellifritt, eds. 1995. *50 Years of Human Engineering: History and Cumulative Bibliography of the Fitts Human Engineering Division.* Wright-Patterson Air Force Base, OH: Crew Systems Directorate, Armstrong Laboratory, Air Force Materiel Command. This book is a presentation of the unclassified publications of the Human Engineering Division of the U.S. federal government. Technical reports are the most common form of publication but some of the research and development work is published in journals and proceedings of scientific and technical societies, or as chapters in books, handbooks, military specifications, and standards or special reports.

McClellan, J. E. and H. Dorn. 2006. Guide to resources. In *Science and Technology in World History: An Introduction,* 2nd ed. Baltimore, MD: John Hopkins University Press. This chapter includes all references the authors used in this well-respected basic textbook. It provides numerous entries for undergraduate students in the history of engineering.

National Museum of History and Technology. 1978. *Guide to Manuscript Collections in the National Museum of History and Technology.* Washington, D.C.: Smithsonian Institution Press. Volume 3 of the series *Archives and Special Collections of the Smithsonian Institution.* The Museum of History and Technology holds important archival collections. One can find in their collections both fundamental source materials and illustrative material not found elsewhere. Most of the entries in this guide concern personal papers, business records, or document files; however, one can also find graphic material, trade literature, and information and reference files.

Niskern, D. 2001. *The History of Technology.* Washington, D.C.: Science Reference Section, Science, Technology and Business Division, Library of Congress. This tracer bullet published by the Library of Congress lists sources for the history of science, invention, medicine, and technology in colonial America. It provides a variety of materials in the collections of the Library of Congress helpful in researching the science and technology of eighteenth-century America. It includes journals, government publications, conference proceedings, dissertations, bibliographies, biographies, and books.

Rider, K. J. 1970. *History of Science and Technology: A Select Bibliography for Students,* 2nd ed. London: Library Association. Although it has been 40 years since it was published by the Library Association of England, this selective bibliography is still considered a valuable source. It is divided into two sections, the first being the history of science and the second being the history of technology. The time periods covered are Ancient, Mediaeval and Renaissance, and Modern (anything after 1600 to the present). Subjects covered are agriculture, building, chemical industries, clocks, engineering (civil, electrical, and mechanical), firearms, machinery, metallurgy, papermaking, photography, printing, textiles, transportation, and woodworking.

Sterling, C. H., and G. Shiers. 2000. *History of Telecommunications Technology: An Annotated Bibliography.* Lanham, MD: Scarecrow Press. A selective guide to materials of historical importance. It records the history of major telecommunication technologies going back almost 200 years. It contains more than 2,500 annotated bibliographic entries of items published in English. Included are general reference works, serial publications, general surveys, institutional and company histories, biographies, plus name and title indexes.

Turner, R., ed. 2002. *Guide to the History of Science*, 9th ed. Chicago: History of Science Society. This guide was written by the History of Science Society. It is the world's oldest and largest society dedicated to understanding the historical context of science, technology, medicine, and society. It includes a membership directory, graduate programs, research centers, libraries, museums, societies and organizations, journals and newsletters, inactive publications, and an index. It is supplemented by an online site at http://www.hssonline.org/profession/index.html (accessed November 10, 2010).

ENCYCLOPEDIAS AND DICTIONARIES

Eugene S. Ferguson's *Bibliography of the History of Technology*, which has been annotated in this chapter, includes an excellent section on older encyclopedias and dictionaries from the 1700s up to the early 1900s, which provides significant sources of information pertaining to the early developments of modern technology. Encyclopedias listed in Sarton (1952) complement the list by Ferguson. The focus of this section is to present some more contemporary works. It is recommended to consult recent encyclopedias about specific engineering fields because they often include a section about historical events.

Day, L., and I. McNeil, eds. 1996. *Biographical Dictionary of the History of Technology*. New York: Routledge. This dictionary includes about 1,500 of the most influential people that have made significant contributions to new technological advances through history. The coverage is worldwide and from ancient times to the twentieth century.

Davidson, P., and K. Lusk-Brooke. 2006. *Building the World: An Encyclopedia of the Great Engineering Projects in History*. Westport, CT: Greenwood Press. It presents the description of forty-two major engineering projects since ancient times; for example the aqueducts of Rome, the Federal highway system; and the channel tunnel. It includes the original documentation of the projects and listings of additional readings.

Jones, W. R. 1996. *Dictionary of Industrial Archaeology*. Phoenix Mill, UK: Sutton Publishing. This dictionary contains over 2,600 terms with a strong focus on the industrial history of Great Britain. Its major emphasis is on the developments that occurred between 1750 and 1850.

Knight, E. H. 1876. *Knight's American Mechanical Dictionary: Being a Description of Tools, Instruments, Machines, Processes, and Engineering; History of Inventions: General Technological Vocabulary; and Digest of Mechanical Appliances in Science and the Arts.* (3 vols.) Boston, MA: Houghton, Mifflin. This dictionary describes topics in the alphabetical order of their names and its general scope is presenting the history of inventions from earliest times until about 1876. Entries are located in the most systemic order so that any detail could be readily reached at a moment's notice. This informative work boasts upwards of 7,000 engravings. Subject matter indexes are found throughout the three volumes and a list of the principal ones follows the preface.

Langmead, D., and C. Garnaut. 2001. *Encyclopedia of Architectural and Engineering Feats*. Santa Barbara, CA: ABC-CLIO. This work gives an overview of the architectural and engineering marvels and innovations that helped improve community living through each successive generation. The many differing factors that go into the progress of a community are discussed in this book. The feats in question are arranged alphabetically and a glossary is provided for the definitions of terms used. There are citations at the end of each article for further reading.

McNeil, I., ed. 1996. *An Encyclopaedia of the History of Technology*. London: Routledge. Edited by a former executive secretary of the Newcomen Society for the History of Engineering, this one-volume encyclopedia is a compilation of articles written by selected members of the history of engineering community. Divided into 22 chapters, this work covers the history of technology from its earliest time to the present. It is an excellent introduction to a very complex and fascinating area. At the end of each chapter is listed other publications for further reading.

A Pictorial History of Science and Engineering; The Story of Man's Technological and Scientific Progress from the Dawn of History to the Present, Told in 1,000 Pictures and 75,000 Words. 1957. New York: Year, Inc. This book is a general history of science and technology. Because no one can be a subject specialist in every field, this work gives us the outline of scientific and technical progress through the ages. It begins in ancient times and proceeds to the twentieth century. There are 1,000 pictures collected from numerous museums and prestigious organizations.

Selin, H., ed. 1997. *Encyclopaedia of the History of Science, Technology, and Medicine in Non-Western Cultures*. Boston: Kluwer Academic. The encyclopedia contributes to the cultural diversity of the sciences by legitimizing the study of other cultures' science and technology. The goal is to engage in a mutual exchange of ideas about both the Western culture and the non-Western cultures. At the beginning of this sizable volume is a list of entries. In the following section, the entries are alphabetically arranged. There are references at the end of each entry for further study, a list of authors, and an index at the end of the book.

Trinder, B., ed. 1992. *The Blackwell Encyclopedia of Industrial Archaeology*. Oxford, U.K.: Blackwell. This encyclopedia contains sizable information about the industrial societies that emerged in the West from the mid-eighteenth to the twentieth century. Scholars from 31 countries have contributed articles that also consider the preindustrial period of each country. It is a guide to the monuments, settlements, and museums that house artifacts from the mid-eighteenth century. It includes pictures, maps, bibliography, index, and an appendix.

HANDBOOKS, MANUALS, AND GUIDES

Handbooks, manuals, and guides are good reference sources because they provide information about how things work, and often include diagrams, pictures, graphics, and tables to help visualize the topic or object being studied. In this section, some titles listed are highly recognized, such as the work of Singer; some others are basic in scope and clearly written for the general public. Nevertheless, these types of materials, whether scholarly or popular in scope, are excellent sources in which to demonstrate the marvels of engineering creativity. Technical-oriented handbooks often have a chapter on the history of that particular field; they are also worth investigating.

Berlow, L. H. 1998. *The Reference Guide to Famous Engineering Landmarks of the World: Bridges, Tunnels, Dams, Roads, and Other Structures*. Phoenix, AZ: Oryx. Gives brief but informative entries for over 600 of the most well-known engineering achievements of the last 5,000 years. Entries are cited for additional information and many are illustrated. Appendix A demonstrates bridge and truss designs, Appendix B illustrates a portfolio of Ohio covered bridges, and Appendix C cites the tallest, longest, and highest for 1997. There is a section for biographies, a section for chronology, a bibliography, a geographical index, and a subject index.

Bunch, B. H., and A. Hellemans. 1993. *The Timetables of Technology: A Chronology of the Most Important People and Events in the History of Technology*. New York: Simon & Schuster. This work devoted to the history of technology is divided into seven time frames: the Stone Ages (2,400,000–4000 BCE), the Metal Ages (4000 BCE–1000 CE), the Age of Water and Wind (1000–1732), the Industrial Revolution (1733–1878), the Electric Age (1879–1946), the Electronic Age (1947–1972), and the Information Age (1973–1993). Included are a name index and a subject index.

Daumas, M., ed. 1969–1979. *A History of Technology and Invention: Progress Through the Ages*. (3 vols.) New York: Crown Publishing. This encyclopedic handbook is the English translation (by E. B. Hennessy) of *Histoire Générale des Techniques*. It is an extensive presentation of "the history of the methods that man has discovered and utilized to improve the conditions of his existence" (from the Preface). It covers prehistoric times to 1860. It includes detailed diagrams and pictures of hundreds of technological methods.

Newhouse, E. L., M. G. Dunn, and E. Lanouette. 1992. *The Builders: Marvels of Engineering*. Washington, D.C.: National Geographic Society, Book Division. This is a popular chronicle of the engineering wonders of the world. It includes famous roads, canals, bridges, railroads, pipelines, towers, tunnels, skyscrapers, sport arenas, exposition halls, domes, cathedrals, and wind, solar, and electric power stations. It has over 400 photographs, detailed diagrams, and drawings.

Oleson, J. P. 2008. *The Oxford Handbook of Engineering and Technology in the Classical World*, Oxford, U.K./New York: Oxford University Press. This handbook covers technological advances from 800 BCE to 500 CE in the Greek and Roman dominated geographical areas. It discusses most technical advances such as machinery, food processing, metallurgy, stoneworking, hydraulic systems, tunnels, channels, weights and measures, and land transport. It is a good basic source for the understanding of technology during the covered period.

Pfammatter, U. 2008. *Building the Future: Building Technology and Cultural History from the Industrial Revolution until Today*. Munich/New York: Prestel. This is a historical handbook dedicated to building engineering and its related fields. It covers the inventions introduced from the Industrial Revolution to the present time.

Singer, C. J., E. J. Holmyard, and A. R. Hall, eds. 1954–1978. *A History of Technology*. (7 vols.) London: Oxford University Press. Written in the form of an encyclopedic handbook, this work is an extensive treatise of the history of technology; it includes thousands of text figures and hundreds of plates. Each chapter of its seven volumes is written by an expert, and covers early society to modern times. It represents one of the contemporary major accounts of the history of technology. Its supplemental includes places and names lists, and a subject index.

Wikander, O., ed. 2000. *Handbook of Ancient Water Technology*. Boston: Brill. This volume presents water management and hydrotechnology from Mesopotamia, Iran, and the Indus Valley to the Atlantic Ocean from Prehistoric times to the sixth or seventh century CE. Topics covered include water supply, urban use, irrigation and rural drainage, larger hydraulic infringements on nature, water power, water as an aesthetic and recreational element, water legislation in the ancient world, and historical context.

MONOGRAPHS AND TEXTBOOKS

Monographic publications are a very important form in the literature of the history of engineering. Some books listed in this section are about a specific area of engineering, such as civil or chemical engineering. Other books deal with social aspects of engineering and its contributions to society. This list is intended to be representative of the topics being published as books in this field. Readers can find a more extensive list of monographs in the bibliographies listed above, searching the indexes and databases listed in this chapter, and by searching library online catalogs. A rich collection of monographic documents also are available at institutional repositories (IRs) of libraries and museums. Some of those institutions are listed in this chapter, but it is recommended to consult a specialized librarian when looking for antiquarian materials.

Addis, W. 2007. *Building: 3000 Years of Design Engineering and Construction*. London: Phaidon Press. A history of construction covering methods and building instruments that has a number of illustrations.

Armytage, W. H. G. 1976. *A Social History of Engineering*, 4th ed. Boulder, CO: Westview Press. The goals of this book are to timeline technological developments as they refer to Britain, and to shed light on how these developments impacted upon the social life of the era and on how certain needs of a community at a certain time gave impetus to achieving technological developments; further, to present the origins of innovations and institutions. It begins at the Stone Age and carries through into the twentieth century. Selected bibliographic references are given for each chapter.

Auyang, S. Y. 2004. *Engineering: An Endless Frontier*. Cambridge, MA: Harvard University Press. The author's purpose in this book is to explain how engineers create technology by presenting a broad picture of modern engineering in both its physical and human dimensions. Topics included are technology takes off, engineering and information, engineers in society, innovation by design, science of useful systems, and leaders who are engineers. Each chapter includes an extended bibliography. Also included are two appendixes: Statistical Profiles of Engineers and U.S. Research and Development.

Belanger, D. O. 1998. *Enabling American Innovation: Engineering and the National Science Foundation*. West Lafayette, IN: Purdue University Press. This work in the *History of Technology* series was assembled by a history of technology scholar. It contains a large amount of research: document files, a comprehensive bibliography, synopses of oral history interviews, summaries, analyses of various topics, charts and graphs, and formatted endnotes. Sponsored by the National Science Foundation, the goal of this book is to provide a historical analysis of the foundation's policies and its support of certain disciplines, showing both the success and failure of particular funded programs and projects.

Billington, D. P. 2006. *Power, Speed, and Form: Engineers and the Making of the Twentieth Century.* Princeton, NJ: Princeton University Press. Inventions covered include the telephone, oil refining, airplane, radio, automobile, electric power, and more within the perspective of their role in transforming our world.

Burstall, A. F. 1965. *A History of Mechanical Engineering.* Cambridge, MA: The M.I.T. Press. This book was intended for mechanical engineering students. The author presents the most important events in the science of mechanical engineering from prehistoric times until the twentieth century. This work was divided into nine chapters, each for a different period in history. Every chapter includes materials available to the mechanical engineer: tools, mechanisms and machines for the mechanical transmission of power and motion, fluid machinery, and heat engines, The final chapter gives a revealing review of the progress of these achievements.

Carter, D. V. 1961. *History of Petroleum Engineering.* Dallas, TX: Boyd Printing Company. The American Petroleum Institute assembled a board of scholars to provide this history of the birth and growth of petroleum engineering. Topics such as the percussion-drilling system; the hydraulic rotary-drilling system; cementing; logging, sampling, and testing; completion methods; production equipment; production techniques and control; reservoir engineering; fluid injection; handling oil and gas in the field; evaluation; research; conservation; unitization; and standardization of oil-field equipment are found. An index is included.

Cutcliffe, S. H. and T. S. Reynolds. 1997. *Technology and American History. A Historical Anthology from Technology and Culture.* Chicago: University of Chicago Press. This is a collection of essays published in *Technology and Culture.* The 15 articles presented are an attempt to cover the history of American technology from late 1700s to the end of the twentieth century.

Furter, W. F., ed. 1980. *History of Chemical Engineering: Based on a Symposium Co-sponsored by the ACS Divisions of History of Chemistry and Industrial and Engineering Chemistry at the ACS/CSJ Chemical Congress, Honolulu, Hawaii, April 2–6, 1979.* Washington, D.C.: American Chemical Society. "The central theme of the book is the historical identification and development of chemical engineering as a profession in its own right, distinct not only from all other forms of engineering, but particularly from all forms of chemistry including applied chemistry and industrial chemistry" (from the Preface). The history of chemical engineering in industry and in education is cited as well as the origins in Canada, Britain, Germany, Japan, Italy, India, and the United States.

Garrison, E. G. 1999. *A History of Engineering and Technology: Artful Methods,* 2nd ed. Boca Raton, FL: CRC Press. The purpose of this book is to help us discover the persons, concepts, and events that helped make engineering what it is today. This book covers the history of engineering from the time of early stone tools studied by archaeologists to the present, with additional insights into what may lie ahead. It includes bibliographical references and an index. It contains information from authentic and highly regarded sources.

Harms, A. A., B. Baetz, and R. Volti. 2004. *Engineering in Time: The Systematics of Engineering History and its Contemporary Context.* London: Imperial College Press. The author presents an innovative approach of the history of engineering by introducing concepts and methods in a chronological time-frame. It is a textbook for engineering undergraduate students.

Hill, D. 1984. *A History of Engineering in Classical and Medieval Times.* LaSalle, IL: Open Court Publishing Company. This work records the major engineering achievements of the peoples of Europe and Western Asia in the period from 600 BCE to 1450. It is concerned with origins of these achievements and their dispersion, paying attention to descriptions of techniques and machines. Research is brought to bear on irrigation and water supplies; dams, bridges, roads; building construction; surveying; water-raising machines; power from water and wind; instruments; automata; and clocks. A bibliography and index are placed at the end of the volume.

Hughes A. C. and T. P. Hughes. eds. 2000. *Systems, Experts, and Computers: The Systems Approach in Management and Engineering, World War II and After.* Cambridge, MA: MIT Press. A systems approach to solving complex problems and managing complex systems came into being after World War II. Thus, the "systems engineer" was born and so too was the field of "systems engineering." This book studies the origins of the systems approach, the organizations and individuals involved, the different systems and projects, the experts and their expertise, computers, the applications of all these innovations in the United States and the world over. Bibliographies are included at the end of each chapter. Notes on the contributors are given at the end of the book.

Kirby, R. S., S. Withington, A. B. Darling, and F. G. Kilgour. 1956. *Engineering in History.* New York: McGraw-Hill. Engineers and historians working together have written this book on the history of engineering. It is intended to present the development of engineering in Western civilization from its origins into the twentieth century; it is a general introduction to this pursuit rather than a definitive history. The interaction between the activities of the engineers and the activities of the people of the communities in which they lived are also discussed because progress in this science cannot come in any other manner. A bibliography is included at the end of the book, and selected bibliographies are provided at the end of most chapters.

McClellan, J. E. and H. Dorn. 2006. *Science and Technology in World History: An Introduction,* 2nd ed. Baltimore, MD: John Hopkins University Press. It is considered to be an excellent textbook for undergraduate students. The authors traced the development of technology from the ancient civilizations to Europe and the United States.

Nebeker, F. 2009. *Dawn of the Electronic Age: Electrical Technologies in the Shaping of the Modern World, 1914–1945.* Piscataway, NJ: IEEE. Details the rapid advancement in electronics in the early twentieth century.

Neuburger, A. 2003. *The Technical Arts and Sciences of the Ancients.* New York: Columbia University Press. This is a comprehensive and well-regarded book on ancient technical science concerning daily life up until the end of the fifth century. Topics covered include mining, metal extractions, and metal working; woodworking; the treatment and preparation of leather; agriculture; fermentation; the production of oils and perfumes; preservation and mummification; ceramic arts, glass, yarns and textiles; dyes, painting, mechanics, lighting, and heating; town planning; fortification; houses, building methods and materials; water supply and drains; and roads, bridges, and shipbuilding.

Rae, J. and R. Volti. 2001. *The Engineer in History,* rev. ed. New York: Peter Lang. This is a collection of details that exemplify engineers in their lives, such as their education and working methods, their influence and relationship with their employers, their perceived value in society, and their power in the socio-political arena. The historical coverage of this book is from antiquity to the end of the Industrial Revolution.

Straub, H. 1964. A *History of Civil Engineering: An Outline from Ancient to Modern Times.* Cambridge, MA: The M.I.T Press, (translated by E. Rockwell). The author's keen interest in the history of his own profession prompted him to share just how modern construction techniques and present-day engineering have gradually developed from two distinct sources: (1) the science of mechanics and (2) the creative craft of building. The book is divided into time frames starting with the ancient world and going forward into the twentieth century.

JOURNALS AND MONOGRAPHIC SERIES

Since the publication of the first scientific journal *Philosophical Transactions of the Royal Society of London* in 1665, journals have had a very important role in the archiving and dissemination of technical and scientific information. From the historical point of view, the archives of technical and engineering journals are themselves valuable sources for the history of engineering. Searching a field of engineering in a given period of time (e.g., metallurgy in the late 1800s) will show the state of that field during that time. In this section, we are presenting journals related to the history of engineering; most of these journals started to publish in the 1900s. Before this time, articles about the history of engineering were published in subject-related engineering or technology journals. This is a representative list. Reference sources with an extensive list of history journals include *The History Journals Guide Periodicals Directory,* a Web site maintained by Stefan Blaschke; *the Bibliography of the History of Technology* by Eugene S. Ferguson; the list of journal titles indexed in *Isis Current Bibliography* http://www.ou.edu/cas/hsci/isis/website/thesaurus/journals.html (accessed November 10, 2010); and in the 2002 *Guide to the History of Science*, 9th ed., edited by Roger Turner. All of these sources are annotated in this chapter.

American Heritage of Invention & Technology. 1985–. New York: American Heritage (8756-7296).

Berichtezur Wissenschaftsgeschichte: Organ der Gesellschaft fuer Wissenschaftsgeschichte. 1978–. Weinheim, Germany: Wiley–VCH Verlag GmbH & Co. KgaA. (0170-6233).

Blaetterfuer Technikgeschichte. Vienna, Austria: Technisches Museum Wien mit Österreichischer Mediathek. 1932–2003. (0170-6233). Originally: Blaetter für Geschichte der Technik (1932–1938).

British Journal for the History of Science. 1962–. Cambridge, U.K.: Cambridge University Press (0007-0874).

Historia Scientiarum: International Journal of the History of Science Society of Japan. 1962–. Tokyo: Nihon Kagakushi Gakkai (History of Science Society of Japan) (0285-4821). Originally called: *Japanese Studies in the History of Science* (1962–1980).

History and Technology: An International Journal. 1983–. London: Routledge (0734-1512).

History of Science: Review of Literature and Research in the History of Science, Medicine and Technology in its Intellectual and Social Context. 1962–. Cambridge, U.K.: Science History Publications Ltd. (0073-2753).

History of Technology. 1976–. London: Mansell Publishing Ltd. (0307-5451).

IA: The Journal of the Society for Industrial Archeology. 1976–. Houghton, MI: Society for Industrial Archeology (0307-5451).

Icon: Journal of the International Committee for the History of Technology. 1995–. Paris: International Committee for the History of Technology (1361-8113).

IEEE Annals of the History of Computing. 1979–. Los Alamitos, CA: IEEE Computer Society (1058-6180).

Indian Journal of History of Science. 1966–. New Delhi: Indian National Science Academy (0019-5235).

International Journal for the History of Engineering & Technology. 2009–. Leeds, U.K.: Maney Pub. Formerly: Transactions of the Newcomen Society (1758-1206).

Isis: International Review Devoted to the History of Science and its Cultural Influences. 1912.–. Chicago: University of Chicago Press, Journals Division (0021-1753).

Jahrbuch fuer Wirtschaftsgeschichte. 1960–. Berlin: Akademie Verlag GmbH (0075-2800).

Journal of the Association for History and Computing. 1998–. Plainview, TX: American Association for History and Computing (1937-5905).

Journal of the Society for Army Historical Research, 1921–. London: Society for Army Historical Research (0037-9700).

Journal of Transport History. 1953–. Manchester, U.K.: Manchester University Press (0022-5266).

Kagakushi Kenkyu (Journal of History of Science, Japan). 1941–. Tokyo: Nihon Kagakushi Gakkai (History of Science Society of Japan) (0022-5266).

Kultur und Technik. 1977–. Munich: Verlag C. H. Beck oHG (0344-5690).

Kwartalnik Historii Nauki i Techniki. 1956–. Warsaw: Polska Akademia Nauk, Instytut Historii Nauki (0023-589X).

Notes and Records of the Royal Society of London. 1938–. London: Royal Society of London (0035-9149).

Osiris: A Research Journal Devoted to the History of Science and its Cultural Influences. 1936–. Chicago University of Chicago Press, Journals Division (0369-7827).

Proceedings of ICE: Engineering History and Heritage. 2009–. London: Institution of Civil Engineers (1757-9449).

Revista da Sociedade Brasileira de Historia da Ciencia. 1985–. Campinas, SP, Brasil: Universidade Estadual de Campinas, Grupo de História e Teoria da Ciência (0103-7188).

Revue D'histoire de la Culture Matérielle. 1991–. Sydney, Nova Scotia, Canada: Cape Breton University Press (1183-1073).

Revue D'histoire des Sciences. 1972–. Paris: Presses Universitaires de France (0151-4105).

Studies in History and Philosophy of Science, Part A. 1970–. Oxford, U.K.: Elsevier Science (0039-3681).

Technikgeschichte. 1909–. Berlin: Kiepert GmbH und Co. KG (0040-117X).

Technology and Culture. 1960–. Baltimore, MD: The Johns Hopkins University Press, Journals Publishing Division (0040-165X).

Transactions of the Newcomen Society. 1921–2008. London: Newcomen Society for the Study of the History of Engineering and Technology (0372-0187).

Voprosy Istorii Estestvoznaniya i Tekhniki. 1956–. Moscow: Izdatel'stvo Nauka (0205-9606).

Zhongguo Ke ji Shi Za Zhi (*The Chinese Journal for the History of Science and Technology*). 2005–. Beijing: Zhongguo Ke Xue Ji Shu Chu Ban She. Originally: *Zhongguo Keji Shiliao* (*China Historical Materials of Science and Technology*). 1980–2004. (1000-0798).

CONFERENCES AND SYMPOSIA

Conferences, symposia, and seminars are very common places for scientists, engineers, and historians interested in the history of engineering to present the results of their research. Some of these gatherings are sponsored by professional societies on the history of science, technology, or engineering while others are sponsored by subject-specific societies like the American Chemical Society, and many other programs are sponsored by academic departments, research centers, and museums. The sources listed in this chapter under Indexes and Abstracts are recommended in order to search for the proceedings of these meetings. Also, the institutions' sites listed in this chapter under Societies and Research Institutions will help the reader to identify oncoming programs.

MUSEUMS AND SPECIAL COLLECTIONS

Museums and specials collections represent sources of importance for researchers. Special collections held at universities, public libraries, and at other institutions provide works published sometimes several centuries back or their reproductions and, in many instances, also collections of graphic works and objects. Most museums have the purpose of preserving large collections of objects, graphics, and other materials; they also provide educational programs for the public. Museums and special collection departments also are involved in research programs. With the development of Web 2.0 technology and better multimedia applications, libraries and museums are uploading extensive collections of virtual objects and documents in archives called *repositories,* making accessible online a wealth of information to researchers. In this section special collections are listed first followed by a list of museums.

LIBRARIES

Case Western Reserve University. Special Collections: History of Science and Technology, http://library.case.edu/ksl/collections/special/historyscience/index.html (accessed September 21, 2010). "In the History of Science collection, one finds early editions of major works, such as *De Fabrica* by Versalius and *Opticks* by Newton. Materials on the History of Technology can be found in the technical papers of the Cleveland firms of Warner and Swasey and Charles F. Brush. In addition, the collections contain important early German, French, English, and American journals. The Natural History collection includes 220 plates of Audubon's Birds of America; Catesby's The Natural History of Carolina, Florida, and the Bahama Islands; and Travels of Lewis and Clark" (from Web page).

Cornell University Library. History of Science Collections, Division of Rare Books and Manuscript Collections http://rmc.library.cornell.edu/collections/scienceandtech.html (accessed September 21, 2010). "In the history of technology, the collection's strengths include the Hollister Collection (300 vols.), which focuses on civil engineering, and the Cooper Collection on American railroad bridges, which includes the original blueprints for many structures that no longer survive. Notable among the technology holdings is Gustave Eiffel's *La tour de trois cents mètres*, a scarce work that includes extraordinarily detailed plans of the Eiffel Tower, photographs of the construction process, and facsimile signatures of the dignitaries who were first to ascend what was then the world's tallest manmade structure" (from Web page).

The Dibner Library of the History of Science http://www.sil.si.edu/libraries/Dibner/ (accessed September 21, 2010). The Dibner Library is the Smithsonian's collection of rare books and manuscripts relating to the history of science and technology. "Contained in this world-class collection of 25,000 rare books and 2,000 manuscript groups are many of the most important works dating from the fifteenth to the nineteenth centuries in the history of science and technology including engineering, transportation, chemistry, mathematics, physics, electricity, and astronomy" (from Web page).

Hagley Museum and Library, The. Research: Manuscripts and Archives Department, http://www.hagley.lib.de.us/library/about/deptdescmanu.html (accessed September 21, 2010). "Hagley's manuscript and archival collections contain the records of more than 1,000 firms as well as the personal papers of the entrepreneurs, inventors, designers, and managers who helped build these businesses. The companies represented range from the mercantile houses of the late eighteenth century to the multinational corporations of the twenty-first" (from Web page).

Huntington Library, Art Collections, and Botanical Gardens, The http://www.huntington.org (accessed September 21, 2010). The library's rare books and manuscripts have collections in electricity from the works of Franklin, Taisnier, Gilbert, Musschenbroek, Beccaria, Galvani, Volta, Ampere, Oersted, Ohm, and Faraday. In civil engineering, it holds collections on early printed books; early English and American surveying books; eighteenth to nineteenth-century American and English engineering materials; Maritime technology and engineering; nineteenth to twentieth-century railroad books and records; Historical photograph collection; California and the American West: mining, geology, architecture, and civil engineering papers. In transportation, it has holdings of printed and manuscript materials, especially British and American railroads, sea transportation, and pre-Wright brothers aeronautics.

Iowa State University. Special Collections: Archives and Manuscripts, http://www.lib.iastate.edu/spcl/collections/index.html (accessed September 21, 2010). The primary areas of selection and preservation for engineering include agricultural engineering, civil engineering and transportation, and energy and electrical power management (i.e., energy management, consumption, efficiency, production, conversion and transmission, primarily of a nonnuclear nature). Also, the Archives of Women in Science and Engineering is maintained.

Library of Congress. Science, Technology and Business Division http://www.loc.gov/rr/ scitech/ (accessed January 30, 2005). The highlights of the special collections holdings of this division include Thomas Jefferson's personal library, 40,000 volumes from the original collection of the Smithsonian Institute; a large collection of manuscripts of major American scientists, inventors, engineers, explorers, and business pioneers, and several special collections in the history of aeronautics. In addition, there is a significant rare book collection in the history of computers and data processing, and several special collections of technical reports, standards, and gray literature in the sciences, technology, and engineering.

Linda Hall Library of Science, Engineering & Technology http://www.lindahall.org/ (accessed January 30, 2005). "The library's major emphasis in engineering is on collecting important current research material. In addition to civil engineering, applied mechanics and related disciplines are the most extensively covered subjects in the library's collection. The collection includes meeting papers from many engineering societies, an extensive collection of individual papers collected by AIAA as well as NASA papers, a variety of government documents and technical reports; the library owns an impressive collection of engineering standards and specifications issued by industry and government. The collection contains not only current standards and specifications, but a large collection of historical documents as well" (from Web page).

Rutgers, The State University of New Jersey. Thomas A. Edison Papers http://edison.rutgers. edu/ (accessed January 30, 2005). "The Thomas A. Edison Papers is a documentary editing project sponsored by Rutgers, The State University of New Jersey, the National Park Service, the Smithsonian Institution, and the New Jersey Historical Commission. Here one can view 180,000 document images, search a database of 121,000 document records and 19,250 names or keyword search 4,000 volume-and-folder descriptions" (from Web page).

Stanford University Libraries. History of Science and Technology Collections http://www-sul.stanford.edu/depts/hasrg/histsci/scihome.html (accessed September 21, 2010). "The History of Science and Technology Collections in the Stanford University Libraries support research and instruction in history of science and technology and related fields. The History of Science and Technology Collection covers the History of Computing; Rare Books in the History of Science; Civil and Electrical Engineering, Classical and High-Energy Physics, Computer Science, Industrial Revolution and Heavy Industry, Military Science, Photography, and Telecommunication" (from Web page).

Wright State University Libraries, Special Collections and Archives http://www.libraries. wright.edu/special (accessed September 21, 2010). "Dunbar Library's emphasis on aviation history and Fordham Library's emphasis on aerospace medicine and human factors engineering combine to make Wright State a nationally known repository for the documentation of some of the twentieth century's most dramatic technologies. The other major focus for both libraries is the local and regional history of the Miami Valley area of Ohio. Together the libraries offer a comprehensive historical perspective on the region, aviation, and aerospace medicine" (from Web page).

Museums

Caltech Institute Archives, California Institute of Technology, The http://archives.caltech. edu/ (accessed September 21, 2010). "The institute archives serve as the collective memory of Caltech by preserving the papers, documents, artifacts and pictorial materials that tell the school's history, from 1891 to the present. Researchers will also find here a wealth of sources for the history of science and technology worldwide, stretching from the time of Copernicus to today" (from Web page).

Canada Science and Technology Museum http://www.sciencetech.technomuses.ca/english/index.cfm (accessed September 21, 2010). "In the library of the Canada Science and Technology Museum, you will find a wealth of material documenting the history and development of science and technology, with special reference to Canada, and the museum's collection. The library collection focuses on the museum's special subject areas: agricultural technology, communications, energy and mining, forestry, graphic arts, land and marine transportation, industrial technology, physical sciences, space and museology" (from Web page).

Deutsches Museum http://www.deutsches-museum.de/ (accessed September 21, 2010). "The archives of the Deutsches Museum are the leading collections for the history of science and technology in Europe. They hold currently 4.3 kilometers of archival material, sources, and documents. The archival collections focus on transportation, aeronautics, and astronautics, the history of physics and chemistry. The library offers literature that primarily relates to the history of the natural sciences and technology, ranging from nonfiction titles to scientific manuals or specialist essays and papers from the invention of the letterpress to the present day" (from Web page).

Franklin Institute Science Museum, The http://www2.fi.edu (accessed September 21, 2010). The "institute features the largest collection of Franklin materials—original works of art, documents, and artifacts—ever assembled, as well as interactive, multimedia installations. The Franklin Institute Award Case Files are a unique repository in the history of science. They are filled with stories of scientific enterprise and social circumstances. In addition, the Franklin Institute Science Museum has the largest collection of artifacts from the Wright brothers' workshop" (from Web page).

Institute and Museum of History of Science: Florence http://www.museogalileo.it/en/index.html (accessed September 21, 2010). Istituto e Museo di Storia della Scienza (IMSS) http://galileo.imss.firenze.it/indice.html. The scientific instrument collection of the Istituto e Museo di Storia della Scienza includes several thousand instruments dating from the sixteenth to the early twentieth century. The archival resources include the Archive of the Royal Museum of Physics and Natural History (ARMU), and the Congresses of Italian Scientists Archive (IRIS). This institute is also listed in this chapter under Societies and Research Institutes.

Museo Nacional de Ciencia y Tecnología http://www.educacion.es/mnct/ (accessed September 21, 2010). This museum has a collection of more than 12,000 objects representing the scientific and technological heritage of Spain.

Museum of the History of Science, Oxford, U.K. http://www.mhs.ox.ac.uk/ (accessed September 21, 2010). "Covers almost all aspects of the history of science, from antiquity to the early twentieth century. Particular strengths include the collections of astrolabes, sundials, quadrants, early mathematical instruments generally (including those used for surveying, drawing, calculating, astronomy, and navigation), and optical instruments (including microscopes, telescopes, and cameras), together with apparatus associated with chemistry, natural philosophy, and medicine. In addition, the museum possesses a unique reference library for the study of the history of scientific instruments that includes manuscripts, incunabula, prints, printed ephemera, and early photographic material" (from Web page).

National Museum of American History, Behring Center, Smithsonian Institution http://americanhistory.si.edu/ (accessed September 21, 2010). "The museum is interested in how objects are made, how they are used, how they express human needs and values, and how they influence society and the lives of individuals. NMAH's natural focus is on the history of the United States of America, including its roots and connections with other cultures. As sources for research, the museum not only offers the historical object, but also significant collections of oral histories, prints, photographs, business Americana and trade literature, and engineering drawings" (from Web page).

National Technical Museum in Prague. History of Science and Technology http://www.ntm.cz (accessed September 21, 2010). A large collection of objects from the National Technical Museum covers the broad spectrum of engineering. These include acoustics, the food industry, metallurgy, and mining as well as the major disciplines of chemical, civil, electrical, industrial, mechanical, and transportation engineering.

Smithsonian Institution, Archives and Special Collections of the Smithsonian Institution http://www.si.edu/ (accessed September 21, 2010). "National Archives, and Archives II located just west of the University of Maryland College Park campus, contains records of many federal agencies involved in science and technology, including Atomic Energy Commission, Coast and Geodetic Survey, Geological Survey, Manhattan Project, National Aeronautics and Space Administration, National Bureau of Standards, National Science Foundation, Office of Naval Research, Office of Scientific Research and Development [World War I], and Weather Bureau" (from Web page).

Smithsonian Institution, The Jerome and Dorothy Lemelson Center for the Study of Invention and Innovation http://invention.smithsonian.org/home/ (accessed September 21, 2010). "The Lemelson Center's oral and video history projects add historical documentation on invention and innovation in the United States. The National Museum of American History Archives Center has vast collections relating to technology, invention, and innovation in the nineteenth and twentieth centuries" (from Web page).

Société du Musée des Sciences et de la Technologie du Canada http://technomuses.ca/index_f. asp (accessed September 21, 2010). "The collection of the Canada Science and Technology Museum encompasses a broad cross section of Canadian scientific and technological heritage. National in scope, this unique collection consists of artifacts, photographs, technical drawings, trade literature, and rare books, all of which are complemented and supported by library holdings of monographs and serials" (from Web page).

INTERNET RESOURCES

In this section, a selected list of Web sites that contain information about the history of engineering are presented. Other resources not listed here, but that can be useful to the readers include specialized search engines; repositories; discussion lists, newsletters, and Weblogs. Although many popular Web sites and blogs can be found by using a search engine, such as Google or Yahoo, most Internet resources of scholarly credibility are supported by professional societies, private corporations, government agencies, and universities; therefore, it is advisable to explore Web resources offered by these types of institutions. The resources mentioned in the section on Societies and Research Institutions, and on Museums and Special Collections, may be helpful in identifying other Web resources.

ASCE History and Heritage of Civil Engineering. Resource Guide http://www.asce.org/ProgramProductLine.aspx?id=15125 (accessed October 20, 2010). This guide hosted by the American Society of Civil Engineers lists resources pertinent to the history of civil engineering.

Carnegie Mellon University Libraries, History: History of Technology and Science http://www.andrew.cmu.edu/user/sc24/History/hotsprimary.html#audio (accessed November 10, 2010). Users find in this Web page lists of primary sources that can be found at Hunt Library, Carnegie Mellon; Hillman Library, the University of Pittsburgh; and the Carnegie Library of Pittsburgh. There is also a link to the Archives Resources.

ECHO: Exploring and Collecting History Online —Science, Technology, and Industry http://echo.gmu.edu/index.php (accessed October 20, 2010). "Echo's research center catalogues, annotates, and reviews sites on the history of science, technology, and industry. You can browse our database of over 5,000 sites by topic, time period, publisher, or content" (from Web page).

Greatest Engineering Achievements of the 20th Century http://www.greatachievements. org/ (accessed October 20, 2010). "This site is sponsored by the National Academy of Engineering. This project is a collaboration of the American Association of Engineering Societies, National Engineers Week, with 27 other professional engineering societies. The overarching criterion used was that those advancements had made the greatest contribution to the quality of life in the past 100 years. Even though some of the achievements, such as the telephone and the automobile, were invented in the 1800s, they were included because their impact on society was felt on the twentieth century" (from Web page).

History of Engineering at Yale University http://www.seas.yale.edu/about-history.php (accessed October 20, 2010). "This Web site exists to collect historical resources relating to the story of Engineering at Yale. Resources available: History of Sheffield Scientific School; The Palmer Models—Exquisite models of nineteenth-century railroad engines and a naval cruiser were crafted by Charles L. Palmer and bequeathed to the Sheffield School in 1908. They are available for private viewing on request; Bibliography—Partial list of significant books and articles about our history" (from Web page).

History of Science Links http://www.clas.ufl.edu/users/ufhatch/pages/10-HisSci/links/ (accessed October 20, 2010). This site is maintained by Robert A. Hatch, University of Florida; it has numerous interesting links. It includes a list on List Serves and Chat Pages.

History of Science Society Guide to U.S. Graduate Study in History of Science http://www. depts.washington.edu/hssexec/hss_gradproglist.html (accessed October 20, 2010). This Web site contains links to every graduate program in the history of science and technology offered in the United States.

HOST http://www.kcl.ac.uk/depsta/iss/library/speccoll/host/about.html (accessed October 20, 2010). This Web site offers information about the HOST program of the United Kingdom; a collaborative retrospective conversion and conservation program for materials on the history of science and technology, 1801–1914.

I.A. Recordings http://www.iarecordings.org/ (accessed October 20, 2010). "Industrial Archaeology Recordings is dedicated to recording past and present industry on film and video. Industrial Archaeology usually deals with disused buildings and machines, but it is also vital to record the processes of an industry while it is still working. This site provides over 560 links to other Industrial Archaeology Pages" (from Web page).

Internet Resources for History of Science and Technology http://www2.lib.udel.edu/subj/hsci/ internet.htm (accessed October 20, 2010). This Web page from the University of Delaware library lists: bibliographies and research guides, databases and information sources, museums, libraries and archives, exhibits, societies, associations, and other organizations, university departments, programs, and centers, and regional resources.

La Storia e la Filosofia della Scienza, della Tecnologia e della Medicina. The History and Philosophy of Science, Technology and Medicine http://www.imss.fi.it/~tsettle/ (accessed October 20, 2004). This is a selection of Web sites and other sources maintained by Thomas B. Settle, of the Istituto e Museo di Storia della Scienza (IMSS) and the Polytechnic University, Florence, Italy.

SOCIETIES AND RESEARCH INSTITUTIONS

In 1920, the first learned society devoted to the history of engineering and technology, the Newcomen Society, was founded in London. In the United States, the Society for the History of Technology was created in 1956. Both of these societies have had a very positive impact on the development of the field of the history of engineering. Many other societies that traditionally have been studying the history of sciences also have become valuable resources for the scholar of the history of engineering. In this section, a number of research institutions that have provided significant support in this field are additionally included. This is not a comprehensive list and other sources found in this chapter can help to identify others.

SOCIETIES, ASSOCIATIONS, AND ORGANIZATIONS

American Association for History and Computing http://www.theaahc.org/ (accessed October 20, 2010). "The American Association for History and Computing (AAHC) is dedicated to the reasonable and productive marriage of history and computer technology. To support and promote these goals, the AAHC sponsors a number of activities, including an annual meeting, annual prizes, an electronic journal (the *Journal of the American Association for History and Computing (JAHC)*), a continuing publication series, and a variety of summer workshops" (from Web page).

American Philosophical Society http://www.amphilsoc.org/ (accessed October 20, 2010). The American Philosophical Society "promotes useful knowledge in the sciences and humanities through excellence in scholarly research, professional meetings, publications, library resources, and community outreach. The American Philosophical Society, this country's first learned society, has played an important role in American cultural and intellectual life for over 250 years" (from Web page).

British Society for the History of Science http://www.bshs.org.uk/ (accessed October 20, 2010). "BSHS is the largest U.K. body dealing with all aspects of the history of science, technology, and medicine." The society publishes the *British Journal for the History of Science* (BJHS) (from Web page).

History of Science Society (HSS) http://www.hssonline.org/ (accessed October 20, 2010). "The History of Science Society is the world's largest society dedicated to understanding science, technology, medicine, and their interactions with society in historical context. Over 3,000 individual and institutional members across the world support the society's mission to foster interest in the history of science and its social and cultural relations" (from Web page).

International Committee for the History of Technology http://www.icohtec.org/ (accessed October 20, 2010). "The International Committee for the History of Technology (ICOHTEC) was founded as the Cold War was being waged with particular bitterness between the nations of the Eastern and Western worlds. The intent was to provide a forum where scholars of both sides might meet and communicate about matters of mutual interest in the history of technology. It provides a new international journal that is essential reading for scholars and researchers in the field of the history of technology and they also sponsor world congresses" (from Web page).

Istituto e Museo di Storia della Scienza (IMSS) http://www.museogalileo.it/en/index.html (accessed October 20, 2010). "The Istituto e Museo di Storia della Scienza is one of the foremost international institutions in the History of Science, combining a noted museum of scientific instruments and an institute dedicated to the research, documentation, and dissemination of the history of science in the broadest senses" (from Web page).

Newcomen Society for the Study of the History of Engineering and Technology, The http://www. newcomen.com/ (accessed October 20, 2010). "The Newcomen Society is the world's oldest learned society devoted to the study of the history of engineering and technology. The society is based in London and is concerned with all branches of engineering: civil, mechanical, electrical, structural, aeronautical, marine, chemical, and manufacturing" (from Web page).

Scientific Instrument Society http://www.sis.org.uk/ (accessed October 20, 2010). "The Scientific Instrument Society (SIS) was formed in April 1983 to bring together people with a specialist interest in scientific instruments, ranging from precious antiques to electronic devices only recently out of production. Collectors, the antiques trade, museum staff, professional historians, and other enthusiasts will find the varied activities of SIS suited to their tastes. The society has an international membership" (from Web page).

Society for the History of Technology (SHOT) http://www.shot.jhu.edu (accessed October 20, 2010). "The Society for the History of Technology was formed in 1958 to encourage the study of the development of technology and its relations with society and culture. An inter-disciplinary organization, SHOT is concerned not only with the history of technological devices and processes, but also with the relations of technology to science, politics, social change, the arts and humanities, and economics" (from Web page).

RESEARCH INSTITUTIONS

American Chemical Society, Division of History of Chemistry http://www.scs.uiuc. edu/~mainzv/HIST/ (accessed October 20, 2010). "The objects of the division shall be to stimulate study and research in the history of chemistry, to offer an opportunity for presen-tation of the results of such specialized study and research, and to encourage a wide general interest among all chemists in the historical phases of their science" (from Web page).

American Institute of Physics. Center for History of Physics http://www.aip.org/history/ (accessed October 20, 2010). "A division of the American Institute of Physics, The Center for History of Physics is the oldest and best-known institution dedicated to the history of a scien-tific discipline. ... Our mission is to preserve and make known the history of modern physics and allied fields including astronomy, geophysics, optics, and the like" (from Web page).

ASCE History and Heritage of Civil Engineering (HHCE) http://www.asce.org/ ProgramProductLine.aspx?id=12884903157 (accessed October 20, 2010). "HHCE's mis-sion is to enhance the knowledge and appreciation of our history and heritage. ... Nearly 200 National and International Historic Civil Engineering Landmarks are featured within this Web site, along with biographies of 43 notable civil engineers" (from Web page).

Australian Science and Technology Heritage Centre of the University of Melbourne http:// www.austehc.unimelb.edu.au/ (accessed October 20, 2010). "The purpose of the centre is the development of projects relating to the history and heritage of science, technology, and medicine in Australia and to provide access to information resources in this area, such as the former Australian Science Archives Project" (from Web page).

Centre for the History of Science, Technology and Medicine, University of Manchester http:// www.chstm.manchester.ac.uk/ (accessed October 20, 2010). "It is located in Manchester, U.K. The Centre for the History of Science, Technology and Medicine (CHSTM) was founded in 1986 to bring together the university's interest in the history of science and medicine. It also includes the National Archive for the History of Computing, a major resource for research in the history and culture of informatics" (from Web page).

Dibner Institute for History of Science and Technology http://dibinst.mit.edu (accessed October 20, 2010). "The Dibner Institute is an international center for advanced research in the history of science and technology and located on the campus of MIT. Each year the institute hosts senior, postdoctoral, and graduate student fellows, as well as symposia, conferences, lectures, and workshops" (from Web page).

IEEE History Center http://www.ieee.org/organizations/history_center/ (accessed October 20, 2010). "IEEE is the Institute of Electrical and Electronics Engineers, Inc. The IEEE established the IEEE History Center in 1980 and, in 1990, the center moved to the cam-pus of Rutgers University, which became a cosponsor. The center maintains many useful resources for the engineer, for the historian of technology, and for anyone interested in the development of electrical and computer engineering and their role in modern society" (from Web page).

Imperial College London—Centre for the History of Science and Technology http://www. imperial.ac.uk/historyofscience/ (accessed October 20, 2010). It is recognized as one of the top research institutes in the United Kingdom. The center's areas of research include: science and technology in ancient Greece and Rome, early modern mathematics and mechanics, historiography of technology, science policy, war, and technology and the British State.

Landmarks Roster, Volunteer Center, History Resources. ASME History, History Links http://www.asme.org/history/index.html (accessed October 20, 2010). "Through its History and Heritage program, ASME encourages public understanding of mechanical engineering, fosters the preservation of this heritage, and helps engineers become more involved in all aspects of history" (from Web page).

The Sidney M. Edelstein Center for the History and Philosophy of Science, Technology, and Medicine (Research Center). Hebrew University of Jerusalem http://sites.huji.ac.il/edelstein/ (accessed October 20, 2010). "The Sidney M. Edelstein Center at The Hebrew University of Jerusalem was established in 1980 to encourage advanced research in the history and philosophy of science, technology, and medicine. In particular, the center fosters research based on the resources of the Einstein Archive, the Quantum Archive, the Edelstein collections on the history of dyeing and chemical technology, and the Yahuda Theological Collection of Isaac Newton" (from Web page).

CONCLUSION

The history of engineering has become an important field of study, originally included with the history of science; the history of engineering has evolved into a distinct subject domain. The history of engineering is part of the history of technology; most publications and professional organizations either use both terms or only use the term *technology*. The number of publications produced is substantial and there are a number of well-designed bibliographic databases that cover the field. Because engineering has an important place in the development of human society, it is expected that, as the different areas of engineering continue making contributions to mankind, the history of engineering will continue to be a healthy and an intriguing field of study.

REFERENCES

Auyang, S. Y. 2004. *Engineering: An endless frontier.* Cambridge, MA: Harvard University Press.

Bell, T. E, and D. Dooling. 2000. *Engineering tomorrow: Today's technology experts envision the next century.* Piscataway, NJ: IEEE Press.

Daumas, M. ed. 1969–1979. *A history of technology and invention: Progress through the ages.* (3 vols.) (translated by E. B. Hennessy). New York: Crown Publishing.

Engineering and materials. Historical review of engineering and technology. 2010. In *Access science. McGraw-Hill encyclopedia of science & technology online,* 10th ed. New York: McGraw-Hill (accessed October 8, 2010).

Garrison, E. G. 1999. *A history of engineering and technology: Artful methods,* 2nd ed. Boca Raton, FL: CRC Press.

Kirby, R. S., S. Withington, A. B. Darling, and F. G. Kilgour. 1956. *Engineering in History.* New York: McGraw-Hill.

McGraw-Hill encyclopedia of science & technology, 10th, ed. 2010. New York: McGraw-Hill.

Sarton, G. 1952. *A guide to the history of science: A first guide for the study of the history of science, with introductory essays on science and tradition.* New York: Ronald Press.

The Oxford English dictionary, 2nd ed. 1989. Oxford, U.K.: Oxford University Press.

14 Industrial Engineering

Nestor L. Osorio and Andrew W. Otieno

CONTENTS

Introduction 341
Searching the Library Catalog 344
    Keywords 345
    LC Subject Headings 345
    LC Call Numbers 346
Abstracts and Indexes 346
Bibliographies and Guides to the Literature 347
Directories 348
Dictionaries and Encyclopedias 350
Handbooks and Manuals 351
Monographs and Textbooks 355
Journals 358
Web Resources 360
Societies, Associations, and Organizations 363
Conferences 366
Conclusions 368
References 368

INTRODUCTION

Industrial Engineering (IE) is concerned with the efficiency of the functioning of operating systems, such as manufacturing, supply, service, and transport. The American Institution of Industrial Engineers defines IE as an area of engineering "… concerned with the design, improvement, and installation of integrated systems of men, materials equipment, and energy. It draws upon specialized knowledge and skill in the mathematical, physical, and social sciences together with principles and methods of engineering analysis and design to specify, predict, and evaluate the results to be obtained from such systems" (Martin-Vega, 2011, p.1.11).

In the early 1960s, IE was focused primarily on work simplification and improvement, using quantitative, science-based tools and techniques. Over the years, IE has evolved to include both design and integration of operating systems. The history of IE dates back to the Industrial Revolution of the mid-1700s when the steam engine and the initial machine tool were developed. The concepts of specialization of labor by Adam Smith, and Eli Whitney's development of interchangeability of parts (Martin-Vega, p. 1.4–1.5) led to the progression of IE as it is today. Notable pioneers also include Frederic Taylor, whose experiments culminated in the standardization of work, and Frank and Lillian Gilbreth, whose studies of human motions formed the basis of work study, human factors, and ergonomics. In 1912, at the annual meeting of the American Society of Mechanical Engineers (ASME), educators, consultants, engineers, and other pioneers, including Taylor and the Gilbreths, formed the Management Division of the ASME, from which a unique new profession of Industrial Engineering was created.

During World War I and II, the need for more rigorous techniques in the planning of complex systems in military operations led to the development of a set of techniques now called operations research (OR). By 1943, the functions of IE had been expanded to include manufacturing engineering, cost control, budgeting, and wages and salary administration. The ASME Management Division continued to expand and, in 1948, the American Institution of Industrial Engineers was formed.

The IE discipline as it has evolved today is sometimes referred to as industrial and systems engineering (Sink et al., 2001, p. 4–10). The discipline is based on a core engineering curriculum, other core courses being mathematics, statistics, engineering economics, and psychology. It comprises four key areas: (1) operations management (OM), (2) human factors engineering, (3) management systems, and (4) manufacturing engineering.

OM is primarily concerned with techniques that focus on the complexity of managing large organizations requiring the effective use of money, materials, equipment, and people. It involves the application of analytical methods from mathematics, science, and engineering that provide an organization with alternatives for the best allocation of resources using OR tools. OM functions include planning, forecasting, resource allocation, performance measurement, scheduling, the design of production facilities and systems, supply chain management, pricing, transportation and distribution, and the analysis of large databases. Human factors engineering, on the other hand, focuses on how efficiently human resources can be utilized. A generic term used interchangeably with this is *ergonomics*. Ergonomics is concerned with the relationship between human factors and how these influence workplace design (ANSI, 1989). The Institute of Ergonomics and Human Factors defines ergonomics as "the application of scientific information concerning humans to the design of objects, systems, and environment for human use" (Institute of Ergonomic and Human Factors, 2004). Human factors, on the other hand, involve the coordination and application of physical movements, and cognitive components, such as information processing and decision making to perform a given task (Marmaras and Kontoginis, 2001). Areas also covered include safety, work measurement and simplification, methods study, and the use of anthropometric data in work systems design. Work measurements and methods study aid in determining optimum cycle times and most efficient methods of performing work-related activities. Anthropometry covers the study of human beings in terms of their physical dimensions, with emphasis on how this can be used for work methods and work place design (ANSI, 1989). Current research in ergonomics focuses more on designs that promote safety of workers, while research in work methods and work simplification continues to focus on better time standards and work methods.

The management function of the IE profession is primarily concerned with aspects of cost control, budgeting, and engineering economics. The manufacturing systems engineering function of IE includes facilities planning and design, layout planning of physical facilities, simulation, quality control, inventory management, and manufacturing technologies. Plant location and layout problems have evolved traditionally from being implemented by analytical methods to the use of more powerful simulation models. As the world's economies continue to evolve, production of goods has become more competitive, hence, there is more emphasis on quality and liability issues. This has led to the development of more stringent quality standards, changing the face of manufacturing into an integrated product design and development activity. Future trends in management suggest that organizations will lay more emphasis on continuous quality improvement. Niche techniques have since evolved into IE, such as lean manufacturing, design for assembly and manufacture, and green manufacturing practices. To support the manufacturing systems, inventory management has moved into a platform of the electronic world, now undertaken by a more rigorous approach of supply chain management. The role of the industrial engineer also includes designing manufacturing processes, equipment, automation and robotics, and materials handling systems.

The term *manufacturing engineering* is usually used interchangeably with others such as *production engineering*, and *manufacturing technology*. The term *manufacturing* is used to describe a set of activities that change a raw material into a more useful product. The Society of Manufacturing Engineers (SME), founded in 1932, defines manufacturing engineering as "that specialty of professional engineering that requires such education and experience as is necessary to understand, apply, and control engineering procedures in manufacturing processes and practices of manufacturing, to research and develop tools, processes, machines, and equipment, and to integrate facilities and systems for producing quality products with optimal expenditures" (Soska, 1984, p. 223). A manufacturing engineer performs duties that include design of the product, selecting appropriate materials for the product, and selection and defining the progression of processes and resources needed for the manufacture of the product (Kalpakjian and Schmid, 2010).

The history of manufacturing engineering as a discipline is closely linked to how manufacturing has evolved over the years. Organized manufacturing can be traced back 6,000 years when manual forming processes began with the production of various articles made of wood, ceramic, stone, and metal. The history can be examined from two perspectives: how materials were discovered, developed, and used to make things, and how systems of production were developed (Groover, 2002, p. 3).

The first materials used for making household utensils and ornamental objects included metals, such as gold, copper, and iron. Whereas gold was found in its pure state by early man, copper was most likely the first metal to be extracted successfully. The production of steel is said to have started about 600 to 800 CE. The periods that followed saw the development of a wide variety of ferrous and nonferrous metals. To date, there have been significant developments in engineering materials. Common engineering materials are now classified to include engineered materials, such as ceramics and reinforced composite materials, alloys of various types, and nanomaterials.

The advent of modern manufacturing had its significance in the Industrial Revolution, which began in England in the 1750s and lasted until the 1830s. The economies of the world were heavily dependant upon agricultural produce. With the invention of Watt's steam engine between 1763 and 1765 and steam power replacing wind and other forms of energy, together with Wilkinson's invention of the water-powered boring machine, modern mechanization began in England and spread quickly to other countries. In the mid-1800s to early 1900s there was expansion in transportation industry, especially the railroad system, necessitating an increase in the need for steel. It is during this period that many consumer products were also developed. The need for more efficient production methods became a very important issue in manufacturing, hence, the second Industrial Revolution. This was characterized by mass production, scientific management techniques in manufacturing, electrification (first electric power generating plant in 1881), and assembly lines (Henry Ford's assembly line in 1913) (Groover, 2002, p. 4).

The two world wars separately played significant roles in continuing to improve manufacturing techniques. During these wars, an increased need for transport, battle ships, and guns resulted in the improvements in fabrication techniques and the development of the arc welding process, which is the most commonly used process in welding today (Lincoln Electric Company, 1994, p. 1.1–1.5). After World War II, the Cold War that ensued created an enormous arms race. More sophisticated aircraft had to be developed and these required more complex parts to be machined or formed. Joint efforts by the U.S. Air Force, Massachusetts Institute of Technology (MIT), and John Parsons resulted in the first numerical control systems being developed (Chang, et al., 1998, p. 316). Since then, and with the development of microprocessors and increased use of computers, major milestones have been established in all aspects of manufacturing.

The evolution of manufacturing engineering over the past decade has made it an integrated concurrent activity where manufacturing engineers consider a product from its design stage, simultaneously with the manufacturing stage. Manufacturing today is characterized by highly automated systems utilizing robotics and integrated enterprise control. Not only is product design very automated through computer-aided design (CAD) systems, but the control of process is equally automated through computer-aided manufacturing (CAM) systems (Benhabib, 2003, p. 9-13). The *Handbook of Manufacturing Engineering* (Walker, 1996) identifies four major areas of the discipline: product and factory development, factory operations, materials, and assembly process. The current trends in product design emphasize simultaneous engineering and multidisciplinary approach in the discipline. Niche technologies that replace traditional lengthy product development cycles and prototyping, such as rapid prototyping, are now used to quickly convert designs from computer representations directly into solid objects. As the optimal design is achieved, the manufacturing engineer is faced with the challenge of designing the production facilities, and, like the industrial engineer, today's facilities designs are achieved through analytical techniques and simulation.

Factory operations functions include manufacturing cost control and estimation, process planning and control, work methods, and quality control. With the evolution of computers in manufacturing, techniques used in today's factory operations include computer-aided process planning and computer-aided testing.

Manufacturing engineering must consider materials used in manufacturing and methods of manufacturing parts. Over the past decades there have been major developments in engineering materials with an emphasis on producing lighter but stronger materials, materials with better corrosion resistance, and with better manufacturing properties. Traditional manufacturing methods that existed were best used for the more conventional metals and alloys, plastics, and ceramics. Newer nontraditional techniques have been developed for the more difficult-to-form or difficult-to-machine alloys, ceramics, and composite materials. Forming and fabrication techniques have become more automated, and machining has been integrated with CAD and CAM systems into computer numerical control (CNC) machining.

The final function of manufacturing engineering focuses on analyzing and defining processes and machines, and the entities of assembly automation. This area also includes selection of materials-handling systems, robotics, assembly lines, and all other components of manufacturing automation.

Over the past two decades, advances in modern manufacturing processes such as lithography, etching, electrochemical machining, ultrasonic machining and micromachining, have led to the possibility of producing very small parts through micro- and nanomanufacturing. Traditionally, manufacturing was restricted to parts whose smallest general dimensions were about 0.5 mm and above (macromanufacturing). Today, with new developments in the technologies mentioned, it is possible to make parts in the range of 1 nm to a few hundred um. Micromanufacturing pertains to the production of parts in the range of 100 nm to u100 m, whereas in nanomanufacturing, the range is anywhere from 1 to about 100 nm (Kalpakjian and Schmid, 2010). Typical applications for micro- and nanomanufacturing have been in the production of devices for microelectromechanical systems (MEMS), micropumps, micro-gears for miniature actuators and devices, and microelectronic circuits, to name a few.

Current trends in quality control began with the development of statistical sampling techniques in the 1920s. Statisticians at the Bell Telephone laboratories were pioneers in the development of what is now dubbed Statistical Quality Control (SQC). The practices of quality control were, until the mid-1980s, closely guarded by individual manufacturers. Stiff competition from manufacturers in other countries, such as Japan, especially in consumer electronics, compelled the rest of the world to reassess quality control practices beyond just SQC. The concept of Total Quality Management (TQM) soon became the norm in the circles of quality control (Hradesky, 1995). Thus, TQM now focuses on consumer satisfaction by including SQC, employee involvement, quality function deployment, and continuous improvement. To ensure competitiveness, many companies have become ISO 9000 (quality management standards) certified.

Industrial and manufacturing engineering are very closely related areas with interlinked skills and functions. Both fields are evolving very rapidly today with a shift from macro to micro and nanoscale manufacturing. Researchers, therefore, must continue to focus on manufacturing practices that are cost-effective and efficient, at a macro level, and on new and innovative methods used to produce micro and nano products. In this chapter, the critical resources for industrial and manufacturing engineering will be outlined.

SEARCHING THE LIBRARY CATALOG

Libraries catalog books and other documents, regardless of their format, according to certain cataloging guidelines. In the United States, these guidelines are established by the Library of Congress.

Libraries outside the United States and Canada may follow some other cataloging systems and users from other countries should become familiar with their local libraries, metadata standards. Materials found in the online catalog include: books, journals, technical reports, dissertations, films, audio and video formats, computer software, specifications, maps, graphics, and reference materials, such as encyclopedias and dictionaries. According to the design of a library online catalog, several search entry avenues are available; the most common being the author, title, and subject term searches. It is also possible to limit the search to certain parameters and to combine several bibliographic fields.

KEYWORDS

Searching the online catalog by using keywords is an effective way to find library materials. Most online catalog interfaces allow the use of keywords in different fields, such as title, subject, or source. The following is a list of keywords that can be used: assembly automation, assembly processes, automated production, automation and robotics, CAD/CAM, computer simulation, control of production, controllers, cost estimating, ergonomics, fabrication process, factory design, factory requirements, finishing, green design, human factors, industrial safety, inventory management, maintenance management, manufacturing process, materials characteristics, micromanufacturing, motion systems, optimization, organization and work design, probabilistic models and statistics, process planning, product design, product development, quality control, reliability, risk analysis, safety, scheduling, statistical process control, storage and warehousing, tooling and equipment, and work measurement. This is not a comprehensive list, but some of the major areas related to manufacturing and industrial engineering are presented.

LC SUBJECT HEADINGS

The Library of Congress Subject Headings are the official terms used in libraries to catalog materials in a specific field. These subject terms may be used in the subject field of a search engine. Examples of Library of Congress Subject Fields include:

Assembly-Line Methods
Automation
Computer Integrated Manufacturing Systems
Engineering Economics
Flexible Manufacturing Systems
Human Engineering
Industrial Engineering
Industrial Engineering–Statistical Methods
Industrial Project Management
Industrial Safety
Manufacturing Processes
Manufacturing Processes–Automation
Manufacturing Resource Planning
Methods Engineering
Operations Research
Plant Layout
Production Control
Production Engineering
Psychology, Industrial
Quality Control
Robots, Industrial

Standardization
Systems Engineering
Work Environment
Work Measurement

LC CALL NUMBERS

Library materials cataloged under the Library of Congress systems can be found under several call number sections. Most materials are in the T, TA, TJ, and TS sections. Other sections with additional sources are RC and HF. The following is a list of some major call numbers:

T55-55.3 Industrial safety
T57.35-57.5 Quantitative methods
T57.6-57.77 Operations research. Systems analysis
T60-T60.8 Work measurement. Methods engineering
TA177.4-185 Engineering economy
TA329-348 Engineering mathematics
TJ 212-225 Control engineering. Automatic control systems
TS155-155.8 Production management. Operations management
TS156-170 Control of production systems
TS171-176 Product design
TS177-182 Manufacturing engineering. Process engineering
TS183 Manufacturing processes
TS184-193 Plant engineering
TS 195-199 Packaging
TS200-2301 Specific industries, i.e., metals, textiles, etc.
RC963-967 Industrial hygiene
HF5548.8 Industrial psychology

Due to the interdisciplinary nature of manufacturing engineering, the readers will find materials in other sections depending on the topics included in the research.

ABSTRACTS AND INDEXES

The literature of industrial and manufacturing engineering is similar to other technical areas; it includes: articles from journals, papers published in conference proceedings, monographs, technical reports, and other documents. Abstracting and indexing services adopt the role of collating, organizing, and summarizing thousands of these documents that are published every year. One of the most important indexes in this field is *Compendex* and it should be the starting point for searching for information about a topic. Moreover, due to the interdisciplinary nature of industrial and manufacturing engineering, it is advisable to search other databases when looking for additional information. Technical reports can be found in *NTIS*; theses in *ProQuest Digital Dissertations*, and patents on the U.S. Patents and Trademark Office's Web site. In addition, other databases related to this field are: *Inspec, Ingenta, Computer & Information Systems Abstracts, Electronics & Communications Abstracts, PsycINFO, Science Citation Index, Google Scholar,* and *ABI/INFORM*. All the databases mentioned in this section are described in appropriate chapters of this work. Two additional sources are *Ergonomics Abstracts* and *International Abstracts in Operations Research* described next. Finally, publisher's databases of technical online journals also have search engines capable of producing bibliographic listing; among them are Elsevier's *Scopus, SpringerLink*, and *Wiley Online Library*.

Ergonomics Abstracts. 1969–. London: Taylor & Francis. Sponsored by the Ergonomics Information Analysis Center at the University of Birmingham, U.K. This publication covers the international literature of ergonomics, human factors, human machine systems, physical environmental influences in the workplace, and work design as well as psychological, physiological, and biomechanical aspects of work.

International Abstracts in Operations Research. 1961–. Basingstoke, Hants, U.K.: Palgrave Macmillan. This index covers about 180 journals in operations research, industrial management, decision sciences, information systems, industrial engineering, and other fields. This publication is sponsored by the International Federation of Operational Research Societies.

BIBLIOGRAPHIES AND GUIDES TO THE LITERATURE

Bibliographies and guides to the literature of a technical subject are produced in several different ways. Books are a common format in which these types of documents are published, although not too many have been published recently. A better way of finding technical bibliographies is by searching in engineering databases such as *Compendex, Inspec, Ergonomics Abstracts,* and the others mentioned in the previous section. These bibliographies are published as articles in journals. The readers also will find them in Web sites and as electronic publications; searching in Google or Google Scholar will produce results. In addition, published "review articles" and "survey articles" of a specific subject normally include extensive lists of references that can be considered subject bibliographies. Finally, some bibliographies are written as technical reports. Because most bibliographies are found as articles in journals, as "review articles," or "survey articles," this section only includes a few examples of those in book format.

Cleland, D. I., G. Rafe, and J. Mosher. 1998. *Annotated Bibliography of Project and Team Management.* Newtown Square, PA: Project Management Institute. This is a comprehensive bibliography covering key topics in project management that spans from 1956 to 1998. It includes annotations from articles and books and there is also a list of dissertations and periodicals. The section on applications covers publications about several industries, such as chemical, aerospace, and nuclear power.

Green, R. J. and H. C. Self. 1995. *50 Years of Human Engineering: History and Cumulative Bibliography of the Fitts Human Engineering Division.* Wright-Patterson Air Force Base, OH: Springfield, VA: Crew Systems Directorate, Armstrong Laboratory, Air Force Materiel Command. The historical activities of this laboratory from 1945 to 1994 are presented with an extensive bibliography. Topics included are human factors, ergonomics, control systems, display systems, work stations, visual perception, design criteria, and control panels. http://contrails.iit.edu/history/50years/index.html (accessed November 2, 2010).

Hanel, N. L. N. 1997. *Selective Guide to Literature on Statistical Information for Engineers.* Washington, D.C.: American Society for Engineering Education. Statistical applications in manufacturing and industrial engineering are numerous. This guide gives a list of basic resources for engineers. It covers indexes and abstracts, bibliographies, Internet resources, encyclopedia and dictionaries, handbooks and manuals, books, proceedings, and journals. http://depts.washington.edu/englib/eld/fulltext/StatInfo.pdf (accessed November 2, 2010).

Jürgensen, A., N. Khan, and W. H. Vanderburg. 2001. *Sustainable Production: An Annotated Bibliography.* Lanham, MD/London: Scarecrow Press. This is a by-product of a project at the Center for Technology and Social Development at the University of Toronto on sustainable engineering. It includes over 250 annotations from books and articles.

Osorio, N. L. 2002. *Selective Guide to Literature on Industrial Ergonomics.* Washington, D.C.: American Society for Engineering Education. This bibliography focuses on industrial ergonomics and has 36 pages of resources, such as indexes and abstracts, literature guides, Internet resources, encyclopedias and dictionaries, handbooks and tables, periodicals, proceedings, books, and industrial standards. http://depts.washington.edu/englib/eld/fulltext/ Ergonomics.pdf (accessed November 2, 2010).

Stein, E. S. and E. Buckley. 1994. *Human Factors at the FAA Technical Center Bibliography, 1958-1994.* Atlantic City International Airport, NJ: Federal Aviation Administration Technical Center. It represents the work done on human engineering for more than 30 years at this facility. The citations are in alphabetical order and are indexed by numbers associated with subject matter categories. It is an archival list of work done by psychologists, engineers, systems analysts, computer scientists, and others. http://en.scientificcommons. org/18523114 (accessed November 2, 2010).

DIRECTORIES

Directories provide valuable information to practitioners and students in the form of practical leads to information about industries and manufacturers. They give, for example, details about how to contact industrial suppliers of components and materials. Systematically arranged, they usually include listings of products and services, plants and equipment, and instrumentation and testing equipment.

There are thousands of industrial directories in the marketplace. Originally published in paper, most of these publications are now in Web format. Gale's *Directories in Print* (*DiP*), a comprehensive compilation of directories for all areas, is listed in the General Section of this book. *DiP* provides the user with a list of directories on a very specific industry, such as the chemical industry. The directories listed in this section are examples of directories and guides of products and services related to manufacturing and industrial engineering. Lists of manufacturers from specific states are not included.

ASTM International Directory of Testing Laboratories. West Conshohocken, PA: American Society of Testing Materials http://www.astm.org/LABS/search.html (accessed November 2, 2010). This online database replaced the former paper product: *International Directory of Testing Laboratories.* 1992–2003. The search engine has three options: Search Text; the Specialized Searches for Lab Name, Testing Services, Lab Services, and Product Tested. Also has a Limit Search to specific geographical region.

Bittence, J. C., ed. 1994. *Guide to Engineering Materials Producers.* Materials Park, OH: ASM International. This is a condensed source for finding manufacturers of metallic and nonmetallic materials. It includes a list of over 900 manufacturers. The second section of this book includes nearly 75 categories of materials.

CatalogXpress. 1900–. Englewood, CO: Information Handling Services, IHS. *CatalogXpress* is an index to product and component information from over 16,000 manufacturers' catalogs and other sources. It also provides general information about an additional half million manufacturers. Available to readers in subscribing institutions at: http://parts.ihs.com/products/procurement/catalogxpress-vendors/ (accessed November 2, 2010). *CatalogXpress* is a subsection of IHS' extensive line of products of engineering information.

DIAL Electronics Weekly Buyers' Guide http://www.dialindustry.co.uk/ (accessed November 2, 2010). *DIAL Electronics Weekly Buyers' Guide* is the fusion from *DIAL Electronics* and *Electronics Weekly,* the two major suppliers in the United Kingdom. It is available free on the Internet and is searchable by product/service, company, and trade name. To gain full access to details about U.K. engineering companies, users need to subscribe.

EUmanufacturer.com http://www.eumanufacturer.com/ (accessed November 2, 2010). This database includes a directory of products and services available from European companies. Although it does not have a comprehensive coverage, it is an example of products and services guides for this part of the world.

GlobalSpec, The Engineering Search Engine http://www.globalspec.com/ (accessed November 2, 2010). It includes a section on Products and Supplies covering 20 industries and a section on Services and Consultants for 10 specific areas. The Tools and Links section has a Directory of Companies and a Part Number Search engine. *GlobalSpec* is open access, but subscription is required.

Industrial Directory http://www.industrialdirectory.com/ (accessed November 2, 2010). This is a manufacturers and suppliers product guide. From a list of over 350 general product categories, the user has access to a directory of specific products; this further leads the user to a directory of companies for that product.

Industry Search Australia and New Zealand http://www.industrysearch.com.au/ (accessed November 2, 2010). This is an information resource for Australia and New Zealand's manufacturing, mining, engineering, transport, construction, farming, electronics, and other industries. It is an example of a regional database for industrial products and services.

ISA Directory of Automation Products and Services; 2009/2010. Research Triangle Park, NC: Instrument Society of America. This is guide to measurement and control products; it includes manufacturers, services, and representatives. The online edition: ISA Directory of Automation http://isadirectoryofautomation.org/ (accessed November 2, 2010) is an open access site with information about companies and products.

Manufacturing.Net http://www.manufacturing.net/ (accessed November 2, 2010). Manufacturing.Net includes a comprehensive database of manufacturers, suppliers, and products. In addition, manufacturing professionals can find in this site a wealth of information for their enterprises. Free registration is required.

Member Product Directory. McLean, VA: AMT–The Association for Manufacturing Technology http://www.amtonline.org/directory/AMTdirectory.cfm (accessed November 2, 2010). This directory provides 18 groups of products from about 700 companies. Browsing also can be done by company name or by geographical location. The online version replaced the paper edition published since 1993 until the early 2000s.

Plant Engineering–New Products. Middletown, OH: Reed Business Information http://www.plantengineering.com/new-products.html (accessed November 2, 2010). The *New Products* section of this journal reviews products and software packages related to plant engineering; and lists manufacturers and services providers. Examples of recent new products include electrical products, compressed air, tools and welding, instrumentation and control, software and components, heating, safety, lubrication, power transmission, and environmental products. Formerly known as *Plant Engineering–Update.*

Skelly, K. J. and E. M. T. Skelly. 1997. *Directory of Safety Standards, Literature and Services.* New York: Van Nostrand Reinhold. This is a reference tool that provides sources for available safety standards, guidelines, and other information. This directory contains eight sections: emergency safety contacts, federal agencies, federal safety standards, societies, organizations and societies, industry safety standards, safety periodicals and directories, safety literature, state agencies and safety standards.

Thomas Register of American Manufacturers. 1905–. New York: Thomas Pub. Co. It is a primary source for finding information about U.S. manufacturers. The paper edition is a multivolume set arranged by products. Available also online at http://www.thomasnet.com/ (accessed November 2, 2010). In the online version users can place orders, view and download CAD drawings, and view thousands of full-text company catalogs. Registration is required.

Top 400 Contractors Sourcebook. 1998–. New York: McGraw-Hill, Engineering News-Record, annual. Provides rankings and overviews of the top 400 contractors for eight major industry sectors: building, transportation, manufacturing, industrial processes, petroleum, power, environmental, and telecommunications. Database version at http://enr.construction.com/toplists/Contractors/001-100.asp (accessed November 2, 2010).

The Top 500 Design Firms Sourcebook. 1998–. New York: McGraw-Hill, annual. Lists the major engineering firms from these sectors: building, transportation, manufacturing, industrial processes, petroleum, power, environmental, and telecommunications. Database version at http://enr.construction.com/toplists/DesignFirms/001-100.asp (accessed November 2, 2010). It is also a supplement to *Engineering News Record (ENR).*

DICTIONARIES AND ENCYCLOPEDIAS

Specialized technical dictionaries and glossaries are necessary because they are recognized reference sources for providing the acceptable terminology in a subject. Compiled by experts, they include term definitions and, in some cases, abbreviations, synonyms, symbols, and cross references. The definitions also might have an explanation of the meaning of the term and usage. An encyclopedia is usually a more extensive treaty in the form of articles or short entries; the purpose of an encyclopedia is to systematically summarize the most important aspects of a field. Encyclopedias define and interpret the topics covered and include valuable bibliographies. Dictionaries and encyclopedias are reference sources to give students and practitioners background information about a topic.

Confer, R. G. and T. R. Confer. 1999. *Occupational Health and Safety: Terms, Definitions, and Abbreviations,* 2nd ed. Boca Raton, FL: Lewis Publishers. This covers about 5,000 terms in the areas of industrial hygiene, safety, occupational medicine, and other related areas in acoustics, chemistry, physics, and biology. Terms on bacteriology, environmental health, epidemiology, illumination, mathematics, and microscopy are also included.

Gass, S. I. and C. M. Harris, ed. 2001. *Encyclopedia of Operations Research and Management Science,* 3rd ed. Berlin: Springer-Verlag. This two-volume encyclopedia provides a comprehensive review of the theoretical and applied areas of operations research and management science. There are 893 major topics covered, and each entry has a list of reference sources. Over 200 experts contributed to this work, which is useful for the professional and students of these fields.

Industrial Engineering Terminology: A Revision of ANSI Z94.0-1989: An American National Standard, Approved 2000, rev. ed. 2000. Norcross, GA: Industrial Engineering and Management Press. This book contains more than 15,000 terms, acronyms, and abbreviations. It is divided into 17 sections related to industrial engineering. Definitions are listed alphabetically within each section. Each definition is an official standard of the American National Standards Institute.

Lee, J., ed. 2002. *Dictionary of Industrial Administration.* (3 vols.) Bristol, U.K.: Thoemmes. This encyclopedic dictionary of industrial management has been published with the contributions of over 100 experts in the field. It is a basic source for quick information on all managerial topics; most entries are from a half to two pages long.

McKenna, T. and R. Oliverson. 1997. *Glossary of Reliability and Maintenance Terms.* Houston, TX: Gulf Publishing. This glossary's goal is creating a common terminology among practitioners; it covers the areas of reliability, process engineering, plant operations, repair, and maintenance technologies. More than 1,000 terms are defined.

Philippsborn, H. E. 1994. *Elsevier's Dictionary of Industrial Technology: In English, German, and Portuguese.* Amsterdam/New York: Elsevier. This multilingual dictionary covers an extensive vocabulary of industrial activity, from agricultural machines to wood products. It is one example of multilingual dictionaries usable in industrial engineering and manufacturing.

Rogelberg, S. G. 2007. *Encyclopedia of Industrial and Organizational Psychology.* (2 vols.) Thousand Oaks, CA: Sage. This encyclopedia includes more than 400 entries about relevant topics in the industrial setting. It includes most of the major issues affecting the workplace, such as employment and staffing, individual differences, training and evaluation, productivity and employee behavior, motivation on the job, management, working groups, organizational structures, and historical aspects of industrial psychology.

Ruggeri, F., R. S. Kenett, and F. W. Faltin, eds. 2007. *Encyclopedia of Statistics in Quality and Reliability*. Sussex, U.K.: John Wiley & Sons. This four-volume set written by leading researchers is a comprehensive treatment of quality and reliability. The summarized information presented originated from hundreds of technical papers not always accessible to the practitioners.

Soroka, W. G. and P. J. Zepf, eds. 1998. *The IoPP Glossary of Packaging Terminology*. Herndon, VA: Institute of Packaging Professionals. This glossary includes nearly 8,000 terms of current acceptance in the packaging industry. The five appendices cover information on weights and measures, container dimensions and tests, package components, terminology, notes on plastics, wooden pallets and pallet bins, and are of great value to the practitioners.

South, D. W. 1994. *Encyclopedic Dictionary of Industrial Automation and Computer Control*. Englewood Cliffs, NJ: Prentice Hall PTR. It is a comprehensive reference source for production automation. Some topics covered are flow line production, numerical control, industrial robotics, material handling, group technology, FMS, automated inspection, process control, and CIM. Most definitions of terms include applications.

Stellman, J. M., ed. 1998. *Encyclopedia of Occupational Health and Safety*, 4th ed. (4 vols.) Geneva, Switzerland: International Labour Office. It is divided into four volumes covering many important areas of occupational safety and health. Major sections in each volume are as follows: Vol. 1: the body and health care; prevention, management, and policy; tools and approaches. Vol. 2: hazards. Vol. 3: chemicals; and industries and occupations. Vol. 4: indexes and guides.

Swamidass, P. M., ed. 2000. *Encyclopedia of Production and Manufacturing Management*. Boston: Kluwer Academic. It covers the field of production and manufacturing management by presenting about 100 articles and more than 1,000 short entries on the most recent technical and strategic innovations in the field. The strategic and technological perspectives of this tool are representative of the competitive nature of today's manufacturing.

Ward, J. L. 2000. *Dictionary of Project Management Terms*, 3rd ed. Arlington, VA: ESI International. This glossary summarizes the technical and nontechnical terminology used in the field of project management. It contains nearly 3,400 terms, phrases, and acronyms. It is arranged alphabetically and contains multiple cross references.

Yam, K. L., ed. 2009. *The Wiley Encyclopedia of Packaging Technology*, 3rd ed. New York: Wiley-Interscience. This encyclopedia has 250 articles covering modern packaging techniques and materials. It offers the users details on all packaging machinery and equipment, and current information about changes in materials, processes, technologies, and regulations that have occurred in recent years.

HANDBOOKS AND MANUALS

There are specific handbooks for industrial engineering and manufacturing engineering, but there are also a wide variety of other handbooks and manuals that complement them by providing relevant technical information related to industrial and manufacturing processes. The following is a list of some of the most relevant handbooks and manuals.

Ayers, J. B. 2006. *Handbook of Supply Chain Management*, 2nd ed. New York: Taylor & Francis. It introduces students and researchers to a large number of new concepts, methods, and tools in SCM. Each of the five SMC essential tasks is covered in separate chapters; numerous case studies are also presented.

Badiru, A. B. 2006. *Handbook of Industrial and Systems Engineering.* Boca Raton, FL: Taylor & Francis. This is a basic handbook for industrial and systems engineering students and researchers. It is divided into 38 major topics, each written by experts.

Badiru, A. B. 2011. *Handbook of Industrial Engineering Equations, Formulas, and Calculations.* Boca Raton, FL: CRC Press. This specialized handbook covers mathematical topics often used by industrial engineers: Computational foundations of industrial engineering, Basic mathematical calculations, Statistical distributions, Methods, and applications, Computations with descriptive statistics, Computations for economic analysis, Industrial production calculations, Forecasting calculations, Six Sigma and lean, Risk computations, Computations for project analysis, Product shape and geometrical calculations, and General engineering calculations.

Bralla, J. G. 1998. *Design for Manufacturability Handbook,* 2nd ed., New York: McGraw Hill. A comprehensive guide to the principles and procedures of design for manufacturability (DFM). Sections include general principles and historical perspectives of DFM, materials, formed metal components, machined components, castings, nonmetallic parts, assemblies, finishes, and new developments.

Bralla, J. G. 2007. *Handbook of Manufacturing Processes: How Products, Components and Materials Are Made.* New York: Industrial. This handbook is divided in two sections: Section I, Manufacturing processes; and Section II, How products, components, and materials are made.

Brewer, A. M. et al. 2001. *Handbook of Logistics and Supply-Chain Management.* Oxford, U.K.: Pergamon. Topics included in this handbook cover most elements of modern transport logistics occurring in domestic and international trade.

Charlton, S. G. and T. G. O'Brien, eds. 2002. *Handbook of Human Factors Testing and Evaluation,* 2nd ed. Mahwah, NJ: Lawrence Erlbaum Associates. Useful resource for obtaining information on human factors research and engineering, ergonomics, and experimental psychology, general principles, techniques, and specific applications of human factors tests and evaluation.

Christopher, W. F. and C. G. Thor, eds. 1993. *Handbook for Productivity Measurement and Improvement.* Cambridge, MA: Productivity Press. This reference provides managers and engineers with some of the most recent literature on topics related to productivity improvement. It covers topics on the evolving theory and specific practices of world-class organizations, quality, and productivity. It also provides thorough coverage on the most advanced methods for the measurement and improvement of quality and productivity.

Cleland, D. I. and B. Bidanda, eds. 1990. *The Automated Factory Handbook: Technology and Management.* Blue Ridge Summit, PA: TAB Professional and Reference Books. A general reference on automated manufacturing and management, it includes chapters on design, planning and control of manufacturing processes, implementation of automation, computer integrated manufacturing, and personnel issues.

Crowson, R., ed. 2006. *Handbook of Manufacturing Engineering,* 2nd ed. Boca Raton, FL: CRC/Taylor & Francis. Completely revised and updated, this new edition contains four volumes: Factory operations: planning and instructional methods; Parts fabrication: principles and process; Parts fabrication: principles and process; and Assembly processes: finishing, packaging, and automation.

Dorf, R. C. and A. Kusiak, eds. 1994. *Handbook of Design, Manufacturing and Automation.* New York: Wiley-Interscience. This handbook provides wide-ranging information on the theory and applications of computer-integrated manufacturing technologies. Coverage includes design and operation of manufacturing processes, fixtures in automation, packaging, costs, robotics, inspection, and manufacturing control.

Gattorna, J., ed. 2003. *Gower Handbook of Supply Chain Management,* 5th ed. Aldershot, Hants, U.K.: Gower Publishing. Chapters are written by experts in supply chain management with case studies in Asia, Europe, and North America. They include supply chains in the context of customers and strategy, operational excellence, supply chain integration and collaboration, virtual supply chains, regional and global supply chains, and other practical considerations in supply chains.

Gilleo, K. 2002. *Area Array Packaging Handbook: Manufacturing and Assembly.* New York: McGraw Hill. This handbook focuses on the fast expanding field of electronics and micro-electronics packaging. It presents the basics of array packaging. Specifically it describes the use of Ball Grid Array, Chip Scale Package, and Flip Chip technologies. It also shows the applicability of each technology with varying applications.

Greene, J. H., ed. 1997. *Production and Inventory Control Handbook,* 3rd ed. New York: McGraw-Hill. This reference book was written in conjunction with the American Production and Inventory Control Society (APICS). It is an essential reference for production and inventory control personnel involved in APICS certification.

Harper, C. A., ed. 2005. *Electronic Packaging and Interconnection Handbook,* 4th ed. New York: McGraw-Hill. This handbook gives a comprehensive coverage of electronic packaging from development and design, to manufacturing, facilities, and testing. It is one of the most commonly used references in electronics packaging. There is also broad coverage on manufacturing for all the major types of electronic products.

Harper, C. A., ed. 2001. *Handbook of Materials for Product Design,* 3rd ed. New York: McGraw-Hill. This reference provides materials, data, information, and guidelines for designers, manufacturers, and users of electromechanical products, as well as those who develop and market materials useful for these products. Contains an extensive list of property and performance data.

Humphreys, K. K. and L. M. English, eds. 2005. *Project and Cost Engineers' Handbook,* 4th ed. New York: Marcel Dekker. A ready reference for cost engineering covering all the basics, cost accounting and estimating, profitability, cost control, project management, operations research, and computer applications in cost engineering.

Hundal, M. S. 2002. *Mechanical Life Cycle Handbook: Good Environmental Design and Manufacturing.* New York: Marcel Dekker. Environmentally friendly product design is a major objective of modern industries. This handbook offers a basic introduction to the issues related to good environmental design.

Ireson, W. G., C. F. Coombs, Jr., and R. Y. Moss, eds. 1996. *Handbook of Reliability Engineering and Management,* 2nd ed. New York: McGraw Hill. A resourceful reference for engineers on all aspects of product reliability. Chapters include management roles in reliability, design for reliability, failure, fault-tree analysis, reliability specification, concurrent engineering, data collection and analysis, testing, failure, maintainability and reliability, quality assurance, mathematical and statistical methods. It also contains data, tables, and charts.

Juran, J. M. and A. B. Godfrey. 2010. *Juran's Quality Handbook*: *The Complete Guide to Performance Excellence,* 6th ed. New York: McGraw-Hill. One of the most essential and widely used reference tools in quality management and engineering. The chapters include quality planning process, quality control process, quality improvement process, process management, costs, measurement and decision making, computer applications to quality systems, ISO 9000, benchmarking, strategic deployment, total quality management, human resources and training, product development, inspection and testing.

Kjell, B. and K. B. Zandin, eds. 2001. *Maynard's Industrial Engineering Handbook,* 5th ed. New York: McGraw-Hill. A ready and very exhaustive reference for industrial engineers and managers. The handbook is organized in several sections with chapters. The scope of each section provides a comprehensive overview of the industrial engineering profession. Many illustrations, tables, and references are included.

Kutz, M., ed. 2002. *Handbook of Materials Selection.* New York: John Wiley & Sons. A general reference for materials properties and application including quantitative methods for selection, major materials, corrosion, material information management, testing and inspection, failure analysis, and manufacturing.

Kutz, M. 2006. *Mechanical Engineers' Handbook; V.3: Manufacturing and Management,* 3rd ed. New York: John Wiley & Sons. From the perspective of mechanical engineering, this handbook provides in 24 chapters a comprehensive view of critical issues in manufacturing.

Laha, D. 2008. *Handbook of Computational Intelligence in Manufacturing and Production Management.* Hershey, PA: Information Science Publishers. The 26 chapter of this handbook written by experts presents the most current advances in manufacturing systems.

McMillan, G. K. and D. M. Considine, eds. 1999. *Process/Industrial Instruments and Controls Handbook,* 5th ed., New York: McGraw Hill. This reference contains the latest methods for increasing process efficiency, production rate, and quality.

Mobley, K., ed. 2001. *Plant Engineer's Handbook.* Boston: Butterworth-Heinemann. This reference is concerned with industrial operations or maintenance. Coverage includes the basics of plant engineering, layout and location, contracts and specifications, energy and water supply, HVAC, safety and health, maintenance, mechanical and electrical equipment, and statistical applications.

Mobley, R. K., L. R. Higgins, and D. J. Wickoff, eds. 2008. *Maintenance Engineering Handbook,* 7th ed. New York: McGraw-Hill. This useful reference tool for engineers and technicians performing maintenance covers organization of maintenance, computer applications in maintenance, costs and control, plant, mechanical equipment and electrical equipment maintenance, instrumentation and reliability, lubrication, welding, and corrosion.

Mulcahy, D. E. 1999. *Materials Handling Handbook.* New York: McGraw-Hill. Reference for latest technologies for design, operation, and maintenance of materials handling systems. Coverage includes basics of product movement, facilities layout, vertical and horizontal systems, unit loads, transportation, and economics.

Nof, S. Y., ed. 1999. *Handbook of Industrial Robotics,* 2nd ed. New York: John Wiley & Sons. Useful as a resource for students, engineers, and managers working robotics. Chapters include historical perspectives of industrial robots, machine intelligence, mechanical design, nanorobotics, robot control, programming, economic aspects, robotics in CIM, ergonomics, design and integration, terminology, and applications.

Nof, S. Y., ed. 2009. *Springer Handbook of Automation.* Berlin: Springer. This handbook covers development and impacts of automation; automation theory and scientific foundations; automation design: theory, elements, and methods; automation design: theory and methods for integration; and automation management. It also has sections on automation in industries, infrastructure and service, medical and healthcare systems, home, office, and businesses. It includes hundred of graphics, tables, and a fully searchable DVD.

Norton, R. L. 2009. *CAM Design and Manufacturing Handbook,* 2nd ed. New York: Industrial. This is an updated edition of this essential reference book for students of CAM design. Topics covered include unacceptable cam curves; double-dwell cam curves, single-dwell cam curves; spline functions; critical path motion (CPM) cam curves; cam size determination; dynamics of cam systems-modeling fundamentals; dynamics of cam systems-force, torque, vibration; modeling cam-follower systems; residual vibrations in cam-follower systems; failure of cam systems-stress, wear, corrosion; cam profile determination; cam materials and manufacturing; lubrication of cam systems; measuring cam-follower performance; case studies; and cam design guidelines.

Peach, R. W., ed. 2003. *The ISO 9000 Handbook,* 4th ed. New York: McGraw-Hill. A ready reference for implementing ISO 9001:2000 standards, also contains CD-ROM with text of ISO 9001:2000, ISO 9000:2000, and ISO 9004:2000 standards.

Pham, H., ed. 2003. *Handbook of Reliability Engineering.* New York: Springer. Provides fundamental and applied work in systems reliability engineering, including methodologies for quality, maintainability, and dependability. It also focuses on methods to find creative reliability solutions and to improve processes.

Rosaler, R. C., ed. 2002. *Standard Handbook of Plant Engineering,* 3rd ed. New York: McGraw-Hill. Chapters include maintenance management, facilities management, plant operations, and power generation. It includes metric conversion tables.

Rushton, A., Oxley, J. and P. Croucher. 2010. *The Handbook of Logistics and Distribution Management,* 4th ed. London: Kogan Page. Chapters include concepts of logistics and distribution, planning for logistics, procurement and inventory decisions, warehousing and storage, freight transport, information and supply chain management, outsourcing, security and safety, environmental aspects, and new concepts in logistics.

Sage, A. P. and W. B. Rouse. eds. 2009. *Handbook of Systems Engineering and Management,* 2nd ed. New York: John Wiley & Sons. Chapters include basics of system engineering, life cycles, tests and evaluation, planning and marketing, systems engineering management, risk management, configuration management, cost management, total quality management, concurrent engineering project management, systems design, human interaction with complex systems, and operations research.

Salvendy, G. ed. 2001. *Handbook of Industrial Engineering: Technology and Operations Management,* 3rd ed. New York: John Wiley & Sons. This is an essential reference for industrial engineers. Organized into parts and chapters, its coverage is very broad and thorough, and includes all major areas in industrial engineering. It includes numerous illustrations, tables, graphs, and references.

Salvendy, G., ed. 2006. *Handbook of Human Factors and Ergonomics,* 3rd ed. New York: John Wiley & Sons. It is a necessary handbook for industrial engineering practitioners. Sections and chapters cover human factors, system design, motion analysis, job design, workplace and environmental design including the biomechanical aspects, health and safety, human–computer interaction, and selected applications.

Stapelberg, R. F. 2009. *Handbook of Reliability, Availability, Maintainability and Safety in Engineering Design.* London: Springer. This handbook provides a compilation of methods and techniques that are essential to guarantee the integrity of processes for engineering design. The four major topics included (availability, maintainability, reliability, and safety) are methodically covered.

Taguchi, G. 2005. *Taguchi's Quality Engineering Handbook.* New York: John Wiley & Sons. This is a good compilation of Taguchi's quality technologies. It is a recommended reference for students and practitioners of industrial quality control.

Tang, C. S. 2008. *Supply Chain Analysis: A Handbook on the Interaction of Information, System and Optimization.* New York: Springer. This handbook is written with a field-practical approach. It presents several case studies demonstrating how the theoretical methods are applied to real situations.

Woodrow, W. 2010. *Sustainable Communities Design Handbook: Green Engineering, Architecture, and Technology.* Burlington, MA: Elsevier Butterworth Heinemann. Written for environmental engineers, civil engineers, architects, mechanical engineers, and energy engineers, this handbook covers the basic technologies and methods for sustainable development.

Wu, B. 2002. *Handbook of Manufacturing and Supply Systems Design: From Strategy Formulation to System Operation.* New York: Taylor & Francis. This handbook provides an organizational framework with guidelines and worksheets to assist engineers who design and manage manufacturing and supply systems.

MONOGRAPHS AND TEXTBOOKS

The production of books in manufacturing engineering is an active part of this profession. Books summarize the accepted knowledge in a field, providing valuable information to the researcher and practitioner. This list offers a selective number of textbooks and treatises covering both research-oriented and practical treatments. In order to find appropriate titles on a specific topic, it is recommended to use a library online catalog, search a database (e.g., *Books in Print*), search a bookstore Web site (e.g., Amazon.com), search *Google Books*, and search a publisher online catalog. In addition, many professional journals have book review sections where readers can find out about new books in their areas. In recent years, most technical books are published in paper and online.

Benhabib, B. 2003. *Manufacturing: Design, Production, Automation and Integration*. New York: Marcel Dekker. This book is divided into three sections: engineering design, discrete parts manufacturing, and automatic control in manufacturing. The first part covers from concept to prototyping, the second presents examples of fabrication processes. In the third section, instrumentation, production control, systems control, and quality control are presented.

Bussmann, S., N. R. Jennings, and M. Wooldridge. 2004. *Multiagent Systems for Manufacturing Control: A Design Methodology*. Berlin/New York: Springer. This book presents the Designing Agent-Based Control Systems (DACS) methodology that is a new concept based on the use of semiautonomous decision makers in the production of products. It gives an overview of agent technologies, and provides case studies to illustrate the DACS methodology. This book is of interest to researchers and practitioners.

Cooper, C. L. and E. A. Locke, eds. 2000. *Industrial and Organizational Psychology Linking Theory with Practice*, Oxford, U.K./Malden, MA: Blackwell Business. This book contains a set of contributing papers written by experts in the field of industrial and organizational psychology. The theoretical and practical approaches presented allow for a blend of practical-based and research-oriented discussions that would benefit students and professionals. It covers most of the relevant topics in the field.

Crowson, R., ed. 2006. *Product Design and Factory Development,* 2nd ed. Boca Raton, FL: CRC Press. This book corresponds to volume 1 of Crowson's *Handbook of Manufacturing Engineering* (2006), which can be purchased as a single book. The seven chapters include: product development, product design, the deductive machine troubleshooting process, factory requirements, factory design, human elements in a factory, and computers and controllers.

Curry, G. L. and R. M. Feldman. 2009. *Manufacturing Systems Modeling and Analysis*. Berlin: Springer-Verlag. Among the topics covered in this book are processing time variability, multiple-stage single-product factory model, multiple products factory models, WIP limiting control strategies, and serial limited buffer models.

Fiset, J. 2009. *Human–Machine Interface Design for Process Control Applications*. Research Triangle Park, NC: Instrumentation, Systems, and Automation Society. Written for designers and industrial control personnel, this book discusses three basic human–machine interfaces: distributed control systems (DCSs), supervisory control and data acquisition system (SCADA), and stand-alone units.

Groover, M. P. 2008. *Automation, Production Systems, and Computer-integrated Manufacturing,* 3rd ed. Upper Saddle River, NJ: Pearson Education. This book provides basic and fundamental coverage of topics in manufacturing systems, such as flow line production, numerical control, industrial robotics, materials handling, flexible manufacturing systems, automated inspection, and process control. It is a textbook for advanced engineering students and includes many examples and exercises.

Groover, M. P. 2010 *Fundamentals of Modern Manufacturing: Materials, Processes, and Systems,* 4th ed. Hoboken, NJ: John Wiley & Sons. Covers the three building blocks of manufacturing with a major coverage of manufacturing processes and secondary coverage of engineering materials and production systems. Materials included are metals, ceramics, polymers, composites, and silicon. It is an introductory treatment of the subject.

Kock, N. F., ed. 2009. *Virtual Team Leadership and Collaborative Engineering Advancements: Contemporary issues and Implications*. Hershey, PA: Information Science Reference. This book presents the combined concepts of collaborative engineering with virtual interaction. It contains both theoretical foundations and real-world applications. Emerging issues, virtual team leadership, and collaborative production are some of the main topics covered.

Lee, E. S., and H. Shih. 2001. *Fuzzy and Multi-Level Decision Making: An Interactive Computational Approach*. London/New York: Springer. This is a theoretical approach to decision-making processes found in hierarchical organizations. The first part of the book covers multilevel programming algorithms, and the second part discusses knowledge representation and fuzzy decision making.

Lee, M. H., and J. J. Rowland, eds. 1995. *Intelligent Assembly Systems*. Singapore/Hackensack, NJ: World Scientific. Written for robotics, mechanical, and systems engineers, this book covers topics like software tools and administration, diagnosis systems and error handling, and sensor–actuator integration methods—all essential for the development of flexible automation systems and robotics.

Levitt, J. 2008. *Lean Maintenance*. New York: Industrial Press. This book presents a practical approach to the components of lean maintenance. It offers a step-by-step analysis for successful lean projects.

Levy, P. E. 2009. *Industrial/Organizational Psychology: Understanding the Workplace,* 3rd ed. New York: Worth Publishers. This is an introduction to industrial psychology. It includes chapters on organizational communication, organizational development, stress management, group behavior, and employee motivation.

Qin, Y. 2010. *Micromanufacturing Engineering and Technology*. Amsterdam: Elsevier Science. It is a well-received book about one of the emerging topics of manufacturing. It covers the basic technical issues of micromanufacturing and presents numerous applications, such as micromechanical-cutting, laser machining, and microforming.

Raid, A., O. Ulgen, and E. William. 2010. *Process Simulation Using WITNESS: Including Lean and Six-Sigma Applications*. Oxford, U.K.: Wiley-Blackwell. A basic text for WITNESS, a simulation software widely used in industry. It includes applications for a manufacturing setting using lean operations and Six-Sigma approaches.

Ravindran, A., ed. 2009. *Operations Research Applications*. Boca Raton, FL: CRC Press. After introducing the readers to basic chapters on quality control, reliability, project management, energy, and production systems, this book provides applications to specific industries: airlines, e-commerce, military, finance, water resources, and supply chain management. Each chapter is written by experts.

Rehg, J. A. 2003. *Introduction to Robotics in CIM Systems,* 5th ed. Upper Saddle River, NJ: Prentice Hall. Written for students and industrial practitioners, this book covers the essentials of automated manufacturing, including design, development, implementation, and support for automated production systems. It also includes information about hardware and software that are needed for the implementation of these systems.

Ridley, J. and J. Channing, eds. 2008. *Safety at Work,* 7th ed. Oxford, U.K.: Elsevier Butterworth Heinemann. Based on the premise that safety is an important function of industrial supervisors and management, this book gives comprehensive coverage of safety standards. It provides guidance about how to comply with regulations and how to achieve a high level of safety in a workplace.

Robinett, R. D., C. R. Dohrmann, G. R. Eisler, J. T. Feddema, G. G. Parker, D. G. Wilson, and D. Stokes, 2002. *Flexible Robot Dynamics and Controls*. New York: Kluwer Academic/ Plenum Publishers. This text represents research and development efforts done by a team of experts at Sandia National Laboratory. Written for graduate courses, it is also a reference source for engineers in the field. It contains many real-world examples with a strong emphasis on hardware solutions.

Smith, R. and B. Hawkins. 2004. *Lean Maintenance: Reduce Costs, Improve Quality, and Increase Market Share*. Amsterdam/Boston: Elsevier Butterworth Heinemann. This book covers the area of lean maintenance (LM) in detail for those in production management, manufacturing processes, and just-in-time systems. The book presents many examples, methodologies, checklists, and explanations of the processes involved in the implementation of LM projects.

JOURNALS

Journals are important publications for engineers, researchers, and students in the field of industrial and manufacturing engineering because they present the results of current research, methods, and developments. Articles published in journals are usually indexed in databases like *Compendex*, *Inspec*, and *Ergonomics Abstracts*. Most articles are now available in electronic format, but some pre-1990s journals are still only in paper form. The list of titles included in this section is a sample of important journals in this area. This list is limited to scholarly publications and, therefore, does not include trade publications.

Advances in Human Factors—Ergonomics. 1984–. Bingley, U.K.: Emerald Group Publishing (0921-2647).

Applied Ergonomics: Human Factors in Technology and Society. 1969–. Oxford, U.K.: Elsevier Science (0003-6870).

Assembly Automation. 1980–. Bingley, Yorkshire, U.K.: Emerald Group Publishing (0144-5154).

CIRP Annals-Manufacturing Technology. 1975–. Bern, Switzerland: Technische Rundschau Edition Colibri (0007-8506).

Computers & Industrial Engineering. 1977–. Oxford, U.K.: Pergamon-Elsevier (0360-8352).

Computers & Operations Research. 1974–. Oxford, U.K.: Pergamon-Elsevier Science (0305-0548).

Concurrent Engineering–Research and Applications. 1993–. London: Sage Publications (1063-293X).

Ergonomics: An International Journal of Research and Practice in Human Factors and Ergonomics. 1957–. Abingdon, Oxon, U.K.: Taylor & Francis (0014-0139).

European Journal of Industrial Engineering, 2007–. Geneva,: Inderscience Enterprises (1751-5254).

European Journal of Operational Research. 1977–. Amsterdam: Elsevier BV (0377-2217).

Human Factors. 1958–. Thousand Oaks, CA: Sage Publications (0018-7208).

Human Factors and Ergonomics in Manufacturing. 1991–. Hoboken, NJ: John Wiley & Sons (1090-8471).

IEEE–ASME Transactions on Mechatronics. 1996–. Piscataway, NJ: Institute of Electrical Electronics Engineers (1083-4435).

IEEE Transactions on Automation Science and Engineering. 2004–. Piscataway, NJ: Institute of Electrical and Electronics Engineers (1545-5955).

IEEE Transactions on Engineering Management. 1963–. Piscataway, NJ: Institute of Electrical Electronics Engineers (0018-9391).

IEEE Transactions on Robotics. 1985–. Piscataway, NJ: Institute of Electrical and Electronics Engineers (1552-3098).

IEEE Transactions on Systems, Man and Cybernetics, Part A: Systems & Humans. 1971–. Piscataway, NJ: Institute of Electrical and Electronics Engineers (1083-4427).

IIE Transactions on Healthcare Systems Engineering. 2010–. Philadelphia: Taylor & Francis (1948-8300).

IIE Transactions: Industrial Engineering Research and Development. 1969–. Philadelphia: Taylor & Francis (0740-817X).

Industrial Engineering and Management. 1966–. Mumbai, India: Chary Publications (0019-8242).

Industrial Engineer: Engineering and Management Solutions at Work. 1949–. Norcross, GA: Institute of Industrial Engineers (1542-894X).

Industrial Management & Data Systems. 1980–. Bingley, Yorkshire, U.K.: Emerald Group Publishing (0263-5577).

International Journal of Advanced Manufacturing Technology. 1985–. Guildford, U.K.: Springer London (0268-3768).

International Journal of Computer Integrated Manufacturing. 1988–. Abingdon, Oxon, U.K.: Taylor & Francis (0951-192X).

International Journal of Flexible Manufacturing Systems: Design, Analysis and Operation of Manufacturing and Assembly Systems. 1988–. Dordrecht, Netherlands: Elsevier (0920-6299).

International Journal of Green Energy. 2004–. Philadelphia: Taylor & Francis (1543-5075).

International Journal of Industrial Engineering: Theory, Applications and Practice. 1994–. El Paso, TX: International Journal of Industrial Engineering (1943-670X).

International Journal of Industrial Ergonomics. 1986–. Amsterdam: Elsevier BV (0169-8141).

International Journal of Machine Tools & Manufacture. 1987–. Oxon, U.K.: Elsevier Science (0890-6955).

International Journal of Materials & Product Technology (IJMPT). 1986–. Bucks, U.K.: Interscience Publishers (0144-3577).

International Journal of Operations and Production Management. 1980–. Yorkshire, U.K.: Emerald Group Publishing Ltd (0144-3577).

International Journal of Production Economics. 1991–. Amsterdam: Elsevier Science BV (0925-5273).

International Journal of Production Research, 1961–. Abingdon, Oxon, U.K.: Taylor & Francis (0022-7543).

Journal of Advanced Manufacturing Systems. 2002–. Singapore: World Scientific Publishing (0219-6867).

Journal of Intelligent Manufacturing. 1990–. Dordrecht, Netherlands: Springer (0956-5515).

Journal of Manufacturing Science and Engineering, Transactions of The ASME. 1996–. New York: American Society Mechanical Engineering (1087-1357).

Journal of Manufacturing Systems. 1982–. Oxford, U.K.: Elsevier (0278-6125).

Journal of Manufacturing Technology Management. 1982–. Yorkshire, U.K.: Emerald Group Publishing (1741-038X).

Journal of the Operational Research Society. 1978–. Basingstoke, Hants, U.K.: Palgrave Macmillan (0160-5682).

Journal of Quality in Maintenance Engineering. 1995–. Yorkshire, U.K.: Emerald Group Publishing (1355-2511).

Journal of Quality Technology, 1969–. Milwaukee, WI: American Society of Quality Control (0022-4065).

Proceedings of the Institution of Mechanical Engineers, Part B: Journal of Engineering Manufacture. 1989–. Westminster, U.K.: Professional Engineering Publishing (0954-4054).

Production Planning and Control. 1990–. London: Taylor & Francis (0953-7287).

Project Management Journal. 1970–. Hoboken, NJ: John Wiley & Sons (8756-9728).

Psychometrika: A Journal Devoted to the Development of Psychology as a Quantitative Rational Science. 1936–. New York: Springer (0033-3123).

Quality and Reliability Engineering International. 1985–. Sussex, U.K.: John Wiley & Sons (0748-8017).

Reliability Engineering and System Safety. 1980–. Oxford, U.K.: Elsevier (0951-8320).

Renewable and Sustainable Energy Reviews. 1997–. Oxford, U.K.: Pergamon-Elsevier Science (1364-0321).

Robotics and Computer-Integrated Manufacturing. 1984–. Oxford, U.K.: Pergamon-Elsevier Science (0736-5845).

WEB RESOURCES

The ability of people, organizations, federal and state agencies, and corporate bodies to produce their own documents and make them accessible on their own Web sites has increased since the 1990s. Many of these Web sites are providing valuable technical and commercial information at no cost to users. Sophisticated search engines like *Scirus* or a portal like The National Digital Library make access to this kind of documentation more effective. Users of the Internet must be aware that in such an open environment, it is important to evaluate the quality of the sources found. In this section, a selective list of Web resources is included; it is not comprehensive, but rather represents examples of Web sites with quality information.

Best Manufacturing Practices http://www.bmpcoe.org/ (accessed October 27, 2010). This is a good source for identifying current manufacturing research, and the promotion of exceptional manufacturing practices, methods, innovative technologies, and procedures.

Center for Integrated Manufacturing Studies (CIMS) http://www.cims.rit.edu/ (accessed October 27, 2010). CIMS uses the expertise of its members from academic, industry, and government resources to improve manufacturing through applied technology and training. This Web site provides information on CIMS's Center for Excellence in Lean Enterprise, Center for Remanufacturing and Resource Recovery, Systems Modernization and Sustainment Center, and Sustainable Systems Research Center.

Commercial Technologies for Maintenance Activities (CTMA) http://ctma.ncms.org (accessed October 27, 2010). This is collaboration between the National Center for Manufacturing Sciences and the Department of Defense. Organizational members benefit from the use of new manufacturing technologies created under CTMA activities. This Web site gives details about CTMA projects and services, and a newsletter is also available.

Consortium for Advanced Manufacturing–International (CAM-I) http://www.cam-i.org/ (accessed October 27, 2010). This organization's main areas of interest or projects are cost management systems (CMS), next generation manufacturing systems (NGMS), robust quality engineering, standards (STD), and enterprise integration program (EIP). The Web site offers details about these programs, forums, and other resources.

Engineering Research Center for Reconfigurable Machining Systems, http://erc.engin.umich.edu/ (accessed October 27, 2010). Based at the University of Michigan, Ann Arbor, this center is dedicated to developing reconfigurable manufacturing systems that will allow rapid and cost-effective applications in industry. This Web site provides a general overview of the center.

ErgoWeb–Resource Center http://www.ergoweb.com/resources/reference/ (accessed October 27, 2010). On this Web site, readers can find information about ergonomics, such as an ergonomics glossary, a history of ergonomics, concepts, case studies, and reference materials. Included in this last section are books and journals listings with some abstracts, over 3,000 bibliographic references, and links to various professional organizations in ergonomics.

European Agency for Safety and Health at Work http://osha.europa.eu/en/front-page (accessed October 27, 2010). There is a considerable amount of information on this Web site about this organization in the form of annual reports, newsletters, magazines, reports, factsheets, forums, conference proceedings, job opportunities, press releases, and events.

Georgia Tech Manufacturing Research Center http://www.marc.gatech.edu/ (accessed October 27, 2010). This Web site describes the facilities in manufacturing processes, applications, and technological solutions to manufacturing problems at the MARC center.

Imaging Products Laboratory (IPL) http://www.ipl.rit.edu (accessed October 27, 2010). Information from Rochester Institute of Technology's Center for Integrated Manufacturing Studies.

Intelligent Manufacturing Systems http://www.ims.org/ (accessed October 27, 2010). IMS' objective is to develop the next generation of manufacturing and processing technologies. This Web site provides information about the activities of nearly 300 companies and 200 research institutions participating in this industrial consortium.

Laboratory for Manufacturing and Productivity (LMP) http://web.mit.edu/lmp/ (accessed October 27, 2010). The LMP is an MIT laboratory engaged in conducting engineering research in manufacturing and in the development of the fundamentals of manufacturing science.

Lean Product Development Initiative (LPDI) http://lpdi.ncms.org/ (accessed October 27, 2010). This is a program of the National Center for Manufacturing Sciences under its program of Commercial Technologies for Maintenance Activities. Users will learn more about LPDI products, services, and contact sources.

Manufacturing.net http://manufacturing.net/ (accessed October 27, 2010). Offers a wealth of practical information to people interested in these areas: aerospace, automotive, chemical and petroleum, electrical and electronics, food and beverage, heavy machinery, medical, metals, pharmaceuticals and biotech, plastics and rubber, and other manufacturing.

Manufacturing Science and Technology Center http://www.sandia.gov/mst/ (accessed October 27, 2010). This division of Sandia National Laboratories focuses on the development and application of advanced manufacturing processes in four areas: the manufacturing of engineering hardware, the production of weapon components, the development of robust manufacturing and processes, and the design and fabrication of unique production equipment. The Web site provides an overview of its operation and contacting resources.

Manufacturing Systems Integration (MISD)–NIST http://www.nist.gov/el/msid/index.cfm (accessed October 27, 2010). The MSID's purpose is to develop technologies and standards that can be used to implement information-intensive manufacturing systems. Through this link, readers can find their products and tools, publications, services, and a description of their major research areas.

NASA–Ames Human Systems Integration Division http://human-factors.arc.nasa.gov/ (accessed October 27, 2010). In addition to giving information about their main areas of research: human-centered design, operations of complex aerospace systems, human performance, and human–automation interaction, this site also has an extensive library of documents on human factors in full text.

NASA–Cognition Lab http://human-factors.arc.nasa.gov/ihi/cognition/ (accessed October 27, 2010). This laboratory is a part of the Human Systems Integration Division at the NASA Ames Research Center. The activities of the lab, their research projects, and contacting information are found on this site.

National Center for Manufacturing Sciences http://www.ncms.org/ (accessed October 27, 2010). This site provides information about the programs and services of this consortium of manufacturers. Among these programs are government partnerships, the Commercial Technologies for Maintenance Activities (CTMA), environmental health and safety, forums, and the InfraGard Manufacturing Industry Association (IMIA).

National Center for Remanufacturing and Resource Recovery, NC3R http://www.reman.rit.edu (accessed October 27, 2010). "NC3R's mission is to deliver to industry advanced technologies and tools for efficient and cost-effective remanufacturing and the design of products that have no negative environmental impacts" (from Web page).

National Coalition for Advanced Manufacturing (NACFAM) http://www.nacfam.org/ (accessed October 27, 2010). The NACFAM is involved in several research activities related to public policy recommendations. This site gives information about membership, policy initiatives, research, upcoming events, and publications.

National Institute for Occupational Safety and Health (NIOSH) http://www.cdc.gov/niosh/ (accessed October 27, 2010). NIOSH's main purpose is to provide programs and policies that prevent work-related illness, injury, disability, and death. This Web page contains details about NIOSH's goals, contacts, organization, research, accomplishments, programs, services, activities, and publications.

NIST Manufacturing Portal http://www.nist.gov/manufacturing-portal.cfm (accessed October 27, 2010). This portal is a gateway to numerous types of support the National Institute of Standards and Technology (NIST) offers to industry. This Web site connects users with NIST programs on energy, green manufacturing, lean manufacturing, metrology, nano-manufacturing, ontologies, process improvement, product data, robotics, simulation, supply chain, sustainable manufacturing, and systems integration.

Occupational Safety and Ergonomics Excellence (OSEE) Program http://www.cims.rit.edu/OSEE/about.aspx (accessed October 27, 2010). This program provides support to companies in all aspects related to ergonomics and health and safety issues.

Occupational Safety and Health Administration (OSHA) http://www.osha.gov/ (accessed October 27, 2010). Extensive information about OSHA is found here, including full-text documents of industrial safety standards, programs, and policies.

PNR Product Realization Network at Stanford (PNR) http://prn.stanford.edu/ (October, 27, 2010). This Web site is an initiative of Stanford University's Graduate School of Business, the School of Engineering, and industry with the purpose of developing new products with practical applications in society.

Quality Management Principles http://www.iso.org/iso/qmp (accessed October 27, 2010). This ISO Web page presents the quality management principles on which the ISO 9000 standards for quality management system are based. The ISO 9000 series is the ISO's most widely known standard in industry.

Royal Society for the Prevention of Accidents (RoSPA) http://www.rospa.com/ (accessed October 27, 2010). It provides access to information about this charitable organization on topics like conferences and events, products and services, training and consultancy, publications, and a list of related links.

Sustainable Systems Research Center (SSRC) http://www.ssrc.rit.edu/ (accessed October 27, 2010). Site is "dedicated to enhancing the environmental and economic performance of products and processes. The center develops and applies tools for sustainable design, a holistic approach to creating high quality products with minimal environmental impact" (from Web page).

Systems Modernization and Sustainment Center (SMS) http://www.sms.rit.edu/ (accessed October 27, 2010). "SMS develops technologies for optimal life-cycle design, management, and modernization of large equipment systems. This is accomplished through maximizing the value of existing systems and designing advanced capabilities in new systems while smartly managing life-cycle costs" (from Web page).

The World Wide Web Virtual Library: Industrial Engineering http://www2.isye.gatech.edu/www-ie/ (accessed October 27, 2010). Housed at Georgia Tech University, this Web page provides links to academic programs, publications, courses, conferences, databases, professional societies, software, commercial entities, other links of interest, and related virtual libraries in industrial engineering.

Zentrum Mensch-Maschine-Systeme–Center of Human–Machine–Systems http://www.tu-berlin.de/zentrum_mensch-maschine-systeme/ (accessed October 27, 2010). Housed at the Technische Universität in Berlin, the center supports interdisciplinary research and system development for effective and efficient human–machine–systems. Besides to form proposals for industrial research partnerships, users can also find a list of related publications.

Listservs, newsgroups, newsletters, Weblogs, and forums are some of the other services available on the Internet. The main purpose of these services is to create a communication platform where professionals working on the same kind of engineering problems form groups to exchange ideas.

There are many of these services on the Internet, and one way of finding them is by searching in databases like USENET, Catalist, JISCmail, Liszt, Google Groups, and Yahoo Groups. The most highly recognized of these services are hosted by professional societies or universities, e.g., the Consortium for Advanced Manufacturing has a forum on Cost Management Systems (CMS). This organization also has a forum on Next Generation Manufacturing Systems (NGMS). It is advisable to become familiar with the Web sites of professional organizations, academic departments, federal and state research, and technical agencies of a specific field in order to identify these kinds of Internet services that can be valuable to the advancement of the reader's career. In many instances, the services are limited to members of the organization, but others do not require membership in the organization to participate. Social networking Web services like Facebook are also being utilized by groups for sharing information.

SOCIETIES, ASSOCIATIONS, AND ORGANIZATIONS

Societies have been the most widely utilized forums of communicating and networking among professionals in industrial and manufacturing engineering. Through publications from various types of journals and proceedings, conferences, expositions, trade shows, seminars, and workshops, societies still remain the pivotal point for exchange and dissemination of research and technical information. All the major societies in industrial and manufacturing engineering maintain Web sites that have been included in this section. Through these Web sites, researchers and professionals can obtain information about the society, their focus and activities, and can often access technical information and publications either directly or through purchases. Other information available on most Web sites also includes calendars of events, conferences, trade shows, job listings, and resources for K–12 students interested in careers in the field of industrial and manufacturing engineering.

American Society for Precision Engineering (ASPE) http://www.aspe.net/ (accessed November 7, 2010). ASPE promotes the advancement of precision design, manufacturing, and measurement by providing a forum to encourage and enable exchange of ideas and information between industry, academia, and government laboratories, and by providing annual and topical meetings, continuing professional education and training, and publications. Membership is wide and diverse.

American Society for Quality (ASQ) http://www.asq.org/ (accessed November 7, 2010). The ASQ, founded in 1946, originally as the American Society for Quality Control, is one of the world's leading authorities on quality issues. Membership is very diverse. It covers quality from engineering to manufacturing, medical, aerospace, and many other areas. The ASQ promotes quality worldwide by advancing learning, quality improvement, and knowledge exchange to improve business results, provides networking and resources for improving quality, and works with the media to promote quality related matters. It also organizes various conferences annually worldwide.

American Society of Mechanical Engineers International http://www.asme.org/ (accessed November 7, 2010). For more details about this society, see Chapter 16 in this book.

American Society of Safety Engineers (ASEE) http://www.asse.org/ (accessed November 7, 2010). "Founded in 1911, ASSE is the oldest and largest professional safety organization. Its more than 30,000 members manage, supervise, and consult on safety, health, and environmental issues in industry, insurance, government, and education. ASSE is guided by a 16-member Board of Directors, which consists of 8 regional vice presidents, 3 council vice presidents, society president, president-elect, senior vice president, vice president of finance, and executive director. ASSE has 12 practice specialties, 150 chapters, 56 sections, and 64 student sections" (from Web page).

American Society of Sanitary Engineering for Plumbing and Sanitary Research http://www.asse-plumbing.org/ (accessed November 7, 2010). Founded in 1906, largely by the efforts of Henry Davis, the chief plumbing Inspector for the District of Columbia, this society has the mission of continuously improving the performance, reliability, and safety of plumbing systems. The society is comprised of a wide variety of membership in seven regions and one international chapter. It has several committees that develop standards for plumbing and drainage, organizes conferences and meetings, and professional development courses.

Association for Facilities Engineering (AFE) http://www.afe.org (accessed November 7, 2010). "AFE provides education, certification, technical information, and other relevant resources for plant and facility engineering, operations, and maintenance professionals worldwide" (from Web page).

Association for Manufacturing Technology (AMT) http://www.mfgtech.org/ (accessed November 7, 2010). "Founded in 1902, AMT (The Association For Manufacturing Technology) represents and promotes the interests of American providers of manufacturing machinery and equipment. Its goal is to promote technological advancements and improvements in the design, manufacture, and sale of members' products in those markets and act as an industry advocate on trade matters to governments and trade organizations throughout the world" (from Web page). The AMT organizes and sponsors the IMTS (International Manufacturing Technology Show), every even year. IMTS is one of the largest trade shows in the Americas.

Human Factors and Ergonomics Society (HFES) http://www.hfes.org/ (accessed November 7, 2010). Founded in 1957, HFES has grown into an internationally recognized society whose mission is "to promote the discovery and exchange of knowledge concerning the characteristics of human beings that are applicable to the design of systems and devices of all kinds" (from Web page). The society has 21 technical groups with local and student chapters; it organizes annual meetings and has several periodical publications.

IEEE Components, Packaging and Manufacturing Technology (CPMT), The http://www.cpmt.org (accessed November 7, 2010). "CPMT society is the leading international forum for scientists and engineers engaged in the research, design, and development of revolutionary advances in microsystems packaging and manufacture. It provides technical assistance, professional development, and international networking opportunities, through its journals, conferences and workshops, committee activities, local chapter events, and educational programs and awards" (from Web page).

IEEE Engineering Management Society (EMS) http://www.ewh.ieee.org/soc/ems/ (accessed November 7, 2010). "The IEEE Engineering Management Society (EMS) directs its efforts toward advancing the practice of engineering and technology management as a professional discipline, encouraging theory development for managing organizations with a high engineering or technical content, and promoting high professional standards among its members" (from Web page).

IEEE Robotics and Automation Society (RAS) http://www.ieee-ras.org/ (accessed November 7, 2010). The IEEE–RAS was established in 1989 as one of the 36 societies and councils sponsored by the Institute of Electrical and Electronics Engineers. The IEEE–RAS serves the robotics and automation interests in the IEEE.

IEEE Systems, Man, and Cybernetics Society http://www.ieeesmc.org/ (accessed November 7, 2010). This society is concerned with the integration of communications, control, cybernetics, stochastics, optimization, and system structure toward the formulation of a general theory of systems, development of systems engineering technology, and human factors engineering. The society also organizes conferences and sponsors various publications.

Institute for Operations Research and the Management Sciences (INFORMS) http://www.informs.org/ (accessed November 7, 2010). "The Institute for Operations Research and the Management Sciences (INFORMS) serves the scientific and professional needs of OR/MS investigators, scientists, students, educators, and managers, as well as the institutions they serve, by such services as publishing a variety of journals that describe the latest OR/MS methods and applications and by organizing professional conferences. The Institute also serves as a focal point for OR/MS professionals, permitting them to communicate with each other and to reach out to other professional societies and to the varied clientele of the profession's research and practice" (from Web page).

Institute of Ergonomic and Human Factors–U.K. http://www.ergonomics.org.uk/ (accessed November 7, 2010). Founded in 1949 as the Ergonomics Society, the Institute of Ergonomics and Human Factors is the only professional society in the United Kingdom dedicated to ergonomists and those interested in ergonomics. It has a diverse membership ranging from students to professionals. It has regional and special interest groups. It provides a forum of networking for professionals in this field, job placements, conferences, journals, workshops, and training.

Institute of Industrial Engineers (IIE) http://www.iienet2.org/ (accessed November 7, 2010). "IIE is the world's largest professional society dedicated solely to the support of the industrial engineering profession and individuals involved with improving quality and productivity. Founded in 1948, IIE is an international, nonprofit association that provides leadership for the application, education, training, research, and development of industrial engineering" (from Web page). The society, which identifies itself as "the Global Association of Productivity and Efficiency Professionals," has chapters worldwide, and membership spans undergraduate and graduate students, engineering practitioners, and consultants in all industries, engineering managers, and engineers in education, research, and government. It organizes an annual conference in addition to various workshops, trainings, and seminars. IIE has two societies: Health Systems (SHS) and Engineering and Management Systems (SEMS). In addition, IIE has eight divisions (Applied ergonomics, Computer and Information Systems, Construction, Engineering economy, Lean, Operations research, Process industries, and Quality control and reliability engineering) and numerous divisions and interest groups.

Institute of Operations Management http://www.iomnet.org.uk/ (accessed November 7, 2010). "The Institute of Operations Management is the professional body for persons involved in operations and production management in manufacturing and service industries in the United Kingdom" (from Web page).

Institution of Engineering and Technology U.K.–Manufacturing Sector, The http://kn.theiet.org/manufacturing/ (accessed November 7, 2010). Formerly known as the Institution of Electrical Engineers, U.K.–Manufacturing Sector. For more details about this society, see Chapter 10 in this book.

Institution of Mechanical Engineers, U.K. http://www.imeche.org/home (accessed November 7, 2010). For more details about this society, see Chapter 16 in this book.

International Ergonomics Association (IEA) http://www.iea.cc/ (accessed November 7, 2010). "The International Ergonomics Association is the federation of ergonomics and human factors societies from around the world. The mission of IEA is to elaborate and advance ergonomics science and practice, and to improve the quality of life by expanding its scope of application and contribution to society" (from Web page).

International Society for Automation (ISA) http://www.isa.org (accessed November 7, 2010) "Founded in 1945, the International Society of Automation is a leading, global, nonprofit organization that is setting the standard for automation by helping over 30,000 worldwide members and other professionals solve difficult technical problems, while enhancing their leadership and personal career capabilities" (from Web page). ISA provides certification and training for automation professionals, develops standards, and also organizations conferences and workshops, in addition to publishing technical papers and books.

Operational Research (OR) Society http://www.orsoc.org.uk (accessed November 7, 2010). "Founded over 50 years ago in succession to the Operational Research Club, which was set up in 1948, the OR society is the world's oldest established learned society catering to the Operational Research (OR) profession, and one of the largest in the world, with 3,000 members in 53 countries" (from Web page). The society also provides information about operational research to interested members of the general public. Membership is very diverse. The society published articles in various journals in addition to organizing training events, conferences, and workshops.

Production and Operations Management Society (POMS) http://www.poms.org/ (accessed November 7, 2010). Founded in 1989, POMS is an international society representing and promoting the interests of professionals in the field of production and operations management through integration and dissemination of knowledge, education, conferences, and journals.

Society for Industrial and Applied Mathematics (SIAM) http://www.siam.org/ (accessed November 7, 2010). SIAM membership is broad and includes students and professionals in applied and computational mathematics, numerical analysis, statistics, and engineers working in industry and academia. SIAM strives to advance the application of mathematics and computational science to engineering, industry, science, and society through membership activities, publications (books and journals), and conferences.

Society of Manufacturing Engineers (SME) http://www.sme.org/ (accessed November 7, 2010). Formed in 1932, SME is a professional organization providing leadership and resources in manufacturing. It is an international society that brings together engineers, companies, educators, students, and others in their need to advance manufacturing in the world. The SME has various technical associations that focus on specific areas, such as automated manufacturing, materials, forming and fabrication, machining, manufacturing education, product and process design, rapid technologies, robotics, and manufacturing research.

Technology Transfer Society (TTS), The http://www.t2society.org/ (accessed November 7, 2010). Founded in 1975, TTS promotes knowledge and opportunities required to achieve technology transfer. Membership is diverse, and includes professionals and students actively involved in technology transfer. TTS organizes conferences and meetings, in addition to publishing a number of technology transfer journals.

CONFERENCES

Conferences and symposiums are usually organized or sponsored by professional engineering societies. The main purpose of a technical conference is to bring students and experts together to report on the most current results of their projects. Therefore, these kinds of meetings allow for the sharing of ideas and for networking. Presentations made at a conference are usually published as the proceedings of the conference in book, CD-ROMs or Web-based documents. Sometimes the proceedings are published as an issue of a journal. Conferences and symposiums may sometimes change their titles or meeting locations and also may have irregularly scheduled meetings. Thousands of conferences are organized every year; this section shows a representative list of recent conferences in the field of industrial and manufacturing engineering.

Annual Reliability and Maintainability Symposium: The International Symposium on Product Quality and Integrity, RAMS. Sponsored by the Reliability Society IEEE, American Institute of Aeronautics and Astronautics (AIAA), Institute of Environmental Sciences and Technology (IEST), Society of Automotive Engineers Inc. (SAE), Society of Reliability Engineers.

ASME International Mechanical Engineering Congress & Exposition. Sponsored by the ASME.

DAAAM International Symposium Intelligent Manufacturing & Automation: Theory, Practice & Education. Sponsored by the Association for International Scientific and Academic Cooperation.

Design for Manufacturability through Design-Process Integration. Sponsored by the SPIE, The International Society for Optical Engineering.

Design for Manufacturing and the Life Cycle Conference (DFMLC). Sponsored by the Technical Committee on Design for Manufacturing, ASME Design Engineering Division.

Environmentally Conscious Manufacturing VI: Disassembly, Modeling Environmental Problems, Reverse and Closed Loop Supply Chains, Sustainability and Industrial Ecology. Sponsored by the SPIE, The International Society for Optical Engineering.

European Annual Conference on Human Decision Making and Manual Control (EAM09). Sponsored by the Université de Reims Champagne–Ardenne, France; National Center for Scientific Research, France.

Human Factors and Ergonomics Society Annual Meeting. Sponsored by the Human Factors and Ergonomics Society, Inc.

IEEE International High Assurance Systems Engineering Symposium (HASE). Sponsored by IEEE Computer Society TCDP, Reliability Society.

IEEE/RSJ International Conference on Intelligent Robots and Systems (IROS 2010). Sponsored by the IEEE Robotics and Automation Society, IEEE Industrial Electronics Society, Robotics Society of Japan, Society of Instruments and Control Engineers, New Technology Foundation.

IFAC Symposium on Information Control Problems in Manufacturing. Sponsored by International Federation of Automatic Control (IFAC).

International Conference on Agile Manufacturing. Sponsored by the ISAM, International Society for Agile Manufacturing.

International Conference on Frontiers of Design and Manufacturing (ICFDM). Sponsored by the National Natural Science Foundation of China (NSFC); Shien-Ming Wu Foundation, USA; Chinese Mechanical Engineering Society (CMES), American Society of Mechanical Engineers (ASME), National Science Foundation, USA (NSF).

International Conference on Mechanical, Industrial, and Manufacturing Technologies (MIMT). Sponsored by the International Association of Computer Science and Information Technology (IACSIT), American Society of Mechanical Engineers.

International Conference on TQM and Human Factors–Towards Successful Integration. Sponsored by the Centre for Studies of Humans, Technology and Organization, Linkoping, Sweden.

International Product Development Management Conference (IPDMC). Sponsored by the European Institute for Advanced Studies in Management (EIASM).

Institute of Ergonomics and Human Factors Annual Conference. Sponsored by The Institute of Ergonomics & Human Factors.

Intelligent Systems in Design and Manufacturing. Sponsored by the SPIE, The International Society for Optical Engineering.

Managing Engineering, Technology, and Innovation for Growth. Sponsored by the IEEE Engineering Management Society.

Manufacturing IT Forum: Connecting IT and Automation. Sponsored by the Instrumentation, Systems, and Automation Society (ISA).

Micro- and Nanotechnology: Materials, Processes, Packaging, and Systems. Sponsored by the SPIE, The International Society for Optical Engineering.

MicroManufacturing Conference & Exhibits. Sponsored by the Society of Manufacturing Engineers.

NanoManufacturing Conference & Exhibits. Sponsored by the Society of Manufacturing Engineers.

Optical Measurement Systems for Industrial Inspection. Sponsored by the SPIE Europe, The International Society for Optical Engineering.

Trends in Manufacturing, Second International CAMT Conference (Centre for Advanced Manufacturing Technologies). Sponsored by the European Commission Centre of Excellence (COFEXC).

United by a Common Vision: 41st Annual Conference of the Association of Canadian Ergonomists. Sponsored by the Association of Canadian Ergonomists (ACE).

WSEAS International Conference on Automatic Control, Modelling & Simulation (ACMOS). Sponsored by the World Scientific and Engineering Academy and Society.

Wireless Solutions for Manufacturing Automation: Insights for Technology and Business Success. Sponsored by the ISA Telemetry and Communications Division, ISA Computer Technology Division.

CONCLUSIONS

Industrial and manufacturing engineering disciplines have expanded dramatically over the past century. With continued global competition for consumer and other products, and the need for high quality at low costs, new technologies have continued to be developed. Traditional design and manufacturing techniques have been transformed into computer- assisted technologies, especially with the development of faster and cheaper microprocessors. With the evolution of the Internet, product design has been drastically transformed into an integrated process involving multidisciplinary approaches, done concurrently with process development, and mostly on a collaborative basis in real time. The integration of enterprise data has changed the face of industrial and manufacturing engineering through niche techniques such as virtual prototyping, design automation, supply chain management, automation technologies, and new management techniques. Research in these disciplines continues to focus on newer and leaner manufacturing processes, environmentally friendly designs and processes, zero defects in production, higher productivities, new materials, especially in micro- and nanotechnology, and alternative processes that use less or nonconventional energies. As the SME suggests, "... innovation, productivity, flexibility, and continuous improvement are key ingredients to success in the constantly evolving world of manufacturing" (SME, 2004). Manufacturing is a key economic activity for every nation as it is one of the leading sources of wealth. Therefore, it is important that industrial and manufacturing engineers be accorded key literature in their discipline in order to keep the practice competitive and innovative.

REFERENCES

American National Standards Institute (ANSI). 1989. *Standard Z94.0: Industrial engineering terminology.* Norcross, GA: Institute of Industrial Engineers.

Benhabib, B. 2003. *Manufacturing—Design, production, automation and integration.* New York: Marcel Dekker.

Chang, T. C., R. A. Wysk, and H. P. Wand. 1998. *Computer-aided manufacturing,* 2nd ed. Upper Saddle River, NJ: Prentice Hall.

Groover, M. P. 2002. *Fundamentals of modern manufacturing—Materials, processes and systems.* New York: John Wiley & Sons.

Hradesky, J. 1995. *Total quality management handbook.* New York: McGraw-Hill.

Institute of Ergonomics and Human Factors of U.K. 2004. http://www.ergonomics.org.uk/ (accessed November 7, 2010).

Kalpakjian, S., and S. R. Schmid. 2010, 6th ed. *Manufacturing engineering and technology.* Upper Saddle River, NJ: Prentice Hall.

Lincoln Electric Company, The. 1994. *The procedure handbook of arc welding,* 13th ed. Cleveland, OH: Lincoln Electric Company.

Marmaras, N. and T. Kontoginis. 2001. Cognitive tasks. In *Handbook of industrial engineering—Technology and operations management,* ed. G. Salvendy, pp. 1013–1040. New York: John Wiley & Sons.

Martin-Vega, L. A. 2001. The purpose and evolution of industrial engineering. In Maynard's *Industrial engineering handbook,* ed. K. B. Zandin, pp. 1.3–1.19. New York: McGraw Hill.

Sink, D. S., D. F. Poirier, and G. L. Smith. 2001. Full potential utilization of industrial and systems engineering in organizations. In *Handbook of industrial engineering—Technology and operations management,* ed. G. Salvendy, p. 3–25. New York: John Wiley & Sons.

SME. 2010. About SME http://www.sme.org/cgi-bin/getsmepg.pl?/html/about.htm&&&SME& (accessed November 7, 2010).

Soska, G. V. 1984. The five disciplines of manufacturing engineering education. Paper presented at the *Proceedings of the Second Annual International Robot Conference*. Wheaton, IL: Tower Conference Management, pp. 223–226.

Walker, J. M. 1996. *The handbook of manufacturing engineering*, New York: Marcel Dekker.

15 Materials Science and Engineering

Leena N. Lalwani and Sara Samuel (based on the first edition by Godlind Johnson)

CONTENTS

Introduction 371
The Literature of Materials Science and Engineering 372
Searching the Library Catalog 372
Indexes and Abstracts 373
    Other Indexes 374
Bibliographies and Guides to the Literature 374
Encyclopedias and Dictionaries 375
Handbooks, Manuals, and Material Property Databases 377
Monographs and Textbooks 384
Journals 384
Patents and Standards 387
Web Sites 387
Associations, Organizations, and Societies 388
Reference 393

INTRODUCTION

Human beings have engaged in some form of "materials engineering" since the beginning of life on Earth. We have even named eras after the materials that were predominantly used in those historic times (e.g., Stone Age, Bronze Age). However, materials science and engineering evolved into a distinct engineering discipline only in the 1960s. The Materials Research Society, founded in 1973, defines the field as: "Materials science and engineering encompasses the study of the structure and properties of any material, as well as using this body of knowledge to create new materials and to tailoring the properties of a material for specific uses" (from MRS Web site).

The backgrounds of today's materials science researchers and practitioners range from physics and chemistry to Earth sciences, biology, medicine, and also art and architecture. Elaborating on the definition above, one can outline four distinct but interrelated activities that materials scientists engage in: (1) determining the structure of materials (e.g., by microscopy, powder diffraction), (2) measuring and describing the properties of materials (both mechanical and physical), (3) devising ways of processing materials (i.e., creating new materials, transforming existing materials, and making things out of them), and (4) matching materials with applications or improving performance of materials that are already being used for an application. While any type of material may be the subject of research, the field grew out of metallurgy and ceramics and now includes metals, polymers, semiconductors, electromaterials, composites, biomaterials, and nanomaterials. Their structural, mechanical, electrical, thermal, magnetic, or optical properties, the so-called "engineering properties," are characterized (measured and described). The many methods of characterization that are employed include powder diffraction, electron microscopy, corrosion studies, and more.

THE LITERATURE OF MATERIALS SCIENCE AND ENGINEERING

Because materials science is such a broad and interdisciplinary field, it is not surprising that the information sources are also many and varied. Handbooks and property guides are especially important. In this guide, the most recent sources will be discussed; for slightly older materials, the reader should refer to guides such as *Guide to Information Sources in Engineering* by Charles R. Lord (2000). In addition, the emphasis lies on information sources specific to materials science and engineering, excluding those of the related sciences, physics and chemistry, and general engineering. Because information seeking has become largely an online activity, there is a focus on e-resources.

SEARCHING THE LIBRARY CATALOG

Considering how this discipline draws on so many different fields and is rapidly developing new areas of research and applications, the best approach to searching the library catalog is to start with keyword searches, combining specific concepts with the terms *materials* or *engineering* (e.g., materials and testing, analysis, microscopy). Once an appropriate title is found, the Subject Headings listed for it may lead to additional possible terms. The following selection of useful Subject Headings also reflects the breadth of subject matter that belongs to the field of materials science:

Acoustical Materials	Materials–Thermal Properties
Biomedical Materials	Microstructure
Building Materials	Nanocomposite Materials
Bulk Solids	Nanodiamonds
Characterization of Materials	Nanofibers
Chromic Materials	Nanophase
Coatings	Nanosilicon
Composite Materials	Nanostructured Materials
Electric Conductors	Nanotubes
Fibers	Nanowires
Foamed Materials	Nonmetallic Materials
Friction Materials	Optical Materials
Geosynthetics	Polymers
Granular Materials	Porous Materials
Hard Materials	Road Materials
Heat Resistant Materials	Sintering
Inhomogeneous Materials	Smart Materials
Laser Materials	Strategic Materials
Magnetic Materials	Super Lattices Materials
Manufacturing Processes	Surfaces (Technology)
Materials	Viscoelastic Materials
Materials at High Temperatures	Welding
Materials at Low Temperatures	

The Library of Congress classification system used in most academic libraries places the core of materials science literature in the TA400s, but many other Ts as well as QC; QD call numbers are also relevant:

QC176 Solid state physics
QC350–467 Optics and light
QD Chemistry of materials
R856–857 Biomedical engineering
T174.7 Nanotechnology
TA329–348 Engineering mathematics
TA401–492 Materials engineering and construction. Mechanics of materials
TN Mining Engineering (including Metallurgy)
TP155–156 Chemical engineering
TP785–869 Clay industry, ceramics, glass
TP1080–1185 Polymers and polymer manufacture

INDEXES AND ABSTRACTS

Indexes and abstracts are the tools to discover a body of research or information about a subject, discipline, or topic. They may cover a large interdisciplinary field or a very narrow and specialized field. These indexes give access to the output of research by indexing the content of a carefully selected group of sources. Aspects of materials science are covered in the large physics (*Inspec*), chemistry (*Chemical Abstracts/SciFinder*), and biology (*Biosis*) indexes, in the general engineering indexes (*Applied Science and Technology Index,* and *Compendex*), and in a number of very specialized tools that will be included here. Unless noted specifically, the indexes listed here exist in electronic form and require subscriptions to be accessed. The *CSA Materials Database with METADEX* includes the most important indexes and abstracts for this discipline. Technical reports also reside within some of these resources.

> *Aerospace and High Technology Database.* 1962–. Bethesda, MD: Cambridge Scientific Abstracts. "Provides bibliographic coverage of basic and applied research in aeronautics, astronautics, and space sciences. The database also covers technology development and applications in complementary and supporting fields, such as chemistry, geosciences, physics, communications, and electronics. In addition to periodic literature, the database also includes coverage of reports issued by NASA, other U.S. government agencies, international institutions, universities, and private firms. This database provides comprehensive international coverage as well as numerous nonserial publications." Included in the *CSA Materials Databases with METADEX* (from Web site).
>
> *Aluminum Industry Abstracts.* 1972–. Bethesda, MD: Cambridge Scientific Abstracts. "Provides comprehensive coverage of the world's technical literature on aluminum, production processes, products, applications, and business developments." Included in the *CSA Materials Database with METADEX* (from Web site).
>
> *Ceramic Abstracts.* 1975–. Staffs, U.K.: CERAM Research Ltd; Bethesda, MD: Cambridge Scientific Abstracts. "This is a comprehensive database for the ceramics industry, providing international coverage on the manufacture, processing, applications, properties, and testing of traditional and advanced ceramics." Includes World Ceramic Abstracts and is included in *CSA Materials Database with METADEX* (from Web site).
>
> *Corrosion Abstracts.* 1980–. Bethesda, MD: Cambridge Scientific Abstracts. "Provides the world's most complete source of bibliographic information in the area of corrosion science and engineering. International sources of literature are scanned and abstracted in the areas of general corrosion, testing, corrosion characteristics, preventive measures, materials construction and performance, and equipment for many industries." Included in *CSA Materials Database with METADEX* (from Web site).
>
> *CSA Materials Research Database with METADEX.* 1966–. Bethesda, MD: Cambridge Scientific Abstracts. Contains "specialist content on materials science, metallurgy,

ceramics, polymers, and composites used in engineering application. Everything from raw materials and refining through processing, welding, and fabrication to end uses, corrosion, performance, and recycling is covered in depth for all metals, alloys, polymers, ceramics, and composites." Includes subfiles: *Aluminum Industry Abstracts, Ceramic Abstracts/ World Ceramics Abstracts, Copper Technical Reference Library, Corrosion Abstracts, Engineered Materials Abstracts, Materials Business File, METADEX* (from Web site).

Engineered Materials Abstracts. 1986–. Bethesda, MD: Cambridge Scientific Abstracts. "The growing importance of polymers, ceramics, and composites in a variety of structural and other advanced applications requires the in-depth coverage provided by *Engineered Materials Abstracts.* Citations taken regard the research, manufacturing practices, properties, and applications of these materials. Begun in 1986, *Engineered Materials Abstracts* is an electronic database containing Ceramics, Composites, and Polymers subfiles. EMA is specifically designed to serve materials sciences researchers, engineers, and scientists." Included in *CSA Materials Database with METADEX* (from Web site).

Materials Business File. 1985–. Bethesda, MD: Cambridge Scientific Abstracts. "Materials Business File covers industry news, product and process developments, plant development and construction, international trade data, government regulations, and management issues related to industries covered by other CSA Technology Research Database (TRD) subfiles. Begun in 1985 as an alerting service, the database now serves also as a historical business/ industry reference." Included in *CSA Materials Database with METADEX* (from Web site).

METADEX. 1968–. Bethesda, MD: Cambridge Scientific Abstracts. "METADEX is the only comprehensive source for information on metals and alloys: their properties, manufacturing, applications, and development. Begun in 1966, METADEX contains over 1,425,000 references and is the database equivalent of *Metals Abstracts, Metals Abstracts Index*, and *Alloys Index.*" Included in the *CSA Materials Database with METADEX* (from Web site).

World Ceramics Abstracts (see *Ceramic Abstracts* above).

OTHER INDEXES

All of the following indexes are useful in materials science and are described fully in other chapters.

> *CAplus* (*Chemical Abstracts*)
> *Compendex*
> *Inspec*
> *National Technical Information Service (NTIS)*
> *Science Citation Index* and *Web of Knowledge*
> *Scopus*

BIBLIOGRAPHIES AND GUIDES TO THE LITERATURE

While chapters devoted to materials science literature may be found in the more comprehensive guides treated in Chapter 2 in this book, very few specific guides or bibliographies have been published recently. The most useful and up-to-date guides may be found on the Web sites of many of the major engineering libraries, for which some examples are listed below:

MIT Libraries Materials Science and Engineering Research Guide http://libguides.mit.edu/ materials (accessed September 29, 2010).

Stanford University Library Materials Science and Engineering Guide http://lib.stanford.edu/ engineering-library/materials-science-engineering (accessed September 29, 2010).

University of California San Diego Science and Engineering Library Materials Science Guide http://libraries.ucsd.edu/sage/subject?subject=10 (accessed September 29, 2010).

University of Michigan Library Materials Science and Engineering Research Guide http://guides.lib.umich.edu/materials (accessed September 29, 2010).

University of Wisconsin Libraries Materials Science and Engineering Research Guide http://researchguides.library.wisc.edu/mse (accessed September 29, 2010).

ENCYCLOPEDIAS AND DICTIONARIES

There are very few new encyclopedias and dictionaries compared to what appeared in the previous edition of this book, possibly due to an increase of information moving online. The following resources should be useful for students looking for an in-depth overview and for researchers needing information about topics outside their own areas of expertise. Dictionaries were chosen for the quality of their definitions and currency.

Bever, M. B., ed. 1986. *Encyclopedia of Materials Science and Engineering.* (8 vols. and 2 suppl.) Oxford, U.K.: Pergamon; Cambridge, MA: MIT Press. This is still the basic, standard reference work of the field, updated periodically by supplementary volumes and by the *Encyclopedia of Advanced Materials.* It also spawned many concise encyclopedias in the early 1990s (e.g., *Concise Encyclopedia of Composites*).

Bloor, D. and R. W. Khan. 1994. *Encyclopedia of Advanced Materials.* (4 vols.) Oxford, U.K./Tarrytown, NY: Pergamon Press. Advanced materials are artificially produced to meet the requirements of particular applications. The rapid developments in this field are covered in more depth than was possible in the standard *Encyclopedia of Materials Science and Engineering* (1986) and its supplements. Examples of topics covered include: aerogel catalysts, biomedical polymers, color in ceramics, fast ion conduction in ceramics, hydrogels, metal matrix composites, and oligomers. There is an excellent subject index and a wealth of illustrations.

Buschow, K. H. J., ed. 2001. *Encyclopedia of Materials: Science and Technology.* (10 vols.) Amsterdam/New York: Elsevier; available online through publisher (updated) and *Knovel* (not updated). Over 2,000 articles written by internationally recognized experts are included in this comprehensive encyclopedia. It includes extensive tables, diagrams, and bibliographies.

Cahn, R. W. and E. Lifshin, eds. 2004. *Concise Encyclopedia of Materials Characterization,* 2nd ed. Amsterdam/Oxford, U.K.: Elsevier. A collection of 116 concise and authoritative articles on most techniques available to characterize materials, ranging from the well-established to the very latest, including: atomic force microscopy, confocal optical microscopy, gamma ray diffractometry, thermal wave imaging, x-ray diffraction, and time-resolved techniques. It is well illustrated and includes useful references.

Cheremisinoff, N. P. 2001. *Condensed Encyclopedia of Polymer Engineering Terms.* Boston: Butterworth-Heinemann. Provides an overview of commercially available polymers.

Davis, J. R., ed. 1992. *ASM Materials Engineering Dictionary.* Metals Park, OH: ASM International. This dictionary covers all material types with detailed definitions, processes, charts, and tables.

Gooch, J., ed. 2011. *Encyclopedic Dictoinary of Polymers,* 2nd ed. (2 vols.) New York: Springer. This resource includes names of chemicals, processes, formulae, and analytical methods that are frequently used in the polymer and engineering fields.

Guston, D. H. 2010. *Encyclopedia of Nanoscience and Society.* (2 vols.) Thousand Oaks, CA: Sage Publications. A reference source that delves into the ethics, economy, funding, and social aspects of nanotechnology.

Keller, H. and U. Erb. 2004. *Dictionary of Engineering Materials.* Hoboken, NJ: John Wiley & Sons. In over 40,000 entries, the authors provide succinct information, including type of material, composition, properties, and use of traditional and new materials, such as dendrimers, fullerenes, and photonics.

Kelly, A. and C. Zweben, eds. 2000. *Comprehensive Composite Materials.* (6 vols.) Amsterdam/New York: Elsevier; available online. Topics covered in the six volumes are fiber reinforcements and general theory of composites, polymer matrix composites, metal matrix composites, carbon/carbon, cement, and ceramic matrix composites, test methods, nondestructive evaluation, smart materials, and design and applications.

Kelly, A., W. Cahn, and M. B. Bever. 1994. *Concise Encyclopedia of Composites,* rev. ed. Oxford, U.K./Tarrytown, NY: Pergamon Press. Culled and updated from the *Encyclopedia of Materials Science and Engineering.*

Li, D., ed. 2008. *Encyclopedia of Microfluidics and Nanofluidics.* (3 vols.) Boston: Springer-Verlag; available online. Over 750 entries covering all aspects on transport phenomena on the nano- and microscale.

Mark, H. F., ed. 2004. *Encyclopedia of Polymer Science and Engineering,* 3rd ed. (12 vols.) New York/Chichester, U.K.: John Wiley & Sons; available online. This third edition is a completely new version. The new edition brings the state-of-the-art up to the twenty-first century, with coverage of nanotechnology, new imaging and analytical techniques, new methods of controlled polymer architecture, biomimetics, and more.

Martin, J., ed. 2009a. *Concise Encyclopedia of Materials for Energy Systems,* Amsterdam/ New York: Elsevier. An up-to-date reference that is inclusive of a variety of energy systems including turbines, renewable power, fuel cells, nuclear power generation, and the transmission and storage of the energy produced. Includes a wide range of materials.

Martin, J., ed. 2009b. *The Concise Encyclopedia of Materials Processing.* Amsterdam/New York: Elsevier. A useful reference source for scholars and professionals with information gathered from *Encyclopedia of Materials: Science and Technology.*

Nalwa, H. S. 2004. *Encyclopedia of Nanoscience and Nanotechnology.* (10 vols.) Stevenson Ranch, CA: American Scientific Publishers; available online. This extensive work brings together the current knowledge about all aspects of nanotechnology. In over 410 chapters, it provides an introduction and overview of most recent advances and emerging new aspects of nanotechnology spanning science to engineering to medicine. Extensive bibliographic references and illustrations.

Novikov, V. 2003. *Concise Dictionary of Materials Science: Structure and Characterization of Polycrystalline Materials.* Boca Raton, FL: CRC Press; available online. Entries for about 1,400 terms cover the description, development, and characterization of polycrystalline materials. The author states in the preface that this dictionary is intended to bridge the gap between textbooks and professional literature. The book also contains a three-page list of symbols, two pages of acronyms, and a two-page list of literature for further reading. An English–German/German–English glossary is included as well.

Rosato, D. V. 2000. *Concise Encyclopedia of Plastics.* Boston: Kluwer Academic. The world of plastics is explained in over 25,000 terms. In addition, lists of abbreviations, conversions, industry associations, Web sites, and more make this dictionary comprehensive and useful.

Salamone, J. C., ed. 1996. *Polymeric Materials Encyclopedia.* (12 vols.) Boca Raton, FL: CRC Press. A comprehensive reference work on the synthesis, properties, and applications of polymeric materials.

Salamone, J. C., ed. 1999. *Concise Polymeric Materials Encyclopedia.* Boca Raton, FL: CRC Press. The most widely applicable articles were culled from the 12-volume *Polymeric Materials Encyclopedia* (see above).

Schramm, L. L. 2008. *Dictionary of Nanotechnology, Colloid and Interface Science.* Weinheim, Germany: Wiley-VCH. A successor to *Dictionary of Colloid and Interface Science.* "Manageably sized dictionary covers theory, experiment, industrial practice and applications for nanotechnology, colloid, and interface science, as well as much of what is now termed materials science" (from publisher's Web site).

Schwartz, M. M., ed. 2002a. *Encyclopedia of Materials, Parts, and Finishes,* 2nd ed. Boca Raton, FL: CRC Press; available online. This second edition of the encyclopedia covers the new materials that have been invented or modified in recent years (including matrix composites, nanostructures, smart piezoelectric materials, shape memory alloys, and intermetallics), and updates information on basic materials as well. Many tables and figures, but no references.

Schwartz, M. M., ed. 2002b. *Encyclopedia of Smart Materials.* (2 vols.) New York: John Wiley & Sons; available online through publisher and *Knovel.* This indexed two-volume set brings together much of the information known about smart materials; smart materials may be defined as materials that are designed to respond to changes in their environment; applications may be found in almost every area of modern engineering, from biomedical to optical and structural. Each article features materials properties, photographs, charts and tables, and a bibliography for further reading.

Schwarz, J. A., C. I. Contescu, and K. Putyera, eds. 2004. *Dekker Encyclopedia of Nanoscience and Nanotechnology.* (5 vols.) New York: Marcel Dekker; available online through *CRCnetBASE.* Another comprehensive encyclopedia recording the state of the art of all aspects of the emerging field of nanoscience.

Schweitzer, P. A., ed. 2004. *Encyclopedia of Corrosion Technology,* 2nd rev. and enlarged ed. New York: Marcel Dekker; available online through *CRCnetBASE.* Provides information on all materials affected by corrosion. References for further reading are included with many entries and the encyclopedia is amply illustrated.

Ullmann's Encyclopedia of Industrial Chemistry, 6th ed. 2003. (40 vols.) New York: Wiley-VCH; available online. State-of-the-art reference work detailing science and technology in all areas of industrial chemistry. Fully international in scope and coverage. With nearly 10,000 tables, 30,000 figures, and innumerable literature sources and cross references, *Ullmann's Encyclopedia* offers a wealth of comprehensive and well-structured information on all facets of industrial chemistry.

HANDBOOKS, MANUALS, AND MATERIAL PROPERTY DATABASES

Handbooks may be the most valuable resource for the practicing engineer as well as students and researchers because they present collections of fundamental facts, formulas, tables, and, for materials science, the properties of materials. An attempt was made to find at least one recent and comprehensive handbook for each subdiscipline of materials science, either by type of material or process or application. We have listed one or two vendors where online access is available, but many of these books may be available through other vendors, such as ebrary, NetLibrary, and MyiLibrary. Most of the volumes listed here are available through either *Knovel* or *CRCnetBASE.*

ACerS-NIST Phase Equilibria Diagram CD-ROM Database Version 3.3. 2011. Westerville, OH: American Ceramics Society. Online version is also available for subscription. The *Phase Equilibria Diagrams* CDROM Database Version 3.3 contains more than 23,000 diagrams published in 21 phase volumes produced as part of the ACerS-NIST Phase Equilibria Diagrams Program: volumes I through XIII; Annuals 91, 92, and 93; High Tc Superconductor I & II; Zirconium and Zirconia Systems; and Electronic Ceramics I. This new version offers full commentary text display in addition to diagram display. This makes the CD the same as the printed volumes with many additional features, including high-quality printing and export capability. http://ceramics.org/publications-and-resources/phase-equilibria-diagrams/ (accessed December 15, 2010).

Amelinckx, S., ed. 1996. *Handbook of Microscopy: Applications in Materials Science, Solid-state Physics and Chemistry.* (3 vols.) New York: Wiley-VCH. A comprehensive treatment of microscopy methods used in the characterization of materials (vols. 1 and 2) and how these techniques can be applied successfully for a given material (vol. 3). Available on *Wiley Online Library.*

Andrews, D., G. Scholes, and G. Wiederrecht, eds. 2011. *Comprehensive Nanoscience and Technology.* (5 vols.) San Diego, CA: Academic Press. An extensive collection of information related to nanomaterials and nanotechnology.

ASM International. 2002–. *ASM Handbooks Online.* Materials Park, OH: ASM International http://products.asminternational.org/hbk/index.jsp (accessed December 15, 2010). Twenty-five volumes of the *ASM Handbook* series plus the desk editions of *Engineered Materials Handbook* and *Metals Handbook* are presented together in an online format: vol. 1, Properties and Selection: Irons, Steels, and High Performance Alloys; vol. 2, Properties and Selection: Nonferrous Alloys and Special-Purpose Materials; vol. 3, Alloy Phase Diagrams; vol. 4, Heat Treating; vol. 5, Surface Engineering; vol. 6, Welding, Brazing, and Soldering; vol. 7, Powder Metal Technologies and Applications; vol. 8, Mechanical Testing and Evaluation; vol. 9, Metallography and Microstructures; vol. 10, Materials Characterization; vol. 11, Failure Analysis and Prevention; vol. 12, Fractography; vol. 13 (A, B, and C) Corrosion; vol. 14 (A and B), Forming and Forging; vol. 15, Casting; vol. 16, Machining; vol. 17, Nondestructive Evaluation and Quality Control; vol. 18, Friction, Lubrication, and Wear Technology; vol. 19, Fatigue and Fracture; vol. 20, Materials Selection and Design; vol. 21, Composites; vol. 22 (A and B) Metals Processing; ASM Desk Editions. Available online on *Knovel.*

ASM Ready Reference Collection. 1997–. Materials Park, OH: ASM International. This collection of comprehensive handbooks includes: properties and units for engineering alloys, electrical and magnetic properties of metals, thermal properties of metals.

ASM Specialty Handbooks. 1993–. Materials Park, OH: ASM International. A nine-volume set that complements and expands on the corresponding *ASM Handbooks.* The set includes Aluminum and Aluminum Alloys, Carbon and Alloy Steels, Cast Irons, Copper and Copper Alloys, Heat Resistant Materials, Magnesium and Magnesium Alloys, Nickel, Cobalt and Their Alloys, Stainless Steel and Tool Materials. Some volumes available online on *Knovel.*

Asmussin, J., and D. K. Reinhard. 2002. *Diamond Films Handbook.* New York: Marcel Dekker. Detailed information on the science, processes, and application of low-pressure diamond deposition, which has been developed over the past two decades to synthesize a form of diamond that is metastable with respect to graphite, as distinguished from the high-pressure synthesis of diamond that is the stable form of carbon. Available online on *CRCnetBASE.*

Bhushan, B. 1998. *Handbook of Micro/Nanotribology,* 2nd ed., Boca Raton, FL: CRC Press. The state-of-the-art of this rapidly evolving field is presented in 16 chapters. Relevant topics range from AFM instrumentation, characterization of solid surfaces, measurement techniques and applications to theoretical modeling of interfaces, and design and construction of magnetic storage devices. Five hundred illustrations and 50 tables are included. Available online on *CRCnetBASE.*

Bhushan, B., ed. 2010 *Springer Handbook of Nanotechnology,* 3rd ed. New York: Springer. Contains information regarding nanofabrication, nanodevices, nanomechanics, and more. The up-to-date third edition includes an additional section that includes information on biomimetics.

Black, J. and G. W. Hastings. 1998. *Handbook of Biomaterial Properties.* London/New York: Chapman and Hall. This is a compilation of data on natural tissues and fluids and how various implantable materials interact with them. The materials covered range from stainless steels, CoCr-based alloys, titanium, and titanium alloys, to thermoplastic polymers and oxide bioceramics. Biocompatibility is discussed with each material and in a chapter on general concepts of biocompatibility. Available online on *Knovel.*

Brady, G. S., ed. 2002. *Materials Handbook: An Encyclopedia for Managers, Technical Professionals, Purchasing and Production Managers, Technicians, and Supervisors,* 15th ed. New York: McGraw-Hill. This classic handbook covers essential information for over 15,000 materials. In the latest edition dozens of material families have been updated extensively, including adhesives, activated carbon, fullerenes, heat-transfer fluids, nanophase materials, nickel alloys, olefins, silicon nitride, stainless steels, and thermoplastic elastomers. Property information is embedded in the encyclopedic entry. Available online through various services.

Brandrup, J. and E. H. Immergut, eds. 2004, 1999. *Polymer Handbook,* 4th ed. (2 vols.) New York: Wiley-Interscience. This is the standard reference for polymer data necessary for theoretical and experimental polymer research. The publisher's online version, *Wiley Database of Polymer Properties,* features added functionality and is updated quarterly. Available online on *Knovel.*

Bunshah, R. F., ed. 2001. *Handbook of Hard Coatings: Deposition Technologies, Properties and Applications.* Park Ridge, NJ: Noyes Publications/Norwich, NY: William Andrews Publications. A comprehensive resource for information on the fabrication, characterization, and applications in the field of hard coatings and wear-resistant surfaces. Available online on *Knovel* and other sources.

Cardarelli, F. 2008. *Materials Handbook: A Concise Desktop Reference,* 2nd ed. London/ New York: Springer-Verlag. This book is divided into 14 families of materials (e.g., metals and alloys, semiconductors, superconductors, magnetic and electrical materials, ceramics). Physico-chemical properties are supplied for each class of materials in tabular form. Emphasis is given to the most common industrial materials in each class. A bibliography is included. Available online on *SpringerLink.*

Chanda, M. and S. K. Roy. 2007. *Plastics Technology Handbook,* 4th ed. *Plastic Engineering,* vol. 72, Boca Raton: CRC Press. Covers processes and property data of industrial polymers. Appendices cover trade names and abbreviations for industrial polymers; formulations; commercial polymer blends and alloys; properties of polymers, rubber compounds, and textile fibers. Available online on *CRCnetBASE.*

CINDAS Materials Properties Data Files, Purdue University, West Lafayette, IN. Contains 3 databases: *Aerospace Structural Metals Database (ASMD), Thermophysical Properties of Matter Database (TPMD),* and *Microelectronics Packaging Materials Database (MPMD).* Contains 6,000 materials and 145,000 data curves. Available online at CINDAS http:// www.cindasdata.com (accessed December 15, 2010).

Composite Materials Handbook—MIL 17. 2002. (5 vols.) Conshohocken, PA: ASTM International. Authoritative source for data on polymer matrix, metal matrix, and ceramic matrix composite materials, updated regularly. Three volumes deal with polymer matrix composites, one with metal matrix composites, and one with ceramic matrix composites. Available online on *Knovel.*

Corrosion Survey Database. Houston, TX: NACE International. Accessed through *Knovel.* "This database is a collection of the results from numerous literature references reporting the effects of exposing 87 metal and nonmetal materials to over 1,500 different exposure media at various temperatures and concentrations resulting in 28,000 pairs of exposed material and medium. The results are presented in searchable tables, and can be browsed by material or exposure medium" (from *Knovel* Web site).

Engineered Materials Handbook. 1987–. (4 vols.) Materials Park, OH: ASM International. The concise versions are included in the *ASM Handbooks Online.* Vol. 1, Composites; vol. 2, Engineering plastics; vol. 3, Adhesives and Sealants; vol. 4, Ceramics and Glasses. Available online on ASM and *Knovel.*

Francombe, M. H., ed. 2000. *Handbook of Thin Film Devices.* (5 vols.) San Diego, CA: Academic Press. Topics covered in this comprehensive handbook range "from basic device physics and design, through growth and device fabrication, performance characteristics, and applications and integration into subsystems." Each volume is devoted to a class of devices: vol. 1, Hetero-structures for High Performance Devices; vol. 2, Semiconductor Optical and Electro-optical Devices; vol. 3, Superconducting Film Devices; vol. 4, Magnetic Thin Film Devices; vol. 5, Ferroelectric Film Devices (from back cover of the book). Available online on *ScienceDirect*.

Frick, J., ed. 2000. *Woldman's Engineering Alloys,* 9th ed. Materials Park, OH: ASM International. "This standard reference book has again been revised with thousands of new records added or updated to include the latest alloys and the most current manufacturing status. The book also continues to list alloys that are no longer produced, but are frequently found in older references and specifications, thus helping you to identify substitutes" (from ASM Web site). Available online on *Knovel*.

Gale, W. F. and T. C. Totemeier, eds. 2004. *Smithells Metals Reference Book,* 8th ed. Engineering Materials Selector Series. Oxford, U.K./Burlington, MA: Elsevier Butterworth-Heinemann. This essential reference work for metal researchers and metallurgists contains essays and data on all aspects of metallurgy. In this new edition chapters were added on: Nonconventional and emerging materials (including) micro/ nano-scale materials; Techniques for the modeling and simulation of metallic materials; Supporting technologies for the processing of metals and alloys; and an extensive bibliography of selected sources of further metallurgical information, including books, journals, conference series, professional societies, metallurgical databases, and specialist search tools. Available online on *Knovel* and *ScienceDirect*.

Gupta, R. K., E. Kennel, and K. Kim. eds. 2009. *Polymer Nanocomposites Handbook*. Boca Raton: CRC Press. Expert authors discuss the development of nanocomposites and their uses and applications.

Habashi, F., ed. 1997. *Handbook of Extractive Metallurgy*. (4 vols.) Weinheim, Germany/New York: Wiley-VCH. A comprehensive compilation of data on metals, their extraction, their alloys, and their most important inorganic compounds. The four volumes are organized by families of metals according to their industrial importance: vol. 1, The Metal Industry Ferrous Metals; vol. 2, Primary Metals, Secondary Metals, Light Metals; vol. 3, Precious Metals, Refractory Metals, Scattered Metals, Rare Earth Metals; vol. 4, Ferroalloy Metals, Alkali Metals, Alkaline Earth Metals.

Harper, C. A., ed. 2001a. *Handbook of Ceramics, Glasses and Diamonds*. Boston/London: Irwin/McGraw-Hill; electronic version: McGraw-Hill, and netLibrary. Emphasis is on innovative uses of these materials; in addition to a wealth of materials properties, processes, and requirements data, selection and design guidelines are included.

Harper, C. A., ed. 2001b. *Handbook of Materials for Product Design,* 3rd ed. New York: McGraw-Hill; also as an electronic book. This is a compendium of materials data needed for selection or creation of new materials. Information is arranged for easy comparison of options in tables, charts, graphs, and illustrations of data on design, testing, specifications, standards, recyclability, and biodegradability. Available online on *Knovel*.

Kaufman, J. G. 2004, 2009. *Aluminum Alloy Database*. Norwich, NY: Knovel; an electronic book on *Knovel*. Originally published in 2004 and updated in 2009. Contains a comprehensive summary of physical, mechanical, and chemical properties of both wrought and cast aluminum alloys. It has interactive searchable and sortable tables.

Kreysa, G. and M. Schutze, eds. 2008. *Dechema Corrosion Handbook,* 2nd ed. Frankfurt am Main: Dechema. Compendium of corrosion data describing the corrosion and chemical resistance of all technically important metallic, nonmetallic, inorganic, and organic materials in contact with aggressive media. Available online on *Knovel*.

Kutz, M., ed. 2002. *Handbook of Materials Selection*. New York: John Wiley & Sons. This is a comprehensive resource for property data as well as selection and evaluation techniques for all groups of materials in use today: metals, ceramics, composites, and plastics. The seven chapters deal with Quantitative methods, Major materials, Finding and managing materials information and data, Testing and inspection, Failure analysis, Manufacturing, Applications and uses. It is heavily illustrated with photos, graphics, and tables, and an index is included. Available online on *Wiley Online Library* and *Knovel*.

Lemaitre, J. 2001. *Handbook of Materials Behavior Models*. (3 vols.) San Diego, CA: Academic Press. "Gathers together 117 models of behavior of materials. ... Presents each model's domain of validity, a short background, its formulation, a methodology to identify the materials parameters, advise on how to use it in practical applications as well as extensive references. Covers all solid materials: metals, alloys, ceramics, polymers, composites, concrete, wood, rubber, geomaterials, such as rocks, soils, sand, clay, biomaterials, etc. Concerns all engineering phenomena: elasticity, viscoelasticity, yield limit, plasticity, viscoplasticity, damage, fracture, friction, and wear" (from publisher's Web site). Available online on *ScienceDirect*.

Lobo, H. and J. V. Bonilla. 2003. *Handbook of Plastics Analysis* (Plastics Engineering), New York: Marcel Dekker; also a netLibrary book. Describes the latest analytical developments, technologies, and equipment used to characterize and examine plastic and composite materials.

Mark, J. E. 2009. *Polymer Data Handbook,* 2nd ed. New York: Oxford University Press; electronic version available. Concise information on the syntheses, structures, properties, and applications of the 218 most important polymeric materials is presented in a tabular format. Available online on *Knovel*.

Narlikar, A. V. and Y. Y. Fu, eds. 2010. *Oxford Handbook of Nanoscience and Technology.* (3 vols.) New York: Oxford University Press. This three-volume set covers the field of nanoscience and technology from various angles including physics, chemistry, materials science, engineering, and more.

Nalwa, H. S., ed. 2003. *Handbook of Organic-Inorganic Hybrid Materials and Nanocomposites.* (2 vols.) Stevenson Ranch, CA: American Scientific Publishers. Brings together up-to-date knowledge on the underlying science, and processing and fabrication, of hybrid materials (covered in vol. 1) and nanocomposites (covered in vol. 2). Some of the topics covered include: different synthetic routes allowing multifunctionality with a wide range of composition, sol-gel chemistry, fibers, xerogels, spectroscopic characterization, mechanical, thermal, electronic, optical, catalytic, and biological properties, polymer–metal interfaces, and their potential commercial applications. Each chapter ends with extensive references.

Nalwa, H. S. 2002a. *Handbook of Nanostructured Materials and Nanotechnology.* (5 vols.) San Diego, CA: Academic Press. Provides comprehensive coverage of the dominant technology of this century. The topics of the five volumes include: synthesis and processing; spectroscopy and theory; electrical properties; optical properties; and organics, polymers, and biological materials. A one-volume concise edition was published in 2002, focusing on synthesis and fabrication as well as the electrical and optical properties of nanoscale materials.

Nalwa, H. S., ed. 2002b. *Handbook of Thin Film Materials.* (5 vols.) San Diego, CA: Academic Press. This is a comprehensive reference work on thin film science and technology, covering semiconductors, superconductors, ferroelectrics, nanostructured materials, and magnetic materials. Sixty-five articles contain state-of-the-art knowledge and data on a vast number of thin film materials, their deposition, processing, and fabrication techniques, spectroscopic characterization, optical characterization probes, physical properties, and structure–property relationships. Ample illustrations and extensive references are included: vol. 1, Deposition and Processing of Thin Films; vol. 2, Characterization and Spectroscopy of Thin Films; vol. 3, Ferroelectric and Dielectric Thin Films; vol. 4, Semiconductor and Superconductor Thin Films; vol. 5, Nanomaterials and Magnetic Thin Films. Index in each volume. *Handbook of Thin Film Devices* is a companion volume.

Plastics Design Library Handbook Series. 1990–. Norwich, NY: William Andrews, Inc. in partnership with Knovel Corporation. So far over 40 volumes have been published in this comprehensive handbook series providing data on all kinds of plastics, their evaluation, characterization, processing, and manufacturing. All volumes exist in hard copy as well as online on *Knovel*.

Poole, C. P. 1999. *Handbook of Superconductivity.* San Diego, CA: Academic Press. This reference book draws together much of the information that has been published in the journal literature, including: thermal, electrical, magnetic, and mechanical properties, phase diagrams, spectroscopic crystallographic structures for many superconductor types, and coherence length and depth tabulations. Traditional superconductors are treated as well as high temperature superconductors. Available online on *ScienceDirect*.

Pruett, K. M. 2000. *Chemical Resistance Guide for Plastics: A Guide to Chemical Resistance of Engineering Thermoplastics, Fluoroplastics, Fibers, and Thermoset Resins.* La Jolla, CA: Compass.

Ramachandran, V. S. and J. J. Beaudoin. 2001. *Handbook of Analytical Techniques in Concrete Science and Technology.* Building Materials Science Series. Norwich, NY: Noyes Publications. A handbook of analytical techniques for new types of concrete and related materials. The techniques range from chemical and thermal analysis to IR and nuclear magnetic resonance spectroscopy to scanning electron microscopy, x-ray diffraction, computer modeling, and more. Available online on *Knovel*.

Revie, R. W. and H. H. Uhlig, eds. 2011. *Uhlig's Corrosion Handbook,* 3rd ed. New York: John Wiley & Sons. After 50 years, this classic corrosion reference book has been completely updated with the latest information, including the development of many nonmetallic materials, their corrosion behavior, and engineering approaches to their corrosion control. Available online on *Wiley* and *Knovel*.

Roberge, P. R. 2000. *Handbook of Corrosion Engineering.* New York: McGraw-Hill. Similar content as *Uhlig's,* but deals with metals only. Appendices include corrosion economics, chemical compositions of engineering alloys, and thermodynamics data and e-pH diagrams. Available online on *Knovel*.

Rosato, D. V., ed. 2001. *Plastics Engineering, Manufacturing and Data Handbook.* (2 vols.) Boston: Kluwer Academic, for the Plastics Institute of America. This comprehensive reference work is presented in two volumes: vol. 1, Fundamentals and Processes; vol. 2, Design, Testing, Marketing, and Regulations. The interrelationships of materials-to-processes are explained for technical and nontechnical readers alike. Information is included on static properties (tensile, flexural), dynamic properties (creep, fatigue, impact), and physical and chemical properties. The text is illustrated by 1,060 figures and 415 tables. Available online on *Knovel*.

Sattler, K. D., ed. 2010. *Handbook of Nanophysics.* Boca Raton, FL: Taylor & Francis. A seven-volume set that has coverage of the physics of nanoscale materials and applications. It "presents an up-to-date overview of the application of nanotechnology to molecular and biological processes, medical imaging, targeted drug delivery, and cancer treatment. Each peer-reviewed chapter contains a broad-based introduction and enhances understanding of the state-of-the-art scientific content through fundamental equations and illustrations, some in color" (from publisher's Web site).

Saunders, N. 1998. *CALPHAD (Calculation of Phase Diagrams): A Comprehensive Guide.* Oxford, U.K./New York: Pergamon. This is an in-depth treatment of the method of computer coupling of phase diagrams and thermochemistry, which makes it possible to calculate the phase behavior of multicomponent materials. See also the journal *CALPHAD* that publishes new developments quarterly. Available online through various services.

Seshan, K. 2002. *Handbook of Thin-Film Deposition Processes and Techniques: Principles, Methods, Equipment and Applications,* 2nd ed. Norwich, NY: Noyes Publications. Brings together information on physical vapor deposition techniques. Available online on *Knovel.*

Shackelford, J. F. and W. Alexander, eds. 2001. *CRC Materials Science and Engineering Handbook,* 3rd ed. Boca Raton, FL: CRC Press. Possibly the most comprehensive materials science handbook containing property data for the major groups of materials (metals, alloys, polymers, ceramics and glass, composites), as well as a section of comparative tables for selecting materials for specific properties. Data have been verified by professional societies, such as ASM International and the American Ceramic Society. Available online on *CRCnetBASE.*

Shah, V. 2007. *Handbook of Plastics Testing and Failure Analysis,* 3rd ed. New York: John Wiley & Sons. Six chapters are devoted to mechanical, thermal, electrical, weathering, optical, and chemical properties, and respective testing. The remainder of the book deals with other plastic testing issues and techniques. Includes tables, illustrations, and extensive references. Available online on *Wiley Online Library.*

Sinha, A. K. 2003. *Physical Metallurgy Handbook.* New York: McGraw-Hill. Enlarged edition of *Ferrous Metallurgy Handbook* (1989). Provides engineers, researchers, and designers with a complete understanding of how various metals perform under a wide range of conditions. Includes properties, forms, and application of ferrous and nonferrous materials.

Smithells, C. J., E. A. Brandes, and G. B. Brook, eds. 1998. *Smithells Light Metals Handbook.* Oxford, U.K./Boston: Butterworth-Heinemann. Drawing on the data within *Smithells Metals Reference Book*, the editors have created a new book dedicated to aluminum, magnesium, and titanium, the most commonly used light metals. An extensive section on binary phase diagrams is included, as well as standards and international materials specifications.

Springer Materials: The Landolt Bornstein Database. Available online at http://www.springermaterials.com (accessed December 15, 2010). "*SpringerMaterials* is based on the *Landolt-Börnstein New Series*, the unique, fully evaluated data collection in all areas of physical sciences and engineering. *SpringerMaterials* also comprises *the Dortmund Data Bank Software and Separation Technology, A Database on Thermophysical Properties* and the *Linus Pauling Files*, a *Database on Inorganic Solid Phases* and chemical safety data" (from Web page). It contains 250,000 substances and material systems, 3,000 properties, and 1,200,000 literature citations.

ThermoDex: An Index of Selected Thermodynamic Data Handbooks: http://www.lib.utexas.edu/thermodex/ (accessed September 30, 2010). Austin, TX: Mallet Chemistry Library, University of Texas at Austin. This database contains records for selected printed and Web-based compilations of thermochemical and thermophysical data. Searching for properties linked to types of compounds will return lists of sources that may contain the data.

Utracki, L. A., ed. 2002. *Polymer Blends Handbook.* (2 vols.) Dordrecht, The Netherlands/Boston: Kluwer Academic. Covers the fundamental principles and technology of thermoplastic blends. Appendices include lists of miscible commercial polymer blends. Available online on *Knovel.*

Van Recum, A. 1998. *Handbook of Biomaterials Evaluation: Scientific, Technical, and Clinical Testing of Implant Materials,* 2nd ed. Philadelphia: Taylor & Francis. A valuable reference work for researchers and practitioners of working on testing and evaluating biocompatible materials. Available online on *CRCnetBASE.*

Villars, P. 1997. *Pearson's Handbook: Crystallographic Data for Intermetallic Phases,* 3rd ed. (2 vols.) Materials Park, OH: ASM International. This is the standard reference work for crystal data, representing over 27,000 compounds.

Waterman, N. A., and M. F. Ashby, eds. 1997. *The Materials Selector,* 2nd ed. London: Chapman and Hall. Information about properties, performance, and processability of metals, plastics, carbon and graphite, glasses, ceramics, polymerics, and composites.

Wessel, J. K. 2004. *Handbook of Advanced Materials.* Hoboken, NJ: John Wiley & Sons. This book brings together the most recent information about advanced materials and their properties. Materials range from polymer composites, continuous fiber ceramic composites to intermetallics and light metal alloys. Chapters on standards and codes, nondestructive evaluation, and rapid prototyping are included. Available online on *Wiley Online Library.*

Whitehouse, D. J. 2003. *Handbook of Surface and Nanometrology,* rev. ed. Bristol, U.K.: Institute of Physics. Explains the physics of surface metrology and nanometrology in nine chapters: General philosophy, Surface characterization, Processing, Instrumentation, Traceability, Surface metrology in manufacture, Surface geometry, Nanometrology, Conclusion and overview. Chapters are illustrated with practical examples and include references. Available online on *CRCNetBASE.*

Wiederrecht, G., ed. 2009. *Handbook of Nanofabrication.* Amsterdam: Elsevier BV. This handbook provides an overview of the methods and challenges of fabricating nanostructures.

MONOGRAPHS AND TEXTBOOKS

From the countless books that have been and are being published relating to materials science, a few classic textbooks were chosen to be listed here as well as the most important monographic and conference proceedings series. The Subject Headings or call numbers outlined in the introduction should be used to discover materials in library catalogs.

Ashby, M. 2010. *Materials Selection in Mechanical Design,* 4th ed. Boston: Butterworth-Heinemann.

Askeland, D. R., P. P. Fulay, and W. W. Wright. 2010. *The Science and Engineering of Materials,* 6th ed. Stamford, CT: Cangage Learning.

Cahn, R.W., P. Haasen, and E. J. Kramer, eds. 1991–. *Materials Science and Technology: A Comprehensive Treatment.* (19 vols.) New York: VCH Publishers. An in-depth, topic-oriented publication devoted to this enormous interdisciplinary field.

Callister, W. D. and D. G. Rethwisch, 2010. *Materials Science and Engineering: An Introduction,* 8th ed. Hoboken, NJ: John Wiley & Sons.

Dieter, G. E. and L. C. Schmidt. 2009. *Engineering Design,* 4th ed. New York: McGraw-Hill.

Gersten, J. I. and F. W. Smith. 2001. *The Physics and Chemistry of Materials.* New York: John Wiley & Sons.

Hosford, W. F. 2008. *Materials for Engineers.* New York: Cambridge University Press.

Materials Research Society Symposia Proceedings. 1980s–. Pittsburgh: Materials Research Society. Proceedings of the society's meetings are published in book form annually. Also available online.

Myers, M. A. and K. K. Chawla. 2010. *Mechanical Behavior of Materials,* 2nd ed. Cambridge, U.K.: Cambridge University Press.

Treatise on Materials Science and Technology. 1972–. (vols. 1–33, then unnumbered), San Diego, CA: Academic Press. Each volume is an in-depth treatment of a topic, such as structural ceramics, metal matrix composites, and auger electron spectroscopy.

JOURNALS

The journal literature is the most important vehicle for scholarly communication for the discipline of materials science. For the selection of the journals listed below, ISI's *Journal Citation Reports* were consulted. The list is limited to English-language titles and excludes trade journals. Journals that are open access are marked with an asterisk (*).

ACS Nano. 2007–. American Chemical Society. (1936-0851).

Acta Biomaterialia. 2005–. (1742-7061) Elsevier BV.

Acta Materialia. 1995–. (1359-6454) Pergamon.

Advanced Composite Materials. 1991–. (0924-3046) VSP.

Advanced Functional Materials. 1992–. (1616-301X) Wiley–VCH Verlag GmbH & Co. KGaA.

Advanced Materials. 1989–. (0935-9648) Wiley–VCH Verlag GmbH & Co. KGaA.

Advances in Applied Ceramics. 1971–. (1743-6753) Maney Publishing.

Annual Review of Materials Research. 1971–. (1531-7331) Annual Reviews.

Applied Composite Materials. 1994. (0929-189X) Springer, Netherlands.

Applied Surface Science. 1985–. (0169-4332) Elsevier BV, North-Holland.

Biointerphases, 2006–. (1559-4106) AVS.

Biomaterials. 1980–. (0142-9612) Elsevier BV.

Biomedical Materials. 2006–. (1748-6041) Institute of Physics Publishing Ltd.

Ceramics International. 1975–. (0272-8842) Pergamon.

Ceramics-Silikaty. 1990. (0862-5468) Vysoka Skola Chemicko Technologicka v Praze, Ustav Skla a Keramiky.

Chemistry of Materials. 1989–. (0897-4756) American Chemical Society.

Composites Science and Technology. 1968–. (0266-3538) Pergamon.

Composite Structures. 1983–. (0263-8223) Elsevier Ltd.

Corrosion Science. 1961–. (0010-938X) Pergamon.

Critical Reviews in Solid State and Materials Sciences. 1980–. (1040-8436) Taylor & Francis Inc.

European Cells and Materials. 2001. (1473-2262) Swiss Society for Biomaterials.

IEEE Transactions on Nanotechnology. 2002–. (1536-125X) Institute of Electrical and Electronics Engineers.

Intermetallics. 1993–. (0966-9795) Elsevier Ltd.

International Journal of Applied Ceramic Technology 2004–. Wiley-Blackwell (1546-542X).

International Materials Reviews. 1956–. (0950-6608) Maney Publishing.

JOM: Journal of the Minerals, Metals and Materials Society. 1949–. (1047-4838) Springer New York LLC.

Journal of Alloys and Compounds. 1991–. (0925-8388) Elsevier BV.

Journal of Applied Polymer Science. 1959–. (0021-8995) John Wiley & Sons.

Journal of Bioactive and Compatible Polymers. 1986–. (0883-9115) Sage Publications Ltd.

Journal of Biomaterials Applications. 1986–. (0885-3282) Sage Publications Ltd.

Journal of Biomaterials Science—Polymer Edition. 1989–. (0920-5063) Technomic Publications.

Journal of Biomedical Materials Research, A, B. 1967–. Part A (1549-3296) Part B (1552-4973) John Wiley & Sons (Official journal of the Society for Biomaterials).

Journal of Ceramic Processing Research. 2000–. (1229-9162) Hanyang University Ceramic, Processing Research Center.

Journal of Composite Materials. 1967–. (0021-9983) Sage Publications Ltd.

Journal of Electroceramics. 1997–. (1385-3449) Springer New York LLC.

Journal of Electronic Materials. 1972–. (0361-5235) Springer New York LLC.

Journal of Engineering Materials and Technology. 1973–. (0094-4289) ASME International.

Journal of Materials Chemistry. 1991–. (0959-9428) Royal Society of Chemistry.

Journal of Materials Research. 1986–. (0884-2914) Materials Research Society.

Journal of Materials Science. 1966–. (0022-2461) Springer New York LLC.

Journal of Materials Science: Materials in Electronics. 1990–. (0957-4522) Springer New York LLC.

Journal of Mechanical Behavior of Biomedical Materials. 2008–. (1751-6161) Elsevier BV.

Journal of Micromechanics and Microengineering. 1990–. (0960-1317) Institute of Physics Publishing Ltd.

Journal of Nanoscience and Nanotechnology. 2001–. (1533-4880) American Scientific Publishers.

Journal of Non-Crystalline Solids. 1968–. (0022-3093) Elsevier BV, North-Holland.

Journal of Phase Equilibria and Diffusion. 2004–. (1547-7037) Springer New York LLC; Continues *Journal of Phase Equilibria.*

Journal of Polymer Science, Part A, B. 1986–. *Part A, Polymer Chemistry* (0887-624X); *Part B, Polymer Physics* (0887–6266) John Wiley & Sons.

Journal of Sandwich Structures and Materials. 1999–. (1099-6362) Sage Publications Ltd.

Journal of the American Ceramic Society. 1918–. (0002-7820) Wiley-Blackwell Publishing, Inc.

Journal of the Ceramic Society of Japan. 2007–. (1882-0743) Nippon Seramikkusu Kyokai.

Journal of the Electrochemical Society. 1948–. (0013-4651) Electrochemical Society, Inc.

Journal of the European Ceramic Society. 1989–. (0955-2219) Elsevier Ltd.

Journal of the Mechanics and Physics of Solids. 1952–. (0022-5096) Pergamon.

Journal of Thermal Spray Technology. 1992–. (1059-9630) Springer New York, LLC.

Journal of Vacuum Science & Technology. 1964–. Part A (1553-1813) Part B (1071-1023) American Institute of Physics.

Macromolecular Bioscience. 2001–. (1616-5187) Wiley–VCH Verlag GmbH & Co. KGaA.

Materials Characterization. 1968–. (1044-5803) Elsevier Inc.

Materials Chemistry and Physics. 1983–. (0254-0584) Elsevier SA.

Materials Science and Engineering, A, B. 1988–. Part A (0921-5093) Part B (0921-5107) Elsevier SA.

Materials Science and Engineering, C. 1998–. (0928-4931) Elsevier SA.

Materials Science and Engineering R: Reports. 1986–. (0927-796X) Elsevier SA.

Materials Today. 1998–. (1369-7021) Elsevier BV.

Mechanics of Advanced Materials and Structures. 2002–. (1537-6494) Taylor & Francis Inc.

Mechanics of Time-Depending Materials. 1997–. (1385-2000) Springer Netherlands.

Metallurgical and Materials Transactions A. 1994–. (1073-5623) Springer New York LLC; formerly *Metallurgical Transactions A.*

MRS Bulletin. 1976–. (0883-7694) Materials Research Society.

Nano Letters. 2001–. (1530-6984) American Chemical Society.

Nanoscale and Microscale Thermophysical Engineering. 2006–. (1556-7265) Taylor & Francis Inc.

Nanotechnology. 1990–. (0957-4484) Institute of Physics Publishing Ltd.

Nano Today. 2006–. (1748-0132) Elsevier Ltd.

Nature Materials. 2001–. (1476-1122) Nature Publishing Group.

Nature Nanotechnology. 2006–. (1748-3387) Nature Publishing Group.

Physical Mesomechanics. 1998–. (1029-9599) Elsevier Inc.

Polymer. 1960–. (0032-3861) Elsevier Ltd.

Polymer Composites. 1980–. (0272-8397) John Wiley & Sons.

Polymer Testing. 1980–. (0142-9418) Elsevier Ltd.

Polymers for Advanced Technologies. 1990–. (1042-7147) John Wiley & Sons.

Progress in Crystal Growth and Characterization of Materials. 1990. (0960-8974) Pergamon.

Progress in Materials Science. 1961–. (0079-6425) Pergamon.

Scripta Materialia. 1996–. (1359-6462) Pergamon; continues Scripta.

Semiconductor Science and Technology. 1986–. (0268-1242) Institute of Physics Publishing Ltd.

Smart Materials and Structures. 1992–. (0964-1726) Institute of Physics Publishing Ltd.

Surface and Coatings Technology. 1986–. (0257-8972) Elsevier SA.

Thin Solid Films. 1967–. (0040-6090) Elsevier SA.

PATENTS AND STANDARDS

As in other engineering disciplines, standards are essential for designing new materials and applications, and the body of intellectual property literature is increasing rapidly and contains a wealth of technical information.

Most standard developing organizations host their own Web sites and sell their standards online; therefore, only the organization most relevant to materials science, ASTM, is listed here. You can also use some aggregator databases like SAI Global (formerly ILI Infodisk), IHS Standards and Specifications, and Techstreet. You can subscribe to some organizations through these databases or purchase standards on a pay per view basis. Information about these can be found in Chapter 2 in this book.

> *ASTM Standards* http://www.astm.org/ (accessed October 1, 2010). West Conshohocken, PA: American Society for Testing and Materials. This is the primary source of standards relating to materials science. Also published in the *Annual Book of ASTM Standards*.

Patents can be searched in databases mentioned in Chapter 2 of this book. USPTO and EPO databases are the most important ones. A step-by-step patent search tutorial is available at http://www.uspto.gov/web/offices/ac/ido/ptdl/CBT/ (accessed October 1, 2010). In addition to the patent databases mentioned above, several of the databases listed under the Indexes and Abstracts section in this chapter include relevant patents. *Chemical Abstracts/SciFinder* is especially useful for finding materials-related patents.

WEB SITES

Many of the sources listed so far are proprietary in some way and have to be paid for by a library or by an individual user. In this section, Web sites were selected that are freely accessible and are either gateways to a host of information on the Internet or are themselves information-rich sites. Most society Web sites are treasure troves of information; they are listed in the following section. University departments of materials science and engineering and government agencies and offices often host information-rich Web sites as well.

> Biomaterials Network http://www.biomat.net/ (accessed October 1, 2010). Maintained by Biomaterials Laboratory, Instituto de Engenharia Biomédica (INEB), University of Porto, Portugal. The major goals of Biomat.net consist of providing an organized and meaningful biomaterials communication resource for scientists, researchers, members of the business community, government, academia, and the general public, and acting as a resource center to disclose resources, organizations, research activity, educational initiatives, scientific events, journals, books, articles, funding opportunities, industrial developments, market analyses, jobs, and every other initiative related to biomaterials science and associated fields, such as tissue engineering.
>
> Center for Computational Materials Science http://cst-www.nrl.navy.mil/ (accessed October 1, 2010). Created and maintained by the National Research Laboratory (NRL). Included are the Crystal Lattice Structures and the Tight-Binding Parameters databases. Current research area summaries and other scientific resources also may be found on this Web site.
>
> Composite Materials http://composite.about.com/ (accessed October 1, 2010). This comprehensive Web site is a complete source to find information and resources for composite materials and the composites industry.
>
> Crystallography Online http://ww1.iucr.org/cww-top/crystal.index.html (accessed October 1, 2010). Provides links to current information concerning crystallography and other topics of interest to crystallographers. This site is part of the International Union of Crystallography.

Department of Energy (DOE) http://www.energy.gov/ (accessed October 1, 2010). Provides energy-related information from the Department of Energy. Covers full-text DOE documents and regulations.

Intute–Engineering http://www.intute.ac.uk/engineering/ (accessed October 1, 2010). Intute is a service created and run by a consortium of seven universities in the United Kingdom. Its mission is to provide access to quality networked information in a variety of subjects. The section "Materials Science and Engineering" lists over 1,000 relevant Web sites.

MatWeb: Material Property Data http://www.matweb.com/ (accessed October 1, 2010). This searchable database covers properties of over 79,000 metals, plastics, ceramics, and composites. Searchable by material name, category, property, and manufacturer.

MEMSnet http://www.memsnet.org/material/ (accessed October 1, 2010). This information resource, hosted by the MEMS and Nanotechnology Exchange, includes a materials index with property information culled from the *CRC Materials Science and Engineering Handbook*.

NIST Data Gateway http://www.nist.gov/materials-science-portal.cfm/ (accessed October 1, 2010). This is the home of the National Institute of Standards and Technology's over 80 databases covering a broad range of substances and properties from many different disciplines. Many consist of data relevant to materials scientists (e.g., Ceramic WebBook and Phase Diagrams and Computational Thermodynamics). The databases may be searched individually or collectively.

Plastics.com http://www.plastics.com/ (accessed October 1, 2010). This resource "features technical advice, blogs, calendars, resources, articles, news, forums and a marketplace" (from Plastics.com Web site).

The Plastics Web http://www.ides.com/ (accessed October 12, 2010). Directory with news, information about plastics, tools and resources, marketplace and pricing information. Especially useful is the Plastic Chemical Resistance data at http://www.ides.com/resources/plastic-chemical-resistance.asp (accessed October 12, 2010).

Platinum Group Metals (PGM) Database http://www.platinummetalsreview.com/jmpgm/index.jsp (cited August 26, 2010). A free database of information on more than 400 alloys, including diagrams and graphs with numerical data points, and phase diagrams of platinum group metals, alloy systems. The database "can display a datasheet of material properties in a user-friendly view, giving a comprehensive picture of a material and its capabilities." The database is sponsored by Anglo Platinum, Johnson Matthey, and the International Platinum Group Metals Association.

ASSOCIATIONS, ORGANIZATIONS, AND SOCIETIES

Professional societies provide an extremely important forum for the exchange of ideas and the dissemination of research results and, thereby, are a major factor in the advancement of a discipline. Each of the societies listed produces a variety of publications (journals, books, technical reports) and holds regular meetings, conferences, or workshops, and all have extensive Web sites providing access to a wealth of information, including job openings and access to experts.

Abrasive Engineering Society (AES) http://www.abrasiveengineering.com (accessed August 12, 2010). Founded in 1957, AES is "dedicated to promoting technical information about abrasives minerals and their uses including abrasives grains and products, such as grinding wheels, coated abrasives, and thousands of other related tools and products that serve manufacturing and the consumer" (from Web page).

American Ceramic Society (ACerS) http://www.acers.org/ (accessed August 12, 2010). "Since 1898, ACerS has been the hub of the global ceramics community and one of the most trusted sources of ceramic materials and applications knowledge. More than 9,500 scientists, engineers, researchers, manufacturers, plant personnel, educators, students, marketing and sales professionals from more than 60 countries make up the members of The American Ceramic Society" (from Web page).

American Concrete Institute (ACI) http://www.concrete.org/ (accessed August 12, 2010). "Founded in 1904 and headquartered in Farmington Hills, Michigan, the American Concrete Institute is advancing concrete knowledge by conducting seminars, managing certification programs, and publishing technical documents. The American Concrete Institute currently has 98 chapters and 20,000 members spanning 108 countries" (from Web page).

American Foundry Society (AFS) http://www.afsinc.org (accessed August 12, 2010). "The leading U.S.-based metalcasting society that assists member companies (metalcasting facilities, diecasters, and industry suppliers) and individuals to effectively manage all production operations, profitably market their products and services and to equitably manage their employees. The American Foundry Society also promotes the interests of the metalcasting industry before the legislative and executive branches of the federal government. With the direction of its volunteer committee structure, the professional staff of the American Foundry Society provides support in the areas of technology, management, and education to further activities that will enhance the economic progress of the metalcasting industry" (from Web page). The Cast Metal Institute is AFS's educational arm.

American Institute for Mining, Metallurgical and Petroleum Engineers (AIME) http://www.aimeny.org/ (accessed August 12, 2010). "AIME was founded in 1871 by 22 mining engineers in Wilkes-Barre, Pennsylvania. As one of the first national engineering societies established in the United States, and along with ASCE (civil), ASME (mechanical), IEEE (electrical), and AIChE (chemical), it is known as an Engineering Founder Society. Together, the Founder Societies form the United Engineering Foundation, Inc. Today, the memberships of the AIME member societies total over 127,000 and includes some of the most important, influential, and innovative figures in the engineering and scientific communities" (from Web page).

American Iron and Steel Institute (AISI) http://www.steel.org (accessed August 12, 2010). AISI began in 1855 as the American Iron Association and "serves as the voice of the North American steel industry in the public policy arena and advances the case for steel in the marketplace as the preferred material of choice. AISI also plays a lead role in the development and application of new steels and steelmaking technology. AISI is comprised of 24 member companies, including integrated and electric furnace steelmakers, and 138 associate and affiliate members who are suppliers to or customers of the steel industry. AISI's member companies represent approximately 75% of both U.S. and North American steel capacity" (from Web page).

American Society for Testing of Materials (ASTM) International http://www.astm.org (accessed August 12, 2010). "ASTM International is one of the largest voluntary standards development organizations in the world—a trusted source for technical standards for materials, products, systems, and services. Known for their high technical quality and market relevancy, ASTM International standards have an important role in the information infrastructure that guides design, manufacturing, and trade in the global economy" (from Web page).

American Vacuum Society (AVS) http://www.avs.org (accessed August 12, 2010). "AVS is a nonprofit organization that promotes communication, education, networking, recommended practices, research, and the dissemination of knowledge on an international scale, in the application of vacuum and other controlled environments to understand and develop interfaces, new materials, processes, and devices through the interaction of science and technology" (from Web page).

American Welding Society (AWS) http://www.aws.org (accessed August 12, 2010). "The American Welding Society (AWS) was founded in 1919 as a multifaceted, nonprofit organization with a goal to advance the science, technology, and application of welding and related joining disciplines. From factory floor to high-rise construction, from military weaponry to home products, AWS continues to lead the way in supporting welding education and technology development to ensure a strong, competitive, and exciting way of life for all Americans" (from Web page).

ASM International http://www.asminternational.org/ (accessed August 12, 2010). "ASM is Everything Material, the society dedicated to serving the materials science and engineering profession. Through our network of 36,000 members worldwide, ASM provides authoritative information and knowledge on materials and processes, from the structural to the nanoscale" (from Web page).

Association for Iron and Steel Technology (AIST) http://www.aist.org (accessed August 12, 2010). "The Association for Iron and Steel Technology (AIST) is a nonprofit organization that advances the technical development, production, processing, and application of iron and steel. AIST membership is comprised of over 15,000 individuals worldwide and includes iron and steel producers, suppliers, academics, and students" (from Web page).

Cryogenic Society of America (CSA) http://www.cryogenicsociety.org/ (accessed August 12, 2010). "CSA is a nonprofit technical society serving all those interested in any phase of cryogenics, the art and science of achieving extremely low temperatures—almost absolute zero" (from Web page).

Electrochemical Society (ECS) http://www.electrochem.org (August 12, 2010). "ECS was founded in 1902 as an international nonprofit, educational organization concerned with a broad range of phenomena relating to electrochemical and solid-state science and technology. The Electrochemical Society has more than 8,000 scientists and engineers in over 70 countries worldwide who hold individual membership as well as roughly 100 corporations and laboratories that hold corporate membership" (from Web page).

Electronic Device Failure Analysis Society (EDFAS) http://edfas.asminternational.org/ (accessed September 17, 2010). "The world's foremost community of professionals working for technology advancement and improved performance of components in their use, manufacturing, and production" (from EDFAS online brochure). An affiliate of ASM.

European Society for Biomaterials (ESB) http://www.esbiomaterials.eu/ (accessed September 17, 2010). "The European Society for Biomaterials (ESB) is nonprofit scientific society whose main objective is to encourage progress in the field of biomaterials in all its aspects, including research, teaching, and clinical applications as well as to foster any other related activity" (from Web page).

Institute of Materials, Minerals, and Mining, U.K. (IOM3) http://www.iom3.org (accessed August 12, 2010). "The Institute of Materials, Minerals and Mining (IOM3) is a major U.K. engineering institution whose activities encompass the whole materials cycle, from exploration and extraction, through characterization, processing, forming, finishing, and application to product recycling and land reuse. It exists to promote and develop all aspects of materials science and engineering, geology, mining and associated technologies, mineral and petroleum engineering, and extraction metallurgy, as a leading authority in the worldwide materials and mining community" (from Web page).

International Association of Nanotechnology (IANT) http://ianano.org (accessed September 17, 2010). "The International Association of Nanotechnology (IANT) is a nonprofit organization with the goals of fostering scientific research and business development in the area of nanoscience and nanotechnology for the benefit of society" (from Web page).

International Metallographic Society (IMS) http://www.internationalmetallographicsociety. org (accessed August 16, 2010). IMS was founded in 1967. Its mission is "to benefit the art and science of metallography and materials characterization. IMS members are involved in material characterization, performance, behavior, and fabrication. Of special interest to members are techniques and the equipment needed for microstructural examination, analysis, and evaluation. Some techniques of interest include light, electron and acoustical microscopy; quantitative/computer-aided microstructural analysis; metallography, ceramography, resinography; and allied sciences for physical and chemical analysis" (from Web page).

International Society of Coating Science and Technology (ISCST) http://www.iscst.org (accessed August 16, 2010). "The ISCST is a technical society formed by an international group of academic and industrial engineers to encourage the use of scientific principles to help understand the processes associated with the application of wet coatings to a substrate. As such, ISCST serves as an interface between scholarly research and industrial practice; both are welcome, but it is the interplay between the two that makes ISCST unique" (from Web page).

International Thermal Spray Association (ITSA), The http://www.thermalspray.org/ (accessed August 16, 2010). "The International Thermal Spray Association is closely interwoven with the history of thermal spray development in this hemisphere. Founded in 1948, and once known as Metallizing Service Contractors, the association has been closely tied to almost all major advances in thermal spray technology, equipment and materials, industry events, education, standards, and market development in North and South America" (from Web page).

International Union of Crystallography (IUCr) http://www.iucr.org/ (accessed August 16, 2010). "The IUCr is an International Scientific Union. Its objectives are to promote international cooperation in crystallography and to contribute to all aspects of crystallography; to promote international publication of crystallographic research; to facilitate standardization of methods, units, nomenclatures, and symbols; and to form a focus for the relations of crystallography to other sciences" (from Web page).

International Union of Materials Research Societies (IUMRS) http://www.iumrs.org (accessed August 16, 2010). "The International Union of Materials Research Societies was established in 1991 as an international association of technical groups or societies that have an interest in promoting interdisciplinary materials research" (from Web page).

Materials Research Society (MRS) http://www.mrs.org (accessed August 16, 2010). "Founded in 1973, MRS now consists of over 15,100 members from the United States as well as nearly 70 other countries. The society is different from that of single discipline professional societies because it encourages communication and technical information exchange across the various fields of science affecting materials" (from Web page).

Metal Powder Industries Federation (MPIF) http://www.mpif.org (accessed August 16, 2010). "The Metal Powder Industries Federation (MPIF) is a not-for-profit association formed by the PM industry to advance the interests of the metal powder producing and consuming industries" (from Web page). MPIF is a federation of six member associations.

NanoAssociation for Natural Resources and Energy Security (NANRES) http://nanres.org/ (accessed September 17, 2010). "NanoAssociation for Natural Resources and Energy Security (NANRES) is a membership industry association founded to promote nanotechnology's role in energy and the environment. With a strong, cross-sector industry base that shares a collective interest and voice in advancing nanotechnology in the U.S. marketplace, NANRES is poised to address the pressing natural resource and energy security challenges facing the United States in the twenty-first century" (from Web page).

National Association for Surface Finishing (NASF) http://www.nasf.org/ (accessed August 12, 2010). "The National Association for Surface Finishing (NASF) is a 501 (c)(6) trade association formed in 2007 as a means to better serve the surface finishing industry. The NASF was formed by reorganizing and bringing together the members of the American Electroplaters and Surface Finishers Society (AESF), the Metal Finishing Associations (MFSA), and the National Association of Metal Finishers (NAMF)" (from Web page).

National Association of Corrosion Engineers (NACE) http://www.nace.org (accessed August 16, 2010). "NACE International is the leader in the corrosion engineering and science community and is recognized around the world as the premier authority for corrosion control solutions. From just 11 engineers, our association has grown to more than 22,000 members in over 100 countries. Built upon decades of knowledge and expertise from dedicated members all around the world, NACE International is involved in every industry and area of corrosion prevention and control, from chemical processing and water systems to transportation and infrastructure protection" (from Web page).

Society for Biomaterials http://biomaterials.org/ (accessed September 17, 2010). "The Society for Biomaterials is a professional society that promotes advances in biomedical materials research and development by encouragement of cooperative educational programs, clinical applications, and professional standards in the biomaterials field. Biomaterials scientists and engineers study cells, their components, complex tissues and organs and their interactions with natural and synthetic materials and implanted prosthetic devices, as well as develop and characterize the materials used to measure, restore, and improve physiologic function, and enhance survival and quality of life" (from Web page).

Society for the Advancement of Material and Process Engineering (SAMPE) http://www.sampe.org (accessed August 16, 2010). "The Society for the Advancement of Material and Process Engineering, SAMPE®, an international professional member society, provides information on new materials and processing technologies through chapter technical presentations, two journal publications, symposia and commercial expositions in which professionals can exchange ideas and air their views. As the only technical society encompassing all fields of endeavor in materials and processes, SAMPE provides a unique and valuable forum for scientists, engineers, designers, and academicians" (from Web page).

Society for Mining, Metallurgy, and Exploration (SME) http://www.smenet.org (accessed August 16, 2010). "The Society for Mining, Metallurgy, and Exploration (SME) is an international society of professionals in the minerals industry. A member society of the American Institute of Mining, Metallurgical and Petroleum Engineers (AIME), SME's roots date back to 1871 when a handful of mining engineers founded AIME. Since its inception, SME has continued to evolve over the years to stay abreast of industry changes and to reflect the ever-broadening interests of its members" (from Web page).

Society of Glass Technology (SGT) http://www.sgt.org (accessed August 16, 2010). "The objects of the Society of Glass Technology are to encourage and advance the study of the history, art, science, design, manufacture, after treatment, distribution, and end use of glass of any and every kind. These aims are furthered by meetings, publications, committees, the maintenance of a library, and the promotion of association with other interested persons and organizations" (from Web page).

Society of Plastics Engineers (SPE) http://www.4spe.org (accessed August 16, 2010). "The objective of the society is to promote the scientific and engineering knowledge relating to plastics. SPE is the only place where people from all parts of the industry can come together around important issues and technologies. SPE's contribution to the plastics industry for over 60 years has made a significant difference to the technologies and innovations the industry enjoys today" (from Web page).

Steel Founders' Society of America (SFSA) http://www.sfsa.org (accessed August 16, 2010). The mission of SFSA is "to serve the steel casting industry in North America" (from Web page).

Surface Analysis Society of Japan (SASJ) http://www.sasj.jp/eng-index.html (accessed August 17, 2010). "SASJ promotes national and international collaborative researches, holds symposia, and publishes related documents in order to improve the authenticity and the standardization of surface analysis techniques. Besides, SASJ promotes construction of database for surface analysis" (from Web page).

The Minerals, Metals and Materials Society (TMS) http://www.tms.org (accessed August 17, 2010). "Headquartered in the United States, but international in both its membership and activities, The Minerals, Metals and Materials Society (TMS) is a rare professional organization that encompasses the entire range of materials and engineering, from minerals processing and primary metals production to basic research and the advanced applications of materials" (from Web page).

The Thermal Spray Society: ASM International (TSS) http://tss.asminternational.org (accessed August 17, 2010). "Founded in 1994, the ASM Thermal Spray Society has grown to a membership of approximately 1,500 individuals around the globe representing over 500 leading companies, research institutions, and universities. The Thermal Spray Society operates according to our vision statement: 'To be the leading global source for thermal spray information—TSS, Your Thermal Spray Information Partner'" (from Web page).

REFERENCE

Lord, C. R. 2000. *Guide to information sources in engineering*, Englewood, CO: Libraries Unlimited.

16 Mechanical Engineering

Aleteia Greenwood and Mel DeSart

CONTENTS

Introduction 395
Searching the Catalog 397
Article Indexes 398
Abstracts 398
Databases and Data Sets 399
Bibliographies and Guides to the Literature 399
Directories 400
Dictionaries and Encyclopedias 400
    Dictionaries: English 400
    Dictionaries: Multilingual 401
Encyclopedias 401
Handbooks and Manuals 402
    General 402
    Automotive 404
    Energy, Heat Transfer, HVACR, Fluid Mechanics/Dynamics/Flow 405
    Hydraulics and Pneumatics 407
    Machinery and Machining 408
    Miscellaneous 409
    Noise, Shock, Vibration 411
    Ocean, Marine, Ship 412
    Pressure Vessels and Piping 412
    Pumps and Pumping 414
    Tribology and Lubrication 415
Monographs and Textbooks 416
Journals 418
Patents and Standards 422
Search Engines and Important Web Sites 422
    Meta Sites 422
    Individual Subject Sites 423
Associations, Organizations, and Societies 424
Acknowledgments 425

INTRODUCTION

If something moves because work (defined as force times distance) is done, chances are very good that mechanical engineering concepts are involved. Mechanical engineering is the branch of engineering that encompasses the generation and application of heat and mechanical power and the design, production, and use of machines and tools. One of the oldest of the engineering disciplines, mechanical engineering concepts have been in use since prehistoric times. From the discovery and use of the most basic tools (lever, wheel and axle, inclined plane (or wedge), etc.) to the most complex machines of the modern world, mechanical engineering concepts impact virtually everything we do.

From those tools and the concepts associated with them came some of the first simple machines, e.g., the pulley, bellows, screw, and the first use of gears. Those were followed by the water wheel, windmill, mechanical clock, lathe, the first pumps, and the printing press with movable type, among hundreds of others.

The Newcomen steam engine, invented in 1712, is regarded as one of the greatest inventions in the history of mechanical engineering and led directly to the Industrial Revolution. But the steam engine that Newcomen invented, Watt, in turn, transformed into the tool that led to much of the industrial development that followed. While railways or tramways existed prior to the steam engine, the development of the high pressure steam engine led directly to the rapid expansion of rail transportation. Other important advances during the Industrial Revolution included the refining of the centrifugal pump, the application of power to spinning and weaving textiles, and, in turn, the development of the factory system.

The late 1800s brought far more varied developments with the invention of the bicycle, the steam turbine, the internal combustion engine, the automobile, the first power station to produce electricity, and the techniques for making steel. The ability to make steel, in turn, increased the capabilities of thousands of other machines. During this period, the foundations of fluid mechanics also appeared, including the discovery of the two different modes of motion of moving fluids: laminar and turbulent. Investigations of drag or resistance to motion in both water and air fed directly into later aerodynamic and hydrodynamic design considerations.

The early twentieth century was the period when other engineering disciplines began to spring from the work of mechanical engineers. The aerodynamic studies of the late 1800s led to the development of gliders and, hence, to the most important invention of the first half of the twentieth century, the airplane. The helicopter and autogyro also were developed during this period. The improvements in the automobile and, in particular, to its assembly, led directly into the blossoming of manufacturing engineering as a discipline. While the power take off was invented in the late 1800s, its expanded use led to the popularity of the tractor in agriculture and the development of many other more specialized powered agricultural implements, from whence came agricultural engineering as a discipline.

A number of other important developments of the first half of the twentieth century focused on heavy machinery, as earth-moving equipment, such as the bulldozer, excavator, road grader, etc., came into being, as well as a number of the tools of war, including the creation of specialized ships (including landing craft) and various forms of armored vehicles, the most mass-produced being the tank.

The 1940s to the present date has seen the invention of the jet engine, robots and robotics, the development of nuclear power and construction of power reactors, the merging of electrical and mechanical concepts in the creation of control systems, and the development of the rocket engine, which allowed humans to venture beyond the boundaries of their home planet for the first time.

For a more detailed discussion of the history of mechanical engineering, the authors offer the following suggested works.

American Society of Mechanical Engineers. History and Heritage Committee. 1997. *Landmarks in Mechanical Engineering.* West Lafayette, IN: Purdue University Press. While the Semler book (below) focuses as much or more on the people behind the achievements, this work from ASME highlights the landmark developments or inventions of mechanical engineering, organized by broad topical area.

Burstall, A. F. 1965. *A History of Mechanical Engineering.* Cambridge, MA: MIT Press. The most comprehensive of the works listed, the volume is broken into nine chapters, from "The Prehistoric Period, before 3000 B.C." up through "Dawn of the Nuclear Age and Space Travel, 1940–1960."

Cressy, E. 1937. *A Hundred Years of Mechanical Engineering*. New York: Macmillan. Although purporting to cover only one hundred years, this work by Cressy, which focuses on the production, distribution, and applications of power, begins with a discussion of the engines of Newcomen and Watt and ends with developments from the early 1930s. Contains over 150 photographs, illustrations, and cutaway drawings.

Lawton, B. 2004. *The Early History of Mechanical Engineering*. Leiden, Boston: Brill. A two-volume set on the early history of mechanical engineering from prehistory to the beginnings of industrialization. Vol. 1 *Power Generation and Transport*, vol. 2 *Manufacturing and Weapons Technology*.

Semler, E. G. 1963–1966. *Engineering Heritage: Highlights from the History of Mechanical Engineering*. London: Heinemann, on behalf of the Institution of Mechanical Engineers. Two volumes comprised of short pieces on particular people or inventions that shaped the history of mechanical engineering.

See also the *Bibliography on the History of Mechanical Engineering and Technology*, compiled by staff of the American Society of Mechanical Engineers http://www.asme.org/Communities/History/Resources/Bibliography_History.cfm (accessed October 10, 2010).

SEARCHING THE CATALOG

One particularly wonderful aspect of online catalogs is the ability to search by keyword. Usually searching by keyword in a library catalog will search the title, author, publisher, subject heading, series, notes, and table of contents fields. Library of Congress subject headings also can be searched. The following Library of Congress subject headings and corresponding classification areas are useful for mechanical engineering.

Biomechanics WE 103
Bearings (Machinery) TJ 1061–1073
Control engineering systems. Automatic machinery (General) TJ 211.5–225
Deformations (Mechanics) TA 417–418
Diesel engines TJ 795–799
Dynamics TJ 170–173, TL 243
Electronic control TJ 163.12
Electromechanical devices TJ 163
Energy conservation TJ 163.26–163.5
Energy storage TJ 165
Engineering design T 353, TA 174
Engineering graphics T 351, T 353
Fasteners TJ 1320–1340
Fluid dynamics TA 357
Fluid mechanics (applied) TA 357–359
Fracture mechanics TA 409
Friction TJ 1075
Gears and Gearing TJ 184–202
Heat engineering TJ 255–265
Heat transmission TJ 260, TJ 265
Hydraulic machinery TJ 836–927
Internal combustion engines TJ 782–793

Locomotives TJ 603–695
Lubrication and lubricants TJ 1075–1081
Machine design TJ 230
Machine parts TJ 243
Machine tools and machining TJ 1180–1201
Machinery, dynamics of TJ 170
Machinery, kinematics of TJ 175
Mass transfer TA 357, TA 418
Materials fatigue TA 418.38, TA 409
Mechanical engineering TJ 144–TJ 159
Mechanical movements TJ 181–TJ 209, TJ 1075
Mechanical wear TJ 170, TJ 1075
Mechanics TA 350, TA 352
Mechanics, Applied TA 350, TA 405
Mechatronics TJ 163.12
Microelectromechanical devices TJ 163
Pneumatic machinery TJ 95–1030
Power and power transmission TJ 1045–1051
Power (Mechanics) TJ 163.6–163.9
Power plants TJ 164

Power resources TJ 163.13–163.25
Prime movers TJ 250
Pumps and pumping engines TJ 899–927
Railroad engineering TF 200
Renewable energy sources TJ 807–830
Rheology TA 418.14
Robots (General) TJ 210–211.49
Servomechanisms TJ 214
Solar energy TJ 809–812
Statics, Applied TA 351
Steam boilers TJ 281–393
Steam engineering TJ 268–280

Steam engines TJ 461–567
Steam power plants. Boiler plants. TJ 395–444
Strains and stresses TA 351, TA 405, TA 407
Strength of materials TA 405
Thermodynamics TJ 265
Tribology TJ 1075–1081
Turbines. Turbomachines TJ 266–267
Vacuum technology TJ 940–949
Vibration TA 355–TA 356, TJ 177, TL 246
Viscosity TA 357.5 V6, TJ 1077
Wind power TJ 820–828

ARTICLE INDEXES

Compendex. Hoboken, NJ: Elsevier Engineering Information, Inc., see Chapter 2, General Engineering Resources in this book.

Engineering Index. 1884–. Hoboken, NJ: Elsevier Engineering Information, Inc., see Chapter 2, General Engineering Resources in this book.

Indexes to ... Publications: Transactions of the ASME. Society Records. 1965?–1981. Easton, PA: American Society of Mechanical Engineers. This annual publication contains indexes to all ASME papers and publications.

ISMEC Bulletin. 1973–1987. Bethesda, MD: Cambridge Scientific Abstracts. This print bibliography/index of mechanical engineering literature was succeeded by ISMEC Mechanical Engineering Abstracts.

ISMEC Mechanical Engineering Abstracts. 1988–1992. Bethesda, MD: Cambridge Scientific Abstracts. This bimonthly print index, with annual cumulated indexes, was succeeded by *Mechanical Engineering Abstracts.*

Mechanical Engineering Abstracts. Bethesda, MD: Cambridge Scientific Abstracts, and Hoboken, NJ: Engineering Information, Inc. This is a "closed" database, with content from 1981 through 2002, when *Mechanical Engineering Abstracts* was succeeded by *Mechanical & Transportation Engineering Abstracts.* The database contains 227,382 records as of its closure at the end of 2002. Also published in paper from 1993 to 2002 under the same title. The print edition was published bimonthly with an annual index. Succeeded *ISMEC Mechanical Engineering.*

ABSTRACTS

Mechanical & Transportation Engineering Abstracts. Ann Arbor, MI: Proquest. This database contains over 400,000 records drawn from the indexing of over 2,600 journals and other publications. As with many former CSA databases, indexed publications are designated either core, priority, or selective with the percentage of content indexed decreasing from core down through selective. Coverage is primarily from 1966 to date, although some older records exist in the database; 30,000 to 35,000 records are added to the database annually in bimonthly updates.

NTIS. Springfield, VA: National Technical Information Service. The NTIS database contains records for over 2,000,000 government technical reports received by NTIS since 1964. Most other databases that cover mechanical engineering content do not cover government technical reports, making NTIS a complementary resource to the other subject-specific indexes and abstracts listed here. See Chapter 2, General Engineering Resources for more information.

SAE Publications and Standards Database. Warrendale, PA: Society of Automotive Engineers. Covers the current and past publications of SAE, including its technical papers (from 1906 forward), magazine articles, books, standards, specifications, and research reports. Supports all aspects of mobility, including mechanics of engines, lubrication, engine design and technology, vehicle body engineering, and fluid mechanics in engines.

Seventy-Seven Year Index, Technical Papers, 1880–1956. New York: American Society of Mechanical Engineers (ASME). A three-part index of the papers published in ASME publications. Separate indexes are included for the *Mechanical Engineering, Journal of Applied Mechanics*, and to the *ASME Transactions*. A 60-year index, covering 1880–1939, was also published.

Technology Research Database. Ann Arbor, MI: Proquest. See Chapter 2, General Engineering Resources for more information.

DATABASES AND DATA SETS

CRCnetBASE: Engineering Collection. Boca Raton, FL: CRC Press. Electronic collection of handbooks and other titles published by CRC Press. Contains dozens of titles in mechanical engineering and related areas. See Chapter 2, General Engineering Resources for more information.

ENGnetBASE. See *CRCnetBASE: Engineering Collection.*

Knovel. Norwich, NY: Knovel Corporation. Electronic collection of books and databases, some enhanced with productivity tools. Contains dozens of titles in mechanical engineering and related areas. See Chapter 2, General Engineering Resources for more information.

SAE Digital Library. Warrendale, PA: Society of Automotive Engineers. Database offers customizable content by the purchaser. Options include the SAE technical papers from 1998 to date, Aerospace Standards (AS), Aerospace Materials Specifications (AMS), Ground Vehicle Standards (J-reports), and the SAE Global Mobility Database (GMD).

BIBLIOGRAPHIES AND GUIDES TO THE LITERATURE

For the most part, the most recent guides in print, devoted specifically to mechanical engineering literature resources, appeared in the late 1980s. Chapters specific to mechanical engineering can be found in: *Scientific and Technical Information Sources* (The MIT Press, 1987) by Ching-chih Chen, *Reference Sources in Science, Engineering, Medicine, and Agriculture* (Technical Communication, 1996) by H. Robert Malinowsky, *Encyclopedia of Physical Sciences and Engineering Information Sources* (Thomson Gale, 1997) by Steven Wasserman, Martin A. Smith, and Susan Mottu, and *Information Sources in Engineering,* 4th ed. (Saur, 2005) edited by Roderick A. MacLeod and Jim Corlett. There are a multitude of bibliographies and guides to the literature accessible through the Internet. Usually these sites are specific to the institutions for which they are made, and point to resources available at or through those institutions. Internet sites that cover a wider range of resources have been included in the section on Search Engines and Internet Sites.

Veilleux, D. 2003. *The Automotive Bibliography: 13,000 Works in English, Czech/Slovak, Danish, Dutch, Finnish, French, German, Italian, Norwegian, Polish, Portuguese, Slovenian, Spanish and Swedish.* Jefferson, NC: McFarland and Company. Comprised of books on motor vehicles and motorization published to 2000. A wide range of items are catalogued, including monographs, theses, biographies, encyclopedias, other reference books, such as handbooks, company and government publications, and buyers', collectors', spotters', and identification guides.

DIRECTORIES

GradSchools.com: The most comprehensive online source of graduate school information http://www.gradschools.com/listings/menus/mech_eng_menu.html (accessed 10 October 2010). This resource provides access to mechanical engineering graduate school programs that offer master's and PhD degrees, and certificate programs. Coverage is international. Provides access to universities by Subject, Program Type, Region, and Country. Provides links directly to the mechanical engineering department of universities. See Chapter 11, Engineering Education in this book.

Peterson's.com http://www.petersons.com/ (accessed 10 October 2010). Provides the ability to search for either college or graduate school programs. Coverage is international with a focus on North America. Searches can be limited by Subject Area, Program, Location, and Degree Level. Information about institutions and their programs includes student body, financial aid, programs and degrees, number of faculty, and research areas. See Chapter 11, Engineering Education in this book.

Thomas Register of American Manufacturers and Thomas Register Catalog File. 1905–2006. New York: Thomas Publishing Co. Current content now available electronically as ThomasNet. See Chapter 14, Industrial Engineering in this book.

Who's Who in Science and Engineering. 1992–. Wilmette, IL: Marquis Who's Who. Look in the Professional Area Index for Engineering: Mechanical to find Mechanical Engineers in North America and internationally. Arranged alphabetically by state, then alphabetically by country. See Chapter 2, General Engineering Resources in this book.

Worldwide Automotive Supplier Directory, 9th ed. 2007/2008. Warrendale, PA: Society of Automotive Engineers. Includes over 10,000 companies and more than 250 products. Locate suppliers geographically or by major product categories, such as body (exterior and interior), body (seating and passenger), chassis, engineering design, manufacturing equipment, powertrain, and testing/instrumentation. Also available to members online through SAE http://www.sae.org/wwsd/ (accessed 10 October 2010). This type of resource has largely been replaced by free online directories and portals.

DICTIONARIES AND ENCYCLOPEDIAS

DICTIONARIES: ENGLISH

Goodsell, D. 1995. *Dictionary of Automotive Engineering,* 2nd ed. Warrendale, PA: Society of Automotive Engineers. A concise work, useful for automotive engineers in the industry as well as engineering students. Includes definitions for over 3,000 terms as well as detailed drawings for 100 of the entries.

Nayler, G. H. F. 1996. *Dictionary of Mechanical Engineering,* 4th ed. Warrendale, PA: Society of Automotive Engineers. Includes definitions to over 4,500 terms. Entries cover traditional as well as newer areas in mechanical engineering, such as micromachining and nanotechnology. Some entries accompanied by illustrations.

Perrot, P. 1998. *A to Z of Thermodynamics.* New York: Oxford University Press. A clear, comprehensive guide with entries that describe the fundamental concepts, principles, and applications of thermodynamics. Entries can include short etymological considerations, historical notes, and discussion of classical paradoxes. Accessible to undergraduate students.

South, D. W., and R. H. Ewert. 1995. *Encyclopedic Dictionary of Gears and Gearing.* New York: McGraw-Hill. Contains alphabetically arranged, clear, concise definitions for common terms, trade names, and abbreviations. Covers the fundamentals of gearing as well as advanced concepts. Some terms are accompanied by illustrations.

DICTIONARIES: MULTILINGUAL

de Coster, J. and O. Vollnhal. 2003. *Dictionary for Automotive Engineering: English–French–German: With Explanations of French and German Terms,* 5th rev. ed. Munich: Saur. The format of this dictionary is English entries to French and German translations. French–English and German–English indexes toward the end of the work refer back to the English language entries. An Index to Illustrations and a brief bibliography complete the volume. Unfortunately, the page numbering for the illustrations in the Index of Illustrations is consistently incorrect.

Ei-Nichi-Chu Kogyo Gijutsu Daijiten, Ei-Nichi-Chu Kikai Yogo Jiten = English–Japanese–Chinese Mechanical Engineering Dictionary. 1988. Tokyo: Kogyo Chosakai. The translation in this dictionary, which encompasses roughly the first half of the volume, is entirely English to Chinese and Japanese. The latter half of the volume contains Chinese and Japanese indices that refer to page numbers where the corresponding English entries appear. All front matter is non-English.

Federation of European Heating and Air Conditioning Associations; Representatives of European Heating and Air Conditioning Associations. 1994. *The International Dictionary of Heating, Ventilating, and Air Conditioning,* 2nd ed. London/New York: Spon Press. Comprised of two major parts. The main part is made up of approximately 4,000 alphabetically arranged terms in English, with translations. The second part of the book contains alphabetical indexes for the other 11 languages included in the dictionary: French, German, Italian, Danish, Finnish, Dutch, Spanish, Swedish, Hungarian, Polish, and Russian. Each alphabetical index is linked to serial numbers that refer the user to the translation in the main part.

Schellings, A. 1998. *Elsevier's Dictionary of Automotive Engineering in English, German, French, Dutch, and Polish.* Amsterdam: Elsevier. A unique resource that comprises technical terms relating to vehicles, materials, assemblies, and components parts. All terms are listed in English. Under each English word are the corresponding other language terms, German, French, Dutch, and Polish, labeled as to which language.

Voskoboinikov, B. S. and V. L. Mitrovich. 2001. *English–Russian Dictionary of Mechanical Engineering and Industrial Automation,* 2nd stereotype ed. Moscow: Russo. Contains over 100,000 terms with content entirely English to Russian. All front matter is in Russian only. Also contains an acronyms dictionary near the end of the volume. A supplemental work by the first author, with an additional 15,000 terms, was published in 2003.

Wunsch, F. 2004. *Wörterbuch Maschinenbau und Tribologie: Deutsch–Englisch, Englisch–Deutsch = Dictionary Machine Engineering and Tribology: German–English, English–German.* Berlin: Springer. Although Machine Engineering leads the title of this work, the content focuses on tribology. Contains over 75,000 terms, plus a brief list of abbreviations and an appendix of British and American weights and measures.

ENCYCLOPEDIAS

Cheremisinoff, N. P., ed. 1986–1990. *Encyclopedia of Fluid Mechanics.* (Suppl.) Houston: Gulf Publishing.
  v. 1. Flow Phenomena and Measurement
  v. 2. Dynamics of Single-Fluid Flows and Mixing
  v. 3. Gas–Liquid Flows
  v. 4. Solids and Gas-Solids Flows
  v. 5. Slurry Flow Technology

v. 6. Complex Flow Phenomena and Modeling
v. 7. Rheology and Non-Newtonian Flows
v. 8. Aerodynamics and Compressible Flows
v. 9. Polymer Flow Engineering
v. 10. Surface and Groundwater Flow Phenomena
Supplement 1. Applied Mathematics in Fluid Dynamics
Supplement 2. Advances in Multiphase Flow
Supplement 3. Advances in Flow Dynamics
Comprehensive coverage of all aspects of fluid mechanics and dynamics. Clearly written
with extensive use of diagrams and reference lists at the end of each entry.

Hewitt, G. F., G. L. Shires, and Y. V. Polezhaev. 1997. *International Encyclopedia of Heat
and Mass Transfer.* Boca Raton, FL: CRC Press. Entries are written by approximately 300
contributors from around the world, including previously unreachable authors from Russia
and Eastern European countries. Contributors are from the academic and industrial sectors. The encyclopedia is intended to support an undergraduate as well as an experienced
practitioner. Entries are highly specialized where necessary, and accessible by the general
reader where possible. Terms are cross referenced as well as linked by the terms "following
from" and "leading to."

*THERMOPEDIA: A-to-Z Guide to Thermodynamics, Heat & Mass Transfer, and Fluids
Engineering.* Redding, CT: Begell House http://www.thermopedia.com/ (accessed
10 October 2010). An online resource based on an updated version of the *International
Encyclopedia of Heat and Mass Transfer,* but that is significantly broader in scope. In addi-
tion to the encyclopedia content, some entries in *THERMOPEDIA* are drawn from other
Begell House publications.

HANDBOOKS AND MANUALS

Many of the works in the following section will be available in electronic form, whether singly from
the publisher or vendor or as part of publisher-based or vendor-based packages.

GENERAL

Avallone, E. A., T. Baumeister, III, and A. M. Sadegh. 2007. *Mark's Standard Handbook
for Mechanical Engineers,* 11th ed. New York: McGraw-Hill. Arguably the best known
handbook in mechanical engineering. *Mark's* covers both theory and applications, with
a slightly greater focus on materials than many other general ME handbooks. Extensive
index at the end of the volume.

Carvill, J. 1994. *Mechanical Engineers Data Handbook.* Boston: Butterworth-Heinemann.
Per the author's preface, this work is designed to be a "… compact but comprehensive
source of information." The title is large format, contains many illustrations in support of
the concepts presented, and still contains nearly 350 pages of data. Text is kept to a minimum and the focus is on equations and data.

Dubbel, H., W. Beitz, and K.-H. Kuttner, eds. 1994. *Handbook of Mechanical Engineering
Dubbel,* 1st English ed., London: Springer-Verlag. English translation (edited by B. J. Davies
and translated by M. J. Shields) of a classic German mechanical engineering handbook.
Presents comprehensive coverage of all mechanical engineering areas, but in a very dense
format (two columns per page, very small print).

Grote, K.-H. and E. K. Antonsson, eds. 2009. *Springer Handbook of Mechanical Engineering.*
New York: Springer. At nearly 1,600 pages, this handbook covers most major areas of
mechanical engineering and beyond (chapters on mathematics, enterprise organization
and optimization, and electrical engineering). Graphically rich, with over 1,800 figures

and 400 tables. The print version of the book also contains an accompanying DVD (for Windows, Mac, or Linux) containing the entire content of the book.

Hicks, T. G., ed. 2006. *Handbook of Mechanical Engineering Calculations,* 2nd ed. New York: McGraw-Hill. While Hicks' pocket guide that follows focuses on listing formulas for specific settings and applications, this volume is a much more extensive, step-by-step guide to what formulas and calculations should be used in dozens of specific applications. Contains hundreds of pages of supporting text.

Hicks, T. G. 2003. *Mechanical Engineering Formulas Pocket Guide.* New York: McGraw-Hill. As the title suggests, the focus of this small format guide is on formulas used by mechanical engineers in various applications. Explanatory text is kept to a minimum.

Kreith, F. and D. Y. Goswami, eds. 2005. *CRC Handbook of Mechanical Engineering,* 2nd ed. Boca Raton, FL: CRC Press. The mechanical engineering entry in the successful CRC Press handbooks series. At roughly 2,500 pages, perhaps the largest and most extensive of all mechanical engineering handbooks. Covers areas such as transportation, communication and information systems, and patent law not covered by most other ME handbooks.

Kutz, M., ed. 2006. *Mechanical Engineers' Handbook,* 3rd ed. New York: John Wiley & Sons. Wiley's entry into mechanical engineering handbooks. Expanded to a four-volume set for the third edition. Volume one covers materials and mechanical design; volume two deals with instrumentation, systems, control, and MEMS; volume three addresses manufacturing and management; and volume four encompasses energy and power.

Lindeburg, M. R. 2006. *Mechanical Engineering Reference Manual for the PE Exam,* 12th ed. Belmont, CA: Professional Publications. Although containing much of the same basic mechanical engineering information as other handbooks in this section, the focus of this work is on those concepts needed for successful completion of the U.S. Professional Engineering (PE) exam in mechanical engineering. Includes examples with detailed solutions.

Marghitu, D. B., ed. 2001. *Mechanical Engineers' Handbook.* San Diego: Academic Press. Almost a combination of handbook and textbook, the presentation is perhaps more at a novice level than that of some of the other works included in this section. Also distinguishing this volume is that many chapters offer the reader problems and worked solutions.

Matthews, C. 2005. *ASME Engineer's Data Book,* 2nd ed. New York: ASME Press.

Matthews, C. 2006. *Engineer's Data Book,* 3rd ed. reprinted with corrections. London/Bury St. Edmund, U.K.: Professional Engineering Publishing. While both titles are pocket-sized, written by the same author, and focus on mechanical engineering equations and data, each title is surrounded by background material and supporting documentation pertinent to its country of origin.

Pope, J. E., ed. 1997. *Rules of Thumb for Mechanical Engineers: A Manual of Quick, Accurate Solutions to Everyday Mechanical Engineering Problems* Houston: Gulf Publishing Co. Large format and soft cover, this volume is very applications-oriented. The focus, across 16 chapters, is on shortcuts, calculations, and practical methods for use in different areas of mechanical engineering.

Smith, E. H., ed. 1998. *Mechanical Engineer's Reference Book,* 12th ed. Oxford, U.K./Boston: Butterworth-Heinemann. Hicks' work (above) is the pocket guide supported by IMechE, while this volume edited by Smith is their full handbook entry. Extensive, with a definite U.K. focus in much of the supporting material.

Timings, R. 2005. *Newnes Mechanical Engineer's Pocket Book,* 3rd ed. Oxford, U.K.: Elsevier/Newnes. Written with a very hands-on, practical approach, and primarily for a British audience, this title covers the basic mathematics and formulae of mechanical engineering, statics, dynamics, fastenings, power transmission, materials, and more. Refers heavily to appropriate British standards.

AUTOMOTIVE

Robert Bosch GmbH. 2007. *Automotive Electrics/Automotive Electronics,* 5th ed. Stuttgart: Robert Bosch. Contains a compilation of content on automotive electrical and electronics systems culled from the German-published Bosch Technical Instruction series.

Robert Bosch GmbH. 2007. *Bosch Automotive Handbook,* 7th ed. Stuttgart: Robert Bosch. A pocket-sized reference work on all aspects of automotive technology, including relevant physics and math concepts. Focused primarily on passenger cars and commercial vehicles. The seventh edition is the first to address hybrids.

Caines, A. J., R. F. Haycock, and J. E. Hillier. 2004. *Automotive Lubricants Reference Book,* 2nd ed. Warrendale, PA: Society of Automotive Engineers. As the title suggests, this work covers content on lubrication that in any way impacts the automotive industry. That includes typical automotive oils, greases, and lubricating fluids, but also includes gas turbine oils, railroad oils, and industrial lubricants used in automotive plants. Contains 15 appendices on various topics, including an extensive glossary.

Cebon, D. 1999. *Handbook of Vehicle–Road Interaction.* Lisse, Exton, PA: Swets & Zeitlinger Publishers. This title is where mechanical and civil engineering meet in the automotive world. This work covers the interaction between vehicles and the surfaces they move upon. Of particular importance to the automotive engineer is how aspects of vehicle dynamics affect stability, performance, and damage to the vehicle or road surface.

Challen, B. and R. Baranescu. 1999. *Diesel Engine Reference Book,* 2nd ed. Woburn, MA: Butterworth-Heinemann. Covers all aspects of the design and performance of diesel engines of all sizes. Includes an extensive section on environmental aspects.

Fenton, J. 1998. *Handbook of Automotive Body and Systems Design.* London: Professional Engineering. Covers numerous aspects of vehicle design, first in general areas, such as aerodynamics, body trim and fittings, and ergonomics, then addresses numerous aspects of electrical and related systems design. Covers road cars, specialist (racing) cars, trucks, and more specialized vehicles, such as tankers, refrigerated vehicles, light vans, ambulances, and others.

Fenton, J. 1998. *Handbook of Automotive Body Construction and Design Analysis.* London: Professional Engineering. Contains detailed descriptions of body fabrication and assembly techniques, panel cutting and forming, finishing, various materials used in automotive construction, and significant content on passenger safety concerns.

Fenton, J. 1998. *Handbook of Automotive Powertrain and Chassis Design.* London: Professional Engineering. The focus here is two-fold: (1) what powers a vehicle and (2) the framework on which that power plant resides. Various types of power plants and designs are discussed, including gasoline, diesel, natural gas, electric, and hybrids. Then content switches to discussions of suspensions, steering and handling, wheels and braking, drive axles and drive trains, and transmissions.

Fitch, J. W. 1994. *Motor Truck Engineering Handbook,* 4th ed. Warrendale, PA: Society of Automotive Engineers. This work focuses specifically on motor trucks— those trucks used in all areas of the trucking industry.

Genta, G. and L. Morello, 2009. *The Automotive Chassis: Volume 1: Components Design, Volume 2: System Design.* New York: Springer. What the Van Busshuysen and Schaefer work (referenced below) does for the internal combustion engine, this pair of volumes does for the chassis (everything that is not the engine or the body). Volume one addresses all of the individual components that go into chassis design, while the second volume covers how a chassis is designed, and how those individual components are merged into an integrated and smoothly functional whole. Treatment is both mathematical and from a hands-on engineering perspective. Both volumes are illustrated. Each volume has its own symbols list immediately following the table of contents, plus individual bibliographies and indices at the end of the volume, while volume two alone contains a series of five appendices.

Jurgen, R., ed. 1999. *Automotive Electronics Handbook,* 2nd ed. New York: McGraw-Hill. This work covers every aspect of electrical systems as they apply to automobiles, from lights and displays to the sensors and actuators that control and monitor the functioning of most systems within the modern automobile. Most chapters have their own glossary.

Mollenhauer, K. and H. Tschöke, eds. 2010. *Handbook of Diesel Engines.* Berlin: London: Springer. A translation of a German work originally published in 2007. Divided into five broad areas (diesel engine cycle, engineer, operation, environmental pollution, and how and where diesel engines are used). A list of pertinent standards, primarily German (DIN), ISO, and SAE, precedes the index at the end of the volume.

SAE Handbook. 2010. New York: Society of Automotive Engineers. The definitive reference work for anyone working in the automotive industry in the United States. Contains most of SAE's automotive-related standards in CD format. Contains numeric and subject indices. No longer produced in print.

Van Basshuysen, R. and F. Schaefer, 2004. *Internal Combustion Engine Handbook: Basics, Components, Systems, and Perspectives*, New York: SAE International and Professional Engineering Publishing. At over 900 pages, this work covers the latest in the theory and practice of internal combustion engine components, systems, and development. Heavily illustrated, and containing nearly 700 bibliographical references, this book addresses the function and design of an internal combustion engine and its individual components as well as the fluids (fuels, oils, and associated filtering) that interact with those engine parts.

Energy, Heat Transfer, HVACR, Fluid Mechanics/Dynamics/Flow

ASHRAE Handbook. Fundamentals. 2009. Atlanta, GA: American Society of Heating, Refrigerating and Air-Conditioning Engineers.

ASHRAE Handbook. Heating, Ventilating and Air-Conditioning Applications. 2007. Atlanta, GA: American Society of Heating, Refrigerating and Air-Conditioning Engineers.

ASHRAE Handbook. Heating, Ventilating and Air-Conditioning Systems and Equipment. 2008. Atlanta, GA: American Society of Heating, Refrigerating and Air-Conditioning Engineers.

ASHRAE Handbook. Refrigeration. 2010. Atlanta, GA: American Society of Heating, Refrigerating and Air-Conditioning Engineers. This set of four handbooks is a comprehensive work on the various facets of heating, ventilating, air-conditioning and refrigerating. Each volume comes in both SI and inch-pound editions, and one volume of the four volume set is updated annually.

Bejan, A. and A. D. Krauss, eds. 2003. *Heat Transfer Handbook.* New York: John Wiley & Sons. In 19 chapters, this title covers every major type of heat transfer: from the basics, such as conduction, convection, and radiation, through boiling and evaporation, and on to more specific topics, such as heat transfer in electronic equipment and in manufacturing and materials processing.

Burton, T., D. Sharpe, D. Jenkins, and E. Bossanyi. 2001. *Wind Energy: Handbook.* New York: John Wiley & Sons. Although beginning with a chapter on wind as a basic resource, the primary focus of this book is on wind turbine aerodynamics, design, installation, and control.

Cheremisinoff, N. P., ed. 1986–1990. *Handbook of Heat and Mass Transfer.* Houston: Gulf Publishing Co. A four-volume set: volume one covers heat transfer operations, volume two mass transfer and reactor design, volume three addresses catalysis, kinetics, and reactor engineering, and volume four covers advances in reactor design and combustion science.

Hewitt, G. F., ed. 2008. *HEDH: Heat Exchanger Design Handbook,* rev. ed. New York: Begell House. This five-volume set is divided into broad topical areas: theory, fluid mechanics, and heat transfer, thermal and hydraulic design of heat exchangers, mechanical design of heat exchangers, and physical properties. This last section is particularly data-focused.

Johnson, R. W., ed. 1998. *Handbook of Fluid Dynamics.* Boca Raton, FL: CRC Press. Covers all aspects of fluid dynamics, but with a particular emphasis on theoretical, computational, and experimental aspects. Chapters related to applications are a secondary focus of this work, not a primary one.

Kreith, F., ed. 2000. *CRC Handbook of Thermal Engineering.* Boca Raton, FL: CRC Press. This handbook focuses on specific current topics in thermal engineering rather than on fundamentals and theory.

Kreith, F. and D. Y. Goswami, eds. 2008. *Energy Management and Conservation Handbook.* Boca Raton, FL: CRC Press. This work takes both broad and specific looks at energy conservation and management, from examining U.S. energy consumption and economics to building and equipment-specific examination of how to improve energy efficiency and conservation.

Kreith, F. and D. Y. Goswami, eds. 2007. *Handbook of Energy Efficiency and Renewable Energy.* Boca Raton, FL: CRC Press. Similar in some ways to the title above, this volume addresses in more detail energy efficiencies of various kinds, and covers more specific engineering topics, such as co-generation, energy storage, and heat exchangers not covered in the previous work. Also included are 10 chapters on various renewable energy technologies.

Kuppan, T. 2000. *Heat Exchanger Design Handbook.* New York: Marcel Dekker. Discusses design, construction, and appropriate modes of operation for various types of heat exchangers. Offers examples of applications in a number of different industries, and addresses appropriate standards and codes.

Landa, H. C., J. M. L. Sanders, and D. C. Landa. 2005. *The Solar Energy Handbook*, 6th ed. Wauwatosa, WI: Film Instruction Co. of America. Beginning with chapters on theory and data, this handbook then moves into various solar energy applications, focusing in particular on heating, ventilating, and air conditioning, but also covering electrical conversion, plus industrial and agricultural applications.

Miller, R. W., ed. 1996. *Flow Measurement Engineering Handbook,* 3rd ed. New York: McGraw-Hill. As the name suggests, the focus here is on flow measurement. A series of 12 appendices make up over a quarter of the volume.

Minkowycz, W. J., E. M. Sparrow, and J. Y. Murthy, eds. 2006. *Handbook of Numerical Heat Transfer,* 2nd ed. New York: John Wiley & Sons. This handbook is divided into two primary sections, Fundamentals and Applications. Within Fundamentals, chapters are devoted to different heat transfer calculation methods (finite difference, finite-element, boundary element, etc.), while Applications focuses on situations where those various methods would and should be used.

Rohsenow, W. M., J. P. Hartnett, and Y. I. Cho, eds. 1998. *Handbook of Heat Transfer,* 3rd ed. New York: McGraw-Hill. Focuses heavily on the physics of heat transfer rather than on applications.

Saleh, J. M., ed. 2002. *Fluid Flow Handbook.* New York: McGraw-Hill. Focusing on fluid dynamics and fluid flow, this work offers practical examples to illustrate various theoretical principles. Useful in a number of engineering disciplines.

Schetz, J. A. and A. E. Fuhs, eds. 1996. *Handbook of Fluid Dynamics and Fluid Machinery.* New York: John Wiley & Sons. Volume one of this three-volume set focuses on the fundamentals of fluid dynamics, volume two on experimental and computational fluid dynamics, and volume three on applications. The emphasis on machinery and the size and coverage of the work set it off from others in the field.

Stoeker, W. F. 1998. *Industrial Refrigeration Handbook.* New York: McGraw-Hill. This volume covers different types of refrigeration systems, particularly multistate systems. Equipment, such as reciprocating and screw compressors, evaporators, chillers, and condensers, are described in detail, as are appropriate applications. Piping, vessels, valves, and controls appropriate to industrial refrigeration are described, as are the types and properties of various refrigerants.

Thumann, A. and D. P. Mehta, 2008. *Handbook of Energy Engineering,* 6th ed. Lilburn, GA: Fairmont Press; Boca Raton, FL: CRC Press. Although focused primarily on power generation, chapters dealing with HVAC systems and equipment will be of interest to some in mechanical engineering.

Tropea, C., A. L. Yarin, and J. F. Foss, eds., 2007. *Springer Handbook of Experimental Fluid Mechanics.* Berlin: Springer. This work is divided into four sections covering basic experiments in fluid mechanics, measurement of various quantities/properties, specific experimental environments and techniques, and data processing and analysis. Accompanied by a DVD with "key data for daily use."

Turner, W. C. and S. Doty, eds. 2009. *Energy Management Handbook,* 7th ed. Lilburn, GA: Fairmont Press; Boca Raton, FL: CRC Press. Chapters on various systems used to generate energy, on HVAC, and on energy storage systems, and in particular the focus on economics, highlight the mechanical engineering aspects of this work.

HYDRAULICS AND PNEUMATICS

Barber, A. 1997. *Pneumatic Handbook,* 8th ed. Oxford, U.K.: Elsevier Science. Comprehensive work on all aspects of compressed air. Unlike most handbooks in this chapter, the *Pneumatic Handbook* contains dozens of advertisements from companies working in the compressed air industry.

Hunt, T. M., N. Vaughn, and R. H. Warring. 1996. *Hydraulic Handbook,* 9th ed. Oxford, U.K.: Elsevier Advanced Technology. Focuses primarily on the mechanical engineering aspects of hydraulics rather than those of concern to the civil engineer. As with a number of other Elsevier-published handbooks, this work also contains advertisements from companies in the hydraulics industry.

McCulloch, D. M. 2003. *Compressed Air and Gas Handbook,* 6th ed. Englewood Cliffs, NJ: Prentice Hall. This work on applications and uses of compressed air and other gases is culled from the accumulated experiences of member companies of the Compressed Air and Gas Institute.

Parr, E. A. 2006. *Hydraulics and Pneumatics: A Technician's and Engineer's Guide,* 3rd ed. Oxford, U.K./Boston: Butterworth-Heinemann. A relatively brief overview of hydraulics and pneumatics. Covered are hydraulic pumps and pressure regulation, air pressure, treatment and regulation, control valves of various types, actuators, hydraulic and pneumatics accessories, process control pneumatics, and safety, fault-finding, and maintenance.

Totten, G. E. and V. J. de Negri, eds. 2009. *Handbook of Hydraulic Fluid Technology,* 2nd ed. Boca Raton: FL: CRC Press. Covers the narrowly focused area of hydraulic fluids and associated technology in great detail. A brief discussion on hardware leads off the volume, followed by chapters on fluid properties, compatibility issues, and testing and analysis. The last half of the volume contains chapters on various hydraulic fluids. Eleven appendices are included.

Yeaple, F. D. 1996. *Fluid Power Design Handbook,* 3rd ed. New York: Marcel Dekker. This volume is very simply and cleanly laid out. Major sections address: hydraulics, including fluids, equipment, and applications; pneumatics, focusing on equipment and applications; accessories, such as filters, tubing, piping hoses, fittings, and seals; control and performance issues; and system design.

MACHINERY AND MACHINING

Bleier, F. P. 1998. *Fan Handbook: Selection, Application and Design*. New York: McGraw-Hill. The basics of stationary and moving air are defined, followed by detailed descriptions and application information about various types of fans, ventilators, and blowers. Appropriate standards are covered, as is information on installation, safety, maintenance, trouble-shooting, and problem solving.

Boyce, M. P. 2006. *Gas Turbine Engineering Handbook,* 3rd ed. Boston: Gulf Professional Publishing. Provides an overview of gas turbine design, including discussions of the major components, such as compressors, turbines, and combustors, along with auxiliary components and accessories, such as bearings, seals, and gears. Installation, operation, and maintenance are also addressed.

Boyes, W. E. and R. Bakerjian, eds. 1989. *Handbook of Jig and Fixture Design,* 2nd ed. Dearborn, MI: Society of Manufacturing Engineers. A handbook that deals not with the materials being machined or the process of machining, but rather with the devices that position, support, and hold an item during machining or other processes (fixtures) and the devices that provide additional guidance to a cutting or machining tool (jigs).

Brown, Jr., T. H. 2005. *Mark's Calculations for Machine Design*. New York: McGraw-Hill. Heavy on formulae and calculations, this work is divided into two main sections. Part one addresses strength of machines and focuses on loading, stresses, buckling, and fatigue. Part two addresses applications and covers assembly, energy, and motion.

Drozda, T. J. and C. Wick, et al., eds. 1983–1998. *Tool and Manufacturing Engineers Handbook* 4th ed. Dearborn, MI: Society of Manufacturing Engineers. This nine-volume set, released over a period of 16 years, covers machining; forming; materials, finishing, and coating; quality control and assembly; manufacturing management; design for manufacturability; continuous improvement; plastic part manufacturing; and material and part handling in manufacturing. Also available as a multi-CD set.

Giampaolo, T. 2009. *The Gas Turbine Handbook,* 4th ed., Lilburn, GA: Fairmont Press. This work examines the various types of gas turbines, including microturbines, and their applications, operation, and troubleshooting. Turbine theory, control systems, inlets and exhaust, accessories, acoustics, and noise control and prevention are also addressed. Four case studies offer real world examples of various issues related to gas turbines.

Lingaiah, K. 2003. *Machine Design Databook,* 2nd ed. New York: McGraw-Hill. One of a number of related handbooks from McGraw-Hill. This work, as its name suggests, focuses on formulas, calculations, and other data needed for various machine design applications. Background or supporting text is kept to a bare minimum.

Logan Jr., E. and R. Roy, eds. 2003. *Handbook of Turbomachinery* 2nd ed., revised and expanded. Boca Raton, FL: CRC Press. This work covers the design and operation of many different types of turbomachines, from gas and steam turbines through turbines used in rocket propulsion and hydraulics, plus automobile superchargers and turbochargers.

Machinability Data Center. 1980. *Machining Data Handbook,* 3rd ed. Cincinnati, OH: The Machinability Data Center. A classic work in the field, this two-volume set is a compendium of data for the machining and finishing of materials into various types of machine parts, both simple and complex.

Machinery's Handbook, 28th ed. 2008. New York: Industrial Press, Inc.

Machinery's Handbook Guide, 28th ed. 2008. New York: Industrial Press, Inc. The *Handbook* covers the properties, testing, and manufacturing of tools and basic types of machine parts. The *Guide* explains in detail how to use the tables and formulas in the *Handbook*. The 28th edition of both the *Handbook* and *Guide* are available in print and on CD-ROM. On CD, the *Guide* is included on the same disc as the *Handbook*, but requires a separate purchase of a *Guide* "key" in order to access that content.

Metal Cutting Tool Handbook 7th ed. 1989. New York: Industrial Press, for the Metal Cutting Tool Institute. Addresses all forms of mechanical metal cutting, but does not address any other forms of cutting (laser, water jet, etc.).

Parmley, R. O., ed. 2005. *Machine Devices and Components Illustrated Sourcebook.* New York: McGraw-Hill. Different than most of the handbooks listed in this chapter, this work offers detailed illustrations of various mechanical parts, their assembly, and their uses in both conventional and innovative applications.

Rothbart, H. A., and T. H. Brown Jr., eds. 2006. *Mechanical Design Handbook; Measurement, Analysis, and Control of Dynamic Systems,* 2nd ed. New York: McGraw-Hill. This work focuses on the fundamentals and design considerations that need to be taken into account by anyone interested in designing and constructing machines of various types.

Sawyer, J. W., and D. Japikse, eds. 1985. *Sawyer's Gas Turbine Engineering Handbook,* 3rd ed. Norwalk, CT: Turbomachinery International Publications. This three-volume work addresses theory and design, selection and various applications, and accessories and support for those in the gas turbine industry.

Sclater, N. and N. P. Chironis. 2001. *Mechanisms & Mechanical Devices Sourcebook,* 4th ed. New York: McGraw-Hill. Similar in format to the Parmley title above; while that title focuses on mechanical components, this work focuses on more complex creations—the combination of components that result in a mechanism or other more complex mechanical devices.

Shigley, J. E., C. R. Mischke, and T. H. Brown, eds. 2004. *Standard Handbook of Machine Design,* 3rd ed. New York: McGraw-Hill. While the McGraw-Hill work by Lingaiah is a databook with the emphasis on data, this work is a handbook with much more explanatory text and background material. Subject area coverage, on the other hand, is quite similar.

Suchy, I. 2006. *Handbook of Die Design,* 2nd ed. New York: McGraw-Hill. Covers various types of dies and their functions and construction. Describes various types of metalworking machinery and operations, including blanking, piercing, bending, forming, and drawing. Significant attention is given to materials, their properties, and surface finishing.

Walsh, R. A. and D. R. Cormier. 2006. *McGraw-Hill Machining and Metalworking Handbook,* 3rd ed. New York: McGraw-Hill. This handbook covers all typical areas of interest to machinists or metalworkers, but also covers standards and standardizing organizations, and safety. Contains a brief list of societies and other organizations with specific connections to various machining and metalworking areas.

MISCELLANEOUS

Ahmad, A., ed. 1997. *Handbook of Optomechanical Engineering.* Boca Raton, FL: CRC Press. Optomechanical engineering addresses where and how mechanical systems interact and impact with modern optics. Topics include optical mounts, adjustment systems, thermal and thermoelastic issues, materials and their properties, and manufacturing.

Bishop, R. H., ed. 2008. *The Mechatronics Handbook,* 2nd ed. Boca Raton, FL: CRC Press. Mechatronics is the integration of mechanical, electrical, and computer systems. This volume details the design and materials of mechatronic systems and their various applications in microelectromechanical systems, sensors and actuators, controls, etc.

Cho, H., ed. 2003. *Opto-mechatronic Systems Handbook: Techniques and Applications.* Boca Raton, FL: CRC Press. Mechatronics is the area where mechanical, electrical, and computer technologies blend and merge. Opto-mechatronics adds optics to mechantronics. This volume addresses the burgeoning new field of opto-mechantronics and applications of those technologies.

Dudley, D. W. 1994. *Handbook of Practical Gear Design,* rev. ed. Lancaster, PA: Technomic Publishing Co. This work not only deals with the design and manufacturing of various types of gears, it also branches into the design of tools to make gears and has an extensive section on gear failures.

Flitney, R. K. and M. W. Brown. 2007. *Seals and Sealing Handbook,* 5th ed. Oxford, U.K.: Elsevier Advanced Technology. Addresses all aspects of seals and sealing technology. As with many other Elsevier handbooks, contains advertisements from companies in the industry.

Gad-el-Hak, M., ed. 2006. *The MEMS Handbook,* 2nd ed. Boca Raton, FL: CRC Press. Covers one of the newest areas of research for mechanical and other engineers: microelectrome-chanical systems. Included are discussions of scaling, mechanical properties, fabrication, and lubrication of MEMS as well as many current and projected applications.

Hoffman, D. M., B. Singh, and J. H. Thomas III, eds. 1998. *Handbook of Vacuum Science and Technology.* San Diego, CA: Academic Press. Addresses the materials, design, and construction of pumps and related systems for the creation of vacuum. Also covers measurement techniques and applications.

Jousten, K., ed. 2008. *Handbook of Vacuum Technology.* Weinheim, Germany: Wiley-Blackwell. Beginning with chapters on the history of vacuum science and technology and the basics of modern vacuum technology, this work then addresses particular applications, such as in condensers and pumps. Materials, components, seals, and leak detection and control are also covered.

Kurfess, T. R., ed. 2005. *Robotics and Automation Handbook.* Boca Raton, FL: CRC Press. This title begins with a history of robotics and moves quickly into how to design, construct, enable, and control various types of robots. Examples are offered of the use of robots in specific industries, such as welding, surgery, assembly, etc.

Nesbitt, B., ed. 2007. *Valves Manual International: Handbook of Valves and Actuators.* Oxford, U.K.: Elsevier. Led by a 40-page glossary of definitions and abbreviations in Chapter 1, this work covers valves and actuators of all types and for virtually any application. Most chapters end with a set of "useful references." The table of contents contains advertisements from valve and actuator manufacturers and suppliers, and the volume closes with a Buyer's Guide.

Parmley, R. O., ed. 1997. *Standard Handbook of Fastening and Joining,* 3rd ed. New York: McGraw-Hill. This work addresses the multitude of processes by which materials or components can be connected. Included are fasteners, such as threads; pins; rings; pipe connections; expansion joints; various types of welding; concrete, lumber, and steel connections; locks; electrical connections; adhesives; rivets; rope and cable connections; shafts and couples; seals; and even rope splicing and tying.

Piotrowski, J. 2007. *Shaft Alignment Handbook,* 3rd ed. Boca Raton: CRC Press. This title deals with the narrow field of shaft alignment, a field crucial to the proper functioning of pumps, gears, turbines, fans, engines, compressors, etc. Particular emphasis is placed on measuring proper function and on techniques to correct misalignments.

Sharpe Jr., W. N., ed. 2008. *Springer Handbook of Experimental Solid Mechanics.* New York: Springer. A revision and expansion of Kobayashi's *Handbook on Experimental Mechanics,* last published in 1993. More than half of the chapters in this work cover areas that have arisen or matured in the 15 years since the last edition. The volume is divided into four sections: solid mechanics topics, contact methods, noncontact methods, and applications.

Siciliano, B. and O. Khatib, eds. 2008. *Springer Handbook of Robotics.* Berlin: Springer. This award-winning handbook is divided into seven parts (Robotics Foundations, Robot Structures, Sensing and Perception, Manipulation and Interfaces, Mobile and Distributed Robotics, Field and Service Robotics, Human-Centered and Life-Like Robotics) with up to 14 chapters per section. Contains a section on further readings at the end of each chapter.

Skousen, P. L. 2004. *Valve Handbook,* 2nd ed. New York: McGraw-Hill. While the Smith and Zappe title (below) focuses solely on valves, this volume covers both valves and actuators. Valve selection criteria are detailed, as are the various types of valves and actuators. Sizing, purchasing issues, and common problems are also discussed.

Smith, P. and R. W. Zappe. 2004. *Valve Selection Handbook,* 5th ed. Houston: Gulf Publishing Co. This work is very applications-oriented. Initial pages are devoted to fundamentals, but the majority of the volume is devoted to detailed descriptions of various valve types, how they are to be installed, and appropriate uses for each. Appendices on ASME safety codes, properties of fluids, and various standards pertaining to valves are included.

Stokes, A. 1992. *Gear Handbook: Design and Calculations.* Oxford, U.K.: Butterworth-Heinemann. The emphasis on this work is on the "calculations" portion of the title. This relatively brief volume contains all the formulae necessary to design, manufacture, and inspect a wide variety of gear types. Produced in association with the Society of Automotive Engineers.

Taylor, J. I. 2000. *The Gear Analysis Handbook.* Tampa, FL: Vibration Consultants, Inc. Narrow in focus, this work describes methods of gear analysis and appropriate tools, data collection, gear physics and specifications, and various gear problems and their causes.

Taylor, J. I. and D. W. Kirkland. 2004. *The Bearing Analysis Handbook.* Tampa, FL: Vibration Consultants, Inc. Bearing defects, problem analysis and data collection, and case histories documenting various types of defects highlight this volume.

Townsend, D. P., ed. 1992. *Dudley's Gear Handbook,* 2nd ed. New York: McGraw-Hill. Broader in scope than the title by Stokes, the emphasis here is on the manufacture, functions, and problems associated with different gear types.

Walsh, R. A. 2000. *Electromechanical Design Handbook,* 3rd ed. New York: McGraw-Hill. This volume addresses designing when the designer must incorporate some combination of mechanical, electrical, and electronic components into the finished product. Emphasis is heavily toward the practical and less to theory.

NOISE, SHOCK, VIBRATION

Barber, A., ed. 1992. *Handbook of Noise and Vibration Control,* 6th ed. Oxford, U.K.: Elsevier Advanced Technology. Includes sections on defining types and causes of vibration and noise, ways of measuring them, and suggestions for remediation. Unlike many Elsevier handbooks, it does not contain advertisements.

Crocker, M. J. 2007. *Handbook of Noise and Vibration Control.* Hoboken, NJ: John Wiley & Sons. This work addresses noise and vibration on multiple fronts, not just from an engineering perspective. Beyond the fundamentals, chapters cover human hearing and speech and the effects of noise, blast, vibration, and shock on people as well as how these effects are generated and can be controlled.

de Silva, C. W., ed., 2005. *Vibration and Shock Handbook.* Boca Raton, FL: CRC Press. With nearly 50 contributors and at nearly 1,900 pages, this title covers virtually every aspect of noise, shock, vibration, and acoustics. Divided into nine sections, each with multiple chapters, this work covers fundamentals, computer techniques, instrumentation and testing, monitoring and diagnosis, suppression and control, and more. Contains an extensive glossary at the end of the volume.

Piersol, A. G., T. L. Paez, and C. M. Harris, eds. 2010. *Harris' Shock and Vibration Handbook,* 6th ed. New York: McGraw-Hill. For any machine with relatively high-speed moving parts, vibration is a serious consideration. This work addresses all aspects of vibration and its control, some of which are directly pertinent to the mechanical engineer.

Taylor, J. I. 2003. *The Vibration Analysis Handbook,* 2nd ed. Tampa, FL: Vibration Consultants, Inc. While the other two books with Taylor as author or co-author focus on vibration in specific locations (gears and bearings), this work is more general in nature. It discusses machinery vibration in general, time and frequency analysis techniques, how to evaluate the condition of machinery, plus chapters on gears, bearings, and press roll and nip problems.

OCEAN, MARINE, SHIP

El-Hawary, F., ed. 2001. *Ocean Engineering Handbook.* Boca Raton, FL: CRC Press. Geared primarily to electrical engineers, also covered are topics such as vehicle control and positioning and thrust control systems, both relevant to mechanical engineering.

Hunt, E. C., et al., eds. 1999–2002. *Modern Marine Engineers Manual,* 3rd ed. Centreville, MD: Cornell Maritime Press. This two-volume set covers those topics of concern to the marine engineer, with many of those topics tied directly to mechanical engineering. Included are engineering materials, steam power plants, bearings and lubrication, steam generation, steam and gas turbines, transmission systems, heat exchangers, piping systems, engines, HVAC and refrigeration systems, and vibration analysis.

Lamb, T., ed. 2003–2004. *Ship Design and Construction.* New York: Society of Naval Architects and Marine Engineers. Volume one addresses the shipbuilding industry and all the basic considerations that go into it (costs, human factors, safety, materials, etc.). Volume two addresses issues specific to the design and construction of particular types of ships, such as cargo ships, passenger ships, tankers, fishing vessels, naval ships (including submarines), and even hovercraft.

PRESSURE VESSELS AND PIPING

Bednar, H. H. 1986. *Pressure Vessel Design Handbook,* 2nd ed. Malabar, FL: Krieger Publishing Company. Describes design considerations for many types and shapes of pressure vessels. Particular consideration is given to various types of stresses and loading, including localized stresses from supports, attachments, etc. Contains a brief glossary.

Bernstein, M. D. and L. W. Yoder. 1998. *Power Boilers: A Guide to Section 1 of the ASME Boiler and Pressure Vessel Code.* New York: American Society of Mechanical Engineers. Focuses entirely on design considerations related to power boilers and their attachments. Materials, piping, fabrication and welding, testing, inspection, certification, quality control, safety, and potential causes of damage or failure are all covered.

Dickenson, T. C., ed. 1999. *Valves, Piping and Pipelines Handbook,* 3rd ed. Oxford, U.K.: Elsevier Advanced Technology. As the title suggests, the handbook focuses entirely on valves and pipes and does not address pressure vessels. Covers over 20 types of valves, various metal and plastic pipes, and contains a large section on determining pipe flow performance.

Ellenberger, J. P. 2010. *Piping and Pipeline Calculations Manual: Construction, Design, Fabrication, and Examination.* Amsterdam, Boston: Butterworth-Heinemann/Elsevier. Written by an author with 40 years of personal experience as an engineer and instructor, this handbook focuses on the calculations, codes and standards related to piping system design, construction and maintenance. Dozens of worked examples support the text.

Frankel, M. L. 2010. *Facility Piping Systems Handbook—For Industrial, Commercial, and Healthcare Facilities,* 3rd ed. New York: McGraw-Hill. Beginning with appropriate codes and standards, this volume covers virtually every type of piping in use in a commercial or industrial setting. Appendices cover pipe distribution systems, a glossary and abbreviations, and unit conventions and conversions, including those for pipe sizes.

Helguero, M. V. 1986. *Piping Stress Handbook,* 2nd ed. Houston: Gulf Publishing Co. As the title suggests, this book addresses specifically the various types and sources of stresses that can affect pipes and piping systems, including internal pressure, corrosion, weight, pressure from external supports, etc. Focus is on data rather than text.

Kohan, A. L. 1998. *Boiler Operator's Guide,* 4th ed. New York: McGraw-Hill. Covers different types of boilers and boiler systems, appropriate construction materials and their properties, fabrication and repair methods, and connections, appurtenances, and auxiliaries. Operational issues, such as combustion, fuels, burners and controls, are addressed, as are inspections, problems, repairs, and maintenance. Each chapter ends with a series of questions and answers.

Malek, M. A. 2007. *Heating Boiler Operator's Manual: Maintenance, Operation, and Repair.* New York: McGraw-Hill. Unlike the Rayaprolu work (below) that addresses all types of boilers, this handbook by Malek focuses on boilers used specifically for heating applications. Various boiler types, fuel systems, and instrumentation and controls are covered.

Matthews, C. 2001. *Engineer's Guide to Pressure Equipment.* London: Professional Engineering. Another pocket guide from Matthews (see General section above for two more). This work focuses on pressure equipment of all kinds, not just vessels or containers. Thus, pipes, flanges, valves, nozzles, and other fittings are considered. Contains a list of pertinent web sites.

McAllister, E. W., ed. 2009. *Pipe Line Rules of Thumb Handbook: Quick and Accurate Solutions to Your Everyday Pipe Line Problems,* 7th ed., Houston: Gulf Publishing Co. Very "how-to" in focus, this handbook is designed primarily for those working with, rather than designing, piping systems. Format poses questions or concerns and then supplies answers.

McCauley, J. F. 2000. *Steam Distribution Systems Deskbook.* Lilburn, GA: Fairmont Press. Unlike some of the boiler handbooks, which cover steam generation, this work focuses largely on steam distribution. Included are discussions of various types of piping systems, insulation, valve systems, and controls.

Megyesy, E. F. 2008. *Pressure Vessel Handbook,* 14th ed. Tulsa, OK: Pressure Vessel Publishing, Inc. Although called a handbook, this work is probably better described as a databook. Text is at a minimum throughout the work, with the focus on data elements for design and construction, geometry and layout, measures and weights, etc.

Menon, E. S. 2005. *Piping Calculations Manual.* New York: McGraw-Hill. Nine chapters cover different types of piping systems: water, fire protection, wastewater and stormwater, steam, compressed air, oil, gas, fuel gas, and cryogenic and refrigeration. Appendices cover units and conversions, pipe properties, and viscosity corrected pump performance.

Moser, A. P. and S. L. Folkman. 2008. *Buried Pipe Design,* 3rd ed. New York: McGraw-Hill. Focuses entirely on considerations related to buried pipe systems. Those include loading and other external forces, pressure considerations, designs of various types of piping systems, choice of materials, and installation.

Moss, D. R. 2004. *Pressure Vessel Design Manual: Illustrated Procedures for Solving Major Pressure Vessel Design Problems,* 3rd ed. Amsterdam/Boston: Elsevier, Gulf Professional Publishing. Designed as more of a "how-to" work than Bednar's book, this work is heavily illustrated and is procedurally oriented. Contains over a dozen appendices on various topics.

Nayyar, M. L. 2002. *Piping Databook.* New York: McGraw-Hill. While Nayyar's handbook (below) is heavily text-oriented, this work is virtually all data related to piping and the piping industry. Included are units and conversions, materials, and various organization specifications for types of pipe, fittings, flanges, and joints. Much more content is devoted to nonmetallic piping than what appears in the handbook.

Nayyar, M. L., ed. 2000. *Piping Handbook,* 7th ed. New York: McGraw-Hill. At over 2,400 pages, this work covers components, materials, codes and standards, manufacturing/fabrication, installation, dozens of design considerations, and piping systems for various applications, primarily related to metal piping. Contains a small section on nonmetallic piping, and 10 appendices, most of which are data tables.

Rayaprolu, K. 2009. *Boilers for Power and Process.* Boca Raton, FL: CRC Press. Divided into three sections (Boiler Fundamentals, Boiler Part and Auxiliaries, Boilers and Firing), this work addresses boiler heat and fluid flow basics, addresses the basic components and attachments of a typical boiler system, and discusses various ways of firing boiler systems.

Sixsmith, T. and R. Hanselka. 1997. *Handbook of Thermoplastic Piping System Design.* New York: Marcel Dekker. This work covers in detail the advances in plastic pipe and piping systems and applications where modern plastics can be used in place of more traditional metal pipe. Includes over 70 pages of basic reference data and nearly 200 pages of chemical resistance data.

Williams Natural Gas Company, Engineering Group. 1996. *Pipe Characteristics Handbook.* Tulsa, OK: Pennwell Books. Although billed as a handbook, virtually the entire work is two sets of data tables: 200 pages on maximum allowable working pressure as a percent of specified minimum yield, and less than 20 pages on properties of pipe.

Pumps and Pumping

Dickenson, T. C., ed. 1995. *Pumping Manual,* 9th ed. Oxford, U.K.: Elsevier Advanced Technology. Covers pump performance and characteristics, 20 different types of pumps, appropriate materials, construction and operation, pump ancillaries (engines, motors, seals, bearing, etc.) and a variety of pumping applications. Contains advertisements from companies in the pumping industry.

Gülich, J. F. 2010. *Centrifugal Pumps,* 2nd ed. Berlin/New York: Springer. This handbook starts with an introduction to fluid mechanics, then moves immediately into pump types, characteristics, applications, operation, and performance. Also addressed are pump noise, vibration, and wear.

Karassik, I. J., ed. 2008. *Pump Handbook,* 4th ed. New York: McGraw-Hill. Covers fewer types of pumps than the Dickenson title, but in greater detail. Contains additional material on pump drivers, controls and valves, pumping systems, intakes and suction piping, and a variety of pumping applications.

Mackay, R. 2004. *The Practical Pumping Handbook.* Oxford, U.K.: Elsevier. As the title suggests, this is a very practical treatment of pumping, piping, and seals. Unlike some other similar works, Mackay's work explores pump selection and purchasing, troubleshooting, and maintenance, in addition to more standard content on the topic. Contains conversion tables and formulae and a brief bibliography at the end of the volume.

Miller, R., M. R. Miller, and H. L. Stewart. 2004. *Pumps and Hydraulics,* 6th ed. Indianapolis: John Wiley & Sons. A basic handbook on pumps and hydraulics. Sectioned into three parts covering basic concepts and properties, pumps, hydraulics, and associated fluids, lines and fittings.

Nesbitt, B., ed. 2006. *Handbook of Pumps and Pumping.* Oxford, Burlington, MA: Elsevier, in association with Roles & Associates Ltd. This work takes a practical approach to pumps and pump-related topics, including classifications, materials, seals, installation, commissioning and maintenance. Seals, couplings and ancillary equipment are addressed, and a listing of worldwide manufacturers and suppliers is provided.

Rayner, R., ed. 1995. *Pump Users Handbook*. Oxford, U.K.: Elsevier Advanced Technology. As the title suggests, this work is very applications-oriented. Criteria for pump selection are covered as well as for installation, testing, and start-up. Different types of pumps and drives are documented, followed by various pumping applications.

Rishel, J. B., T. H. Durkin, and B. L. Kincaid. 2006. *HVAC Pump Handbook,* 2nd ed. New York: McGraw-Hill. Covers all aspects of HVAC pumps and pumping. Includes design basics, selection criteria, installation and operation, performance issues and measures, and the use of pumps in open and closed HVAC cooling systems and in HVAC hot water systems.

TRIBOLOGY AND LUBRICATION

Association of Iron and Steel Engineers. 2007. *Lubrication Engineers Manual,* 3rd ed. Warrendale, PA: Association for Iron and Steel Technology. General manual for anyone involved with lubrication or maintenance. Included are sections on fundamentals, test methods, statistics, performance, and others.

Bhushan, B., ed. 2001. *Modern Tribology Handbook*. Boca Raton, FL: CRC Press. Volume one of this two-volume set covers the principles of tribology at both the macro and micro levels in 20 chapters, while volume two covers tribological materials and coatings, plus their many and varied applications in industry.

Bhushan, B. and B. K. Gupta. 1991. *Handbook of Tribology: Materials, Coatings and Surface Treatments*. New York: McGraw-Hill. Although this work touches on the basics of friction, wear, and lubrication, the vast majority of the text deals with different types of coatings and their application/deposition and with treating surfaces by various methods.

Bloch, H. P. 2009. *Practical Lubrication for Industrial Facilities,* 2nd ed. Lilburn, GA: Fairmont Press. This updated edition by Bloch describes categories of lubricants and their testing, addresses in individual chapters specific types of lubricants (natural and synthetic oils, greases, waxes, etc.), then delves into the appropriate industrial uses of those lubricants.

Booser, E. R., ed. 1983–1994. *CRC Handbook of Lubrication: (Theory and Practice of Tribology)*. Boca Raton, FL: CRC Press. This three-volume set from CRC is somewhat odd in design in that volumes one and two were issued in 1983 and 1984, respectively, then volume three was issued in 1994 and covers new developments and updates to the field in the decade since publication of the original two volumes. Volume one focuses on application and maintenance, volume two on theory and design, and volume three on monitoring, materials, synthetic lubricants, and applications.

Booser, E. R., ed. 1997. *Tribology Data Handbook*. Boca Raton, FL: CRC Press. Designed as a follow-up or companion volume to the CRC set listed immediately above, the focus of this work is on data. Among the topics covered are properties of various lubricants, lubrication specifications for different types of equipment, material properties, industrial application processes, and friction, wear, and surface characterization.

Neale, M. J., ed. 1995. *Tribology Handbook,* 2nd ed. Oxford, U.K.: Butterworth-Heinemann. This work is very applications-directed and touches only marginally on theory. Performance characteristics are primary in presentation. Various types of bearings, drives, seals, lubricants and lubrication systems are addressed as well and failures, environmental effects, and maintenance and repair.

Totten, G. E., ed. 2006– . *Handbook of Lubrication and Tribology,* 2nd ed. Boca Raton, FL: CRC Press. The first volume in a revision of the *CRC Handbook of Lubrication* (see one of the Booser entries above) is all that has been published to date. The focus of volume one is applications and maintenance. Significantly revised and expanded, it covers topics such as biotribology and the tribology of hard disk drives, among others, not addressed in the first edition.

Vižintin, J., M. Kalin, K. Dohda, and S. Jahanmir., eds. 2004. *Tribology of Mechanical Systems: A Guide to Present and Future Technologies.* New York: ASME Press. Addresses lubrication and appropriate lubricants for all types of mechanical systems. Included are typical metal-based systems, such as engines, but also covered are materials, such as ceramics, polymers and composites.

MONOGRAPHS AND TEXTBOOKS

There are hundreds of textbooks covering mechanical engineering and its subdisciplines. The textbooks included here represent an overview of some of the textbooks currently used in classrooms. Monographs were chosen as representative of classical works in the field of mechanical engineering. Both textbooks and monographs were chosen based on book reviews in *Books in Print* and in consultation with mechanical engineering professors. Major publishers for monographs and textbooks include Dover (for reprints of seminal works), as well as Cambridge University Press, John Wiley & Sons, McGraw-Hill Higher Education, Oxford University Press, Prentice Hall, and Princeton University Press.

Beer, F. P., E. R. Johnston, Jr., J. T. DeWolf, and D. Mazurek. 2009. *Mechanics of Materials,* 5th ed. New York: McGraw-Hill. Covers mechanics of materials in general, as well as stress and strain, torsion, bending, stresses under specific loading conditions, deflection of beams, and energy methods. Contains new and revised problems, and homework sets with answers at the back of the book. Presents concepts clearly and in detail, and includes illustrations and diagrams to explain concepts.

Beer, F. P., E. R. Johnston, Jr., E .R. Eisenberg, W. E. Clausen, and G. H. Staab. 2009. *Vector Mechanics for Engineers: Statics and Dynamics,* 9th ed. New York: McGraw-Hill. Designed for beginning courses in statics and dynamics. Each chapter contains sample problems with solutions, end-of-chapter problems, and computer problems as well as a review and summary. Almost 40% of the problems are new in this edition.

Budynas, R., J. K. Nisbett, and J. E. Shigley. 2011. *Shigley's Mechanical Engineering Design,* 9th ed. Boston: McGraw-Hill. A classic intended for beginning students, this text covers the significant machine components and design decisions encountered in machine design in a clear and concise manner. Presented in four parts (basics, failure prevention, design of mechanical elements, and analysis tools), emphasis is on developing good design and problem-solving skills. Includes new applications and problem sets, as well as an Online Learning Center to support both students and instructors.

Den Hartog, J. P. 1985. *Mechanical Vibrations,* 4th ed. New York: Dover Publications. Reprint. Originally published 1956. New York: McGraw-Hill. Covers the fundamentals of mechanical vibrations. Can be used by practicing engineers as well as for classroom instruction. An elementary knowledge of dynamics and calculus is necessary, but differential equations are explained in detail. Examples have been drawn from real-life experiences of the author and his friends. Each chapter includes problems that illustrate typical practical situations with answers included at the back.

Den Hartog, J. P. 1987. *Advanced Strength of Material.* New York: Dover Publications. Reprint. Originally published 1952. New York: McGraw-Hill. Intended for the intermediate student studying strength of materials. Covers torsion, rotating disks, membrane stresses in shells, bending of flat plates, two-dimensional theory of elasticity and buckling. Includes problem sets and answers.

Dorf, R. C., and R. H. Bishop. 2010. *Modern Control Systems,* 12th ed. Upper Saddle River, NJ: Prentice Hall. Introduces control systems, including a brief history of control theory and practice. Provides coverage of classical control and modern control methods including mathematical models of systems, state variable models, feedback control system characteristics, performance of feedback control systems, stability of linear feedback systems, the root locus method, frequency response methods, stability in the frequency domain, design of feedback, state variable feedback, robust control, and digital control systems. Students have the opportunity to apply theory and analysis using the many examples provided throughout the book. Exercises and problems at the end of each chapter. Incorporates computer-aided design and analysis using MATLAB® and LabVIEW MathScript, updated.

Fermi, E. 1956. *Thermodynamics*. New York: Dover. Based on lectures given by Fermi in 1936, this book is about pure thermodynamics. Fermi describes the state of a system and its transformations, the first and second laws of thermodynamics, thermodynamic potentials, and thermodynamics of dilute solutions. This is an elementary treatment of thermodynamics, but the reader should be familiar with the fundamentals of thermometry and calorimetry.

Gere, J. M. 2004. *Mechanics of Materials,* 6th ed. London: Brooks/Cole. Describes the fundamentals of mechanics of materials. Principal topics are analysis and design of structural members subjected to tension, compression, torsion, and bending as well as stress, strain, elastic behavior, inelastic behavior, and strain energy. Transformations of stress and strain, combined loadings, stress concentrations, deflections of beams, and stability of columns are also covered. Includes many problem sets with answers in the back.

Hibbeler, R. C. 2011. *Mechanics of Materials,* 8th ed. Upper Saddle River, NJ: Prentice Hall. Clear and comprehensive, this text includes stress, strain, mechanical properties of materials, axial load, torsion, bending, stress and strain transformations, design and deflection of beams and shafts, buckling of columns, and energy methods. Includes a photorealistic art program that helps students visualize concepts.

Incropera, F. P. and D. P. DeWitt. 2006. *Fundamentals of Heat and Mass Transfer,* 6th ed. New York: John Wiley & Sons. Provides the basics of heat and mass transfer. This edition includes areas of current interest such as fuel cells and alternative energy devices, electronics cooling, microscale heat transfer, and biological as well as bioheat transfer. Contains an extensive collection of new, revised, updated examples and homework problems illustrating real life engineering processes.

Lewis, E. V., ed. 1988. *Principles of Naval Architecture,* 2nd ed. Jersey City, NJ: Society of Naval Architects and Marine Engineers. This comprehensive three-volume set, SNAME's leading reference work, covers load modeling, flows around hulls, propeller design, vibration, stability and control, and motion in waves. Vol. 1 Stability and Strength; vol. 2 Resistance, Propulsion; vol. 3 Seakeeping and Controllability.

Meirovitch, L. 2001. *Fundamentals of Vibrations*. Boston: McGraw-Hill. Covers all basic concepts at a level appropriate for undergraduate engineering students and is excellent for introductory vibrations courses. Meirovitch, one of the leading authorities in the field of vibration, emphasizes analytical developments and computational solutions. Includes physical explanations, problems at the end of each chapter, worked-out examples, and illustrations.

Mott, R. L. 2007. *Machine Elements in Mechanical Design,* 5th ed. Upper Saddle River, NJ: Prentice Hall. Provides a current, thorough, practical approach to designing machine elements in the context of complete mechanical design. Includes some of the primary machine elements, such as belt drives, chain drives, gears, shafts, keys, couplings, seals, and rolling contact bearings. It also covers plain surface bearings, linear motion elements, fasteners, springs, machine frames, bolted connections, welded joints, electric motors, controls, clutches, and brakes.

Mott, R. L. 2008. *Applied Strength of Materials,* 5th ed. Upper Saddle River, NJ: Prentice Hall. Useful for undergraduate, introductory level courses. Coverage is comprehensive in the key areas of strength of materials. Includes basic concepts, design properties, direct stress, deformation, and design, torsional shear stress and torsional deformation, shearing forces and bending moments in beams, stress due to bending, shearing stresses in beams, deflection of beams, combined stresses, columns, pressure vessels, and riveted and bolted connections.

Munson, B. R., D. F. Young, T. H. Okiishi, and W. W. Huebsch. 2008. *Fundamentals of Fluid Mechanics,* 6th ed. New York: John Wiley & Sons. Intended as an undergraduate textbook on fluid statics, dynamics, and kinematics. This edition includes a CD-ROM/ DVD with 150 video segments designed to aid student comprehension and new problem types to engage students such as case studies from news stories, as well as 1,300 homework problems.

Nise, N. S. 2010. *Control Systems Engineering,* 6th ed. Hoboken, NJ: John Wiley & Sons. Focuses on the practical application of control systems engineering. Explains how to analyze and design real-world feedback control systems. Real world examples demonstrate the design process. This edition includes 20% new end-of-chapter problems that highlight biomedical, aerospace, and other engineering applications. Additional skill assessment exercises, in-chapter examples, review questions, and problems reinforce key concepts.

O'Hanlon, J. F. 2003. *A User's Guide to Vacuum Technology,* 3rd ed. New York: John Wiley & Sons. A detailed, practical guide to current vacuum technology that emphasizes the fundamentals while covering operation procedures, understanding, and selection of equipment used in semiconductor, optics, packaging, and related coating technologies.

Timoshenko, S. P. 1983. *Strength of Materials, Part 1 and Part 2,* 3rd ed. Reprint. Malabar, FL: R. E. Krieger Publishing Co. Part 1: Elementary Theory and Problems covers the fundamentals of strength of materials, such as stress, deformation, and strain. Part 2: Advanced Theory and Problems includes later developments that are of practical importance in the fields of strength of materials, and theory of elasticity. Both books include problems and solutions. Many of the problems illustrate how to apply the theory to solve practical design problems.

Timoshenko, S. P., and D. H. Young. 1968. *Elements of Strength of Materials,* 5th ed. Princeton, NJ: Van Nostrand. Timoshenko, a pioneer in the field of applied mechanics, influenced the teaching of mechanics throughout the world through his research and textbooks. This book covers the fundamentals of the elements of strength of materials, such as stress, deformation, and strain.

Ullman, D. G. 2010. *The Mechanical Design Process,* 4th ed. Boston: McGraw-Hill. Primarily appropriate for mechanical engineering senior design courses. Moves through the design process from beginning to end, from why study the design process to wrapping up the process and supporting the finished product. Includes properties of the 25 most commonly used materials in mechanical design, safety as a design variable, and human factors in design. Also includes case studies and examples from industry along with 20 online templates.

Van Ness, H. C. 1983. *Understanding Thermodynamics.* New York: Dover. Covers the basic concepts of thermodynamics. Not intended to cover thermodynamics in the same way as a textbook, meant more as a supplement to assist students through the difficult early stages of a beginning course in thermodynamics. Topics covered include the first law of thermodynamics, the concept of reversibility, heat engines, power plants; the second law of thermodynamics; and statistical mechanics. Written in an engaging, informal style.

White, F. M. 2011. *Fluid Mechanics,* 7th ed. Boston: McGraw-Hill. Covers fluid mechanics fundamentals from physical concepts to engineering applications. Includes pressure distribution in a fluid, integral relations for a control volume and differential relations for fluid flow, viscous flow in ducts, compressible flow, open-channel flow, as and turbomachinery. Includes modern day examples such as a flying car and kite-driven ships and new material on microflow concepts.

JOURNALS

Mechanical engineering journals cover a broad range of topics from heating and refrigeration to automobiles and thermodynamics. To determine the most significant journals in the mechanical engineering field, the following resources were consulted: *Journal Citation Reports* (using the categories: Mechanics; and Engineering, Mechanical); *Guide to Information Sources in Engineering* (Libraries Unlimited, 2000) by Charles Lord; *Scientific and Technical Information Sources* by Ching-chih Chen; *Encyclopedia of Physical Sciences and Engineering Information Sources* by Steven Wasserman, Martin Smith, and Susan Mottu; and *Core List of Books and Journals in Science and Technology* (Onyx, 1987) by Russell Powell and James Powell Jr.

Acta Mechanica. 1965–. Vienna: Springer (0001-5970).

Advances in Applied Mechanics. 1948–. Amsterdam: Elsevier (0065-2156).

Annual Review of Fluid Mechanics. 1969–. Palo Alto, CA: Annual Reviews (0066-4189).

Applied Mechanics Reviews. 1948–. New York: ASME International (0003-6900).

Applied Thermal Engineering. 1980–. Amsterdam: Elsevier (1359-4311).

ASHRAE Journal. 1959–. New York: American Society of Heating, Refrigerating and Air-Conditioning Engineers (0001-2491).

Automotive Engineering International. 1998–. Warrendale, PA: Society of Automotive Engineers (1543-849X).

Combustion and Flame. 1963–. Amsterdam: Elsevier (0010-2180).

Computers and Fluids. 1973–. Amsterdam: Elsevier (0045-7930).

Energy Conversion and Management. 1980–. Amsterdam: Elsevier (0196-8904).

Experimental Thermal and Fluid Science, 1988–. Amsterdam: Elsevier (0894-1777).

Experiments in Fluids. 1983–. New York: Springer (0723-4864).

Heat Transfer Engineering. 1979–. Philadelphia: Taylor & Francis (0145-7632).

IEEE/ASME Transactions on Mechatronics. 1996–. Piscataway, NJ: Institute of Electrical and Electronics Engineers (1083-4435).

International Journal of Engine Research. 2000–. London: Professional Engineering Publications (1468-0874).

International Journal of Fatigue. 1979–. Amsterdam: Elsevier (0142-1123).

International Journal of Heat and Fluid Flow. 1979–. New York: Elsevier (0142-727X).

International Journal of Heat and Mass Transfer. 1960–. Amsterdam: Elsevier (0017-9310).

International Journal of Impact Engineering. 1983–. Amsterdam: Elsevier (0734-743X).

International Journal of Machine Tools and Manufacture. 1987–. Amsterdam: Elsevier (0890-6955).

International Journal of Mechanical Sciences. 1960–. Amsterdam: Elsevier (0020-7403).

International Journal of Multiphase Flow. 1974–. Amsterdam: Elsevier (0301-9322).

International Journal of Plasticity. 1985–. Amsterdam: Elsevier (0749-6419).

International Journal of Refrigeration/Revue Internationale du Froid. 1978–. Amsterdam: Elsevier (0140-7007).

International Journal of Thermal Science. 1999–. Amsterdam: Elsevier (1290-0729).

Journal of Applied Mechanics. Transactions of the ASME. 1933–. New York: ASME International (0021-8936).

Journal of Biomechanical Engineering. Transactions of the ASME. 1977–. New York: ASME International (0148-0731).

Journal of Computational and Nonlinear Dynamics. Transactions of the ASME. 2006–. New York: ASME International (1555-1415).

Journal of Computing and Information Science in Engineering. Transactions of the ASME. 2001–. New York: ASME International (1530-9827).

Journal of Dynamic Systems Measurement and Control. Transactions of the ASME. 1971–. New York: ASME International (0022-0434).

Journal of Electronic Packaging. Transactions of the ASME. 1989–. New York: ASME International (1043-7398).

Journal of Energy Resources Technology. Transactions of the ASME. 1979–. New York: ASME International (0195-0738).

Journal of Engineering for Gas Turbines and Power. Transactions of the ASME. 1984–. New York: ASME International (0742-4795).

Journal of Engineering Materials and Technology. Transactions of the ASME. 1973–. New York: ASME International (0094-4289).

Journal of Engineering Mechanics. 1983–. New York: American Society of Civil Engineers (0733-9399).

Journal of Fluid Mechanics. 1956–. Cambridge: Cambridge University Press (0022-1120).

Journal of Fluids Engineering. Transactions of the ASME. 1973–. New York: ASME International (0098-2202).

Journal of Fuel Cell Science and Technology. Transactions of the ASME. 2004–. New York: ASME International (1550-624X).

Journal of Heat Transfer. Transactions of the ASME. 1959–. New York: ASME International (0022-1481).

Journal of Manufacturing Science and Engineering. Transactions of the ASME. 1996–. New York: ASME International (1087-1357).

Journal of Mechanical Design. Transactions of the ASME. 1990–. New York: ASME International (1050-0472).

Journal of Mechanisms and Robotics. Transactions of the ASME. 2009–. New York: ASME International (1942-4302).

Journal of Medical Devices. Transactions of the ASME. 2007–. New York: ASME International (1932-6181).

Journal of Microelectromechanical Systems. 1992–. Piscataway, NJ: Institute of Electrical and Electronics Engineers (1057-7157).

Journal of Micromechanics and Microengineering. 1991–. Bristol, U.K.: Institute of Physics (0960-1317).

Journal of Nanotechnology in Engineering and Medicine. Transactions of the ASME. 2010–. New York: ASME International (1949-2944).

Journal of Non-Newtonian Fluid Mechanics. 1976–. Amsterdam: Elsevier (0377-0257).

Journal of Offshore Mechanics and Arctic Engineering. Transactions of the ASME. 1987–. New York: ASME International (0892-7219).

Journal of Pressure Vessel Technology. Transactions of the ASME. 1974–. New York: ASME International (0094-9930).

Journal of Rheology. 1957–. New York: Society of Rheology (0148-6055).

Journal of Solar Energy Engineering. Transactions of the ASME. 1980–. New York: ASME International (0199-6231).

Journal of Sound and Vibration. 1964–. Amsterdam: Elsevier (0022-460X).

Journal of Statistical Mechanics—Theory and Experiment. 2004–. Bristol, U.K.: Institute of Physics (1742-5468).

Journal of Strain Analysis for Engineering Design. 1976–. London: Professional Engineering Publishing (0309-3247).

Journal of Thermal Science and Engineering Applications. Transactions of the ASME. 2009–. New York: ASME International (1948-5085).

Journal of Tribology. Transactions of the ASME. 1984–. New York: ASME International (0742-4787).

Journal of Turbomachinery. Transactions of the ASME. 1986–. New York: ASME International (0889-504X).

Journal of Vibration and Acoustics. Transactions of the ASME. 1990–. New York: ASME International (1048-9002).

Mechanical Engineering. 1906–. New York: American Society of Mechanical Engineers (0025-6501).

Mechanical Systems and Signal Processing. 1987–. Amsterdam: Elsevier (0888-3270).

Mechanics of Materials. 1982–. Amsterdam: Elsevier (0167-6636).

Mechanism and Machine Theory. 1972–. Amsterdam: Elsevier (0094-114X).

Nanoscale and Microscale Thermophysical Engineering. 1997–. Philadelphia: Taylor & Francis (1556-7265).

Nonlinear Dynamics. 1990–. New York: Springer (0924-090X).

Numerical Heat Transfer. Part A, Applications. 1989–. Philadelphia: Taylor & Francis (1040-7782).

Numerical Heat Transfer. Part B, Fundamentals. 1989–. Philadelphia: Taylor & Francis (1040-7790).

Probabilistic Engineering Mechanics. 1986–. Amsterdam: Elsevier (0266-8920).

Proceedings of the Combustion Institute. 2000–. Amsterdam: Elsevier (1540-7489).

Proceedings of the Institution of Mechanical Engineers. Part A, Journal of Power and Energy. 1990–. London: Professional Engineering Publishing (0957-6509).

Proceedings of the Institution of Mechanical Engineers. Part B, Journal of Engineering Manufacture. 1989–. London: Professional Engineering Publishing (0954-4054).

Proceedings of the Institution of Mechanical Engineers. Part C, Journal of Mechanical Engineering Science. 1983–. London: Professional Engineering Publishing (0954-4062).

Proceedings of the Institution of Mechanical Engineers. Part D, Journal of Automobile Engineering. 1989–. London: Professional Engineering Publishing (0954-4070).

Proceedings of the Institution of Mechanical Engineers. Part E, Journal of Process Mechanical Engineering. 1989–. London: Professional Engineering Publishing (0954-4089).

Proceedings of the Institution of Mechanical Engineers. Part F, Journal of Rail and Rapid Transit. 1989–. London: Professional Engineering Publishing (0954-4097).

Proceedings of the Institution of Mechanical Engineers. Part G, Journal of Aerospace Engineering. 1989–. London: Professional Engineering Publishing (0954-4100).

Proceedings of the Institution of Mechanical Engineers. Part H, Journal of Engineering in Medicine. 1989–. London: Professional Engineering Publishing (0954-4119).

Proceedings of the Institution of Mechanical Engineers. Part I, Journal of Systems and Control Engineering. 1991–. London: Professional Engineering Publishing (0959-6518).

Proceedings of the Institution of Mechanical Engineers. Part J, Journal of Engineering Tribology. 1994–. London: Professional Engineering Publishing (1350-6501).

Proceedings of the Institution of Mechanical Engineers. Part K, Journal of Multi-body Dynamics. 1999–. London: Professional Engineering Publishing (1464-4193).

Proceedings of the Institution of Mechanical Engineers. Part L, Journal of Materials Design and Applications. 1999–. London: Professional Engineering Publishing (1464-4207).

Proceedings of the Institution of Mechanical Engineers. Part M, Journal of Engineering for the Maritime Environment. 2002–. London: Professional Engineering Publishing (1475-0902).

Proceedings of the Institution of Mechanical Engineers. Part N, Journal of Nanoengineering and Nanosystems. 2004–. London: Professional Engineering Publishing (1740-3499).

Proceedings of the Institution of Mechanical Engineers. Part O, Journal of Risk and Reliability. 2006–. London: Professional Engineering Publishing (1748-006X).

Proceedings of the Institution of Mechanical Engineers. Part P, Journal of Sports, Engineering and Technology. 2008–. London: Professional Engineering Publishing (1754-3371).

Progress in Energy and Combustion Science. 1976–. Oxford: Elsevier (0360-1285).

Rheologica Acta. 1958–. New York: Springer (0035-4511).

SAE Transactions. 1893–. Warrendale, PA: Society of Automotive Engineers (0096-736X).
Tribology International. 1968–. Amsterdam: Elsevier (0301-679X).
Tribology Letters. 1995–. New York: Springer (1023-8883).
Wear. 1957–. Amsterdam: Elsevier (0167-6636).
Wind Energy. 1998–. New York: John Wiley & Sons (1095-4244).

PATENTS AND STANDARDS

There are no databases or other resources specifically geared toward patents in mechanical engineering (see Chapter 2, General Engineering Resources in this book for patent information). However, there are a number of organizations that produce standards in ME and related areas. General standards resources that offer mechanical engineering standards are listed in Chapter 2. Standardizing organizations producing codes and standards specifically related to mechanical engineering are listed below.

> *ASME Standards.* New York: American Society of Mechanical Engineers. With over 600 codes and standards being currently produced and distributed, ASME is the preeminent standardizing organization in the world for mechanical engineering and has been producing codes and standards for nearly a century. Of particular note is the *ASME Boiler and Pressure Vessel Code.* Produced since 1914, the *Code* is now adopted by 49 of the 50 states in the United States and by all the provinces in Canada.
>
> *ASTM Standards.* West Conshohocken, PA: American Society for Testing and Materials. With over 12,000 standards in 15 sections published in over 70 volumes, the *Annual Book of ASTM Standards* is an invaluable resource for mechanical engineers. Available in print, CD-ROM, or online.
>
> SAE Standards. Warrendale, PA: Society of Automotive Engineers. SAE has produced over 8,300 standards, primarily in two broad areas: aerospace and ground vehicle. Aerospace content is, in turn, divided primarily between Aerospace Materials Specifications and Aerospace Standards. Ground vehicle standards are primarily known as J-Reports, with most published annually in the *SAE Handbook.* Both aerospace and ground vehicle standards are available in print, on CD-ROM, or online.

See also Chapter 3, Aeronautical and Aerospace Engineering, and Chapter 20, Transportation Engineering in this book.

SEARCH ENGINES AND IMPORTANT WEB SITES

Topical Web sites can readily be divided into two groups: meta sites, which collect links to more specific topic-specific sites, and those topic-specific sites. Representative meta sites are listed immediately below, followed by a select list of topic-specific sites.

META SITES

The following sites were accessed October 28, 2010.

> Australasian Virtual Engineering Library: Mechanical Engineering http://avel.library.uq.edu.au/browse/browse_tree_204.html
> BUBL Links
> > Energy Research http://www.bubl.ac.uk/link/e/energyresearch.htm
> > Energy Technology http://www.bubl.ac.uk/link/e/energytechnology.htm

Fluid Mechanics http://www.bubl.ac.uk/link/f/fluidmechanics.htm

Mechanical Engineering http://bubl.ac.uk/link/m/mechanicalengineering.htm

Pneumatics http://www.bubl.ac.uk/link/p/pneumatics.htm

Eng-Links http://www.shender4.com/eng-links.htm (accessed November 3, 2010)

Galaxy Directory: Mechanical Engineering (plus more than a dozen subdisciplines) http://www.galaxy.com/galaxy/Engineering-and-Technology/Mechanical-Engineering.html (accessed October 28, 2010)

icademic: Mechanical Engineering http://www.icademic.org/97445/Mechanical-Engineering (accessed November 3, 2010)

iCrank.com http://www.icrank.com (accessed October 28, 2010)

Intute: Engineering (click on "Mechanical Engineering and Related Industries") http://www.intute.ac.uk/engineering/ (accessed October 28, 2010)

Open Directory Project: Science: Technology: Mechanical Engineering http://www.dmoz.org/Science/Technology/Mechanical_Engineering/ (accessed October 28, 2010)

ViFaTec (Virtuelle Fachbibliothek Technik) http://vifatec.tib.uni-hannover.de/fit/index.php3?L=e (click "browse" next to the mechanical engineering category) (accessed October 28, 2010)

Yahoo! Directory: Engineering: Mechanical Engineering http://dir.yahoo.com/Science/engineering/mechanical_engineering/ (accessed October 28, 2010)

INDIVIDUAL SUBJECT SITES

Animated Engines http://www.keveney.com/Engines.html (accessed October 28, 2010). As the title suggests, this site provides access to animations of how nearly 20 different types of engines function.

Automotive Intelligence http://www.autointell.com/ (accessed October 28, 2010). One of a number of sites collecting news and information on the automotive industry.

E-fluids.com http://www.efluids.com (accessed October 28, 2010). A meta site in its own right, e-fluids.com brings together a variety of resources related to flow engineering, fluid mechanics research, education, and related topics.

Leonardo Centre for Tribology/University of Sheffield http://www.tribology.group.shef.ac.uk/index.html (accessed October 28, 2010). Based at the University of Sheffield, this site offers both information on the research and teaching of the tribology team at Sheffield as well as tools and information resources in tribology and lubrication.

MEMS and Nanotechnology Clearinghouse http://www.memsnet.org/ (accessed October 28, 2010). A clearinghouse site for information on micro-electro-mechanical (microelectro-mechanical) systems (MEMS) and nanotechnology. Offers background information via a Beginner's Guide and a glossary, as well as content for the engineering student or engineer, including industry news, a conferences and workshops calendar, list of job openings, etc.

Shock and Vibration Information Analysis Center http://www.saviac.org/ (accessed October 28, 2010). This site, supported by the U.S. Department of Defense and managed by the U.S. Army Engineer Research and Development Center, is focused on research and analysis in the field of shock and vibration technology.

Thermal Connection http://www.tak2000.com/ThermalConnection.htm (accessed October 28, 2010). Thermal Connection collects tools, data, and links relevant to anyone working in heat transfer and related fields.

Vibrationdata.com http://www.vibrationdata.com (accessed October 28, 2010). A subscription-based site offering tutorials, software, and other resources related to acoustics, shock and vibration, signal processing, and other areas.

ASSOCIATIONS, ORGANIZATIONS, AND SOCIETIES

Acoustical Society of America http://asa.aip.org (accessed October 28, 2010). "The premier international scientific society in acoustics, dedicated to increasing and diffusing the knowledge of acoustics and its practical applications" (from Web site).

American Bearing Manufacturers Association http://www.americanbearings.org/ (accessed October 28, 2010). "The American Bearing Manufacturers Association (ABMA) is a non-profit association consisting of American manufacturers of antifriction bearings, spherical plain bearings, or major components thereof. The purpose of ABMA is to define national and international standards for bearing products and maintain bearing industry statistics" (from Web site).

American Gear Manufacturers Association http://www.agma.org (accessed October 28, 2010). AGMA's membership is largely companies/corporations involved in some aspect of the gear, coupling, or mechanical power transmission industries.

American Society for Nondestructive Testing http://www.asnt.org (accessed October 28, 2010). "ASNT exists to create a safer world by promoting the profession and technologies of nondestructive testing … through publication, certification, research, and conferencing" (from Web site).

American Society of Heating, Refrigerating, and Air-Conditioning Engineers http://www.ashrae.org (accessed October 29, 2010). "ASHRAE will advance the arts and sciences of heating, ventilation, air conditioning, refrigeration, and related human factors to serve the evolving needs of the public and ASHRAE members" (from Web site).

American Society of Mechanical Engineers http://www.asme.org (accessed October 29, 2010). "Founded in 1880 as the American Society of Mechanical Engineers, today's ASME is a 120,000-member professional organization focused on technical, educational and research issues of the engineering and technology community. ASME conducts one of the world's largest technical publishing operations, holds numerous technical conferences worldwide, and offers hundreds of professional development courses each year. ASME sets internationally recognized industrial and manufacturing codes and standards that enhance public safety" (from Web site).

American Welding Society http://www.aws.org (accessed October 29, 2010). "The American Welding Society (AWS) was founded in 1919 as a multifaceted, nonprofit organization with a goal to advance the science, technology, and application of welding and related joining disciplines" (from Web site).

Association of Energy Engineers http://www.aeecenter.org (accessed October 29, 2010). "AEE is your source for information on the dynamic field of energy efficiency, utility deregulation, facility management, plant engineering, and environmental compliance" (from Web site).

Combustion Institute http://www.combustioninstitute.org (accessed October 29, 2010). "The Combustion Institute is an educational nonprofit, international, scientific society whose purpose is to promote and disseminate research in combustion science" (from Web site).

Compressed Gas Association http://www.cganet.com (accessed October 29, 2010). "Since 1913, the Compressed Gas Association has been dedicated to the development and promotion of safety standards and safe practices in the industrial gas industry" (from Web site).

Institution of Mechanical Engineers http://www.imeche.org/Home (accessed October 29, 2010). "Established in 1847, the Institution of Mechanical Engineers (IMechE) is the leading body for professional mechanical engineers. With a worldwide membership now in excess of 75,000 engineers, the IMechE is the United Kingdom's qualifying body for Chartered and Incorporated mechanical engineers" (from Web site).

International Union of Theoretical and Applied Mechanics http://www.iutam.net (accessed October 29, 2010). "The International Union of Theoretical and Applied Mechanics (IUTAM) was formed in 1946 with the object of creating a link between persons and national or international organizations engaged in scientific work (theoretical or applied) in solid and fluid mechanics or in related sciences" (from Web site).

NACE International http://www.nace.org (accessed October 29, 2010). "NACE International was originally known as 'The National Association of Corrosion Engineers' when it was established in 1943. With more than 60 years of experience in developing corrosion prevention and control standards, NACE International has become the largest organization in the world committed to the study of corrosion" (from Web site).

National Fluid Power Association http://www.nfpa.com (accessed October 29, 2010). NFPA's mission is to "advance hydraulic and pneumatic motion control technology" and "to act for and represent, on a worldwide basis, companies active in the U.S. fluid power industry" (from Web site).

SAE International: the Society of Automotive Engineers http://www.sae.org (accessed October 29, 2010). "The Society of Automotive Engineers has more than 84,000 members—engineers, business executives, educators, and students from more than 97 countries—who share information and exchange ideas for advancing the engineering of mobility systems. SAE is your one-stop resource for standards development, events, and technical information and expertise used in designing, building, maintaining, and operating self-propelled vehicles for use on land or sea, in air or space" (from Web site).

Society for Experimental Mechanics http://www.sem.org (accessed October 29, 2010). "The Society for Experimental Mechanics is composed of international members from academia, government, and industry who are committed to interdisciplinary application, research and development, education, and active promotion of experimental methods to: (a) increase the knowledge of physical phenomena; (b) further the understanding of the behavior of materials, structures and systems; and (c) provide the necessary physical basis and verification for analytical and computational approaches to the development of engineering solutions" (from Web site).

Society of Naval Architects and Marine Engineers http://www.sname.org (accessed October 29, 2010). "The Society of Naval Architects and Marine Engineers is an internationally recognized nonprofit, technical, professional society of individual members serving the maritime and offshore industries and their suppliers. SNAME is dedicated to advancing the art, science and practice of naval architecture, shipbuilding, and marine engineering" (from Web site).

Society of Tribologists and Lubrication Engineers http://www.stle.org (accessed October 29, 2010). STLE's mission is "to advance the science of tribology and the practice of lubrication engineering in order to foster innovation, improve the performance of equipment and products, conserve resources and protect the environment" (from Web site).

Vibration Institute http://www.vibinst.org (accessed October 29, 2010). "A nationally recognized not-for-profit organization dedicated to the exchange of practical information about vibration and condition monitoring" (from Web site).

ACKNOWLEDGMENTS

The authors wish to thank Bonnie Osif for the opportunity to contribute to the 2nd edition. Aleteia wishes to thank Mel DeSart because once again, it was a pleasure collaborating. Mel wishes to thank Aleteia for taking the lead on this revision and for her energy, good humor, and willingness to co-author … again, and to thank reference assistants Nathalie Wargo and Talea Anderson for doing some URL checking.

17 Mining Engineering

Jerry Kowalyk

CONTENTS

Introduction and Scope 427
History of Mining Engineering 428
History of Mining Engineering 428
    History of Mining Web Sites 429
Information Sources and Guides to the Literature of Mining Engineering and Technology 429
Searching the Library Catalog 430
Abstracts and Indexes 430
Journals 434
    Selected Mining Journals 434
    Journal Titles Related to Mining Environmental Management 436
Handbooks and Manuals 436
Dictionaries 438
Encyclopedias 438
Directories 439
Government Ministries 440
    Australia 440
    Canada 440
    United States 441
Associations, Organizations, and Societies 443
    Australia 443
    Canada 444
    International 445
    United Kingdom 445
    United States 445
Metal Prices: Indexes 447
Industry Profiles and Yearbooks 447
Statistics 448
Educational Subject Guides to Mining and Mining Engineering on Selected Web Sites 449
Conclusion 451

INTRODUCTION AND SCOPE

Mining involves the recovery of solid, liquid, or gas mineral substances from the Earth for direct use or conversion into a useable product. Recovery of energy minerals in the form of natural gas and petroleum has developed into a separate industry with its own technology. That aspect of mining engineering is discussed in Chapter 19, Petroleum Engineering and Refining, in this book.

This guide points to information resources available in either print or online for mining engineering. Where available, URLs have been provided to readers for further exploration of the pertinent Internet resource. Further mineral processing through refinement and fabrication techniques are the property of metallurgical, material, and chemical engineering, and are not within the scope of this chapter.

HISTORY OF MINING ENGINEERING

Mining ranks as one of the basic industries of early civilization and, after early hunting, gathering, and agriculture is very nearly concurrent with the development of civilization. From the gathering of surface deposited raw materials during the Paleolithic era, 300,000 years ago in Africa, to the first underground mines of the Stone Age 40,000 years ago, and on to contemporary underground excavations, mining has supplied the basic resources used by human civilization.

In ancient history, mining work was performed predominantly by slaves (prisoners of war, criminals, or political prisoners). As surficial deposits became exhausted, the late Roman Empire and early Middle Ages saw a more complicated extension of mines underground. Slaves were replaced by skilled artisans and guild organizations. Ownership and the financial support of mines through investment funding encouraged the formation of organized company enterprises, with guilds taking on the form of modern trade labor organizations.

The discovery and refinement of metallurgical processes in the eighteenth century made possible the transformation of labor to more efficient factory systems. With the Industrial Revolution of the late eighteenth century, the quest for mineral resources became internationalized as nations projected economic, political, and military power abroad to satisfy an increasing mechanized demand for raw minerals and sources of energy. As globalization increased, it may be argued that the history of many major wars of the twentieth century and the early decades of this century may be attributable to national governments identifying their own national and economic security with resource security.

Mining is not pursued as an isolated activity. Minerals discovered through prospecting and exploration must first submit to geological investigation and economic evaluation before mining engineers can plan, develop, and proceed with resource recovery and extraction itself. Following exhaustive mineral exploitation, environmental engineers conclude the mining process with closure and reclamation of the mine site with a view to sustainable future uses for the land, such as wildlife refuges, real estate development, or waste disposal sites.

Within the realm of mining activity proper, the mining engineer is concerned with the methods, tools, and processes for recovering mineral resources from the relative near surface of the Earth or extracting those resources from underground excavations. The extraction process must be executed in a logical set of unit operations consisting of organized drilling, blasting, loading, and hauling in a continuous, productive cycle. Throughout the production cycle a myriad detailed supplementary (auxiliary) operations are carried out for successful production cycles. Power, pumping, lighting, ventilation, air conditioning, water, communication, lighting, and waste disposal need to be supplied and maintained to sustain the entire operation up to the day of its closure.

This chapter has been compiled to help bring an awareness of the published and unpublished sources of information in mining engineering that. An exhaustive list of resources is not provided. Business, economic, and financial components of the petroleum industry are not part of the scope of this guide, though some titles listed herein may include discussion of those aspects.

HISTORY OF MINING ENGINEERING

Standard monographic works in English, which outline the history of the human mining experience include:

Craddock, P. and J. Lang, eds. 2002. *Mining and Metal Production through the Ages*. London: British Museum.
Davies, O. 1935. *Roman Mines in Europe*. Oxford, U.K.: Clarendon Press.
Dibner, B. 1958. *Agricola on Metals*. Norwalk, CT.: Burndy Library.
Gregory, C. E. 1980. *A Concise History of Mining*. Oxford, U.K.: Pergamon.
Lynch, M. 2002. *Mining in World History*. London: Reaktion.
Rickard, T. A. 1932. *A History of American Mining*. New York: McGraw-Hill.

Rickard, T. A. 1932. *Man and Metals: A History of Mining in Relation to the Development of Civilization.* New York: McGraw-Hill.

Shepherd, R. 1993. *Ancient Mining.* London: Elsevier Applied Science.

Spence, C. C. 1993. *Mining Engineers & the American West; the Lace-boot Brigade, 1849–1933.* Moscow, ID: University of Idaho.

Temple, J. 1972. *Mining: An International History.* New York: Praeger.

HISTORY OF MINING WEB SITES

Bir Umm Fawakhir: Insights into Ancient Egyptian Mining http://www.tms.org/pubs/journals/JOM/9703/Meyer-9703.html (accessed May 6, 2010). An archaeological survey of the site at Bir Umm Fawakhir in Egypt.

Bronze Age http://www.encyclopedia.com/topic/Bronze_Age.aspx (accessed May 6, 2010). Encyclopedia.com, a free Web-based encyclopedia, provides users with millions of articles from the *Columbia Encyclopedia, Oxford's World Encyclopedia*, and the *Encyclopedia of World Biography.* The article on the Bronze Age includes imbedded hyperlinks to related topics as well as to other "prehistory" discussions in the encyclopedia about the Copper Age, Coal Mining, Iron Age, etc.

Copper: The Red Metal http://www.unr.edu/sb204/geology/copper2.html (accessed May 6, 2010). A thorough scholarly discussion about the early history of copper mining, smelting, metallurgy, and mining technique. Geographic themes abound with discussions about copper in Sub-Saharan Africa; the Bronze Age in Asia Minor, Europe, and Asia; the Bronze Age to the Fall of Rome, copper in the Middle Ages and Renaissance, in the Western Hemisphere, and copper mining in the United States. The article includes references.

History of Mining: The Miner's Contribution to Society http://www.dmtcalaska.org/course_dev/intromining/01history/notes01.html (accessed May 6, 2010).

History of Parys and Mona Copper Mines http://www.amlwchhistory.co.uk/parys/history.htm (accessed May 6, 2010). A useful starting point for visitors is the site index to listing of hyperlinks to topics like history (Bronze Age, Roman, Dark Ages), geology processes (open cast, shaft working, drainage, ore preparation, calcining, smelting, precipitation, manufacturing, brass production, transport, and work conditions) as well as photographs and research links to maps and technical reports.

Mining History Network http://projects.exeter.ac.uk/mhn/ (accessed May 6, 2010). The Mining History Network is hosted at Exeter University in the United Kingdom. The Web site includes an A to Z listing of mining historians, bibliographies of British and North American mining history, and links to related Web sites. Details of a number of mining history organizations are also available including Exeter Mining History Research Group, Australian Mining History Association, Northern Mine Research Society, Japan Mine Research Society, Early Mining Research Group, and the Peak District Mines Historical Society.

Pictorial Walk Through the 20th Century, A. "Honoring the U.S. Miner." U.S. Department of Labor, Mine Safety and Health Administration (MSHA) http://www.msha.gov/century/century.htm (accessed May 6, 2010).

INFORMATION SOURCES AND GUIDES TO THE LITERATURE OF MINING ENGINEERING AND TECHNOLOGY

Anthony, L. J., ed. 1988. *Information Sources in Energy Technology.* London/Boston: Butterworths.

Chen, C.-C. 1987. *Scientific and Technical Information Sources,* 2nd ed. Cambridge, MA: MIT Press.

Hurt, C. D. 1998. *Information Sources in Science and Technology,* 3rd ed. Englewood, CO: Libraries Unlimited, Inc.

Lord, C. R. 2000. *Guide to Information Sources in Engineering.* Englewood, CO: Libraries Unlimited.

Macleod, R. A. and J. Corlett. 2005. *Information Sources in Engineering,* 4th ed. New York: Bowker-Saur.

Malinowsky, H. R. 1994. *Reference Sources in Science, Engineering, Medicine, and Agriculture.* Phoenix: Oryx.

Malinowsky, H. R. 1980. *Science and Engineering Literature: A Guide to Reference Sources,* 3rd ed. Littleton, CO: Libraries Unlimited.

Mildren, K. W. and P. J. Hicks, eds. 1996. *Information Sources in Engineering,* 3rd ed. London: Bowker-Saur.

Mount, E. 1976. *Guide to Basic Information Sources in Engineering.* New York: Halsted Press.

Mount, J. Books about Mining http://members.cox.net/pmount/miningbooks3.html (accessed July 20, 2010). An alphabetical list of books about mining, prospecting, mines, miners, mining history, economic minerals, mining engineering, and some geological, geotechnical, and petroleum engineering that are currently in print. Compiled by a retired paleontologist/geologist and geosciences librarian, Jack Mount. Each title links to additional information and availability for purchase at Amazon.com. There is a separate section devoted to books on mining history.

New York Public Library, Engineering Societies Library. 1975. *Bibliographic Guide to Technology.* Boston: G. K. Hall.

Wasserman, S., D. E. Wilt, and J. B. Erickson, eds. 1997. *Encyclopedia of Physical Sciences and Engineering Information Sources,* 2nd ed. Detroit: Gale.

Wood, D. N., J. E. Hardy, and A. P. Harvey, eds. 1989. *Information Sources in the Earth Sciences,* 2nd ed. London/New York: Bowker-Saur.

SEARCHING THE LIBRARY CATALOG

When looking for books on mining engineering, search by Library of Congress (LC) subject or keyword in your local Library Catalog. If you know of a particular title or author you would like to find, search by that specific field in the catalog. Examples of relevant subject headings for Mining Engineering include:

Mining Engineering; Blasting; Tunneling

To browse for books on mining engineering in a library that shelves according to the Library of Congress Classification scheme, the following ranges of call numbers can lead to the correct area of the Library.

Metallurgy; Mining engineering: TN 1-997
Mineral deposits. Prospecting: TN 263-271
Practical mining operations; Safety: TN 275-325
Mine transportation, haulage and hoisting; Mining machinery: TN 331-347
Mining of particular metals: TN 400-580
Mineral and metal industries: HD 9506-9624

ABSTRACTS AND INDEXES

Effective use of bibliographic databases requires an understanding of search strategy development and techniques. Online help and user guides normally do accompany databases.

Key databases for mining engineering include:

AESIS, Australia's Geoscience, Minerals, and Petroleum Database was developed by the Australian Mineral Foundation (AMF), Geoscience Australia, Commonwealth Scientific and Industrial Research Organisation (CSIRO), the National Library of Australia, the Australian Geoscience Information Association, and private industry. The Australian Mineral Foundation is available on Thomson Dialog (Bluesheet File 105). *AESIS* indexed all available Australian information related to mines and mining. Published and unpublished resources cover blasting, drilling, extractive metallurgy, mineral exploration, mineral industry, mineral processing, mineral resources, mineralogy, geology, and petroleum subject areas. The database specialized in material written by Australians, or about Australia or Papua New Guinea covering the period from 1851 through 2001, and contained nearly 200,000 references in the English language to worldwide technical literature. *AESIS* is no longer updated. AESIS is no longer updated, but remains available on Informit as an archive database. Subscription information may be found on the Informit site at http://www.informit.com.au/indexes_AESIS.html (accessed March 26, 2011).

AusGeoRef may be viewed as the new successor and continuation of *AESIS*. In 2008, the American Geological Institute http://www.agiweb.org/georef/ausgeoref/index.html (accessed March 27, 2011) and Geoscience Australia http://www.ga.gov.au/library/ausgeoref/index.jsp (accessed March 27, 2011) collaborated to expand coverage of *AusGeoRef* to include references drawn from the *AESIS* database. *AusGeoRef* includes more than 180,000 bibliographic references on Australia, with more than 74,000 new references incorporated from AESIS. The references are drawn from the journal literature, meeting proceedings and abstracts, books, reports, and maps. The database is updated on a weekly basis.

General categories covered in *AusGeoRef* are mineralogy, geochemistry (isotopes, hydrochemistry), geochronology (includes radiometric and relative age determination); petrology (includes igneous, metamorphic, and sedimentary rocks); oceanography; paleontology (includes invertebrates, vertebrates, plants); geologic maps; geophysics (includes geophysical surveys, well-logging, seismology); hydrogeology (includes groundwater and hydrologic cycle); environmental science (includes soil and water pollution, reclamation); geomorphology; soils; economic geology (includes metals, industrial minerals, petroleum, mining); and engineering geology.

American Mineralogist Crystal Structure Database. 2003–. http://www.minsocam.org/MSA/Crystal_Database.html. Compiled by Bob Downs and Paul Heese of the University of Arizona. It includes every structure published in the *American Mineralogist, The Canadian Mineralogist*, and *The European Journal of Mineralogy*. A complete description of the database is available at http://www.minsocam.org/MSA/ammin/toc/abstracts/2003_Abstracts/Jan03_Abstracts/Downs_p247_03.pdf (accessed August 3, 2010). The database is maintained by the Mineralogical Society of America and the Mineralogical Association of Canada, and financed by the National Science Foundation.

Applied Science and Technology Abstracts. 1983–. (Thomson Dialog Bluesheet File 99). Subject coverage includes civil engineering, energy resources and research, environmental engineering, geology, metallurgy, mineralogy, mining engineering, and petroleum and gas engineering.

Arctic and Antarctic Regions. mid-1800s–. http://www.ebscohost.com/thisTopic.php?marketID=1&topicID=868 (accessed July 29, 2010). Formerly known as *NISC* and now owned by EBSCOhost, *Arctic and Antarctic Regions* is the world's largest collection of international polar databases covering a time period from the mid-1800s to present. It includes references to scientific periodicals, monographs, proceedings of conferences and symposia, government reports, theses, dissertations, pipeline documents, consultants' reports, and monographs. Subject areas covered are diverse, but those generally relevant to mining engineering are cold regions engineering, exploration, geosciences (climate, geography, geology), health and natural sciences.

Arctic Science and Technology Information System (ASTIS) http://www.arctic.ucalgary.ca/ index.php?page=astis_database (accessed July 20, 2010). ASTIS contains over 70,000 records describing publications and research projects about northern Canada. Records contain abstracts, detailed subject and geographic indexing terms, and links to 16,000 publications that are available online. *ASTIS* is maintained by the Arctic Institute of North America http://www.ucalgary.ca/aina/ (accessed March 27, 2011) at the University of Calgary.

Cambridge Scientific Abstracts Databases http://www.csa.com/ (accessed June 29, 2010). Mining engineering subject coverage emphasis is on the environmental management of mining activity with databases like *Pollution Abstracts* (1981–), *Water Resources Abstracts* (1967–), and *TOXLINE* (1999–). Articles in toxicology include chemicals, pharmaceuticals, pesticides, environmental pollutants, mutagens, and teratogens.

Canada Energy Minerals and Metals Information Centre (EMMIC) http://www2.nrcan.gc.ca/ es/msd/emmic/web-en-emmic/index.cfm?fuseaction=home.welcome (accessed June 29, 2010). EMMIC is a section within the Energy Sector's Management Services Division of Natural Resources Canada. EMMIC provides library and information services to the Energy Sector, Minerals and Metals Sector, Corporate Services Sector, and Executive offices as well as information to other libraries via interlibrary loan and to outside institutions and the public (from Web page). For subject guides and library/information resources at EMMIC on mining and mineral processing go to http://www2.nrcan.gc.ca/es/msd/ emmic/web-en-emmic/index.cfm?fuseaction=subjects.subjects (accessed March 27, 2011). As an additional bonus, EMMIC users may access the ETDE's *World Energy Base* database free of charge. ETDE's *Energy Database* contains the world's largest collection of energy literature of more than 4 million records with coverage from 1995 to the present. The database contains bibliographic references and abstracts to journal articles, reports, conference papers, books and theses.

Canadian Research Index (Microlog). 1982–. http://www.proquest.com/en-US/products/ brands/pl_mm.shtml (accessed March 27, 2011). *Canadian Research Index* provides access to Canadian government publications at the municipal, provincial, and federal levels. It includes policy papers, scientific and technical reports, statistical reports, annual reports, etc. The *Microlog* number accompanying each record can be used to locate the documents, which are available in the *Microlog* microfiche collection. Indexes all depository Canadian government publications at the municipal, provincial, and federal levels. Includes scientific and technical reports, statistical reports, policy papers, and annual reports.

Environment Abstracts on LexisNexis Environmental. 1975–. http://www.lexisnexis. com/academic/1univ/envir/default.asp (accessed: July 20, 2010). Content from mining related journals includes titles such as: *African Mining Monitor, Coal Week, Coal Week International, Engineering and Mining Journal, Metals Week, Mining Annual Review, Mining Journal, Mining Magazine, Waste News,* and *Waste Treatment Technology News.* The principal focus of this database is American Law, Environmental Codes and Regulations, Case Law, Regulatory agency decisions, and Case Law and Agency Actions.

GEOBASE. Bibliographic database for the Earth, geographical, and ecological sciences. An Earth science database of bibliographic information and abstracts covering Earth sciences, ecology, geology, geography, geomechanics, human geography, mines and mining oceanography, and related disciplines. Material includes refereed scientific papers; trade journals and magazine articles; product reviews, directories, books, conference proceedings, and reports. GEOBASE is currently available worldwide via Dialog, OCLC, and *ScienceDirect.*

GeoBase http://www.geobase.ca/geobase/en/ (accessed June 29, 2010). *GeoBase* is a Canadian federal, provincial, and territorial government initiative overseen by the Canadian Council on Geomatics (CCOG) to ensure the provision and access to quality geospatial data with free and unrestricted use. *GeoBase* comes with full metadata and provides references, and contexts to a wide variety of thematic data for government, business, and individual applications. The data address concerns such as sustainable resource development (forestry, mines, energy, water, etc.); public safety, protection, and sanitation (emergency response, disaster management, tracking SARS); and environmental protection (greenhouse gas effects, global warming, natural risks, flooding, etc.). It is a free, online geospatial database.

GeoRef. 1785–present. Compiled by The American Geological Institute http://www.agiweb. org/georef/index.html (accessed June 30, 2010). *GeoRef* offers international bibliographic coverage of all aspects of Earth sciences, geology, and geophysics from 1785 to the present including archaeology, economics, education, engineering and environmental geology, expeditions, fossils, geologic hazards (avalanches, hurricanes, sunspots, volcanoes, etc.), instruments, land use, landfills, magnetic fields, and maps, medical and military geology, and soils. The database contains over 2.5 million references to books, theses, journals, reports, maps, conferences, and government documents in all areas of the Earth sciences. GeoRef includes pure and applied research in applied geology, exploration geophysics, mineral exploration, petroleum and environmental organic chemistry, Earth science, and economic and engineering geology.

Geoscan http://sst.rncan.gc.ca/esic/geoscan_e.php (accessed June 30, 2010). A free database of 40,000 publications from the Geological Survey of Canada issued since 1986. Some are free, others may be ordered for a fee as photocopies.

IMMAGE (commercial database of IMM Information Services). The *IMMAGE* database is compiled by the Institution of Mining and Metallurgy in the United Kingdom and consists of 105,000 references (expanding by up to 3,000 per year), to published papers, books, reports, etc., covering technical research and development and operating practices of the international minerals industry. Subject coverage includes tunneling and underground excavation, including rock mechanics, instrumentation and automation, economic geology, mining technology, and extraction processing in nonferrous metals and industrial mineral fields. Web access is available through The Institute of Materials, Minerals and Mining at http://www.iom3.org/content/immage-reference-database-international-minerals-industry (accessed March 27, 2011). Free access to *IMMAGE* online is available to IOM3 members as a new member benefit. Nonmembers can request trial access to *IMMAGE* online.

METADEX (*Metals Abstracts*) CSA Internet Database Service http://www.csa.com/fact-sheets/metadex-set-c.php (accessed March 27, 2011). *METADEX* contains over 950,000 references from 2,000 journals, patents, dissertations, reports, books, and conference proceedings (1966–current) covering all aspects of metallurgical science, including corrosion, metallography, mechanical and physical properties, testing and quality control, physical properties, extraction and refining, machining, welding, metal matrix composites, powder technology, surface finishing, heat treatment, testing and quality control, alloy production and development, thermal treatment, and application, and end uses.

MINABS Online (The Mineralogical Society) http://www.minabs.com/ (accessed March 27, 2011). *Mineralogical Abstracts* was an essential research tool for 70 years with abstracts covering mineralogy, petrology, geochemistry, crystallography, gemmology, meteoritics, and allied subjects. In January 2004, this printed journal had been replaced with an electronic journal known as *MINABS Online*. Like its print predecessor, *MINABS Online* was a specialized abstracting publication covering international research in mineralogy, crystallography, geochemistry, petrology, environmental mineralogy, and related topics in Earth science. According to The Mineralogical Society Web site regarding the availability of *MINABS Online* in the future, "… With effect from December 31, 2008, this journal has been discontinued. The journal continued to be available until May 2010 from www. minabs.com. This service has now been closed. Later in 2010, all of the abstracts from 1982 to 2008 will be made available, free of charge to all internet users, under auspices of

GeoRef ..." The proposed link will be http://www.minersoc.org/pages/e_journals/minabs. html (accessed July 05, 2010).

Mineral and Energy Technology (CANMET) http://canmetenergy-canmetenergie.nrcan-rncan.gc.ca/eng/ (accessed June 29, 2010). Database subject coverage includes agriculture, energy, environment, fisheries, forestry, life sciences, natural resources, Earth sciences, physical sciences, technology, and telecommunications.

ScienceDirect. 1995–. http://www.sciencedirect.com/ (accessed March 27, 2011). This general science database contains articles from journals published by Elsevier. Examples of Elsevier mining journals indexed include *International Journal of Coal Geology, International Journal of Rock Mechanics and Mining Sciences, International Journal of Rock Mechanics and Mining Science & Geomechanics Abstracts, Mining Science and Technology,* and *Minerals Engineering.*

JOURNALS

Journal literature (magazines, periodicals, serials, and trade literature publications) all help researchers to keep abreast of trends, current issues, and techniques in the industry. Some prominent mining engineering journal titles, ranked high by their impact factor over the past three years in ISI's *Journal Citation Reports* are listed below. Scope notes for these Subject Categories have been quoted also, in full from the *JCR* Web site: http://thomsonreuters.com/products_services/science/science_products/a-z/journal_citation_reports/ (accessed March 27, 2011).

The Description fields of information were quoted from *Ulrich's Periodical Directory* http://www. ulrichsweb.com/ulrichsweb/ (accessed March 27, 2011), an excellent tool for locating journals in a subject area. It is useful to note the definitions they use for subject pertinent to this chapter. In Ulrich's Periodical Directory, subject categories assigned to these periodical titles were a variant combination of Engineering, Geological, Metallurgy and Metallurgical Engineering, or Mining and Mineral Processing.

Engineering, Geological: "Engineering, Geological includes multidisciplinary resources that encompass the knowledge and experience drawn from both the geosciences and various engineering disciplines (primarily civil engineering). Resources in this category cover geotechnical engineering, geotechnics, geotechnology, soil dynamics, earthquake engineering, geotextiles and geomembranes, engineering geology, and rock mechanics."

Metallurgy and Metallurgical Engineering: "Metallurgy and Metallurgical Engineering includes resources that cover the numerous chemical and physical processes used to isolate a metallic element from its naturally occurring state, refine it, and convert it into a useful alloy or product. Topics in this category include corrosion prevention and control, hydrometallurgy, pyrometallurgy, electrometallurgy, phase equilibria, iron-making, steel-making, oxidation, plating and finishing, powder metallurgy, and welding."

Mining and Mineral Processing: "Mining and Mineral Processing includes resources on locating and evaluating mineral deposits; designing and constructing mines; developing mining equipment; supervising mining operations and safety; and extracting, cleaning, sizing, and dressing mined material. Relevant topics in this category include exploration and mining geology, rock mechanics, geophysics, and mining science and technology."

SELECTED MINING JOURNALS

E&MJ: Engineering & Mining Journal. New York: McGraw-Hill. (0095-8948) http://www. mining-media.com (accessed March 27, 2011). Covers exploration, deposit discoveries and development, milling, smelting, refining, and other extractive processing of metals and nonmetallics. Issues include regional news coverage from the United States and Canada, Latin America, Africa, Asia, and Australia as well as an international coal section. World Mining Equipment sections deal with operating strategies, suppliers, reports, processing solutions, and the latest technology issues for mine operators.

Canadian Mining Journal. Southam Business Communication Inc. (0008-4492) http://www. canadianminingjournal.com (accessed March 27, 2011). The leading mining and exploration journal in Canada. Covers mineral exploration trends, metal prices, and new geological models, underground mine developments and operating performances, new technology and services.

CIM BULLETIN. Canadian Institute of Mining, Metallurgy and Petroleum (0317-0926) http://www.cim.org/mainEn.cfm (accessed March 27, 2011). Technical data, papers, and information on mining practice, and research on mineral engineering subjects to promote the technological interests of people involved in the development of the industry in Canada, with special feature articles on new mining developments and technologies. Content also includes monthly features, career opportunities, industry equipment, and society and division newsletters.

Engineering Geology: An International Journal. Elsevier BV (0013-7952) http://www. elsevier.com/wps/find/journaldescription.cws_home/503330/description#description (accessed March 27, 2011). Contains original studies, case histories, and comprehensive reviews in the field of engineering geology. Special features include bibliographies, charts, illustrations, and book reviews.

International Journal of Mineral Processing. Elsevier Science BV (0301-7516) http://www. elsevier.com/locate/minpro (accessed March 27, 2011). Covers all aspects of the processing of solid-mineral materials, such as metallic and nonmetallic ores, coals, and other solid sources of secondary materials, etc. Special features in the journal include bibliographies, charts, illustrations, and book reviews.

International Journal of Rock Mechanics and Mining Sciences. Pergmon Elsevier Science Ltd. (1365-1609) http://www.elsevier.com/locate/ijrmms (accessed March 27, 2011). Original research, new developments, and case studies in rock mechanics and rock engineering for mining and civil applications. Special features include advertising, abstracts, charts, illustrations, and book reviews.

International Journal of Surface Mining, Reclamation and Environment. A A Balkema (1389-5265) http://www.balkema.nl (accessed March 27, 2011) is incorporated as part of the larger Taylor & Francis Group. The journal examines all aspects of surface mining technology and waste disposal systems relating to coals, oil sands, industrial minerals, and metalliferous deposits. Includes computer applications and automation processes.

Mining Journal. Mining Journal Ltd. (0026-5225) http://www.mining-journal.com/ (accessed March 27, 2011). Weekly international coverage of political, financial, and technical news affecting mining industry activity. Special features include advertising, illustrations, and book reviews.

SME 2010 Guide to Minerals & Materials Science Schools. Society for Mining, Metallurgy and Exploration (978-0-87335-232-1) http://www.smenet.org/store/mining-books.cfm/ SME-2010-Guide-to-Minerals-%26-Materials-Science-Schools/18-EMSME (accessed March 27, 2011). This publication identifies educational institutions offering programs in mining related environmental engineering, geological engineering, geophysical engineering, materials science and engineering, metallurgical engineering, mineral engineering, mineral processing engineering, mining engineering, and mining engineering technology.

Transactions of the Institution of Mining and Metallurgy Section A: Mining Technology. Maney Publishing (0371-7844) http://maney.co.uk/index.php/journals/mnt/ (accessed July 6, 2010).

Transactions of the Institution of Mining and Metallurgy Section B: Applied Earth Science. Maney Publishing (0371-7453) http://maney.co.uk/index.php/journals/aes/ (accessed July 6, 2010).

Transactions of the Institution of Mining and Metallurgy Section C, Mineral Processing and Extractive Metallurgy. Maney Publishing (1743-2855) http://maney.co.uk/index.php/journals/mpm/ (accessed July 6, 2010).

The above three journals cover practical scientific and engineering aspects of ore genesis, metal properties, mineral processing, product preparation, separation, extraction, purification, and post-treatment by physical and chemical methods. The scope of the journals also includes discussion of the economic aspects of mine management and finance, mineral resources management, planning and mining law, mining machinery and technology, effluents, and associated environmental issues.

JOURNAL TITLES RELATED TO MINING ENVIRONMENTAL MANAGEMENT

IEA Coal Research Newsletter. 1999–. International Energy Agency (1747-2687). Available in PDF format at http://www.iea-coal.org.uk/site/2010/home (accessed March 27, 2011).

International Journal for Numerical and Analytical Methods in Geomechanics (0363-9061) http://interscience.wiley.com/jpages/ (accessed March 27, 2011).

International Journal of Earth Sciences. Springer–Verlag. (1437-3254) http://www.springer.com/earth+sciences/geology/journal/531 (accessed March 27, 2011).

International Journal of Geomechanics. American Society of Civil Engineers (1532-3641) http://ascelibrary.org/gmo (accessed March 27, 2011).

Ironmaking and Steelmaking. Maney Publishing (0301-9233) http://maney.co.uk/index.php/journals/irs/ (accessed March 27, 2011).

Journal of Geotechnical and Geoenvironmental Engineering. ASCE (1090-0241) http://www.asce.org/Journal.aspx?id=2147486639 (accessed March 27, 2011).

Mine Water and the Environment, Journal of International Mine Water Association. Formerly *International Journal of Mine Water.* Springer Scientific Germany (1025-9112) http://www.springer.com/earth+sciences/geology/journal/10230 (accessed March 27, 2011).

Minerals and Metallurgical Processing Journal. Society for Mining, Metallurgy and Exploration (0747-9182) http://www.smenet.org/minerals-and-metallurgical-processing-journal/ (accessed March 27, 2011).

Northern Miner Online (0029-3164) http://www.northernminer.com/ (accessed March 27, 2011). Covers the worldwide mining activities of mining companies based in North America or listed on North American stock exchanges.

HANDBOOKS AND MANUALS

Handbooks and manuals are a valuable source for quickly needed information like facts, figures, technical specifications, data, properties, design and testing criteria, explosive formulas and calculations, recommended procedures along the production cycle, and log interpretation criteria.

ASM Handbook. ASM International http://products.asminternational.org/hbk/index.jsp (accessed March 27, 2011). The *ASM Handbook* comprises 21 volumes providing major reference information to industry about mining extraction, metallurgical processing and fabrication of metals; testing, inspection, and failure analysis; microstructural analysis and materials characterization; corrosion, and wear phenomena in machinery and equipment, some of it used in the mining industry.

Bickel, J. O., et al. 1996. *Tunnel Engineering Handbook,* 2nd ed. New York: Chapman Hall. A comprehensive, state-of-the-art overview written for practicing engineers on the classification types, planning, design, layout, construction methods, and rehabilitation of tunnels above and below ground. (Out of print; hard to find.)

Canadian Minerals Yearbook: Review and Outlook. Compiled by the Minerals and Metals Sector, Natural Resources Canada. Ottawa: Public Works and Government Services Canada, http://www.nrcanrncan.gc.ca/mms-smm/busi-indu/cmy-amc-eng.htm (accessed March 27, 2011) Compiled annually from 2006- by the Minerals and Metals Sector, Natural resources Canada. Ottawa: Public Works and Government Services Canada. The *Yearbook* contains invaluable information about Canada's mineral production, world developments in regard to a particular mineral, health and safety aspects of the mineral, market information, the outlook for the industry, and many statistical reports.

Caterpillar Performance Handbook. 1988–. Peoria, IL.: Caterpillar Tractor Co. Handbook and manual of earthmoving machinery, industrial power trucks, and farm tractors.

Fuerstenau, M. C. and K. N. Han, eds. 2003. *Principles of Mineral Processing.* Littleton, CO: Society for Mining, Metallurgy, and Exploration, Inc. Examines all aspects of minerals processing from the handling of raw materials to separation strategies to the remediation of waste products. The book relates recent developments in engineering, chemistry, computer science, and environmental science to explain how these disciplines contribute to the production of minerals and metals efficiently and economically from ores.

Gertsch, R. E., et al. 1998. *Techniques in Underground Mining: Selections from Underground Mining Methods Handbook.* Littleton, CO: Society for Mining, Metallurgy and Exploration. Sections in the book review support methods used in mine planning and mining including pillars, room-and-pillar methods in open-stope mining, sublevel caving, and block and panel caving.

Hartman, H. L. 1992. *SME Mining Engineering Handbook.* Littleton, CO: Society for Mining, Metallurgy and Exploration. Also available on CD-ROM. This is the one, indispensable, essential reference work anyone in the mining industry must have in their collection, even if it is the only one. A distillation of every aspect of mining engineering. Provides professional practitioners with an authoritative reference and design source. The book covers all branches of mining (metal, coal, and nonmetal), all locales of mining (surface, underground, and hybrid) with a concentration on mining in the United States. Numerous references are also devoted to international practices.

International Society of Explosives Engineers. 2000. *ISEE Blasters Handbook,* 17th ed. Cleveland: ISEE. A manual describing explosives and standard, practical methods of use. (Formerly the *ETI* and, prior to that, *DuPont Blasters Handbook.*)

Karassik, I. J., et al. 2007. *Pump Handbook,* 4th ed. New York: McGraw-Hill. Provides practical data on the design, application, specification, purchase, operation, troubleshooting, and maintenance of pump technology used in every type of engineering application throughout industry, including mining. This is an essential reference and guide.

Marcus, J. J. 1997. *Mining Environmental Handbook: Effects of Mining on the Environment and American Environmental Control.* Edge River, NJ: World Scientific. This handbook examines the effects of mining on the environment and environmental laws that deal with mining in the United States both currently and in future perspective.

Minerals Handbook: Statistics and Analyses of the World's Minerals Industry. 1982–. London: Macmillan. Incorporates information for each of over 50 metals and minerals; an essential reference for minerals industry analysts and economists and for any mining or exploration company. Discusses world reserves and reserve base, production capacity, adequacy of reserves, consumption, end-use patterns, value of contained metal, substitutes, technical possibilities, and marketing arrangements.

Mining Annual Review (CD-ROM). 2000–2005. London: Aspermont, UK. This annual review documented some of the latest trends in mining and looked at the industry in most countries around the world. It provided a thumbnail sketch of the state of mining internationally.

Stack, B. 1982. *Handbook of Mining and Tunneling Machinery.* New York: John Wiley & Sons. The author describes the history, evolutionary development, and interrelationship of tunneling machines employed in Australia, the United Kingdom, United States, and Germany from earliest times to the present. Coverage is given for the entire range of associated drilling machines, including "horizontal and vertical, raise drills and shaft-boring machines, shields, mechanized shields, slurry machines, rock tunneling machines … reaming machines, incline-boring machines and the various boom-type machines, such as road-headers and mounted impact breakers, etc." (from Preface, xxvii).

Taggart, A. F. 1954. *Handbook of Mineral Dressing: Ores and Industrial Minerals.* New York: John Wiley & Sons. A reference work and guide to industrial minerals and cement. Covers crushing, grinding, screen sizing, classification, washing and scrubbing, gravity concentration, flotation and electrical methods of concentration, dewatering, filtration, drying, storage and mill transport, sampling and testing, design, and construction of ore-treatment plants; with 70 pages of mathematical and other tables.

DICTIONARIES

Dictionaries and encyclopedias provide definitions and clarify concepts. The terminology peculiar to mining is set out in a number of works. Samples of standard references to this unique vocabulary may be found in the following sources:

American Geological Institute, compilers. 2000. *Dictionary of Mining, Mineral, and Related Terms,* 2nd ed. Alexandria, VA: American Geological Institute. This dictionary focuses solely on mining and geological terminology related to mining, including industrial minerals, pollution, and environmental mining terminology. The CD-ROM version of this dictionary was compiled and edited by the staff of the U. S. Bureau of Mines, Washington, D.C. 1996.

Blackburn, W. H., et al. 1997. *Encyclopedia of Mineral Names.* Ottawa, Ontario, Canada: Mineralogical Association of Canada/Association Mineralogique du Canada.

Clark, A. 1993. *Hey's Mineral Index: Mineral Species, Varieties and Synonyms,* 3rd ed. London/New York: Chapman & Hall. An alphabetical listing of all known minerals, their chemical composition, and type of locality. Also includes an additional finding index arranged by mineral chemical composition.

Dorian, A. F. 1993. *Elsevier's Dictionary of Mining and Mineralogy in English, French, German and Italian.* Amsterdam/New York: Elsevier, French and European Publications, Inc.,

Glossary of Mining Terms http://www.portal.state.pa.us/portal/server.pt/community/training/14001/glossary__mining_terms/589150 (accessed July 9, 2010). Compiled by the Pennsylvania Department of Environmental Protection, Bureau of Mine Safety, the online glossary contains thousands of terms related mainly to coal mining.

Neuendorf, K. K. E., J. P. Mehl, Jr., and J. A. Jackson, eds. 2005. *Glossary of Geology,* 5th ed. Alexandria, VA: American Geologic Institute. Rock blasting terms and symbols: A dictionary of symbols and terminology in rock blasting and related areas like drilling, mining, and rock.

ENCYCLOPEDIAS

There is no encyclopedia devoted specifically to mining engineering. As a process technology, mining engineering applies the lessons learned from many disciplines, such as the geosciences, geochemistry, materials engineering, environmental sciences, and applied automotive mechanics.

Carr, D. and N. Herz. 1989. *Concise Encyclopedia of Mineral Resources.* Oxford, U.K./ New York: Pergamon Press; Cambridge, MA: MIT Press. "Based on material from the *Encyclopedia of Materials Science and Engineering,* first published in 1986, with revisions and updated material." This work is both a dictionary of terminology and encyclopedic discussion of concepts in mining and mineral resources, nonmetallic minerals, and advances in materials science and engineering.

Dasch, E. J., ed. 1996. *Macmillan Encyclopedia of Earth Sciences.* (2 vols.) New York: Macmillan Reference.

Frye, K., ed. 1982. *The Encyclopedia of Mineralogy.* London: Springer, London Limited. A basic mineralogical dictionary, and volume 4B from the series title: *Encyclopedia of Earth Sciences.*

Marshall, C. P. and R. W. Fairbridge, eds. 2006. *Encyclopedia of Geochemistry.* London: Springer.

Middleton, G. V., ed. 2003. *Encyclopedia of Sediments & Sedimentary Rocks.* Dordrecht/ Boston: Kluwer Academic.

Roberts, W. L., T. J. Campbell, and G. R. Rapp, eds. 1990. *Encyclopedia of Minerals,* 2nd ed. New York: Van Nostrand Reinhold.

DIRECTORIES

Directories provide access to information about the petroleum industry and the location of people, companies, products, services, manufacturers, and suppliers of components to the industry. In recent years, online access has proven to be of greater use in providing up-to-date information on a timely basis than have traditional hard copy publications.

American Mines Handbook. 1989–2003. Toronto: Business Information Group. (0840-8610). (Merged/incorporated into *Canadian and American Mines Handbook* (1712-4042) published 2004–. Lists information on public mining companies registered in the United States. Includes mines and projects summary, five-year stock range table, and statistical tables.

CAMESE. *Compendium of Canadian Mining Suppliers.* Don Mills, Ont.: Canadian Mining Journal in co-operation with the Canadian Association of Mining Equipment and Services for Export (1997–) (1485-8401). Canadian Association of Mining Equipment and Services for Export—"CAMESE is a national, sectoral export trade association. It exists to support Canadian mining suppliers in global marketing, and to assist foreign buyers in finding Canadian sources for mining equipment and services" (from Web page). Browse Supplier information is also available online at http://www.camese.org (accessed July 9, 2010).

Canadian and American Mines Handbook. 2004–. Toronto: Business Information Group. Formed by union of *Canadian Mines Handbook* and *American Mines Handbook*, this new annual handbook and directory lists over 1,200 Canadian and 300 U.S. and foreign mining companies with updates on mines, mining area maps, and historic stock prices.

C.I.M. Directory. 1990–. Montréal: Canadian Institute of Mining, Metallurgy & Petroleum. The directory lists the names, addresses, and companies of all members of the Canadian Institute of Mining, Metallurgy and Petroleum with access to thousands of active metallurgists and materials engineers throughout Canada and the world.

Financial Times Business. *Global Mining Directory.* 2001–. London: FT Business. Provides details of the world's nonferrous metals industry with in-depth operational analyses, ownership, and subsidiary information, plus three-year financial results to over 300 exploration and mining companies. Includes lists of over 700 service and supply companies as well as over 4,000 key industry contact personnel.

International Mining Directory. 1984–. London: Kogan Page. compendium of detailed information about mining and mine-equipment companies, consultants and service companies worldwide, as well as the largest, most comprehensive equipment buyers' guide in the mining industry. At the end of the directory is a brief, five-page glossary containing definitions of mining terminologies and technology.

Mining-technology.com—the Web site for the mining industry http://www.mining-technology.com/ (accessed July 9, 2010). Net Resources International, a trading division of SPG

Media Limited (U.K.). Web site provides international news coverage by geographical region of current mining industry projects extracting energy minerals, ferrous, base and precious metals, industrial minerals, and gemstones. Included is a comprehensive Products and Services catalog and directory to mining industry equipment worldwide; an A to Z company index listing mining contractors and suppliers alphabetically as well as industry news releases about the latest mining technology products, and the job market.

Natural Resources Canada. *Mining Associations/Companies in Canada.* http://www.nrcan-rncan.gc.ca/mms-smm/busi-indu/cmy-amc/mac-eng.htm (accessed July 9, 2010). Directory of Canadian links to federal, provincial, and territorial government agencies and companies responsible for the mining industry within their particular jurisdiction.

Walker, S. 2000. *Major Coalfields of the World,* 2nd ed. London: IEA Coal Research. (Out of print; hard to find.) Provides a clear and comprehensive analysis of the world's coalfields considering the geology, structure, stratigraphy, quality, industry structure and performance, transport infrastructure, production costs, and market potential of each major coalfield.

GOVERNMENT MINISTRIES

Begin searches for government information resources on mining and minerals by visiting the home page Web site of the jurisdiction in question. There, search for the relevant departmental agencies with stated responsibility for energy and natural resources. Government sites often include valuable statistical and commodity pricing data.

AUSTRALIA

Australian Government—Geoscience Australia http://www.agso.gov.au/ (accessed July 11, 2010). Originally the Bureau of Mineral Resources, Geology and Geophysics, this site has been created by a Commonwealth government department with a broad range of responsibilities in the area of geosciences. An especially useful map resource is the link to MapConnect at http://www.ga.gov.au/mapconnect/ (accessed April 5, 2011), which permits users to download topographic and geological map data or to make their own map. An excellent, all inclusive, up-to-date Web site linking visitors to governmental institutions and private industries participating in Australian mining may be found at the Web site of the next entry.

Australia Mining and Exploration http://www.reflections.com.au/MiningandExploration/index.html (accessed July 11, 2010). "A portal about Australasia's Mining and Exploration Industry … covers all aspects of the mining industry from exploration through to mining, processing, and transport including information on mining and exploration companies announcements and reports, commodities, calendar of events and conferences, and details of government and mining related organizations, publications, drilling companies, geological and mining consultants, mining plant and equipment, assayers, financiers, share brokers, accountants, etc. Also includes sections on the stock market with daily market reports, stock prices, and charts on indices, commodities, and companies" (from Web site).

CANADA

In Canada, information about all mining and mineral matters at the federal government level is compiled under one agency, Natural Resources Canada http://www.nrcan.gc.ca (accessed July 11, 2010) and is accessible through their basic Web root directory. The principal goal of providing electronic resources through NRCAN is to increase Canadian competitiveness for attracting industrial investments in mineral and energy exploration and development sustained through timely knowledge transfers. Government information about all aspects of Canadian mining activity (acts and regulations, statistics, geodetic surveys, scientific and technological innovation, cartographic, geological, environmental, business investment opportunities, Investment Tax Credit information for Exploration, Mining Legislation and Regulations, Standards, Online Maps, and educational resources) may be obtained.

Mapping Federal–Provincial–Territorial Mining Knowledge http://mmsd.mms.nrcan.gc.ca/stat-stat/map-car/index-eng.aspx (accessed July 11, 2010). Provides information on major exploration projects and producing mines in Canada. The site maps exploration projects, producing mines, and resource-dependent communities in Canada. All plug-ins required for viewing infrastructures of maps for producing mines, exploration projects, and deposits on this site are included for downloading. Maps products are searchable by commodity. Also available are country profile reports of Gross Domestic Product (GDP), inflation rates, economic profiles, and resource commodities produced in a country of interest.

The Map Image Rendering Database for Geoscience (MIRAGE) http://gdr.nrcan.gc.ca/mirage/index_e.php (accessed July 11, 2010). MIRAGE (Map Image Rendering DAtabase for GEoscience) is a digital image library of Geological Survey of Canada maps. Begun in July 1998, the aim of the project is to make ESS Geoscience maps accessible through the Internet. Thousands of paper maps were scanned and image linked to their metadata records. Users can search this metadata to discover geoscience maps that they can view and download. The scanned images are stored as both MrSID compressed images and as Adobe® Acrobat files.

UNITED STATES

Bureau of Mines Minerals Yearbook. 1932–1993. http://minerals.usgs.gov/minerals/pubs/usbmmyb.html (accessed July 11, 2010). Links to the image scan, full text contents of the *Minerals Yearbooks* from 1932–1993. Courtesy of the University of Wisconsin Ecology and Natural Resources Collection http://digicoll.library.wisc.edu/EcoNatRes/ (accessed July 11, 2010).

Metal Industry Indicators http://minerals.usgs.gov/minerals/pubs/mii/ (accessed July 11, 2010). *The Metal Industry Indicators* (*MII*) is a monthly newsletter in PDF format that analyzes and forecasts the economic health of five metal industries: primary metals, steel, copper, primary aluminum, and aluminum mill products.

Mine and Mineral Processing Plant Locations http://minerals.usgs.gov/minerals/pubs/mapdata/ (accessed July 11, 2010). Provides the locations of 1,879 coal mines and facilities, 8 uranium mines, and 1,965 mines and processing plants for 74 types of nonfuel minerals and materials shown on large lithologic maps. The localities account for most of the fuel and nonfuel minerals and materials produced in the United States in 1997 other than crushed stone, sand and gravel, and common clay.

Mine Safety and Health Administration (U.S. Department of Labor) http://www.msha.gov/ and MSHA's Mine Data Retrieval System (DRS) http://www.msha.gov/drs/drshome.htm (accessed July 11, 2010). MSHA provides mine overviews, accident histories, violation histories, inspection histories, inspector and operator dust samplings, and employment/production data.

Mineral Commodity Summaries http://minerals.usgs.gov/minerals/pubs/mcs/ (accessed July 11, 2010). "Published on an annual basis, the reports are the earliest government publication to furnish estimates covering nonfuel mineral industry data. Data sheets contain information on the domestic industry structure, government programs, tariffs, and five-year salient statistics for over 90 individual metallic and nonmetallic minerals and materials" (from Web page).

Mineral Industry Surveys http://minerals.usgs.gov/minerals/pubs/commodity/mis.html (accessed July 11, 2010). "Mineral Industry Surveys (MIS) are periodic online statistical and economic publications designed to provide timely statistical data on production, distribution, stocks, and consumption of significant mineral commodities. These publications are issued monthly, quarterly, or annually" (from Web page).

Mineral Resource Data System (MRDS) http://tin.er.usgs.gov/mrds/ (accessed July 11, 2010). May be formally cited as U.S. Geological Survey, 2005, Mineral Resources Data System: U.S. Geological Survey, Reston, VA. MRDS contains over 300,000 sites with variable-length records of metallic and nonmetallic mineral resources of the world. A record contains descriptive information about mineral deposits and mineral commodities. The types of information in the database include deposit name, location, commodity, deposit description, geologic characteristics, production, reserves, potential resources, and references. A description of the primary "metallic and nonmetallic mineral resources throughout the world" [and the United States] including information by "deposit name, location, commodity, deposit description, geologic characteristics, production, reserves, potential resources, and references" (from Web site). MRDS also includes a featured link to Mineral Resources Online Spatial Data http://mrdata.usgs.gov/ (accessed July 11, 2010). The site functions as an enhanced data download facility to provide users the choice to select data by geographic area rather than download an entire dataset. At present this enhanced data selection is available for the geochemical data, MRDS, MAS/MILS, active mines, and radiometric ages. All of the data are available free of charge. Additionally, graphic visualizations of nonrenewable mineral resource production, consumption, import, and export can be seen by using the U.S. Minerals Databrowser at http://www.mazamascience.com/Minerals/USGS/. The Databrowser graphs US/World data presented in charts and tables from the U.S. Geological Survey Minerals dataset at http://minerals.usgs.gov/ds/2005/140/ (accessed July 11, 2010).

Minerals Yearbook. 1994–. http://minerals.usgs.gov/minerals/pubs/myb.html (accessed July 11, 2010). "*The Minerals Yearbook* is an annual publication that reviews the mineral and material industries of the United States and foreign countries; contains statistical data on minerals, including economic and technical trends; chapters list approximately 90 commodities and over 175 countries" (from Web page).

National Environmental Publications Internet Site (NEPIS) http://www.epa.gov/nepis/ (accessed July 11, 2010). Keyword searchable index and full text site for over 35,000 U.S. digital titles of EPA books, reports, and documents. EPA's print publications are available through the National Service Center for Environmental Publications (NSCEP), and EPA's digital publications are stored in the National Environmental Publications Internet Site (NEPIS) database.

National Library for the Environment: Congressional Research Service Reports http://ncseonline.org/nle/crs/ (accessed July 11, 2010). Index and full text site for reports on environmental or resource issues prepared for the U.S. Congress. A subject search on the term "Mining" leads searchers to a full text collection of informational research reports about issues relating to legislative acts and natural conservation regulations for members of the House and Senate.

National Institute for Occupational Safety and Health (NIOSH) at http://www.cdc.gov/niosh/ (accessed July 11, 2010) and the Centers for Disease Control and Prevention (CDC) at http://www.cdc.gov/ (accessed July 11, 2010). Within both are hyperlinks to a number of free databases and search engines leading to full bibliographic citations for thousands of publications about mine safety, mine construction and stability, hazardous materials handling, equipment and instrumentation, mining byproducts, and training programs for safety and health.

NIOSH (National Institute for Occupational Safety and Health–Mining Site Index) http://www.cdc.gov/niosh/mining/default.htm (accessed July 11 2010). A bibliographic database of occupational safety and health publications, documents, grant reports, and other communication products related to mining and supported in whole or in part by the National Institute for Occupational Safety and Health (NIOSH). Also includes statistics on mining accidents, fact sheets on minerals and coal, and materials on injury prevention, ergonomics, hearing loss, fire prevention, ventilation, explosives, dust, aging miners, and many other topics.

NIOSHTIC-2 http://www2a.cdc.gov/nioshtic-2/ (accessed July 11, 2010). "NIOSHTIC-2 is a bibliographic database of 35,455 occupational safety and health citations, documents, grant reports, and other communication products supported by the National Institute for Occupational Safety and Health" (from Web page).

Nonmetallic Mineral Products Industry Indexes http://minerals.usgs.gov/minerals/pubs/imii/ (accessed July 11, 2010). The USGS has prepared leading and coincident indexes for the Nonmetallic Mineral Products Industry (NAICS 327). The former name for this industry was the Stone, Clay, Glass, and Concrete Products Industry (SIC 32) under the Standard Industrial Classification (SIC) system. The SIC has been replaced by the North American Industry Classification System (NAICS). These indexes are similar to the ones in Metal Industry Indicators. The latest report for these indexes, November 2004, is available in PDF format. Historical data for these new indexes are available back to 1948. See also the section on Statistics for a discussion on the Statistical Compendium (U.S. Bureau of Mines) and the USGS Commodity Statistics and Information.

United States Geological Survey (USGS) http://www.usgs.gov/ (accessed July 11, 2010). Self-described as the "Federal source for science about the Earth, its natural and living resources, natural hazards, and the environment." Current topics of national and international concern are exemplified in the links to the "USGS Responds to Deepwater Horizon Oil Spill" at http://www.usgs.gov/deepwater_horizon/ (accessed March 27, 2011) and the eruption of the Iceland Volcano, Eyjafjallajökull at http://volcanoes.usgs.gov/publications/2010/iceland.php (accessed March 11, 2011).

USGS Minerals Information http://minerals.usgs.gov/minerals/ (accessed July 11, 2010) provides subscription information and full text links to the following major publications of the USGS.

ASSOCIATIONS, ORGANIZATIONS, AND SOCIETIES

Generally the Web sites listed offer a single convenient location where mission statements, scope, and objectives of the associations and their organizational divisions, including statements of benefits and reasons for joining, membership forms, membership directories, by-laws, newsletters, lists of monographic and serial publications produced by the association in support of scientific research, technology or legal aspects of the mining industry, and its impact on economies and the environment, may be found. As a bonus, all Web sites will generally include valuable resource links to member companies, related organizations and associations, government agencies, schools and universities, commodities information, production statistics, and links useful to the mining industry or in support of that industry. Society Web sites may provide readers with updated discussions of current issues and events that affect the mining industry, or issues where an association is engaging representation and support on behalf of its constituents. The following lists of major associations, institutes, and societies are selective samples at best.

AUSTRALIA

The Australasian Institute of Mining and Metallurgy http://www.ausimm.com.au/ (accessed July 12, 2010). "The Australasian Institute of Mining and Metallurgy represents the interests of member professionals in all areas of mining, exploration, and minerals-processing industries located in a network of branches throughout the Asia Pacific region. Within its member base of private enterprise, government, research, education, and the support sector, including consultancies, manufacturers, and suppliers, the AusIMM promotes continuing programs of research and educational development" (from Web site).

Australian Mineral Industries Research Association (AMIRA) http://www.amira.com.au/ (accessed July 12, 2010). AMIRA International is an industry association and not-for-profit, private sector company that manages collaborative research for more member companies in the minerals industry in Australasia, Asia, Europe, Africa, and North and South America. The association provides a forum for the minerals industry to meet, network, and cooperate in areas of common interest. AMIRA develops and manages jointly funded research projects on a fee-for-service basis on behalf of its members.

CSIRO http://www.csiro.au/csiro/channel/_ca_dch2t.html (accessed July 12, 2010). CSIRO is Australia's Commonwealth Scientific and Industrial Research Organization. The CSIRO Division of Exploration and Mining supplies R&D to the Australian exploration and mining industry; undertakes major research initiatives in designing mines and mining equipment; and offers technologically assisted prospecting techniques and resource development in socially, economically, and environmentally sustainable and acceptable ways. In the Search window, type "mining" to obtain approximately 3,300 search results on that topic.

CANADA

The following list is selective. Comprehensive listings of Canadian mining associations, institutes, mining companies, and metal producers may be found under links to the Minerals and Metals Sector: http://www.nrcan-rncan.gc.ca/mms-smm/index-eng.htm (accessed April 5, 2011) at Natural Resources Canada.

Canadian Institute of Mining, Metallurgy and Petroleum–CIM/ICM http://www.cim.org/ mainEn.cfm (accessed July 12, 2010). The Canadian Institute of Mining, Metallurgy and Petroleum is the result of individuals in the mining industry seeking a vehicle for lobbying for safety laws and workers' protection as well as a method of ensuring the communication of ideas. CIM has three main objectives: (1) the facilitation of exchange of knowledge and technology, (2) fraternity, and (3) the recognition of excellence through conferences, publications, and awards. The most recent 2006–2008 Strategic Plan stated as one of its goals "world-class professional development, networking, and knowledge sharing" (from Web page). The CIM Standing Committee on Reserve Definitions, the Guidelines for the Estimation of Mineral Resources and Mineral Reserves Committee, the Mineral Property Valuation Committee, and the CIM/CSA Working Committee are continuing to participate in the development of global standards for resource and reserve definition.

Coal Association of Canada http://www.coal.ca/ (accessed July 12, 2010). "Headquartered in Calgary, Alberta, the Coal Association of Canada represents companies engaged in the exploration, development, use, and transportation of coal. Its members include major coal producers and coal-using utilities, the railroads and ports that ship coal, and industry suppliers of goods and services. The Coal Association of Canada provides a forum to discuss and coordinate the views of its members on matters of common interest. It provides a respected voice that enhances the viability of the industry by advocating the clean use of coal through technology development and communication with members, government, and public-sector stakeholders" (from Web page).

Mineralogical Association of Canada http://www.mineralogicalassociation.ca/ (accessed July 12, 2010). "The Mineralogical Association of Canada (MAC) was formed in 1955 as a nonprofit scientific organization to promote and advance the knowledge of mineralogy and the allied disciplines of crystallography, petrology, geochemistry, and mineral deposits" (from Web page).

The Mining Association of Canada http://www.mining.ca/www/ (accessed July 12, 2010). The Mining Association of Canada (MAC) is the national organization of the Canadian mining industry. It comprises companies engaged in mineral exploration, mining, smelting, refining, and semifabrication. Member companies account for the majority of Canada's output of metals and major industrial materials. The primary role of MAC is the presentation of industry information and views to the federal government and parliamentary committees. MAC provides information about the mining industry to the media and to schools, libraries, and other public institutions. Most companies with producing mines and plants are members.

INTERNATIONAL

International Council on Mining & Metals (based in London) http://www.icmm.com/ (accessed July 12, 2010). ICMM is made up of representatives from private industry, national, and international mining associations. "The International Council on Mining and Metals (ICMM) was established in 2001 to act as a catalyst for performance improvement in the mining and metals industry. Today, the organization brings together 19 mining and metals companies as well as 30 national and regional mining associations and global commodity associations to address the core sustainable development challenges faced by the industry" (from Web page).

Institute of Materials, Minerals and Mining http://www.iom3.org/ (accessed July 12, 2010). "The Institute of Materials, Minerals and Mining (IOM3) is a major U.K. engineering institution whose activities encompass the whole materials cycle, from exploration and extraction, through characterization, processing, forming, finishing, and application, to product recycling and land reuse. It exists to promote and develop all aspects of materials science and engineering, geology, mining, and associated technologies, mineral, and petroleum engineering and extraction metallurgy, as a leading authority in the worldwide materials and mining community" (from Web page).

UNITED KINGDOM

Mining History Network http://projects.exeter.ac.uk/mhn/ (accessed July 12, 2010). The Mining History Network is both an information resource for mining historians and a focus to improve opportunities for continuing communication between researchers.

UNITED STATES

In addition to the selective listing below, more exhaustive lists of U.S. regional and state mining associations as well as international mining associations may be found at http://www.miningusa.com/associat/associ.asp (accessed July 12, 2010).

American Institute of Mining, Metallurgical, and Petroleum Engineers (AIME) http://www.aimeny.org/ (accessed July 12, 2010). Founded in 1871, the American Institute of Mining, Metallurgical, and Petroleum Engineers (AIME) with headquarters in Littleton, Colorado, is according to its mission statement "a nonprofit corporation organized and operated to advance and disseminate, through the programs of member societies, knowledge of engineering and the arts and sciences involved in the production and use of minerals, metals, energy sources and materials." AIME activities involve sponsorship of regularly scheduled member meetings/events, conferring of awards/scholarships, providing library research services, maintaining mining-related industry links, and supporting charitable giving. AIME is composed of five separately incorporated units, AIME Institute headquarters, and four autonomous member societies: Society for Mining, Metallurgy, and Exploration (SME); The Minerals, Metals, & Materials Society (TMS); Association for Iron and Steel Technology (AIST); and Society of Petroleum Engineers (SPE). There is also a Woman's Auxiliary to AIME (WAAIME) (from Web page).

American Society of Appraisers (ASA) http://www.appraisers.org/ASAHome.aspx (accessed July 12, 2010). ASA is a multidisciplinary organization of accredited appraisal professionals recognized by the U.S. Congress as a source of appraisers and appraisal standards. The Web site provides a directory to valuation expertise for mine and quarry machinery and specialty equipment, oil and gas machinery and equipment, business valuation, and real and personal property.

American Society of Mining and Reclamation (ASMR) http://dept.ca.uky.edu/asmr/W/ (accessed July 12, 2010). The primary goal of the society is to expand opportunities for technology transfer to meet the needs of reclamationists worldwide. To this end, the society seeks to keep their membership informed about the latest developments in reclamation technology both nationally and internationally. Another goal for the society is to provide written and verbal technology transfer between members of similar organizations and interests in Canada, China, Australia, Great Britain, and the United States through the formation of the International Affiliation of Land Reclamationists (IALR). In future, the society plans on using its Web site to keep members informed of its activities, and as a supplement to its regularly published newsletter.

Mining and Metallurgical Society of America http://www.mmsa.net/ (accessed July 12, 2010). "The Mining and Metallurgical Society of America (MMSA) is a professional organization dedicated to increasing public awareness and understanding about mining and why mined materials are essential to modern society and human well-being. MMSA delivers this message to two different public sectors: (1) public policy makers including elected officials and regulatory agencies, and (2) to all levels of the educational system." In support of the latter activity, MMSA provides teachers at all levels, from elementary through university, classes with teaching materials and funded programs to improve public education about mining and minerals (from Web page).

The Society of Economic Geologists, Inc. http://www.segweb.org/ (accessed March 11, 2010). Society of Economic Geologists, Inc. (SEG) is an international organization with interests in the field of economic geology. Membership includes representatives from industry, academia, and government institutions. The objectives of the society are "… to advance the science of geology through the scientific investigation of mineral deposits and mineral resources and the application thereof to exploration, mineral resource appraisal, mining and mineral extraction. … To disseminate information arising from investigations of mineral deposits and mineral resources through SEG publications, meetings, symposia, conferences, field trips, short courses, workshops and lecture series" (from Web page).

Society for Mining Metallurgy and Exploration http://www.smenet.org/ (accessed July 12, 2010). The Society for Mining, Metallurgy, and Exploration (SME) is an international society of professionals in the minerals industry. A member society of the American Institute of Mining, Metallurgical and Petroleum Engineers (AIME), the SME is organized into five distinct divisions: coal and energy, environmental, industrial minerals, mineral and metallurgical processing, and mining and exploration. The range of programs and services offered to SME members facilitates professional development and information exchange through society publications, professional registration, peer-review of technical papers, college accreditation programs, meetings and exhibits, public education, and short courses.

United States National Mining Association http://www.nma.org/ (accessed July 12, 2010). The National Mining Association (NMA) is a 501(c) 6, nonprofit trade association and represents the political presence of over 325 mining corporations of the American mining industry in Washington, D.C. before Congress, the administration, federal agencies, the judiciary, and the media. Membership includes coal, metal, and industrial mineral producers, mineral processors, equipment manufacturers, state associations, bulk transporters, engineering firms, consultants, financial institutions, and other companies that supply goods and services to the mining industry.

METAL PRICES: INDEXES

Average World Steel Transaction Prices ($US/ton) http://www.meps.co.uk/world-price.htm (accessed July 12, 2010). The latest monthly steel price updates are available to MEPS subscribers through an annual subscription.

Metal Bulletin Prices and Data (London) http://www.metalbulletin.com/Login.aspx?stubId =2338 (accessed July 12, 2010). With data available through annual subscription, the MetalBulletin.com Web site covers international prices for nonferrous metals, iron and steel, and scrap and secondary with the latest news on mergers, acquisitions, tenders, financials, and import/export news. The prices archive contains over 800 prices series, going back to 1992, allowing a researcher to track price trends and create graphs online.

Metal Price Charts on InfoMine–Precious Metals, Base Metals, etc. http://www.infomine. com/investment/charts.aspx (accessed July 12, 2010). The InfoMine Web site provides, aggregates, integrates, and syndicates access to worldwide mining and mineral exploration information for investors in the mining and exploration industry. Interactive commodity and currency charts provide prices of metals in the currencies of countries that mine the commodities over a selectable 15 day to 3 year range. The revenue model for the InfoMine Web site is based on the dual system of subscription and advertising. While there is considerable free content on InfoMine, to get the most out of the site, users may need to subscribe (from Web page).

Mineral Industry Indicators http://minerals.usgs.gov/minerals/pubs/mii/ (accessed July 12, 2010). Monthly newsletter in PDF file format that analyzes and forecasts the economic health of five U.S. metal industries: primary metals, steel, copper, primary aluminum, and aluminum mill products.

INDUSTRY PROFILES AND YEARBOOKS

Canadian & American Mines Handbook. Published by Business Information Group. Published annually in October, the handbook provides concise snapshots of over 2,400 Canadian and U.S. mining companies, mines, and associated organizations. such as smelters, refineries, and industry associations.

Canadian Minerals Yearbook: Review and Outlook. 1994–. http://www.nrcan-rncan.gc.ca/ mms-smm/busi-indu/cmy-amc-eng.htm (accessed July 12, 2010). Provides full text access to Canadian mineral statistics from 1886–2008. Each year, the Minerals and Metals Sector (MMS) of Natural Resources Canada does a comprehensive review of developments and outlook for the mineral industry.

Metals and Minerals Annual Review. 1990–1998. London: Mining Journal Ltd. Reviews the state of metals markets and industry trends in the industry. Provides detailed coverage about precious metals and minerals, steel industry metals, industrial and energy minerals. Continues as *Mining Annual Review* below.

Mining Annual Review. 2001–. London: Mining Journal Ltd. http://www.mining-journal. com/ (accessed July 12, 2010). An annual report (also called *Mining Journal*) on global mining industry activity through over 130 country reports, 70 commodity reports plus technical articles written by experts on developments in technology in mineral exploration, surface mining, underground mining, and in mineral and coal processing.

STATISTICS

Australian Bureau of Statistics (ABS) http://www.abs.gov.au/ (accessed July 12, 2010). In the Search window, type terms such as "mining" or "mining statistics" to obtain a list of search results linking to statistical resources on mineral and petroleum exploration in Australia. The ABS site provides key performance indicators, publications, and articles about the mining industry with data sourced from the ABS and statistical publications of mining related international organizations.

Canadian Minerals Yearbook. 1994–. http://www.nrcan-rncan.gc.ca/mms-smm/busi-indu/cmy-amc-eng.htm (accessed July 12, 2010). Provides full text access to Canadian mineral statistics from 1886–2003. Each year, the Minerals and Metals Sector (MMS) of Natural Resources Canada does a comprehensive review of developments in the mineral industry. Mineral statistics are provided from 1886 and a Chronological Record of Mining Events 1604–1943.

Commodity Statistics and Information http://minerals.usgs.gov/minerals/pubs/commodity/ (accessed July 11, 2010). Statistics and information on the worldwide supply of, demand for, and flow of minerals and materials essential to the U.S. economy, the national security, and protection of the environment. Includes an alphabetical index to publications, contacts, and links to information about dozens of specific commodities.

Crowson, P. 2006. *Minerals Handbook: Statistics and Analyses of the World's Minerals Industry, (1996–1997).* New York: Macmillan Publishers. (Out of print but still available for order from Macmillan Distribution Limited.) Intended as a comprehensive introductory guide for the nonspecialist, the book provides basic data on 52 minerals and metals to permit informed debate on mining and mineral policies. Figures provided list domestic production, trade and consumption, geographical sources of net imports, shares of world production and consumption, historic growth of consumption, end-use patterns, and estimates on world reserves and reserve bases.

Energy, Mines, and Resources Canada. 1977–. *Statistical Review of Coal in Canada.* Ottawa: Energy, Mines and Resources Canada. Annually reviews production, consumption, and export statistics of Canadian metallurgical grade bituminous coal and subbituminous coal used for domestic electricity generation. Includes statistics on coal imported into Canada from the United States and South American export markets.

Gold Institute, The http://www.goldinstitute.org/ (accessed July 12, 2010). The institute's Web site offers links to anything and everything related to the gold industry in particular or mining in general. The information is divided into a number of categories covering American, Canadian, and Australian associations and organizations, international exchanges, and gold price data and statistics on world gold mine production, with supplementary links to mining technology related Web sites.

Mineral Industry Surveys and Minerals Yearbook (US): (*Vol. I–Metals & Minerals*) http://minerals.usgs.gov/minerals/pubs/myb.html (accessed July 12, 2010). U.S. Geological Survey Statistical data on worldwide production, distribution, stocks, and consumption of mineral commodities.

Natural Resources Canada. *Minerals and Mining Statistics Online* http://mmsd.mms.nrcan.gc.ca/stat-stat/index-eng.aspx (accessed July 12, 2010). "The Minerals and Mining Statistics Division (MMSD) is part of the Minerals and Metals Sector (MMS) of Natural Resources Canada. MMSD is responsible for the collection, analysis and dissemination of comprehensive statistics on the Canadian minerals industry describing ore reserves, exploration, and mineral production and use, and for the conduct, in collaboration with Statistics Canada and the provincial and territorial governments, of various surveys of mine complex development and production" (from Web page).

Statistics Canada. *Coal and Coke Statistics* http://dsp-psd.pwgsc.gc.ca/Collection-R/ Statcan/45-002-XIB/45-002-XIB-e.html (accessed July 12, 2010). "Production, imports, exports, stocks, and disposition of coal by province together with supply and disposition of coke in Canada are provided in this publication. Documents are available full text in Adobe® Acrobat PDF format" (from Web page).

Statistics Canada. *Primary Iron and Steel* http://dsp-psd.pwgsc.gc.ca/Collection-R/ Statcan/41-001-XIB/41-001-XIB-e.html (accessed July 12, 2010). "This online publication provides current data on the Canadian iron and steel industry. It includes monthly production and shipment statistics in metric tons for pig iron, steel primary forms, and steel castings at the Canada level. It also includes monthly shipment data in metric tons on rolled steel products broken down by product and consuming industry, again at the Canada level. The December issue includes a list of reporting firms" (from Web page).

U.S. Bureau of Mines. *Statistical Compendium* http://minerals.usgs.gov/minerals/pubs/stat/ (accessed July 12, 2010). "This Statistical Compendium provides a consistent set of official, long-term (20 years or more) data series through 1990 for selected commodities to facilitate analysis of long-term trends in these mineral sectors. The Web site provides an index list of mineral commodity data published in the *Statistical Compendium*" (from Web page).

U.S. Geological Survey. Mineral Resources Program http://minerals.usgs.gov/ (accessed July 12, 2010). "The Mineral Resources Program funds science to provide and communicate current information on the occurrence, quality, quantity, and availability of mineral resources." Links are provided to announcements of USGS Mineral research grants, database, and documentation for the National Geochemical Survey, historical statistics for mineral commodities across the United States, current commodity statistics, and quantitative global mineral resource assessments with links to mineral resource spatial data online. Some of the documents on this Web site are in PDF format and require the free Adobe Acrobat Reader (from Web page).

EDUCATIONAL SUBJECT GUIDES TO MINING AND MINING ENGINEERING ON SELECTED WEB SITES

BUBL LINK Catalog of Internet Resources. Mining http://www.bubl.ac.uk/link/m/mining. htm (accessed July 12, 2010). BUBL is a free Internet-based information service for the U.K. higher education community, but is widely used worldwide by nonlibrarians as well. BUBL is located and run from the Centre for Digital Library Research of the University of Strathclyde, Glasgow, Scotland. The BUBL subject guide to Mining includes hundreds of links to Internet resources and services; descriptions are searchable and can be browsed by alphabetical order or Dewey Decimal Classification.

Country Mine http://www.infomine.com/countries/ (accessed July 12, 2010). Complete mining information on the world mining industry by country. Includes mines and mining properties in the country, a snapshot of the country's mining activities, links to geological and other maps.

Earth Sciences Information Resources–Internet Links, Branner Earth Sciences Library and Map Collections at Stanford University http://library.stanford.edu/depts/branner/research_ help/index.html (accessed July 12, 2010). An excellent meta-collection of Internet resources for the Earth Sciences, Geographic Information Systems (GIS), and the Branner Earth Sciences Library and Map Collection, including links especially of interest to mining engineers, such as professional associations and societies, government agencies of the United States (USGS), the European continent, Asia and the Pacific, Canada, Mexico, and the

Americas, links to Global Databases (including the poles), software tools and utilities, and academic institutions with programs in mining, petroleum engineering, and the geological sciences.

The Future of Mining, Deloitte Touche Tohmatsu http://www.deloitte.com/view/en_CA/ca/index.htm (accessed July 19, 2010) is a Canadian professional financial services firm, providing accounting, audit, tax advisory, and consulting services. The mining industry over the past few years has shown extreme price volatility, market uncertainty, cycles of boom and bust, and resource limitations. This has not been an environment conducive for mining companies to be seeking investment for mineral exploration, extraction, and processing. *Tracking the Trends 2010* by Deloitte provides an excellent summary and overview of the top 10 issues and trends emerging in the global mining sector. The document is available for download in PDF format at http://www.deloitte.com/assets/Dcom-Canada/Local%20Assets/Documents/EandR/ca_en_energy_Tracking_the_trends_2010_121109.pdf (accessed July 19, 2010).

Galaxy Engineering and Technology Mining http://www.galaxy.com/dir40993/Mining.htm (accessed: July 12, 2010). Galaxy is a searchable Internet directory of contextually relevant information. Information is researched, compiled, classified, and organized by human Internet librarians who are master's of library sciences graduates with experience in specific topic areas resulting in subject depth and concentrated, relevant coverage. Mining-related links are given to lists of mining companies, industry employment opportunities, geological surveys, government agencies, industry information, investment vehicles, trade magazines, organizations, and prospecting sites.

Intute http://www.intute.ac.uk/ (accessed July 12, 2010). Intute is a free online search service that finds the best Web resources for academic study and research. Intute is created by a consortium of seven universities in the United Kingdom, working together with a host of partners. A description of each resource is provided with direct access to the resource itself. In the Search Web Resources window, type a word or short phrase such as "mining" to retrieve hundreds of Web links to mineral mining sites and research.

Mindat.org http://www.mindat.org/ (accessed July 12, 2010), The site states that it is "the largest mineral database and mineralogical reference Web site on the Internet." The site contains worldwide data on mineral properties, localities, and other mineralogical information worldwide. Visitors are invited to search minerals by properties, by chemistry; to browse alphabetical and chemical indexes of minerals, or to link to mineral dealers. The site has 38,322 mineral names (including synonyms, varieties, etc.) and 210,495 different localities (from Web page).

Mines, Mining, and Mineral Resources–Library of Congress Science Reference Services. Science Tracer Bullets Online. Tracer Bullet 94-2 http://www.loc.gov/rr/scitech/tracer-bullets/minestb.html (accessed July 12, 2010). This guide reviews the literature on mines, mining, and mineral resources in the collections of the Library of Congress with the exception of materials on coal, uranium, petroleum, metallurgy, alloys, and gemstones.

Mining Education. Mining Internet Services http://www.miningusa.com/ (accessed July 12, 2010). Advertising itself as providing information to the mining industry as a courtesy, the Web site includes links to articles on what coal miners do, the history of coal, a glossary of coal mining terminology, map and GIS resources, directories and addresses of journals, publishers, associations, and companies in the transportation, coal, and metal mining industries.

NRCan Library–Energy, Minerals, and Management http://www2.nrcan.gc.ca/es/msd/emmic/web-en-emmic/ (accessed July 12, 2010). Links to Canadian publications, mineral and metal sector policies, dictionaries and other reference works, lists of mining associations, institutes, mining companies and metal producers, electronic journals, statistics and literature databases, conferences, and meetings.

CONCLUSION

Mining engineering is a specialized subfield requiring the integration of very specialized knowledge across disciplines, and from a variety of sources specific to the field, including aspects of chemistry, chemical, civil, environmental, mechanical, mining, and petroleum engineering. Pressed by a future of persistent global economic uncertainty and with increases in the number of coal mining accidents, growing tailings ponds, and the horrific scale of BP's Gulf of Mexico oil spill, governments and societies are demanding that corporations shift their focus to the research and development of innovative clean energy and mineral processing techniques in an environmentally safe, responsible, and sustainable manner.

18 Nuclear Engineering

Mary Frances Lembo

CONTENTS

Introduction, History, and Scope of the Discipline 453
Searching the Library Catalog 454
Abstracts and Indexes 454
Bibliographies and Guides to the Literature 456
Directories 456
Encyclopedias and Dictionaries 457
Handbooks and Manuals 458
Monographs and Textbooks 460
Journals and Series 462
Patents and Standards 463
Search Engines and Important Web Sites 464
Associations, Organizations, and Societies 465
Conclusion 466
Acknowledgments 467
References 467

INTRODUCTION, HISTORY, AND SCOPE OF THE DISCIPLINE

On December 2, 1942, the first successful self-sustaining atomic chain reaction was achieved at the University of Chicago, thus ushering in the Nuclear Age and the discipline of nuclear engineering. Rather than being a theoretical endeavor of physicists, "nuclear fission propelled the subject into the military arena, leading to the enormous Manhattan Project in the United States that produced the Hiroshima and Nagasaki bombs (McKay, 1984, preface)". The demands of an undertaking on the scale of the Manhattan Project presented new and unique challenges to the engineering industry. Developing processes to create and refine plutonium and uranium, constructing buildings and reactors to facilitate these processes, and designing nuclear weapons were all immediate challenges made even more imperative by the military threat of Germany and Japan during World War II.

After the nuclear bombs dropped on Hiroshima and Nagasaki effectively ended World War II, scientists were in a position to devote their energies to nonmilitary uses for nuclear energy. Atomic technology, originally developed for its destructive uses, was now being tapped for peaceful applications, such as producing electricity and developing medical isotopes. However, nuclear production also resulted in the challenge of cleaning up wartime facilities and safe storage of waste by-products as well as maintaining active nuclear power plants. The field of nuclear engineering covers the entire range of nuclear energy production including: design, construction, operation and maintenance of nuclear power and naval propulsion reactors, reactor safety, development of nuclear weapons, disposal of radioactive wastes, and the production of radioisotopes (*Encyclopaedia Britannica*, 1993, pp. 423–424).

During the Nuclear Age of the 1950s and 1960s, in the excitement of this new technology, many texts were published, which are still relevant today. As the interest in nuclear power waned, fewer current books on the topic were available in print. However, with the renewed interest in nuclear power as a viable alternative to fossil fuels, newer materials are again being published. This chapter includes both "classic" items as well as current references.

As in other engineering disciplines, the journal literature plays a major role for nuclear engineers, as do handbooks and standards. The technical report literature also plays an important role. Significant publishing agencies key to nuclear engineering include the American Nuclear Society in the United States and the International Atomic Energy Agency worldwide.

SEARCHING THE LIBRARY CATALOG

The following is a list of useful Library of Congress Subject Headings in nuclear engineering and related fields. Using the phrase "Nuclear engineering" in combination with other keyword terms should provide a targeted topic search.

Nuclear engineering:

Nuclear energy
Nuclear facilities
Nuclear fission
Nuclear fuel
Nuclear fuel elements
Nuclear fuels
Nuclear industry
Nuclear power plants
Nuclear reactors
Radioactive decontamination
Radioactive waste disposal
Radioactive wastes
Spent nuclear fuels

The Library of Congress (LC) call numbers classify the nuclear literature into two main areas: the TK9000+ range, which emphasizes the applied aspect of nuclear engineering, and the QC770+ range, which demonstrates the nuclear physics aspect of the discipline.

TK9001-9401 Nuclear engineering. Atomic power
QC770-798 Nuclear and particle physics. Atomic energy. Radioactivity

In addition, health effects of nuclear radiation are covered in RA1231.

ABSTRACTS AND INDEXES

The discipline of nuclear engineering is covered in the general indexing and abstracting services, such as *Compendex* and *Inspec*. The resources listed below are services that are more specific to nuclear engineering. The following indexes and abstracts will assist users in locating literature in nuclear engineering covered in journal publications, conference proceedings, and technical reports.

Energy Citations Database (ECD). 1948–. Oak Ridge, TN: Office of Scientific and Technical Information, U.S. Department of Energy. *Energy Citations Database* contains bibliographic records for energy and energy-related research from the U.S. Department of Energy and its predecessor agencies. This database is publicly available at http://www.osti.gov/energycitations/ (accessed August 10, 2010). *ECD* includes the information found in *Information Bridge* as well as *Nuclear Science Abstracts* and continues *Energy Research Abstracts*.

Energy Research Abstracts, 1977–1995. Oak Ridge, TN: Technical Information Center, U.S. Department of Energy. Continues *ERDA Energy Research Abstracts* and *Nuclear Science Abstracts.* Energy Research and Development Administration (ERDA) was the predecessor agency to the U.S. Department of Energy. In addition, the scope was expanded to include broader coverage of nonnuclear energy information.

Energy Science and Technology. 1974–. Oak Ridge, TN: Office of Scientific and Technical Information, U.S. Department of Energy. Available online from Dialog and STN, but use is restricted to users located in the United States and to the International Energy Agency (IEA) Energy Technology Data Exchange (EDTE) member countries. The database corresponds, in part, to *Energy Research Abstracts* and *INIS AtomIndex.* Continues *Nuclear Science Abstracts,* but with broader coverage to include nonnuclear energy information.

Information Bridge. 1991–. Oak Ridge, TN: Office of Scientific and Technical Information, U.S. Department of Energy. This source contains the full text of Department of Energy (DOE) research and development reports generally from 1991 onward, but reports are being added retrospectively as they become available in electronic format. The database allows for full-text searching and is publicly available at http://www.osti.gov/bridge/ (accessed August 10, 2010). The full-text documents and bibliographic citations found in *Information Bridge* are also incorporated into the more comprehensive *Energy Citations Database.*

INIS (*International Nuclear Information System Database*). 1970–. Vienna: International Atomic Energy Agency (IAEA). Supersedes *INIS AtomIndex.* The *INIS* database contains worldwide literature on the peaceful uses of nuclear science and technology. Beginning in 1992, the database also includes the economic and environmental aspects of all nonnuclear energy sources. Access is via the Internet at http://inisdb.iaea.org/ (accessed August 10, 2010). Information from *INIS* is also included in *Energy Science and Technology Database,* but is restricted to the International Energy Agency (IEA) Energy Technology Data Exchange (EDTE) member countries.

Nuclear Science Abstracts. 1948–1976. Oak Ridge, TN: Oak Ridge Directed Operations, Technical Information Division. *Nuclear Science Abstracts* covers the nuclear science and technology literature, and includes scientific and technical reports of the predecessor agencies of the U.S. Department of Energy (the Atomic Energy Commission and the Energy Research and Development Administration). The information from this database is also included in *Energy Citations Database.*

Power Reactor Information System (PRIS). 1970–. Vienna: International Atomic Energy Agency (IAEA). *PRIS* provides specific data, such as location, operator, owner, suppliers, and design characteristics, on operating reactors, reactors under construction, and reactors in the process of being decommissioned in IAEA Member States as well as Taiwan and China. Additional data includes energy production and operational events. http://nucleus.iaea.org/sso/NUCLEUS.html?exturl=http://www.iaea.or.at/programmes/a2/ (accessed August 11, 2010).

Research Reactor Database (RRDB). 1970?–. Vienna: International Atomic Energy Agency (IAEA). *RRDB* contains technical data on over 670 research reactors in 69 countries and the European Union. Each facility updates the data within the database, which is then reviewed by IAEA staff. http://nucleus.iaea.org/RRDB/RR/ReactorSearch.aspx?rf=1 (accessed August 11, 2010).

Waste Management Research Abstracts (WMRA). 1975?–2005. Vienna: International Atomic Energy Agency (IAEA). WMRA is a collection of research summaries dealing with the topic of radioactive waste, including environmental restoration and decommissioning of nuclear facilities. The collection of these abstracts began in the late 1960s. Originally published annually, there are 30 printed volumes in this collection. Links to volumes 23–30 (1999–2005) are available at http://www-pub.iaea.org/MTCD/publications/seriesMain.asp (accessed August 18, 2010). List is alphabetical.

BIBLIOGRAPHIES AND GUIDES TO THE LITERATURE

Although no bibliography on nuclear engineering has been published recently, the following provide solid background and historical references for the field.

Anthony, L. J. 1966. *Sources of Information on Atomic Energy, International Series of Monographs in Library and Information Science, Vol. 2.* Oxford/New York: Pergamon Press. While dated, this book provides an introduction to atomic energy, information sources from a variety of countries, and a discussion of published sources in the areas of atomic energy, high energy physics, nuclear power and engineering, ionizing radiation and radioisotopes, and controlled nuclear fusion and plasma physics. A subject and author index is included along with an index to organizations and to periodicals.

Information International Associates, Inc., Oak Ridge National Laboratory. 1980–1997. *Nuclear Facility Decommissioning and Site Remedial Actions: A Selected Bibliography.* (18 vols.) ORNL/EIS-154, Oak Ridge, TN: Martin Marietta Energy Systems. This multivolume bibliography contains citations and abstracts of documents relevant to environmental restoration, nuclear facility decontamination and decommissioning, uranium mill tailings management, and site remedial actions. Volumes 1, 2–6, and 8–10 can be found in full-text PDF format from *Energy Citations Database.*

International Atomic Energy Agency, Commission of the European Communities, OECD Nuclear Energy Agency. 1991. *Water Cooled Reactor Technology: Safety Research Abstracts*, No.1. IAEA/WCRT/SRA/01. Vienna: International Atomic Energy Agency (IAEA). This collection of abstracts focuses on the subject areas related to nuclear safety, such as pressurized light water-cooled and moderated reactors, boiling light water-cooled and moderated reactors, light water-cooled and graphite moderated reactors, pressurized heavy water-cooled and moderated reactors, and gas-cooled graphite moderated reactors.

Lester, R. and A. J. Walford. 2005. *The New Walford's Guide to Reference Resources: Science, Technology and Medicine, Vol. 1.* London: Facet Publishing. This new guide, now organized by subject, contains references to a full variety of resources, including journal articles, online database services, journal articles, encyclopedias, dictionaries, handbooks, and so on. In addition, a special emphasis is placed on new digital resources, such as blogs, RSS feeds, and Web sites. Within the section on technology are several pages devoted to nuclear energy. The physics section also contains resources on nuclear and particle physics. The 2005 edition is the first volume of a series that succeeds the original *Walford's Guide to Reference*, which was published in eight editions between 1959 and 2000.

United Nations. Department of Political and Security Council Affairs. Atomic Energy Commission Group. 1949–1953. *An International Bibliography on Atomic Energy.* Lake Success, NY: Atomic Energy Commission Group, Department of Security Council. This bibliography comes in two volumes and is international in scope. Volume 1 deals with political, economic, and social aspects and includes two supplements. It tends to be more selective rather than comprehensive. Volume 2 deals with scientific aspects and includes one supplement. This volume focuses on the peaceful purposes of atomic energy.

DIRECTORIES

Changes in personnel, location, or other data cannot be reflected quickly in print format; however, the following resources can act as a starting point for locating directory-type information. Societies and professional organizations often have membership directories that may be available via their Web sites.

American Nuclear Society. 2009. *2009–2010 World Directory of Nuclear Utility Management*, 21st ed. La Grange Park, IL: American Nuclear Society. This directory, compiled by the editors of *Nuclear News*, contains listings for important personnel at nuclear plant sites, such as plant managers, maintenance superintendents, radioactive waste managers, and public relations contacts. The plant listings are arranged by country and include status information (operating, under construction, or decommissioned). International organizations and contact information are also included. Available in a paperback print version or a print/CD-ROM combination.

Business Press International. 2010. *World Nuclear Industry Handbook*. Sutton, Surrey, U.K.: Business Press International. Issued as a supplement to the November issue of *Nuclear Engineering International*, this publication provides up-to-date contact information as well as products and services information.

International Atomic Energy Agency. 1959–1976. *Directory of Nuclear Reactors*. Vienna: International Atomic Energy Agency (IAEA). This 10-volume set discusses electric power and research reactor projects. Information includes name and location of reactor, data on reactor physics, core, fuel element, reflector and shielding, safety and containment, and a brief bibliography for each reactor.

International Atomic Energy Agency. 1998. *Directory of Nuclear Research Reactors*. Vienna: International Atomic Energy Agency (IAEA). This reference contains technical information on research reactors known to the IAEA at the end of October 1998. IAEA collected the information compiled in this directory by sending questionnaires to the reactor operators. Data that IAEA could not verify is listed in Part III of the directory.

International Atomic Energy Agency. 2009. *Nuclear Fuel Cycle Information System: A Directory of Nuclear Fuel Cycle Facilities*, IAEA-TECDOC-1613. Vienna: International Atomic Energy Agency (IAEA). This publication provides data on civilian nuclear fuel cycle facilities with an international perspective. The information for this directory was obtained by questionnaires to IAEA member states and from additional published sources. Information includes name and location of facility, start-up and shutdown dates, and so on. An introduction to the nuclear fuel cycle industry and processes, and a glossary of terms and references are also included. Available at http://www-pub.iaea.org/MTCD/publications/PDF/te_1613_web.pdf (accessed August 16, 2010).

International Atomic Energy Agency. 2010. *Nuclear Power Reactors in the World, Reference Data Series No. 2*, 30th ed. Vienna: International Atomic Energy Agency (IAEA). This resource is a collection of tables providing specific reactor data, such as nuclear power reactors in operation and under construction, December 31, 2009, scheduled construction starts during 2009, nuclear share of electricity generation (as of 2009), and more. Information is gathered by questionnaires to member states. Available at http://www-pub.iaea.org/MTCH/publications/PDF/iaea-rds-2-30-web.pdf (accessed March 22, 2011).

Longman Publishing Co. 1996. *World Energy and Nuclear Directory: Organizations and Research Activities in Atomic and Non-atomic Energy*, 5th ed. Harlow, Essex, U.K.: Longman; Detroit, MI: Distributed in the United States and Canada by Gale Research. This directory provides contact information as well as budget, staff size, and summaries of research interests for research and development centers around the world in the areas of energy and nuclear technology. The entries are arranged alphabetically by the language of that country.

ENCYCLOPEDIAS AND DICTIONARIES

Encyclopedias and dictionaries provide useful background information for researchers who are beginning in nuclear engineering or who need to get up to speed on a topic outside their area of expertise. Multilanguage dictionaries provide assistance to researchers working with documents in foreign languages or with teams from other countries.

Alter, H. 1986. *Glossary of Terms in Nuclear Science and Technology*. La Grange Park, IL: American Nuclear Society. Prepared by ANS-9, the American Nuclear Society Standards

Subcommittee on Nuclear Terminology and Units. This concise glossary provides brief definitions of nuclear terms, including nuclear and reactor physics, health physics, and regulatory terms. Also includes an appendix on a classification system for reactors.

Atkins, S. E. 2000. *Historical Encyclopedia of Atomic Energy*. Westport, CT: Greenwood Press. This encyclopedia provides an alphabetical listing and brief explanations of key events, people, organizations, and sites significant in the history of nuclear energy. A chronology of atomic energy, an index, and a selected biography are also included.

Clason, W. E. 1986. *Elsevier's Dictionary of Nuclear Science and Technology: In English– American (with Definitions), French, Spanish, Italian, Dutch, and German*, 5th ed. Amsterdam/New York: Elsevier Publishing. This polyglot dictionary lists words in English, defines them, and then gives the translation for the five languages listed in the title. The dictionary covers a variety of nuclear engineering fields, such as nuclear physics and chemistry, metallurgy, and ionizing radiation. It also has a section for each language referring back to the entry number on the chart. A brief bibliography is included at the end.

International Atomic Energy Agency. 2002. *IAEA Safeguards Glossary,* 2001 ed. International Nuclear Verification Series No. 3. Vienna: International Atomic Energy Agency (IAEA). This glossary is divided into 13 parts, each of which covers a different area related to IAEA safeguards. Definitions and explanations are given for each term or concept. There is also a translation section where the terms are translated into the six official IAEA languages (Arabic, Chinese, English, French, Russian, and Spanish) as well as German and Japanese. Available at http://www-pub.iaea.org/MTCD/publications/PDF/nvs-3-cd/PDF/NVS3_scr. pdf (accessed August 17, 2010).

International Atomic Energy Agency. 2003. *Radioactive Waste Management Glossary*. Vienna: International Atomic Energy Agency (IAEA). This concise publication provides definitions of terms used in radioactive waste management. http://www-pub.iaea.org/ MTCD/publications/PDF/Pub1155_web.pdf (accessed August 17, 2010.)

International Atomic Energy Agency. 2007. *IAEA Safety Glossary: Terminology Used in Nuclear Safety and Radiation Protection*. Vienna: International Atomic Energy Agency (IAEA). The glossary provides definitions and explanations of technical terms used in IAEA safety standards and safety publications. http://www-pub.iaea.org/MTCD/publications/PDF/Pub1290_web.pdf (accessed October 24, 2010).

Office of Intelligence. U.S. Department of Homeland Security, Office of Science and Technology. 2003. *Nuclear Terms Handbook*. Washington, D.C.: Department of Homeland Security. This concise handbook contains not only a glossary of nuclear terms, but also includes frequently used acronyms and abbreviations, a materials description and uses list, and a world list of nuclear power plants with related maps. A bibliography is included at the end of the book.

HANDBOOKS AND MANUALS

Handbooks and manuals are essential to the study of nuclear engineering because they provide key facts, equations and formulas, and fundamental information in a concise and easy-to-use format for the engineer. The following are a list of reference handbooks important to the study of nuclear engineering.

Argonne National Laboratory. 1963. *Reactor Physics Constants, ANL-5800,* 2nd ed. Washington, D.C.: U.S. Atomic Energy Commission. This handbook contains data on fission properties, selected cross sections, constants for thermal homogeneous reactors, and shielding constants. A subject index is available at the end of the book. Available at http:// www.osti.gov/bridge/purl.cover.jsp?purl=/4620873-seNtgm/ (accessed August 18, 2010).

Baum, E. M., H. D. Knox, and T. R. Miller. 2010. *Chart of the Nuclides,* 17th ed. New York: Knolls Atomic Power Laboratory, Lockheed Martin. Available as either a wall chart or a textbook version, this publication shows the key nuclear properties of the known stable and radioactive forms of the elements. In chart format, the nuclides are arranged with the atomic number along the vertical axis and the neutron number along the horizontal axis. Descriptive information includes a history of the development of the periodic table, descriptions of the type of data on the chart, and unit conversion factors and fundamental physics constants.

Cacuci, D. G. 2010. *Handbook of Nuclear Engineering.* New York: Springer. This four-volume e-book provides thorough coverage of all areas of nuclear engineering. Volume 1 focuses on the fundamentals of nuclear engineering; volumes 2, 3, and 4 discuss nuclear reactor analysis, power reactors, experimental reactors, and the nuclear fuel cycle, respectively. Other topics include radioactive waste disposal, safeguards, and nuclear nonproliferation.

Etherington, H. 1958. *Nuclear Engineering Handbook.* New York: McGraw-Hill. This classic publication presents data used in all aspects of nuclear engineering. Divided into 14 sections, the book contains information regarding mathematical data and tables, nuclear data, experimental techniques, radiological protection, fluid and heat flow, reactor materials, and isotopes.

Frisch, O. R. 1958. *The Nuclear Handbook.* Princeton, NJ: Van Nostrand. This book is intended as a desk reference source for those interested in the science and technology of nuclear physics. Material is presented in a concise format. Topics covered include radiation effects and protection, elements and isotopes, natural radioactivity, nuclear materials, particle accelerators, and nuclear reactors.

Kok, K. D. 2009. *Nuclear Engineering Handbook.* Boca Raton, FL: CRC Press. Organized in three sections (Nuclear Power Reactors, Nuclear Fuel Cycle Processes and Facilities, and Engineering and Analytical Applications), this handbook provides an introduction to nuclear engineering intended for engineers and engineering students. Aspects of nuclear engineering covered include design, licensing, construction, and operation of nuclear power plants, as well as the fuel cycle process and the decontamination/decommissioning of facilities.

Nero, A. V. 1979. *A Guidebook to Nuclear Reactors.* Berkeley, CA: University of California Press. The first part of this text provides a general introduction to nuclear power plants, including basic reactor design features, environmental interactions, and nuclear power plant emissions. Part 2 discusses commercial nuclear power plants. Uranium resources and other nuclear materials are discussed in Part 3. Part 4 looks into advanced reactor systems. Appendices include abbreviations and units, reactions, and the nuclear fuel cycle. A glossary and an index are included.

Poenaru, D. N. and W. Greiner. 1996. *Handbook of Nuclear Properties.* Oxford Studies in Nuclear Physics 17. Oxford, U.K./New York: Oxford University Press. This handbook begins with information on atomic masses and shell model interpretations of nuclear masses. Additional information includes nuclear deformations and nuclear stability. Tables on fundamental constants, energy conversion factors, particle properties, alpha-particle emitters, and a table of nuclides are also included.

U.S. Atomic Energy Commission. 1960–1964. *Reactor Handbook,* 2nd ed. New York: Interscience Publishers. This four-volume set provides data on the materials, fuel processing, and physics of nuclear engineering. Volume 1 discusses fuel materials, cladding and structural materials, control, moderator, coolant, and shielding materials. Volume 2 provides background on fuel processing including aqueous and nonaqueous separations, reconversions, and radioactive waste disposal. Volume 3 presents the physics of nuclear reactors, including the theory of neutron transport and reactor dynamics. In addition, this volume discusses shielding information, including neutron attenuation, heat generation in shields, and sources of neutrons and gamma rays. Volume 4 presents the engineering of nuclear reactors with discussions on fluid flow and heat transfer, reactor operations and safety, and types of reactors.

Wick, 0. J. 1980. *Plutonium Handbook: A Guide to the Technology*. LaGrange Park, IL: American Nuclear Society. The *Plutonium Handbook* takes a multidisciplinary approach to examining the element plutonium. It contains basic information on the physics, metallurgy, engineering, chemistry, and health and safety of plutonium. An author and subject index is included at the end of the book.

MONOGRAPHS AND TEXTBOOKS

The following is a list of monographs and textbooks that provide background for a basic understanding in the areas of nuclear engineering, nuclear waste issues, reactor physics theory, and power plant design.

Aasen, A. and P. Olsson, eds. 2009. *Nuclear Reactors, Nuclear Fusion and Fusion Engineering*. New York: Nova Science Publishers. Current critical research in both fission and fusion energy production and reactor technology is presented in this publication.

Bell, G. I. and S. Glasstone. 1970. *Nuclear Reactor Theory*. New York: Van Nostrand Reinhold. This book serves as an introduction to nuclear reactor theory for physicists, mathematicians, and engineers, and explains the physical and mathematical concepts used in nuclear reactors. The chapters include references and exercises. An appendix at the end includes selected mathematical functions.

Bryan, J. C. 2008. *Introduction to Nuclear Science*. Boca Raton, FL: CRC Press. *Introduction to Nuclear Science* covers nuclear chemistry and physics, current applications and health issues. A strong emphasis is placed on providing relevant mathematical concepts applicable to the field of nuclear science. In addition, information of use to medical professionals is provided.

Duderstadt, J. J. and L. J. Hamilton. 1976. *Nuclear Reactor Analysis*. New York: John Wiley & Sons. Designed for the nuclear engineering student, this book provides the basic scientific principles of nuclear fission chain reactions and applications in nuclear reactor design. References and problems are included at the end of each chapter along with appendices, which include selected nuclear data, selected mathematical formulas, and nuclear power reactor data.

Glasstone, S. and A. Sesonske. 1994. *Nuclear Reactor Engineering,* 4th ed. New York: Chapman & Hall. This two-volume publication covers the principles of nuclear engineering and applications for the design and operation of nuclear power plants. The first volume discusses the fundamentals of nuclear engineering, including nuclear fission, radioactivity, neutron transport behavior, and the basics of nuclear design. Volume 2 focuses on the applications and other advanced topics including energy transport and fuel management as well as environment, health, and safety concerns. Innovations in plant design and challenges to the nuclear power industry are also discussed. References and exercises are included at the end of most chapters.

Jevremovic, T. 2009. *Nuclear Principles in Engineering,* 2nd ed. New York: Springer Science+Business Media. This book covers the basic principles in nuclear physics, including safe reactors, medical imaging, food technology radiation shielding, and nuclear space applications with emphasis on engineering applications. There is a focus on descriptive models and analogies with simplified mathematics.

Knief, R. A. 2008. *Nuclear Engineering: Theory and Technology of Commercial Nuclear Power,* 2nd ed. La Grange Park, IL: American Nuclear Society (reprint of the 1992 edition). This textbook, intended for undergraduate and graduate nuclear engineering courses, was originally based on a class the author taught at the University of New Mexico. Practical aspects of commercial nuclear power are discussed. Descriptions of nuclear reactor designs and fuel-cycle steps are included as well as all facets of nuclear plant operation from startup, power operation to shutdown modes. There is also an increased focus on nuclear reactor safety.

Lamarsh, J. R. 1972. *Introduction to Nuclear Reactor Theory.* Reading, MA: Addison Wesley Publishing. This publication is based on a one-year course in nuclear reactor theory at Cornell and New York universities. The book gives readers a fundamental understanding of the principles regarding the operation of a nuclear reactor. References and exercises are included at end of each chapter. Appendices cover conversion factors, selected constants, isotopes of importance in nuclear engineering, and properties of selected molecules.

Lamarsh, J. R. and A. J. Baratta. 2001. *Introduction to Nuclear Engineering,* 3rd ed. Upper Saddle River, NJ: Prentice Hall. This book, now in its third edition, was originally based on the class notes and lectures of the late Dr. John R. Lamarsh. It contains an overview of the field of nuclear engineering and discusses the basics of atomic and nuclear physics, nuclear reactor theory, and reactor design, and both U.S. and non-U.S. nuclear reactor design. Information on nuclear reactor safety is also included. References and practice exercises are included at the end of most chapters.

Murray, R. L. 2008. *Nuclear Energy: An Introduction to the Concepts, Systems and Applications of Nuclear Processes,* 6th ed. New York: Elsevier Science and Technology Books. This publication, written for the undergraduate engineering student, provides an overview of the field of nuclear energy and its uses. It focuses on the processes, uses, and future outlook of nuclear energy, and nuclear nonproliferation. New trends in the industry are discussed including probabilistic safety analysis (PSA), electric power industry deregulation, performance characteristics of nuclear power plants, storage and disposal of radioactive wastes, and developments in decontamination and decommissioning. The basic concepts of energy, nuclear reactions, and effects of nuclear energy on humans are discussed. With the sixth edition, new coverage includes information on the concerns of nuclear safety and terrorism as well as nuclear plant security and the detection of nuclear materials. Facts on nuclear waste management including the Yucca Mountain Repository and new information on the outlook for new power reactors are also included. Exercises are included at the end of each chapter and a list of selected references is given at the end of the book. Internet resources and a student Web site are included. An instructor Web site with exercise solutions and premade PowerPoint® slides is also available. The author studied under J. Robert Oppenheimer at UC Berkeley and made contributions to the uranium separation process for the Manhattan Project at Berkeley and at Oak Ridge.

Murray, R. L. and K. L. Manke, eds. 2003. *Understanding Radioactive Waste,* 5th ed. Columbus, OH: Battelle Press. Written for the general public, this publication provides the reader with the background information necessary to understand the complexities of radioactive waste and its relationship to the political climate.

Olander, D. R. 1976. *Fundamental Aspects of Nuclear Reactor Fuel Elements*, TID-27611-P1. Oakridge, TN: Technical Information Center. Energy Research and Development Administration. This book was based on lectures in graduate courses in the Department of Nuclear Engineering, University of California, Berkeley. Part 1 reviews selected aspects of statistical thermodynamics, crystallography, chemical thermodynamics, and physical metallurgy as it applies to nuclear reactor fuel elements. Part 2 shows how these principles apply to issues of nuclear fuel elements. Each chapter includes a section on nomenclature, references, and problems. Solutions of problems, in handwritten format, are available at http://www.osti.gov/bridge/product.biblio.jsp?query_id=0&page=0&osti_id=7290222 (accessed August 18, 2010).

Rust, J. H. 1979. *Nuclear Power Plant Engineering.* Buchanan, GA: Haralson Publishing. This text focuses on the technological and engineering aspects of nuclear power plant design and nuclear energy. This book is based on the author's lectures on nuclear power at the University of Virginia and the Georgia Institute of Technology. Chapters include references and exercises. Appendices at the end cover conversion factors, properties of reactor coolants, and materials.

Shultis, J. K. and R. E. Faw. 2007. *Fundamentals of Nuclear Science and Engineering,* 2nd ed. Boca Raton, FL: CRC Press. Acting as an introduction to the fundamentals of nuclear science and engineering, this text provides the basic foundation to understand the field of nuclear science. An examination of particle accelerators, nuclear fusion reactions, and nuclear medical diagnostic technology is included.

Stacey, W. M. 2007. *Nuclear Reactor Physics,* 2nd ed. Weinheim, Germany: Wiley-VCH. This book is useful as a textbook as well as a comprehensive reference on the topic of nuclear reactor physics. Part 1 of the book introduces the reader to the basic principles and computational methodology needed to understand the subject matter. Part 2 provides more advanced development of the principles of nuclear reactor physics and is intended for those becoming specialists in the field.

Waltar, A. E. and A. B. Reynolds. 1981. *Fast Breeder Reactors.* New York: Pergamon Press. This book discusses the basic principles and methods as well as design features of fast breeder reactors. Each chapter ends with references and problems to be solved. Appendices include fast reactor data, comparison of homogeneous and heterogeneous core designs, and a list of symbols.

JOURNALS AND SERIES

The journal literature plays a significant role in the communication of research and development in nuclear engineering. The following is a list of key journals in the field of nuclear energy and engineering. The list has been compiled by consulting with the Institute for Scientific Information's (ISI) *Journal Citation Reports* and by investigating the collection of the Pacific Northwest National Laboratory Technical Library. All of the journals listed below are available in electronic format.

Advances in Nuclear Science and Technology. 1962–. Plenum Press. Monographic series (0065-2989).

Annals of Nuclear Energy. 1975–. Pergamon Press. Continues Annals of Nuclear Science and Engineering (0306-4549).

Annals of the ICRP. 1977–. International Commission on Radiological Protection. Pergamon Press. (0146-6453).

Atomic Data and Nuclear Data Tables. 1973–. New York: Academic Press. Merges Atomic Data and Nuclear Data Tables (0092-640X).

Atomic Energy/Atomnaia Energiia. 1993–. Consultants Bureau. Translated from the Russian. Continues Soviet Atomic Energy (1063-4258).

Fusion Engineering and Design. 1987–. North-Holland Publishing. Continues Nuclear Engineering and Design/Fusion (0920-3796).

Fusion Science and Technology: An International Journal of the American Nuclear Society. 2001–. American Nuclear Society. Continues Fusion Technology (1536-1055).

IEEE Transactions on Nuclear Science. 1963–. Institute of Electrical and Electronics Engineers. Continues IRE Transactions on Nuclear Science (0018-9499).

Journal of Nuclear Materials. Journal des Materiaux Nucleaires. 1959–. North Holland Publishing. Articles in English, French, or German (0022-3115).

Journal of Nuclear Materials Management (JNMM). 1986–. Institute of Nuclear Materials Management. Summaries in English and Japanese. Continues Nuclear Materials Management (0893-6188).

Journal of Nuclear Science and Technology. 1964–. Nihon Genshiryoku Gakkai. Atomic Energy Society of Japan (0022-3131).

Kerntechnik. 1987–. C. Hanser. Continues Atomkernenergie/Kerntechnik (0932-3902).

Nuclear Engineering and Design: An International Journal Devoted to the Thermal, Mechanical and Structural Problems of Nuclear Energy. 1965–. Elsevier. Continues Nuclear Structural Engineering (0029-5493).

Nuclear Engineering International. 1968–. London: Heywood-Temple Industrial Publications. Continues Nuclear Engineering (London) (0029-5507).

Nuclear Fusion. Fusion Nucleaire. 1960–. Vienna: International Atomic Energy Agency (0029-5515).

Nuclear Future: Journal of the Institution of Nuclear Engineers and the British Nuclear Energy Society. 2005–. London: Thomas Telford. Formed by the merger of Nuclear Energy (United Kingdom) and The Nuclear Engineer (United Kingdom) (1745-2058).

Nuclear News. 1959–. La Grange Park, IL: American Nuclear Society (0029-5574).

Nuclear Plant Journal. 1987–. Glen Ellyn, IL: EQES. Continues Nuclear Plant Safety (0892-2055).

Nuclear Science and Engineering: the Journal of the American Nuclear Society. 1956– La Grange Park, IL: American Nuclear Society (0029-5639).

Nuclear Technology. 1971–. La Grange Park, IL: American Nuclear Society. Continues Nuclear Applications and Technology (0029-5450).

Progress in Nuclear Energy. 1977–. New York: Elsevier (0149-1970).

Radiation Measurements. 1977–. Oxford: Pergamon Press. Continues Nuclear Tracks and Radiation Measurements (1350-4487).

Radiation Protection Dosimetry. 1981–. Ashford, Kent, U.K.: Nuclear Technology Publishing (0144-8420). *Radiochimica Acta.* 1962–. New York: Academic Press (0033-8230).

Transactions of the American Nuclear Society. 1958–. La Grange Park, IL: American Nuclear Society (0003-018X).

Waste Management: Proceedings of the Symposium on Waste Management. 1979–. Tucson: Arizona Board of Regents. (0275-6196).

PATENTS AND STANDARDS

Patents and, in particular, standards play a major role in the nuclear engineering literature. For information on searching patents, see the patent section in Chapter 2 of this book. Look for significant patents from the following companies: Areva, Babcock & Wilcox, British Nuclear Fuels, Combustion Engineering, DuPont, Framatome, General Atomics, General Electric Company, Japan Atomic Energy Research Institute, Siemens, and Westinghouse Electric Corporation.

The following associations include standards pertinent to nuclear engineering:

American Nuclear Society (ANS) Standards http://www.ans.org/standards/ (accessed August 25, 2010). ANS has over 90 standards specific to the nuclear industry.

American Society of Civil Engineers issues the following standards for nuclear facility construction:

Seismic Analysis of Safety-Related Nuclear Structures and Commentary, ASCE 4-98. 2000. Reston, VA: American Society of Civil Engineers. This publication provides the requirements for the analysis of new nuclear structure designs or the evaluation of existing structures to determine their reliability in the event of an earthquake.

Seismic Design Criteria for Structures, Systems and Components in Nuclear Facilities, ASCE 43-05. 2005. Providing criteria for the design of nuclear facilities, this standard will be useful to those designing new nuclear structures.

A Summary Description of Design Criteria, Codes, Standards, and Regulatory Provisions Typically Used for the Civil and Structural Design of Nuclear Fuel Cycle Facilities. 1988. Reston, VA: American Society of Civil Engineers. Although this title is not specifically a standard, it is useful in understanding the government regulations and industry codes and standards for those who are constructing a new nuclear facility. It brings together information on the requirements from a variety of different resources.

American Society for Testing and Materials (ASTM) Section 12 - Nuclear, Solar, and Geothermal Energy. This section of the ASTM standards contains over 300 active standards in the area of nuclear energy.

American Society of Mechanical Engineering. 2004. *ASME Boiler and Pressure Vessel Code*. The *ASME Boiler and Pressure Vessel Code* provides requirements for the design and fabrication of boilers and pressure vessels and nuclear power plant components during construction. The Code is available in both print and electronic format. The print version of The Code is regularly updated by addenda.

IEEE Nuclear Standards http://www.ieee.org/publications_standards/index.html (accessed August 25th, 2010). The IEEE has over 130 standards related to nuclear and plasma science, accelerator technology, fusion, nuclear instruments, plasma science and applications, radiation effects, and reactor instruments and controls.

International Atomic Energy Agency (IAEA) http://www.iaea.org/Publications/Standards/index.html (accessed August 25, 2010). The IAEA publishes standards in the areas of nuclear safety, radiation protection, radioactive waste management, transport of radioactive materials, safety of nuclear fuel cycle facilities, and quality assurance.

U.S. Department of Energy (DOE) Technical Standards http://www.hss.energy.gov/nuclearsafety/ns/techstds/standard/standard.html (accessed August 26, 2010). Within the Department of Energy's list of online approved DOE standards are several standards and handbooks applicable to nuclear engineering.

U.S. Nuclear Regulatory Commission. *Regulatory Guide*. Washington, D.C.: U.S. Nuclear Regulatory Commission. Office of Standards Development. The Nuclear Regulatory Commission (NRC) Guides provide the requirements that bind those people or organizations who receive a license from the NRC to operate nuclear facilities. The Guides are issued in 10 broad divisions with individual titles for each division (Power reactors, Research and test reactors, Fuels and materials facilities, Environmental and siting, Materials and plant protection, Products, Transportation, Occupational health, Antitrust and financial review, and General). An electronic version is available at http://www.nrc.gov/reading-rm/doc-collections/reg-guides/ (accessed August 25, 2010).

SEARCH ENGINES AND IMPORTANT WEB SITES

Most scholarly publications are not freely available via the Internet; however, the field of nuclear engineering has a slight advantage. Due to the efforts of the Department of Energy's (DOE) Office of Science and Technology Information (OSTI), much of the government-sponsored, publicly available DOE research is obtainable on the Internet. See the abstracts and indexes sections for links to *Information Bridge* and *Energy Citations Database*. In addition, society Web sites may contain pertinent information and the major societies in nuclear engineering host Web sites that may be found in the Societies section.

International Atomic Energy Agency (IAEA) http://www.iaea.org/ (accessed August 25, 2010). The IAEA was founded in 1957 to promote the peaceful use of atomic energy. It is an organization within the United Nations and is very active in safeguards, verification, and monitoring of nuclear materials. In working with other organizations, IAEA supports research and development in nuclear technology that will aid food, environmental, and health issues facing developing countries. IAEA publications, technical documents and reports may be found at http://www.iaea.org/Publications/index.html (accessed August 25, 2010).

Lawrence Berkeley National Laboratory (LBNL), Isotopes Project http://ie.lbl.gov (accessed August 25, 2010). The purpose of the Isotopes Project is to gather, evaluate, and disseminate data regarding neutron capture and radioactive decay as well as developing new techniques to distribute the data.

National Nuclear Data Center (NNDC) http://www.nndc.bnl.gov/index.jsp (accessed August 25, 2010). The NNDC is a worldwide resource for data on nuclear physics, nuclear research, and applied nuclear technologies. In addition, the NNDC focuses on data compilation and evaluation for nuclear information.

New Mexico's Digital Collections. University of New Mexico, Centennial Science and Engineering Library. Nuclear Engineering Wall Charts http://econtent.unm.edu/cdm4/browse.php?CISOROOT=%2Fnuceng (accessed August 27, 2010). This site contains wall charts of nuclear reactors around the world from the World's Reactors series, originally published by, and used by permission from Nuclear Engineering International.

Nuclear Age Timeline http://www.em.doe.gov/publications/timeline.aspx (accessed August 25, 2010). The Timeline, created by the Department of Energy's Office of Environmental Management, gives an historical outlook to the development of the nuclear age from the development of the x-ray to the clean-up of the nuclear weapons complexes (1895–1993). Links to references and selected resources are provided as well.

Nuclear Regulatory Commission (NRC) http://www.nrc.gov (accessed August 25, 2010). The NRC is an independent agency tasked with regulating the civilian use of nuclear materials. The site includes an Electronic Reading Room at http://www.nrc.gov/reading-rm.html (accessed August 25, 2010), which provides access to NRC documents and reports. The NRC also contains a docket collection, which provides specific licensing information for power and research reactors in the United States. For assistance in accessing this information, contact the NRC Public Document Room.

ASSOCIATIONS, ORGANIZATIONS, AND SOCIETIES

Societies and associations play a vital role in the discipline of nuclear engineering. The following societies create standards and technical documents as well as publishing conference proceedings and journals to disseminate nuclear engineering literature. Society Web sites are another useful source of information where you can discover additional details about the society, purchase materials, or view their publications. Upcoming conference dates are often listed on these pages, as are professional development opportunities and employment announcements.

American Nuclear Society (ANS) http://www.new.ans.org/ (accessed August 25, 2010). The ANS was founded in 1954 and facilitates the exchange of information with members in the development and safe application of nuclear science and technology.

Canadian Nuclear Association–Association Nucleaire Canadienne (CNA) http://www.cna.ca/ (accessed August 26, 2010). The CNA was established in 1960 to promote the peaceful use of nuclear technology within Canada.

Electric Power Research Institute (EPRI) http://my.epri.com (accessed August 31, 2010). EPRI, an independent, nonprofit organization, brings together scientists, engineers, and other experts in the field of energy production to focus on research and development involving power generation and delivery for the public. EPRI is also involved in policy and economic analyses for energy planning.

European Nuclear Society (ENS) http://www.euronuclear.org/ (accessed August 20, 2010). Established in 1975, the ENS is an association of 26 nuclear societies from 25 countries. Its goals are to promote the peaceful use of nuclear energy through the promotion of science and nuclear engineering.

IEEE Nuclear and Plasma Sciences Society (NPSS) http://ewh.ieee.org/soc/nps/ (accessed August 20, 2010). When the society was first established in 1949, it was known as the Professional Group on Nuclear Science within the Institute of Radio Engineers (IRE). When the Institute of Electrical and Electronic Engineers (IEEE) was formed in 1963 with the merger of the American Institute of Electrical Engineers (AIEE) and the IRE, the professional group became the Nuclear Science Group. In 1972, Plasma Science was added and the group was promoted to a society. Their goal is to encourage the advancement of nuclear and plasma sciences.

Institute of Nuclear Materials Management (INMM) http://www.inmm.org/ (accessed August 26, 2010). INMM provides guidance and professional development opportunities in the management of nuclear materials.

Institute of Nuclear Power Operations (INPO) http://www.inpo.info/ (accessed August 31, 2010). Established in 1979, INPO is a not-for-profit organization that focuses on safety and reliability in the operation of commercial nuclear power plants.

International Commission on Radiation Units and Measurements (ICRU) http://www.icru.org/ (accessed August 26, 2010). Established in 1925, this organization has developed recommendations on "(1) quantities and units of radiation and radioactivity; (2) procedures suitable for the measurement and application of these quantities in diagnostic radiology, radiation therapy, radiation biology, and industrial operations; and (3) physical data needed in the application of these procedures, the use of which tends to assure uniformity in reporting" (from Web page).

International Commission on Radiological Protection (ICRP) http://www.icrp.org/ (accessed August 31, 2010). ICRP is a not-for-profit organization based in the United Kingdom. It was founded in 1928 as the International X-ray and Radium Protection Committee. They act as an advisory committee by providing recommendations and guidance on radiation protection.

National Council on Radiation Protection and Measurements (NCRP) http://www.ncrponline.org/ (accessed August 26, 2010). The NCRP provides information, guidance, and recommendations on radiation protection and measurements. The council also facilitates cooperation among organizations concerned with the scientific and technical aspects of radiation protection and measurements.

Nuclear Energy Agency/Agence pour l'énergie nucléaire (NEA) http://www.nea.fr/ (accessed August 26, 2010). As an agency within the Organization for Economic Cooperation and Development (OECD), the NEA assists member countries in maintaining and developing technologies for safe, environmentally sound, and peaceful use of nuclear energy.

Nuclear Institute http://www.bnes.com/ibis/Nuclear Institute/Home (accessed August 25, 2010). The Nuclear Institute was established in January of 2009 with the merger of the British Nuclear Energy Society (BNES) and the Institution of Nuclear Engineers. Between them, these institutions have been serving the nuclear community since the late 1950s.

World Nuclear Association (WNA) http://www.world-nuclear.org/ (accessed August 26, 2010). The aims of the WNA are to promote the peaceful use of nuclear energy. In addition, the WNA focuses on all facets of the nuclear fuel cycle from mining to fuel fabrication to the safe disposal of spent nuclear fuel.

CONCLUSION

The discipline of nuclear engineering has traditionally had close ties with physics and chemistry. As the understanding of the physics of nuclear reactors advanced, and the knowledge of nuclear physics moved beyond the theoretical, the discipline of nuclear energy began to grow on its own. Although no new nuclear reactors have been built in the United States since 1979, this is changing with the renewed interest in nuclear energy as a viable alternative to fossil fuels for energy generation. In fact, from 2007 to 2010, the U.S. Nuclear Regulatory Commission has received 17 applications for new nuclear power plants and is expecting five additional ones in 2011 (U.S. NRC, 2010). In the meantime, existing reactors continue to function and need to be kept in working order. Legacy issues of cleaning up wartime facilities and waste by-products are still challenges for today's nuclear engineer. Finding new technologies to deal with these past issues is as crucial as continuing to search for safer and cleaner methods of producing nuclear energy. To move forward, nuclear engineers must have access to the literature of the discipline, including both current and historical publications.

As this book goes to press, the world's eye is turned toward a crisis in Japan. On March 11, a 9.0 magnitude earthquake and subsequent tsunami devastated the northern region of Japan and wreak havoc on the Fukushima Nuclear Power Plant. Radiation workers struggle to get electricity restored to the units so that the spent fuel can be cooled to prevent a catastrophic radiation release. Will this recent nuclear incident sour the United States and other countries from developing nuclear energy as an alternative to fossil fuels? This places a strong emphasis on the need to identify and locate pertinent information on this topic and other engineering subjects as efficiently as possible.

ACKNOWLEDGMENTS

Many thanks to Patricia Cleavenger, Nancy Doran, David J. Senor, and John M. Tindall for their support and assistance in writing this chapter.

REFERENCES

Encyclopaedia Britannica, Inc. 1993. Nuclear Engineering. In *The new encyclopaedia Britannica: Macropaedia–Knowledge in depth*. Chicago: Encyclopaedia Britannica, Inc.

McKay, A. 1984. *The making of the atomic age*. New York: Oxford University Press.

U.S. Nuclear Regulatory Commission. *Expected new nuclear power plant applications*. Updated June 21, 2010. http://www.nrc.gov/reactors/new-reactors/new-licensing-files/expected-new-rx-applications.pdf (accessed August 27, 2010).

19 Petroleum Engineering and Refining

Randy Reichardt

CONTENTS

Introduction and Scope of the Chapter 469
History of Petroleum Engineering 470
Searching the Catalog 472
Abstracts and Indexes 472
Bibliographies and Guides to the Literature 474
Dictionaries..475
    Dictionaries: Multilingual 476
Directories and Yearbooks 476
Encyclopedias 479
Government Resources 480
Handbooks and Manuals...480
Maps and Atlases 483
Monographs and Textbooks 484
Journals and Magazines 486
    Scholarly Journals..486
    Trade and Business Journals 487
Patents 489
Search Engines and Portals...489
Associations, Societies, and Organizations...490
Standards..491
Statistics and Data..493
References...495

INTRODUCTION AND SCOPE OF THE CHAPTER

The petroleum industry includes both upstream and downstream activities. Petroleum engineers work upstream, responsible for the exploration and development of oil and gas fields and wells. Petroleum engineers' work focuses on three areas: drilling, reservoir, and production engineering, in the search for unprocessed crude oil, also known as petroleum.

Petroleum refining, processing, and distribution are considered downstream activities and generally fall within the subject scope of the chemical engineer. Crude oil, transported from the ground to the refinery, is used to produce many products, including petroleum gas, gasoline, kerosene, heavy oil (fuel oil), coke, asphalt, tar, and other byproducts.

This chapter will highlight key resources that provide coverage in either upstream, downstream, or both disciplines. A thorough list of resources for each category is provided. The business, economic, and financial components of the petroleum industry are not part of the scope of this guide, although selected titles listed in this chapter do cover some of these aspects.

HISTORY OF PETROLEUM ENGINEERING

As noted in the American Petroleum Institute's *All about Petroleum*, the use of petroleum by the Sumerians, Mesopotamians, and Egyptians dates back thousands of years. Asphalt, bitumen, and pitch were used to inlay mosaics in walls, line water canals, and grease chariots. The Chinese first discovered underground deposits of oil, and built bamboo pipelines to transport oil and natural gas. By 1500 CE, the Chinese were drilling wells greater than 2,000 feet deep (American Petroleum Institute, Web site).

In one sense, petroleum engineering is an offshoot of mining engineering. Mining of the Earth's resources dates back millennia and has always involved the extraction and preparation of coal, minerals, and metals. Petroleum engineering involves extraction as well, but of oil and gas only, and is a relatively new profession.

The petroleum industry began to emerge in the United States in the mid-1800s. The first American oil company, Pennsylvania Rock Oil Company, was formed in 1854 (American Petroleum Institute). The first well in the United States drilled specifically to recover oil became known as the Drake Well, and was created on August 28, 1859, at Oil Creek, near Titusville, Pennsylvania (American Petroleum Institute 1961, pp. 170–175). The Drake Well heralded the beginning of the petroleum engineering industry in the United States.

However, the first commercial oil well in North America was dug and began operations one year earlier, in Canada in 1858. Many consider James Miller Williams to be the father of the North American petroleum industry, having dug a well in the Enniskillen Township in Ontario in August 1858, specifically searching for oil (Morritt, 1993, pp. 25–34; Cronin, 1955, pp. 15–20). Williams set up two pumps to bring the oil to the surface, and soon the well was producing between 5 and 100 barrels a day. The Oil Museum of Canada now resides on the site of North America's first commercial oil well, in Oil Springs, Ontario (Oil Museum of Canada, 2001). A number of titles have appeared recently, covering historical aspects of the petroleum industries in the United States and Canada (Finch, 2007; Grey, 2005; Lucier, 2008).

The development of the industry soon followed in Russia in the 1870s, and by 1900, had expanded to the Middle East and other parts of Asia (Giebelhaus, 1996). In the twentieth century, the petroleum engineering and refining industries expanded to most parts of the planet. In North America, the industry grew in Canada, the United States, and Mexico, as new oil fields and wells were discovered. The British Petroleum (BP) Company was formed in 1901. Elsewhere in the world, discoveries of oil deposits led to the introduction of petroleum engineering operations in Africa, Asia, the Middle East, and Latin America. In 1960, countries from these regions joined to form OPEC, the Organization of Petroleum Exporting Countries. OPEC member nations in 2009 were Algeria, Angola, Ecuador, Iran, Iraq, Kuwait, Libya, Nigeria, Qatar, Saudi Arabia, the United Arab Emirates, and Venezuela. In 2009, OPEC countries accounted for 79.6% of world-proven crude oil reserves by region (OPEC, 2010).

A very detailed history of petroleum engineering in the United States was published in 1961 by the American Petroleum Institute (American Petroleum Institute, 1961). A newer edition has not been issued. Published in 1975, *Trek of the Oil Finders: A History of Exploration for Petroleum*, written by Edgar Wesley Owen, is primarily of interest to petroleum geologists. However, the first chapter of the book offers a brief but concise history of oil and petroleum, from the ancient record through to the Drake Well (Owen, 1975). *Petrochemicals: The Rise of an Industry* was published in 1988. The author, Peter H. Spitz, notes that the petrochemical industry is an American phenomenon, created by oil and chemical companies in the 1930s and early 1940s, but drew much of its early technology from the Germans, who had built a substantial chemical industry since the early 1800s (Spitz, 1988). The American Oil & Gas Historical Society offers many petroleum history resources, and publishes the newsletter, *The Petroleum Age* (American Oil & Gas Society, Web site; American Oil & Gas Historical Society, *The Petroleum Age,* Web site).

The offshore petroleum industry has evolved into a distinct branch of the oil and gas industry over the past few decades. The history of the offshore industry, from its beginnings in the late nineteenth century through to the mid-1960s, has been documented by F. Jay Schempf in the book, *Pioneering Offshore: The Early Years*, and features in-depth interviews with over 125 important contributors to its advancement and development into a distinct technology (Schempf, 2007).

In 2010, the offshore petroleum industry dealt with the worst ecological disaster in U.S. history. On April 20, 2010, the explosion of the Deepwater Horizon semisubmersible drilling unit in the Gulf of Mexico resulted in a sea floor gusher, and was estimated to be releasing at least 42,000 barrels of crude oil per day, and possibly much more (Davies, 2010). The explosion claimed the lives of 11 people and injured 17 others. The BP oil spill released both oil and methane gas into the deep waters of the Gulf of Mexico. The impacts and ramifications of the spill could potentially be felt for years, affecting marine plant and animal life (Joye and MacDonald, 2010; Martin, 2010) as well as the tourism and fishing industries of the Gulf States. As of July 15, 2010, BP had successfully halted the flow of oil from its broken wellhead using a containment cap (*CBC News*, 2010) The long-term environmental consequences have yet to be determined or confirmed, but there is little doubt the offshore petroleum industry will change as a result of the Deepwater Horizon accident.

In addition to the publications referenced above and listed at the end of this chapter, the following titles and Web site cover various aspects and time periods of the history of the petroleum and petrochemical industries.

Bamberg, J. H. 1994. *The History of the British Petroleum Company, v.2: The Anglo-Iranian Years, 1928–1954*. Cambridge, U.K.: Cambridge University Press.

Cavnar, Bob. 2010. *Disaster on the Horizon: High Stakes, High Risks, and the Story Behind the Deepwater Well Blowout*. White River Junction, VT: Chelsea Green Pub.

Chapman, K. 1991. *The International Petrochemical Industry: Evolution and Location*. Oxford, U.K.: Blackwell.

Chastko, P. 2004. *Developing Alberta's Oil Sands: From Karl Clark to Kyoto*. Calgary, Alberta: University of Calgary Press.

Ferrier, R. W. 1982. *The History of the British Petroleum Company, Vol. 1: The Developing Years, 1901–1932*. Cambridge, U.K.: Cambridge University Press.

Freudenburg, William R. 2011. *Blowout in the Gulf: the BP Oil Spill Disaster and the Future of Energy in America*. Cambridge, MA: MIT Press.

Jones, G. 1981. *The State and Emergence of the British Oil Industry*. London: Macmillan.

McCoy, Margaret A., and Judith A Salerno. 2010. *Assessing the Effects of the Gulf of Mexico Oil Spill on Human Health: A Summary of the June 2010 Workshop*. Washington, DC: National Academies Press.

Pees, S. T. Oil History. http://www.petroleumhistory.org/OilHistory/OHindex.html (accessed July 23, 2010).

Shah, S. 2004. *Crude: The Story of Oil*. New York: Seven Stories Press.

Skupien, Phyllis Lipka. 2010. *2010 Gulf Coast Oil Disaster: Litigation and Liability*. Wayne, PA: Andrew Publications.

Stackenwalt, F. Michael. 1998. Dmitrii Ivanovich Mendeleev and the Emergence of the Modern Russian Petroleum Industry, 1863–1877. *Ambix: The Journal of the Society for the Study of Alchemy and Early Chemistry* 45 (2): 67–84.

United States. National Commission on the BP Deepwater Horizon Oil Spill and Offshore Drilling. 2011. *Deep Water: The Gulf Oil Disaster and the Future of Offshore Drilling: Report to the President*. [Washington, DC]: National Commission on the BP Deepwater Horizon Oil Spill and Offshore Drilling: For sale by the Supt. of Docs., U.S. G.P.O.

Williamson, H. F. and A. R. Daum. 1959. *The American Petroleum Industry: The Age of Illumination 1859–1899*. Evanston, IL: Northwestern University Press.

Williamson, H. F. and A. R. Daum. 1959. *The American Petroleum Industry: The Age of Illumination 1859–1899*. Evanston, IL: Northwestern University Press.

Williamson, H. F., R. L. Andreano, A. R. Daum, and G. C. Klose. 1963. *The American Petroleum Industry: The Age of Energy 1899–1959*. Evanston, IL: Northwestern University Press.

Yergin, D. 1991. *The Prize: The Epic Quest for Oil, Money and Power*. New York: Simon & Schuster.

SEARCHING THE CATALOG

Most online library catalogs have a keyword search function, allowing the user to type in any word or phrase, when searching for material in the library system. Keyword searching generally covers titles, subject headings, notes, tables of contents, series, publisher, and author fields. When searching a library catalog, the following Library of Congress Subject Headings are among the most useful and relevant to both petroleum engineering and petroleum refining and processing:

Bitumen	Gas Reservoirs
Coal–Tar Products	Heavy Oil
Enhanced Oil Recovery	Hydrocarbons
Natural Gas	Petroleum Engineering
Natural Gas Pipelines	Petroleum Geology
Oil Field Flooding	Petroleum Industry and Trade
Oil Fields	Petroleum Pipelines
Oil Reservoir Engineering	Petroleum Products
Oil Sands	Petroleum Refineries
Oil Well Drilling	Petroleum: Refining
Oil Well Pumps	Petroleum Reserves
Oil Wells	Petroleum, Synthetic
Petroleum	Secondary Recovery of Oil
Petroleum Chemicals	Thermal Oil Recovery
Petroleum Coke	

The Library of Congress Classification System includes a number of locations within its schedules for petroleum engineering and petroleum refining. The main classification ranges include the following:

HD 9560-9579	Petroleum Industry and Trade
HD 9581	Pipelines. Petroleum Pipelines
HD 9581-9582	Natural Gas
TN 850	Bitumen
TN 860-879.5	Petroleum. Petroleum Engineering
TN 871	Enhanced Oil Recovery
TN 879.5-879.6	Petroleum Pipelines.
TP 690-692.5	Petroleum Refining. Petroleum Products

ABSTRACTS AND INDEXES

The literature of petroleum engineering and refining can appear in many different formats: scholarly and trade journal articles, conference papers, patents, Web sites, standards, handbooks and manuals, monographs, government reports, theses and dissertations, gray literature, encyclopedias and dictionaries, articles in newspapers and magazines, and so forth. In 1987, Pearson and Ellwood listed 15 indexing and abstracting services (in print), and 67 databases, divided into bibliographic, textual, and databases containing only numbers and/or data (Pearson and Ellwood, 1987). No attempt has been made to provide similar coverage in this section. Only core resources are listed, together with a number of related but important titles. The most up-to-date and comprehensive list of petroleum industry databases can be found in each issue of the *Gale Directory of Databases*. In the 2008, Part 1, edition of the directory, 68 databases are listed in the category: Petroleum and petroleum technology (Hall and Romaniuk, 2008).

The following abstracting and indexing services provide good coverage of publications in petroleum engineering and/or refining and processing. Databases offering related coverage are also listed.

EnCompassLIT and *EnCompassPAT*. 1964–. http://www.ei.org/encompasslit_pat (accessed July 23, 2010). Previously known as *ApiLiT* and *ApiPAT*, these databases are now available via Elsevier Engineering Information, as part of the Engineering Village platform. Subject coverage is the downstream petroleum, petrochemical, natural gas, energy, and allied industries. *EnCompassLIT* is a technical literature database that began in 1964, containing over 870,000 records as of July 2010. Entries are selected from over 1,500 scientific journals, trade magazines, conference papers, and technical reports. *EnCompassPAT* also began in 1964, and covers the patent literature of these related industries. Over 450,000 patents were indexed as of July 2010. Patents are selected from 40 patenting authorities across the globe. *EnCompassLIT* and *EnCompassPAT* are also available via Dialog (as Ei EnCompassLit and Ei EnCompassPat, and via.STN as *ENCOMPLIT* and *ENCOMPPAT*.

OnePetro http://www.onepetro.org/mslib/app/search.do (accessed July 23, 2010). Launched in March 2007, formerly known as the SPE (Society of Petroleum Engineers) e-Library. Primary coverage is of the oil and gas exploration and production industry, with peripheral coverage of other subjects based on the papers available from participating societies. Contains full text of over 85,000 documents as of July 2010, including all papers from the SPE e-Library back to 1927. In addition to the SPE, the database includes selected publications from the following organizations: API (American Petroleum Institute), ARMA (American Rock Mechanics Association), ASSE (American Society of Safety Engineers), NACE International (National Association of Corrosion Engineers), OTC (The Offshore Technology Conference), SWPLA (Society of Petrophysicists and Well Log Analysts), SPEE (Society of Petroleum Evaluation Engineers), and WPC (World Petroleum Council).

Petroleum Abstracts. 1965–. http://www.pa.utulsa.edu/ (accessed July 23, 2010). Provides coverage of oil and gas exploration, production, transportation storage, environment, safety and health, relevant to the upstream sector of the petroleum industry. Entries are created from over 300 journals, 200 conference proceedings, patents, government documents, books, reports, dissertations, maps, and other sources. Over 900,000 entries as of July 2010, with 35,000 added yearly. *PA* is available in a number of formats, including via vendors, such as Dialog, Ovid, EBSCO/EBSCOhost, Questel/Orbit, and STN, where it is also known as *TULSA*. It is available via the Internet as the *Petroleum Abstracts Discovery Database*. The *Petroleum Abstracts Bulletin* is published 50 times a year, and is sent by e-mail, with 400–600 abstracts per issue.

Other relevant databases include: *Chemical Abstracts* http://www.cas.org/ (accessed July 23, 2010). *Chemical Abstracts* is also known as *SciFinder* on the Web, CA Search: *Chemical Abstracts* (Dialog), and *CAPlus* (STN). While mentioned in Chapter 7, Chemical Engineering, of this book, it is important to note that CA provides good coverage related to petroleum technology, including fossil fuels and derivatives, natural gas exploration, reservoirs, petroleum refining and processing, and hydrocarbons.

Compendex http://www.ei.org/compendex (accessed July 23, 2010). Covered in more detail in Chapter 2, General Engineering, of this book, *Compendex* provides solid coverage of the literature of the upstream and downstream petroleum industries.

Energy Information Platform of Alberta (EIPA) http://eipa.alberta.ca/ (accessed July 23, 2010). This is a publicly accessible database of over 52,000 citations to public and private research on hydrocarbon resources in Alberta, which holds the largest reserves of oil sands in the world. EIPA was previously known as the *Alberta Oil Sands Information Services* (*AOSIS*), which had developed the *Heavy Oil, Enhanced Recovery, and Oil Sands* (*HERO*) database.

Fuel and Energy Abstracts. Available via *ScienceDirect*, also published bi-monthly in paper copy. Provides summaries of world literature on scientific, technical, environmental, and commercial aspects of fuel and energy. Covers over 800 international publications, monographs, conference proceedings, reports, surveys, and statistical analyses. Number of abstracts per issue can vary from 500 to 800. The online indexing is inconsistent. Individual abstracts are indexed in most issues. However, in newer issues, the citations are buried under headings, such as Liquid Fuels, Heat Pumps, and Fuel Science and Technology, and cannot be retrieved without a secondary keyword search of the actual subject index of a particular issue.

GeoRef. American Geological Institute. 1785–. Over 2.2 million records on North America from 1669, and other areas of the world since 1933. While primarily a geosciences database, good coverage is provided in areas including: petroleum engineering, exploration, geology, maps, products, reserves and resources, reservoir properties, oil and gas fields, and pipelines.

Databases of importance to the field of petroleum engineering and refining, which are discussed elsewhere in this book, include: *NTIS* (U.S. National Technical Information Service), *Dissertation Abstracts* (*ProQuest Dissertations & Theses: Full Text*), *Scopus*, and *Web of Science*.

Databases covering government resources that would include material on petroleum engineering and refining, which are freely accessibly to anyone:

science.gov http://www.science.gov/ (accessed July 23, 2010). science.gov searches more than 42 databases and over 2,000 selected Web sites from 14 U.S. federal agencies.

WorldWideScience.org http://worldwidescience.org/ (accessed July 23, 2010). A gateway to approximately 70 national science portals from over 60 countries.

Information about other databases related to petroleum and petroleum technology can be found using the *Gale Directory of Databases*.

BIBLIOGRAPHIES AND GUIDES TO THE LITERATURE

The most recent guides devoted specifically to petroleum literature resources appeared in the late 1980s. None that could be identified have been published since that time. The 193-page guide by Pearson and Ellwood was thorough and detailed in its coverage, with over 420 entries in numerous categories, together with extensive lists of publishers and associations (Pearson and Ellwood, 1987). Charles R. Lord's *Guide to Information Sources in Engineering* was published in 2000, but did not place significant emphasis on information resources in petroleum engineering (Lord, 2000). In 2005, the fourth revised edition of *Information Sources in Engineering* (Macleod and Corlett, 2005) was published with a 12-page chapter devoted to petroleum and offshore engineering. In 2006, the first edition of *Using the Engineering Literature* was published, and included the first version of this chapter. As such, this revised version must be considered the most current guide to the literature of petroleum at the time of publication.

The titles listed below, despite being outdated, remain important for historical reasons and for identifying earlier editions of important titles, as well as titles that are now out of print or have ceased publication.

Anthony, L. J., ed. 1988. *Information Sources in Energy Technology.* London: Butterworths. Covering information sources in a number of energy resources, this title is divided into three distinct components: energy in general, fuel technology, and specific energy resources. Petroleum and oil are covered in the chapter on liquid fuels, and natural gas in the chapter on gaseous fuels. Rather than present key resources in alphabetical, numbered, or bulleted lists, titles in each section are identified and discussed in paragraphs, rendering the book somewhat less than user-friendly.

Myers, A., et al., ed. 1993. *Petroleum and Marine Technology Information Guide: A Bibliographic Sourcebook and Directory of Services.* London: E & FN Spon. This title is the fourth edition of a work that began as a bibliography of offshore technology in 1981. The third edition was renamed and expanded coverage to include marine technology, and online databases in the oil and gas industry. The fourth edition was renamed and revised again, adding sources on petroleum exploration, reservoirs, and production to its bibliography. Primary focus of the work is on the offshore and marine industry. Included are a bibliography by subject, organizations providing information services, online databases and CD-ROMs, and an industry directory relevant to petroleum and marine technology.

Pearson, B. C. and K. B. Ellwood. 1987. *Guide to Petroleum Reference Literature.* Littleton, CO: Libraries Unlimited Inc. An exhaustive survey of the literature of petroleum at the time of its publication. Coverage included guides to the literature, bibliographies, indexing and abstracting services, dictionaries, encyclopedias and yearbooks, handbooks, manuals, basic texts, directories, statistical sources, databases (including bibliographic, data and full text), periodicals, professional and trade associations, and publishers. In 2010, this title was digitized by the Hathi Trust, but Full View was not available due to copyright restrictions. http://catalog.hathitrust.org/Record/000847675 (accessed July 23, 2010).

Stark, M., ed. 1988. *Bibliography of Petroleum Information Resources.* Washington D.C.: Petroleum and Energy Resources Division, Special Libraries Association. This bibliography was created by at least eight members of the Petroleum and Energy Resources Division of SLA. Coverage included handbooks, databases, indexes and abstracts, directories, journals, maps and atlases, standards, legislation and regulations, journals, and theses and dissertations.

DICTIONARIES

Association of Desk and Derrick Clubs, comp. 2007. *D & D Standard Oil & Gas Abbreviator,* 6th ed. Tulsa: PennWell. An indispensable tool in the oil and gas industry, useful in deciphering over 11,500 abbreviations and acronyms. Divided into the following sections: abbreviations with definitions, definitions with abbreviations, abbreviations for logging tools and services, federal environmental acronyms, pipe coating terminology and definitions, mnemonics, abbreviations for companies, associations and organizations, and miscellaneous information and symbols. The fifth edition includes a searchable CD-ROM with the full text of the book, with an additional chapter on universal conversion factors, and a Michigan stratigraphic chart.

Dictionary for the Oil & Gas Industry, A, rev. ed. 2005. Austin, TX: Petroleum Extension Service, Continuing & Extended Education, University of Texas at Austin. Includes over 11,000 definitions of terms used in all aspects of the industry including petroleum geology, exploration, drilling, production, pipelines, refining and processing, and management. Includes contact information for industry associations and government agencies.

Hyne, N. J. 1991. *Dictionary of Petroleum Exploration, Drilling & Production.* Tulsa: PennWell. Comprehensive dictionary of terminology used in upstream petroleum. Features include graphs, charts, diagrams, and photos. Includes an appendix with additional information, such as drilling and completion records, detailed diagrams of a rotary drilling rig, cable tool drilling rig, and a crank counterbalanced beam pumping unit, geological time scale, giant oil and gas fields, geological features, drill stem test symbols, flow sheet symbols, and U.S. land subdivisions.

Langenkamp, R. D. 1994. *Handbook of Oil Industry Terms & Phrases,* 5th ed. Tulsa: PennWell. Over 4,200 entries, with 1,000 new entries since the fourth edition. It appears that Langenkamp has coordinated the publication of this title with his related work, *The Illustrated Petroleum Reference Dictionary*: entries in both current, 1994 editions are identical, with the latter publication featuring over 200 illustrations.

Langenkamp, R. D. 1994. *Illustrated Petroleum Reference Dictionary,* 4th ed. Tulsa: PennWell. Over 4,200 entries, including 1,280 new entries since the third edition, along with over 200 illustrations. Covers both technical and slang terminology. Included, as "adjuncts" to the dictionary, are two other complete works: *The D&D Standard Oil Abbreviator*, and *Universal Conversion Factors*, compiled and edited by Stephen Gerolde. The contents of the dictionary are identical to the fifth edition of Langenkamp's *Handbook of Oil Industry Terms & Phrases*, the difference being the inclusion of illustrations.

Vassiliou, M. S. 2009. *Historical Dictionary of the Petroleum Industry.* Lanham, MD: Scarecrow Press. Covers the history of the petroleum industry from premodern times to the twenty-first century. Includes a chronology and over 400 cross-referenced entries covering topics including companies, technologies, places, people, events, and phenomena related to the industry. Note that this book was originally published in 2009 with the title, *The A to Z of the Petroleum Industry.*

DICTIONARIES: MULTILINGUAL

Bernadiner, M. and G. V. Chilingar. 2004. *Practical English–Russian–English Petroleum Dictionary.* Coral Springs, FL: Lumina Press. Includes approximately 10,000 English and Russian terms and phrases, covering topics in petroleum engineering and geology. Coverage includes exploration, seismic survey, logging, drilling, completion, modeling, operation, and production.

Kedrinskiĭ, V.V. 2004. *Anglo-russkiĭ slovar' po khimii i pererabotke nefti: okolo 60,000 terminov (English-Russian dictionary of petroleum chemistry and processing)*, 6-e izd., stereotipnoe. Mosvka: Russo.

Moureau, M., and G. Brace. 2008. *Dictionnaire du Pétrole et Autres Sources d'Energie: Anglais–Français, Français–Anglais (Comprehensive Dictionary of Petroleum and Other Energy Sources: English–French, French–English).* Paris: Éditions Technip. Coverage of oil and natural gas and related energy sources; 80,000 terms translated or explained, with 10,000 definitions or clarifications in both languages.

Proubasta, M.D. 2006. *Glossary of the Petroleum Industry: English–Spanish and Spanish–English (Glosario de la Industria Petrolera)*, 4th ed. Tulsa: PennWell. Updated to include more than 25,000 technical terms, covering the oil and gas industry as well as related fields. "The fourth edition draws heavily from the editorial content of *Oil and Gas Journal Latinoamérica* as a source of new technical vocabulary and of actual usage in different Spanish-speaking countries" (from lib.store.yahoo.net/lib/pennwell/ProbastaNBI.pdf (accessed October 19, 2010).

DIRECTORIES AND YEARBOOKS

Directories provide immediate access to information about the petroleum industry, including people, companies, products, services, manufacturers and suppliers of components, and cover both the upstream and downstream components of the petroleum industry. The advent of online access has helped alleviate the problem of information in print going out of date quickly, especially in directories that are published annually. Directories are also included in many of the sites listed in the section, Search Engines and Portals.

Alberta Oil & Gas Directory. Annual. Edmonton, Alberta: Armadale Publications. Includes section called CANADA-Z Oil Gas Mining. Available online at http://www.global-serve. net or http://www.albertaoilandgas.com/ (accessed July 23, 2010). This publication doubles as a directory for the Alberta and Canada-wide petroleum industry. Most of the Canadian petroleum and natural gas resources are found in Alberta, which is why the guide focuses on Alberta, and then the rest of Canada.

Arab Oil & Gas Directory, 36th ed. 2010. Paris: Arab Petroleum Research Center. Available in print and electronic versions; http://www.arab-oil-gas.com/indexns.htm (accessed July 23, 2010). Covers all aspects of the oil, gas, and petrochemical industries in Algeria, Bahrain, Egypt, Iran, Iraq, Jordan, Kuwait, Lebanon, Libya, Mauritania, Morocco, Oman, Qatar, Saudi Arabia, Sudan, Syria, Tunisia, United Arab Emirates, and Yemen, as well as OAPEC and OPEC. Provides surveys, information, data, maps, statistics, addresses, information on projects and activities of foreign oil companies operating in the Middle East and North Africa, and financial and business analyses.

Canadian Oil Register. Annual. Edmonton, Alberta: JuneWarren-Nickle's Energy Group. Available at http://www.canadianoilregister.com (accessed July 23, 2010), requires a subscription. Previously known as *Nickle's Canadian Oil Register*. Provides listings of more than 3,200 companies related to the Canadian energy industry, including oil and gas producers, explorers and developers, service and supply companies, consultants, engineers, pipeline contractors, designers, construction and fabricators, geophysical data brokers and contractors, drilling contractors, pipeline companies, transportation, and oilfield construction companies. Includes an Internet index of Web sites.

Canadian Oilfield Service and Supply Directory. Annual. Edmonton, Alberta: JuneWarren-Nickle's Energy Group http://www.cossd.com/ (accessed July 23, 2010). This annual publication is divided into six sections: white, all companies alphabetically; grey, by location; blue, oil company producers, developers, and explorers; orange, hotels and motels; green, over 160 environmental categories; and yellow, over 1,820 specialized oilfield categories. Maps are also included. Thirtieth edition (2010) also available in GPS format.

FSU Oil and Gas Statistic Yearbook. 2010. Moscow: RPI. Provides comprehensive historical and 2009 year-end information on upstream, downstream, and midstream operations in the major oil and gas producing countries and companies in the former Soviet Union.

Global Oil & Gas Directory. Abingdon, U.K.: Achilles Group Limited http://www.oildir.com/ (accessed 23 July 2010). A directory of suppliers based on a network of existing databases produced by Achilles. For a supplier to appear in this directory, it must appear in at least one of the Achilles/First Point databases: FPAL–U.K. and The Netherlands; JQS–Norway and Denmark; RPP–Venezuela; SICLAR–Argentina.

Gulf Coast Oil Directory. 2010. Houston: Atlantic Communications. A directory of the petroleum industry in the U.S. Gulf Coast region. Subject coverage includes blowout and firefighting, drilling contractors, geologists, geophysical and seismographic services, petroleum and consulting engineers, pipeline contractors and operators, well testing, completion and wireline services, and a buyer's guide to oilfield products. More information at http://www.oilonline.com/Directory.aspx (accessed March 29, 2011).

Gulf States Petroleum Directory. Houston: Hart Energy Publishing, LP. An online directory of offshore operations in the Gulf of Mexico, and onshore activity stretching from the Permian Basin and South Texas to the Anadarko Basin, ArkLaTex, Oklahoma, and Florida. Industry sectors covered include exploration, production, land, pipelining, refining, gas processing, drilling, well servicing, geophysical contractors and data brokers, equipment and supplies, field and nonfield services, information technology, and offshore equipment and supplies. Buyers' Guide is included. More information available at http://www.harten-ergy.com/directories/gspd.php (accessed July 23, 2010).

Houston/Texas Oil Directory. 2010. Houston: Atlantic Communications. A directory of the petroleum industry for Texas and Oklahoma. Subject coverage is identical to the *Gulf Coast Oil Directory.* More information at http://www.oilonline.com/Directory.aspx (accessed March 29, 2011).

Industry Canada. *Oil and Gas Industry Company Directories* http://www.ic.gc.ca/eic/site/ogt-ipg.nsf/eng/h_og00003.html / (accessed July 23, 2010). Links to Canadian pipeline, onshore and offshore oil and gas company directories, associations, and equipment and services.

Marcellus Directory. 2010. Houston: Hart Energy Publishing. Covers the oil and gas activity in the Marcellus Shale Natural Gas Field Formation, which extends through Pennsylvania, New York, Ohio, and West Virginia, part of the Devonian Black Shale Field. Provides the same coverage as all Hart directories. More information at http://www.hartenergy.com/directories/marcellus.php (accessed July 23, 2010).

North American Petroleum Directory. Houston: Hart Energy Publishing http://www.northamericanpetroleumdirectory.com/ (accessed July 23, 2010). An online database of companies, personnel, and contact information for the oil and gas industry in the United States and Canada.

Oil & Gas Directory, 40th ed. 2010. Houston: Information Services http://informationser-vices.com/oil_and_gas_directory/index.htm (accessed July 23, 2010). The directory provides American and worldwide coverage of exploration, well drilling, and produc-tion. Geographic areas include beyond the United States include Canada, Latin America, Europe, Africa, Middle East, and Asia-Pacific. The fortieth edition (2010) included list-ings for 13,059 companies: 9,959 United States, 1,026 Canadian, 2,074 outside of North America. Thirty supply and service sections are listed, and there is a separate section for oil and gas companies. Regional indexes are also included.

PC Directory. 2001–. Tulsa: Midwest Publishing Company http://www.midwestpub.com/index.html (accessed July 23, 2010). Midwest Publishing Company of Tulsa, Oklahoma, has produced directories for the energy industries since 1943. In 2001, it ceased publishing the directories in print and began offering them as part of a software package called *PC Directory,* described as "a custom, self-contained software that resides directly on your desktop or laptop computer." The software now is available as an online product only. Directories are updated monthly. Directories available include pipeline transmission and gathering, refining and gas processing, petrochemical, drilling and well servicing, offshore and international exploration and production, and the North American exploration and production industry.

Petroleum Equipment Directory. Annual. Tulsa: Petroleum Equipment Institute http://www.pei.org/search/default.asp (accessed July 23, 2010). Provides listings of Canadian and U.S. Institute members, with contact and product information, covering the petroleum market-ing and liquid handling equipment industry. Available in print and CD-ROM and search-able online at the above URL.

Petroleum Supply Americas. Semiannual. Lakewood, NJ: OPIS/STALSBY http://www.opisnet.com/directories/psa.asp (accessed July 23, 2010). Available in print, CD-ROM, online. Provides coverage of the petroleum supply industry for North, Central, and South America. Divided into sections by company, personnel, products handled, and country-state-city. Types of industries listed include banking/financial, consultants, crude oil, ethanol production facilities, feedstocks/intermediates, natural gas liquids, oil refiners, petrochemicals, and transportation/storage.

Petroleum Supply Europe. Lakewood, NJ: OPIS/STALSBY http://www.opisnet.com/directo-ries/pse.asp (accessed July 23, 2010). Available in print and CD-ROM. Provides coverage of the petroleum supply industry for Europe, Asia, Africa, Australia, and the Middle East. Layout similar to *Petroleum Supply Americas.*

Petroleum Terminal Encyclopedia. Annual. Lakewood, NJ: OPIS Directories http://www. opisnet.com/directories/pte.asp (accessed July 23, 2010). Available in print, standard CD-ROM, geocoded CD-ROM. Covers North America, Central America, South America, Australia, Europe, and the Far East, with indexes by state/country, company, and type of terminal, for over 1,800 bulk liquid storage facilities worldwide. Information for each terminal may include all of the following: address, physical location, corporate head-quarters, terminal type, methods used for supply and outloading, outloading features, pipelines used by terminal, high and low water depths and berth length, total storage capacity, and individual storage capacity by product. Despite its title, this is a directory, not an encyclopedia.

Renewable Fuels Supply Americas. Semiannual. Lakewood, NJ: OPIS Directories http:// www.opisnet.com/directories/renewablefuels.asp (accessed July 23, 2010.) Available in print and CD-ROM. Provides coverage of the renewable fuels supply industry, such as ethanol and biodiosel, and includes topics such as carbon footprint /life-cycle issues and biodiesel feedstocks.

Rocky Mountain Petroleum Directory. Houston: Hart Energy Publishing, LP. A directory of offshore operations in the Rocky Mountain states from New Mexico to Canada, and in Kansas, Nebraska, and Oklahoma. Coverage identical to Hart's Gulf States Petroleum Directory, except for offshore. More information available at http://www.hartenergy.com/ directories/rmpd.php (accessed July 23, 2010).

Subsea Oil & Gas Directory. Subsea.Org http://www.subsea.org/ (accessed July 23, 2010). Online directory and industry catalog for the subsea oil and gas industry. Also provides indus-try news, links, and conference information as well as an overview of the subsea industry.

Who's Who in Natural Gas and Power. Annual. Lakewood, NJ: OPIS/STALSBY http://www. opisnet.com/directories/wwnatgas.asp (accessed July 23, 2010). Available in print and CD-ROM. Similar in layout to other OPIS directories, this publication covers both the natural gas and electric power and utility companies. Detailed listings cover all segments of the natural gas industry, including consulting, equipment, processor, producer, training and education, and transportation.

ENCYCLOPEDIAS

There is no "traditional" encyclopedia devoted specifically to petroleum engineering and refining technical topics. As Pearson and Ellwood noted, the *International Petroleum Encyclopedia* func-tions as a yearbook, with a focus on statistics, data, and maps. Good coverage of processes related to downstream petroleum activities can be found in encyclopedias such as *Ullmann's Encyclopedia of Industrial Chemistry* and the *Kirk–Othmer Encyclopedia of Chemical Technology* (both covered in Chapter 7, Chemical Engineering, of this book). Given the extensive coverage of the upstream and downstream petroleum industry provided throughout many handbooks and manuals as well as the aforementioned encyclopedias, perhaps the industry does not see the need for a full-blown encyclopedia covering petroleum engineering and refining topics.

Hilyard, J., ed. 2010. *International Petroleum Encyclopedia.* Tulsa: PennWell. Published annu-ally by PennWell. Provides world coverage by geographic regions of the year's activities in petroleum and natural gas. Figures for each country include refining capacity, oil produc-tion, oil reserves, and gas reserves. Key statistics are provided, and statistical tables cover oil production and consumption, oil refining, natural gas, and petroleum prices. Historical figures are included. Maps are also provided, showing locations of oil fields, oil sands, gas fields, crude, natural gas and products pipelines, refineries, and tanker terminals. This is an important resource for historical and current data, and for the latest information on developments in the petroleum industry.

Wiley Critical Content: Petroleum Technology. 2007. New York: John Wiley & Sons. For a library unable to afford the *Kirk–Othmer* or *Ullmann's* encyclopedias, this book will fill that gap in its reference collection. The two-volume set features over 50 articles from both encyclopedias arranged thematically as follows: (1) exploration, production, and refining; (2) refined products and fuels; and (3) petrochemicals.

GOVERNMENT RESOURCES

When searching for government resources on petroleum and energy, the recommendation is to begin with the home page or Web site of the jurisdiction in question, and search for the relevant department, such as one dealing with energy, natural resources, petroleum, natural gas, and the like. For example, if you were looking for federal government information in North America on petroleum, oil and/or natural gas, your search could begin with the following federal departments:

Canada: Natural Resources Canada. Energy Sector http://nrcan.gc.ca/eneene/index-eng.php (accessed July 23, 2010). Responsible for the Canadian federal energy policy and all renewable and nonrenewable energy sources, including natural gas, petroleum, crude oil, and offshore oil and gas. The National Energy Board http://www.neb-one.gc.ca/clf-nsi/index.html (accessed July 23, 2010) exists to regulate pipelines, energy development, and trade in the Canadian public interest.

United States: Department of Energy http://www.energy.gov (accessed July 23, 2010). Includes links to the Office of Fossil Energy, the Office of Oil and Natural Gas, and the National Energy Technology Laboratory. Statistical information from the Energy Information Administration is covered elsewhere in this chapter.

Individual states and provinces will have equivalent departments or ministries. In Texas, the Bureau of Economic Geology at the University of Texas, Austin, began operations in 1909. In addition to functioning as a research unit at the University of Texas, Austin, it is also the State Geological Survey and the Regional Lead Organization for the Petroleum Technology Transfer Council http://www.beg.utexas.edu/ (accessed July 23, 2010).

Alberta is the major oil producing province in Canada, and information relating to its oil- and natural gas-based activities is available from its Department of Energy, as well as the Energy Resources Conservation Board, an independent, quasijudicial agency of the Government of Alberta that regulates the development of Alberta's energy resources: oil, natural gas, oil sands, coal, and pipelines. Additional information is available at their Web sites: http://www.energy.gov.ab.ca and http://www.ercb.ca/ (accessed July 23, 2010).

Government sites often include valuable statistical and pricing data. See the Statistics section in this chapter for more information. For databases covering government resources, please see the section Abstracts and Indexes.

HANDBOOKS AND MANUALS

Handbooks and manuals are valuable to the petroleum and refining engineer, supplying figures, data, properties, design and testing criteria, formulas and calculations, refining processes, well log interpretation criteria, and other much-needed quick reference information. Of note is that most of the titles listed here, and dozens of others relating to narrower topics in petroleum engineering and refining, are available both in print and electronically either on the publishers' sites or via full-text monograph databases, such as *Knovel, CRCnetBASE, Referex, eBrary, AccessEngineering,* and others.

Ahmed, T. 2010. *Reservoir Engineering Handbook,* 4th ed. Boston: Gulf Professional Publishers. This book "explains the fundamentals of reservoir engineering and their practical applications in conducting a comprehensive field study" (from Preface, p. xv). Divided into 17 chapters., the coverage includes reservoir fluid behavior and properties, fundamentals of reservoir fluid flow, oil and gas well performance, oil recovery mechanisms, and methods for the prediction of oil reservoir performance.

American Society of Mechanical Engineers. Shale Shaker Committee. *Drilling Fluids Processing Handbook.* 2005. Amsterdam: Elsevier. An updated version of an earlier work, *Shale Shakers and Drilling Fluids Systems,* published in 1999. The revised work expands on the earlier title to include many other aspects of drilling control, and was written by 21 experts in drilling and drilling fluids. Divided into 20 detailed chapters, and includes an extensive glossary and index. Coverage includes drilling fluids, tank arrangement, shale shakers, mud cleaners, and other pieces of equipment used in the field. Each chapter is divided into subsections, which are, in turn, subdivided, providing for fast access to any part of the book.

Cholet, H., ed. 2008. *Well Production Practical Handbook,* new ed., expanded. Paris: Editions Technip. Intended as a reference guide for oilfield operators and petroleum engineers, but also provides solutions to practical problems. Includes guidelines, recommendations, formulas, and charts. Topics covered: well technology, well productivity evaluation, and control, stimulation, horizontal and multilateral wells, and production improvement. Information in the book is based on a wide variety of sources, including papers published by the Society of Petroleum Engineers.

Fink, J. K. 2003. *Oil Field Chemicals.* Amsterdam: Gulf Professional Publishing. A comprehensive and detailed summary of chemicals used in the oil field. The compilation was prepared by critically examining over 20,000 references from the literature, primarily from Petroleum Abstracts and patent databases. Only materials that are publicly accessible have been included. The author notes that the information presented is not complete, and that developments from the past 10 years have been screened for inclusion. The book is divided into 22 chapters, organized according to applications of parallel job processes, beginning with drilling, through to demulsifiers. Two indices, one chemical and one general, are provided.

Gabolde, G. and J.P. Nguyen. 2006. *Drilling Data Handbook,* 8th ed., Paris: Editions Technip. Provides data for drilling and production, divided into 13 tabbed sections. Data are primarily in tabular form, with some graphs and diagrams. Topics covered include drill string standards, casing, tubing, and line pipe standards, drilling bits and downhole motors, drilling mud, cementing, direction drilling, and more.

Grace, R. D. 2003. *Blowout and Well Control Handbook.* Amsterdam: Gulf Professional Publishing. This handbook covers blowout containment and well control procedures, for the drilling or petroleum engineer. Coverage includes equipment in well control operations, classic pressure control procedures while drilling or tripping, special conditions, problems, and procedures in well control, fluid dynamics and special services in well control, relief well design and operation, and contingency planning. A case study of the 1985 blowout of the E. N. Ross No. 2 well, in Rankin County, Mississippi, is presented. The book closes with a brief history and overview of the Al-Awda Project: the oil fires of Kuwait.

Guo, B., K. Sun, and A. Ghalambor. 2008. *Well Productivity Handbook: Vertical, Fractured, Horizontal, Multilateral, and Intelligent Wells.* Houston: Gulf Publishing. Provides information and guidance for modeling oil and gas production wells. Covers petroleum fluid properties, reservoir deliverability, wellbore flow performance, and productivity of intelligent well systems.

Jones, D. S. J. and P. R. Pujadó. 2006. *Handbook of Petroleum Processing*. Dordrecht, The Netherlands: Springer. A reference work designed for both researchers and practicing engineers who will be working in the petroleum processing industry. Provides an introduction to crude oils, detailed coverage of the refining processes, off-sites and utilities, environmental and safety aspects, refinery planning, and equipment needs.

Khan, M. I. and M. R. Islam. 2007. *The Petroleum Engineering Handbook - Sustainable Operations*. Houston: Gulf Publishing Co. Focuses on the "greening" of all practices in the petroleum industry from exploration and extraction through to refining and gas processing. An extensive set of references are included.

Lake, L. W., ed. 2006–2007. *Petroleum Engineering Handbook*. Richardson, TX: Society of Petroleum Engineers. First major revision since 1987 of the Society of Petroleum Engineers' important reference work. Expanded from one to seven volumes, the handbook covers these topics: general engineering, drilling, facilities and construction, production operations, reservoir engineering and petrophysics, and emerging and peripheral technologies.

Lapeyrouse, N. J. 2002. *Formulas and Calculations for Drilling, Production, and Workover*, 2nd ed. Amsterdam: Gulf Professional Publishing. The primary purpose of this book is to serve as a reference source to those who do not routinely use formulas and calculations. It is divided into five sections: basic formulas, basic calculations, drilling fluids, pressure control, and engineering calculations. Conveniently coil-bound to open flat on a desk or in the oilfield.

Lyons, W. C., B. Guo, and F. A. Seidel. 2001. *Air and Gas Drilling Manual*, 2nd ed. New York: McGraw-Hill. The second edition of this work has been written as an engineering practice book for both engineers and Earth scientists working in air and gas drilling. Dividing it into three sections, the authors cover basic technology, air and gas drilling fundamentals, and deep well operations. Five appendices are included.

Lyons, W. C. and G. J. Plisga, eds. 2005. *Standard Handbook of Petroleum & Natural Gas Engineering*, 2nd ed., Burlington, MA/Oxford, U.K.: Gulf Professional Publishing. A revised edition of the two-volume set published in 1994. The 1994 edition began as a project to revise and rewrite the fifth edition of the *Practical Petroleum Engineer's Handbook*, published in 1970. When the authors (27 in total) realized that revisions would be inadequate, they chose to write a new handbook in the style of handbooks of other major engineering fields. As such, the initial chapters cover basic mathematics, general engineering and science, and auxiliary equipment. Specific petroleum engineering chapters follow, covering drilling and well completions, reservoir engineering, production engineering, and petroleum economics. The second edition features contributions from 75 authors, most of whom are practicing engineers in industry. Published information from the American Petroleum Institute and the Society of Petroleum Engineers was used in preparation of this handbook.

McAllister, E. W., ed. 2009. *Pipeline Rules of Thumb Handbook*, 7th ed. Tulsa: Pennwell Books. The seventh edition of the classic work features 30% new and updated content. Extensive coverage includes construction, pipe design, control valves, corrosion, gas and liquid pipelines, pumps, measurement, instrumentation, leak detection, tanks, maintenance, economics, and rehabilitation and risk evaluation.

Meyers, R. A., ed. 2005. *Handbook of Petrochemicals Production Processes*. New York: McGraw-Hill. Mentioned in this chapter because it is a companion volume to *Handbook of Petroleum Refining Processes*, and shares a similar format. Provides detailed descriptions of 53 petrochemical production processes, divided into 18 technologies. The petrochemicals described within represent the most economically important petrochemicals, including ethylbenzene, ethylene, phenols and acetone, propylene and light olefins, polyethylene, polypropylene, and polystyrene.

Meyers, R. A., ed. 2004. *Handbook of Petroleum Refining Processes,* 3rd ed., New York: McGraw-Hill. Detailed coverage of 61 licensable petroleum refining processes, divided into 15 technologies, such as catalytic cracking, hydrotreating, and visbreaking and coking. A valuable reference tool for engineering students working on term projects involving process design.

Mian, M. H. 1991–1992. *Petroleum Engineering Handbook for the Practicing Engineer.* (2 vols.) Tulsa: PennWell. Volume 1 covers engineering economics, basic rock and fluid properties, well log interpretation, reservoir engineering and evaluation, and secondary oil recovery. Volume 2 covers principles of transient test analysis, transient testing of oil and gas wells, drilling technology, and production technology.

Mokhatab, S., W. A. Poe, and J. Speight. 2006. *Handbook of Natural Gas Transmission and Processing.* Burlington, MA: Gulf Professional Publishing. Provides detailed coverage of natural gas fundamentals, energy pricing, raw gas transmission, processing, recovery, plant controls and automation, simulation, environmental aspects, maximizing profitability, and gas project plant management.

Muhlbauer, W. K. 2004. *Pipeline Risk Management Manual,* 3rd ed., Oxford, U.K.: Gulf Professional Publishing. The new, expanded third edition of this title now includes offshore pipelines and distribution systems and cross-country liquid and gas transmission lines. Divided into three sections, covering risk evaluation at a glance, customizing the basic risk assessment model, and risk management. Over 50 examples are included to help illustrate the concepts and models.

Parkash, S. 2010. *Petroleum Fuels Manufacturing Handbook: Including Speciality Products and Sustainable Manufacturing Technologies.* New York: McGraw-Hill. Divided into two parts, the first few chapters covering petroleum fuels, and the remaining chapters dealing with petroleum specialty products.

Parkash, S. 2003. *Refining Processes Handbook.* Amsterdam: Gulf Professional Publishing. Provides an overview of operations and processes involved with refining of crude oil into products.

Samuel, G. R., and X. Lio. 2009. *Advanced Drilling Engineering: Principles and Design.* Houston: Gulf Publishing. This handbook covers elaborate drilling processes and engineering well design aspects. Begins with essential topics such as well trajectory and wellbore positioning, and also provides in-depth coverage of well-path planning for directional and extended-reach wells.

Speight, J. G. 2007. *Natural Gas: A Basic Handbook.* Houston: Gulf Publishing Co. A practical handbook designed for introductory engineers, managers and analysts. Divided into two sections covering the origin and properties, and the recovery and processing of natural gas.

Speight, J. G. 2008. *Synthetic Fuels Handbook: Properties, Process and Performance.* New York: McGraw-Hill. Covers the production and properties of fuels made from natural gas and hydrates, petroleum and heavy oil, tar sand bitumen, coal, oil shale, and related nonpetroleum based products, such as wood, biomass, and domestic and industrial waste.

MAPS AND ATLASES

Maps and atlases are important in petroleum engineering as reference tools and can cover different topics, such as well locations, pipelines, oil fields, gas fields, refineries, gas plants, oil sands deposits, sour gas, and tanker terminals. In addition to the selected sources listed below, petroleum-related maps and atlases are often available from the appropriate state or provincial agency or department responsible for energy or economic geology.

American Association of Petroleum Geologists http://bookstore.aapg.org (accessed July 23, 2010). The AAPG publishes a number of maps and atlases, including some in digital/GIS format. One example is the Digital GIS Atlas of O&G Fields, a four-module product covering the U.S. Gulf Coast (100 fields), east South America and western Africa (53 fields), and Southeast Asia and Oceania (32 fields).

GEOSCAN Database. Geological Survey of Canada http://geoscan.ess.nrcan.gc.ca/starweb/geoscan/servlet.starweb?path=geoscan/geoscan_e.web (accessed July 23, 2010). A bibliographic database of over 40,000 records covering GSC publications. Maps are included, and searches can be restricted to maps about petroleum-related features.

JuneWarren-Nickle's Energy Group. JWN Maps & Charts http://www.junewarren-nickles.com/page.aspx?id=maps (accessed July 23, 2010). A set of 24 Canadian maps and charts distributed over a 12-month period, and available to subscribers of *Oilweek, Oilsands Review*, and *Oil & Gas Inquirer.* Maps include *in situ* oil sands, emergency locator map, oil sands lease map, and Western Canadian Sedimentary Basin gas plant locator. Also available every second December is the *Canadian Oilfield Gas Plant Atlas*, which includes over 145 color maps that cover over 12,500 gas plant facilities. The eighth edition, 2010–2011, is the most recent version available.

MAPSearch http://www.mapsearch.com/home.cfm (accessed July 23, 2010). Provides maps, databases, and GIS products on oil, gas, petroleum, and electric power in the United States and Canada. MAPSearch products are published as GIS (Geographical Information Systems) data, or in wall map, state map, or atlas format. The most detailed references available are five major U.S./Canada atlases on crude oil, LPG/NPL (liquefied petroleum gas/natural gas liquid), natural gas, refined products, and petrochemicals/olefins. Each atlas contains information on the pipelines and facilities for the commodity covered. State maps and systems maps are also published.

National Geologic Map Database. United States Geological Survey http://ngmdb.usgs.gov/ (accessed July 23, 2010). Included in the database is a geoscience map catalog on over 67,000 maps, data, and related products from over 300 publishers, covering both U.S. state and territory areas. The advanced search function allows for limiting a search by resources, including petroleum and coal.

Oilfield Publications Limited (OPL) http://www.oilpubs.com/ (accessed July 23, 2010). OPL publishes atlases, maps, GIS mapping, and wall charts of interest to the international offshore oil and gas industry. OPL also publishes vessel registers, books, online data, databases, and CD-ROMs. OPL publishes the *World Deepwater Atlas*, the fourth edition of which appeared in 2005.

Petroleum Economist, The, http://www.petroleum-economist.com/ (accessed July 23, 2010). Publishes large, detailed maps that are included with a subscription to the journal, and are suitable for wall mounting. Coverage is either world or by selected geographic region. In September 2008, the journal published the *World Energy Atlas 2009* (London: Petroleum Economist), which includes maps of all major oil and gas fields, and pipelines, gas processing and storage facilities, deepwater fields, major oil refineries by capacity, liquid natural gas facilities, and tanker terminals, together with enhanced topographic information throughout.

MONOGRAPHS AND TEXTBOOKS

Where to begin, if one wanted to build a solid collection of monographs on "All Things Petroleum." Rather than list a few dozen key titles to build or upgrade a petroleum engineering and refining collection, the significant publishers whose output is dedicated primarily to the petroleum industry are presented and discussed.

Editions Technip http://www.editionstechnip.com/ (accessed 2010). Based in Paris, Editions Technip was founded in 1956 by Institut Français du Pétrole (IFP), and publishes on all aspects of the oil and gas industry, as well as related disciplines, such as geology, hydrocarbon chemistry, and economics. In the early years, Editions Technip authors came only from the IFP, but are now chosen in addition from among industry specialists, and French universities and research centers. Over 1,000 titles were available as of 2010, including original works in English, English translations of its top French titles, and international symposia. Editions Technip publishes the journal, *Oil & Gas Science and Technology, Revue de l'Institut Français du Pétrole.*

Gulf Publishing Company http://www.gulfpub.com/ (accessed March 29, 2011). Founded in 1916, Gulf Publishing of Houston, Texas, publishes a variety of technical books and manuals in a wide range of energy topics. Petroleum-related titles cover drilling, exploration, gas processing, offshore, petrochemicals, pipelines and flow, process engineering, petroleum engineering, production, and refining. Gulf also publishes two major trade journals: *World Oil* and *Hydrocarbon Processing.*

Oilfield Publications Limited (OPL) http://www.oilpubs.com/ (accessed July 23, 2010). In addition to maps, atlases, and wall charts, OPL also publishes books covering issues related to the offshore oil and gas industry.

PennWell http://www.pennwell.com/ (accessed March 29, 2011). Based in Tulsa, Oklahoma, PennWell publishes a wide range of titles in business, well logging, drilling, exploration, refining and processing, reservoir engineering, offshore, production, and pipelines and storage, as well a series of subject-related maps, and atlases of the continental United States and Canada. A very important reference title published annually by PennWell, and mentioned elsewhere in this chapter, is the *International Petroleum Encyclopedia.* PennWell's set of 14 "nontechnical" titles on petroleum and natural gas topics is a must for any library collection focusing on the petroleum industry. Examples in the series include: *Oil & Gas Production in Nontechnical Language* (Raymond and Leffler, 2006), *Petrochemicals in Nontechnical Language* (Burdick and Leffler, 2010), and *Petroleum Refining in Nontechnical Language* (Leffler, 2008).

PETEX: The Petroleum Extension Service, Continuing and Extended Education, University of Texas at Austin http://www.utexas.edu/ce/petex/ (accessed March 24, 2011). PETEX develops, produces, and delivers training courses, publications, and digital training material designed for field-level personnel in the petroleum industry. Coverage includes drilling, personnel and rig management, offshore technology, well servicing, pipelines, and production.

PetroSkills http://www.petroskills.com/index.aspx (accessed July 23, 2010). Formerly known as OGCI, Oil and Gas Consultants, Inc., PetroSkills is a training and consulting company, and publishes a number of important titles used in their courses, covering many basic aspects of petroleum engineering including: reservoir engineering, petroleum geology, decision analysis, production operations, risk and decision analysis, and structural styles in petroleum exploration.

Society of Petroleum Engineers http://www.spe.org/ (accessed March 24, 2011). The Society of Petroleum Engineers publishes titles in four series: monograph, textbook, reprint, and conference proceedings. Topics include completions, drilling, economics, enhanced oil recovery, health, safety and environment, formation evaluation, management, production and facilities, reservoir, as well as a set of reference titles.

Other publishers issuing selected titles in petroleum engineering, refining, processing, and related topics include CRC Press, Butterworth-Heinemann, American Association of Petroleum Geologists, American Institute of Chemical Engineers, Professional Engineering Publishing, Elsevier Science,

McGraw-Hill, Taylor & Francis, Oxford University Press, Blackwell, A.A. Balkema, Academic Press, and the International Association of Drilling Contractors.

Monographic series consist of individual titles sharing a broader subject theme or topic and can number in the hundreds of volumes. The following monographic series include titles that would be of interest to the petroleum engineering and refining professional. It should be noted that titles within chemical engineering monographic series about petroleum refining and processing also will be of interest.

> *AAPG Memoir.* Tulsa: American Association of Petroleum Geologists (0271-8510).
> *AAPG Studies in Petroleum Geology.* Tulsa: American Association of Petroleum Geologists (0271-8510).
> *Developments in Petroleum Science.* Amsterdam: Elsevier (0376-7361).

JOURNALS AND MAGAZINES

A scholarly journal will contain articles of interest written by and for experts and researchers in a specific field or discipline. An article in a scholarly journal reports in-depth research, and includes a detailed literature review of the topic in question, to that point in time. Each article is peer-reviewed, meaning it has been examined by one or more subject experts in the field before it is accepted for publication. The peer-review process is anonymous: the editor of the journal, before accepting the article, submits it to one or more scholars, who vet the manuscript, checking it for accuracy, validity, etc. The scholars examining the paper are not provided with the author's identity nor is the author made aware of the names of the reviewers.

The following peer-reviewed titles are recommended for collections in petroleum engineering, refining, and processing. The list includes journals that appear in the 2009 *ISI Journal Citation Reports*, under the heading Engineering, Petroleum as well as in *Ulrich's Web Global Serials Directory,* under the heading Petroleum and Gas. Titles selected from *Ulrich's* are ones for which indexing and abstracting is provided by at least one abstracting and indexing service, and which were tagged as Refereed. The journals in the list publish some or all articles in English.

Scholarly Journals

> *AAPG Bulletin.* 1917–. Tulsa: American Association of Petroleum Geologists (0149-1423, 1558-9153).
> *Applied Energy.* 1975–. Oxford, U.K.: Pergamon (0306-2619, 1872-9118).
> *Biofuels, Bioproducts and Biorefining.* 2007–. Chichester, U.K.: John Wiley & Sons (1932-104X, 1932-1031).
> *Bulletin of Canadian Petroleum Geology.* 1953–. Calgary, Alberta: Canadian Society of Petroleum Geologists (0007-4802).
> *Chemical and Petroleum Engineering.* 1965–. New York: Springer New York LLC (0009-2355, 1573-8329). Translation of: *Khimicheskoe i Neftegazovoe Mashinostroenie* (1029-8770).
> *Chemistry and Technology of Fuels and Oils.* 1965–. New York: Springer New York LLC (0009-3092, 1573-8310). Translation of *Khimiya i Tekhnologiya Topliv i Masel* (0023-11690).
> *Energy & Fuels.* 987–. Washington, D.C.: American Chemical Society (0887-0624, 1520-5029).
> *Energy Sources. Part A. Recovery, Utilization, and Environmental Effects.* 2006–. Philadelphia: Taylor & Francis (1556-7036, 1556-7230).
> *Energy Sources. Part B. Economics, Planning, and Policy.* 2006–. Philadelphia: Taylor & Francis (1556-7249, 1556-7257).
> *Fuel: The Science and Technology of Fuel and Energy.* 1922. New York: Elsevier (0016-2361, 1873-7153).
> *Fuel Processing Technology.* 1978–. New York: Elsevier (0378-3820, 1873-7188).
> *GeoArabia.* 1996–. Manama, Bahrain : Gulf PetroLink (1025-6059, 1819,169X).

International Journal of Offshore and Polar Engineering. 1991–. Golden CO: International Society of Offshore and Polar Engineers (1053-5381).

International Journal of Oil, Gas and Coal Technology. 2008–. Olney, U.K.: Inderscience (1753-3309, 1753-3317).

Journal of Canadian Petroleum Technology. 1962–. Calgary, Alberta: SPE Canada (0021-9487).

Journal of Petroleum Geology. 1978–. Beaconsfield, U.K.: Scientific Press (0141-6421, 1747-5457).

Journal of Petroleum Science and Engineering. 1987–.New York: Elsevier (0920-4105, 1873-4715).

Journal of the Energy Institute. 1927–. London: Maney Publishing (0144-2600, 1746-0220).

Journal of the Japan Petroleum Institute (Sekiyu Gakkaishi). 1958–. Japan Petroleum Institute (1346-8804, 1349-273X).

Journal of Tribology. New York: American Society of Mechanical Engineers (0742-4787, 1528-8897).

Lubrication Science. 1988–. New York: John Wiley & Sons (0954-0075, 1557-6833).

Marine and Petroleum Geology. 1971–. New York: Elsevier (0264-8172, 1873-4073).

Natural Resources Research. 1992–. New York: Kluwer (1520-7439, 1573-8981).

Oil & Gas Science & Technology: Revue de l'Institut Francais du Petrole. 1946–. Paris: Editions Technip (1294-4475).

Oil Shale. 1988–. Tallinn, Estonia: Estonian Academy Publishers (1736-7492, 0208-189X).

Open Petroleum Engineering Journal, The. 2008–. Hilversum, The Netherlands: Bentham (1874-8341).

Petroleum Chemistry. 1962–. Moscow: IAPC "Nauka/Interperiodica" (0965-5441, 1555-6239). Translation of *Neftekhimiya* (0028-2421).

Petroleum Geoscience. 1995–. Bath, U.K.: Geological Society Publishing House (1354-0793, 2041-496X).

Petroleum Science and Technology. 1983–. Philadelphia: Taylor & Francis (1091-6466, 1532-2459).

Petrophysics. 1962–. Houston: Society of Petrophysicists and Well Log Analysts (1529-9074).

SPE Drilling & Completion. 1961–. Richardson, TX: Society of Petroleum Engineers (1064-6671, 1930-0204).

SPE Economics & Management. 2009–. Richardson, TX: Society of Petroleum Engineers (2150-1173).

SPE Journal. 1996–. Richardson, TX: Society of Petroleum Engineers (1086-055X, 1930-0220).

SPE Production & Operations. 1961–. Richardson, TX: Society of Petroleum Engineers (1930-1855, 1930-1863).

SPE Projects, Facilities & Construction. 2006–. Richardson, TX: Society of Petroleum Engineers (1942-2431).

SPE Reservoir Evaluation & Engineering. 1998–. Richardson, TX: Society of Petroleum Engineers (1094-6470, 1930-0212).

TRADE AND BUSINESS JOURNALS

Petroleum, oil, and gas trade journals and magazines provide subscribers and society members with the most current information, news, and developments in the field. Trade journals can provide weekly and monthly data, prices, and statistics, information about forthcoming conferences, the latest society or association news, news regarding personnel movements and promotions, often with accompanying biographical information, and reviews of the newest products and services available to the industry. A search on *Ulrich's Web Global Series Directory* for titles under the heading Petroleum and Gas, and restricted to Serial type: Trade, returned 725 titles. Many of the trade publications listed are region, state, or country specific. What follows is a very selective list of what might be considered more general trade journals for the petroleum industries.

AAPG Explorer. 1979–. http://www.aapg.org/explorer/ (accessed July 23, 2010). Tulsa: American Association of Petroleum Geologists (0195-2986).

Alberta Oil. 2005–. http://www.albertaoilmagazine.com (accessed July 23, 2010). Edmonton, Alberta: Venture Publishing (1912-5291).

American Gas. 1918–. http://www.aga.org/Newsroom/magazine/Pages/default.aspx (accessed July 23, 2010). Washington D.C.: American Gas Association (1043-0652).

American Oil & Gas Reporter. 1958–. http://www.aogr.com/ (accessed July 23, 2010). Haysville, KS: National Publishers Group Inc. (0145-9198).

Asian Oil and Gas. 1980–. http://www.oilonline.com/Magazines/AsianOilGas.aspx (accessed July 23, 2010). Houston: Atlantic Communications (1026-6461).

Biodiesel Magazine. 2004–. http://www.biodieselmagazine.com (accessed July 23, 2010). Grand Forks, ND: BBI International Media (1935-7621).

CIM Magazine. 2006–. http://www.cim.org/bulletin (accessed July 23, 2010). Montreal PQ: Canadian Institute of Mining, Metallurgy & Petroleum (1718-4177).

Drilling Contractor. 1944–. http://www.drillingcontractor.org (accessed July 23, 2010). Houston: International Association of Drilling Contractors (0046-0702).

E&P. 1973–. http://www.epmag.com/ (accessed July 23, 2010). Houston: Hart Energy Publishing (1527-4063).

Energy Processing Canada. 1908–. http://www.northernstar.ab.ca/default.asp?id=27 (accessed July 23, 2010). Calgary, Alberta: Northern Star Communications (0319-5759).

Hydrocarbon Processing. 1922–. http://www.hydrocarbonprocessing.com (accessed July 23, 2010). Houston: Gulf Publishing (0887-0284).

JPT: Journal of Petroleum Technology. 1949–. http://www.spe.org/jpt (accessed July 23, 2010). Richardson, TX: Society of Petroleum Engineers (0149-2136).

LNG Journal. 1996–. http://www.lngjournal.com (accessed July 23, 2010). London: Nelton Publications (1365-4314).

New Technology Magazine. 1937–. http://www.newtechmagazine.com/ (accessed July 23, 2010). Calgary, Alberta: Junewarren-Nickle's Energy Group (1480-2147).

North American Pipelines http://www.napipelines.com/ (accessed July 23, 2010). Peninsula, OH: Benjamin Media Inc. (2150-9190).

NPN—National Petroleum News. 1909–. http://www.npnweb.com (accessed July 23, 2010). Park Ridge, IL: M2Media360 (0149-5267).

Offshore. 1954–. http://www.offshore-mag.com/index.html (accessed July 23, 2010). Tulsa: PennWell (0030-0608).

Offshore Engineer. 1975–. http://www.oilonline.com/Magazines/OffshoreEngineer.aspx (accessed July 23, 2010). Houston: Atlantic Communications (0305-876X).

Oil & Gas Inquirer. 1989–. http://www.oilandgasinquirer.com/ (accessed July 23, 2010). Edmonton, Alberta: JuneWarren-Nickles' Energy Group (1204-4741).

Oil & Gas Journal. 1902–. http://www.ogj.com/index.cfm (accessed July 23, 2010). Tulsa: PennWell (0030-1388).

Oilsands Review. 2006–. http://www.oilsandsreview.com/ (accessed July 23, 2010). Edmonton, Alberta: JuneWarren-Nickles'Energy Group (1912-5305).

Oilweek. 1948–. http://www.oilweek.com/ (accessed July 23, 2010). Edmonton, Alberta: JuneWarren-Nickles' Energy Group (1200-9059).

OPEC Bulletin. 1967–. http://www.opec.org/opec_web/en/publications/76.htm (accessed July 23, 2010). Vienna: Organization of the Petroleum Exporting Countries (0474-6279).

Petroleum Review. 1914–. http://www.energyinst.org.uk/index.cfm?PageID=9 (accessed July 23, 2010). London: Energy Institute (0020-3076).

Pipeline & Gas Journal. 1970–. http://www.pipelineandgasjournal.com/ (accessed July 23, 2010). Houston: Oildom Publishing (0032-0188).

PipeLine and Gas Technology. 2002–. http://www.pipelineandgastechnology.com/ (accessed July 23, 2010). Houston: Hart Energy Publishing (1540-3688).

Pipelines International Magazine. 2009–. http://pipelinesinternational.com/magazine/ (accessed March 24, 2011). Melbourne, Australia: Great Southern Press (1837-1167).

Upstream. 1996–. http://www.upstreamonline.com (accessed July 23, 2010). Oslo: Norges Handels og Sjoefartstidende (0807-6472).

World Oil. 1916–. http://www.worldoil.com/ (accessed July 23, 2010). Houston: Gulf Publishing (0043-8790).

PATENTS

Patents are critical to all engineering disciplines. For many reasons, patent searching is important to the engineer. For example, 80% of patent data is not published elsewhere, and will contain copious amounts of detail, research data and results, and technical drawings of inventions: www.patex.ca/html/ip_information/about_patents.html (accessed October 19, 2010). In the publication cycle, patents will precede conference papers, which will in turn precede publication in scholarly journals. As such, the newest technologies, unless kept proprietary by their inventors, are first disclosed in published patents and patent applications. Searching the patent literature can save the engineer time and money. When planning to develop new technology, a patent search may reveal that the technology already exists, thus avoiding duplication of effort. Examining the patent literature can lead to new ideas and challenges in research and development, allow for the tracking of the work of competitors, and even predict forthcoming hot areas of R&D.

Databases that index patents include *Petroleum Abstracts* and *SciFinder.* Please see Chapter 2, General Engineering Resources, in this book for a full list of patent searching resources.

SEARCH ENGINES AND PORTALS

This is a selective set of Web sites that provide links to resources on petroleum engineering and refining, such as buyer's guides, product showcases, auctions, job openings, discussion forums, company listings, industry news, etc. It is by no means exhaustive.

Most, if not all, major petroleum industry players have Web sites, as do petroleum engineering departments at academic and technical institutions and colleges, and publishers of print and electronic resources, including monographs, encyclopedias, scholarly and trade journals, newsletters, and statistics. A number of the sites below include membership fees or subscription only access to either part of or the entire site. Finally, while browsing throughout each portal or directory listed below, users will, at some point, discover that some links have changed or disappeared; such is the life of a Web site.

Canadian Wellsite http://www.canadian-wellsite.com/ (accessed July 23, 2010). Designed specifically for the Canadian oilpatch. Some sections require subscription access. Free access is available to the oilfield directory, classifieds, events, software, and discussion group.

FCC Network, The, http://www.thefccnetwork.com/ (accessed July 23, 2010). An online network covering fluid catalytic cracking and related petroleum refining processes. Registration is free, and includes access to a monthly newsletter, technical papers and catalyst reports, tips and techniques for troubleshooting, tools and strategies for FCC performance optimization, and submission of questions to an advisory panel.

Hydrocarbon Online http://www.hydrocarbononline.com/ (accessed July 23, 2010). Provides links to services, suppliers, markets for buying and selling, career development, news and community for the hydrocarbon processing industry.

Oil and Gas Insight http://www.oilandgasinsight.com/ (accessed 23 July 2010). Fee-based service specializing in oil and gas market intelligence, trend analysis, company profiles, and forecasts for the oil and gas industry on a country-by-country basis across the United States, Canada, Latin America and the Caribbean, Central, Eastern, Southeast, and Western Europe, Asia, the Middle East, and Africa.

Oil and Gas International http://www.oilandgasinternational.com/ (accessed July 23, 2010). Fee-based site covering worldwide exploration and production news, information, and analysis for the upsteam petroleum industry.

Oil and Gas Online http://www.oilandgasonline.com (accessed July 23, 2010). From VertMarkets, the same company that produces Hydrocarbon Online, with the same features as that site. Coverage of the international upstream petroleum industry.

Oil.com http://www.oil.com/ (accessed July 23, 2010). One of many sites produced by the WorldNews Network. Provides links to sites covering oil prices, news and industry sites, refineries, and related energy sites.

Oilfield Directory http://www.oilfielddirectory.com/ (accessed July 23, 2010). Established in 1996, the Oilfield Directory provides extensive coverage of the petroleum industry, including links to news feeds and oil prices, a global product and services directory, an equipment section, job bank, and discussion forum. Companies interested in more detailed listings of their services can pay an annual membership fee for additional access.

OilOnline http://www.oilonline.com/ (accessed July 23, 2010). One of the oldest oil and gas portals on the Web, having been online since December 1, 1995. Links are divided into industry news, key indicators, careers, industry information, equipment and services, and an online store. This site also hosts the trade journals *Offshore Engineer* and *Asian Oil & Gas*.

Refining Online http://www.refiningonline.com/ (accessed July 23, 2010). A Web portal for the petroleum refining industry. Registration is free. Features include an interactive Q&A, a refining industry search engine, a technical knowledge base, events calendar, and membership directory.

RIGZONE http://www.rigzone.com/ (accessed July 23, 2010). The "gateway to the oil & gas industry," RIGZONE focuses on the upstream oil and gas industry. The information available on the site is categorized into the following zones: news and analysis, insight and expertise, oil and gas directory, forthcoming events, meetings, and conferences, a data center on offshore rig activity, career center, and equipment market. Members can subscribe to four newsletters. For a subscription fee, members can subscribe to the RIGZONE News Professional Edition. The industry directory is perhaps the best available online at the time of publication, providing detailed subject access by company, region, product, service, e-commerce, government and education, associations and societies, and news and information. RIGZONE is a well-designed, easy-to-use site.

ASSOCIATIONS, SOCIETIES, AND ORGANIZATIONS

Below is a selective list of major associations, institutes, and societies relevant to petroleum engineering and refining. There are many more than can be listed here. For example, RIGZONE lists 108 societies and associations, and Oil Online lists over 125.

American Association of Drilling Engineers http://www.aade.org/ (accessed on December 12, 2010).

American Association of Petroleum Geologists (AAPG) http://www.aapg.org/ (accessed on December 12, 2010).

American Gas Association http://www.aga.org/ (accessed on December 12, 2010).

American Institute of Chemical Engineers (AIChE) http://www.aiche.org/ (accessed on December 12, 2010).

American Petroleum Institute (API) http://www.api.org (accessed on December 12, 2010).

American Society of Gas Engineers http://www.asge-national.org/ (accessed on December 12, 2010).

Canadian Association of Petroleum Producers (CAPP) http://www.capp.ca/ (accessed on December 12, 2010).

Canadian Institute of Mining, Metallurgy and Petroleum (CIM) http://www.cim.org (accessed on December 12, 2010).

Canadian Petroleum Products Institute http://www.cppi.ca/ (accessed on December 12, 2010).

Energy Institute (U.K.) http://www.energyinst.org/home (accessed on December 12, 2010).

Gas Technology Institute http://www.gastechnology.org/ (accessed on December 12, 2010).

Independent Petroleum Association of America http://www.ipaa.org/ (accessed on December 12, 2010).

Institut Français du Pétrole (IFP), The, http://www.ifp.fr/ (accessed on December 12, 2010).

Instituto Mexicano del Petróleo (IMP) http://www.imp.mx/ (accessed on December 12, 2010).

International Association of Drilling Contractors http://www.iadc.org/ (accessed on December 12, 2010).

International Association of Oil & Gas Producers http://www.ogp.org.uk/ (accessed on December 12, 2010).

International Energy Agency http://www.iea.org/ (accessed on December 12, 2010).

International Society of Offshore & Polar Engineers http://www.isope.org/ (accessed on December 12, 2010).

NACE International—The Corrosion Society http://www.nace.org (accessed on December 12, 2010).

National Petroleum Council http://www.npc.org/ (accessed on December 12, 2010).

National Petrochemical & Refiners Association http://www.npradc.org/ (accessed on December 12, 2010).

North America Energy Standards Board http://www.naesb.org/ (accessed on December 12, 2010).

Offshore Engineering Society (U.K.) http://www.oes.org.uk/ (accessed on December 12, 2010).

Society of Petroleum Engineers (SPE) http://www.spe.org/ (accessed on December 12, 2010).

Society of Petrophysicists and Well Log Analysts http://www.spwla.org/ (accessed on December 12, 2010).

STANDARDS

Standards are an essential part of an engineer's work. A standard establishes parameters for design, capacity, or property characteristics, which permit interchangeability of parts and materials. Engineering companies cannot compete locally or in the global marketplace without integrating standardization into every phase of their operations. "thinkstandards.net" (IEEE Web site) is an example of a good Web site with basic information about standards, including definitions, a brief history, the benefits and return on investment (ROI) of standards, and a list of links to standards resources.

Standards of importance to the petroleum industry include those produced by the following:

American Petroleum Institute http://www.api.org (accessed July 23, 2010). API is the primary standards development and issuing body for petroleum engineering and refining. API publishes hundreds of standards, recommended practices, specifications, codes, technical publications, reports, and studies, for both the upstream and downstream areas. Mandatory for any petroleum engineering collection.

American Society for Testing and Materials. *ASTM Annual Book of Standards* http://astm.org/ (accessed July 23, 2010). ASTM publishes the *Annual Book of Standards*, a 70-volume set containing over 12,000 standards. Section 05 of the set is Petroleum Products and Lubricants, over 630 standards, including methods to measure properties of natural and liquefied petroleum, crude petroleum, and pure light hydrocarbons. Also available online by subscription.

International Standards Organization http://www.iso.org/ (accessed July 23, 2010). ISO publishes approximately 120 standards dealing with various aspects of the petroleum and natural gas industry, including offshore structures, steel pipe for pipelines, pipeline transportation systems, subsea production systems, drilling and production equipment, drilling fluids, casing, and tubing.

Standards applicable to aspects of petroleum engineering, refining, processing, and testing are also issued by these organizations and associations:

AFNOR: Association Française de Normalisation
AIChE: American Institute of Chemical Engineers
ASME: American Society of Petroleum Engineers
AWS: American Welding Society
BSI: British Standards Institute
GPA: Gas Processors Association
ISA: The Instrumentation, Systems and Automation Society
NFPA: National Fire Protection Association
TEMA: Tubular Exchange Manufacturers Association

Special mention is made of three important compilations of methods, procedures, and correlations involving petroleum and gas.

Engineering Data Book, 12th ed. 2004. Tulsa: Gas Processors Suppliers Association. Available in SI (International System of Units) and FPS (foot-pound-second) versions. Published in two loose-leaf binders to allow for updates. First published in 1935. Useful for the field or plant engineers who are determining operating and design parameters. Also of use to design engineers as a general reference tool for accepted engineering practice in estimating, preliminary design, feasibility studies, and onsite operating decisions. Students studying engineering design in areas such as design of processing plants or refineries will find useful and practical information, data, and procedures within. Divided into 26 sections, covering equipment, storage, and processes.

IP Standard Methods for Analysis and Testing of Petroleum and Related Products and British Standard 2000 Parts 2010, 68th ed. 2010. London: Energy Institute. Published annually in three volumes. A compilation of test methods for the analysis and testing of petroleum and petroleum products, using traditional and modern instrumentation techniques. The 2010 edition contains 305 full methods and 20 proposed methods. Joint methods with BSI, EN ISO, and ASTM are included. The methods are used for quality control and are important for national (U.K.) and international trading of petroleum and petroleum products.

Technical Data Book—Petroleum Refining. 1997. Washington, D.C.: API Publishing Services. A loose-leaf, three-volume set providing physical and thermodynamic data and correlations needed by the petroleum refining industry for design of equipment and process evaluation. Divided into 15 chapters, each chapter devoted to a single property or group of related properties, such as hydrocarbon characterization, critical properties of pure hydrocarbons, defined mixtures, natural gases, and petroleum fractions, viscosity, combustion, and 11 others. The sixth edition was the last available in print. The print version of the *Technical Data Book* has been replaced by the eighth edition (2006), an interactive software application that includes over 130 API standard methods and 20 software programs. This edition is produced by EPCON International http://www.epconsoftware. com/ (accessed March 29, 2011). The EPCON site indicates that the printed three-volume set accompanies the electronic version, but it is not clear if the printed edition can be purchased separately from the software package.

STATISTICS AND DATA

Statistics and data are crucial to the petroleum industries. From hourly commodity prices to annual estimates of oil reserves, the need for this kind of information is ongoing and relentless.

American Petroleum Institute. *API Data™* https://accessapi.api.org/accessapi/index.html (accessed July 23, 2010). The American Petroleum Institute provides petroleum industry statistics via API Data™, an online subscription service available to API members and nonmembers. Included in the service are weekly bulletins of refinery inputs, production, imports and inventories, monthly statistical reports, inventories of natural gas liquids and liquefied refinery gases, and imports and exports of crude oil and petroleum products. Pricing for the individual reports is high and designed for the single corporate subscriber.

American Petroleum Institute. *Basic Petroleum Data Book.* Washington, D.C.: American Petroleum Institute (0730-5621). Published twice a year in print, also available online. Covers primarily U.S. statistics, with some world data. Sections include energy, crude oil reserves, exploration and drilling, production, financial, prices, demand, refining, imports, exports, offshore, transportation, natural gas, OPEC, environmental, and miscellaneous. Historical data are included. The API also publishes the *Weekly Statistical Bulletin* and the *Monthly Statistical Report.*

BP Statistical Review of World Energy www.bp.com/statisticalreview (accessed July 23, 2010). London: British Petroleum Co.. Published annually in print and available online. Coverage includes oil and natural gas, and includes statistics for reserves, production, consumption, prices, stocks, refining and trade movements. The online version offers downloads in different formats such as PDF, Excel®, and PowerPoint®, and includes an "energy charting tool," which can create charts and graphs that in turn can be exported for further analysis.

Busby, R. L., ed. *International Petroleum Encyclopedia.* (See: Encyclopedias.)

Canadian Association of Petroleum Producers. *Statistical Handbook for Canada's Upstream Petroleum Industry* http://www.capp.ca/getdoc.aspx?DocId=167463&DT=NTV (accessed July 23, 2010). First published in 1955, and data updated regularly online. Provides historical summaries of the progress of the petroleum industry in Canada. Coverage includes land, exploration, and drilling, reserves, production, expenditures/revenue, prices, demand/consumption, refining, and imports/exports. Sections covering transportation, energy, and world data have been discontinued, but continue to appear in the online edition.

Oil & Energy Trends: Annual Statistical Review. 1979–. London: Wiley-Blackwell Publishing. Gathers together the relevant data on the world oil industry, including oil and gas reserves, active and drilled wells, refinery capacity and production, oil and oil products demand and prices (0953-1033, 1746-9066).

Oil & Gas Journal Online Research Center http://www.pennenergy.com/index/research-and_data/oil-and_gas.html (accessed July 23, 2010). A fee-based service that provides reports, statistics, surveys, GIS data, and industry research on energy, oil, and gas. Oil and gas data include prices, imports, exports, stocks (inventories), demand and consumption, and storage. The *Oil & Gas Journal Energy Database* has over 100,000 data series. The Oil & Gas Journal Online Research Center also publishes a series of electronic energy industry

directories, covering various aspects of the oil and natural gas and electric power industries. These include pipeline, refining and gas processing, petrochemicals, liquid terminals,
and gas utility. The directories are not available online, but designed to be downloaded to
one computer at a time. When connected to the Internet, users can update listings when
made available.

OPEC. *Annual Statistical Bulletin* http://www.opec.org/opec_web/en/ (accessed July 23,
2010). Contains tables, charts, and graphs covering the world's reserves of oil and gas,
crude oil and product output, exports, refining, tankers, and other data. OPEC publications
can be downloaded free of charge from the OPEC Web site. Print copies are available for
a subscription charge.

OPEC. *World Oil Outlook* http://www.opec.org/opec_web/en/ (accessed July 23, 2010). This
annual publication "concentrates on exploring the possible developments of oil supply and
demand." Divided into two sections, one covering oil supply and demand outlook to 2030, the
other covering oil downstream outlook to 2030. Features extensive lists of tables and figures.

Twentieth Century Petroleum Statistics, 64th ed. 2008. Dallas: DeGolyer and MacNaughton.
Published annually. Covers primarily U.S. statistics, with some world data. Despite the title,
it does include twenty-first century data. U.S. statistics include crude reserves, production,
imports by source, producing oil and gas wells, refining capacity, natural gas reserves,
drilling costs, and much more.

United States. Department of Energy. Energy Information Administration http://www.eia.
doe.gov/ (accessed July 23, 2010). Created in 1977 by the U.S. Congress, the Energy
Information Administration (EIA) provides extensive, detailed current and historical
information, statistics and data by geographic area, fuel, sector, and price. Petroleum and
natural gas are included. Some files are available in text, PDF, and Excel® format, which
can be exported for further use by the searcher. The EIA also publishes some of its statistics in data in print titles, such as the *Petroleum Supply Annual* and the *Annual Energy
Review.* Among its annual publications are ones covering the following: coal, energy outlook, energy review, international energy annual, international energy outlook, natural gas,
and petroleum supply.

For reasons unknown, certain publications covering statistics relating to the petroleum industry
are no longer published. They are included here for those who may be interested in obtaining copies
for historical research:

Guide to Petroleum Statistical Sources, 10th ed. 1997–1998. New York: API EnCompass.
The guide was divided into three sections: statistical databases, recurring statistical information in print, and related publications of interest. Use of the guide helped simplify and
expedite the search for a wide variety of petroleum industry statistical information.

Natural Gas Statistics Sourcebook, 7th ed. 2001. Tulsa: PennWell. No volume published since the
seventh edition. Contained monthly, quarterly, and annual data for the important parameters
of segments of the U.S. and worldwide natural gas industry, including reserves, exploration
and drilling, production, imports and exports, prices, processing, demand, and consumption.

Refining Statistics Sourcebook, 6th ed. 1999. Tulsa: Pennwell. No volume published since
the sixth edition. Contained monthly and annual data for the important parameters of segments of the U.S. and worldwide petroleum refining industry, including petroleum product demand, capacity and inputs, refining production, imports and exports, crude oil and
petroleum products stocks, prices, and transportation and petroleum movement.

Data for both sourcebooks were collected from the Oil & Gas Journal Energy Database. Current
data is available from the Oil & Gas Journal Online Research Center.

Statistical Annual. 2000. London: Petroleum Economist. Last known volume is 2000. Primary focus was on production statistics for oil, natural gas, coal, hydro power, and nuclear power. Data were presented in graphs and tables only.

REFERENCES

American Oil & Gas Historical Society http://www.aoghs.org/ (accessed July 23, 2010).

American Oil & Gas Historical Society. *The petroleum age* http://sites.google.com/site/petroleumage/ (accessed July 23, 2010).

American Petroleum Institute. *All about petroleum* http://classroom-energy.org/oil_natural_gas/progress_through_petroleum/petroleum/aboutpetroleum01.html (accessed July 23, 2010).

American Petroleum Institute. 1961. *History of petroleum engineering.* New York: American Petroleum Institute.

Burdick, D. L., and W. L. Leffler. 2010. *Petrochemicals in nontechnical language,* 4th ed. Tulsa: PennWell.

CBC News. 2010. *Gulf oil leak stopped: BP* http://www.cbc.ca/world/story/2010/07/15/bp-oil-cap-test.html (accessed July 23, 2010).

Cronin, F. 1955. North America's father of oil. *Imperial Oil Review* (April): 15–20.

Davies, S. 2010. Deep oil dilemma. *Engineering and Technology,* (June 5) 5 (8): 44–49.

Finch, D. 2007. *Pumped: Everyone's guide to the oil patch.* Calgary, Alberta: Fifth House.

Giebelhaus, A. W. 1996. The emergence of the discipline of petroleum engineering: An international comparison. *ICON: Journal of the International Committee for the History of Technology* 2: 108–122.

Gray, E. 2005. *The great Canadian oil patch: The petroleum era from birth to peak,* 2nd ed. Edmonton, Alberta: JuneWarren Publications.

Hall, L. D., and B. Romaniuk, eds. 2008. *Gale directory of databases.* Vol. 1: Online databases. Detroit: Thompson Gale.

IEEE thinkstandards.net http://thinkstandards.net (accessed July 23, 2010).

Joye, S. and I. MacDonald. 2010. Offshore oceanic impacts from the BP oil spill. *Nature Geoscience* 3, doi:10.1038/ngeo902, http://www.nature.com/ngeo/journal/v3/n7/pdf/ngeo902.pdf> (accessed July 23, 2010).

Leffler, W. L. 2008. *Petroleum refining in nontechnical language.* Tulsa: PennWell Books.

Lord, C. R. 2000. *Guide to information sources in engineering.* Englewood, CO: Libraries Unlimited.

Lucier, P. 2008. *Scientists & swindlers: Consulting on coal and oil in America, 1820–1890.* Baltimore: Johns Hopkins University Press.

Macleod, R. A. and J. Corlett, eds. 2005. *Information sources in engineering,* 4th ed. London: Bowker-Saur.

Martin, C. 2010. Oil spill ramifications pour in. *Current Biology* 20 (10). doi:10.1016/j.cub.2010.05.005, http://download.cell.com/current-biology/pdf/PIIS0960982210005695.pdf (accessed July 23, 2010).

Morritt, H. 1993. *Rivers of oil: The founding of North America's petroleum industry.* Kingston, Ontario: Quarry Press.

Oil Museum of Canada. 2001. Black gold: Canada's oil heritage http://epe.lac-bac.gc.ca/100/205/301/ic/cdc/blackgold/default.htm (accessed July 23, 2010).

OPEC Annual Statistical Bulletin. 2009, 2010. 22. http://www.opec.org/library/Annual%20Statistical%20Bulletin/interactive/2009/FileZ/ASB.pdf (accessed July 23, 2010).

Osif, Bonnie A., ed. 2006 *Using the engineering literature.* London: Routledge.

Owen, E. W. 1975. The earliest oil industry. In *Trek of the oil finders: A history of exploration for petroleum,* pp. 1–14. Tulsa: The American Association of Petroleum Geologists.

PATEX. About patents and patent search http://www.patex.ca/html/ip_information/about_patents.html (accessed July 23, 2010).

Pearson, B. C. and K. B. Ellwood. 1987. *Guide to the petroleum reference literature.* Littleton, CO: Libraries Unlimited, Inc.

Raymond, M., and W.L. Leffner. 2006. *Oil and gas production in nontechnical language.* Tulsa OK: PennWell Corporation.

Schempf, F. J. 2007. *Pioneering offshore: The early years.* Tulsa: Pennwell Custom Publishing.

Spitz, P. H. 1988. *Petrochemicals: The rise of an industry.* New York: John Wiley & Sons.

20 Transportation Engineering

Rita Evans and Kendra K. Levine

CONTENTS

Introduction 498
The Literature of Transportation Engineering 498
Searching the Library Catalog 498
    Keywords, LC Subject Headings, and LC Call Numbers 498
Bibliographic Databases, Indexes, and Abstracts 499
Bibliographies and Guides to the Literature 500
Databases and Datasets 501
Directories 503
Dictionaries and Glossaries 504
Handbooks and Manuals 505
    General 505
    Highways and Traffic 505
    Nonmotorized Transportation 508
    Pipelines 509
    Public Transportation 509
    Railroads 510
    Vehicles 510
Monographs and Textbooks 511
    General 511
    Freight 511
    Highways and Traffic 512
    Nonmotorized Transportation 512
    Pipelines 512
    Railroads 513
    Vehicles 513
Journals 513
    General 514
    Highways and Traffic 515
    Maritime 516
    Pipelines 516
    Railroads 516
    Vehicles 517
Proceedings 517
Patents 519
Standards and Specifications 519
Important Web Sites and Portals 520
Associations, Organizations, and Societies 523
Acknowledgments 525
References 526

INTRODUCTION

Transportation engineering is a subdiscipline of civil engineering. Beginning in the mid-nine-teenth century, industrialization and urbanization generated a need for better infrastructure and an improved transportation system. Waterway and canal development was followed by the growth of rail transportation. The advent of motorized vehicles and the tremendous demand for roads and related facilities made the need for engineers specializing in transportation more urgent.

The dynamics of traffic flow was one of the earliest areas of research in transportation engineer-ing with research activities underway by the 1950s. A catalyst for the emergence of transportation engineering was the establishment in the United States of the federal Department of Transportation in 1967, and the move to multimodalism on the part of state departments of transportation.

Today, transportation engineering encompasses a wide range of activities. Transportation engineers design, construct, and maintain facilities. They are involved in traffic engineering and operations, and logistics. They may focus on a particular mode, such as highways, but intermodal operations are increasingly important. Transportation engineering does not exist in a vacuum, and issues related to city and regional planning overlap with the more technical aspects.

THE LITERATURE OF TRANSPORTATION ENGINEERING

The literature of transportation engineering has developed with the discipline. In 1922, the Advisory Board on Highway Research of the National Research Council began publishing the proceedings of its annual meeting in the *Bulletin of the National Research Council*. In 1963, the Highway Research Board began publishing *Highway Research Record* (renamed the *Transportation Research Record* in 1974). The American Society of Civil Engineers published the *Journal of the Highway Division* beginning in 1957; this became *Transportation Engineering Journal* in 1968 and *Journal of Transportation Engineering* in 1983. The American Association of State Highway Officials pub-lished works on the design of roads and highways in the mid-1950s; this material was used in the design of the Interstate Highway System (Sinha, 2002).

The Highway Research Information Service (HRIS) began operations in 1967. This early effort by the Transportation Research Board (TRB) to provide access to transportation engineering infor-mation involved acquisition, abstracting and indexing, record storage, file storage, and batch and online retrieval. In the mid-1970s, TRIS (Transportation Research Information Services) was intro-duced; the Transportation Division of the Special Libraries Association worked closely with TRB to develop and implement this database.

Today, Web sites are an essential source of information from government agencies, professional organizations, and commercial suppliers. Statistics, standards, patents, and other information that used to be difficult to identify and locate is now often just a few clicks away. Many journals are available in electronic editions, and the *TRIS* database (now *TRID*) is accessed online. Print resources remain important, however. As with any engineering discipline, handbooks, manuals, and textbooks are still critical information sources.

This chapter outlines some of the more important resources for transportation engineering. It is not comprehensive, but does highlight a selected list of resources that encompass the range of types of materials and information.

SEARCHING THE LIBRARY CATALOG

KEYWORDS, LC SUBJECT HEADINGS, AND LC CALL NUMBERS

Searching the library catalog for information on transportation engineering frequently involves searching by mode, such as highway, rail, or marine transportation. Some key Library of Congress (LC) Subject Headings include:

Automobiles	Local transit	Street-railroads
Bridges	Locomotives	Streets
Container ports	Logistics	Subways
Electronic traffic controls	Marine accidents	Traffic accidents
Elevated highways	Marine engineering	Traffic congestion
Express highways	Motor vehicles	Traffic engineering
Ferries	Naval architecture	Traffic flows
Geographic Information Systems	Pavements	Traffic safety
Harbors	Pedestrians	Traffic signs and signals
High speed trains	Ports	Transportation
Highway capacity	Railroad bridges	Transportation engineering
Highway communications	Railroad engineering	Travel time (Traffic engineering)
Highway engineering	Railroad tunnels	Trucking
Highway-Railroad grade crossings	Railroads	Trucks
Intelligent transportation systems	Roads	Tunneling

LC Free-floating subdivisions:

Transportation

LC headings often lag behind new areas of research and are often slow to adopt new terminology. *Street-railroads*, for example, is used rather than the much more common term *light rail*. Identifying the correct LC Subject Headings will result in more useful catalog search results.

The Library of Congress classification system places most transportation engineering-related information in the T schedule and a few topics in the H schedule:

HE331-380	Traffic engineering. Roads and highways. Streets
TA501-625	Surveying
TA800-820	Tunneling. Tunnels
TA1001-1280	Transportation engineering
TF1-1602	Railroad engineering
TE1-450	Highway engineering. Roads and pavements
TG1-470	Bridge engineering
TL1–484	Motor Vehicles. Cycles
TP315–360	Fuel
TP690–692.5	Petroleum refining. Petroleum products
VM1–989	Naval architecture. Shipbuilding. Marine engineering

In addition to LC Subject Headings, research in transportation engineering can be facilitated by the use of the *Transportation Research Thesaurus* (*TRT*) http://trt.trb.org/trt.asp? (accessed August 12, 2010). The *TRT* is used for indexing and retrieval of transportation information. It covers all modes and functions. TRID, the primary bibliographic database for transportation, uses the *TRT*, which allows for much more focused subject searching than LC's Subject Headings.

OCLC's (Online Computer Library Center's) *WorldCat* is the catalog of choice in locating sources of transportation information.

BIBLIOGRAPHIC DATABASES, INDEXES, AND ABSTRACTS

Journal literature, conference proceedings, and technical reports are primary sources of transportation engineering information, and there are several indexing and abstracting services that provide access to this literature. Particularly useful are the indexes listed below, but other indexes, such as *Web of Science, Compendex, Inspec,* and *PsycInfo,* cover aspects of transportation engineering and should be used as a complement to the subject-specific indexes.

TRID (the *TRIS* and *ITRD* Database) is the most comprehensive international source of bibliographic information in transportation. Primary Resource. *TRID* is produced by the Transportation Research Board and the Joint Transport Research Centre of OECD. It is available free of charge to the public on TRB's Web site at http://trid.trb.org/ (accessed April 5, 2011). More than 800,000 records on all modes and disciplines cover transportation systems, roads and highways, traffic, urban transportation, safety, freight and passengers, intermodal transportation, energy, and environmental effects. Sources include technical reports, articles from journals and trade publications, conference proceedings, theses, books, and summaries of research in progress. Each record is available in English, French, German, or Spanish. Most records include abstracts. Electronic links to the full text of many documents and to document suppliers are provided. Records are indexed using the *Transportation Research Thesaurus* and the multilingual ITRD thesaurus.

Automobile Abstracts (0309–0817) http://www.mira.co.uk (accessed September 3, 2010). Motor Industry Research Association (MIRA). Automobile Abstracts, Warwickshire: MIRA; monthly. Electronic journal of abstracts of technical articles from worldwide automotive literature. Covers all aspects of vehicle design and performance as well as fuel, lubricants, materials, production, and environmental aspects.

Research in Progress (RiP) http://rip.trb.org/ (accessed August 17, 2010). Database from the Transportation Research Board tracks current and recently completed government-funded transportation research projects. Projects funded by the U.S. Department of Transportation and state departments of transportation and projects performed by U.S. university research centers are included. International research projects from the ITRD database are included as well as records from the Transportation Association of Canada.

SAE Publications and Standards Database http://store.sae.org/psdb/ (accessed August 17, 2010). Database from the Society of Automotive Engineers provides access to 100,000 SAE technical papers and articles, and ground vehicle and aerospace specifications and standards, dating back to 1906. The database covers engine performance, aerodynamic design, alternative fuels, materials and manufacturing, and safety.

SafetyLit: Injury Prevention Literature Update & Archive Database http://www.safetylit.org/citations/index.php?fuseaction=citations.advancesearch (accessed September 3, 2010). Produced by San Diego State University in collaboration with the World Health Organization, it contains records for more than 100,000 scholarly articles on injury prevention and safety promotion. All users can view abstracts; full text is available to institutional licensees.

TRB Publications Index http://pubsindex.trb.org/ (accessed August 17, 2010) from the Transportation Research Board contains more than 41,000 records with abstracts for all TRB, Highway Research Board (HRB), Strategic Highway Research Program (SHRP), and Marine Board publications from 1923 until the present. All NCHRP and TCRP publications, special reports, and records are included. Each individual paper and article in the *Transportation Research Records, Conference Proceedings, Research Circulars*, and *TR News* is indexed. Records can be browsed by mode.

BIBLIOGRAPHIES AND GUIDES TO THE LITERATURE

Guides produced by the Transportation Division of the Special Libraries Association have been the most comprehensive sources available for transportation engineering. Many bibliographies have been produced for specific topics by librarians, academics, and government agencies. These are best identified through focused searches in library catalogs using appropriate subject headings and the subheading Bibliography.

The National Transportation Library (NTL) publishes a number of subject bibliographies http://ntl.bts.gov/ref/biblio/ (accessed August 17, 2010) including *Sources of Information in Transportation* (1990; 2001). Compiled by members of SLA's Transportation Division, the individual bibliographies in the *Sources* series cover general transportation, aviation, highways, urban transportation, inland waterways, intelligent transportation systems, maritime transportation, trucking, pipelines, and hazardous materials transportation. They include basic references, statistical sources, directories, periodicals, conferences, indexes, abstracts, and electronic resources. NTL staff members have produced bibliographies on more focused topics, such as congestion, transportation planning, and the transportation workforce. Some entries are annotated.

DATABASES AND DATASETS

Most countries gather, analyze, and disseminate statistics regarding transportation of people and goods, and the data are often available electronically. Some organizations and commercial publishers are also sources for statistics.

The Bureau of Transportation Statistics (BTS) http://www.bts.gov/ (accessed September 13, 2010) is a branch of the U.S. Department of Transportation (USDOT) and the Research and Innovative Technology Administration (RITA). BTS is responsible for compiling, presenting, and hosting statistical analysis and abstracts for all of the transportation modes. The main resource offered through their site is TranStats http://www.transtats.bts.gov/ (accessed September 13, 2010), the multimodal transportation database that provides datasets by mode, agency, and subject for download. Some other important publications or data sources from BTS include:

National Transportation Statistics http://www.bts.gov/publications/national_transportation_ statistics/ (accessed September 13, 2010) is an annual digest of transportation statistics nationwide. It covers several aspects of the transportation system including infrastructure, safety, energy, and the environment. Each section is available for download from 2000 to the present in a number of formats; HTML, XLS, and CSV. The entire report is available for download as a PDF.

Transportation Statistics Annual Reports http://www.bts.gov/publications/transportation_ statistics_annual_report/ (accessed September 13, 2010) provides an overview of the status of transportation in the United States with the accompanying statistical datasets. These annual reports also include recommendations for improvements to the transportation system. Topics covered include travel behavior, travel time, costs, transit, emissions, and the economy. Individual tables can be downloaded in HTML, XLS, or CSV, while the whole report is available in PDF.

National Transportation Atlas Database (NTAD) http://www.bts.gov/publications/national_ transportation_atlas_database/ (accessed September 13, 2010) provides geospatial data of transportation infrastructure and facilities in the United States. The data is made available in ESRI shape files. Editions prior to 2009 are available only on CD.

The Energy Information Administration http://www.eia.doe.gov/ (accessed September 13, 2010) in the U.S. Department of Energy (USDOE) publishes several resources covering energy production and consumption domestically and internationally. All of the following publications are available for free online from 1995 onwards.

Annual Energy Review. Annual. Washington, D.C.: U.S. Department of Energy. Energy Information Administration http://www.eia.doe.gov/aer/ (accessed September 13, 2010). This publication provides an overview of energy consumption, creation, and trade going back to 1949. Reports are broken down by energy type as well as sector, with one for transportation. Data is available for download in PDF, HTML, and XLS. The monthly energy reviews are available for download as XLS.

U.S. Department of Energy. Energy Information Administration. 2010–. *Annual Energy Outlook 2004 with Projections to 2035* http://www.eia.doe.gov/oiaf/aeo/ (accessed September 13, 2010). Presents a mid-term forecast and analysis of U.S. energy supply, demand, and prices. The 2010 edition includes additional analysis on alternative energy markets. Some accompanying datasets are available for download in XLS.

International Energy Statistics (IES) http://tonto.eia.doe.gov/cfapps/ipdbproject/ IEDIndex3.cfm (accessed September 13, 2010) is a database that examines international energy consumption and production by type, indicators, and country. Many of the tables are available for download in XLS. This source is the continuation of the *International Energy Annual* (IEA) http://www.eia.doe.gov/iea/contents.html (accessed September 13, 2010).

The primary resource for transportation-related energy statistics is the *Transportation Energy Data Book,* http://cta.ornl.gov/data/index.shtml (accessed September 13, 2010), produced by Oak Ridge National Laboratory as part of the Center for Transportation Analysis, which is sponsored by USDOT and USDOE. An annual compilation of statistics from a number of other sources produced by several different agencies, it contains data on petroleum and energy, emissions, and vehicles. All data tables are available for download in XLS, and the whole document is available as a PDF.

The Office of Highway Policy Information http://www.fhwa.dot.gov/policyinformation (accessed August 12, 2010) in the Federal Highway Administration (FHWA), U.S. Department of Transportation, serves as a repository of highway information and statistics for the agency. The primary resource they offer is the *Highway Statistics Series* http://www.fhwa.dot.gov/policy/ohpi/ hss/index.cfm (accessed August 12, 2010). The series compiles statistics from state, federal, and local agencies, covering topics such as highway infrastructure; travel; driver's licenses and motor vehicle registrations; motor fuel; bridges; revenue, debt, and expenditures; safety; and performance. Full versions of the tables are available from 1999 to present. Archives of the series are available for download in PDF going back to 1945.

Crain Communications' Automotive News Data Center http://www.autonews.com/section/ datacenter (accessed September 14, 2010) provides licensed subscribers with detailed data on production, pricing, sales, and inventory. It also contains vehicle specifications and car cutaways. Tables can be downloaded in XLS or PDF.

European Transport Statistics http://ec.europa.eu/transport/publications/statistics/statistics_ en.htm (accessed September 14, 2010), sponsored by the European Commission, provides transportation statistics by mode for all countries in Europe. Statistics are broken down by performance of infrastructure, mode, and passenger and freight transportation. Tables are available for download in XLS.

The Federal Transit Administration's National Transit Database (NTD) http://www.ntdpro-gram.gov/ntdprogram/ (accessed September 14, 2010) is a primary source for statistics on public transportation in the United States. The NTD site has an area for transit agencies to report their statistics regularly and another for the public to access and download the aggregated tables. Self-extracting XLS files are available to download for the annual, monthly, and safety data. Information about specific agencies can be accessed through the Annual Transit Profiles (Section 15), which provides data points on service, financial information, and performance indicators. The National Transit Summaries and Trends aggregates transit operating statistics by mode, and provides some analysis of changes over the years. These are available for download in PDF from 1998.

The International Road Federation http://www.irfnet.org (accessed September 14, 2010) publishes *World Road Statistics*. The 2009 edition has worldwide data for 2002 to 2007 on roads and vehicles, and presents data by country for road networks, traffic volumes, motor vehicle production, vehicle export/import, vehicle registration, fuels, safety, taxation, and expenditures. Published in hard copy, on CD with XLS files, and in PDF format.

The National Highway Traffic Safety Administration's Fatality Analysis Reporting System (FARS) http://www-fars.nhtsa.dot.gov (accessed September 15, 2010) provides access to data of vehicle crashes on public roadways in the United States that involve fatalities. Queries can be run comparing a number of crash factors, such as the environment, participants, and vehicles. Results can be exported as TXT files.

National Household Travel Survey (NHTS) http://nhts.ornl.gov/index.shtml/ (accessed September 13, 2010) contains survey data about personal travel behavior, such as trip length, type of vehicle, and time, among others. The 2009 NHTS was launched with a new analysis tool that makes it easier for people to create their own tables. The online table designer is available for the 2009, 2001, and 1995 surveys. Frequently used tables are available for immediate download in HTML or XLS. Researchers can access more complete tables after registering with a log in. Earlier versions of the survey are available for download.

The National Safety Council http://www.nsc.org/ (accessed September 14, 2010) compiles and publishes statistics and reports, including *Injury Facts* that compiles annual data on fatal and nonfatal unintentional injuries, including those on streets and highways. Published since 1927, it was previously known as *Accident Facts*. Available in hard copy and on CD.

Railroad Performance Measures from the Association of American Railroads http://www.railroadpm.org/ (accessed September 14, 2010) provides weekly data from the seven major (Class I) U.S. freight railroads.

United Nations Economic Commission for Europe (UNECE) Transport Statistics http://www.unece.org/trans/main/wp6/transstatpub.html (accessed September 14, 2010) collects statistics of transportation indicators, such as investment and number of employees, for different modes within the UNECE region. These compilations are available for download in PDF. The statistical overview *Road Traffic Accidents* enables easy comparison in traffic safety between Europe and North America. These tables are available in XLS.

Ward's Communications http://wardsauto.com/keydata/ (accessed September 14, 2010) compiles extensive statistics on the global and U.S. automobile industries. Licensed subscribers can access statistics including detailed data on car and light truck platforms, dealers, vehicle production, sales, registrations, marketing, and vehicle specifications. Tables can be downloaded in XLS or PDF.

DIRECTORIES

Print directories are no longer the essential tools they were just a few years ago. Many professional societies and organizations have membership directories available online, as do most academic and research institutions and government agencies. A search engine, such as Google, can yield information on organizations and individuals.

Online and print directories still perform a useful function, particularly those that compile information from disparate sources and those that point to vendors and equipment manufacturers; some noteworthy ones include:

ATA's Truck Fleet Directory: Profiles of For-Hire and Private Fleets in the U.S. (updated fourth quarter 2009) Alexandria, VA: American Trucking Associations. One of the few sources of truck fleet information, this directory on CD lists more than 50,000 fleets organized by state. Fleets can be searched by state, size, type (for-hire, private, etc.), commodity type, trailer type, zip code, area code, and SIC code.

Containerisation International Yearbook. Annual. London: Informa U.K. Ltd. This print yearbook contains port and terminal facilities details, equipment manufacturer and service provider information, a vessel register, and entries for industry associations and shipbrokers.

Jane's Information Group http://catalog.janes.com/catalog/public/html/transport.html (accessed August 12, 2010) publishes a number of authoritative annual directories for the transportation industry; all are available in print and electronic versions. Each title includes a directory of manufacturers. Those more important to transportation engineering are the following:

> *High-Speed Marine Transportation* provides detailed specifications for all types of high speed marine craft including component and propulsion systems.
>
> *Marine Propulsion* provides detailed specifications and images of transmissions, propellers, and related systems for all diesel and gas turbine engine and propulsion systems.
>
> *Urban Transport Systems* profiles transportation systems in more than 400 cities worldwide and gives statistics on fleet size, ridership, and route length. Other operational statistics may be provided; some entries include route maps.
>
> *World Railways* covers railway systems in 140 countries and includes system details on organization, finance, operations, staff, planned developments, and rolling stock.

Library and Information Center Directory http://www.transportationresearch.gov/NTL/NTKN/LibraryDirectory2/Forms/Default.aspx (accessed August 12, 2010). Hosted by Bureau of Transportation Statistics, U.S. Department of Transportation. This international directory of more than 350 government, academic, association, and corporate transportation libraries and information centers provides details on collections, services, access, network affiliations, and staff.

Select Transportation Associations, Societies, and Unions http://ntl.bts.gov/associations.html (accessed August 17, 2010). Compiled by the National Transportation Library, this alphabetical directory lists more than 200 organizations and links to their Web sites.

DICTIONARIES AND GLOSSARIES

Manuals and handbooks often include glossaries specific to some part of transportation engineering. Notable transportation dictionaries include:

American Association of State Highway and Transportation Organizations. 2009. *AASHTO Transportation Glossary,* 4th ed. Washington, D.C.: AASHTO. Focus is on highways and bridges, but also defines terms for air, nonmotorized, public and water transportation.

Cavinato, J. L. 2000. *Supply Chain and Transportation Dictionary,* 4th ed. Norwell, MA: Kluwer Academic. Covers terms in supply chain management, transportation, distribution, logistics, material, and purchasing.

Dictionary http://www.bts.gov/dictionary/index.xml (accessed August 17, 2010). Bureau of Transportation Statistics, U.S. Department of Transportation. This online dictionary defines more than 6,000 terms. Definitions were obtained from government agencies, associations, and private organizations.

Dinkel, J. 2000. *Road and Track Illustrated Automotive Dictionary.* Cambridge, MA: Robert Bentley Publishers. Defines basic items as well as complex systems. Many illustrations.

Glossary of Transportation Terms http://web1.ctaa.org/webmodules/webarticles/articlefiles/TransportationGlossary.pdf (accessed August 17, 2010), Community Transportation Association. Provides definitions and context for about 100 terms, agencies, and programs related to public transportation.

Goodsell, D. 1995. *Dictionary of Automotive Engineering*. Warrendale, PA: Society of Automotive Engineers. Defines more than 3,000 terms used worldwide.

National Transit Database Glossary http://www.ntdprogram.gov/ntdprogram/Glossary.htm (accessed September 15, 2010). Federal Transit Administration. Defines hundreds of terms used in a public transit context. Notes when terms are used in NTD annual reporting modules.

Railway Age's Comprehensive Railroad Dictionary, 2nd ed. 2002. Omaha, NE: Simmons-Boardman Books. Defines more than 4,000 terms and includes 240 illustrations.

Terminology http://termino.piarc.org/search.php (accessed August 17, 2010). World Road Association/Association Mondiale de la Route. This online resource provides access to terms on road and traffic engineering from eight sources, including the PIARC Lexicon, which contains a large number of terms and expressions used in road technology and related fields. Users can search in one language and specify results in one or more other languages. Equivalent terms in other languages are identified; some definitions are given. *PIARC Technical Dictionary of Road Terms* (8th ed., 2007) is available in print.

HANDBOOKS AND MANUALS

Handbooks and manuals are essential components of the transportation engineering literature. Facts, formulas, and basic information for specific aspects of transportation engineering are compiled in a manner that facilitates quick access. Most handbooks and manuals are limited to one mode of transportation. Listed below are titles grouped by mode.

GENERAL

Hall, R. W., ed. 2003. *Handbook of Transportation Science,* 2nd ed. Boston, MA: Kluwer Academic. Print and electronic. Focuses on the properties and characteristics common to all modes of transportation. Organized by subject, chapters written by different authors address several broad topics: discrete choice, travel demand, and vehicle operation; flows and congestion, including traffic control and system interactions; spatial models used in network analysis and design and network assignment; and routing and network models. Extensive use of tables, figures, and mathematical models; references at the end of each chapter. The index and table of contents lack detail.

Myer Kutz, M., ed. 2004. *Handbook of Transportation Engineering*. New York: McGraw-Hill. Print and electronic. Provides design techniques, examples of applications, and guidelines. The 38 chapters are organized in five parts: networks and systems; traffic, streets, and highways; safety, noise, and air quality; nonautomobile transportation; and operations and economics.

HIGHWAYS AND TRAFFIC

The American Association of State Highway and Transportation Officials (AASHTO) is a primary publisher of highway and traffic engineering manuals. Most publications are available in print and electronic versions. Among the most-used titles are:

- *A Policy on Geometric Design of Highways and Streets (2004)* (5th ed.). Primary Resource. Known as the Green Book, this is the standard guide for engineers designing the physical layout and dimensions of streets and roads. Cover design controls and criteria; the elements of design; cross-section elements; intersections; and grade separations and interchanges. Local roads and streets, collectors, arterials, and freeways are each addressed in separate chapters. Available in print and electronic versions.

- *AASHTO Maintenance Manual for Roadways and Bridges* (4th ed.) (2007). Provides basic information regarding the processes, methods, and materials used in maintaining roadways and bridges.
- *Guide for Design of Pavement Structures* (1993) and Supplement (1998). Covers the range of design from construction of new pavements to rehabilitation and reconstruction of existing pavements. Includes alternative design procedures.
- *Highway Safety Manual* (2010). Assists practitioners in identifying countermeasures, setting project priorities, comparing alternative treatments, and quantifying and predicting the safety performance of roadway elements. It includes tools for analyzing the effects of measures on highway safety.
- *Hot Mix Asphalt Paving Handbook* (2000) (2nd ed.). Covers paving, plant operations, materials transportation, surface preparation, and quality control.
- *Manual for Bridge Evaluation* (2008) with Interim Revisions (2010). Describes how to determine the physical condition, maintenance needs, and load capacity of highway bridges in the U.S. Provides inspection procedures and load rating practices that meet the National Bridge Inspection Standards (NBIS). Ratings examples are included.
- *Mechanistic-Empirical Pavement Design Guide, Interim Edition: A Manual of Practice* (2008). This design and analysis tool for evaluating pavement structures uses realistic principles to enable engineers to create more reliable pavement designs. Project specific traffic, climate, and materials data for estimating damage accumulation over a specified pavement service life characterize the mechanistic-empirical approach. Applicable to designs for new, reconstructed, and rehabilitated flexible, rigid, and semi-rigid pavements.
- *Model Drainage Manual* (2005). This CD manual compiles design policies and procedures for agencies to use in developing customized manuals.
- *Roadside Design Guide* (2002) (3rd ed.) with Chapter Update (2006). Compiles current information on safety treatments aimed at minimizing injury severity when vehicles leave the roadway.

American Concrete Institute (2004) *ACI Manual of Concrete Practice*, Farmington Hills, MI: The Institute. Contains concrete and masonry codes, specifications, requirements, guides, and reports used in design of structures and pavements and for maintenance and rehabilitation. Available as six-volume set or on CD.

Brockenbrough, R.L. (2009) *Highway Engineering Handbook* (3rd ed.), London: McGraw-Hill. Aimed at practitioners in highway design and construction. Begins with environmental issues and the regulatory framework, and moves on to highway location and geometric and cross-section design. Pavement design incorporates AASHTO specifications and covers preventive maintenance, pavement management, and rehabilitation. Bridge engineering covers bridge types, materials, deck design, and construction, corrosion, and coatings. Some chapters include references.

Chen, W-F. and Duan, L. (ed.) (2003) *Bridge Engineering*, Boca Raton, FL: CRC Press. Covers all major areas of bridge design, construction, maintenance and inspection. Discusses superstructures, substructures, and seismic design. Includes all types of bridges: concrete, steel, box, truss, and suspension. Contains many tables and formulas; each chapter has references.

Currin, T.R. (2001) *Introduction to Traffic Engineering: A Manual for Data Collection and Analysis*, Pacific Grove, CA: Brooks/Cole. Describes step-by-step techniques for observing and analyzing spot speeds, turning movements, saturation flow rates, control delay, parking, trip generation, and platoon ratio. Provides sample data collection and data analysis forms. An appendix includes additional information on statistical analysis.

Fwa, T.F. (ed.) (2006) *Handbook of Highway Engineering*, Boca Raton, FL: CRC Press. Print and electronic. Addresses socioeconomic and environmental aspects as well as functional requirements of highway planning and development; functional and structural design; and construction, maintenance and management. Extensive references are provided for each of 22 individual chapters.

Guide for the Geometric Design of Driveways <http://onlinepubs.trb.org/onlinepubs/nchrp/nchrp_rpt_659.pdf> (accessed August 30, 2010), Washington, DC: Transportation Research Board. NCHRP Report 659. Contains guidelines for driveways which complement information in the AASHTO Green Book.

Highway Capacity Manual (HCM) (2000) Washington, DC: Transportation Research Board. Primary Resource. The standard reference manual for estimating the capacity and determining service levels of highways and other transportation facilities, HCM covers highways and other roadways, intersections, transit facilities, and facilities for cyclists and pedestrians. The thirty-one chapters are organized into five parts: overview; concepts; methodologies; corridor and area-wide analyses; and simulations and other models. The CD version includes tutorials, example problems and video clips. The Highway Capacity Manual Applications Guidebook from TRB (2003) contains case studies illustrating applications of the HCM to analysis of real-world traffic situations.

The Institute of Transportation Engineers publishes a number of traffic engineering handbooks and manuals. Among the more important are:

- *Trip Generation* (2008) (8th ed.), 3 vols. Primary Resource. Provides statistics used in calculating the forecasted trip generation rate for a wide variety of land uses, from housing to schools to multi-use developments. Data is based on more than 4,000 trip generation studies, but caution is urged when dealing with data from small sample sizes. Many municipalities specify use of ITE's figures in their planning processes.
- *Highway Safety Manual* (2010), 3 vols. Provides tools to enable practitioners to select countermeasures, set project priorities, compare alternative treatments, and quantify and predict the safety performance of road elements.
- *Parking Generation: An ITE Informational Report* (2010) (4th ed.). Contains land use descriptions, parking generation rates, equations and data plots; includes updated peak demand rates and quantitative and qualitative information on the influence of numerous factors on parking demand rates.
- *Traffic Control Devices Handbook* (2001). Augments the Manual of Uniform Traffic Control Devices for Streets and Highways (MUTCD). Provides basic information and criteria on traffic control devices to enable small jurisdictions to handle MUTCD requirements. Includes numerous tables and figures and some chapters have extensive references.
- *Traffic Engineering Handbook* (2008) (6th ed.). Provides practitioners with an overview on topics including users, vehicles, safety, traffic flow, operations, traffic control, parking, signs and markings, communications and maintenance.
- *Traffic Signal Maintenance Handbook* (2010). Outlines the impact of the management, installation and operations of traffic control systems on the maintenance of these systems. Describes cost-effective approaches and applications involving traffic signal systems and Intelligent Transportation System (ITS) technologies.
- *Traffic Signal Timing Manual* (2008). Provide a comprehensive collection of traffic signal timing concepts, analytical procedures and applications.
- *Trip Generation Handbook: An ITE Recommended Practice* (2001) (2nd edn). Provides assistance in choosing independent variables and time periods for analysis and methods for conducting studies. Also useful when estimating trip generation for multi-use developments.

Lay, M.G. (2009) *Handbook of Road Technology* (4th ed.) (2 vols), Melbourne: Gordon & Breach. Provides background and practical information on roadway planning, pavements, traffic, and transport. Written from an Australian perspective.

National ITS Architecture <http://www.iteris.com/itsarch/> (accessed August 18, 2010) U.S. Department of Transportation. The National ITS Architecture provides a common framework for planning, defining, and integrating intelligent transportation systems. This site provides access to all architecture documents. Describes functions, subsystems and data flow. Key Concepts of the National ITS Architecture <http://www.iteris. com/itsarch/index.htm> (accessed August 18, 2010) serves as background.

Parke, G., Hewson, N., *ICE Manual of Bridge Engineering* (2nd ed.) (2008) London: Thomas Telford. Presents a comprehensive overview of concept, analysis, design, construction, and maintenance of many types of bridges from a British perspective. Chapters on load distribution, structural analysis, design of specific types of bridges, technical advancements, substructures, maintenance, and monitoring are written by a variety of experts. Makes extensive use of figures and photographs; most chapters have bibliographies.

Roundabouts in the United States(2007) <http://onlinepubs.trb.org/onlinepubs/nchrp/ nchrp_rpt_572.pdf> (accessed August 30, 2010) Washington, DC: Transportation Research Board. NCHRP Report 572. Presents methods of estimating the safety and operational impacts of roundabouts. Provides updated design criteria for roundabouts.

Traffic Signs Manual http://www.dft.gov.uk/pgr/roads/tss/tsmanual/ (accessed August 18, 2010) U.K. Department for Transport. Provides guidance on the use of traffic signs and road markings as prescribed by the U.K. Traffic Signs Regulations.

U.S. Federal Highway Administration (2009) Manual on Uniform Traffic Control Devices (MUTCD) http://mutcd.fhwa.dot.gov/ (accessed September 15, 2010), Washington, DC: The Administration. Primary resource. Defines the standards used in the U.S. for the installation and maintenance of traffic control devices on all streets and highways. MUTCD covers signs, markings, signals, temporary traffic control, and highway-rail and light transit grade crossings.

U.S. Federal Highway Administration (2004) Standard Highway Signs, Washington, DC: The Administration. Contains detailed drawings of standard highway signs (regulatory, warning, guide, emergency management, school) and pavement markings for use by agencies.

In the U.S., each state transportation department publishes manuals and other information related to highway and pavement engineering with specifications and regulations applicable in that particular state. FHWA provides links <http://www.fhwa.dot.gov/webstate.htm> (accessed August 18, 2010) to all state transportation websites.

Nonmotorized Transportation

American Association of State Highway and Transportation Officials. 1999. *Guide for the Development of Bicycle Facilities,* 3rd ed. Washington, D.C.: The Association. Available in print and on CD. Provides information on planning, design, construction, and maintenance of facilities to enhance safe bicycle travel.

American Association of State Highway and Transportation Officials. 2004. *AASHTO Guide for the Planning, Design and Operation of Pedestrian Facilities.* Washington, D.C.: The Association. Describes effective measures for providing pedestrian facilities on public rights-of-way along streets and highways. Also examines the connection between pedestrian mobility and site design.

Bicycle Countermeasures Selection System (BIKESAFE) http://www.bicyclinginfo.org/ bikesafe (accessed August 18, 2010). Presents information on bicycle accident factors and analysis, and selecting facility improvements. Provides tools for selecting appropriate countermeasures or treatments; includes 50 case studies.

Institute of Transportation Engineers. 1998. *Design and Safety of Pedestrian Facilities: A Recommended Practice.* Washington, D.C.: The Institute. Provides guidelines for designs to allow pedestrians safe and efficient opportunities to walk near streets and highways.

Institute of Transportation Engineers. 2010. *Designing Walkable Urban Thoroughfares: A Context Sensitive Approach*. Washington, D.C.: The Institute. Provides guidance and demonstrates how to support walkable, connected neighborhoods, mixed land uses, and easy access for pedestrians and bicyclists. Available in hard copy and in PDF format.

Pedestrian and Bicycle Information Center Case Study Compendium http://katana.hsrc.unc. edu/cms/downloads/COMPENDIUM.web.8-29-2008.pdf (accessed August 30, 2010). Contains brief case studies of successful pedestrian and bicycle projects in the United States and other countries; includes section on engineering.

Pedestrian Road Safety Audit Guidelines and Prompt Lists http://drusilla.hsrc.unc.edu/ cms/downloads/PedRSA.reduced.pdf (accessed August 18, 2010). U.S. Federal Highway Administration. Outlines basic concepts used in conducting pedestrian road safety audits. The guidelines provide detailed descriptions of potential pedestrian safety issues.

Roadway and Pedestrian Facility Design http://www.walkinginfo.org/engineering/roadway.cfm (accessed August 30, 2010). Pedestrian and Bicycle Information Center. Examines elements of roadway design and operations important to pedestrians. Provides detailed information on topics such as bicycle lanes, traffic lane reduction, roundabouts, sidewalks and curb ramps.

Technical Guide for Conducting Pedestrian Safety Assessments in California http://www. techtransfer.berkeley.edu/tse/psa_handbook.pdf (accessed September 3, 2010). University of California, Berkeley: Institute of Transportation Studies, Technology Transfer Program. Provides detailed information on conducting assessments of pedestrian safety and accessibility at existing or future roadways and facilities.

PIPELINES

Ellenberger, P. 2010. *Piping and Pipeline Calculations Manual: Construction, Design Fabrication and Examination*. Burlington, MA; Oxford, U.K.: Butterworth-Heinemann. Provides a quick reference guide to codes and standards and gives practical advice on compliance. Covers codes from ASME and API.

McAllister, E. W. 2009. *Pipeline Rules of Thumb Handbook: A Manual of Quick, Accurate Solutions to Everyday Pipeline Engineering Problems,* 7th ed. Boston, MA: Elsevier Gulf Professional Publishing. Provides practical techniques and quick methods for pipeline design, engineering, and construction.

PUBLIC TRANSPORTATION

Public Transportation Standard Bus Procurement Guidelines RFP. 2010. http://www.apta. com/resources/reportsandpublications/Documents/APTA_Bus_Procurement_Guidelines. doc (accessed August 18, 2010). American Public Transportation Association. Serves as a template for agencies requesting proposals for negotiated bus procurement contracts when vehicles incorporate new technology.

Public Transportation Security Series (TCRP 86). 2005–2006. http://www.trb.org/ Publications/Public/PubsTCRPProjectReportsAll.aspx (accessed August 17, 2010). Transportation Research Board. Transit Cooperative Research Program. This 13-volume series assembles information pertaining to specific transit security issues. Topics include emergency operations, continuity of operations, tunnel safety, passenger security inspections, ferries, and guidelines for emergency training exercises.

Transit Capacity and Quality of Service Manual, 2nd ed. 2006. http://www.trb.org/ Publications/Blurbs/Transit_Capacity_and_Quality_of_Service_Manual_2nd_153590. aspx (accessed August 18, 2010). Transportation Research Board. Transit Cooperative

Research Program. Provides a framework for measuring transit availability and quality of service from the riders' perspective. Provides tools for calculating the capacity of bus, ferry, rail and transit services as well as the capacity of stations, terminals, and stops.

RAILROADS

Three complementary annual publications from the American Railway Engineering and Maintenance Association (AREMA) serve as primary railroad engineering reference materials:

Communications and Signals Manual of Recommended Practice. Contains recommendations and instructions for all aspects of railroad communications and signals.

Manual for Railway Engineering. Compiles engineering reference material in four volumes: Track, Structures, Infrastructure, and Passenger and Systems Management. Thousands of pages document design, specifications, definitions and procedures. Includes a detailed index.

Portfolio of Trackwork Plans. Contains plans and specifications for switches, frogs, turnouts and crossovers, crossings, and rail and special trackwork.

American Railway Engineering and Maintenance Association. 2005. *Practical Guide to Railway Engineering*. Landover, MD: AREMA. Book includes full-text CD. Provides an overview of railway engineering history and details on how a railroad operates, right-of-way management issues, track design and layout, and railway bridge design concepts; includes a primer on signals and communication systems. Includes chapters and course modules on basic track, drainage, environmental permitting, structures, electric traction, and passenger transit.

Iwnicki, S., ed. 2006. *Handbook of Railway Vehicle Dynamics*. Boca Raton, FL: CRC Press. Print and electronic. Covers the factors that influence the dynamic behavior of railway vehicles: wheel–rail interface, suspension and suspension component design, simulation and testing of systems, interaction with the surrounding infrastructure, and noise generation.

Unsworth, J. F. 2010. *Design of Modern Steel Railway Bridges*. Boca Raton, FL: CRC Press. Print and electronic. Complements recommended practices from AREMA regarding design, maintenance, and rehabilitation of steel superstructures.

VEHICLES

Automotive Handbook, 7th ed. 2007. Stuttgart: Bosch. Provides facts and figures for passenger and commercial vehicles. Covers processes, systems, and techniques in automotive technology.

Garrett, T. K. 2001. *The Motor Vehicle*, 13th ed. Warrendale, PA: SAE International. Provides current information on vehicle technology with sections on engines, transmissions, and the carriage unit.

Gillespie, T. D. 1992. *Fundamentals of Vehicle Dynamics*. Warrendale, PA: Society of Automotive Engineers. Explains the performance of an automotive vehicle, with chapters focusing on acceleration performance, braking performance, aerodynamics and rolling resistance, ride, tires, steady-state cornering, suspensions, steering systems, and roll-over.

Robert Bosch GmbH. 2008. *Bosch Automotive Electrics, Automotive Electronics Handbook*, 5th ed. Plochingen, Germany: Bentley Publishers. Covers the topics of electrical and electronic systems in motor vehicles with a particular emphasis on networking.

Wech, L. and U. W. Seiffert. 2007. *Automotive Safety Handbook*, 2nd ed. Warrendale, PA: SAE International. Covers both active and passive safety systems and addresses accident avoidance, occupant protection, biomechanics, and the interrelationships among the occupant, the vehicle, and the restraint system.

MONOGRAPHS AND TEXTBOOKS

Monographs and textbooks provide in-depth discussion of design, control mechanisms, operations, and other aspects of transportation.

How to Develop a Pedestrian Safety Action Plan http://safety.fhwa.dot.gov/ped_bike/ ped_focus/docs/fhwasa0512.pdf (accessed September 3, 2010). U.S. Department of Transportation, Federal Highway Administration. Serves as a reference for pedestrian safety design through safety treatments such as street redesign and countermeasures.

Planning Complete Streets for an Aging America http://assets.aarp.org/rgcenter/ppi/ liv-com/2009-12-streets.pdf (accessed September 3, 2010). AARP. This report outlines actions to incorporate the needs of aging motorists and pedestrians in the design of streets and other transportation facilities. Includes policy, planning, and design recommendations.

GENERAL

Banks, J. H. 2002. *Introduction to Transportation Engineering,* 2nd ed. Boston, MA: McGraw-Hill. Textbook. Print and electronic. Focuses on highway and road design and traffic flow, control, and capacity. Addresses social, political, and economic contexts including environmental impacts, travel demand, planning, and project evaluation.

Fricker, J. D. and R. K. Whitford. 2004. *Fundamentals of Transportation Engineering: A Multimodal Systems Approach.* Upper Saddle River, NJ: Pearson Prentice Hall. This textbook begins with an examination of traffic flow and highway design and moves into planning, safety, pavement design, and freight movement. Makes extensive use of scenarios, figures, illustrations, and includes exercises, references, and glossaries for each chapter.

Papacostas, C. S. and P. D. Prevedouros. 2001. *Transportation Engineering and Planning*, 3rd ed. Upper Saddle River, NJ: Prentice Hall. This undergraduate textbook addresses design and operations; transportation systems, including modes, planning, and forecasting; transportation impacts; and supporting elements, including queuing and simulation. Includes exercises.

Sussman, J. 2004. *Introduction to Transportation Systems.* Boston, MA: Artech House. Provides a graduate-level introduction to transportation systems including system components, freight and passenger transportation, intelligent transportation systems, and modeling.

FREIGHT

Ioannou, P. A. 2008. *Intelligent Freight Transportation.* Boca Raton, FL: CRC Press. Print and electronic. Seventeen chapters address different aspects of intelligent transportation systems applied to freight. Topics include automation, modeling, port choice and competition, inland ports, optimization, and labor and environmental issues.

Jones, E. C. and C. A. Chung. 2008. *RFID in Logistics: A Practical Introduction.* Boca Raton, FL: CRC Press. Print and electronic. Provides a structured description of RFID and how it is used in the supply chain.

Muller, G. 1999. *Intermodal Freight Transportation,* 4th ed. Washington, D.C.: Eno Transportation Foundation. Provides comprehensive overview and discusses terminal operations, transport equipment, containerization, and technology trends.

Murphy, P. R., and D. F. Wood. 2010. *Contemporary Logistics,* 10th ed. Upper Saddle River, NJ: Prentice Hall. Print and electronic. Provides overview of logistics, elements of logistics systems, materials handling, transportation management, inventory, warehousing and supply management, and implementing a logistics system. Stresses transportation as a critical element in the physical system.

Highways and Traffic

Boyce, P. R. 2009. *Lighting for Driving: Roads, Vehicles, Signs, and Signals.* Boca Raton, FL: CRC Press. Print and electronic. Examines standards and technologies for lighting for driving.

Castro, C. and T. Horberry., eds. 2004. *Human Factors of Transport Signs.* Boca Raton, FL: CRC Press. Print and electronic. Examines sign research and new technologies for improving signaling.

Daganzo, C. F. 1997. *Fundamentals of Transportation and Traffic Operations.* New York: Elsevier Science. Introduces the basics of traffic operations, and covers tools and their applications.

Daganzo, C. F. 2010. *Logistic Systems Analysis, 4th ed.* New York: Springer Berlin Heidelberg. Print and electronic. Takes an integrated approach to logistics systems and examines operation and organization. Focuses on finding reasonable, rational solutions. Extensive references.

Dagleish, M. and N. Hoose. 2008. *Highway Traffic Monitoring and Data Quality.* Boston: Artech House. Describes vehicle, traffic, and environmental sensors and applications for capturing data, such as vehicle weight, size, speed, license number, and travel time. Provides statistical techniques for data analysis and quality control.

Garber, N. J. and L. A. Hoel. 2008. *Traffic and Highway Engineering, 4th ed.* Toronto, Ontario: CL-Engineering. Print and electronic. Presents material at an introductory level. Covers standard topics of traffic engineering, safety, traffic flow, intersections, capacity, design, drainage, materials, and pavements.

Ghosh, S. and T. S. Lee. 2010. *Intelligent Transportation Systems: Smart and Green Infrastructure Design,* 2nd ed. Boca Raton, FL: CRC Press. Print and electronic. Reviews developments and implementations of ITS worldwide and examines challenges to its adoption.

Homburger, W. S., J. W. Hall, W. R. Reilly, and E. C. Sullivan. 2007. *Fundamentals of Traffic Engineering,* 16th ed. Berkeley, CA: Institute of Transportation Studies, University of California. Serves as an introduction to the field of traffic engineering; designed for use in a one-week course for practicing professionals, but appropriate for university engineering courses. Addresses traffic studies, capacity, demand, traffic control devices, geometric design, control, operations, and traffic management. Extensive references.

Mannering, F. L., S. S. Wahburn, and W. P. Kilareski. 2008. *Principles of Highway Engineering and Traffic Analysis,* 4th ed. New York: John Wiley & Sons. Print and electronic. Covers fundamental issues in highway engineering and traffic analysis. Makes extensive use of examples and problems.

Wright, P. H. and K. Dixon. 2004. *Highway Engineering,* 7th ed. New York: John Wiley & Sons. Textbook covering the design of roadways and related facilities. Addresses traffic engineering, materials, drainage, pavements, and maintenance.

Nonmotorized Transportation

Kachroo, P. 2009. *Pedestrian Dynamics: Mathematical Theory and Evacuation Control.* Boca Raton, FL: CRC Press. Print and electronic. Derives pedestrian traffic models for multidirectional flow.

Pipelines

Antaki, G. A. 2003. *Piping and Pipeline Engineering: Design, Construction, Maintenance, Integrity, and Repair.* Boca Raton, FL: CRC Press. Print and electronic. Covers technical principles in materials, design, construction, inspection, testing, and maintenance. Includes codes and standards, design analysis, welding and inspection, corrosion mechanisms, failure analysis, and valve selection and application.

Hiltscher, G., W. Mühlthaler, and J. Smits. 2007. *Industrial Pigging Technology: Fundamentals, Components, Applications.* Weinheim, Germany: Wiley-VCH. Print and electronic. Describes equipment used for planning and designing pigging units and includes many practical examples.

Kennedy, J. L. 1993. *Oil and Gas Pipeline Fundamentals,* 2nd ed. Houston: Gulf Publishing. Provides basic information on types of pipelines, design, construction practices, welding techniques and equipment, operations, maintenance, repair, inspection, and safety.

Larock, B. E., R. W. Jeppson, and G. Z. Watters. 1999. *Hydraulics of Pipeline Systems.* Boca Raton, FL: CRC Press. Provides comprehensive treatment of pipeline hydraulics. Includes CD with software for performing complex computations.

Liu, H. 2003. *Pipeline Engineering.* Boca Raton, FL: CRC Press. Print and electronic. Covers the essentials of pipeline engineering. Presents equations used in analyzing pipe flow. Describes nonfluid-mechanics topics such as materials, sensors, and computer controls.

Mohitpour, M., A. Murray, and H. Golshan. 2007. *Pipeline Design and Construction,* 3rd ed. New York: ASME International. Contains practical techniques for the design, construction, commissioning, and assessment of pipelines and related facilities.

Najafi, M. 2004. *Trenchless Technology: Pipeline and Utility Design, Construction, and Renewal.* New York: McGraw-Hill. Covers design guidelines, construction, environmental considerations, and equipment, methods, and materials for minimal surface disruption through the use of trenchless technologies.

Smith, P. 2005. *Piping Materials Guide.* New York: Elsevier Gulf Professional Publishing. Print and electronic. Covers the selection of piping materials, materials and fitting, troubleshooting techniques for corrosion control, and inspections.

RAILROADS

Hay, W. H. 1982. *Railroad Engineering.* New York: John Wiley & Sons. Classic textbook that covers the fundamentals for locating, constructing, operating, and maintaining a railroad. Includes problem examples and an index.

Shabana, A. A., K. E. Zaazaa, and H. Sugiyama. 2007. *Railroad Vehicle Dynamics: A Computational Approach.* Boca Raton, FL: CRC Press. Print and electronic. Provides guidance on developing methods to analyze vibration and stability.

Wickens, A. H. 2003. *Fundamentals of Rail Vehicle Dynamics.* Lisse, The Netherlands: Swets & Zeitlinger. Print and electronic. Focuses on the problem of guidance and stability and its resolution through proper design of the suspension connecting the wheels and car body. Presents models for exploring new suspension and vehicle designs.

VEHICLES

Bishop, R. 2005. *Intelligent Vehicle Technology and Trends.* Norwood, MA: Artech House. Print and electronic. Provides overview of sensing and control systems.

Stone, R. and J. K. Ball. 2004. *Automotive Engineering Fundamentals.* Warrendale: SAE International. Covers the principles involved in designing a vehicle; includes engines, transmissions, steering systems, brakes, tires, and aerodynamics.

JOURNALS

Journal literature is an essential form of information in transportation engineering. While a number of publications address the discipline as a whole, most sources address only one mode. Most of the journals described below are peer-reviewed; a few trade publications have been included. Unless noted otherwise, journals are available in electronic format, but access may require a subscription or license agreement.

GENERAL

Accident Analysis and Prevention (0001–4575) www.elsevier.com/locate/aap (accessed September 15, 2010). Elsevier (bimonthly). Scope: medical, legal, economic, educational, behavioral, theoretical, or empirical aspects of transportation accidents, including pre-injury and post-injury phases.

Journal of the Transportation Research Forum (1046–1469) http://www.trforum.org/journal/ (accessed September 1, 2010). Fargo, ND: Upper Great Plains Transportation Institute; biannual. Publishes papers presented at TRF Annual Forum.

Journal of Transportation Engineering (0733–947X) http://www.asce.org/Journal.aspx?id=2147486764. (accessed September 1, 2010). American Society of Civil Engineers; monthly. Contains technical articles on planning, design, construction, maintenance, and operation of air, highway, and urban transportation as well as pipeline facilities.

Transportation Journal (0041–1612) http://www.astl.org/i4a/pages/index.cfm?pageid=3288 (accessed September 1, 2010). American Society of Transportation and Logistics; quarterly. Covers the supply chain and logistics management and aims for a balance between practical and scholarly information.

Transportation Research Record: Journal of the Transportation Research Board (0361–1981) http://www.trb.org/Finance/Public/TRRJournalOnline1.aspx (accessed September 1, 2010). Transportation Research Board. Primary Resource. The official publication of TRB, the Record contains peer-reviewed technical papers presented at TRB conferences on all aspects and modes of transportation. Each record contains multiple papers focused on a single topic, such as pavement maintenance, highway capacity, rail–highway crossings, bridges, or soil mechanics.

Transportation Research series http://www.elsevier.com/wps/find/journaldescription.cws_home/601350/relatedpublications#relatedpublications (accessed September 1, 2010). Published in six parts. Elsevier. Primary resource.

Transportation Research, Part A: Policy and Practice (0965–8564). Scope: General interest papers on passenger and freight covering all aspects of transportation from various disciplines, including engineering.

Transportation Research, Part B: Methodological (0191–2615). Scope: Development and solution to problems, particularly those such as traffic flow, analysis of transportation networks, and queuing theory that require mathematical analysis.

Transportation Research, Part C: Emerging Technologies (0968–090X). Scope: The implications of technologies from operations research, computer science, electronics, control systems, and telecommunications on the planning, design, control, implementation, management, and rehabilitation of transportation systems.

Transportation Research, Part D: Transport and Environment (1361–9209). Scope: Environmental impact of all modes; focuses on empirical findings and regulatory, planning, technical, or fiscal policy responses with implications for the design and management of transportation systems.

Transportation Research, Part E: Logistics and Transportation Review (1366–5545). Scope: Transport economics, infrastructure, public policy evaluation, empirical studies of management, logistics and supply chain management, and logistics and operations models.

Transportation Research, Part F: Traffic Psychology and Behavior (1369–8478). Scope: behavioral and psychological aspects of traffic and transportation.

HIGHWAYS AND TRAFFIC

Asphalt Technology News (1083–687X) http://www.ncat.us/info-pubs/asphalt-technology. html (accessed September 1, 2010). National Center for Asphalt Technology; biannual. Newsletter aimed at transferring asphalt research findings to practitioners.

Better Roads (0006–0208) http://www.betterroads.com (accessed September 1, 2010). Randall-Reilly Publishing Co. LLC; monthly. Practical articles aimed at highway contractors and state and local government agencies involved in road and bridge engineering and maintenance.

HMAT: Hot Mix Asphalt Technology (not supplied) http://www.naylornetwork.com/nap-nxt/ (accessed September 1, 2010). National Asphalt Pavement Association; bimonthly. Promotes the use of asphalt pavements through success stories, buyers' guides, and information on bituminous materials, mixtures, applications, and maintenance.

ITE Journal (0162–8178) http://www.ite.org/itejournal/index.asp (accessed September 1, 2010). Institute of Transportation Engineers; monthly. Reports on research involving surface transportation systems, particularly traffic engineering.

ITS International (1463–6344) http://www.itsinternational.com/ (accessed September 1, 2010). Swanley, Kent, U.K.: Route One Publishing; bimonthly. Focuses on the deployment of technology for intelligent transportation systems and covers the full range of applications, including mass transit and light rail systems.

Journal of Bridge Engineering (1084–0702) http://scitation.aip.org/beo/ (accessed September 1, 2010). American Society of Civil Engineers; bimonthly. Publishes research about the practice and profession of bridge engineering; covers projects, design, construction, inspection, safety, repair, and rehabilitation.

NCHRP Publications http://www.trb.org/Publications/PubsNCHRPPublications.aspx (accessed September 1, 2010). Transportation Research Board. The National Cooperative Highway Research Program conducts research in areas that affect highway planning, design, construction, operation, and maintenance nationwide. Results are disseminated through the NCHRP Report, Synthesis of Practice, Summary of Progress, and Web-only series of publications.

Public Roads (0033–3735) http://www.fhwa.dot.gov/publications/publicroads/index.cfm (accessed September 1, 2010). Federal Highway Administration, Turner-Fairbank Research Center; bimonthly. Reports on advances and innovations in highway and traffic research and technology with an emphasis on developments in federal highway programs.

Roads and Bridges (8750–9229) http://www.roadsbridges.com (accessed September 1, 2010). Scranton Gillette Communications; monthly. Provides information on advancements in the road and bridge industry, including reviews of products and construction and maintenance equipment. Features many small, successful projects.

Traffic Engineering and Control (0041–0683) http://www.tecmagazine.com/ (accessed September 3, 2010). Hemming Group Ltd.; 11 issues per year. Covers traffic control, traffic planning, and transportation management from a British perspective.

Traffic Injury Prevention (1538–9588) http://www.tandf.co.uk/journals/titles/15389588.asp (accessed September 3, 2010). Taylor & Francis; bimonthly. Focuses on research, interventions, and evaluations within the areas of traffic safety, crash causation, injury prevention, and treatment.

Traffic Technology International (1356–9252) http://www.ukipme.com/mag_traffic.htm (accessed September 3, 2010). Dorking, Surrey, U.K.: UKIP Media and Events; bimonthly. Covers advanced traffic management systems including enforcement, intelligent parking, road pricing, electronic tolls collection, and detection and incident management. Contains many product reviews.

Tunneling and Underground Space Technology (incorporating *Trenchless Technology*) (0886–7798) http://www.elsevier.com/wps/find/journaldescription.cws_home/799/description#

description (accessed September 3, 2010). Elsevier Science; quarterly. Focuses on technical improvements and cost-effective methods for the planning, design, construction, operation, and maintenance of underground structures.

Tunnels and Tunnelling International (1369–3999) http://www.tunnelsonline.info/ (accessed September 3, 2010). World Market Intelligence; monthly. International in coverage, T&T publishes technical profiles of ongoing and completed tunnels, including rail, vehicular, and pipeline projects. Covers techniques and methods, equipment, geologic conditions, safety, and environmental issues.

Maritime

Fast Ferry International (0954–3988) http://www.fastferryinfo.com (accessed September 3, 2010). Fast Ferry International; 10 issues per year. Features articles on new vessels and designs.

International Journal of Maritime Engineering, The (1479-8751) http://www.rina.org.uk/ijme.html (accessed September 3, 2010). The Institution; quarterly. Published as *Part A of the Transactions of The Royal Institution of Naval Architects*. Provides a forum for technical issues related to the design and construction of ships and offshore structures.

Journal of Ship Production (8756–1417) http://www.sname.org/SNAME/SNAME/Publications/JournalofShipProduction/Default.aspx (accessed September 3, 2010). Jersey City, NJ: Society of Naval Architects and Marine Engineers; quarterly. Covers technical aspects of shipyards and the production of merchant and naval ships.

Journal of Ship Research (0022–4502) http://www.sname.org/SNAME/SNAME/Publications/JournalofShipResearch/Default.aspx (accessed September 3, 2010). Jersey City, NJ: Society of Naval Architects and Marine Engineers; quarterly. Publishes technical papers on applied research in hydrodynamics, propulsion, ship motions, structures, and vibrations.

Journal of Waterway, Port, Coastal and Ocean Engineering (0733–950X) http://ascelibrary.aip.org/wwo/ (accessed September 3, 2010). American Society of Civil Engineers; bimonthly. Covers engineering aspects of dredging, floods, pollution, and sediment transport, and the development and operation of ports, harbors, and offshore facilities.

Pipelines

Journal of Pipeline Engineering http://www.j-pipe-eng.com/ (accessed September 3, 2010). Beaconsfield, Bucks, U.K.: Scientific Surveys Ltd; bimonthly. Provides technical information on pipes and pipelines research with a European focus including pipeline design, corrosion, inspection, and data management.

Railroads

Japanese Railway Engineering (0448–8938) Japan Railway Engineers' Association; biannual. Publishes articles in English on railway engineering in Japan.

Quarterly Report of RTRI (0033–9008) http://www.rtri.or.jp/infoce/qr_E.html (accessed September 3, 2010). Railway Technical Research Institute. Provides in-depth coverage of technological developments in the Japanese railroad industry including vehicles, tracks, tunnels, and modeling. Full-text available online from 2001 to present.

RTR–European Rail Technology Review (0079–9548) http://www.eurailpress.de/rtr-archiv.html (accessed September 13, 2010). DVV Media Group GmbH I Eurailpress; quarterly. Publishes detailed information on track infrastructure, local and mainline rolling stock and components, signaling, control and command systems, and modern railway operations.

Railway Track and Structures (RT&S) (0033–9016) http://www.rtands.com/ (accessed September 3, 2010). Omaha, NE: Simmons-Boardman Books; monthly. Focuses on rail engineering, maintenance-of-way, and communications and signals in the United States. Reports on railroad testing done at the Transportation Technology Center, Inc. (TTCI).

VEHICLES

Automotive Engineering International (1543–849X) http://www.sae.org/mags/aei/ (accessed September 3, 2010). SAE International; monthly. Covers new products and technology for the global automotive industry. Focuses on applied vehicle engineering and design concepts.

IEEE Transactions on Vehicular Technology (0018–9545) http://ieeexplore.ieee.org/xpl/ RecentIssue.jsp?reload=true&punumber=25 (accessed September 3, 2010). New York: Institute of Electrical and Electronic Engineers; bimonthly. Focuses on electrical and electronic technology in vehicles and vehicle systems, including mobile communication devices used with such systems.

International Journal of Vehicle Autonomous Systems (1471–0226) http://www.inderscience. com/browse/index.php?journalID=30 (accessed September 3, 2010). Geneva, Switzerland: Inderscience; quarterly. Reports on driver assistance systems, intelligent vehicle systems, collision avoidance, by-wire systems, and new electrical and electronic systems.

International Journal of Vehicle Design: The Journal of Vehicle Engineering and Components (0143–3369) http://www.inderscience.com/browse/index.php?journalID=31 (accessed September 3, 2010). Geneva, Switzerland: Inderscience Publishers; monthly. Publishes articles on engineering design and research into all types of self-propelled vehicles and components.

Vehicle System Dynamics: International Journal of Vehicle Mechanics and Mobility (0042–3114) http://www.ingentaconnect.com/content/tandf/vesd (accessed September 3, 2010). Lisse, The Netherlands: Taylor & Francis; monthly. Publishes research on vehicle dynamics, intelligent highway systems, and related topics.

PROCEEDINGS

Many associations and societies publish proceedings of meetings, seminars, and other gatherings. Some are quite specialized and proceedings dealing with detailed technical aspects of any mode can be identified using catalogs and indexes described previously. Some of the more important published proceedings include the following:

American Association for the Advancement of Automotive Medicine http://www.carcrash. org/publications_proceedings.htm (accessed September 3, 2010). *Proceedings of the Annual Conference.* Des Plaines, IL: The Association. Papers focus on motor vehicle crash injury prevention and control from a multidisciplinary perspective. Available in hard copy and on CD.

American Public Transportation Association http://www.apta.com/mc/Pages/default.aspx (accessed September 14, 2010). *Bus and Paratransit Conference, Rail Conference, and Annual Meeting Proceedings.* Papers address bus and rail transit vehicle design, safety, maintenance, operations, and security.

American Railway Engineering and Maintenance Association http://www.arema.org/ (accessed September 3, 2010). Beginning in 1998, these proceedings superseded the *AREA Proceedings.* Produced by the Communications and Signals and the Tracks and Structures Committees, they contain all technical papers presented at the annual conference and committee reports. The American Railway Engineering Association (AREA) published the *Bulletin* five times a year from 1900 to 1997; in 1997 AREA became part of AREMA. Four issues of the *Bulletin* contain technical papers, proposed manual changes, and committee reports that were then bound together to form AREA's annual proceedings. The fifth issue is a Membership and Committee Directory.

Association for European Transport Conference http://www.aetransport.org/lc_cms/page_view.asp?id=677 (accessed September 13, 2010). Association for European Transport; annual. Papers on topics such as transportation management, modeling, public transportation, road maintenance, climate change, and high speed rail. Content is primarily British with some international coverage. Formerly known as the PTRC Annual Meeting. Beginning with 2001, complete papers are available only on CD.

Association of Asphalt Paving Technologists http://www.asphalttechnology.org/annual-journal.html (accessed September 3, 2010). *Asphalt Paving Technology.* The Association; annual. The proceedings include all papers and reports presented at the annual conference. Covers all aspects of asphalt pavements from a global perspective: materials, mix design, testing, construction, and performance. Published irregularly from 1974 to 1986; annually since 1988.

CODATU (*Cooperation for Urban Mobility in the Developing World*) http://www.codatu. org/ (accessed September 13, 2010). The Association; biennial. Also known as the *Urban Transport Conference*, this international meeting is designed to further the development of urban transportation in developing and industrialized countries.

IEEE/ASME Joint Rail Conference http://www.asmeconferences.org/JRC2010/ (accessed September 13, 2010). ASME Rail Transportation Division, IEEE Vehicular Technology Section, ASCE T&DI Rail Transportation Committee, et al.; annual. Focuses on innovations in railway mechanical and electrical engineering, including high-speed passenger rail. Available online to members and subscribers from 1989 to the present.

IEEE Intelligent Transportation Systems Council (ITSC). *Conference Proceedings.* http://iee-explore.ieee.org/xpl/conhome.jsp?punumber=1000396 and *Intelligent Vehicles Symposium Proceedings* http://ieeexplore.ieee.org/xpl/conhome.jsp?punumber=1000397 (accessed September 3, 2010). Piscataway, NJ: The Institute; annual. Focuses on advanced technologies for vehicles.

Institute of Transportation Engineers (ITE) http://www.ite.org/bookstore/index.asp (accessed September 13, 2010). Primary resource. *Annual Meeting Compendium of Papers* (title varies) and *Technical Conference Compendium of Papers* (title varies), Washington, D.C. Papers on technical topics including traffic engineering, traffic operations, traffic management, safety, mobility, and transit. Proceedings issued by the Institute of Traffic Engineers from 1930 to 1976. Beginning in 1997, available only on CD.

International Congress on Noise Control Engineering http://www.i-ince.org/ (accessed September 13, 2010). *Internoise Proceedings.* Place of publication varies: The International Institute of Noise Control Engineering (INCE); annual. Covers research on aircraft, rail, traffic, and pavement noise. INCE/USA sponsors the annual *Noise-Con Conference* with the Acoustical Society of America.

International Symposium on Transportation and Traffic Theory (ISTTT). Place of publication varies. Triennial from 1959–2005; biannual since 2005. Focuses on theoretical approaches to planning, modeling, management, and operation of traffic and transportation systems.

Society of Automotive Engineers (SAE) http://www.sae.org/events/ (accessed September 13, 2010). SAE sponsors numerous conferences and symposia on all aspects of the design, manufacture, and service life-cycle technology for ground and aerospace vehicles.

Stapp Car Crash Conference Proceedings http://www.stapp.org/pubs.shtml (accessed September 13, 2010). Warrendale, PA: Society of Automotive Engineers; annual. Presents research in impact biomechanics, human injury tolerance, and automotive safety measures and appliances. Papers are published in the *Stapp Car Crash Journal.*

Transportation Association of Canada. *Proceedings of TAC Annual Conference.* Ottawa. Covers topics such as geometric design, pavements, road maintenance, traffic control, and the environment. Published on CD.

Transportation Research Board. *Annual Meeting Compendium of Papers* (title varies), Washington, D.C.: National Academy of Sciences; annual. Primary Resource. An essential resource for research in all aspects of transportation engineering. Papers are published in series by subject. Covers the full range of multimodal transportation topics: pavements, structures, transportation management, traffic engineering, freight transportation, safety, rail transit, security, aviation, logistics, modeling, and so on. Many preprints are revised and published in the *Transportation Research Record* journal. Beginning in 1998, available only on CD. Prior to 1962, published as *Proceedings of the Annual Meeting of the Highway Research Board.*

World Conference on Transport Research (WCTR) http://wctrs.ish-lyon.cnrs.fr/ (accessed September 13, 2010). Proceedings; triennial. Papers on transportation modeling, operation, planning, and control.

World Congress on Intelligent Transport Systems http://www.itsworldcongress.com/ (accessed September 13, 2010). Title varies. An international meeting and exhibition sponsored by ITS America, ITS Japan, and ERTICO-ITS Europe; annual. Proceedings available on CD.

PATENTS

Patents and standards can be key resources for transportation engineers involved in the design of new equipment and processes, particularly in the area of motor vehicles. The development of electronic databases and online resources has greatly enhanced access to patent information. Key patent databases and portals are included in Chapter 2, General Engineering Resources, of this book.

STANDARDS AND SPECIFICATIONS

Many standards and specifications used in transportation engineering are issued by organizations such as ASTM and ISO while those applying to more specialized areas are issued by a variety of associations. Some of the more important issuing bodies and standards are listed below. Additional information on standards can be found in Chapter 2, General Engineering Resources, in this book.

American Association of State Highway and Transportation Officials, Washington, D.C.: AASHTO. AASHTO publishes a number of specifications for bridges and highways:

LRFD Bridge Construction Specifications, 5th ed. 2010. Companion volume to LFRD Bridge Design Specifications focused on construction.

LRFD Bridge Design Specifications, 5th ed. 2010. Specifications are based on the load-and-resistance factor design (LRFD) philosophy. Includes CD.

LRFD Guide Specifications for Design of Pedestrian Bridges, 2nd ed. 2004. Addresses the design and construction of bridges designed primarily for pedestrians, bicyclists, equestrian riders, and light maintenance vehicles.

Specifications for Seismic Isolation Design, 3rd ed. 2010. Provides fundamental requirements for seismic isolation design.

Standard Specifications for Highway Bridges, 17th ed. 2002. Known as *The Bridge Book,* this is the basic source of structural design standards for public and private sector bridge engineers. Includes CD.

Standard Specifications for Transportation Materials and Methods of Sampling and Testing, 30th ed. 2010. Contains more than 400 materials and specifications and test methods commonly used in highway construction. Includes relevant ASTM specifications.

American Society of Testing and Materials. 2010. *Annual Book of ASTM Standards. Section 4, Construction; Volume 4.02, Concrete and Aggregates; Volume 4.03, Road and Paving Materials and Vehicle-Pavement Systems.* Philadelphia: ASTM. Covers road and paving materials. Provides specifications, tests, and practices for field measurements, traffic monitoring, and vehicle-to-roadside communication.

Association of American Railroads. *Manual of Standards and Recommended Practices (MSRP)* http://www.aar.com/aar_standards-publications.htm (accessed September 14, 2010). Washington, D.C.: ARA. The MSRP contains about two dozen sections, most of which have been updated since 2005. Includes regularly adopted specifications, standards, and recommended practices for freight cars, locomotives, and their components.

Code of Federal Regulations. Title 49 Transportation http://www.access.gpo.gov/nara/cfr/cfr-table-search.html (accessed September 14, 2010). Washington, D.C.: Office of the Federal Register, National Archives and Records Service, General Services Administration; annual. Contains the general and permanent rules published in the Federal Register by the U.S. Department of Transportation, the modal administrations, and other federal departments and agencies.

Institute of Transportation Engineers http://www.ite.org/standards/index.asp (accessed September 14, 2010). ITE develops Intelligent Transportation System (ITS) standards under a cooperative agreement with U.S. Department of Transportation. It also issues Recommended Practices for traffic operations, safety, and design, including:
Neighborhood Street Design Guidelines. 2010.
Designing Walkable Urban Thoroughfares. 2010.
Guidelines for the Design and Application of Speed Humps. 2007.
Guidelines for Activation, Modification or Removal of Traffic Control Signals. 2005.
Smart Growth Transportation Guidelines. 2003.

ITS Standards Program http://www.standards.its.dot.gov/ (accessed September 14, 2010). Provides access to intelligent transportation systems standards from AASHTO, IEEE, ITE, NEMA, SAE and other organizations.

Society of Automotive Engineers (SAE) http://standards.sae.org/ (accessed September 14, 2010). SAE develops standards for the automotive, ground vehicle, and commercial vehicle industries. In addition to individual standards, it publishes an annual handbook with more than 2,000 full-text documents on CD. Most standards and standard collections are available in PDF format.

U.S. Federal Highway Administration. 1996. *Standard Specifications for Construction of Roads and Bridges on Federal Highway Projects* http://www.efl.fhwa.dot.gov/design/manual/Fp96.pdf (accessed September 14, 2010). Washington, D.C.: FHWA. Covers the process used for road and bridge construction on federal highway projects from bidding through construction.

IMPORTANT WEB SITES AND PORTALS

Web sites and portals are essential resources for transportation engineering information. There are hundreds of sites with relevant information; this is a selection of the more important ones.

The U.S. Department of Transportation (USDOT) http://www.dot.gov/ (accessed September 14, 2010) is an excellent source of transportation engineering information. Primary Resource. The Web site contains an overview of the department's agenda and efforts.

The Federal Highway Administration http://www.fhwa.dot.gov/ (accessed September 14, 2010). FHWA is a primary source for information on highway engineering for projects receiving federal funding. Extensive online information is available for bridges, pavements, environmental design, safety, and tribal lands; a list of FHWA sites is available at http://www.fhwa.dot.gov/topics.htm (accessed September 14, 2010). Among the important FHWA sites are:

FHWA's Office of Infrastructure http://www.fhwa.dot.gov/infrastructure/ (accessed September 14, 2010) provides technical assistance and program assistance in federal-aid highway programs, asset management, construction, pavements and bridges. It also has information about current technical recommendations, agency sponsored projects, and history.

In the FHWA Office of Bridge Technology http://www.fhwa.dot.gov/bridge/ (accessed September 14, 2010), the National Bridge Inspection Program includes the National Bridge Inspection Standards http://www.fhwa.dot.gov/bridge/nbis.htm (accessed September 14, 2010) and the National Bridge Inventory http://www.fhwa.dot.gov/bridge/nbi.htm (accessed September 14, 2010) of deficient or functionally obsolete bridges, listed by state. Tables are available for download in XLS.

The FHWA Office of Pavement Technology http://www.fhwa.dot.gov/pavement/ (accessed September 14, 2010) includes publications, conferences, training programs, contacts, and links to information on asphalt and concrete pavements.

The Center for Transportation Analysis (CTA) from Oakridge National Lab http://cta.ornl.gov/cta/ (accessed September 15, 2010) performs research sponsored from a number of agencies, such as USDOT and USDOE. Their primary focus is the intersection of transportation, energy, and the environment.

The Construction and Maintenance Web site http://www.fhwa.dot.gov/construction/ in FHWA's Office of Asset Management links to *National Highway Specifications* http://fhwapap04.fhwa.dot.gov/nhswp/index.jsp (accessed September 14, 2010) with a searchable library of highway specifications from throughout the United States. It also links to the *Construction Program Management and Inspection Guide* http://www.fhwa.dot.gov/construction/cpmi04tc.cfm (accessed September 14, 2010) where there are links to topics such as design-build, safety, and state preference for materials. The *Generic Construction Related Review Guidelines* http://wwwcf.fhwa.dot.gov/construction/reviews/ (accessed September 14, 2010) provides examples of process and in-depth reviews performed by field offices on subjects such as pavements, structures, environmental protection, and work zone safety.

In FHWA's Office of Operations http://www.ops.fhwa.dot.gov/index.asp (accessed September 14, 2010), the Workzone Mobility and Safety Program http://ops.fhwa.dot.gov/wz/index.asp (accessed September 14, 2010) links to a set of practitioner's tools including best practices, fact sheets, reports, technologies, and decision support tools.

The Research and Innovative Technology Administration (RITA) http://www.rita.dot.gov/ (accessed September 14, 2010) is the administration responsible for coordinating research within USDOT. Because the focus is on research and the development of new technologies, they are the multimodal arm of USDOT. RITA houses several offices that work to these ends. They include:

The Bureau of Transportation Statistics, http://www.bts.gov/about/ (accessed September 14, 2010), which warehouses many statistics and datasets for transportation at the federal and local level.

The National Transportation Library (NTL) http://ntl.bts.gov/ (accessed September 14, 2010) is a part of BTS and supports transportation research, policy, operations, and technology transfer throughout the United States. The NTL's digital library serves as a repository of technical, research, and policy documents from public, academic, and private organizations.

The Intelligent Transportation Systems Program Office (ITS-PO) http://www.its.dot.gov/about.htm (accessed September 14, 2010) provides information about the development and use of ITS applications. It provides guidelines for ITS architecture and data collection and houses databases for benefits and costs associated with ITS deployment, and lessons learned.

Intelligent Transportation Systems Applications Overview http://www.itsoverview.its.dot.gov/ (accessed August 30, 2010). U.S. Department of Transportation, Research and Innovative Technology Administration. Provides links to intelligent infrastructure and intelligent vehicle applications addressed by the federal Intelligent Transportation Systems program.

The Local Technical Assistance Program (LTAP) Clearinghouse http://www.ltap.org/ (accessed September 14, 2010) connects to LTAP centers in each state and Tribal Technical Assistance Program (TTAP) centers. Centers provide employees of local agencies with cost-effective access to training, information clearinghouses, and technology updates.

Maritime Administration http://www.marad.dot.gov (accessed September 14, 2010) links to information on the Marine Transportation System initiative, security, shipbuilding, and domestic operation of ports.

The National Highway Traffic Safety Administration http://www.nhtsa.gov/ (accessed September 14, 2010) provides data on crashes through databases, such as the Fatality Analysis Reporting System (FARS) and the Crashworthiness Data System.

The National Transportation Safety Board http://www.ntsb.gov/ (accessed September 14, 2010) publishes accident reports for highways, marine transportation, pipelines, and railroads.

The Office of Pipeline and Hazardous Materials Safety Administration http://www.phmsa. dot.gov/pipeline (accessed September 14, 2010) provides information about the construction, maintenance, and safety of pipelines used in the transportation of petroleum, natural gas, and hazardous materials.

The Research Development and Technology Office (RDT) http://www.rita.dot.gov/rdt/ (accessed September 14, 2010) coordinates research across different offices and agencies, which includes overseeing the University Transportation Centers (UTC) program http:// utc.dot.gov/ (accessed September 14, 2010). The UTC program focuses on transportation research and education at centers at more than 30 universities across the United States.

The Turner–Fairbank Highway Research Center (TFHRC) http://www.tfhrc.gov/ (accessed September 15, 2010) performs research for FHWA. Their primary focus is new technologies to promote a safer and more reliable highway system. Two particularly important research areas at the center include:

> The Safety Research Program http://www.tfhrc.gov/safety/index.htm (accessed September 14, 2010) focuses on intersections, pedestrians and bicyclists, roadsides, run-off-road accidents and speed management, and provides access to technical reports, articles, and other resources.

> Operations and Intelligent Transportation Systems http://www.tfhrc.gov/its/its.htm (accessed September 14, 2010) has information on enabling technologies and travel management products and services as well as links to technical reports and articles.

The Volpe National Transportation Center http://www.volpe.dot.gov/ (accessed September 15, 2010) performs systems research for RITA. Their main focus is the development of market-driven innovation ready for deployment.

Two other highway and traffic engineering-related sites of interest are:

The Institute of Transportation Engineers (ITE) http://www.ite.org develops intelligent transportation system (ITS) standards http://www.ite.org/standards/index.asp (accessed September 14, 2010) in a cooperative agreement with the U.S. Department of Transportation. ITE's site also has technical information http://www.ite.org/technical/default.asp (accessed September 14, 2010) and links on topics, such as traffic calming, road safety audits, safety and signal timing.

Urban Mobility Information from the Texas Transportation Institute (TTI) http://mobility.tamu.edu/ (accessed September 15, 2010) provides information and resources about traffic monitoring and congestion, with emphasis on the analysis of traffic data.

Other selected sites with transportation engineering information are:

American Association of Port Authorities (AAPA) http://www.aapa-ports.org/ (accessed September 14, 2010). Provides links to sites with port information for North America.

ARRB http://www.arrb.com.au/home.aspx (accessed September 15, 2010). Provides access to research and information from Australia's main transportation organization.

BIMCO https://www.bimco.org/ (accessed September 14, 2010). Provides members with access to databases of international information on technical aspects of shipping, cargoes, safety, environmental issues, and so on.

Lloyd's Marine Intelligence Unit http://www.seasearcher.com/lmiu/index.htm (accessed September 14, 2010). Provides members with access to detailed databases on vessels, shipping, casualties, and shipping companies.

National Truck Equipment Association (NTEA) http://www.ntea.com/ (accessed September 14, 2010). Provides access to technical reports and articles, safety information, and a hoist database.

Office of Coast Survey, U.S. National Ocean Service, National Oceanic and Atmospheric Administration (NOAA) http://www.nauticalcharts.noaa.gov/ (accessed September 14, 2010). Provides access to databases, charts, surveys, and other navigational products.

Physical Oceanographic Real-Time Systems (PORTS) http://tidesandcurrents.noaa.gov/ports.html (accessed September 14, 2010). From U.S. National Ocean Service, National Oceanic and Atmospheric Administration (NOAA), supplies shipmasters and pilots with water levels, currents, and other data to avoid groundings and collisions.

Railway Track and Structures magazine's Links to Important Railroad Industry Information http://www.rtands.com/links-to-important-rail-industry-information.html (accessed September 14, 2010) provides quick access to the sites of Class I, regional, and short line railroads, associations, regulatory agencies, and information resources.

Transport Canada http://www.tc.gc.ca/en/menu.htm (accessed September 15, 2010). This site provides access to statistics, policies, regulations, and other information about all modes of transportation.

The World Bank's Web site on Transport http://www.worldbank.org/transport (accessed September 14, 2010) covers World Bank-funded projects in developing nations, and contains data about rail and rural accessibility.

ASSOCIATIONS, ORGANIZATIONS, AND SOCIETIES

For transportation engineering, entities such as professional associations and organizations are important sources of information. Most deal with just one mode, but some, such as the Transportation Research Board, are multimodal in focus.

American Association of State Highway and Transportation Officials, The (AASHTO) http://www.transportation.org/ (accessed September 13, 2010). This advocacy association fosters the development, operation, and maintenance of an integrated, intermodal national transportation system. Publishes essential handbooks, standards, and guidelines including the *Manual on Uniform Traffic Control Devices* (with FHWA and ITE), *A Policy on Geometric Design of Highways and Streets,* and bridge standards.

American Public Transportation Association (APTA) http://www.apta.com/ (accessed September 13, 2010). An advocacy group for bus, rapid transit, and commuter rail systems, APTA sponsors conferences on bus and rail transit and applications of intelligent transportation systems. It publishes proceedings, reports, and statistics on fares and fare collection, vehicles, alternative fuels, safety, sustainability, funding sources, and policy. Formerly known as the American Public Transit Association and the American Transit Association.

American Railway Engineering and Maintenance Association www.arema.org/eseries/script-content/index.cfm (accessed September 13, 2010). This association publishes manuals, guides, handbooks, and recommended practices for the design, construction, and maintenance of railway infrastructure. Formed from the 1997 merger of the American Railway Bridge and Building Association, the American Railway Engineering Association, and the Roadmasters and Maintenance of Way Association, along with functions of the Communications and Signal Division of the Association of American Railroads.

American Road and Transportation Builders Association (ARTBA) http://www.artba.org/ (accessed September 13, 2010). This association lobbies for increased funding for transportation projects, conducts public information campaigns, and publishes data on construction material costs and wages and benefits.

American Society of Civil Engineers (ASCE) http://www.asce.org/ (accessed September 13, 2010). The world's largest publisher of civil engineering information, including technical and professional journals, proceedings, and standards, ASCE also sponsors conferences and continuing education programs. Covers highway, bridge, and marine engineering.

American Society of Mechanical Engineers (ASME) http://www.asme.org/ (accessed September 13, 2010). ASME's Rail Transportation Division (RTD) focuses on intercity freight and passenger railroads as well as rail rapid transit. RTD is a partner in the annual IEEE/ASME Joint Rail Conference, which focuses on innovation in railway mechanical and electrical engineering.

American Society of Naval Engineers (ASNE) http://www.navalengineers.org (accessed September 13, 2010). ASNE promotes naval engineering through publications and sponsorship of conferences.

ARRB Group http://www.arrb.com.au/ (accessed September 13, 2010). Formerly the Australian Road Research Board, ARRB conducts a technical research program for the Austroads Council in asset management, bituminous surfaces, pavement technologies, and road safety engineering.

Association of American Railroads (AAR) http://www.aar.org/ (accessed September 13, 2010). AAR represents the major freight railroads in North America and Amtrak. It publishes manuals, handbooks, recommended practices, and technical standards. A subsidiary, Transportation Technology Center, Inc., conducts research on rolling-stock, track components, structures, signals, and safety devices; it also provides training.

Canadian Transportation Research Forum (CTRF) http://ctrf.ca/ (accessed September 14, 2010). CTRF members include transportation professionals in the private sector, government agencies, and universities. It holds an annual meeting and publishes proceedings.

Institute of Transportation Engineers (ITE) http://www.ite.org/ (accessed September 13, 2010). ITE offers education programs and provides certification for professional traffic operations engineers. Twelve areas-of-interest councils develop standards and recommended practices, organize conferences and seminars, and issue briefings and informational reports. ITE's publications include standard works, such as *Transportation Engineering Handbook, Trip Generation* and *Parking Generation.*

Intermodal Association of North America (IANA) http://www.intermodal.org/ (accessed September 13, 2010). IANA is an association of railroads, truckers, port authorities, suppliers, and manufacturers, and promotes intermodal transportation. It administers UIIA, the standard equipment interchange contract.

International Road Federation (IRF) http://www.irfnet.org/ (accessed September 13, 2010). IRF promotes road planning, development, construction, and usage through a wide range of conferences. Compiles and publishes *World Road Statistics* on CD.

International Union of Railways (IUR) http://www.uic.org (accessed September 13, 2010). IUR, primarily a European organization, facilitates international railway transportation by providing standards, regulations, and recommendations, and publishing technical documents to promote safety and interoperability.

National Asphalt Paving Association, The (NAPA) http://www.hotmix.org/ (accessed September 13, 2010). NAPA supports an active research program to explore environmental issues and improve the quality of asphalt pavement mixes and paving techniques. Produces technical, marketing, and educational materials including *HMAT Magazine*.

Railway Technical Research Institute (RTRI) http://www.rtri.or.jp/ (accessed September 13, 2010). RTRI, the research center for the seven Japan railway companies, develops basic technology and research applications, such as the Maglev system, promotes technology transfer, and studies safety.

Society of Automotive Engineers (SAE) http://www.sae.org/servlets/index (accessed September 13, 2010). SAE sponsors meetings, conferences, and symposia on the design, manufacture, and life cycle of motor and aerospace vehicles. It publishes technical papers and develops and publishes standards.

Transportation Association of Canada (TAC) http://dev.tac-atc.ca/english/index.cfm (accessed September 13, 2010). TAC promotes knowledge on technical guidelines and best practices for roadways and urban transportation systems. Publishes manuals and handbooks on design, construction, safety, and operations.

Transportation Research Board, The (TRB) http://www.trb.org/ (accessed September 13, 2010). TRB, a unit of the National Research Council, is the primary transportation research organization in the U.S. Primary Resource. Charged with promoting transportation innovation through research, its interests include all modes of transportation and cover technical and policy aspects. It facilitates information sharing through its annual meeting and an extensive publication program of reports and peer-reviewed technical papers, including its journal, *Transportation Research Record*. TRB manages research programs, including cooperative programs in highway, transit, aviation, and freight research. It produces *TRID*, the online database of transportation information. TRB is supported by state departments of transportation, the U.S. Department of Transportation, and other federal modal agencies.

TRL (Transport Research Laboratory) http://www.trl.co.uk/ (accessed September 13, 2010). TRL is an independent organization in the United Kingdom that provides research and testing and publishes reports. Research areas include safety, sustainability, travel time, and infrastructure management.

ACKNOWLEDGMENTS

We are deeply indebted to our colleagues in the Transportation Division of SLA who contributed to *Sources of Information in Transportation*, particularly Susan Dresley, series editor. Our special thanks to editor Bonnie Osif for giving us the challenge and opportunity of writing this resource guide to transportation engineering.

REFERENCES

Sinha, K. C., ed. 2002. Development of transportation engineering research, education, and practice in a changing civil engineering world. *Journal of Transportation Engineering* 128 (4): 301–313.

Sources of Information in Transportation, 4th ed. 1990. Compiled by members of the Transportation Division, Special Libraries Association, Monticello, IL: Vance Bibliographies.

Sources of Information in Transportation, 5th ed. 2001. Compiled by members of the Transportation Division, Special Libraries Association; Series Editor, Susan C. Dresley, Washington, D.C.: National Transportation Library, Bureau of Transportation Statistics, U.S. Department of Transportation http://www.ntl.bts.gov/ref/biblio/ (accessed September 13, 2010).

Index

50 Years of Human Engineering: History and Cumulative Bibliography of the Fitts Human Engineering Division 324, 347
100 of the World's Tallest Buildings 174
21st-Century Engineer: A Proposal for Engineering Education Reform 266

A

A to Z of Thermodynamics 400
AAPG Bulletin 486
AAPG Explorer 488
AAPG Memoir 486
AAPG Studies in Petroleum Geology 486
Aasen, A. 460
AASHTO Guide for the Planning, Design and Operation of Pedestrian Facilities 508
AASHTO Maintenance Manual for Roadways and Bridges 506
AASHTO Transportation Glossary 504
Abbate, J. 16
Abbot, J. H. 52
Abbott, M. 121
Abbreviations Dictionary 23
Abelson, H. 206
Aberdeen's Concrete Construction 178
ABET (see Accreditation Board for Engineering and Technology)
Abraham, M. A. 197
Abrasive Engineering Society 388
Absher, M. 40
Abstracts in New Technologies and Engineering 17, 282
AccessEngineering 4
Accessibility and the Bus Systems: From Concepts to Practice 160
AccessScience 4, 12
Accident Analysis and Prevention 514
Accreditation Board for Engineering and Technology 259, 273, 275
ACerS-NIST Phase Equilibria Diagram 377
ACHEMA 125
ACHEMASIA 125
ACHEMAMERICA 125
Acid Rain: Overview and Abstracts 285
ACI Manual of Concrete Practice 85, 505
ACI Materials Journal 178
ACI Structural Journal 178
ACM Career Resource Center 233
ACM Digital Library 226, 230, 244
ACM Events and Conferences 222
ACM Journals 224
ACM Portal 263
ACM Transactions on Sensor Networks 252
Acoustical Society of America 304, 424
Acquisitions Streamlining and Standardization Information System 56

Acronyms, Initialisms & Abbreviations Dictionary, 22
ACS Nano 385
ACS Style Guide; Effective Communication of Scientific Information 126
Acta Astronautica 55
Acta Biomaterialia 104, 385
Acta Materialia 385
Acta Mechanica 419
Ada: A Life and Legacy 209
Adair, John 148
Adams, C. P. 26
Addis, W. A. 170, 328
Adhesives in Civil Engineering 177
Adler, D. 172
Adsorption 123
Advanced Buildings Technologies and Practices 179
Advanced Building Systems: A Technical Guide for Architects and Engineers 86
Advanced Composite Materials 385
Advanced Dam Engineering: For Design, Construction and Rehabilitation 188
Advanced Drilling Engineering: Principles and Design 483
Advanced Functional Materials 385
Advanced Materials 385
Advanced Strength of Material 416
Advanced Transit Association 160
Advances in Applied Ceramics 385
Advances in Applied Mechanics 419
Advances in Biochemical Engineering/Biotechnology 123
Advances in Cement Research 178
Advances in Chemistry Series 114
Advances in Computers 224
Advances in Energetic Materials and Chemical Propulsion 54
Advances in Human Factors—Ergonomics 358
Advances in Nuclear Science and Technology 462
Advances in Polymer Technology 123
Advances in Space Research 55
AERADE 57
Aeronautical Engineer's Data Book 53
Aeronautical Journal 55
Aerosol Science and Technology 299
Aerosol Technology: Properties, Behavior, and Measurement of Airborne Particles 121
Aerospace America 55
AeroSpace and Defence Industries Association of Europe 58
Aerospace and High Technology Database 49, 373
Aerospace Engineering Education During the First Century of Flight 59
Aerospace Industries Association 58
Aerospace International 55
Aerospace Materials Specifications 22
Aerospace Structural Metals Handbook 52

AESIS, Australia's Geoscience, Minerals, and Petroleum Database 431
Agar, J. 209
Agardy, F. J. 290, 295
AGRICOLA 63
Agricola on Metals 428
Agricultural Engineering Abstracts/CAB Abstracts 63
Agricultural Engineering Index 63, 65
Agricultural Engineering International: The CIGR Journal of Scientific Research and Development 69
Agricultural Systems 70
Agricultural Water Management 70
Agriculture, Ecosystems & Environment 70
AGRIS 64
Agronomy Journal 70
Ahmad, A. 409
AHS International 58
AIAA Aerospace Design Engineers Guide 53
AIAA Journal 55
Air and Gas Drilling Manual 482
Air and Waste Management Association 311
Air CFRs Made Easy 291
Air Pollution 296
Air Pollution Control Engineering 296
Air Pollution Control Equipment Selection Guide 292
Air Pollution Control Technology Handbook 292
Air Pollution Engineering Manual 292
Airport Engineering 158
Airport Planning and Management 160
Airport Terminals 158
Air Quality 296
Air Quality Control: Formation and Sources, Dispersion, Characteristics and Impact of Air Pollutants– Measuring Methods, Techniques for Reduction of Emissions and Regulations for Air Quality Control 296
Air Quality Control Handbook 291
AISC 175
Akay, M. 100
Alan Turing.net 210
Alberta Oil 488
Albright, L. F. 117
Albright's Chemical Engineering Handbook 117
Alembert, J.'d 45
Alexander, C. 248
Alexander, J. F. 32
Alexander, W. 383
Algorithms and Data Structures 207
All about Petroleum 495
Allard S.K. ix, x
All Conferences.Com 223
Allen, E. 85, 87
Alley, E. R. 185, 291, 292
Alter, H. 457
Aluminum Alloy Database 380
Aluminum Association 176, 180
Aluminium Design and Construction 177
Aluminum Design Manual 176
Aluminum Federation 180
Aluminum Industry Abstracts 373
Almanac of Architecture & Design 83
Alred, G. J. 32
Alvarez-Cohen, L. 295, 315

AMA Agricultural Mechanization in Asia, Africa, and Latin America 70
Ambio: A Journal of the Human Environment 299
Amelinckx, S. 377
American Academy of Environmental Engineers 286, 311
American Association for Aerosol Research 312
American Association for Artificial Intelligence 232
American Association for History and Computing 338
American Association for the Advancement of Automotive Medicine 517
American Association for the Advancement of Science 29, 268
American Association of Drilling Engineers 490
American Association of Engineering Societies 29
American Association of Petroleum Geologists 484, 490
American Association of Port Authorities 523
American Association of State Highway and Transportation Officials 161, 194, 504, 505, 508, 519, 523
American Astronautical Society 58
American Bearing Manufacturers Association 424
American Ceramic Society 180, 388
American Chemical Society 114, 125, 312, 339
    Division of Biochemical Technology 125
    Division of Fuel Chemistry 125
    Division of History of Chemistry 339
    Division of Industrial and Engineering Chemistry 125
    Division of Polymeric Materials 125
American Concrete Institute 21, 180, 194, 389, 506
American Conference of Governmental Industrial Hygienists 312
American Council on Education 265
American Foundry Society 389
American Gas 488
American Gas Association 490
American Gear Manufacturers Association 424
American Geological Institute 141, 438
American Heritage of Invention and Technology 14, 331
American Historical Association's Guide to Historical Literature 324
America: History and Life 318
American Indian Council of Architects and Engineers 40
American Indian Science and Engineering Society 40
American Industrial Hygiene Association 312
American Institute for Medical and Biological Engineering 107
American Institute for Mining, Metallurgical and Petroleum Engineers 388
American Institute of Aeronautics and Astronautics 58
American Institute of Architects 91
American Institute of Chemical Engineers 123, 127, 491, 492
American Institute of Hydrology 190
American Institute of Mining, Metallurgical, and Petroleum Engineers 445
American Institute of Physics. Center for History of Physics 339
American Institute of Steel Construction 85, 180, 194
American Institute of Timber Construction 85, 175, 180
American Iron and Steel Institute 180, 194, 389
American Membrane Technology Association 312
American Mineralogist Crystal Structure Database 431
American Mines Handbook 439

American National Standards Institute 21, 24, 56, 89, 342, 368
American Nuclear Society 456, 463, 465
American Oil & Gas Historical Society 495
American Oil & Gas Reporter 488
American Oil & Gas Society 470
American Petroleum Institute 470, 491, 493, 495
American Philosophical Society 338
American Public Health Association 187
American Public Transit Association 160
American Public Transportation Association 517, 524
American Public Works Association 160
American Railway Engineering and Maintenance Association see American Railway Engineering and Maintenance of Way Association
American Railway Engineering and Maintenance of Way Association 156, 194, 510, 517, 524
American Rock Mechanics Association 141
American Society for Engineering Education 8, 29, 31, 123, 272, 273, 275
  Biomedical Engineering Division 108
American Society for Nondestructive Testing 424
American Society for Precision Engineering 363
American Society for Quality 363
American Society for Testing and Materials 22, 24, 89, 193, 389, 464, 492, 520
American Society of Agricultural and Biological Engineers 61, 77, 78
American Society of Agricultural Engineers 61, 65
American Society of Agronomy 190
American Society of Appraisers 446
American Society of Biomechanics 107
American Society of Civil Engineers 20, 80, 131, 132, 134, 172, 176, 177, 187, 194, 198, 201, 463, 524
  Committee on Sustainability 197
American Society of Civil Engineers' Civil Engineering Database 132
American Society of Gas Engineers 491
American Society of Heating, Refrigerating, and Air Conditioning Engineers 85, 91, 194, 304, 424
American Society of Mechanical Engineers 20, 194, 199, 363, 397, 424, 463, 534
  Bioengineering Division 108
  History and Heritage Committee 396
  Shale Shaker Committee 481
American Society of Mining and Reclamation 446
American Society of Naval Engineers 524
American Society of Safety Engineers 363
American Society of Sanitary Engineering for Plumbing and Sanitary Research 363
American Soil and Foundation Engineers 141
American Technical Education Association 273
American Underground Construction Association 141
American Universities and Colleges 265
American Vacuum Society 389
American Water Resources Association 190, 312
American Water Works Association 190, 194, 284, 304, 312
American Water Works Association Journal 186
American Welding Society 181, 389, 424, 492
Americans with Disabilities Act 44
American Wood Council 180
America's Best Graduate Schools 264
Amstock, J. S. 176

Analog Integrated Circuits and Signal Processing 224
Analysis of Structures: An Integration of Classical and Modern Methods 88
Ancient Mining 429
Anderson, J. D. 45, 59
Anderson, R. 39
Anderson, T. B., 39
Andrews, D. 378
Angelo, J. A. 51
Angus, H. T. 175
Animated Engines 423
Annales de l'Institut Technique du Batiment et Travaux Publics 177
Annals of Biomedical Engineering
Annals of Nuclear Energy 462
Annals of the ICRP 462
Annotated Bibliography of Project and Team Management 347
Annotated Bibliography on the History of Data Processing 208
Annual Book of ASTM Standards 387, 520
Annual Energy Outlook 2004 with Projections to 2035 502
Annual Energy Review 501
Annual Reliability and Maintainability Symposium 366
Annual Review of Biomedical Engineering 104
Annual Review of Environment and the Resources 299
Annual Review of Fluid Mechanics 419
Annual Review of Materials Research 385
Annual Reviews 97
Annual Statistical Bulletin 494
Antaki, G. A. 512
Anthony, L.J. 49, 429, 456
Antonsson, E.K. 402
API Data 493
Applegate, G. 187
Applied and Environmental Microbiology 70
Applied Catalysis 123
Applied Catalysis. B, Environmental 299
Applied Composite Materials 385
Applied Energy 486
Applied Engineering in Agriculture 70
Applied Ergonomics: Human Factors in Technology and Society 358
Applied Mechanics Reviews 419
Applied Science and Technology Index 17, 82, 97, 245, 262, 431
Applied Strength of Materials 417
Applied Surface Science 385
Applied Thermal Engineering 419
Appraisal of Existing Iron and Steel Structures 175
Appraisal of Existing Structures 173
Appropriate Technology 70
APWA Reporter 158
Aqua 186
Aquacultural Engineering 70
Aqualine 184
Aquatic Sciences and Fisheries Abstracts 143, 281, 283
ARCAT Specs 83
Architect 81
Architectural & Engineering News 88
Architectural Design 88
Architectural Design: Integration of Structural and Environmental Systems 7

*Architectural Details: Classic Pages from Architectural
        Graphic Standards* 30
Architectural Engineering Design 85
Architectural Engineering Institute 79, 80, 91
*Architectural Engineering: With Special Reference to
        High Building Construction, Including Many
        Examples of Chicago Office Buildings* 87
*Architectural Expression of Environmental Control
        Systems* 87
Architectural Graphic Standards 44, 89
Architectural Index 82
Architecture in Space Structures 177
Architectural Record 81, 87
Architectural Review 80
Architectural Science Review 88
Architecture Today 80, 88
*Archives of Environmental Contamination and
        Toxicology* 299
Arctic and Antarctic Regions 431
Arctic Science and Technology Information System 432
*Area Array Packaging Handbook: Manufacturing and
        Assembly* 353
Argonne National Laboratory 458
Armstrong-Wright, A. 158
Armytage, W. H. G. 328
Arntzen, C. J. 67
Arora, H. 99
ARRB 523, 524
*Art in Structural Design: An Introduction and
        Sourcebook* 87
Art of Computer Programming 206, 207
Arts and Humanities Citation Index 20
Arup, Ove, and Partners 150
ASABE Membership Roster 66
ASABE Technical Library 64
Asano, T. 297
ASAE Standards 76
ASCE (*see American Society of Civil Engineers*)
ASCE Civil Engineering Database 82
ASCE History and Heritage of Civil Engineering 339
        Resource Guide 336
ASEE Annual Conference Proceedings 272
*ASEE Directory of Engineering and Engineering
        Technology Colleges* 265
*ASEE Directory of Graduate Engineering and Research
        Statistics* 265
*ASEE Directory of Undergraduate Engineering
        Statistics* 265
ASEE Engineering Deans Institute 272
ASEE Prism 257, 271
Ashby, M. 384
Ashford, N. 158
ASHRAE (see American Society of Heating,
        Refrigerating, and Air Conditioning Engineers)
ASHRAE Fundamentals 405
ASHRAE Handbook 89
*ASHRAE Heating, Ventilating and Air-Conditioning
        Applications* 405
*ASHRAE Heating, Ventilating and Air-Conditioning
        Systems and Equipment* 405
ASHRAE Refrigeration 405
ASHRAE Journal 419
Askeland, D. R. 384
Asian Civil Engineering Coordinating Council 134

Aslezova, S. 51
ASME (see American Society of Mechanical Engineers)
ASME Boiler and Pressure Vessel Code 464
ASME Engineer's Data Book 403
ASME History 340
ASME International Mechanical Engineering Congress
        272, 66
ASME International Steam Tables for Industrial Use 119
ASME standards 422
ASM Handbook 85, 436
ASM International 193, 378, 390
ASM Materials Engineering Dictionary 375
ASM Ready Reference Collection 378
ASM Specialty Handbooks 378
Asmussin, J. 378
Asphalt Institute 160
Asphalt Technology News 515
Aspray, W. 209
Assembly Automation 358
Association for Computing Machinery 205, 208, 232, 255
Association for European Transport Conference 518
Association for Facilities Engineering 364
Association for Iron and Steel Technology 390
Association for Manufacturing Technology 364
Association for Planning Supervisors 153
Association for Project Management 154
Association for Specialist Fire Protection 181
Association for the Advancement of Medical
        Instrumentation 107, 108
Association for Unmanned Vehicle Systems
        International 58
Association for Women in Science 40
Association Francaise de Normalisation 492
Association of American Railroads 503, 520, 524
Association of Asphalt Paving Technologists 518
Association of Asphalt Paving Technologists Journal 158
Association of Coastal Engineers 146
Association of Consulting Engineers 153
Association of Directors of Environment, Economy,
        Planning, and Transport 161
Association of Energy Engineers 424
Association of Environmental and Engineering
        Geologists 141, 213
Association of Environmental Engineering and Science
        Professors 312
Association of Iron and Steel Engineers 415
Association of Metropolitan Water Agencies 161
Association of Pedestrian and Bicycle Professionals 161
Association of State and Interstate Water Pollution Control
        Administrators 313
Association of State Dam Safety Officials 190
Association of State Drinking Water Administrators 313
Associations Unlimited 255
ASTM (see American Society for Testing and Materials)
*ASTM Dictionary of Engineering, Science, and
        Technology* 11, 84
ASTM International Directory of Testing Laboratories 348
ASTM Standards 387, 422, 491
Atanasoff: Forgotten Father of the Computer 210
AT&T Labs Fellow Program 42
*ATA's Truck Fleet Directory: Profiles of For-Hire and
        Private Fleets in the U.S.* 503
Atkins, S. E. 458
Atkinson, G. S. D. 196

Atkinson, J. H. 136
Atkinson, K. 156
Atlas Conferences Inc 223
Atmospheric Environment 299
Atmospheric Radiation Measurement Program 282
Atomic Data and Nuclear Data Tables. 462
Atomic Energy/Atomnaia Energiia 462
Atomization and Sprays 55
Audio Engineering Society 255
Auger, C. P. 10, 28, 50
AusGeoRef 431
Australasian Institute of Mining and Metallurgy 443
Australasian Virtual Engineering Library: Mechanical
    Engineering 422
Australian Bureau of Statistics 448
Australian Government—Geoscience Australia 440
Australian Mineral Industries Research Association 444
Australia Mining and Exploration 440
Australian Science and Technology Heritage Centre of the
    University of Melbourne 339
Authoritative Dictionary of IEEE Standards Terms 249
*Automated Factory Handbook: Technology and
    Management* 352
Automatica 55, 252
*Automation, Production Systems, and Computer-
    integrated Manufacturing* 356
Automobile Abstracts 500
*Automotive Bibliography: 13,000 Works in English,
    Czech/Slovak, Danish, Dutch, Finnish,
    French, German, Italian, Norwegian,
    Polish, Portuguese, Slovenian, Spanish and
    Swedish* 399
Automotive Chassis 404
Automotive Electrics/Automotive Electronics 404
Automotive Electronics Handbook 404
Automotive Engineering Fundamentals 513
Automotive Engineering International 419, 517
Automotive Handbook 510
Automotive Intelligence 423
Automotive Lubricants Reference Book 404
Automotive Safety Handbook 510
Autonomous Robots 252
Auyang 318, 328, 340
Avallone, E .A. 53, 402
Average World Steel Transaction Prices 447
Avery Index to Architectural Periodicals 82
Aviation Today 57
Aviation Week and Space Technology 55
AWWA Sourcebook 286
Ayers, J. B. 351

B

Babcock, D. L. 30
*BABEL: A Glossary of Computer Oriented Abbreviations
    and Acronyms* 213
Bachman, L. R. 87
Bacteriology Abstracts 283
Badiru, A. B. 352
Baetz, B. 329
Bagchi, A. 297
Baier, R. 158
Bailey, M.R. 133
Bainbridge, W. S. 214

Baine, C. 35
Bains, W. 67
Baird, G. 87
Baird, J. A. 175
Bakerjian, R. 408
Balachandran, S. 315
Baldwin, Jerry 5
Ball, J.K. 513
Ballast, D. K. 84
Bangash, M. Y. H. 172, 174, 175
Banks, J. H. 511
Banville, D. L. 114
Baratta, A. J. 461
Barber, A. 407, 411
Barbosa-Canovas, G.V. 68
Barbour Index 170
Barker, J. A. 137
Bartlett, A. 150
Barr, D. I. H. 187
Barr, M. 212
Barry, R.L. 31
Bartos, J.M. 164
Baruth, E. E. 297
Basic Engineering Series and Tools (series) 33
Basic Petroleum Data Book 493
Basselm, J.A. 67
Basta, N. 31
Batik, A. L. 22
Bauingenieur 177
Baum, E. M 459
Baumbach, G. 296
Baumeister, T. 53, 402
Bausch, P. 17
Bautechnik 177
Bear, J. 265
Bear, M. 265
Beardmore, N. 183
Bearing Analysis Handbook 411
*Bear's Guide to the Best Computer Degrees by Distance
    Learning* 265
Beaudoin, J.J. 382
Beck, S. 51
Beckham, B. 35
*Beckham's Guide to Scholarships for Black and Minority
    Students* 35
*Becoming Leaders: A Practical Handbook for Women in
    Engineering, Science, and Technology* 36
Bedding, B. 176
Bednar, H. H. 412
Beer, D. F. 32
Beer, F. P. 416
Beim, H. J. 291
Beitz, W. 402
Bejan, A. 405
Belanger, D. O. 328
Bell, G. I. 460
Bell, T.E. 340
Bell Labs Technical Journal 224
Belzer, J. 215
Benchmarks for Science Literacy 268
Bender, D. A. 67
Bender, A. E. 67
Benders' Dictionary of Nutrition and Food Technology 67
Benefits of Flood Alleviation 188

Benhabib 343, 356, 368
Benjamin, N.B.H. 151
Bennett, J. 150
Benvenuto, E. 170
Bepress 220
Bergeson, L.L. 165
*Berichtezur Wissenschaftsgeschichte: Organ der
    Gesellschaft fuer Wissenschaftsgeschichte* 331
Berinstein, P. 13
*Berkshire Encyclopedia of Human-Computer
    Interaction* 214
Berlow, L. H. 172, 327
Bernoulli, D. 45
Bernstein, L. B. 187
Bernstein, M. D. 412
Berthouex, P. M. 294
Best Manufacturing Practices 360
Beton Arme 178
Beton-Kalendar 178
Beton und Stahlbetonbau 178
Better Roads 515
Bever, M. B. 375, 376
BFRL Strategic Goal: Measurement Science for
    Sustainable Infrastructure Materials 200
Bhandari, A. 295
Bhogal, A. 201
Bhushan, B. 415
Bianchina, P. 84
*Bibliografia Historica de la Ciencia y la Tecnica en
    Espana* 323
Bibliografia Italiana di Storia della Scienza 321
Bibliographic Guide to the History of Computer
    Applications, 1950–1990 209
Bibliographic Guide to Technology 430
Bibliographic Guide to the History of Computing,
    Computers and the Information Processing
    Industry 209
Biographical Dictionary of the History of Technology 326
Bibliography for History of Engineering 323
Bibliography of Agricultural Engineering Books 65
*Bibliography of Bibliographies of Agricultural
    Engineering and Related Subjects* 65
Bibliography of British and Irish History 323
Bibliography of the History of Technology 324, 331
*Bibliography on the History of Mechanical Engineering
    and Technology* 397
Bibliography: Underrepresented Students 35
Bickel, J. O. 436
Bickford, J. H. 171
Bicycle Countermeasures Selection System 508
Bidanda, B. 352
Bidgoli, H. 16, 215
Bies, J. D. 13
BIKESAFE 508
Billing, B. 153
Billington, D. P. 170, 329
Bilton, Nick x
BIMCO 523
Bindocci, C. G. 324
Bing x
Biochemical Engineering Journal
BioCommerce Data's Biotechnology Directory 97
Biodegradation 70, 299
Biodiesel Magazine 488

Biodynamics Database 101
BioEnergy Research 70
BioEngineering Abstracts 97
Biofiltration for Air Pollution Control 296
Biofuels 70
Biofuels, Bioproducts, and Biorefining 70, 486
*Biographical Dictionary of American Civil
    Engineers* 133
Biographical Dictionary of Civil Engineers 133
*Biographical Dictionary of Civil Engineers in Great
    Britain and Ireland* 133
Bioinspiration & Biomimetics 252
Biointerphases 385
Biological & Agricultural Index 64
Biological Engineering 71
Biological Wastewater Treatment 207
Biomass & Bioenergy 71
Biomaterials 385
Biomaterials Forum 104
Biomaterials Network 105, 387
BIOMAT-L 106
BIOMCH-L 106
Biomechanics and Modeling in Mechanobiology
*Biomechanics: The Magazine of Body Movement and
    Medicine* 104
BioMed Central 105
Biomedical Engineering and Design Handbook 102
Biomedical Engineering Desk Reference 102
Biomedical Engineering Handbook 100, 101
Biomedical Engineering Network 106
Biomedical Engineering Society 108
Biomedical Materials 385
Biomedical Microdevices 104
Biomedical Technology and Devices Handbook 102
BioMolecular Engineering Research Center 101
BioOne 97
Bioprocess and Biosystems Engineering 71
Bioresource Technology 71
Biosensors & Bioelectronics 71
BIOSIS Previews 64, 97
Biosystems Engineering 71
Biotechnology and Bioengineering 71
Biotechnology and Bioengineering Abstracts 97
Biotechnology for Biofuels 71
Biotechnology from A to Z 67
Biotechnology Research Abstracts 283
Bird, R. B. 120
Birkhauser Construction Manuals 85
*Bir Umm Fawakhir: Insights into Ancient Egyptian
    Mining* 429
Bishop, R. 513
Bishop, R. H. 409, 416
Bisio, A. 286
Biswas, A. K. 183
Bittence, J. C. 348
Bittnar, P. 164
Bitton, G. 288
Black, J. 101, 378
Blackburn, W. H. 438
Blacks in Science and Medicine 36
Blackwell Encyclopedia of Industrial Archaeology 327
*Blaetter fuer Technikgeschichte. Vienna, Austria:
    Technisches Museum Wien mit
    Osterreichischer Mediathek* 331

Blake, L. S. 133
Bland, A.E. 164
Blaschke, S. 331
Blast Effects on Buildings 173
Bleier, F. P. 408
Blicq, R. 33
Bloch, H. P. 415
Blockley, D. 132, 150
Bloor, D. 375
Blow, G. J. 158
Blowout and Well Control Handbook 481
Blue Book of Building and Construction 153
BMES Bulletin 104
BNA 250
Boedecke, K.J.
Boiler Operator's Guide 413
Boilers for Power and Process 414
Bold and the Brave: A History of Women in Science and Engineering 38
Bonilla, J.V. 381
Books in Print 54
Books24x7: ITPro 234
Booser, E. R. 415
Boots, S. 286
Borisenko, V. E. 164
Bosch Automotive Handbook 404, 510
Bossanyi, E. 405
Boulding, J. R. 292
Bovik, A. 248
Bovill, C. 87
Bowlin, G. L. 100
Boyce, M. P. 408
Boyce, P. R. 512
Boyes, W. E. 408
BP Statistical Review of World Energy 493
Brady, G. S. 11, 379
Bralla, J. G. 352
Brandes, E. A. 383
Brandrup, J. 379
Branan, C. 117
Branin, J. 220
BRE 188
BRE Bibliography of Structural Failures 171
Bregman, J. I. 158, 288
Brenner, M. 14
Brewer, A. M. 352
Breyer, D. 175
Brick Development Association 181
Bridge Building: Art and Science 179
Bridge Construction and Engineering 179
Bridge Deck Behaviour 173
Bridge Engineering 505
Bridge Engineering: A Global Perspective 173
Bridge Engineering Handbook 172
Bridge Engineering. Proceedings of the Institution of Civil Engineers 177
Bridge Management 174
Bridge Site 179
Bridging the Gap: Rethinking the Relationship of Architect and Engineer: The Proceedings of the Building Arts Forum 87
Bristow, G. V. 51
British Constructional Steelwork Association Ltd 181
British Geotechnical Association 141

British Humanities Index 321
British Journal for the History of Science 331
British Library, The—Manuscripts Catalogue 323
British Masonry Society 181
British Society for the History of Science 338
British Standards Institute 492
British Standards Institution 22, 194
Brockenbrough, R. L. 156, 175, 506
Brogdon, J. 65
Bronikowski, R. J. 30
Bronzino, J. D. 100, 101
Brook, G.B. 383
Brooks, F. P. 206
Brooks, H. 84
Broome, J. 150
Brown, C.A. 292
Brown, G. 173
Brown, L.C. 294
Brown, M.W. 410
Brown, R. H. 68
Brown, R. W. 137
Brown, T.H. 408, 409
Brownell, F. W. 291
Brownfields: Managing the Development of Previously Developed Land 159
Bronze Age 429
Brusaw, C.T. 32
Bruschi, B.A. 39
Bryan, J. C. 460
BUBL Links 422
    Mining 449
Bucher, W. 12
Budynas, R. 174, 416
Buettner, D. R. 85
Builder 88
Builder.com 235
Building and Environment 299
Building Arts Forum 87
Building a Workforce for the Information Economy 269
Building Research Establishment 181
Building the World: An Encyclopedia of the Great Engineering Projects in History 326
Builder's Comprehensive Dictionary 85
Builders: Marvels of Engineering 327
Building (Construction) References 170
Building Design & Construction 88
BuildingGreen.com 82
Building Law Reports 149
Building Science Abstracts
Building Systems Integration Handbook 86
Building the Future: Building Technology and Cultural History from the Industrial Revolution until Today 328
Building: 3000 Years of Design Engineering and Construction 170, 328
Bulleit, W. M. 33
Bulletin of Canadian Petroleum Geology 486
Bulletin of Environmental Contamination and Toxicology 299
Bulletin of the International Association for Shell and Spatial Structures 177
Bulletin of the Seismological Society of America 138
Bunch, B. H. 327
Bunni, N. G. 151

Bunshah, R. F. 379
Burdick, D. L. 495
Bureau of Mines Minerals Yearbook 441
Bureau of National Affairs 306
Bureau of Transportation Statistics 501, 521
Buried Pipe Design 413
Burke, G. 288
Burke, R. J. 38
Burks, A. R. 209
Burstall, A. F. 329
Burton, F. 298
Burton, G. A. 293
Burstall, A. F. 396
Burton, T. 405
Busby, R. L. 493
Bhushan, B. 378
Buschow, K. H. J. 375
Business Press International 457
Buss, E. W. 248
Bussmann, S. 356
Butler, R. B. 85
By Design: Better Places to Live

C

C10 225
Caballero, B. 67
CACHE 112, 127
Cacuci, D. G. 459
Cadick, J. 248
Cahn, R. W. 375, 384
Cahn, W. 376
Caines, A. J. 404
California Air Resources Board Ambient Air Quality
    Standards 303
California Institute of Technology (CalTech)
California Polytechnic State University–San Luis
    Obispo 80
California State University, Long Beach, Chemical
    Engineering 114
Calishain, T. 17
*Callaham's Russian–English Dictionary of Science and
    Technology* 13
Callen, H. 120
Callister, W. D. 384
*CALPHAD (Calculation of Phase Diagrams):
    A Comprehensive Guide* 382
Caltech Institute Archives 334
Camacho, C. 40
Cambridge Aerospace Dictionary 51
Cambridge Air and Space Dictionary 52
Cambridge Dictionary of Science and Technology 52
Cambridge Dictionary of Space Technology 52
*Cambridge Encyclopedia of Space: Missions,
    Applications, and Exploration* 52
Cambridge Scientific Abstracts 281, 432
CAM Design and Manufacturing Handbook 354
Camenson, B. 32
CAMESE 439
Campbell, T.J. 439
Canada Energy Minerals and Metals Information
    Centre 432
Canada Science and Technology Museum 335
Canadian Agricultural Safety Association 77

Canadian Aeronautics and Space Journal 55
Canadian & American Mines Handbook 438, 447
Canadian Association of Petroleum Producers 491, 493
Canadian Biosystems Engineering 71
Canadian Civil Engineer 134
Canadian Codes Centre 194
Canadian Geotechnical Journal
Canadian Geotechnical Society 138, 141
Canadian Institute of Food Science and Technology 77
Canadian Institute of Mining, Metallurgy and Petroleum
    444, 491
Canadian Journal of Chemical Engineering
*Canadian Journal of Electrical and Computer
    Engineering Revue* 225
Canadian Minerals Yearbook: Review and Outlook 346,
    447, 448
Canadian Mining Journal 435
Canadian Nuclear Association–Association Nucleaire
    Canadienne 465
Canadian Patents Database 26
Canadian Petroleum Products Institute 491
Canadian Research Index 432
Canadian Society for Bioengineering 77
Canadian Society for Civil Engineering 135
Canadian Society for Engineering in Agricultural, Food
    and Biological Systems 77
Canadian Transportation Research Forum 524
Canadian Wellsite 489
Cantu, N. E. 36
CA Patent Index
Capelli-Schellpfeffer, M. 248
CAplus 113, 124
Cardarelli, F. 12, 379
CA Reg File 113
Carnegie Mellon University Libraries Engineering and
    Science Library 218
   History: History of Technology and Science 336
Carr, D. 438
Carter, D. V. 329
Carvill, J. 402
Case for IRs: A SPARC Position Paper 220
Case Western Reserve University Special Collections:
    History of Science and Technology 333
CASreact 113
Cast Iron: Physical and Engineering Properties 175
Castro, C. 512
Catalog of Free Compilers and Interpreters 237
CatalogXpress 348
Catalysis Letters
Catalysis Reviews: Science and Engineering
Catalysis Today
Caterpillar Performance Handbook 437
Cavinato, J. L. 504
Cayley, Sir George 45
CBS News 495
C/C++ Users Journal 225
CDM Regulations Explained 152
Cebon, D. 404
CEH Wallingford 190
Cement and Concrete Aggregates 178
Cement and Concrete Research 178
Cement Minerals Products Association 181
Center for Computational Materials Science 387
Center for Integrated Manufacturing Studies 360

Center for Sustainable Systems 200
Center for the Advancement of Hispanics in Science and
        Engineering Education 40
Centre for the History of Science, Technology and
        Medicine, University of Manchester 339
Center for Transportation Analysis 522
Centre for Nanomaterials Applications in
        Construction 165
Centre for Window and Cladding Technology 181
Centrifugal Pumps 414
Centrum Holst 176
Ceramic Abstracts 373
Ceramics International 385
Ceramics-Silikaty 385
CERAM Research 181
Ceruzzi, P. E. 209
CESMM3 Price Database 151
CFRs Made Easy 288
Challen, B. 404
Chanda, M. 379
Chang, K. 248
Chang, N.-B. 197
Chang, T.C. 369
Channell, D. F. 264, 324
Channing, J. 357
Charles Babbage Institute: Center for the History of
        Information Technology 210
Charlton, S. G. 352
Charlton, T. M. 170
Chartered Institute of Building 151
Chartered Institution of Wastes Management 161
Chartered Institution of Water and Environmental
        Management 190
Chart of the Nuclides 459
Chase, M. W., 53
Chatterton, J.B. 188
ChemCats
ChemFinder.com
ChE Links 114
Chemical Abstracts Service 98
Chemical and Engineering News 123
Chemical and Petroleum Engineering 486
Chemical and Process Thermodynamics 121
Chemical Carcinogenesis Research Information
        System 105
Chemical Engineering 123
Chemical Engineering and Processing 123
Chemical Engineering and Technology 123
Chemical Engineering Communications 123
Chemical Engineering Education 123 263, 264, 271
Chemical Engineering Faculty Directory 265
Chemical Engineering Progress 123, 127
Chemical Engineering Journal 123
Chemical Engineering Research and Design 132
Chemical Engineering Science 123
Chemical Engineer's Resource Page 114
*Chemical Information Mining: Facilitating Literature-
        Based Discovery* 114
Chemical Market Reporter 123
Chemical Properties Handbook 118
*Chemical Resistance Guide for Plastics: A Guide
        to Chemical Resistance of Engineering
        Thermoplastics, Fluoroplastics, Fibers, and
        Thermoset Resins* 382

Chemical Safety Manual 122
Chemical Week 123
Chemie-Ingenieur-Technik 123
Chemistry and Industry 123
Chemistry and Physics of Carbon 123
Chemistry and Technology of Fuels and Oils 486
Chemistry of Materials 385
ChemList 113, 114
Chemosphere 299
Chen, C-C. 399, 429
Chen, J. 165
Chen, W. 248
Chen, W. F. 85, 137, 171, 172, 506
Chen, W.-K. 217, 248
Cheremisinoff, N. P. 51, 286, 288, 291, 293, 375, 401, 405
Chicago Manual of Style 126
Chin, D. A. 297
Chironis, N.P. 409
Cho, H. 409
Cho, Y.I. 406
Choice 248
Cholet, H. 481
Chopey, N. P. 117
Chrest, A. P. 158
Chrimes, M.M. 131, 133, 201, 202
Christiansen, D. 248
Christie, B. A. 36
Christopher, W. F. 352
Chung, C.A. 511
Chung, H. W. 151
CIGR 68, 77
CIGR Handbook of Agricultural Engineering 68
Cimbala, J.M. 296
CIM BULLETIN 435
C.I.M. Directory 439
CIM Magazine 488
CINDAS Materials Properties Data Files 379
Circuits and Filters Handbook 217, 248
Circulation 186
CIRIA 173, 176, 188
CIRP Annals-Manufacturing Technology 358
CiteSeer 221
Civieltechnisch Centrum Uitvoering Research en
        Regelgeving 181
Civil and Offshore Engineering Research 146
Civil Engineering 134
Civil Engineering Construction Contracts 153
Civil Engineering Database 81, 132
Civil Engineering Handbook 85, 133
Civil Engineering Magazine 134
Civil Engineering Materials 173
Civil Engineering of Underground Rail Transport 159
Civil Engineering Procedure 152
*Civil Engineering Sealants in Wet Conditions—Review
        of Performance and Material Guidance in
        Use* 176
Civil Engineering through the Ages 131, 202
Civil Engineer's Reference Book 133
Civil Engineering through the Ages 131, 202
Clark, A. 438
Clark, J. 13
Clark, J. R. 144
Clason, W. E. 458
Clause, H.R. 11

Clausen, W.E. 416
Clean Air Handbook 291
Clean Water Handbook 186, 293
Clean Water Report 299
Cleland, D. I. 347, 352
Cleveland Clinic 94
Climate Research 299
Clinical Engineering Handbook 101
Clinical Gait Analysis 106
Clinical Oral Implants Research 104
ClinicalTrials.gov 105
Cloud, G. S. 65
Clough, R. H. 151
Clough, S.R. 165
CNET 236
CNET Networks 239
Coal Association of Canada 444
*Coastal Engineering: An International Journal for
        Coastal, Harbor, and Offshore Engineers* 145
Coastal Engineering Journal 145
Coastalmanagement.com 145
Coastal Zone Management Handbook 144
Coasts, Oceans, Ports, and Rivers Institute 146
Cobb, G. 173
Cobeen, K.E. 175
CODATU 518
Code of Federal Regulations 22, 307, 520
Code of Practice for Project Management 151
Code of Silence: Ethics of Disasters 31
Coghill, A. M. 126
CoGPrints 221
Cohen, I. B. 209
Cold-Formed Steel Design 175
Cold Regions Science and Technology 299
*Cold War and American Science: The Military-
        Industrial-Academic
Complex and MIT and Stanford* 267
Cole, M. 38
Coles, S. L. 242, 257
Collected Algorithms of the ACM 237
*College blue book: Scholarships, fellowships, grants, and
        loans* 43
Colorado Water Institute 191
Combustion and Flame 55, 122, 419
Combustion Institute 424
Combustion Science and Technology 55, 122
Commercial Technologies for Maintenance Activities 360
Commission Internationale du Genie Rural (see CIGR)
Commission on Professionals in Science and
        Technology 36
Commodity Statistics and Information 448
Communication for Professional Engineers 153
*Communications and Signals Manual of Recommended
        Practice* 510
Compendex 16, 18, 35, 49, 62, 83, 98, 208, 227, 230,
        244, 245, 252, 262, 272, 282, 346, 374, 398,
        473, 500
Compendium of Canadian Mining Suppliers 439
Compendium of On-Line Soil Survey Information 140
*Compilation of Air Pollutant Emission Factors,
        Stationary Point and Area Sources* 291
Compilation of EPA's Sampling and Analysis Methods
        157, 305
Compilers.net 237

CompInfo: The Computer Information Center: Computer
        Standards 222, 236
Complete Guide to Consulting Contracts 152
Complete Guide to Dowsing 187
Composite Materials 387
Composite Materials Handbook—MIL 17 79
Composites Science and Technology 385
Composite Structures 385
*Composite Structures for Civil and Architectural
        Engineering* 87
Compost Science & Utilization 71
Comprehensive Composite Materials 376
*Comprehensive Dictionary of Engineering and
        Technology* 13
Comprehensive Index of Publications 65
Comprehensive Nanoscience and Technology 378
Compressed Air and Gas Handbook 407
Computational Fluid Dynamics 191
Computer Abstracts International 230
Computer Abstracts International Database 228
Computer & Control Abstracts 18
Computer and Information Systems Abstracts 228
Computer Applications in Engineering Education 270
Computer Communications 225
*Computer Desktop Encyclopedia: The Indispensable
        Reference on Computers* 215
Computer Engineering Handbook 217
Computer from Pascal to von Neumann 209
*Computer Glossary: The Complete Illustrated
        Dictionary* 212
Computer Information Systems Abstracts 230, 246
Computer Journal 225
*Computer Networks: The International Journal
        of Computer and Telecommunications
        Networking* 225
Computers and Electrical Engineering 225
Computers and Chemical Engineering 123
Computers and Electronics in Agriculture 71
Computers and Fluids 419
Computers & Industrial Engineering 358
Computers & Operations Research 358
Computers and Structures 177
Computer Science: A Guide to Web Resources 238
Computer Science Handbook 217
Computer Science Index 228
Computer Science Journals 223
Computer Source 228
Computer Standards and Interfaces 225
Computer Systems Science and Engineering 225
Computer Technology 230
Computerworld 225
Computing Before Computers 209
Computing Information Directory 205
Computing Research Association 232
Computing Research Policy Blog: Advocacy and Policy
        Analysis for the Computing
Community 239
Computing Research Repository 221
Computing Reviews 211
Comyns, A .E. 115
*Conceptual Structural Design: Bridging the Gap between
        Architects and Engineers* 87
*Concise Dictionary of Biomedicine and Molecular
        Biology* 99

Concise Dictionary of Materials Science: Structure and Characterization of Polycrystalline Materials 376
Concise Encyclopedia of Composites 376
Concise Encyclopedia of Computer Science 216
Concise Encyclopedia of Materials Characterization 375
Concise Encyclopedia of Materials for Energy Systems 376
Concise Encyclopedia of Materials Processing 376
Concise Encyclopedia of Mineral Resources 438
Concise Encyclopedia of Plastics 376
Concise History of Mining 428
Concise Polymeric Materials Encyclopedia 376
Concrete 178
Concrete Construction 178
Concrete Construction Engineering Handbook 174
Concrete Construction Handbook 174
Concrete International 179
Concrete Reinforcing Steel Institute 181
Concrete Science and Engineering 179
Concrete Society 181
Concurrent Engineering–Research and Applications 358
Condensed Encyclopedia of Polymer Engineering Terms 375
Confer, R. G. 350
Confer, T.R. 350
Conference on Deep Foundations 142
Conference Proceedings Citation Index 20
Confronting the 'New' American Dilemma, Underrepresented Minorities in Engineering: A Data-Based Look at Diversity 43
Conservation of Bridges 172
Considine, D. M. 67
Considine, G. D. 67
Consortium for Advanced Manufacturing–International 360
Constructed Wetlands for Wastewater Treatment: Municipal, Industrial, and Agricultural 297
Constructing the Team 152
Construction and Building Materials 177
Construction and Design of Cable Stayed Bridges 174
Construction and Maintenance Web 521
Construction Best Practice Programme 154
Construction Building Envelope and Interior Finishes Databook 172
Construction Contracting 151
Construction Equipment Guide 151
Construction Glossary: An Encyclopedic Reference and Manual 85
Construction Industry 153
Construction Industry Board 151
Construction Industry Council 151, 154
Construction Industry Law Newsletter 149
Construction Industry Research and Information Association 151
Construction Information Service 171
Construction Law 153
Construction Law Journal 150
Construction Law Reports 150
Construction Management 152
Construction Management and Economics 150
Construction Management Association of America 154
Construction Materials Reference Book 172
Construction Materials: Their Nature and Behaviour 173

Construction Materials: Types, Uses and Applications 86
Construction Metallique 179
Construction of Marine and Offshore Structures 144
Construction: Principles, Materials, and Methods 86
Construction Project Management 153
Construction Review 88
Construction Specifications Handbook 86
Construction Specifications Institute 154
Construction Specifier 88
Construction WebLinks 90
Contaminants of Emerging Concern 295
Contaminated Land 160
Contaminated Rivers: A Geomorphological-Geochemical Approach to Site Assessment and Remediation 298
Containerisation International Yearbook 504
Contemporary and Historical Literature of Food Science and Human Nutrition 65
Contemporary Logistics 511
Contescu, C.I. 377
Contextual Factors in Education: Improving Science and Mathematics Education for Minorities and Women 38
Contract and Fee-Setting Guide for Consultants and Contractors 149
Contractor's Dictionary of Equipment, Tools, and Techniques 12
Control Systems Engineering 418
Cook, N. J. 173
Cook, R. L. 31
Cooke, A. 17
Cookfair, A.S. 26
Coombs, C.F. 353
Cooper, A. R. 156, 291
Cooper, C. L. 356
Cooper, J. 264
Cooperation for Urban Mobility in the Developing World 518
Copper: The Red Metal 429
Corbett, E. C. 151
Corlett, J. 10, 50, 97, 399, 430, 495
Corbitt, R. A. 185, 288
Cormier, D.R. 409
Cornell University Library. History of Science Collections 333
Corrosion Abstracts 373
Corrosion Science 385
Corrosion Survey Database 37
Cortado, J. W. 208, 209, 324
Cosloy, S.D. 99
Cost Benefit Analysis for Engineers
Costruzione Metalliche 179
Coull, A. 174
Council on Forest Engineering 77
Council on Tall Buildings and Urban Habitat 179
Country Mine 449
Countryside Commission 158
Coutts, B.E. 324
Covington, C.A. 212
Covington, M.A. 212
Covington, M.M. 212
Cowan, H. J. 84, 85
Cox, R. C. 133
Craddock, P. 428

Crafts-Lighty, A. 96
Craig, R. F. 138
Crain Communications' Automotive News Data
        Center 502
CRC Dictionary of Agricultural Sciences 68
CRC Handbook of Chemistry and Physics 117
CRC Handbook of Engineering in Agriculture 68
*CRC Handbook of Lubrication: (Theory and Practice of
        Tribology)* 415
CRC Handbook of Mechanical Engineering 403
CRC Handbook of Thermal Engineering 406
CRC Materials Science and Engineering Handbook 383
CRCnetBASE 247, 399
CRC Press 230
Cressy, E. 397
Critical Reviews in Biotechnology 123
*Critical Reviews in Environmental Science and
        Technology* 300
Critical Reviews in Food Science and Nutrition 72
*Critical Reviews in Solid State and Materials
        Sciences* 385
Crocker, D. 51
Crocker, M. J. 411
Croney, D. 158
Croney, P. 158
Cronin, F. 495
Crosbie, M. 30
Cross-Rudkin, P. S. 133
Croucher, P. 355
CROW 161
Crow, R. 220
Crowson, P. 448
Crowson, R. 352, 356
Crop Science 72
CrossRef 17
Cryogenic Society of America 390
Crystallography Online 387
CSA/ASCE Civil Engineering Abstracts 132, 282
CSA Illumina 17, 33
CSA Materials Research Database with METADEX 373
CSIRO 444
CSPress/JohnWiley 230
Current Bibliography in the History of Technology 322
Current Biology 495
Current Contents Connect 284
Current Opinion in Biotechnology 72
*Curriculum Guidelines for Undergraduate Degree
        Programs in Computer Engineering* 204
Currin, T.R. 506
Curry, G. L. 356
Curwell, S. 173
Cussler, E. L. 112, 120, 127
Cutcliffe, S. H. 329
Cybermetics Lab 220

D

DAAAM International Symposium Intelligent
        Manufacturing & Automation 366
Daganzo, C. F. 512
Dagleish, M. 512
Dahl, O-J. 206
Daigger, G. T. 297
Daintith, J. 211

*Damage Tolerant Design Handbook: A Compilation of
        Fracture and Crack Growth Data for High
        Strength Alloys.* 53
Dam decommissioning in France 189
Daniels, K. 86
Danner, R. P. 118
Darling, A.B. 330, 340
Darnay, A. J. 284
Darrin, A. G. 53
Darton, M. 13
Das, B. M. 138
Dasch, E. J. 439
Datasheet Archive 246
Datta, A. K. 69
Daubert, T. E. 118
Daumas, M. 318, 327, 340
Davis, E. B. 97
Davis, J. 173
Davis, J. R. 101, 375
Davidson, P. 326
Davies, O. 428
Davies, S. 495
Davis, L. R. 295
Davis, M. 32, 33, 53
Davis, W. T. 192
Davis' Handbook of Applied Hydraulics 185
*Dawn of the Electronic Age: Electrical Technologies
        in the Shaping of the Modern World,
        1914–1945* 330
Day, D. A. 151
Day, L. 326
Day, R. W. 137
DECHEMA 123
Dechema Corrosion Handbook 380
de Coster, J. 401
Deep Foundations Institute 141
Deepwater Horizon ix, 278
*DEFRA (Annual) Digest of Environmental
        Statistics* 189
*Dekker Encyclopedia of Nanoscience and
        Nanotechnology* 4, 377
Dekosky, R. K. 324
Deliiska, B. 68
Della-Giustina, D. 122
Delleur, J. W. 185
Delphion 27
de Negri, V.J. 407
Den Hartog, J. P. 416
Department for the Environment, Transport, and the
        Regions (see DETR)
DePetro, T.G. 50
Derwent Biotechnology Resource 98
Desai, V.R. 297
Desalination 300
Deshusses, M.A. 296
Design and Construction of Silos and Bunkers 174
Design and Performance of Road Pavements 158
*Design and Safety of Pedestrian Facilities:
        A Recommended Practice* 508
Design Automation for Embedded Systems 225
Design-Build Institute of America 154
*Designer's Guide to Wind Loading of Building
        Structures* 173
Design for Manufacturability Handbook 352

Design for Manufacturability through Design-Process Integration 366

Design for Manufacturing and the Life Cycle Conference 367

Design Guide for Marine Treatment Schemes 186

Designing Walkable Urban Thoroughfares: A Context Sensitive Approach 509, 520

Design Institute for Physical Property Data 118

Design Manual for Roads and Bridges 157

Design of Landfills and Integrated Solid Waste Management 297

Design of Modern Steel Railway Bridges 510

Design of Small Dams 189

Design of Wood Structures 175

Design Paradigms: Case Histories of Error and Judgment in Engineering 174

Design Recommendations for Multi-Storey and Underground Car Parks 156

Designs for Science Literacy 268

de Silva, C. W. 411

Dessauer, Fredrich 94

Detail: Zeitschrift fur Architektur & Baudetail & Einrichtung 88

DETR 151, 159

Deutsches Institut fuer Normung 195

Deutsches Museum 335

Developing a Safety and Health Program 121

Developmental and Reproductive Toxicology 105

Developments in Structural Form 173, 88

Devine, J. 17

Devinny, J. S. 296

DeWitt, D.P. 417

DeWolf, J.T. 416

DevX.com 236

De Zuane, J. 185, 293

Dhir, R.K. 173

DIAL Electronics Weekly Buyers' Guide 348

Dialog 18, 33, 230, 281

Diamond Films Handbook 378

Dibner, B. 428

Dibner Library of the History of Science 333, 339

Dickenson, T. C. 412, 414

Dickson, P. 51

Dictionary 504

Dictionary of Agricultural and Environmental Science 68

Dictionary of Agricultural and Food Engineering 67

Dictionary of Agricultural Sciences 68

Dictionary of Algorithms and Data Structures 213

Dictionary of Architectural and Building Technology 84

Dictionary of Architecture and Construction 12

Dictionary of Automotive Engineering 400, 505

Dictionary for Automotive Engineering: English–French–German: With Explanations of French and German Terms 401

Dictionary of Aviation 51

Dictionary of Bioinformatics and Computational Biology 99

Dictionary of Biomedical Science 99

Dictionary of Bioscience 99

Dictionary of Building 85

Dictionary of Building Preservation 12

Dictionary of Civil Engineering 133

Dictionary of Civil Engineering and Construction (Wiley) 133

Dictionary of Computer and Internet Terms 212

Dictionary of Computer Science, Engineering and Technology 212

Dictionary of Computing 211

Dictionary of Ecodesign 85

Dictionary of Engineering Acronyms and Abbreviations 22

Dictionary of Engineering and Technology, with Extensive Treatment of the Most Modern Techniques and Processes 13

Dictionary of Engineering Materials 84, 375

Dictionary of Environmental & Civil Engineering 12, 85, 288

Dictionary of Environmental and Sustainable Development 196

Dictionary of Environmental Science and Engineering 287

Dictionary of Environmental Science and Technology 287

Dictionary of Food Science and Technology 68

Dictionary of Geotechnics 137

Dictionary of Industrial Archaeology 326

Dictionary of Industrial Administration 350

Dictionary of Mechanical Engineering 400

Dictionary of Mining, Mineral, and Related Terms 438

Dictionary of Nanotechnology, Colloid and Interface Science, 164, 376

Dictionary of Project Management Terms 351

Dictionary of Report Series Codes 28

Dictionary of Soil Mechanics and Foundation Engineering 137

Dictionary of the Space Age 51

Dictionary of Water and Sewage Engineering 185

Dictionary of Water and Waste Management 185, 287

Dictionnaire du Genie Civil 133

Diesel Engine Reference Book 404

Dieter, G. E. 384

Diffusion: Mass Transfer in Fluid Systems 120

Diggers, R. 248

Digital Bibliography and Library Project 221

Digital Dissertations 4

Digital Signal Processing: A Practical Guide for Engineers and Scientists 217

Dijkstra, E. W. 206

Dinkel, J. 504

DIPPR Data Compilation of Pure Compound Properties 18

Directory of Biomedical and Health Care Grants 98

Directory of Engineering Document Sources 22

Directory of Geoscience Departments 266

Directory of Financial Aids for Minorities 36

Directory of Health Organizations 105

Directory of Human Factors/Ergonomics Graduate Programs in the United States and Canada 265

Directory of Nuclear Reactors 457

Directory of Nuclear Research Reactors 457

Directory of Open Access Journals 105, 224

Directory of Published Proceedings 20

Directory of Safety Standards, Literature and Services 349

Directory of Universities and Approved Programs 66

Discipline of Programming 206

Displays 225

Dissertation Abstracts 4

Diversity/Careers in Engineering & Information Technology 38
Dixon, K. 512
Djarova, M. 68
DOAR 220
Dobrowolski, J. A. 174
Document Engineering Company 24
DOE Information Bridge 283, 455
Doepke, M. 261, 275
Dohda, K. 416
Dohrmann, C.R. 357
Doing It Differently: Systems for Rethinking Construction 150
Doing the Right Thing: All Ethics Guide for Engineering Students 31
Donahue, R.L. 68
Dooley, J. H. 62, 78
Dooling, D. 340
Doran, D. K. 172
Dorf, R. C. 13, 248, 352, 416
Dorian, A. F. 438
Dorn, H. 325, 330
Dornfest, R. 17
Downing, D. A. 212
Dr. Dobb's Journal 225
Dresley, Susan 5, 526
Drew, S. W. 67, 78
Drexel University 80
DGRWeb 265
Drilling Contractor 488
Drilling Data Handbook 481
Drilling Fluids Processing Handbook 481
Drozda, T. J. 408
DTIC Online: Public Technical Reports 49
Duan, L. 172, 506
Dubbel, H. 402
Dubbey, J. M. 209
DuBrock, C.P. 39
Duderstadt, J. J. 460
Dudley, D. W. 409
Dudley's Gear Handbook 411
Durkin, T. H. 415
Durrans, S. R. 298
Dutton, H. 176
Dwight, J. 177
Dym, C. L. 263
Dyro, J. 101
Dyrud, M. 264

E

E&MJ: Engineering & Mining Journal 434
E&P 488
Earley, M.W. 248
Early History of Mechanical Engineering 397
Earth Manual 138
Earthquake Engineering Abstracts 136, 282
Earthquake Engineering and Structural Dynamics 138
Earthquake Engineering Handbook 137
Earth Retention Systems Handbook 138
Earthscan 256
Earth Sciences Information Resources–Internet Links 449
EarthWeb 236
Eastlake, C.N. 52

Eaton, A.D. 305
EBSCO Technical Package 226
Eccleston, C. H. 289
ECHO: Exploring and Collecting History Online— Science, Technology, and Industry 336
ECN: Energy Research Centre of the Netherlands 200
Ecodesign: A Manual of Ecological Design 87
Ecology Abstracts 283
Edgar, T. F. 112, 127
Editions Technip 485
Educating the Engineer for the 21st Century 268
Educating the Engineer of 2020: Adapting Engineering Education to the New Century 269
Educational Programs in Agricultural and Biological Engineering and Related Fields—United States, Canada, and Ireland 66
Edwards, J. T. 159
Eekhout, M. 177
Effective Environmental Assessments: How to Manage and Prepare NEPA Eas 289
Efficiency and Sustainability in the Energy and Chemical Industries: Scientific and Case Studies 121
E-fluids.com 423
Egger-Sider, F. 17
Eggleston, B. 151
Ei Compendex (see Compendex)
Ei-Nichi-Chu Kogyo Gijutsu Daijiten, Ei-Nichi-Chu Kikai Yogo Jiten 400
Ei Thesaurus 9
Eisenberg, E.R. 416
Eisler, G..R. 357
Ela, W.P. 295
Electrical & Electronic Abstracts 18
Electrical Engineering: a Pocket Reference 249
Electrical Engineering Handbook 248
Electric Power Distribution Handbook 249
Electric Power Research Institute 254, 465
Electrical Safety Handbook 248
Electrochemical Society 390
Electromagnetic Compatibility Handbook 249
Electromechanical Design Handbook 411
Electronic Device Failure Analysis Society 390
Electronic Engineers Master Catalog 246
Electronic Packaging and Interconnection Handbook 248, 353
Electronics and Communications Abstracts 246
Electronics Handbook 250
Elementary Principles of Chemical Processes 121
Elements of Chemical Reaction Engineering 121
Elements of Ocean Engineering 145
Elements of Programming Style 206
Elements of Strength of Materials 418
El-Hawary, F. 144, 412
Eliasson, J. 159
Ellenberger, J. P. 412
Ellenberger, P. 509
Ellifritt,T.S. 324
Ellwood, K.B. 495
Elsevier 230
Elsevier's Dictionary of Agriculture in English,German, French, Russian and Latin 68
Elsevier's Dictionary of Automotive Engineering in English, German, French, Dutch, and Polish 401

Elsevier's Dictionary of Civil Aviation 51
Elsevier's Dictionary of Engineering 13
Elsevier's Dictionary of Industrial Technology: In English, German, and Portuguese 350
Elsevier's Dictionary of Nuclear Science and Technology: In English–American (with Definitions), French, Spanish, Italian, Dutch, and German 458
Elsevier's Dictionary of Nutrition and Food Processing in English, German, French and Portuguese 68, 84
Elsevier's Dictionary of Soil Mechanics and Geotechnical Engineering 137
Elsevier's Dictionary of Technology: English-Spanish 13
Elsevier's Dictionary of Technology: Spanish–English 13
Elsevier's Dictionary of Water and Hydraulic Engineering 185
EMBASE 98
Embedded Systems Dictionary 212
Embry-Riddle Aeronautical University 57
Emden's Construction Law 132
Emerald's International Civil Engineering Abstracts
Emerson, C.J. 36
Enabling American Innovation: Engineering and the National Science Foundation 328
Encyclopedia of Advanced Materials 375
Encyclopedia of Agricultural, Food and Biological Engineering 68
Encyclopedia of Agricultural Science 67
Encyclopedia of Architectural Technology 84
Encyclopedia of Architecture: Design, Engineering & Construction 85
Encyclopedia of Association and Information Sources for Architects, Designers, and Engineers 84
Encyclopedia of Associations 11
Encyclopedia of Biomaterials and Biomedical Engineering 100
Encyclopedia Britannica 202, 453, 467
Encyclopedia of Architectural and Engineering Feats 326
Encyclopedia of Building Technology 84
Encyclopedia of Chemical Processing and Design 116
Encyclopedia of Chemical Technology 100
Encyclopedia of Computer and Computer History 216
Encyclopedia of Computer Science 216, 223, 249
Encyclopedia of Computer Science and Technology 215
Encyclopedia of Corrosion Technology 377
Encyclopedia of Energy Technology and the Environment 286
Encyclopedia of Environmental Control Technology 286
Encyclopedia of Environmental Analysis and Remediation 287
Encyclopedia of Environmental Information Sources 315
Encyclopedia of Environmental Science and Engineering 287
Encyclopedia of Fluid Mechanics 51, 401
Encyclopedia of Food Engineering 67
Encyclopedia of Food Sciences and Nutrition 67
Encyclopedia of Food Science and Technology 67
Encyclopedia of Genetics, Genomics, Proteomics, and Informatics 100
Encyclopedia of Industrial Biotechnology: Bioprocess, Bioseparation, and Cell Technology 67, 100
Encyclopedia of Industrial and Organizational Psychology 350

Encyclopedia of Industrial Chemistry see Ullmann's Encyclopedia of Industrial Chemistry
Encyclopedia of Life Sciences 100
Encyclopedia of Materials, Parts, and Finishes 377
Encyclopedia of Materials Science and Engineering 375
Encyclopedia of Materials: Science and Technology 375
Encyclopedia of Medical Devices and Instrumentation 100
Encyclopedia of Microfluidics and Nanofluidics 376
Encyclopedia of Mineral Names 438
Encyclopedia of Mineralogy 439
Encyclopedia of Minerals 439
Encyclopedia of Nanoscience and Nanotechnology 100, 376
Encyclopedia of Nanoscience and Society 375
Encyclopedia of Occupational Health and Safety 351
Encyclopedia of Operations Research and Management Science 350
Encyclopedia of Optical Engineering 248
Encyclopedia of Physical Sciences and Engineering Information Sources 399, 430
Encyclopedia of Polymer Science and Engineering 376
Encyclopedia of Production and Manufacturing Management 351
Encyclopedia of Sediments & Sedimentary Rocks 439
Encyclopedia of Separation Science 116
Encyclopedia of Smart Materials 85, 377
Encyclopedia of Software Engineering 215
Encyclopedia of Space Science and Technology 52
Encyclopedia of Statistics in Quality and Reliability 351
Encyclopedia of Technical Aviation 51
Encyclopaedia of the History of Science, Technology, and Medicine in Non-Western Cultures 325
Encyclopaedia of the History of Technology 326
Encyclopedia of Water Science 68
Encyclopedic Dictionary of Gears and Gearing 400
Encyclopedic Dictionary of Industrial Automation and Computer Control 351
Encyclopedic Dictionary of Polymers
Encyclopedic Dictionary of Named Processes in Chemical Technology 115
Energy and Fuels, 123, 496
Energy Citations Database 283, 454
Energy Conversion and Management 418
EnergyFiles: Environmental Sciences, Safety, and Health 282
Energy Information Administration 254
Energy Institute (U.K.) 491
Energy Management and Conservation Handbook 406
Energy Management Handbook 407
Energy Manual; Sustainable Architecture 86
Energy, Mines, and Resources Canada 448
Energy Processing Canada 488
Energy Research Abstracts 455
Energy Science and Technology 282, 455
Energy Sources. Part A. Recovery, Utilization, and Environmental Effects 486
Energy Sources. Part B. Economics, Planning, and Policy 486
Engineer in America: A Historical Anthology from Technology and Culture 268
Engineering (London) 14
Engineered Materials Abstracts 374
Engineered Materials Handbook 379

Engineering: An Endless Frontier 328, 340
Engineering and Materials 318
Engineering and Technology 495
Engineering and Technology Degrees 265
Engineering and Technology Enrollments 266
Engineering Communication: A Practical Guide to Workplace Communications for Engineers 33
Engineering, Construction and Architectural Management 150
Engineering Data Book 492
Engineering Education 263
Engineering Education: Designing an Adaptive System 269
Engineering Education Quality Assurance 267
Engineering Education: Renewing America's Technology 270
Engineering Education: Research and Development in Curriculum and Instruction 260, 267
Engineering Ethics: Balancing Cost, Schedule, and Risk Lessons Learned from the Space Shuttle 31
Engineering Ethics: Concepts and Cases 31
Engineering for Architecture 87
Engineering Formulas 14
Engineering for Sustainable Development: Guiding Principles 201
Engineering Geology 138, 435
Engineering Guide to the Safety of Embankment Dams in the U.K 188
Engineering Handbook 13
Engineering Heritage: Highlights from the History of Mechanical Engineering 397
Engineering Index (see also Compendex) 16, 18, 48
Engineering in Emergencies 173
Engineering in History 330, 340
Engineering in Life Sciences 72
Engineering Institute of Canada 273
Engineering in Time: The Systematics of Engineering History and its Contemporary Context 329
Engineering Journal 179
Engineering Manual: A Practical Reference of Data and Methods in Architectural, Chemical, Civil, Electrical, Mechanical, and Nuclear Engineering 86
Engineering Mathematics Handbook 14
Engineering Meteorology 174
Engineering News-Record (see ENR)
Engineering Program 37
Engineering Project Appraisal 153
Engineering Properties of Foods 69
Engineering Research Center for Reconfigurable Machining Systems 360
Engineering Standard: A Most Useful Tool 22
Engineering Structures 177
Engineering Success: Persistence Factors of African American Doctoral Recipients in Engineering and Applied Science 37
Engineering Tomorrow: Today's Technology Experts Envision the Next Century 340
Engineer of 2020 269, 274, 275
Engineer's Data Book 403
Engineer's Guide to Pressure Equipment 413
Engineer's Guide to Technical Communication 33
Englibrary 29
Eng-Links 423

English, L.M. 353
English–Chinese Dictionary of Engineering and Technology 13
English–Japanese–Chinese Mechanical Engineering Dictionary 401
English–Russian Dictionary of Mechanical Engineering and Industrial Automation 401
EngNetBase 4, 13
Enhancing the Postdoctoral Experience for Scientists and Engineers: A Guide for Postdoctoral Scholars, Advisers, Institutions, Funding Organizations, and Disciplinary Societies 269
ENR 14, 15, 88
ENR Directory of Construction Information Resources 4
Evaluating and Improving Undergraduate Teaching in Science, Technology, Engineering, and Mathematics 269
ENVIROnetBASE 289
Environment Abstracts 283, 283, 432
Environment Agency 160
Environmental and Engineering Geophysical Society 141, 313
Environmental Assessment 157
Environmental Biotechnology: Concepts and Applications 298
Environmental Biotreatment: Technologies for Air, Water, Soil, and Waste 295
Environmental Data Report 285
Environmental Engineering 290, 295
Environmental Engineering Abstracts 282, 283
Environmental Engineering Dictionary 287
Environmental Engineering Dictionary and Directory 84, 287
Environmental Engineering Science 295, 300
Environmental Engineering Selection & Career Guide 286
Environmental Engineer's Mathematics Handbook 290
Environmental Health Perspectives 300
Environmental Hydrology and Hydraulics: Eco-Technological Practices for Sustainable Development 297
Environmental Impact Assessment: A Guide to Procedures 159
Environmental Impact Statements 158, 288
Environment International 300
Environmental Law Handbook 289
Environmentally Conscious Manufacturing 367
Environmental Management 300
Environmental Modelling and Software: With Environment Data News 300
Environmental Monitoring 295
Environmental Pollution 300
Environmental Progress 125
Environmental Regulatory Glossary 288
Environmental Remediation. Cost Data-assemblies 290
Environmental Remediation. Cost Data-unit Price 290
Environmental Science and Pollution Control 297
Environmental Science and Technology 300
Environmental Sciences and Pollution Management 283
Environmental Site Assessment, Phase I: A Basic Guide 156
Environmental Technologies Handbook 288
Environmental Technology Resources Handbook 289
Environment, Development and Sustainability 199

Environanotechnology 164

Envisioning a 21st Century Science and Engineering Workforce for the United States: Tasks for University, Industry, and Government 270

EPA Full-Text Reports–National Environmental Publications Internet Site 283

EPA Publications Bibliography 285

EPA Technical Reports 283

EPA Test Methods Collections 305

E-Print Network: Research Communications for Scientists and Engineers; Computer Technologies and Information Sciences 221, 308

Equity and Excellence in Education 39

Erb, U. 12, 22, 84, 375

Ergonomics Abstracts 347

Ergonomics: An International Journal of Research and Practice in Human Factors and Ergonomics 358

ErgoWeb–Resource Center 360

ERIC 35, 262

Erickson, J.B. 430

Ernst, R. 13

Eryity, V. 176

ESCAP 191

Escarameia, M. 188

Eshbach, O. W. 12

Eshbach's Handbook of Engineering Fundamentals 12

Essence of Geotechnical Engineering: 60 Years of Geotechnique

Essentials of Offshore Structures: Theory and Applications 136, 145

Estimating for Building and Civil Engineering Works 151

Etherington, H. 459

EUCC Coastal Guide 146

Euler, L. 45

EUmanufacturer.com 348

Eurocodes Expert 195

Europaische Gesellschaft fur Ingenieur-Ausbildung 273

European Agency for Safety and Health at Work 360

European Annual Conference on Human Decision Making and Manual Control 367

European Cells and Materials 385

European Journal of Engineering Education 271

European Journal of Industrial Engineering 358

European Journal of Operational Research 358

European Nuclear Society 465

European Patent Office 27

European Society for Biomaterials 108, 390

European Society for Engineering Education 273

European Society of Biomechanics 108

European Sources of Scientific and Technical Information 10

European Space Agency 57

European Transport Statistics 502

Evaluation of Retention Strategies of Texas A&M University's Minority

Extreme Searcher's Internet Handbook: A Guide for the Serious Searcher 17

Everyone's Guide to the Oil Patch 495

Ewert, R.H. 400

Expected New Nuclear Power Plant Applications 467

Experimental Thermal and Fluid Science 419

Experiments in Fluids 419

Expert Systems with Applications 252

Eyjafjallajokul ix

Enzyklopadie der Technischen Chemie 116

F

Facebook 1, 5

Facility Piping Systems Handbook—For Industrial, Commercial, and Healthcare Facilities, 412

Facts on File Dictionary of Biotechnology and Genetic Engineering 99

Facts on File Dictionary of Space Technology 51

Fairbridge, R. W. 439

Faherty, K. F. 176

Fahlbusch, H. 183

Faltin, F.W. 351

Fan, L-S. 127

Fan, M. 112, 164

Fang, H-Y. 137

Fan Handbook: Selection, Application and Design 408

Farkas, D.F. 61, 78

Farrall, A. W. 67

Fast Breeder Reactors 462

Fatality Analysis Reporting System 503

Faulkner, N. 38

Faw, R.E. 462

FCC Network 489

Feddema, J.T. 357

Federal Aviation Administration 22

Federal Communications Commission 22

Federal Register 307

Federal Transit Administration's National Transit Database 502

Federation Internationale de Beton 182

Federation of Civil Engineering Associations of the Mexican Republic 135

Federation of European Heating and Air Conditioning Associations 401

Felder, G.N. 263

Felder, R. 112. 121, 263, 264

Feldman, I. 24

Feldman, R.M. 356

Fenske, R. H. 39

Fenton, J. 404

Ferguson, E. S. 324, 331

Ferguson, H. 131, 202

Fermi, E. 417

Fernandez Troyano, L. 173

Ferrous Metallurgy Handbook 383

FIDIC: A Practical Legal Guide 151

FIDIC Forms of Contract 151

Field, H. L. 69

Financial Times Business 439

Finch, D. 470, 495

Finglas, P. 67

Fink, J. K. 481

Finkelstein, L. 33

Finn, W. E. 100

Fire Protection Handbook 22

First Computers—History and Architectures 209

First Electronic Computer: The Atanasoff Story 209

Fischer, K. 13

Fischer, R. E. 87

Fiset, J. 356

Fiske, E. B. 266
Fiske Guide to Colleges 266
Fitch, J. W. 404
Flexible Robot Dynamics and Controls 357
Flickinger, M. C. 67, 78, 100
Flight International 55
Flitney, R. K. 410
Flood Estimation Handbook 186
Flow Measurement Engineering Handbook 406
Flow, Turbulence, and Combustion 55
Fluid Flow Handbook 406
Fluid Mechanics 418
Fluid Phase Equilibria 123
Fluid Power Design Handbook 407
FOLDOC: Free Online Dictionary of Computing 214
Folger, H. S. 121
Folkman, S. L. 413
Fontana, D. 318
Food Analytical Methods 72
Food and Agriculture Organization of the United
    Nations 191
Food and Bioprocess Technology 72
Food and Bioproducts Processing 72
Food Biotechnology 72
Food Control 72
Food Engineering Reviews 72
Food Manufacturing Efficiency 73
Food Master 66
Foods and Food Production Encyclopedia 67
Food Science and Technology International 73
Forde, M. 172
Forensic Engineering 177
Foresight Institute 108, 163
*Formulas and Calculations for Drilling, Production, and
    Workover* 482
*Formula Handbook for Environmental Engineers and
    Scientists* 288
Forrester Research 235
Forum on the Construction Industry 154
Foss, J.F. 407
Foundation Design and Construction 139
Foundation Engineering Handbook 137
Fox, M. A. 269
Francis 322
Francis, F. J. 67
Francombe, M. H. 379
Frankel, M. L. 412
Franklin Institute Science Museum 335
Franson, M. A. H. 305
Frisch, O. R. 459
Frye, K. 439
*Free Compilers and Interpreters for Programming
    Languages* 238
Freedman, A. 212
Freeman, H. 156
Freeman, R. L. 217
FreeTechBooks 234
Freitag, J. K. 87
Frick, J. P. 23, 380
Fricker, J. D. 511
Fridley, K.J. 175
Friedlander, S. K. 296
Friedman, D. 173
Fritze, R. H. 324

Frize, M. 38
Frize, P.R.D. 38
*From Analysis to Action: Undergraduate Education
    in Science, Mathematics, Engineering, and
    Technology* 270
*From Dits to Bits: A Personal History of the Electronic
    Computer* 210
*From ENIAC to UNIVAC: An Appraisal of the Eckert–
    Mauchly Computers* 210
*From Research to Printout: Creating Effective Technical
    Documents* 33
*From Scarcity to Visibility: Gender Differences in
    the Careers of Doctoral Scientists and
    Engineers* 38
From Theory to Practice in Soil Mechanics 136
Frontiers in Education Conference 270, 272
FSTA: Food Science & Technology Abstracts 64
Fu, Y.Y. 381
Fuchs, M. 85
Fuel 123
Fuel Processing Technology 123, 486
Fuel: The Science and Technology of Fuel and Energy 486
Fuerstenau, M. C. 437
Fuhs, A.E. 54, 406
Fukushima Daiichi ix
Fulay,P.P. 384
Fulltext Sources Online 15
*Fundamental Aspects of Nuclear Reactor Fuel
    Elements* 461
*Fundamentals in Air Pollution: From Processes to
    Modelling* 296
Fundamentals of Environmental Discharge Modeling 295
Fundamentals of Fluid Mechanics 418
Fundamentals of Heat and Mass Transfer 417
Fundamentals of Nuclear Science and Engineering 462
*Fundamentals of Patenting and Licensing for Scientists
    and Engineers* 107
Fundamentals of Ethics for Scientists and Engineers 31
*Fundamentals of Modern Manufacturing: Materials,
    Processes, and Systems* 356
Fundamentals of Rail Vehicle Dynamics 513
Fundamentals of Traffic Engineering 512
*Fundamentals of Transportation and Traffic
    Operations* 512
*Fundamentals of Transportation Engineering:
    A Multimodal Systems Approach* 511
Fundamentals of Vehicle Dynamics 510
Fundamentals of Vibrations 417
Furter, W. F. 329
Fusion Engineering and Design 462
*Fusion Science and Technology: An International Journal
    of the American Nuclear Society* 462
Future of electronic data 220
Future of Mining, Deloitte Touche Tohmatsu 450
*Fuzzy and Multi-Level Decision Making: An Interactive
    Computational Approach* 357
Fwa, T.F. 506

G

Gabolde, G. 481
Gad-el-Hak, M. 410
Gafford, W. R. 83
Galambos, T. V. 175

Galaxy Directory: Mechanical Engineering 423
Galaxy Engineering and Technology Mining 450
Gale, W. F. 380
Gale Directory of Databases 16, 17, 495
Gale Group Computer Database 229
Gallagher, L. M. 186, 293
Gallant, B. 293
Galloway, P. D. 266
Ganic, E. N. 13
Ganssle, J. 212
Garber, N. J. 512
Garland, D.J. 54
Garner, G. O. 32
Garnaut, C. 326
Garrett, T. K. 510
Garrison, E. G. 317, 329, 340
Garson, L.R. 126
Gartner, Inc. 235
Gas Processors Association 492
Gass, S. I. 350
Gas Technology Institute 491
Gas Turbine Engineering Handbook 408
Gas Turbine Handbook 40
Gates Millennium Scholars 42
Gattorna, J. 352
Gauthetrou, J. 138
Gaylord, C.N. 172
Gaylord, E. H. 172
Gear Analysis Handbook 411
Gear Handbook: Design and Calculations 411
Geddes, S. 151
General Motors Women's Retail Network Dealer
        Development Scholarship 43
Genetics Home Reference 105
GENE-TOX 105
Genta, G. 404
GeoArabia 486
GeoEngineer 140
GEOBASE 432
Geoforum.com 140
GeoIndex 140
Geo-Institute 141
German Geotechnical Society 195
GeoRef 136, 284, 433
Georgia Tech Manufacturing Research Center 360
Georges, A. 38
Geoscan 433, 484
GeoScienceworld 136
Geosynthetics International 139
GeoTechLinks 140
Geotechnical Earthquake Engineering Handbook 137
Geotechnical Engineering Investigation Handbook
        137, 139
*Geotechnical Engineering: Proceedings of the
        Institution of Civil Engineers* 139
Geotechnique 139
Geraghty & Miller's Groundwater Bibliography 286
Gere, J. M. 417
Gersten, J. I. 384
Gertsch, R. E. 437
*Getting Out the Gates: Underrepresented Minority
        Students' Search for Success in Introductory
        Chemistry Courses to Continue on the
        STEM Path* 37

Gerwick, B. C. 144
Ghalambor, A. 481
Ghassemi, A. 289
Ghirardi, I. 52
Ghosh, S. N. 297, 512
Giampaolo, T. 408
Gibbons, M. 257
Giebelhaus, A. W. 495
Gieck, K. 14
Gilleo, K. 353
Gillespie, T. D. 510
Gilpin, A. 196
Gladden, S. C. 151
Glass Construction Manual 177
Glasstone, S. 460
Global Change Research Information Office 307
Global Journal of Engineering Education 271
Global Mining Directory 439
Global Mobility Database 284
GlobalSpec, The Engineering Search Engine 349
Glossary of Biotechnology Terms 99
Glossary of Geology 438
Glossary of Mining Terms 438
Glossary of Reliability and Maintenance Terms 350
Glossary of Terms in Nuclear Science and Technology 457
Glossary of Transportation Terms 504
Glover, T. J. 14
Godel, J. B. 83
Godfrey, A.B. 353
Godfrey, K. A. 151
Godfrey, L. E. 28
Godfrey, P. 150
Godish, T. 296
Gogotsi, Y. 164
*Going beyond Google: The Invisible Web in
        Learning and Teaching* 17
Gold Institute 448
Goldstine, H. H. 209
Golshan, H. 513
Golze, A. R. 186
Gooch, J. 375
Goodman, R. E. 202
Good Nano Guide 165
Goodowens, J. B. 151
Goodsell, D. 400, 505
Google ix, x, 245
Google Directory: Aerospace 57
Google Directory: Electrical Engineering 254
Google Hacks 17
Google Patents 27
Google Scholar ix, 1, 18, 262
Gordon, B. 263
Gordon, T. T. 26
Gordon Research Conferences 108
Gosling, P. J. 99
Gosselink, J.G. 298
Goswami, D.Y. 403, 406
Gottlieb, D. W. 289
Government Printing Office 307
Government Reports Announcement and Index 48
Gower Handbook of Project Management 149
Gower Handbook of Supply Chain Management 352
GPS Solutions 252
Grace, R. D. 481

GradSchools.com 233, 266, 400
Graduate Program Directory 66
Graduate Programs in Engineering and Applied
    Sciences 266
Grady, C. P. L. 297
Granqvist, C-G. S. 165
Grant, H. 270
Grasso, D. 267
Grava, S. 156
Gray, E. 495
Gray, P. 267
gray literature 3, 10, 27
Grayson, L. 260, 261, 267, 275
Grazda, E. E. 14
Greatbuildingsonline 179
Great Canadian Oil Patch: The Petroleum Era from
    Birth to Peak 495
Greatest Engineering Achievements of the
    20th Century 337
Great Jobs for Engineering Majors 32
Green, R. J. 347
Greiner, W. 459
Grey, 470
GRDC 189
Green, D.W. 117
Green, R. J. 324
Green, S. 65
Green Book 505
Green Building Council 80, 83
Green Building Materials; A Guide to Product Selection
    and Specification 86
Green Construction Handbook 150
Greene, J. H. 353
Green Guide; The Design, Construction and
    Operation of Sustainable Buildings 85
Green Nanotechnology: Solutions for Sustainability and
    Energy in the Built Environment 165
GreenSource: The Magazine of Sustainable
    Design 89
GreenSpec: Product Directory with Guideline
    Specifications 84
Green Studio Handbook: Environmental Strategies for
    Schematic Design 87
Gregory, C. E. 428
Greider, A. P. 66
grey literature (see gray literature)
Grey Literature Network Services 3
GreyNet 27
Griffin, P. 38
Grimm, N. R. 86
Griskey, R.G. 117
Grondzik, W. T. 87
Groover, M. P. 343, 356, 368
Grote, K.-H. 402
Ground Improvement: Proceedings of the Institution of
    Civil Engineers 139
Ground Water 300
Groundwater Chemicals Desk Reference 293
Groundwater Hydrology 298
Ground Water Monitoring and Remediation 300
Ground-Water Remediation Technologies Analysis
    Centre 189
Grubb, P. 107
Guedes, P. 84

Guidance Manual on Environmentally Sound
    Management of Waste 294
Guidebook to Nuclear Reactors 459
Guide for Design of Pavement Structures 506
Guide for the Development of Bicycle Facilities 508
Guide for the Geometric Design of Driveways 507
Guidelines for Activation, Modification or Removal of
    Traffic Control Signals 520
Guidelines for the Design and Application of Speed
    Humps 520
Guide to Basic Information Sources in Engineering 430
Guide to Computing Literature 205, 208, 223
Guide to Consultants 66
Guide to Engineering Materials Producers 348
Guide to Information Sources in Engineering 10, 50, 315,
    372, 393, 430, 495
Guide to Manuscript Collections in the National Museum
    of History and Technology 325
Guide to Petroleum Statistical Sources 494
Guide to Presenting Technical Information: Effective
    Graphic Communication 33
Guide to Selected Bioinformatics Internet Resources 96
Guide to Stability Design Criteria for Metal
    Structures 175
Guide to Standards 21, 24
Guide to the History of Science 325, 331, 340
Guide to the Petroleum Reference Literature 495
Guide to the Reservoirs Act, 1975 188
Guide to the Structural Use of Adhesives 177
Guide to Writing as an Engineer 32
Guinness Book of Structures 174
Gulf Publishing Company 485
Gulich, J. F. 414
Gunaratne, M. 137
Gunston, B. 51
Guo, B. 481, 482
Gupta, B. K. 415
Gupta, R. K. 380
Gupta, R. S. 295
Guston, D. H. 375

H

Haasen,P. 384
Habashi, F. 380
Hackerman, N. 269
Haden, C. 37
Hagley Museum and Library 333
Haines, R. W. 86
Halford, B. 127
Hall. A. R. 328
Hall, C. W. 65, 67
Hall, J.W. 512
Hall, L. D. 495
Hall, R. W, 505
Halliday, S. 172
Hallman, W. 53
Halpin, D. W. 152
Hamblen, J. 165
Hambly, E. C. 173
Hamilton, L.J. 460
Hammer, D. A. 297
Han, K.N. 437
Hancock, J. M. 99

Handbook for Productivity Measurement and Improvement 352

Handbook of Adhesives and Sealants in Construction 176

Handbook of Advanced Plasma Processing Techniques 121

Handbook of Air Pollution Prevention and Control 291

Handbook of Analytical Techniques in Concrete Science and Technology 82

Handbook of Ancient Water Technology 183, 328

Handbook of Advanced Materials 384

Handbook of Applied Mathematics 14

Handbook of Applied Mathematics for Engineers and Scientists 14

Handbook of Architectural Technology 85

Handbook of Atmospheric Science: Principles and Applications 292

Handbook of Automotive Body Construction and Design Analysis 404

Handbook of Automotive Body and Systems Design 404

Handbook of Automotive Powertrain and Chassis Design 404

Handbook of Aviation Human Factors 54

Handbook of Batteries 249

Handbook of Biomaterial Properties 101, 378

Handbook of Biomaterials Evaluation: Scientific, Technical, and Clinical Testing of Implant Materials 102, 383

Handbook of Bolts and Bolted Joints 171

Handbook of Ceramics, Glasses and Diamonds 380

Handbook of Chemical and Environmental Engineering Calculations 118, 290

Handbook of Chemical Engineering Calculations 117

Handbook of Chemical Risk Assessment: Health Hazards to Humans, Plants, and Animals 285

Handbook of Civil Engineering Calculations 133

Handbook of Computational Intelligence in Manufacturing and Production Management 354

Handbook of Corrosion Engineering 382

Handbook of Dam Engineering 186

Handbook of Design, Manufacturing and Automation 352

Handbook of Die Design 409

Handbook of Diesel Engines 405

Handbook of Dredging Engineering 144

Handbook of Drinking Water Quality 185, 293

Handbook of Electrical Safety in the Workplace 248

Handbook of Electric Motors 249

Handbook of Energy Efficiency and Renewable Energy 406

Handbook of Energy Engineering 407

Handbook of Engineering in Agriculture (CRC) 68

Environmental Engineers' Handbook 289

Handbook of Environmental Design

Handbook of Environmental Engineering 290

Handbook of Environmental Engineering Calculations 289

Handbook of Environmental Management and Technology 288

Handbook of Extractive Metallurgy 380

Handbook of Industrial and Systems Engineering 352

Handbook of Industrial Engineering Equations, Formulas, and Calculations 352

Handbook of Flow Visualization 54

Handbook of Fluid Dynamics 53, 54, 406

Handbook of Fluid Dynamics and Fluid Machinery 54, 406

Handbook of Food Engineering 68

Handbook of Food Engineering Practice 69

Handbook of Geotechnical Investigation and Design Tables 138

Handbook of Glass in Construction 176

Handbook of Groundwater Engineering 185

Handbook of Hard Coatings: Deposition Technologies, Properties and Applications 379

Handbook of Heat and Mass Transfer 405

Handbook of Heat Transfer 406

Handbook of Highway Engineering 505

Handbook of Human Factors and Ergonomics 355

Handbook of Human Factors Testing and Evaluation 352

Handbook of HVAC Design 86

Handbook of Hydraulic Fluid Technology 407

Handbook of Image and Video Processing 248

Handbook of Industrial Engineering: Technology and Operations Management 355

Handbook of Industrial Robotics 354

Handbook of International Safety Practices 248

Handbook of Jig and Fixture Design 408

Handbook of Logistics and Distribution Management 355

Handbook of Logistics and Supply-Chain Management 352

Handbook of Lubrication and Tribology 415

Handbook of Manufacturing Engineering 343, 369

Handbook of Manufacturing Processes: How Products, Components and Materials Are Made 352

Handbook of Materials Behavior Models 381

Handbook of Materials for Medical Devices 101

Handbook of Materials for Product Design 353, 380

Handbook of Materials Selection 353, 380

Handbook of Mathematics for Engineers and Scientists 14

Handbook of Mechanical Engineering Calculations 403

Handbook of Mechanical Engineering Dubbel 402

Handbook of Micro/Nanotribology 38

Handbook of Microscopy: Applications in Materials Science, Solidstate Physics and Chemistry 377

Handbook of Mineral Dressing: Ores and Industrial Minerals 438

Handbook of Mining and Tunneling Machinery 437

Handbook of Nanofabrication 384

Handbook of Nanophysics 380

Handbook of Nanostructured Biomaterials and Their Applications in Nanobiotechnology 102

Handbook of Nanostructured Materials and Nanotechnology 381

Handbook of Natural Gas Transmission and Processing 483

Handbook of Nuclear Engineering 459

Handbook of Nuclear Properties 459

Handbook of Neuroprosthetic Methods 100

Handbook of Noise and Vibration Control 411

Handbook of Numerical Heat Transfer 406

Handbook of Optomechanical Engineering 409

Handbook of Organic-Inorganic Hybrid Materials and Nanocomposites 381

Handbook of Petrochemicals Production Processes 482

Handbook of Petroleum Processing 482

Handbook of Petroleum Refining Processes 483

*Handbook of Physical-Chemical Properties
and Environmental Fate for Organic
Chemicals* 290
Handbook of Plastics Analysis 381
Handbook of Plastics Recycling 157
Handbook of Plastics Testing and Failure Analysis 383
*Handbook of Pollution Control and Waste
Minimization* 289
Handbook of Pollution Prevention Practices 288
Handbook of Practical Gear Design 409
Handbook of Public Water Systems 293
Handbook of Pumps and Pumping 414
Handbook of Railway Vehicle Dynamics 510
Handbook of Reliability Engineering 354
*Handbook of Reliability, Availability, Maintainability
and Safety in Engineering Design* 355
*Handbook of Reliability Engineering and
Management* 353
*Handbook of RF/Microwave Components and
Engineering* 248
Handbook of Road Technology 507
*Handbook of Soil Analysis: Mineralogical, Organic and
Inorganic Methods* 138
Handbook of Solid Waste Management 294
*Handbook of Space Engineering, Archaeology,and
Heritage* 53
Handbook of Space Technology 53
Handbook of Superconductivity 382
Handbook of Supply Chain Management 351
Handbook of Surface and Nanometrology 384
Handbook of Sustainable Development 196
*Handbook of Systems Engineering and
Management* 355
Handbook of Technical Formulas 13
Handbook of Technical Writing 32
Handbook of Theoretical Computer Science 217
*Handbook of Thermoplastic Piping System
Design* 414
*Handbook of Thin-Film Deposition Processes and
Techniques: Principles, Methods, Equipment
and Applications* 383
Handbook of Thin Film Devices 379
Handbook of Thin Film Materials 381
Handbook of Transportation Engineering 505
Handbook of Transportation Science 505
*Handbook of Tribology: Materials, Coatings and Surface
Treatments* 415
Handbook of Turbomachinery 408
Handbook of Vacuum Science and Technology 410
Handbook of Vacuum Technology 410
Handbook of Vehicle–Road Interaction 404
*Handbook of Water and Wastewater Treatment
Technologies* 293
Hanel, N. L. N. 347
Hansen, V. E. 188
Harms, A. A. 329
Harper, C. 248
Haq, G. 160
Hardy, J.E. 430
Harper, C. A. 353, 380
Harris, C. E., Jr. 31
Harris, C. M. 12, 350, 411
Harris, F. 152
Harris' Shock and Vibration Handbook 411

Hart, G. V. 248
Hart, H. 33
Hart, R. D. 85
Hartman, H. L. 437
Hartnett, J.P. 406
Hashagen, U. 209
Hastings, G. 101, 378
HathiTrust Digital Library 323
Harvey, A.P. 430
Hawkins, B. 357
Hawley's Condensed Chemical Dictionary 115
Hay, W. H. 513
Haycock, R.F. 404
Haynes, W. M. 117
Hayward, A, 175
*Hazardous Air Pollutant Handbook: Measurements,
Properties, and Fate in Ambient Air* 292
*Hazardous Building Materials: A Guide to the
Selection of Environmentally Responsible
Alternatives* 173
Hazardous Chemicals Desk Reference 293
Hazardous Substances Data Bank 105, 122
Hazardous Waste Clean-Up Information 160
Hazardous Waste Consultant 301
*Hazardous Waste Operations and Emergency Response
Manual* 293
*Hazardous Waste Management Compliance
Handbook* 157
Hazardous Waste/Superfund Week 301
HCI Bibliography: Human-Computer Interaction
Resources 238
HDR Engineering, Inc. 293
Health and Safety Sciences Abstracts 283
Health Devices International Sourcebase 98
Health Devices Sourcebook 98
Heat and Mass Transfer 123
Heat Exchanger Design Handbook 406
*Heating Boiler Operator's Manual: Maintenance,
Operation, and Repair* 413
Heat Transfer Engineering 419
Heat Transfer Handbook 405
HEDH: Heat Exchanger Design Handbook 406
Hegger, M. 85
Heinsohn, R. J. 296
Heisler, S. I. 12
Heitsch, A. 263
Heldman, D. R. 68, 69
Helguero, M. V. 413
Hellemans, A. 327
Helmholz, H.L.F. 45
Hemmendinger, D.
Henry, K. 165
Herbich, J. 144
Hermond, D. S. 37
Herz, N. 438
Hess-Kosa, K. 156
Hewitt, C. N. 292
Hewitt, G. F. 402, 406
Hewson, N. 508
*Hey's Mineral Index: Mineral Species, Varieties and
Synonyms* 438
Heywood, J. 260, 267, 275
Hibbeler, R. C. 417
Hicks, P.J. 430

Hicks, T.G. 13, 14, 149, 403
Hicks, S. D. 14
High-Performance Buildings Research 180
High-Performance Building 87
High-Speed Marine Transportation 504
High Technology Research Database 18
Hightower, C. 96
Highway Capacity Manual 157, 507
Highway Design and Construction 157, 159
*Highway Design and Traffic Engineering
    Handbook* 157
Highway Engineering 512
Highway Engineering Handbook 505
Highway Maintenance Handbook 156
Highway Research Board 498, 500, 519
Highways 159
Highway Safety Manual 506, 507
Highway Statistics 160, 502
Highway Traffic Analysis and Design 159
Highway Traffic Monitoring and Data Quality 512
Hildebrandt, D. M. 205
Hill, D. 329
Hillier, J.E. 404
Hiltscher, G. 513
Hinds, W. C. 121
Hispanic Scholarship Fund 41
*Historia Scientiarum: International Journal of the
    History of Science Society of Japan* 331
Historical Abstracts 322
Historical Building Construction 173
*Historical Dams: Foundations of the Future Resting on
    the Achievements of the Past* 183
Historical Encyclopedia of Atomic Energy 458
Historic Concrete: Background to Appraisal 175
History and Technology: An International Journal 331
History Journals Guide. Periodicals Directory 323
*History of Aerodynamics and its Impact on Flying
    Machines* 59
History of Chemical Engineering 329
History of Chemical Engineering Web Page 112
*History of Civil Engineering: An Outline from Ancient
    to Modern Times* 330
History of Computing Series 209
History of Computing Technology 210
*History of Engineering and Technology: Artful
    Methods* 329
*History of Engineering in Classical and Medieval
    Times* 329
History of Engineering at Yale University 337
*History of Engineering Science: An Annotated
    Bibliography* 264
History of Hydraulics 183
History of Hydrology 183
History of Mechanical Engineering 329, 396
History of Medicine 105
History of Mining: The Miner's Contribution to
    Society 429
History of Modern Computing 209
History of Parys and Mona Copper Mines 429
History of Petroleum Engineering 329, 495
*History of Programming Languages.: Proceedings of
    the History of Programming Languages
    Conference* 210
History of Science Links 337

*History of Science: Review of Literature and Research
    in the History of Science, Medicine and
    Technology in its Intellectual and Social
    Context* 331
History of Science Society 338
    Guide to U.S. Graduate Study in History of
        Science 337
*History of Science and Technology: A Select Bibliography
    for Students* 325
History of Science, Technology and Medicine 262,
    321, 322
History of Strength of Materials 170
History of Suspension Bridges in Bibliographic Form 171
History of Technology 325, 328, 331
*History of Technology and Invention: Progress through
    the Ages* 327, 340
*History of Telecommunications Technology: An Annotated
    Bibliography* 325
*History of the Internet: A Chronology, 1843 to the
    Present* 17
*History of Theory of Structures in the Nineteenth
    Century* 170
Hitchcock, D. 26
HMAT: Hot Mix Asphalt Technology 515
Hoag, K. 267
Hoare, C.A.R. 206
Hock, R. 17
Hoffman, D. M. 410
Hoffman, E. S. 175
Hoel, L.A. 512
Holgate, A. 87
Holistic Engineering Education 267
Hollister, S. C. 260, 275
Holmberg, R. 138
Holmes, J. D. 173
Holmyard, E.J. 328
Holroyd, T. M. 152
Holz, H. 152
Holzman, A.G. 215
Homburger, W. S. 512
Honda, H. 32
Hoose, N. 512
Hopkin, V.D. 54
Hopkins, J.B. 50
Hopkins, J. P. 50
Hornbostel, C. 86
Horonjeff, R. 159
Hosford, W. F. 384
HOST 337
Hot Mix Asphalt Paving Handbook 506
Household Products Database 105
Howard Aiken: Portrait of a Computer Pioneer 209
How Buildings Work: The Natural Order of Architecture 7
Howell, T. A. 68
How to Develop a Pedestrian Safety Action Plan 511
*How to Find Chemical Information: A Guide for
    Practicing Chemists,Educators, and Students*
*How to Solve Problems: For Success in Freshman
    Physics, Engineering and Beyond* 268
Hoyle, D. 23
Hoyt, M. 165
Hradesky 344, 369
HR Wallingford 191
Huang, C.P. 164

Hudson's Building and Engineering Contracts 153
Huebsch,W.W. 418
Hughes A. C. 330
Hughes T.P. 330
Human Factors 358
Human Factors and Aerospace Safety 55
Human Factors and Ergonomics Society 364
Annual Meeting 367
Human Factors and Ergonomics in Manufacturing 358
Human Factors of Transport Signs 512
*Human–Machine Interface Design for Process Control
    Applications* 356
Humphreys, K. K. 353
Hundal, M. S. 353
Hundred Years of Mechanical Engineering 397
Hung, Y-T. 296
Hunt, E. C. 412
Hunt, R. E. 137
Hunt, T. 186
Hunt, T. M. 407
Huntington Library, Art Collections, and Botanical
    Gardens 333
Hurt, C.D. 11, 430
Hutchinson, A. R. 177
Hutchinson, B. S. 66
HVAC Pump Handbook 415
HVAC Systems Design Handbook 86
Hydrological Sciences Branch 191
Hydrological Sciences Journal 186
Hydraulic Engineering Research Unit 191
Hydraulic Handbook 186, 407
Hydraulic Structures 188
*Hydraulics and Pneumatics: A Technician's and
    Engineer's Guide* 407
Hydraulics of Pipeline Systems 513
Hydrocarbon Online 489
Hydrocarbon Processing 488
Hydrology We 189

I

IAEA Safeguards Glossary 458
*IAEA Safety Glossary: Terminology Used in Nuclear
    Safety and Radiation Protection* 458
I.A. Recordings 337
*IA: The Journal of the Society for Industrial
    Archeology* 331
Ibarz, A. 68
IBM Journal of Research and Development 224, 225
IBM Systems Journal 224
icademic: Mechanical Engineering 423
ICE (see Institution of Civil Engineers)
ICE Bibliography on Prestressed Concrete 1920–1957 170
ICE Conditions of Contract 151
ICE Manual of Bridge Engineering 508
ICE Manual of Structural Materials 172
*ICID Multilingual Technical Dictionary on Irrigation and
    Drainage* 185
iCivilEngineer 90, 134
ICOLD Technical Dictionary on Dams 185
ICONDA—the International Construction Database 132
*Icon: Journal of the International Committee for the
    History of Technology* 331, 495
iCrank.com 423

IDC 234
IEA Coal Research Newsletter 436
IEEE (see Institution of Electrical and Electronic
    Engineers)
IEEE Annals of the History of Computing 208, 210, 331
IEEE/ASME Joint Rail Conference 518
IEEE–ASME Transactions on Mechatronics 358, 419
IEEE Components, Packaging and Manufacturing
    Technology 364
IEEE Computer Society 210, 224, 232
IEEE Computer Society Career Services Center 234
IEEE Computer Society Conferences 222
IEEE Computer Society Digital Library 226, 239
IEEE Engineering in Medicine and Biology Magazine 104
IEEE Engineering in Medicine and Biology Society 108
IEEE Engineering Management Society 364
IEEE History Center 339
IEEE Intelligent Transportation Systems Council 518
IEEE International High Assurance Systems Engineering
    Symposium 367
IEEE Nuclear and Plasma Sciences Society 466
IEEE Nuclear Standards 464
IEEE/Oceanic Engineering Society 146
IEEE Potentials 271
IEEE Power and Energy Magazine 255
IEEE Press 230
IEEE Robotics and Automation Society 364
IEEE/RSJ International Conference on Intelligent Robots
    and Systems 367
IEEE Spectrum 250, 257
IEEE Standards Online 222
IEEE Systems, Man, and Cybernetics Society 364
IEEE thinkstandards.net 495
*IEEE Transactions on Aerospace and Electronic
    Systems* 55
*IEEE Transactions on Automation Science and
    Engineering* 358
IEEE Transactions on Engineering Management 358
IEEE Transactions on Medical Imaging 104
IEEE Transactions on NanoBioscience 256
IEEE Transactions on Nanotechnology 256, 385
*IEEE Transactions on Neural Systems and
    Rehabilitation* 104
IEEE Transactions on Nuclear Science 462
IEEE Transactions on Power Systems 255
IEEE Transactions on Robotics 358
IEEE Transactions on Smart Grid 255
IEEE Transactions on Sustainable Energy 255
*IEEE Transactions on Systems, Man and Cybernetics,
    Part A: Systems & Humans* 358
IEEE Transactions on Vehicular Technology 517
IEEEXplore 18, 245, 251, 263
IESNA Lighting Handbook: Reference and Application 86
IET Renewable Power Generation 255
*IFAC Symposium on Information Control Problems in
    Manufacturing* 367
IFMBE News 104
Ifrah, G. 209
iGoogle 5
IHS 4DOnline Parts Universe 247
*IIE Transactions: Industrial Engineering Research and
    Development* 358
IIE Transactions on Healthcare Systems Engineering 358
ILI 24

ILI Metals Infobase 21
I Live in the Future and Here's How it Works x
Illingworth, J. R. 152
Illinois Institute of Technology 80
Illston, J. M. 173
Illuminating Engineering Society of North America 91
Illustrated Dictionary of Biotechnology 99
*Illustrated Dictionary of Building Materials and
        Techniques* 84
*Illustrated Encyclopedic Dictionary of Building and
        Construction Terms* 84
Imaging Products Laboratory 360
IMMAGE 433
Immergut,E.H. 379
Impact and Explosion 172
Imperial College London—Centre for the History of
        Science and Technology 339
Imperial Oil Review 495
Immune Building Systems Technology 296
*Implementing ISO 14000: A Practical, Comprehensive
        Guide to the ISO 14000 Environmental
        Management Standards* 24
Improvement: Standards and Sustainable Building 199
Ince, S. 183
Incropera, F. P. 121, 417
Independent Petroleum Association of America 491
Index and Directory of Industry Standards 23
Index of NACA Technical Publications 48
*Index of NASA Technical Publications and Technical
        Publication Announcements* 48
Index to EPA Test Methods 305
*Index to Place of Publication of ASME Papers
        1950–1977* 21
Indian Concrete Journal 179
Indian Journal of History of Science 331
Indoor Air Pollution 292
Indoor Air Quality Engineering 296, 297
Indoor Air Quality Handbook 86
*Indoor and Built Environment: The Journal of
        the International Society of the Built
        Environment* 301
Industria Italiana del Cemento 177
Industrial and Applied Microbiology Abstracts 283
Industrial and Engineering Chemistry Research 123
*Industrial and Organizational Psychology Linking Theory
        with Practice* 356
Industrial Biotechnology 73
Industrial Crops and Products 73
Industrial Directory 349
*Industrial Engineer: Engineering and Management
        Solutions at Work* 359
Industrial Engineering and Management 358
*Industrial Engineering Terminology: A Revision of ANSI
        Z94.0-1989: An American National Standard,
        Approved 2000* 350
Industrial Management & Data Systems 359
*Industrial/Organizational Psychology: Understanding the
        Workplace* 357
*Industrial Pigging Technology: Fundamentals,
        Components, Applications* 513
Industrial Refrigeration Handbook 407
Industrial Robot—An International Journal 252
Industry Search Australia and New Zealand 349
Informaworld 33

Information and Technology Centre for Transport and
        Infrastructure (see CROW)
Information Bridge (see DOE Information Bridge)
Information Handling Services 24
Information International Associates 456
Information Sources in Biotechnology 96
Information Sources in Engineering 10, 49, 50, 97, 399,
        430, 495
Information Sources in Grey Literature 10, 28, 50
Information Sources in Patents 26
Information Sources in Science and Technology 430
Information Sources in the Earth Sciences 430
*Information Technology Standards: Quest for the
        Common Byte* 23
InformationWeek 226
InfoTech Trends 235
Infotrac 246
InfoWorld 239
InfraGuide: The National Guide to Sustainable
        Infrastructure 201
Infrastructure Report Card for America's
        Infrastructure 134
Ingenta 33
Ingre, David 33
INIS *(see International Nuclear Information System
        Database)*
Inland Navigation: Locks, Dams, and Channels 144
Innovations in Science and Technology Education 271
Innovative Food Science and Emerging Technologies 73
Innovators: The Engineering Pioneers Who Made
        America Modern 170
Inspec 16, 18, 98, 208, 227, 230, 244, 246, 262, 263
Inspec Classification 18
Inspec List of Journals 18
Inspec Thesaurus 18
Institutional Repositories 220
Institute and Museum of History of Science: Florence 335
Institute for Operations Research and the Management
        Sciences 364
Institute for Scientific Information 16
Institute of Biological Engineering 77, 94, 108
Institute of Environmental Sciences and Technology 313
Institute of Ergonomic and Human Factors–U.K. 365, 368
        Annual Conference 367
Institute of Food Science and Technology 77
Institute of Food Technologists 77
Institute of Hydrology 186
Institute of Industrial Engineers 365
Institute of Marine Engineering, Science and
        Technology 146
Institute of Materials, Minerals, and Mining, U.K.
        390, 445
Institute of Medicine 19
Institute of Nanotechnology 163
Institute of Noise Control Engineering of the USA 313
Institute of Nuclear Materials Management 466
Institute of Nuclear Power Operations 466
Institute of Operations Management 365
Institute of Physics and Engineering in Medicine 109
Institute of Radio Engineers 94
Institute of Transportation Engineers 156, 161, 507, 509,
        518, 520, 522, 524
Institut Francais du Petrole 491
Institution of Agricultural Engineers 77

Institution of Civil Engineers 135, 152, 159, 177, 188
Institution of Electrical and Electronic Engineers 20, 94,
    107, 244, 249, 254, 255
Institution of Engineering and Technology 16, 255
    Manufacturing Sector 365
Institution of Highways and Transportation 161
Institution of Mechanical Engineers, U.K 365
Institution of Structural Engineers 156, 177, 182
Institution Repository Search 220
Instituto Mexicano del Petroleo 491
Instrumentation, Systems and Automation Society 492
Instrument Society of America 94
*Interactive Teaching: Promoting Better Learning
    Using Peer Instruction and Just-in-
    TimeTeaching* 264
*Integrated Buildings: The Systems Basis of
    Architecture* 87
Integrated nanosystems 166
Integrated Risk Information System 105
Integration, The VLSI Journal 225
Intelligent Assembly Systems 357
Intelligent Freight Transportation 511
Intelligent Manufacturing Systems 361
Intellectual Property 26
Intellectual Property Office of Singapore 27
Intelligent Systems in Design and Manufacturing 367
Intelligent Transportation Systems Applications
    Overview 521
Intelligent Transportation Systems Program Office 521
*Intelligent Transportation Systems: Smart and Green
    Infrastructure Design* 512
Intelligent Vehicle Technology and Trends 513
Interior Graphic Standards 30
Intermetallics 385
Intermodal Association of North America 525
Intermodal Freight Transportation 511
*Internal Combustion Engine Handbook: Basics,
    Components, Systems, and Perspectives* 405
International Abstracts in Operations Research 347
*International Academic Programs Agricultural, Food, or
    Biological Engineering Departments* 66
International Aerospace Abstracts 48
International Association for Continuing Education and
    Training 273
International Association for Earthquake Engineering 182
International Association for Shell and Spatial
    Structures 182
International Association of Bridge and Structural
    Engineering 182
International Association of Drilling Contractors 491
International Association of Hydrogeologists 191
International Association of Hydrological Sciences 191
International Association of Nanotechnology 390
International Association of Oil & Gas Producers 491
International Astronautical Federation 58
International Atomic Energy Agency 456, 457, 458, 464
International Bibliography on Atomic Energy 456
International Breakwaters Conferences 147
International Building Code 89, 175
International Civil Aviation Organisation 161
International Code Council 89, 195
International Commission of Agricultural Engineering
    (see CIGR)
International Commission on Irrigation and Drainage 191

International Commission on Large Dams 191
International Commission on Radiation Units and
    Measurements 466
International Commission on Radiological Protection 466
International Committee for the History of
    Technology 338
*International Communications in Heat and Mass
    Transfer* 123
International Concrete Federation 182
International Conference of the International Association
    for Computer Methods and Advances in
    Geomechanics 143
International Conference on Agile Manufacturing 367
International Conference on Coastal Engineering 147
International Conference on Energy Sustainability 199
International Conference on Frontiers of Design and
    Manufacturing 367
International Conference on Mechanical, Industrial, and
    Manufacturing Technologies 367
International Conference on Ocean, Offshore Mechanics
    Arctic Engineering
International Conference on TQM and Human Factors–
    Towards Successful Integration 367
International Congress on Noise Control Engineering 518
International Construction Database 170
International Construction Law Review 150
International Council for Building Research Studies and
    Documentation 182
International Council for Science, Resources:
    Environment/Sustainability 201
International Council on Mining & Metals 445
International Council on Nanotechnology 166
International Critical Tables of Numeric Data 117
*International Dictionary of Heating, Ventilating, and Air
    Conditioning* 401
International Directory of Chemical Engineering
    URLs 126
*International Directory of Engineering Societies and
    Related Organizations* 11
International Electrotechnical Commission 22, 25, 253
*International Encyclopedia of Abbreviations and
    Acronyms in Science and Technology* 23
International Energy Agency 491
International Engineering History and Heritage 201
International Environmental Modelling and Software
    Society 313
International Ergonomics Association 365
International Federation for Medical and Biological
    Engineering 109
International Federation for Information Processing 233
International Federation of Consulting Engineers 154
International Food Information Service 68
International Geosynthetics Society 142
*International Handbook of Earthquake Engineering:
    Codes, Programs,and Examples* 138
International Institute for Environment and
    Development 198
International Institute of Noise Control Engineering 314
*International Journal for Numerical and Analytical
    Methods in Geomechanics* 436
*International Journal for the History of Engineering &
    Technology* 331
*International Journal of Advanced Manufacturing
    Technology* 359

International Journal of Aerospace Innovations 55
International Journal of Agricultural and Biological Engineering 73
International Journal of Applied Ceramic Technology 385
International Journal of Chemical Kinetics 55
International Journal of Chemical Reactor Engineering 123
International Journal of Computer Integrated Manufacturing 39
International Journal of Earth Sciences 436
International Journal of Engineering Education 271
International Journal of Engine Research 419
International Journal of Fatigue 419
International Journal of Flexible Manufacturing Systems: Design, Analysis and Operation of Manufacturing and Assembly Systems 359
International Journal of Food Engineering 73
International Journal of Food Science and Technology 73
International Journal of Geomechanics 436
International Journal of Green Energy 359
International Journal of Forest Engineering 73
International Journal of Heat and Fluid Flow 55, 419
International Journal of Heat and Mass Transfer 55, 123, 419
International Journal of High Performance Computing Applications 225
International Journal of Humanoid Robotics 252
International Journal of Impact Engineering 419
International Journal of Industrial Engineering: Theory, Applications and Practice 359
International Journal of Industrial Ergonomics 359
International Journal of Innovative Computing Information and Control 252
International Journal of Machine Tools & Manufacture 359, 419
International Journal of Maritime Engineering 516
International Journal of Materials & Product Technology 359
International Journal of Mechanical Sciences 419
International Journal of Mineral Processing 435
International Journal of Multiphase Flow 419
International Journal of Nanotechnology 167
International Journal of Offshore and Polar Engineering 145, 487
International Journal of Oil, Gas and Coal Technology 487
International Journal of Operations and Production Management 359
International Journal of Pavement Engineering 158
International Journal of Plasticity 419
International Journal of Production Economics 359
International Journal of Production Research 359
International Journal of Project Management
International Journal of Refrigeration/Revue Internationale du Froid 419
International Journal of Robotics Research 252
International Journal of Robust and Nonlinear Control 55
International Journal of Rock Mechanics and Mining Sciences 139, 435
International Journal of Solids and Structures 177
International Journal of Space Structures 177
International Journal of Spray and Combustion Dynamics 55

International Journal of Surface Mining, Reclamation and Environment 435
International Journal of Sustainable Development and Planning 199
International Journal of Sustainable Energy 199
International Journal of Thermal Science 419
International Journal of Vehicle Autonomous Systems 517
International Journal of Vehicle Design: The Journal of Vehicle Engineering and Components 517
International List of Agricultural and/or Biological Engineering Societies 66
International List of Agricultural Engineering Societies 76
International Materials Reviews 385
International Metallographic Society 390
International Microelectronic and Packaging Society 255
International Mining Directory 439
International Navigation Association 147
International Navigation Congress 148
International Nuclear Information System Database 455
International Organization for Standardization 21, 24, 25, 89, 107
International Product Development Management Conference 367
International Reference Guide to Space Launch Systems 50
International Road Federation 502, 525
International Satellite Directory 50
International Scientific and Technical Gliding Organisation 58
International Society for Automation 365
International Society for Concrete Pavements 161
International Society for Rock Mechanics 142
International Society for Soil Mechanics and Geotechnical Engineering 142
International Society of Biomechanics 109
International Society of Coating Science and Technology 391
International Society of Explosives Engineers 437
International Society of Exposure Science 314
International Society of Indoor Air Quality and Climate 314
International Society of Offshore & Polar Engineers 147, 491
International Symposium on Common Ground, Consensus Building and Continual
International Standards Organization 492
International Symposium on Transportation and Traffic Theory 518
International Telecommunication Union 25, 254
International Thermal Spray Association 391
International Toxicity Estimates for Risk 105
International Transport Research Database
International Transport Research Documentation 155, 156
International Tunneling and Underground Space Association 142
International Union for Physical and Engineering Sciences in Medicine 109
International Union for Conservation of Nature 196
International Union of Crystallography 391
International Union of Food Science & Technology 77
International Union of Materials Research Societies 391
International Union of Railways 195, 525

International Union of Theoretical and Applied
    Mechanics 425
International Water Association 191, 314
International Water Management Institute 191
Internet and American Business 209
Internet and Personal Computing Abstracts 229
Internet.com 236, 240
Internet Encyclopedia 16, 215
Internet Engineering Task Force 222
Internet Ethics 17
Internet Public Library 122
*Internet Resources for History of Science and
    Technology* 337
Internet Scout Project 29
*Internet Technologies Handbook: Optimizing the IP
    Network* 249
*Introduction to Agricultural Engineering Technology:
    A Problem Solving Approach* 69
*Introduction to Architectural Science: The Basis of
    Sustainable Design* 88
*Introduction to Chemical Engineering
    Thermodynamics* 121
Introduction to Engineering Communication 33
Introduction to Electrical Engineering 257
Introduction to Engineering Ethics 31
Introduction to Environmental Engineering 31
*Introduction to Environmental Engineering and
    Science* 295
Introduction to Food Engineering 69
Introduction to Heat Transfer 121
Introduction to the History of Structural Mechanics 170
Introduction to Nuclear Engineering 461
Introduction to Nuclear Reactor Theory 461
Introduction to the OariNZ Project 220
Introduction to Robotics in CIM Systems 357
*Introduction to Traffic Engineering: A Manual for Data
    Collection and Analysis* 505
Introduction to Transportation Engineering 511
Introduction to Transportation Systems 160, 511
Intute 134, 388, 450
Intute: Computer Science 238
Intute: Mechanical Engineering 423
Inventing the Internet 16
*Investigating Minority Student Participation in an
    Authentic Science Research Experience.* 37
*Invisible Web: Uncovering Information Sources Search
    Engines Can't See* 17
Ioannou, P. A. 511
IoPP Glossary of Packaging Terminology 351
Iowa State University 333
*IP Standard Methods for Analysis and Testing of
    Petroleum and Related Products and British
    Standard 2000 Parts 2010* 492
Ireson, W. G. 353
Irish Nanotechnology Association 163
Ironmaking and Steelmaking 436
Irrigation and Drainage 73, 186
Irrigation and Drainage Systems 73, 186
Irrigation Association 192
Irrigation Association of Australia 192
*Irrigation Maintenance and Operations: Learning
    Process* 188
Irrigation Principles 188
Irrigation Science 74

Irwin, J. D. 241, 257
Irvine, D. J. 152
ISA Directory of Automation Products and Services 349
Isaacs, G.W. 61, 78
Isakowitz, S. J. 50
ISEE Blasters Handbook 437
*Isis Current Bibliography of the History of Science and its
    Cultural Influences* 322, 331
*Isis: International Review Devoted to the History of
    Science and its Cultural Influences* 331
Islam, M. R. 482
ISMEC Bulletin 398
ISMEC Mechanical Engineering Abstracts 398
ISO 14000 and ISO 9000 23
ISO 9000 Handbook 354
*ISO 9000 Quality Systems Handbook: Using the
    Standards as a Framework for Business
    Improvement* 23
*ISPRS Journal of Photogrammetry and Remote
    Sensing* 252
Issues in Science and Technology Librarianship 96
*Is There an Engineer Inside You? A Comprehensive Guide
    to Career Decisions in Engineering* 35
Istituto e Museo di Storia della Scienza 338
IstructE 173
ITE Journal 515
ItmWeb 236
ITRD (see also TRID)155, 156, 500
ITS International 515
ITS Standards Program 520
IttoolboxBlogs 240
Iwnicki, S. 510

J

Jackson, A.V. 292
Jackson, J. A. 438
Jackson, N. 173
Jackson, S. A. 270
Jacobs, P.F. 97
Jaffe, A. B. 26
Jahm, R. K. 157
Jahanmir, S. 416
Jahrbuch fuer Wirtschaftsgeschichte 331
Jain, S.K. 290
Jakkula, A. A. 171
Jane's All the World's Aircraft 50
Jane's Information Group 504
Jane's Space Systems and Industry 51
Jane's Urban Transport Systems 157
Jansen, R. B. 188
Japanese Railway Engineering 516
Japan Patent Office 27
Japan Society of Civil Engineers 135
Japikse, D. 409
Java.net 240
JAXA: Japan Aerospace Exploration Agency 57
Jenkins, D. 405
Jensen-Buttler, C. 159
Jeppson, R.W. 513
Jeris, J.S. 118, 290
Jevremovic, T. 460
Johnson, R. W. 53, 406
Johnson Foundation for Medical Physics 94

Johnston, E.R. 416
John von Neumann and the Origins of Modern Computing 209
Joint Task Force on Computer Engineering Curricula 204, 240
JOM: Journal of the Minerals, Metals and Materials Society 385
Jones, D. S. J. 481
Jones, E. C. 511
Jones, N. 173
Jones, W. R. 326
Jordening, H.-J. 298
Journal Citation Reports 16, 20, 70, 224, 251, 486
Journal of Advanced Manufacturing Systems 359
Journal of Aerosol Science 123, 301
Journal of Aerospace Computing, Information, and Communication 55
Journal of Agricultural Safety and Health 74
Journal of Aircraft 55
Journal of Alloys and Compounds 385
Journal of Applied Mechanics. Transactions of the ASME 419
Journal of Applied Polymer Science 123, 385
Journal of Architectural Engineering 89
Journal of Atmospheric and Oceanic Technology
Journal of Atmospheric Chemistry 301
Journal of Bioactive and Compatible Polymers 385
Journal of Biobased Materials and Bioenergy 74
Journal of Biomaterials Applications 385
Journal of Biomaterials Science—Polymer Edition 104, 385
Journal of Biomechanical Engineering. Transactions of the ASME. 419
Journal of Biomechanics 104
Journal of Biomedical Materials Research, Part A, 385
Journal of Bioscience and Bioengineering 74
Journal of Biotechnology 123
Journal of Bridge Engineering 177, 515
Journal of Canadian Petroleum Technology 487
Journal of Catalysis 123
Journal of Ceramic Processing Research 385
Journal of Chemical Engineering Data 123
Journal of Chemical Engineering of Japan 124
Journal of Chemical Technology and Biotechnology 74, 124
Journal of Circuits Systems and Computers 225
Journal of Cold Regions Engineering 139
Journal of Composite Materials 385
Journal of Computational and Nonlinear Dynamics. Transactions of the ASME 19
Journal of Computational Physics 55
Journal of Computer and System Sciences 225
Journal of Computer Science and Technology 225
Journal of Computing and Information Science in Engineering. Transactions of the ASME 419
Journal of Construction Engineering and Management 155
Journal of Contaminant Hydrology 186
Journal of Dynamic Systems Measurement and Control. Transactions of the ASME 419
Journal of Earthquake Engineering 139
Journal of Electroceramics 38
Journal of Electronic Materials 385

Journal of Electronic Packaging. Transactions of the ASME 419
Journal of Energy Resources Technology. Transactions of the ASME 420
Journal of Engineering Education 40, 263, 271
Journal of Engineering for Gas Turbines and Power. Transactions of the ASME 420
Journal of Engineering Materials and Technology. Transactions of the ASME 385, 420
Journal of Engineering Mechanics 177, 420
Journal of Engineering Technology 264, 271
Journal of Environmental Engineering 186, 301
Journal of Environmental Management 301
Journal of Environmental Quality 74
Journal of Environmental Science and Health. Part A, Toxic/Hazardous Substances and Environmental Engineering 301
Journal of Environmental Systems 302
Journal of Field Robotics 252
Journal of Flood Risk Management 186
Journal of Fluid Mechanics 55, 420
Journal of Fluids Engineering 55
Journal of Food Engineering 74
Journal of Food Process Engineering 74
Journal of Food Processing and Preservation 74
Journal of Food Science 74
Journal of Food Science and Technology 75
Journal of Fuel Cell Science and Technology. Transactions of the ASME 420
Journal of Geodesy 252
Journal of Geotechnical and Geoenvironmental Engineering 140, 436
Journal of Green Building 89
Journal of Guidance, Control, and Dynamics 55
Journal of Hazardous Materials 302
Journal of Heat Transfer 55, 420
Journal of High Speed Networks 225
Journal of History of Science, Japan 331
Journal of Hydraulic Engineering 186, 302
Journal of Hydraulic Research 186
Journal of Hydroinformation 186
Journal of Hydrologic Engineering 186
Journal of Hydrology 287, 302
Journal of Information Storage and Processing Systems 225
Journal of Infrastructure Systems 158
Journal of Intelligent Manufacturing 359
Journal of Intelligent Material Systems and Structures 55
Journal of Irrigation and Drainage Engineering 75, 187
Journal of Loss Prevention in the Process Industries 124
Journal of Machine Learning Research 253
Journal of Management in Engineering 150
Journal of Manufacturing Science and Engineering, Transactions of The ASME 359, 420
Journal of Manufacturing Systems 359
Journal of Manufacturing Technology Management 359
Journal of Materials Chemistry 385
Journal of Materials Education 271
Journal of Materials Research 385
Journal of Materials Science 35
Journal of Materials Science: Materials in Electronics 385

Journal of Mechanical Behavior of Biomedical Materials 386
Journal of Mechanical Design. Transactions of the ASME 420
Journal of Mechanisms and Robotics. Transactions of the ASME 420
Journal of Medical Devices. Transactions of the ASME 420
Journal of Membrane Science 124
Journal of Microelectromechanical Systems 420
Journal of Micromechanics and Microengineering 386, 420
Journal of Nanomaterials 167
Journal of Nanomechanics and Micromechanics 168
Journal of Nanoengineering and Nanosystems
Journal of Nanoscience and Nanotechnology 385
Journal of Nanotechnology in Engineering and Medicine. Transactions of the ASME 420
Journal of Network and Computer Applications 225
Journal of Neural Engineering 104
Journal of Non-Crystalline Solids 255, 386
Journal of Non-Newtonian Fluid Mechanics 420
Journal of Nuclear Materials. Journal des Materiaux Nucleaires 462
Journal of Nuclear Materials Management 462
Journal of Nuclear Science and Technology 462
Journal of Offshore Mechanics and Arctic Engineering. Transactions of the ASME 420
Journal of Performance of Constructed Facilities 177
Journal of Petroleum Geology 487
Journal of Petroleum Science and Engineering 487
Journal of Phase Equilibria and Diffusion 386
Journal of Physical Chemistry 56
Journal of Pipeline Engineering 516
Journal of Polymer Science 386
Journal of Pressure Vessel Technology. Transactions of the ASME 420
Journal of Process Control 124
Journal of Professional Issues in Engineering Education and Practice 271
Journal of Propulsion and Power 56
Journal of Quality in Maintenance Engineering 359
Journal of Quality Technology 359
Journal of Quantitative Spectroscopy and Radiative Transfer 56
Journal of Rheology 420
Journal of Sandwich Structures and Materials 386
Journal of Science Education and Technology 271
Journal of Separation Science 124
Journal of Ship Production 516
Journal of Soil and Water Conservation 75
Journal of Solar Energy Engineering. Transactions of the ASME 420
Journal of Sound and Vibration 420
Journal of Spacecraft and Rockets 56
Journal of Statistical Mechanics—Theory and Experiment 420
Journal of Strain Analysis for Engineering Design 420
Journal of Structural Engineering 177, 178
Journal of Supercomputing 225
Journal of Supercritical Fluids 124
Journal of Systems Architecture 225
Journal of the ACM 245

Journal of the Air and Waste Management Association 301
Journal of the American Ceramic Society 385
Journal of the American Helicopter Society 56
Journal of the American Water Works Association 301
Journal of the Association for History and Computing 331
Journal of the Astronautical Sciences 56
Journal of the British Interplanetary Society 55
Journal of the Ceramic Society of Japan 385
Journal of the Electrochemical Society 385
Journal of the Energy Institute 487
Journal of the European Ceramic Society 386
Journal of the Japan Petroleum Institute 487
Journal of the Learning Sciences 271
Journal of the Mechanical Behavior of Biomedical Materials 104
Journal of the Mechanics and Physics of Solids 386
Journal of the Operational Research Society 359
Journal of the Society for Army Historical Research 331
Journal of the Transportation Research Forum 514
Journal of Thermal Science and Engineering Applications. Transactions of the ASME 420
Journal of Thermal Spray Technology 386
Journal of Thermophysics and Heat Transfe 56
Journal of Tissue Engineering and Regenerative Medicine 104
Journal of Transport History 331
Journal of Transportation Engineering 514, 526
Journal of Tribology. Transactions of the ASME 420, 487
Journal of Turbomachinery. Transactions of the ASME 56, 420
Journal of Urban Planning and Development 158
Journal of Vacuum Science & Technology 386
Journal of Vibration and Acoustics. Transactions of the ASME 420
Journal of Water and Climate Change 187
Journal of Water and Health 187
Journal of Water Resources, Planning and Management 187, 302
Journal of Waterway, Port, Coastal, and Ocean Engineering 145, 516
Journal of Women and Minorities in Science and Engineering 38, 39, 40
Journal Storage Project (see JStor)
Journal TOCs 34
Jousten, K. 410
Joyce, R. 152
Joye, S. 495
JPT: Journal of Petroleum Technology 488
JStor 16, 18
Jumper, E. 59
JuneWarren-Nickle's Energy Group 484
Juo, P. S. 99
Juran, J. M. 353
Juran's Quality Handbook: The Complete Guide to Performance Excellence 353
Jurgen, R. 404
Jurgensen, A.

K

Kachroo, P. 512
Kadlec, R. H. 298
Kagakushi Kenkyu 331

Kaiser, K. L. 249
Kalin,M. 416
Kalpakjian 343, 344, 368
Kansas State University 80
Kaplan, S.M. 249
Karnofsky, B. 157
Karassik, I. J. 414, 437
Kaufman, J. G. 380
Keane, C. M. 266
Keeley, S, 286
*Keeping What We've Got: The Impact of Financial Aid on
        Minority Retention in Engineering* 38
Keith, L. H. 157, 305
Keller, H. 12, 22, 84, 375
Kelly, A. 376
Kendrick, P. 159
Kennedy, J. L. 513
Kenett, R.S. 351
Kennel, K.E. 380
Kennish, M. J. 144
Kent, A. 215, 249
Kent, J.A. 115, 118
Kernigham, B. W. 206
Kerns, D. V. 241, 257
Kerntechnik 463
*Keyguide to Information Sources in Agricultural
        Engineering* 66
*Keyguide to Information Sources in Food Science and
        Technology* 65
Khan, M. I. 481
Khan, N. 347
Khan, R.H. 375
Khandan, N. 289
Khatib, J. 198
Khatib, O. 420
Kilareski, W.P. 512
Kilgour, F. G. 330, 340
Kilman, G.B. 249
Kim, D.-H. 87
Kim, K. 380
Kincaid, B.L. 415
Kindred Association 157
King, D. W. 5
King, R. B. 157
Kirby, R. S. 318, 330, 340
Kirchhoff, G.R. 45
Kirk, R. 100
Kirk–Othmer Encyclopedia of Chemical Technology,
        10, 115, 116
Kjell, B. 353
Kline, R. R. 267
Knee, Michael 238, 240
Knief, R. A. 460
Knight, E. H. 12, 326
*Knight's American Mechanical Dictionary: Being a
        Description of Tools, Instruments, Machines,
        Processes, and Engineering; History of
        Inventions; and Digest of Mechanical
        Appliances in Science and the Arts* 12, 326
Knovel 4, 14 , 117, 399
Knox, H.D. 459
Knuth, D. E. 206
Kock, N. F. 356
Kohan, A. L. 413

Kok, K. D. 459
Komendant, A. E. 87
Kontoginis, T. 342, 368
Koo, J. H. 165
Kories, R. 249
Kopp, J. F. 305
Kowalski, W. J. 296
Kramer, E. J. 384
Krauss, A.D. 405
Kreith, F. 294, 403, 406
Kreysa, G. 380
Kruse, K. 30
Kucklick, T. R. 101
Kultur und Technik 332
Kuppan, T. 406
Kurfess, T. R. 410
Kurtz, M. 14, 403
Kurtz, J-P 133
Kutz, M. 102, 353, 354, 380
Kusiak, A. 352
Kuttner, K-H. 402
Kwartalnik Historii Nauki i Techniki 332
Kwok, A. G. 87
Kyle, B. G. 121

L

Laboratory for Manufacturing and Productivity 361
Laboube, R.A. 175
Lackie, J. 99
Lacombe, A. 5
Laha, D. 354
Laidler, D. W. 159
Lake, L. W. 482
Lam, P. C. 40
La Mantia, F. 157
Lamarsh, J. R. 461
Lamb, T. 412
Lambert, R. 173
Lamm, R. 157
Landa, D.C. 406
Landa, H. C. 406
Landen, K. 23
Landis, R. B. 38
Landolt-Börnstein: Substance/Property Index
        118, 120, 383
Landwards 75
Lane, C. N. 285
Langford, D. 17
Langley, Samuel 45
Langmead, D. 326
Lapeyrouse, N. J. 482
LaPlante, P. A. 212
Larock, B. E. 513
Larsen, O. P. 87
Larson, H. 176
Lasers in Medical Science 104
La Storia e la Filosofia della Scienza, della Tecnologia e
        della Medicina 337
Latham, M. 152
Lauffenburger, D. A. 121
Lawrence Berkeley National Laboratory 465
Lawton, B. 397
Lay, M.G. 507

Layton, E. T. 267
Leadership and Management in Engineering
Leadership in Engineering and Environmental Design
    (see LEED)
Lean Construction Institute 154
Lean Maintenance 357
*Lean Maintenance: Reduce Costs, Improve Quality, and
    Increase Market Share* 57
Lean Product Development Initiative 361
Learning to Live with Rivers 188
Lecture Notes in Computer Science 253
Lee, C. C. 287, 289, 293
Lee, E. S. 357
Lee, J. 37, 350
Lee, M. H. 357
Lee, S.C. 290
Lee, T.S. 512
LEED 80
LEED Materials: A Resource Guide to Green Building 84
LEED Reference Manuals 86
Leffler, W.L. 495
Lehr, J. 286
Lehr, J. H. 286
Lemaitre, J. 381
Lemelson Center for the Study of Invention and Innovation
    210, 336
Leonardo Centre for Tribolog 423
Lerum, V. 87
Leslie, S. W. 267
Lester, R. 456
Leto, J. 14
Levine, J., ix, x
Levitt, J. 357
Levy, M. 87, 88
Levy, P. E. 357
Levy, S. M. 172
Lewis, E. E. 32
Lewis, E. V. 417
Lewis, R. A. 68
Lewis, R. J. 115, 157, 293
LexisNexis Academic 35
Ley, W., K. 53
Li, D. 376
Libicki, M. C. 23
Library and Information Center Directory 504
Library Literature and Information Science 1
Library of Congress Subject Headings 9, 207, 240
Science, Technology and Business Division 334
Lide, D.R. 117
Liew, J.Y.R. 85
Life Safety Code 90
Lifshin, E. 375
Lightfoot, E.N. 120
*Lighting for Driving: Roads, Vehicles, Signs, and
    Signals* 512
Light Rail Now 160
Lilienthal, Otto 45
Lim, H.C. 297
Lin, S.D. 289, 293
Lincoln Electric Company 343, 368
Linda Hall Library of Science, Engineering &
    Technology 335
Lindeburg, M. R. 403
Linden, D. 249

Linderman, J.J. 121
Lingaiah, K. 408
Links to Important Railroad Industry Information 523
Linux Journal 226
Lio, X. 483
Liptak, B. G. 289
Literature of Agricultural Engineering 65
Literature of Chemical Technology 114
*Literature Resources for Chemical Process Industries:
    A Collection of Papers* 114
Liu, D. H. F. 289
Liu, H. 513
Liu, M. 88
Lloyd's Marine Intelligence Unit 523
LNG Journal 488
Lnstruct 170
*Load & Resistance Factor Design: Manual of Steel
    Construction* 85
Lobo, H. 381
Local Technical Assistance Program 522
Locke, E.A. 356
Loftin, M.K. 86
Logan, E. 408
Long, S. 38
Look, B. G. 138
LoPresti, P.G. 100
Lord, Charles 9, 10, 50, 240, 315, 372, 393, 430, 495
Lottermoser, B. G. 298
Low Impact Development Center 192
LRFD Bridge Construction Specifications 519
LRFD Bridge Design Specifications 519
*LRFD Guide Specifications for Design of Pedestrian
    Bridges* 519
Lubrication Engineers Manual 415
Lubrication Science 487
Lucier, P. 470, 495
Lukoff, H. 210
Lund, D.B. 68
Lundberg, M. 158
Lusk-Brooke, K. 326
Lyall, S. 88
Lynch, J. M. 287
Lynch, M. 428
Lyons, R. G. 217
Lyons, W. C., 482

M

Ma, K.C. 290
Ma, M. 107
Macaulay, D. 30
Machinability Data Center 408
Machine Design Databook 408
*Machine Devices and Components Illustrated
    Sourcebook* 409
Machine Elements in Mechanical Design 417
Machinery's Handbook 408
Machinery's Handbook Guide 408
Machining Data Handbook 408
Mackay, C. 165
Mackay, D. 29
Mackay, E. B. 152
Mackay, R. 414
Macleod, R. A. 10, 50, 97, 399, 430, 495

Macmillan Dictionary of Measurement 13
Macmillan Encyclopedia of Earth Sciences 439
Macnab, A. 137
Macromolecular Materials and Engineering 123
Macromolecular Bioscience 386
Mader, C. L. 144
Maes 39
Magazine of Concrete Research 179
Mainstone, R. 88, 173
Major Coalfields of the World 440
Maizell, R. E. 114
*Making of an Engineer: An Illustrated History of
    Engineering Education in United States and
    Canada* 267
Making of the Atomic Age 467
*Making Technology Masculine: Men, Women, and
    Modern Machines in America, 1870–1950* 267
Makin' Numbers: Howard Aiken and the Computer 209
Maloney, J.D. 117
*Management Procurement and Law—Proceedings of the
    ICE* 150
*Man and Metals: A History of Mining in Relation to the
    Development of Civilization* 429
Manly, B. F. J. 295
Madrid, C. 12
Maintenance Engineering Handbook 354
Making of America 262
*Making of an Engineer: An Illustrated History of
    Engineering Education in the United States
    and Canada* 260
*Making of the Modern World: The Goldsmiths'–Kress
    Library of Economic Literature* 322
Malek, M. A. 413
Malinowsky, H.R. 10, 107, 110, 399, 430
Mamaril, N. 35, 43
*Management Development in the Construction
    Industry* 152
*Managing Engineering and Technology: An Introduction
    to Management for Engineers* 30
*Managing Engineering, Technology, and Innovation for
    Growth* 367
Managing the Engineering Design Function 30
Manke, K.L. 461
Manly, B. F. J. 290
Mannering, F. L. 512
Manning, R. 183
*Manual 52: Guide for Design of Steel Transmission
    Lines* 172
*Manual 57: Operation and Maintenance of Irrigation and
    Drainage Systems* 187
*Manual 72: Guide for Design of Steel Transmission Pole
    Structures* 172
*Manual 74: Guidelines for Electrical Transmission Line
    Structural Loading* 172
Manual for Bridge Evaluation 506
Manual of Bridge Engineering 172, 508
Manual of Hydrology 183
Manual of Numerical Methods in Concrete 174
Manual of Railway Engineering 156, 510
Manual of Standards and Recommended Practices 520
Manual of Steel Construction, Metric 175
Manual on Uniform Traffic Control Devices 21, 25, 508
*Manufacturing: Design, Production, Automation and
    Integration* 356

Manufacturing IT Forum 367
Manufacturing.net 349, 361
Manufacturing Science and Technology Center 361
Manufacturing Systems Integration 361
Manufacturing Systems Modeling and Analysis 56
Manzhirev, A.V. 14
Map Image Rendering Database for Geoscience 441
Mapping Federal–Provincial–Territorial Mining
    Knowledge 441
Marcellus Shale 278
Marcus, J. J. 437
Marghitu, D. B. 403
MAPSearch 484
Marciniak, J. J. 215
Marine and Petroleum Geology 487
*Marine Engineering: Part of the Proceedings of the
    Institution of Civil Engineers* 145
Marine Propulsion 504
Marine Structures 145
Maritime Administration 522
Maritime Research Institute Netherlands 147
Maritime Technology Society 147
Mark, H. 52
Mark, H. F. 376
Mark, J. E. 381
*Marketing and Selling A/E and Other Engineering
    Services* 151
Mark's Calculations for Machine Design. 408
Marks' Standard Handbook for Mechanical Engineers
    53, 402
Marmaras, N. 342, 368
Marshall, C. P. 439
Martin, C. 495
Martin, J. 376
Martin, T. C. 242, 257
Martin, M. 31
Martindale's Calculators' Online Center/Engineering
    Center 145
Martin-Vega, L.A. 341, 368
Masonry Institute of America 182
Massachusetts Institute of Technology 46
Masters, G. M. 295
*Masterworks of Technology: The Story of Creative
    Engineering, Architecture, and Design,* 8, 32
Materials and Structures 178
Materials Business File 374
Materials Characterization 386
Materials Chemistry and Physics 386
Materials for Engineers 384
Materials Handbook 11
Materials Handbook: A Concise Desktop Reference
    12, 379
*Materials Handbook: An Encyclopedia for Managers,
    Technical Professionals, Purchasing and
    Production Managers, Technicians, and
    Supervisors* 379
Materials Handling Handbook 354
Materials Research Society 109, 391
Materials Research Society Symposia Proceedings. 384
Material Safety Data Sheets 122
Material Safety Data Sheets Online 122
*Materials Science and Technology: A Comprehensive
    Treatment* 384
Materials Science and Engineering 386

Materials Science and Engineering: An Introduction 384
Materials Selection in Mechanical Design 384
Materials Selector 384
Materials Today 386
Mathai, P.K. 298
Mathematical Programming Glossary 214
Mathematical Work of Charles Babbage 209
*Mathematics, Engineering, Science Achievement
    Program* 41
*Mathematics for Mechanical Technicians and
    Technologists: Principles, Formulas,
    Problem Solving* 13
Matheson Gas Data Book 120
MathSciNet 229
MATLAB 112, 127
Matthews, C. 33, 53, 403, 413
MatWeb: Material Property Data 388
Maynard's Industrial Engineering Handbook 353
Mays, G. C. 173, 177
Mays, L. W. 293, 298
Mazur, E. 264, 267
Mazurek, ED. 416
McAllister, E. W. 413, 482, 509
McCaffer, R. 152
McCartney, B. 144
McCarthy, J. F. 86
McCauley, J. F. 413
McClellan, J. E. 325, 330
McClintock, H. 159
McCluskey, J. 159
McCormick, B. 59
McCulloch, D. M. 407
McGowan, M. 30
*McGraw-Hill Concise Encyclopedia of Science and
    Technology* 12
*McGraw-Hill Dictionary of Computing &
    Communications* 212
*McGraw-Hill Dictionary of Electrical and Computer
    Engineering* 212, 249
*McGraw-Hill Dictionary of Scientific and Technical
    Terms* 12, 212, 213
McGraw-Hill Encyclopedia of Engineering 12
*McGraw-Hill Encyclopedia of Networking &
    Telecommunications* 216
McGraw-Hill Encyclopedia of Science and Technology
    12, 318, 340
*McGraw-Hill Machining and Metalworking
    Handbook* 409
McGraw-Hill's Engineering Companion 13
McInnnis, A. 152
McKay, A. 453, 467
McKee, G.D. 305
McKelvey, F.A. 159
McKenna, T. 350
McLeland, L-N. 124
McMillan, G. K. 354
McMurrey, D.A. 32
McNeil, I. 326
McWilliam, R.W. 133
Meadows, D. 86
Means, R.S. 290
Means Costworks 153
Mechanical and Electrical Equipment for Buildings 87

Mechanical and Transportation Engineering Abstracts
    282, 398
Mechanical Behavior of Materials 384
*Mechanical Design Handbook; Measurement, Analysis,
    and Control of Dynamic Systems,* 409
Mechanical Design Process 418
Mechanical Engineering 93, 421
Mechanical Engineering Formulas Pocket Guide 403
Mechanical Engineering Index 398
*Mechanical Engineering Reference Manual for the PE
    Exam* 403
Mechanical Engineers Data Handbook 402
Mechanical Engineers' Handbook 403
*Mechanical Engineers' Handbook; V.3: Manufacturing
    and Management* 354
Mechanical Engineer's Reference Book 403
*Mechanical Life Cycle Handbook: Good Environmental
    Design and Manufacturing* 353
Mechanical Systems and Signal Processing 421
Mechanical Vibrations 416
Mechanics of Advanced Materials and Structures 386
Mechanics of Materials 416, 417, 421
Mechanics of Time-Depending Materials 386
Mechanism and Machine Theory 421
Mechanisms & Mechanical Devices Sourcebook 409
Mechanistic-Empirical Pavement Design Guide 506
Mechatronics Handbook 409
Medical Device R&D Handbook 101
Medical Device Register On-Line 99
Medical Image Analysis 104
Medline 98, 105
Meeting the Challenge 38
Megyesy, E. F. 413
Mehta, D.P. 407
Mehl, J. P. 438
Meinck, F. 185
Meiners, P. A. 35
Meirovitch, L. 417
Meisel, A. 84
Mellon Mays Undergraduate Fellowship Program 43
Member Product Directory 349
Memoirs of a Computer Pioneer 209
*Memories that Shaped an Industry: Decisions Leading to
    IBM System 360* 209
MEMS and Nanotechnology Clearinghouse 423
MEMS Handbook 410
MEMSnet 388
Menon, E. S. 413
*Merck Index: An Encyclopedia of Chemicals, Drugs, and
    Biologicals* 100
Merdinger, C. J. 131, 202
Merkley, G. P. 188
Merritt, F.S. 86, 175
Mesri, G. 139
Metadata Made Simpler: A Guide for Libraries 23
Metadex (see Metals Index)
Metal Bulletin Prices and Data 447
Metal Cutting Tool Handbook 409
Metal Powder Industries Federation 391
Metal Price Charts on InfoMine–Precious Metals, Base
    Metals 447
Metallic Materials Specification Handbook 23
Metallurgical and Materials Transactions A. 386
Metals & Alloys in the Unified Numbering System 21, 23

Index

561
Metals and Minerals Annual Review 447
Metals Index (Metadex) 17, 374, 433
Metal Industry Indicators 441
Metcalf, J. 5
Metcalf and Eddy, Inc. 188, 298
Methods, H. 298
Methods for Chemical Analysis of Water and Wastes 305
Meyers, R. A. 287, 482, 483
Mian, M. H. 483
Micromanufacturing Engineering and Technology 357
Microporous and Mesoporous Materials 123
Microprocessors and Microsystems 225
Microsoft Manual of Style for Technical Publications 33
Middleton, G. V. 439
Meier, H. W. 86
Metric Handbook 172
Micro- and Nanotechnology: Materials, Processes,
    Packaging, and Systems 367
MicroManufacturing Conference & Exhibits 367
Microsoft Computer Dictionary 213
Mildren, K. W. 50, 54, 59, 430
*Milestones in Computer Science and Information
    Technology* 210
Miller, J. R. 298
Miller, M. A. 249
Miller, M. R. 414
Miller, R. 414
Miller, R. W. 406
Miller, S.M.O. 298
Miller, T.R. 459
Milwaukee School of Engineering 80
MINABS Online 433
Mindat.org 450
Mine and Mineral Processing Plant Locations 441
Mineralogical Association of Canada 444
Mineral and Energy Technology 434
Mineral Commodity Summaries 441
Mineral Industry Indicators 447
Mineral Industry Surveys 441, 448
Mineral Resource Data System 442
Mineral Resources Program 140
Minerals and Metallurgical Processing Journal 436
Minerals and Mining Statistics Online 448
*Minerals Handbook: Statistics and Analyses of the
    World's Minerals Industry* 437, 448
Minerals, Metals and Materials Society 393
Minerals Yearbook 442
Mine Safety and Health Administration 441
*Mine Wastes: Characterization, Treatment,
    Environmental Impacts* 298
*Mine Water and the Environment, Journal of
    International Mine Water Association* 436
Mines, Mining, and Mineral Resources–Library of
    Congress Science Reference Services. Science
    Tracer Bullets Online 450
Mining and Metallurgical Society of America 446
Mining and Metal Production through the Ages 428
Mining: An International History 429
Mining Annual Review 437, 447
Mining Association of Canada 445
Mining Associations/Companies in Canada 440
Mining Education. Mining Internet 450
*Mining Engineers & the American West; the Lace-boot
    Brigade, 1849–1933* 429

*Mining Environmental Handbook: Effects of Mining on
    the Environment and American Environmental
    Control* 437
Mining History Network 429, 445
Mining in World History 428
Mining Journal 435
Mining-technology.com 439
Minkowycz, W. J. 406
Minority & Women Doctoral Directory 35
Minority Engineer 39
Minority Science and Engineering Improvement
    Program 41
Minrath, W.R. 14
Minsk, L. D. 157
Miracola, C. L. 37
Mischke, C.R. 409
Missouri University of Science and Technology 80
MIT Libraries Materials Science and Engineering
    Research Guide 374
MIT OpenCourseWare Web, 112, 127, 272
MIT Press: Computer Science and Intelligent Systems 231
Mitrovich, V.L. 401
Mitsch, W. J. 298
Mobile Networks and Applications 225
Mobley, K. 354
Mobley, R. K. 354
Model Drainage Manual 506
*Modeling Tools for Environmental Engineers and
    Scientists* 289
Modelling Transport 159
Modern Construction Management 152
Modern Control Systems 416
Modern Marine Engineers Manual 412
Modern Steel Construction 179
Modern Tribology Handbook 415
Mohitpour, M. 513
Mohle, H. 185
Mokhatab, S. 483
Molecular Biomethods Handbook 102
Molecular modeling, viewing and drawing 127
Molecules to Go 102
Mollenhauer, K. 405
Mollenhoff, C. R. 210
*Money for College: A Guide to Financial Aid for
    African-American Students* 36
Montgolfier, J-E. 45
Montgolfier, J-M. 45
Montgomery, J. H. 293
Moore, J. 102
Morello, L. 404
Moretto, L. 33
Morgan, B. 66
Morgan & Claypool Publishers 231
Morgan Kaufmann: Computing Books 231
Morritt, H. 495
Morse, L.C. 30
Moschovitis, C. J. P. 17
Moss, R.Y. 353
Moser, A. P. 413
Mosher, J. 347
Moss, D. R. 413
Motorway Achievement 159
Motorway Archive Trust 159, 160
Motor Truck Engineering Handbook 404

Motor Vehicle 510
Mott, R. L. 417
Mottu, S. 399
Mount, E. 430
MRS Bulletin 386
Mueller, J.F. 149
Muhlbauer, W. K. 482
Muhlthaler, W. 513
Mulcahy, D. E. 354
Mulder, K. 197
Muller, G. 511
Mulligan, C. N. 295
*Multiagent Systems for Manufacturing Control:
    A Design Methodology* 56
Multilingual Aeronautical Dictionary 52
Mulvany, W. T. 183
Municipal Engineer 158
Munson, B. R 418
Murphy, P. R. 511
Murray, A. 513
Murray, R. L. 461
Murthy, J.Y. 406
Museo Nacional de Ciencia y Tecnologia 335
Museum of the History of Science 335
Myer Kutz, M. 505
Myers, M. A. 384
*Mythical Man-Month: Essays on Software
    Engineering* 206

N

NACE International 425, 491
NACME (see National Action Council for Minorities in
    Engineering)
Najafi, M. 513
Nalwa, H. S. 100, 102, 376, 381
NanoAssociation for Natural Resources and Energy
    Security 391
Nanodot 106
Nanofolio: What is Nanotechnology? 166
Nanoforum: European Nanotechnology Gateway 166
Nano Letters 386
NanoManufacturing Conference & Exhibits 367
Nanomaterials Handbook 164
Nanoscale and Microscale Thermophysical Engineering
    386, 420
Nanotechnology, 168, 386
Nanotechnology and the Environment 165
*Nanotechnology: Basic Calculations for Engineers and
    Scientists* 165
Nanototechnology Knowledge Transfer Network 166
Nanotechnology Institute 163
*Nanotechnology in Construction: Proceedings of the
    NICOM3* 164
Nanotechnology Now 166
NANO: The Magazine for Small Science 168
Nano Today 168, 386
Narins, B. 213
Narlikar, A. V. 381
NASA 273
NASA–Ames Human Systems Integration Division 361
NASA–Cognition Lab 361
Nassas, S. 171
NASA Technical Report Server 48, 49

National Academies 19, 43
National Academy of Engineering 16, 19, 269
National Academy of Sciences 19, 34
National Academy Press Publications 273
National Action Council for Minorities in Engineering
    34, 38, 40, 41, 43
*National Action Plan for Addressing the Critical
    Needs of the U.S. Science, Technology,
    Engineering and Mathematics (STEM)
    Education System* 268
National Aeronautics and Space Administration 57
National Ag Safety Database 65
*National Ambient Air Quality Standards for Criteria
    Pollutants* 304
National Asphalt Paving Association 525
National Association for Surface Finishing 391
National Association of Clean Air Agencies 314
National Association of Corrosion Engineers 392
National Association of County Engineers 161
National Association of Environmental Professionals 314
National Association of Multicultural Engineering
    Program Administrators 41
National Building Specification 154
*National Cataloging Unit for the Archives of
    Contemporary Scientists* 323
National Center for Biotechnology Information 102, 106
National Center for Health Statistics 102
National Center for Manufacturing Sciences 361
National Center for Remanufacturing and Resource
    Recovery 361
National Coalition for Advanced Manufacturing 361
National Coalition of Underrepresented Racial and
    Ethnic Advocacy Groups in Engineering and
    Science 41
National Council of Examiners for Engineers and
    Surveying 31, 32, 273
National Council on Radiation Protection and
    Measurements 466
National Defense Science & Engineering & Graduate
    Fellowship Program 43
National Design Specification for Wood Construction. 175
*National Directory of Foundation Grants for Native
    Americans* 35
National Electric Code 21, 90
National Engineering Handbook 186
National Environmental Methods Index 307
National Environmental Publications Internet Site 442
National Fire Prevention Association 195, 254, 492
National Fluid Power Association 425
National Geologic Map Database 484
National Geophysical Data Center 142
National Ground Water Association 184, 192, 314
National Hazards Review 178
National Highway Traffic Safety Administration 522
Fatality Analysis Reporting System 503
National Household Travel Survey 503
National Information Service for Earthquake Engineering
    141, 142
National Information Standards Organization 21, 23, 25
National Institute for Engineering Ethics 30, 31
National Institute for Farm Safety 78
National Institute for Occupational Safety and Health 22,
    361, 442
    Mining Site Index 442

National Institute for Standards and Technology 22, 23, 25, 195
National Institute of Advanced Industrial Science and Technology 166
National Institute of Biomedical Imaging and Bioengineering 106
National Institutes of Health 94
National Inventors Hall of Fame 27
National Inventory of Dams 189
National ITS Architecture 508
National Joint Utilities Group 161
National Library for the Environment: Congressional Research Service Reports 442
National Library of Medicine 105
National Museum of American History 335
National Museum of History and Technology 325
National Nanotechnology Initiative 162
National Nanotechnology Institute 166
National Nuclear Data Center 465
National Organization for the Professional Advancement of Black Chemists and Chemical Engineers 41
National Petrochemical & Refiners Association 491
National Petroleum Council 491
National Primary Drinking Water Standards and National Secondary Drinking Water Standards 305
National Recommended Water Quality Criteria 306
National Research Council 35
National Research Council Canada 28
Natural Resources Canada 448
Natural Resources Conservation Service 186
Natural Resources Conservation Service Soils 141
National Safety Council 503
National Science Digital Library 105, 273
National Science Foundation 106, 268, 273
National Society of Black Engineers 41
National Society of Professional Engineers 8, 30, 31, 91, 273
National Technical Information Services (see NTIS)
National Technical Museum in Prague 336
National Transit Database Glossary 505
National Transportation Atlas Database 501
National Transportation Library 500, 521
National Transportation Safety Board 522
National Transportation Statistics 501
National Truck Equipment Association 523
National Water Quality Assessment Data Warehouse 285
National Water Resources Association 192
Nation's Building News (see Builder)
Natural Gas: A Basic Handbook 483
Natural Gas Statistics Sourcebook 494
Natural Resources Canada 440
Natural Resources Conservation Service Soils 141
Natural Resources Engineering 69, 440–1
Nature Materials 386
Nature Nanotechnology 386
Navier, C. 45
Navigation Engineering Practice and Ethical Standards 144
Nawy, E. G. 174
Nayler, G. H. F. 400
Naylor, T.E. 50
Nayyar, M. L. 413, 414
Nazaroff, W. W. 295, 315

NCHRP Publications 515
NCSTRL 218–9
Neale, M. J. 415
Neal–Schuman Authoritative Guide to Evaluating Information on the Internet 17
Nebraska Tractor Tests 69
Nebeker, F. 94, 95, 110, 250, 257, 330
NEEDS 273
Neighborhood Street Design Guidelines 520
Neilsen, J. 173
Nelson, P. M. 157
Nemecek, J. 164
Nemerow, N. L. 290, 295
Nero, A. V. 459
Nesbitt, B. 410, 414
Netlingo 214
Networked Computer Science Technical Reference Library 218
Networks 225
Network World 240
Neuendorf, K. K. E. 438
Neuburger, A. 330
Neumayer, E. 196
Neville, A. M. 174
Newberry, C. 59
Newcomen Society for the Study of the History of Engineering and Technology 338
New Concepts for Sustainable Management of River Basins 188
New Engineering Contract 152
New England Journal of Higher Education 263
New Generation Computing 225
Newhouse, E. L. 327
New Mexico Water Resources Research Institute 192
Newnes Electrical Power Engineer's Handbook 249
Newnes: Electronics and Computer Engineering 231
Newnes Mechanical Engineer's Pocket Book 403
New Penguin Dictionary Civil Engineering 132
New Publications of the Geological Survey 285
New Scientist 14, 15
New Steel Construction 179
New Technology Magazine 488
Newton, D. 107, 110
Newton, H. 213
Newton, I. 45
Newton's Telecom Dictionary: Covering Telecommunications, Networking, Information Technology, the Internet, Fiber Optics, RFID, Wireless, and VoIP 213
New Walford's Guide to Reference Resources: Science, Technology and Medicine, Vol. 1 456
New Way Things Work 30
New World Dictionary of Computer Terms 204
New World Encyclopedia 219, 240
New York Public Library, Engineering Societies Library 430
Nguyen, J.P. 481
Nienhuis, P. H. 188
Nill, K. R. 99, 103
NIOSHTIC-2 443
Nisbett, J.K. 416
Nise, N. S. 418
Niskern, D. 325
NIST/ASME Steam Properties Database 119

NIST Chemistry WebBook 118
NIST Data Gateway 388
NIST-JANAF Thermochemical Tables 53
NIST Manufacturing Portal 362
NOAA Coastal Services Center 146
Noise Control Engineering Journal 302
Nof, S. Y. 354
Nonlinear Dynamics 421
Nonmetallic Mineral Products Industry Indexes 443
North America Energy Standards Board 491
North American Industry Classification System 22
North American Pipelines 488
North Carolina A&T 80
Northern Miner Online 436
Norton, R. L. 354
Notes and Records of the Royal Society of London 332
Novak, P. 188
Novikov, V. 376
NPN—National Petroleum News 488
NRCan Library–Energy, Minerals, and Management 450
NSSN 22, 25, 56
NTIS 3, 19, 49, 231, 246, 398
*NTIS Alerts: Biomedical Technology and Human Factors
         Engineering* 104
*Nuclear Energy: An Introduction to the Concepts,
         Systems and Applications of Nuclear
         Processes* 461
Nuclear Energy Agency/Agence pour lenergie
         nucleaire 466
*Nuclear Engineering and Design: An International
         Journal Devoted to the Thermal,
         Mechanical and Structural Problems of
         Nuclear Energy* 463
Nuclear Engineering Handbook 459
Nuclear Engineering International 463
*Nuclear Engineering: Theory and Technology of
         Commercial Nuclear Power* 460
Nuclear Engineering Wall Charts 465
*Nuclear Facility Decommissioning and Site
         Remedial Actions: A Selected
         Bibliography* 456
*Nuclear Fuel Cycle Information System:
         A Directory of Nuclear Fuel Cycle
         Facilities* 457
Nuclear Fusion. Fusion Nucleaire 463
*Nuclear Future: Journal of the Institution of Nuclear
         Engineers and the British Nuclear Energy
         Society* 463
Nuclear Handbook 459
Nuclear Institute 466
Nuclear News 463
Nuclear Plant Journal 463
Nuclear Power Plant Engineering 461
*Nuclear Power Reactors in the World, Reference Data
         Series No. 2* 457
Nuclear Principles in Engineering 460
Nuclear Reactor Analysis 460
Nuclear Reactor Engineering 460
Nuclear Reactor Physics 462
*Nuclear Reactors, Nuclear Fusion and Fusion
         Engineering* 460
Nuclear Reactor Theory 460
Nuclear Regulatory Commission see U.S. Nuclear
         Regulatory Commission

Nuclear Science Abstracts 455
*Nuclear Science and Engineering: the Journal of the
         American Nuclear Society* 463
Nuclear Technology 463
Nuclear Terms Handbook 458
*Numerical Data and Functional Relationships in Science
         and Technology*
Numerical Heat Transfer. Part A, Applications 421
Numerical Heat Transfer. Part B, Fundamentals 421
Numerical Modeling of Water Waves 144
Numerical Recipes Books Online 234
NWIS Web 285
NWS Hydrologic Information Center 189

O

Oakhill, A.C. 175
Oakridge National Lab 522
OARiNZ 220
O'Brien, J. L. 149
O'Brien, T.G. 352
*Occupational Health and Safety: Terms, Definitions, and
         Abbreviations* 350
Occupational Outlook Handbook 8, 259, 266, 275
Occupational Safety and Ergonomics Excellence 362
Occupational Safety and Health Administration 22, 362
Oceanic Abstracts 284
*Ocean Engineering: An International Journal of Research
         and Development* 145
Ocean Engineering Handbook, 144, 412
Ochs, M.A. 61, 78
OCLC FirstSearch 281
Oreovicz, F.S. 268
*Office of Biological and Environmental Research
         Abstracts Database* 283
Office of Bridge Technology 180, 521
Office of Coast Survey 147, 523
Office of Highway Policy Information 502
Office of Infrastructure 521
Office of Operations 521
Office of Pavement Technology 521
Office of Pipeline and Hazardous Materials Safety
         Administration 522
Office of Water Services 192
Offshore 488
Offshore Engineer 488
Offshore Engineering Information Service 146
Offshore Engineering Society (U.K.) 147, 491
Offshore Technology Conference
O'Flaherty, C. A. 159
O'Hanlon, J. F. 418
Oil & Energy Trends: Annual Statistical Review 493
Oil & Gas Journal 488
Oil & Gas Journal Online Research Center 493
Oil and Gas Insight 490
Oil and Gas International 490
Oil and Gas Online 490
Oil and Gas Pipeline Fundamentals 513
*Oil and Gas Production in Nontechnical
         Language* 495
*Oil & Gas Science & Technology: Revue de l'Institut
         Francais du Petrole* 487
Oil.com 490
Oil Field Chemicals 481

Oilfield Directory 490
Oilfield Publications Limited 484, 485
Oil Museum of Canada 495
OilOnline 490
Oilsands Review 488
Oil Shale 487
Oilweek 488
Okiishi, T. H. 418
Oklahoma State University 80
Oklobdzija, V. G. 217
Olander, D. R. 461
Oldenziel, R. 267
O'Leary, B.L. 53
Oleson, J. P. 327
Oliu, W. E. 32
Oliverson, R. 350
Olsen, W.C. 65
Olsson, P. 460
O'Neil, M. J. 100
Online Ethics Center for Engineering and Science 30, 31
Online Mendelian Inheritance 105
OPEC 470, 494
OPEC Bulletin 488
Open Directory: Computers 238
Open Directory Project: Science: Technology: Mechanical
        Engineering 423
Open Petroleum Engineering Journal 487
Operational Research Society 365
Operations and Intelligent Transportation Systems 522
Operations Research Applications 357
Oppenheim, N. 159
Opportunities in Engineering Careers 31
Optical Measurement Systems for Industrial
        Inspection 367
*Opto-mechatronic Systems Handbook: Techniques and
        Applications* 409
O'Reilly, M. 153
O'Reilly Books 234
O'Reilly Network 236
Oreovicz, F. S. 268
Organic Process Research and Development 124
Organisation for Economic Cooperation and
        Development 294
Orlitt, A. 151
Ortuzar, J. D. 158
Osif, Bonnie 5, 495
*Osiris: A Research Journal Devoted to the History of
        Science and its Cultural Influences* 332
Osorio, N. L. 347
Ossicini, S. 164
OSTG: Open Source Technology Group Network 236
Oswalt Institute for Physics 94
Othmer, D.F. 100
*Our Own Devices: The Past and Future of Body
        Technology* 30
OVID 281
Owen, E. W. 495
Oxford Dictionary of Biomedicine 99
Oxford English Dictionary 317, 340
*Oxford Handbook of Engineering and Technology in the
        Classical World* 327
Oxford Handbook of Nanoscience and Technology 381
*Oxford Science Publications: Mathematics, Statistics, and
        Computer Science* 231

Oxford University's Chemical and Other Safety
        Information 122
Oxley, J. 355
Ozelton, E.C. 175

P

Packaging Technology and Science 75
Paez, T.L. 411
Pankratz, T. M. 84, 287
Pansu, M. 138
Papacostas, C. S. 511
Parfitt, K. 88
Parkash, S. 483
Parke, G. 508
Parker, D. H. 188
Parker, G.G. 357
Parker, S. P. 12, 99
Parking: A Handbook of Environmental Design 159
Parking Generation: An ITE Informational Report 507
Parking Structures 158
Parmley, R. O. 409, 410
Parr, E. A. 407
Parry, W.T. 119
Partnering in Design and Construction 151
Partnering in the Team 151
PATEX 495
Patil, A. 267
Patent Fundamentals for Scientists and Engineers 26
Patent Librarian's Notebook 27
Patents as Scientific and Technical Literature 26
*Patents, Citations, and Innovations: A Window on the
        Knowledge Economy* 26
Patents for Chemicals, Pharmaceuticals and
        Biotechnology: Fundamentals of Global Law,
        Practice and Strategy 107
Patent Searching Made Easy 26
*Patent, Trademark, and Copyright Searching on the
        Internet* 26
*Paths to Discovery: Autobiographies from Chicanas
        with Careers in Science, Mathematics, and
        Engineering* 36
Patnaik, P. 290
Patterson, M. E. 61, 78
Paul, M.A. 137
Paz, M. 138
PBS TeacherSource 273
PC Magazine 226
*PCI Design Handbook: Precast and Prestressed
        Concrete* 85
PCI Journal 179
PC World 226
Peach, R. W. 354
Pearson, B. C. 495
Pearson, W. 39
*Pearson's Handbook: Crystallographic Data for
        Intermetallic Phases* 383
Peck, R.B. 139
*Pedestrian and Bicycle Information Center Case Study
        Compendium* 509
*Pedestrian Dynamics: Mathematical Theory and
        Evacuation Control* 512
*Pedestrian Road Safety Audit Guidelines and Prompt
        Lists* 509

Peer Instruction: A User's Manual 267
Penning Rowsell, E. C. 188
Penn State 80, 221
Pennsylvania State University Department of
    Architectural Engineering 86
Pereira, N.C. 296
Perez, D. 35
Performance Evaluation 225
Perl Journal 226
Perrot, P. 400
Perry, J. H. 86
Perry, R. 86, 117
Perry's Chemical Engineers' Handbook 62, 117, 118
Personal and Ubiquitous Computing 25
Persson, P-A. 138
Peschke, M. 23
Peters, Tom
Petersen, J. K. 249
Peterson's.com 400
PETEX: The Petroleum Extension Service 485
Petrochemicals in Nontechnical Language 495
Petrochemicals: The Rise of an Industry 495
Petroleum Age 470, 495
Petroleum Economist 484
Petroleum Engineering Handbook 482
*Petroleum Engineering Handbook for the Practicing
    Engineer* 483
*Petroleum Engineering Handbook – Sustainable
    Operations* 482
*Petroleum Fuels Manufacturing Handbook: Including
    Speciality Products and Sustainable
    Manufacturing Technologies* 483
Petroleum Geoscience 487
Petroleum Review 488
Petroleum Science and Technology 487
Petrophysics 487
Petroski, H. 30, 174
PetroSkills 485
Pfaffenberger, B. 213
Pfammatter, U. 328
Pfafflin, J. R. 287
Pham, H. 354
Philippsborn, H. E. 68, 350
Photovoltaic Design & Installation For Dummies 256
*Physical and Thermodynamic Properties of Pure
    Chemicals: Data Compilation*
Physical Mesomechanics 386
Physical Metallurgy Handbook 383
Physical Oceanographic Real-Time Systems 523
Physics Abstracts 18
Physics and Chemistry of Materials 384
Physics in Medicine and Biology 104
Physics of Fluids 56
*Pictorial History of Science and Engineering; The Story
    of Man's Technological and Scientific Progress
    from the Dawn of History to the Present, Told
    in 1,000 Pictures and 75,000 Words* 326
Pictorial Walk Through the 20th Century, A. "Honoring
    the U.S. Miner 429
Piersol, A. G. 411
Pierzynski, G. M. 298
Pinkus, R. L. B. 31
Pioneering offshore: The Early Years 495
Piotrowski, J. 410

Pipe Characteristics Handbook 414
Pipeline & Gas Journal 488
PipeLine and Gas Technology 489
Pipeline Design and Construction 513
Pipeline Engineering 513
Pipeline Risk Management Manual 483
*Pipe Line Rules of Thumb Handbook: Quick and
    Accurate Solutions to Your Everyday Pipe
    Line Problems* 413, 482, 509
Pipelines International Magazine 489
*Piping and Pipeline Calculations Manual:
    Construction, Design, Fabrication, and
    Examination* 412, 509
*Piping and Pipeline Engineering: Design, Construction,
    Maintenance, Integrity, and Repair* 512
Piping Calculations Manual 413
Piping Databook 413
Piping Handbook 414
Piping Materials Guide 513
Piping Stress Handbook 413
Planner 158
Planning and Design of Airports 159
Planning Complete Streets for an Aging America 511
Planning for Cycling 159
Plant Engineering–New Products 349
Plant Engineer's Handbook 354
Plastics.com 388
Plastics Design Library Handbook Series. 382
*Plastics Engineering, Manufacturing and Data
    Handbook* 82
Plastics Technology Handbook 379
Plastics Web 388
Plate, E. J. 174
Platinum Group Metals (PGM) Database 388
Plauger, P.J. 206
Plisga, G. J. 482
Pluschke, P. 292
Plutonium Handbook: A Guide to the Technology 460
Pneumatic Handbook 407
PNR Product Realization Network at Stanford 362
*Pocket Book of Technical Writing for Engineers and
    Scientists* 33
Pocket Ref 14
Podmore, P.L.J. 173
Podolny, W. 174
Poe, W. A. 483
Poenaru, D. N. 459
Poirier, D.F. 368
Polezhaev, Y.V. 402
Policy on Geometric Design of Highways and Streets 505
Pollution Abstracts 283
Pollution Engineering 302
Polyanin, A. D. 14
Polymer 386
Polymer Blends Handbook 383
Polymer Composites 386
Polymer Data Handbook 381
Polymer Engineering and Science 124
Polymer Handbook 379
Polymeric Materials Encyclopedia 376
Polymer Nanocomposites 165
Polymer Nanocomposites Handbook 380
Polymer Reaction Engineering 124
Polymers for Advanced Technologies 386

Polymer Testing 386
Poole, C. P. 382
Pope, J. E. 403
Port Designer's Handbook 144
Porteous, A. 287
Porter, J.D. 39
Portfolio of Trackwork Plans 510
Portland Cement Association 162, 182
Portraits in Silicon 210
Ports 148
Portugal, F. H. 5
Postharvest Biology and Technology 75
Poston, T. R. 12
Powder Technology 124
*Power Boilers: A Guide to Section 1 of the ASME Boiler
    and Pressure Vessel Code* 412
Power Electronics Handbook 249
Power Reactor Information System 455
*Power, Speed, and Form: Engineers and the Making of
    the Twentieth Century* 329
Practical Design of Timber Structures to Eurocode 5 176
*Practical Environmental Bioremediation: The Field
    Guide* 157
Practical Foundation Engineering Handbook 137
Practical Handbook of Marine Science 144
*Practical Handbook of Soil, Vadose Zone, and Ground-
    Water Contamination: Assessment, Prevention,
    and Remediation* 290
Practical Guide to Railway Engineering 510
Practical Lubrication for Industrial Facilities 415
Practical Pumping Handbook 414
Practical Recycling Handbook 157
*Practical Structural Analysis for Architectural
    Engineering* 87
*Practice Periodical on Structural Design and
    Construction* 178
Precast/Prestressed Concrete Institute 182
Prentice-Hall Professional Technical Reference 231
Pressman, A. 30
Pressure Vessel Design Handbook 412
*Pressure Vessel Design Manual: Illustrated Procedures
    for Solving Major*
Pressure Vessel Design Problems 413
Pressure Vessel Handbook 413
Preston, S. D. 37
Prevedouros, P. D. 511
Price, G. 16, 17, 127
*Principles of Design for Deconstruction to Facilitate
    Reuse and Recycling* 173
Principles of Estimating 152
Principles of Geotechnical Engineering 138
*Principles of Highway Engineering and Traffic
    Analysis* 512
Principles of Mineral Processing 437
Principles of Naval Architecture 417
*Principles of Pollution Abatement: Pollution Abatement
    for the 21st Century* 295
Pritchard, M.S. 31
Probabilistic Engineering Mechanics 420
Proceeding Conference for Industry and Education
    Collaboration 272
*Proceedings of ICE: Engineering History and
    Heritage* 332
Proceedings of TAC Annual Conference 519

*Proceedings of the Annual Meeting of the Highway
    Research Board* 519
*Proceedings of the Architectural Engineering 2003
    Conference* 88
Proceedings of the Combustion Institute 421
Proceedings of the ICE: Engineering Sustainability 199
*Proceedings of the Institution of Mechanical Engineers.
    Part A, Journal of Power and Energy* 421
*Proceedings of the Institution of Mechanical Engineers.
    Part B, Journal of Engineering Manufacture*
    359, 421
*Proceedings of the Institution of Mechanical Engineers.
    Part C, Journal of Mechanical Engineering
    Science* 421
*Proceedings of the Institution of Mechanical Engineers.
    Part D, Journal of Automobile Engineering* 421
*Proceedings of the Institution of Mechanical
    Engineers. Part E, Journal of Process
    Mechanical Engineering* 420
*Proceedings of the Institution of Mechanical Engineers.
    Part F, Journal of Rail and Rapid Transit.* 421
*Proceedings of the Institution of Mechanical Engineers.
    Part G, Journal of Aerospace Engineering*
    56, 421
*Proceedings of the Institution of Mechanical Engineers.
    Part H, Journal of Engineering in Medicine*
    421
*Proceedings of the Institution of Mechanical
    Engineers. Part I, Journal of Systems and
    Control Engineering* 420
*Proceedings of the Institution of Mechanical Engineers.
    Part J, Journal of Engineering Tribology* 421
*Proceedings of the Institution of Mechanical Engineers.
    Part K, Journal of Multi-body Dynamics* 421
*Proceedings of the Institution of Mechanical Engineers.
    Part L, Journal of Materials Design and
    Applications* 421
*Proceedings of the Institution of Mechanical Engineers.
    Part M, Journal of Engineering for the
    Maritime Environment* 421
*Proceedings of the Institution of Mechanical Engineers.
    Part N, Journal of Nanoengineering and
    Nanosystems* 168, 421
*Proceedings of the Institution of Mechanical Engineers.
    Part O, Journal of Risk and Reliability* 421
*Proceedings of the Institution of Mechanical Engineers.
    Part P, Journal of Sports, Engineering and
    Technology* 421
*Proceedings of the 3rd Workshop on Global Engineering
    Education* 268
Process Biochemistry 74, 124
*Process/Industrial Instruments and Controls
    Handbook* 354
*Process Simulation Using WITNESS: Including Lean and
    Six-Sigma Applications* 357
Process Safety and Environmental Protection 124
Process Safety Progress 124
*Procurement Routes for Partnering: A Practical
    Guide* 150
Production and Inventory Control Handbook 353
Production and Operations Management Society 366
Production Planning and Control 359
Professional Handbook of Building Construction 85

Professional Women & Minorities: A Total Human Resources Data Compendium 36
Profiles in Science 105
Programmer's Heaven 238
Programming Languages: History and Fundamentals 207
Progress in Aerospace Sciences 56
Progress in Computational Fluid Dynamics 56
Progress in Crystal Growth and Characterization of Materials 386
Progress in Electromagnetics Research 253
Progress in Energy and Combustion Science 421
Progress in Materials Science 386
Progress in Nuclear Energy 463
Progress in Quantum Electronics 253
Project and Cost Engineers' Handbook, 353
Project Management Journal 359
Project Muse 16, 19
Project 2061 268
Propellants, Explosives, Pyrotechnics 56
Properties of Concrete 174
Properties of Hazardous Industrial Materials 156
Proprietary Trench Support Systems Technical Note 95, 152
ProQuest Direct 35
Protein Data Bank 102
Portland Cement Association
Psychology of Computer Programming 207
PLoS Biology 104
PLoS Computational Biology 104
PLoS Genetics 104
Pruett, K. M. 382
Psychometrika: A Journal Devoted to the Development of Psychology as a Quantitative Rational Science 359
Public Library of Science 104, 109
Public Roads 515
Public Transportation Security Series 509
Public Transportation Standard Bus Procurement Guidelines RFP 509
Public Transport in Third World Cities 158
Public Works 158
PubMed Central 70, 105
Pugh, E. W. 209
Pujado, P. R. 482
Purdue University 379
Putyera, K. 377
Product Design and Factory Development 356
Pump Handbook 414, 437
Pumping Manual 414
Pumping Station Design 188
Pumps and Hydraulics 414
Pump Users Handbook 415
Putnam, R. 85

Q

Quadriennial International Conference on Soil Mechanics and Geotechnical Engineering 143
Quakeline Database 136
Qualitative Examination of the Nature and Impact of Three California Minority Engineering Programs 36

Quantitative Model to Gauge Success for Underrepresented Minorities in the Pennsylvania State University College of Engineering 37
Quality and Reliability Engineering International 359
Quality Criteria for Water 306
Quality Handbook for the Architectural, Engineering, and Construction Community 86
Quality Management Principles 362
Quarterly Report of RTRI 516
Qin, Y. 357

R

Rabins, M.J. 31
Radiation Measurements 463
Radiation Protection Dosimetry 463
Radioactive Waste Management Glossary 458
Rae, J. 330
Rafe,G. 347
Raid, A. 357
Railway Age's Comprehensive Railroad Dictionary 505
Railroad Engineering 513
Railroad Performance Measures 503
Railway Technical Research Institute 525
Railway Track and Structures 517, 523
Railroad Vehicle Dynamics: A Computational Approach 513
Ralston, A. 216
Ramachandran, V. S. 382
RAMP: Risk Appraisal and Management of Projects 152
Ramsey/Sleeper Architectural Graphic Standards 30
Randall, R. E. 145
Rankine, W. 45
Rao, M. A. 69
Rapid Guide to Hazardous Air Pollutants 291
Rapley, R. 102
Rapp, G. R. 439
Rastegar, S. 95, 110
Ratay, R. T. 149
Ratner, B. D. 102
Rauhut, B. 268
Ravenet, J. 174
Ravindran, A. 357
Rayaprolu, K. 414
Rayleigh, Lord (John William Strutt) 45
Raymond, M. 495
Rayner, R. 415
RCRA Orientation Manual 294
Rea, M. S. 86
Reactor Physics Constants, ANL-5800 458
Reactor Handbook 459
Read, G. C. 188
Reader's Advisor, Vol. 5: The Best in Science, Technology, and Medicine 11
ReadWriteWeb 29
Real People Working in Engineering 8, 32
Receptors, Models for Binding, Trafficking and Signalling 121
Recycling in America: A Reference Handbook 157
Reddy, D. V. 145
Reddy, T.B. 249
Redei, G. P. 100
Redman, H. 28

Reference Guide to Famous Engineering Landmarks of the World 172, 327

Reference Manual for Telecommunications Engineering 217

Reference Sources in Science, Engineering, Medicine, and Agriculture 10, 110, 399, 430

Reference Sources in History: An Introductory Guide 324

Referex 247

Refining Online 490

Refining Processes Handbook 483

Refining Statistics Sourcebook 494

Regenerative Medicine 104

Registry of Open Access Repositories 220

Regulatory Guide 464

Rehabilitation of Drains and Sewers 188

Rehg, J. A. 357

Reichert, M. 40

Reilly, E. D. 210, 216

Reilly, W.R. 512

Reimbert, A.M. 174

Reimbert, M. L. 174

Reinhard, D.K. 378

Reinforced Concrete Designers Handbook 174

Reliability Engineering and System Safety 359

Remarkable Structures: Engineering Today's Innovative Buildings 88

Remediation: The Journal of Environmental Cleanup Costs, Technologies and Techniques 302

Remote Sensing of Environment 253

Renewable Agriculture and Food Systems 75

Renewable and Sustainable Energy Reviews 359

Renewable Energy Focus 255

Rennison, R.W. 133

Rensselaer Polytechnic Institute 260

Reporting Results: A Practical Guide for Engineers and Scientists 33

Report on Ship Channel Design. Manual of Practice 107, 144

Research and Innovative Technology Administration 521

Research Development and Technology Office 522

Research Doctorate Programs in the United States: Continuity and Change 266

Research in Higher Education 39

Research in Progress 500

Research on Institutional Repositories 220

Research Portfolio Online Reporting Tools 98

Research Reactor Database 455

Reservoir Engineering Handbook 481

Resource: Engineering & Technology for a Sustainable World 75

Resources, Conservation, and Recycling 302

Resources in Science and Engineering Education 112

Retention by Design: Achieving Excellence in Minority Engineering Education 38

Retention of Underrepresented Students in Engineering Degree Programs: An Evaluation Study 37

Rethinking Construction 151

Rethinking Construction Implementation Toolkit 151

Rethwisch, D. G. 384

Revie, R. W. 382

Reviews in Chemical Engineering 123

Revista da Sociedade Brasileira de Historia da Ciencia 332

Revolt of the Engineers: Social Responsibility and the American Engineering Profession 267

Revue D'histoire de la Culture Materielle 332

Revue D'histoire des Sciences 332

Reynolds, A.B. 462

Reynolds, C. E. 174

Reynolds, J. P. 118, 290

Reynolds, J.S. 87

Reynolds, T. S. 268, 275, 329

RFID in Logistics: A Practical Introduction 511

Rheologica Acta 421

Ricci, P. 23

Rice, P. 176

Rickard, T. A. 428, 429

Ricketts, J. T. 86

Rider, K. J. 325

Ridley, J. 357

Riegel, E. R. 118

Riegel's Handbook of Industrial Chemistry 115, 118

RIGZONE 490

RILEM 182

Rippen, A. L. 67

Rishel, J. B. 415

Risk Abstracts 283

Risk and Reliability Analysis: A Handbook for Civil and Environmental Engineers 290

Ritter, A. 88

Ritter, E.M. 67, 112

Ritter, S.K. 127

Rizzo, J. A, 294

River and Channel Revetments: A Design Manual 188

Rivers of oil: The founding of North America's petroleum industry 495

Rizvi, S.S.H. 69

Road and Track Illustrated Automotive Dictionary 504

Road Pricing in Urban Areas 159

Road Pricing, the Economy and the Environment 159

Roads and Bridges 515

Roadside Design Guide 506

Roadway and Pedestrian Facility Design 509

Roadwork, Theory and Practice 159

Roark's Formulas for Stress and Strain 174

Roberge, P. R. 382

Robert Bosch GmbH. 404, 510

Roberts, W. L. 439

Robinett, R. D. 357

Robinson, P. 264

Robotics and Automation Handbook 410

Robotics and Autonomous Systems 253

Robotics and Computer-Integrated Manufacturing 253, 359

Rock Blasting and Explosives Engineering 138

Rogelberg, S. G. 350

Rogers, M. 153

Rohert Maillart and the Art of Reinforced Concrete 170

Rohsenow, W. M. 406

Rojas, R. 209, 216

Roman Mines in Europe 428

Ronald E. McNair Postbaccalaureate Achievement Program 41

Rosaler, R. C. 86, 355

Rosato, D. V. 376, 382

Roscosmos: Russian Federal Space Agency 57

Ross, R. B. 23

Rosser, S. V. 268, 275
Roosa, S. A. 197
Rothbart, H. A. 409
Rothery, B. 23
Rotstein, E. 69
Roundabouts in the United States 508
Rouse, H. 183
Rouse, W.B. 355
Rousseau, R. 121, 206
Rowland, J.J. 357
Roy, R. 408
Roy, S.K. 379
Royal, K. 35, 43
Royal Academy of Engineering Visiting Professors
        Scheme in Engineering Design for Sustainable
        Development 201
Royal Aeronautical Society 58
Royal Society for the Prevention of Accidents 362
RSS (real simple syndication) 5, 239
RTR–European Rail Technology Review 516
Ruddock, C. 133
Rugarcia, A. 264
Ruggeri, F. 351
*Rules of Thumb for Chemical Engineers: A Manual of
        Quick, Accurate Solutions to Everyday Process
        Engineering Problems* 117
*Rules of Thumb for Mechanical Engineers: A Manual
        of Quick, Accurate Solutions to Everyday
        Mechanical Engineering Problems* 403
*Rural Routes and Networks: Creating and Preserving
        Routes That Are Sustainable, Convenient,
        Tranquil, Attractive, and Safe* 158
Rush, R. D. 86
Rushton, A. 355
Rust, J. H. 461
Rutgers 334
Ryall, M. J. 172, 174

S

Sadegh. A.M. 53, 402
SAE (see Society of Automotive Engineers)
SAE Dictionary of Aerospace Engineering 52
SAE Digital Library 399
SAE Handbook 405
SAE International: the Society of Automotive
        Engineers 425
SAE Publications and Standards Database. 399, 500
SAE Standards 422
SAE Transactions 422
Safarian, S. S. 174
Safari Books Online 235, 251
Safety at Work 357
Safety Evaluation of Existing Dams 189
*SafetyLit: Injury Prevention Literature Update & Archive
        Database* 500
Safety of Nano-Materials Interdisciplinary Research
        Centre 167
Safety Research Program 522
Sage, A. P. 355
Salamone, J. C. 376
Saleh, J. M. 406
Salter, R. J. 159
Salvadori, M. G. 88

Salvato, J. A. 290, 295
Salvendy, G. 355
Sammet, J. E. 207
Samet, J.M. 86
Sammons, V. O. 36
Samuel, G. R. 483
Sanders, J.M.L. 406
Sankaranarayanan, K. 121
Sanks, R. L. 188
Sarma, M.S. 242, 257
Sarte, S. B. 198
Sarton, G. 340
Sattler, K. D. 382
Saunders, N. 382
Sawyer, J. W. 409
Sax's Dangerous Properties of Industrial Materials 157
Scalzi, J.B. 174
Scarl, D. 268
Schenk, M. T. 11
Schaefer,F. 405
Schellings, A. 401
Schempf, F. J. 495
Schetz, J. A. 54, 406
Schiechtl, H. M. 188
Schifftner, K. C. 292
Schinzinger, R. 31
Schittich, C. 177
Schlachter, G. 36
Schmid 343, 344
Schmidt, D. 97
Schmidt, L.C. 384
Schmidt, R. 268
Schmidt-Walter, H. 249
Schnelle, K. B. 292
Scholes, G. 378
Schramm, L. L. 164, 376
Schutze, M. 380
Schwartz, M. 85, 377
Schwarz, J. A. 377
Schweitzer, P. A. 377
Science 14, 15
Science Abstracts 18
Science Accelerator 28
Science and Engineering Indicators 2010 270
*Science and Engineering Literature: A Guide to Reference
        Sources* 430
Science and Engineering of Materials 84
*Science and Technology in World History: An
        Introduction* 325, 330
Science Citation Index 16, 20, 221
ScienceDirect 19, 33, 244, 246, 434
Science Education 264
Science.gov 19, 105, 231, 307
Science News 15
Science of the Total Environment 302
Scientific and Technical Aerospace Reports 48
Scientific and Technical Information Sources 399, 429
*Scientific and Technical Literature: An Introduction to
        Formal Communication* 11, 399
Scientific Instrument Society 338
Scientific Papers and Presentations 33
*Scientists & Swindlers: Consulting on Coal and Oil in
        America* 495
Scifinder 113, 124

Scifinder Registry File 115
Scirus 19, 219, 238
SciTech Book News 248
Sclater, N. 409
SCOPE 104
Scopus 19, 229, 246
Scott, J. S. 85, 185, 287
Scott, W. 153
Scripps Institution of Oceanography 147
Scripta Materialia 386
Seals and Sealing Handbook 410
Second Bibliographic Guide to the History of Computing,
        Computers, and the Information Processing
        Industry 324
Second Life 1
Seebauer, E. G. 31
Seidel, F.A. 482
Seiffert, U.W. 510
Seismic Analysis of Safety-Related Nuclear Structures
        and Commentary 462
Seismic Design Criteria for Structures, Systems and
        Components in Nuclear Facilities 463
Seismological Society of America 142
Select Transportation Associations, Societies, and
        Unions 504
Selected Office of Water Methods and Guidance 305
Selective Guide to Literature on Agricultural
        Engineering 65
Selective Guide to Literature on Aerospace
        Engineering 50
Selective Guide to Literature on Computer Science 206
Selective Guide to Literature on Industrial
        Ergonomics 347
Selective Guide to Literature on Statistical Information
        for Engineers 347
Self, H.C. 324, 347
Selin, H. 326
Sellers, K. 165
Semiconductor Science and Technology 386
Seminumerical Algorithms 206
Semler, E. G. 397
Sensing and Instrumentation for Food Quality and
        Safety 75
Sensor Technology Handbook 250
SE Online: Software Engineering Online 236
Separation and Purification Reviews 124
Separation and Purification Technology 124
Separation Science and Technology 124
Seshan, K. 383
Sesonske, E. 460
Seven Pillars of Partnering 150
Seventy-Seven Year Index, Technical Papers 399
Sewer Rehabilitation Manual 186
Sewers 188
Sewers: Replacement and New Construction 188
Seymour, E. 264
Shabana, A. A. 513
Shackelford, J. F. 383
Shaft Alignment Handbook 410
Shah, V. 383
Shaping the Future. New Expectations for
        Undergraduate Education in Science,
        Mathematics, Engineering, and
        Technology 270

Shaping the Future. Volume II: Perspectives on
        Undergraduate Education in Science,
        Mathematics, Engineering and Technology 270
Sharpe, C. C. 26
Sharpe, D. 405
Sharpe, W.N. 410
Sheldon, T. 216
Shell Bitumen Handbook 157
Shenson, H. I. 149
Shepherd, R. 429
Sherman, C. 16, 17
Sherwood, B. 160
Shigley, J. E. 409, 416
Shiers, G. 325
Shigley's Mechanical Engineering Design 416
Shih, H. 357
Ship Design and Construction 412
Shires, G.L. 402
Shock and Vibration Information Analysis Center 423
Short, T. A. 249
Shul, R. J. 121
Shultis, J. K. 462
Siciliano, B. 410
Sidney M. Edelstein Center for the History and Philosophy
        of Science, Technology, and Medicine 340
Silevitch, M. 263
Silos 173, 174
Simmons, H. L. 86
Simon, T. M. 37
Sims, J.T. 298
Singer, C. J. 328
Singh, A. 36
Singh, B. 410
Singh, B.R. 288
Singh, R.P. 69
Singh, V. P. 290
Sinha, A. K. 383
Sinha, K.C. 526
Sink, D.S. 369
Site Management for Engineers 152
Sixsmith, T. 414
Skelly, E.M.T. 349
Skelly, K. J. 349
Skempton, A. 133
Skill Development for Engineers: An Innovative Model for
        Advanced Learning in the Workplace 267
Skousen, P. L. 411
Skvarenina, T. L. 249
Skogerboe, G. V. 188
Slater, R. 210
Smart Computing 226
Smart Growth Transportation Guidelines 520
Smart Materials and Structures 6, 386
Smart Materials in Architecture, Interior Architecture and
        Design 88
SME Mining Engineering Handbook 437
SME 2010 Guide to Minerals & Materials Science
        Schools 435
Smilauer, J. 164
Smith, E. H. 403
Smith, F.W. 384
Smith, G. B. 165
Smith, G.L 368
Smith, J .M. 121

Smith, M.A. 399
Smith, P. 411, 513
Smith, P.D. 173
Smith, P. G. 185, 287
Smith, P.R. 84
Smith, R. 357
Smith, R.J.H. 152
Smith, S. W. 217
Smithells, C. J. 383
Smithells Light Metals Handbook 383
Smithells Metals Reference 380
Smithsonian Institution 210, 336
Smithsonian Physical Tables
Smits, J. 513
Smoke, Dust, and Haze: Fundamentals of Aerosol Dynamics 296
Smoltczyk, U. 138
Snell, M. 153
Snow and Ice Control Manual for Transportation Facilities 157
Social History of Engineering 328
Social Sciences Citation Index 20
Societe du Musee des Sciences et de la Technologie du Canada 336
Society for Biomaterials 109, 392
Society for Experimental Mechanics 425
Society for Industrial and Applied Mathematics 233, 253, 366
Society for Industrial Microbiology 109
Society for Mining, Metallurgy, and Exploration 392, 446
Society for the Advancement of Chicanos and Native Americans in Science 42
Society for the Advancement of Material and Process Engineering 392
Society for the History of Technology 274, 339
Society of Automotive Engineers 20, 56, 304, 518, 520, 525
Society of Construction Law 155
Society of Economic Geologists 446
Society of Environmental Toxicology and Chemistry 314
Society of Glass Technology 392
Society of Hispanic Professional Engineers 42
Society of Manufacturing Engineers 342, 366, 369
Society of Mexican American Engineers and Scientists 42
Society of Naval Architects and Marine Engineers 147, 425
Society of Petroleum Engineers 485, 491
Society of Petrophysicists and Well Log Analysts 491
Society of Plastics Engineers 392
Society of Tribologists and Lubrication Engineers 425
Society of Women Engineers 29, 42, 274
Software Development Times 226
Software Patent Institute Database of Software Technologies 222
Soil & Tillage Research 75
Soil Construction and Geotechnics 138
Soil Dynamics and Earthquake Engineering 140
Soil Mechanics 138
Soil Mechanics in Engineering Practice 139
Soils and Environmental Quality 298
Soils and Foundations 140
Soil Science Society of America 74, 76, 142
Soil Science Society of America Journal 76, 142
Soil Testing Manual: Procedures, Classification Data and Sampling Practices 137

Soil Use and Management 76
Solar Power Your Home For Dummies 256
Solar Domestic Water Heating 256
Solar Energy Handbook 406
Solar Energy Materials and Solar Cells 255
Solie, J.B. 69
Somerville, S. H. 137
Sorby, S. A. 33
Soroka, W. G. 351
Soska, G. V. 369
Sourbes-Verger, I. 52
Source materials for architectural engineering 83
Sources of Construction Information 83
Sources of Information in Transportation 526
Sources of Information on Atomic Energy, International Series of Monographs in Library and Information Science 456
Sorting and Searching 206
South, D. W. 351, 400
Southeastern Consortium for Minorities in Engineering 42
Spaceflight 56
SpaceRef.com 57
Sparrow, E.M. 406
Special Libraries Association 274
SPE Drilling & Completion 487
SPE Economics & Management 487
SPE Journal 487
SPE Production & Operations 487
SPE Projects, Facilities & Construction 487
SPE Reservoir Evaluation & Engineering 487
Specifications for Seismic Isolation Design 519
Speight, J. G. 483
Spellman, F. R. 290
Spence, C. C. 429
Spengler, J. D. 86
Spicer, C. W. 292
Spiegel, R. 86
Spier, R.E. 67, 78
Spitz, P. H. 495
Spon's Civil Engineering and Highway Works Price Book 153
Sportisse, B. 296
Spreadsheets for Structural Engineering 180
Spreadsheet Solutions for Structural Engineers 180
SpringerAlerts 34
Springer: Computer Science 231
Springer Handbook of Automation 354
Springer Handbook of Experimental Solid Mechanics 410
Springer Handbook of Mechanical Engineering 402
Springer Handbook of Nanotechnology 378
Springer Handbook of Robotics 410
SpringerMaterials 119, 383
Staab, G.H. 416
Stacey, W. M. 462
Stack, B. 437
Stafford-Smith, B. 174
Stahl, D. 23
Stahlbau 179
Stallmeyer, J. E. 172
Stand-Alone Solar Electric Systems 256
Standards Activities of Organizations in the United States 23
Standard Handbook for Aeronautical and Astronautical Engineers 53

Standard Handbook for Civil Engineers 86, 133
Standard Handbook of Architectural Engineering 86
Standard Handbook of Consulting Engineering Practice 149
Standard Handbook of Environmental Engineering 288
Springer Handbook of Experimental Fluid Mechanics 407
Standard Handbook of Engineering Calculations 14
Standard Handbook of Environmental Engineering 185
Standard Handbook of Fastening and Joining 410
Standard Handbook of Hazardous Waste Treatment and Disposal
Standard Handbook of Heavy Construction 149
Standard Handbook of Machine Design 409
Standard Handbook of Petroleum & Natural Gas Engineering 482
Standard Handbook of Plant Engineering 355
Standard Highway Signs 508
Standard Industrial Classification 22
Standard Methods for the Examination of Water and Wastewater 197, 294, 305
Standard Periodical Directory 14, 15
Standards: A Resource and Guide for Identification, Selection, and Acquisition 23
Standard Specifications for Construction of Roads and Bridges on Federal Highway Projects 520
Standard Specifications for Highway Bridges 519
Standard Specifications for Transportation Materials and Methods of Sampling and Testing 519
Standards Engineering Society 22, 25
Standard Z94.0: Industrial engineering terminology 368
Standing Committee on Structural Safety 182
Stanford University Libraries. History of Science and Technology Collections 334
Stanford University Library Materials Science and Engineering Guide 374
Stanford University Libraries Swain Chemistry and Chemical Engineering Library 114
Stanford University Earth Sciences Library and Map Collections 449
Stapelberg, R. F. 355
Stapp Car Crash Conference Proceedings 518
Stark, T. 86
State University of New York at Albany Library 220
Statistical Annual 495
Statistical Handbook for Canada's Upstream Petroleum Industry 493
Statistical Handbook on Technology 13
Statistical Record of the Environment 284
Statistical Review of Coal in Canada 448
Statistics Canada 449
Statistics for Environmental Engineers 294
Statistics for Environmental Science and Management 290, 295
Steam Distribution Systems Deskbook 413, 413
Stedman, T. L. 99
Stedman's Medical Dictionary 99
Steedman, J.C. 174
Steel Bridges in the World 180
Steel, Concrete and Composite Design of Tall Buildings 174
Steel Construction Institute 175, 182
Steel Designers Manual 175
Steel Detailer's Manual 175

Steel Founders' Society of America 392
Stein, B. 87
Stein, D 188, 209
Stein, J. S. 85
Steinberg, M. L. 99
Steinmetz: Engineer and Socialist 267
Stellman, J. M., 351
Stensel, H.D. 298
Stephens, J. H. 174
Sterling, C. H. 325
Stern, A. C. 296
Stern, N. B. 210
Stern, R. 188
Stewart, B. A. 68
Stewart, H.L. 414
STN 19, 281
Stoeker, W. F. 407
Stokes, A. 411
Stokes, D. 357
Stokes, G. 45
Stone, R. 513
StormwaterAuthority.org 308
Stormwater Collection Systems Design Handbook 293
Stormwater Conveyance Modeling and Design 298
Stormwater Effects Handbook: A Toolbox for Watershed Managers, Scientists, and Engineers 293
Stony Brook Algorithm Repository 237
Story of Electricity 257
Strassen und Plaetze Neu Gestaltet 158
Strategies for National Sustainable Development: A Handbook for Their Planning and Implementation 196
Straub, H. 330
Streeter, V. L. 54
Strength of Materials 418
Strong, D. L. 157
Structurae 180
Structural Alloys Handbook 54
Structural Concrete 179
Structural Design for Physical Security 172
Structural Design Guide to the AISC (LRFD) Specification for Buildings 175
Structural Design in Architecture 88
Structural Design of Tall Buildings 178
Structural Detailing in Steel 175
Structural Engineer 177
Structural Engineering Handbook 171, 172
Structural Engineering Institute 91
Structural Engineering International 178
Structural Engineering: The Nature of Theory and Design 170
Structural Engineer's Pocket Book 173
Structural Glass 176
Structural Impact 173
Structural Plastics Design Manual 176
Structural Plastics Selection Manual 176
Structural Safety 178
Structural Steel Designer's Handbook 175
Structural Use of Glass in Buildings 177
Structure and Interpretation of Computer Programs 206
Structured Programming 206
Structures and Building 177

Strutt, John William 45
Studies in History and Philosophy of Science 332
Suchy, I. 409
Sugiyama, H. 513
Sullivan, E.C. 512
Sullivan, T. F. P, 288
Summary Description of Design Criteria, Codes,
*   Standards, and Regulatory Provisions*
*   Typically Used for the Civil and*
*   Structural Design of Nuclear Fuel Cycle*
*   Facilities* 463
Sun, K. 481
Sunley, J. 176
SUPERLCCS: Gale's Library of Congress Classification
*   Schedules* 9
Supply Chain Analysis: A Handbook on the Interaction of
*   Information, System and Optimization* 355
Supply Chain and Transportation Dictionary 504
Surface Analysis Society of Japan 392
Surface and Coatings Technology 386
Surveyor 158
Sussman, G. J. 206
Sussman, J. 160, 206, 511
Sustainability and Adaptability in Infrastructure
*   Development* 159, 197
Sustainability of Construction Materials 198
Sustainability Science Abstracts 283
Sustainability Science and Engineering 197
Sustainability: Science, Practice, and Policy 200
Sustainable Buildings Industry Council 198
Sustainable Development for Engineers: A Handbook and
*   Resource Guide* 197
Sustainable Development Handbook 197
Sustainable Communities Design Handbook:
*   Green Engineering,Architecture, and*
*   Technology* 355
Sustainable Construction 172
Sustainable Construction: Company Indicators 151
Sustainable Development 199
Sustainable Development International 201
Sustainable Drainage Systems: Hydraulic, Structural and
*   Water Quality Advice* 188
Sustainable Engineering Practice: An Introduction 197
Sustainable Infrastructure: The Guide to Green
*   Engineering and Design* 198
Sustainable Production: An Annotated Bibliography 347
Sustainable Systems Research Center 362
Sutherland, R. J. M 175
Swamidass, P. M. 351
SWE 39
Sweet's Catalog File 84
Synthetic Fuels Handbook: Properties, Process and
*   Performance* 483
Systems Analysis for Sustainable Engineering Theory and
*   Applications* 197
Systems, Experts, and Computers: The Systems Approach
*   in Management and Engineering, World War II*
*   and After* 330
Systems Modernization and Sustainment Center 362
Synthesis: The Digital Library of Engineering and
    Computer Science 105
Szokolay, S. V. 88
Szycher, M. 99
Szycher's Dictionary of Medical Devices 99

T

Tables for the Hydraulic Design of Pipes, Sewers and
*   Channels* 187
Taggart, A. F. 438
Taguchi, G. 355
Taguchi's Quality Engineering Handbook 355
Tall Building Structures 174
Tang, C. S. 355
Tapley, B. D. 12
Taranath, B. S. 174
Taschenbuch der Technischen Formeln 13
Task Committee of the Waterways and Navigation
    Engineering Committee of the Coasts,
    Oceans, Ports, and Rivers Institute of
    ASCE 144
Task Committee of the Waterways Committee 144
Task Committee on Underwater Investigations 144
Tau Beta Pi Association 274
Taylor, J. I. 411, 412
Tchobanoglous, G. 294, 298
Teacher's Domain: Multimedia Resources for the
*   Classroom and Professional Development* 273
Teaching Engineering 268
Teaching the Majority: Breaking the Gender Barrier in
*   Science, Mathematics, and Engineering* 268
TechBookReport 211
TechCrunch 29
Technical Arts and Sciences of the Ancients 330
Technical Data Book—Petroleum Refining 492
Technical Data for Professional Practice 30
Technical Guide for Conducting Pedestrian Safety
    Assessments in California 509
Technical Report Archive & Image Library 28, 219
Technically-Write! 33
Techniques in Underground Mining: Selections from
*   Underground Mining Methods Handbook* 437
Technikgeschichte 332
Technology and American History. A Historical
*   Anthology from Technology and Culture* 329
Technology and Culture 332
Technology and Culture of the Society for the History of
*   Technology* 322
Technology Mathematical Handbook 14
Technology Research Database 399
Technology Review 14, 15
Technology Transfer Society 366
Technorati 29
Technos 271
TechRepublic 237
Techstreet 25
TechTarget Network 237
TechWeb 237
TechXtra: Engineering, Mathematics, and Computing
    134, 239
TecTrends 229
Telecommunications Illustrated Dictionary 249
Telliard, W. A. 305
Tellus. Series B, Chemical and Physical Meteorology 302
Temple, J. 429
Temporary Works: Their Role in Construction 152
TenLinks.com: Ultimate Civil Engineering Directory 134
Tenner, E. 30
Tennessee State University 80

Tenopir, Carol ix, x, 5
Terminology 505
Terzaghi, K. 136, 139
Test Methods for Evaluating Solid Waste: Physical/
    Chemical Methods 305
Texas Transportation Institute 523
Theodore, L. 118, 165
Threlfall, A.J. 174
Theoretical Foundations Chemical Engineering 124
Theory of Wing Sections Including a Summary of Airfoil
    Data 52
Thermodex: An Index of Selected Thermodynamic and
    Physical Property Resources 119
Thermodynamics and an Introduction to Themostatics 120
Theodore, L. 288, 290
Thermal Connection 423
Thermal Spray Society 393
ThermoDex 383
Thermodynamics 417
THERMOPEDIA: A-to-Z Guide to Thermodynamics,
    Heat & Mass Transfer, and Fluids
    Engineering 402
Thesaurus of Engineered Materials, 9
Thesaurus of Sanitary and Environmental
    Engineering 281
Thinking Like an Engineer 32
Thin Solid Films 255, 386
Thomann, A. E. 13
Thomas, J.H. 410
Thomas A. Edison Papers 334
ThomasNet 11, 67
Thomas Register of American Manufacturers 11, 349, 400
Thompson, D.C. 24
Thor, D.G. 352
Thoresen, C. A. 144
Thumann, A. 407
Tibor, T. 24
Tidal Power Plants 187
Tilly, G. 172
Timber Construction Manual 85, 175
Timber Designers Manual 175
Timber Engineering: STEP 1 and STEP 2 176
Timber in Construction 176
Timber Research and Development Association 183
Time-Saver Standards 87
Time-Saver Standards for Architectural Design Data 30
Timetables of Technology: A Chronology of the Most
    Important People and Events in the History of
    Technology 327
Timings, R. 403
Timoshenko, S. P. 170, 418
Tissue Engineering and Regenerative Medicine 104
To Engineer Is Human: The Role of Failure in Successful
    Design 30
To Recruit and Advance: Women Students and Faculty in
    Science and Engineering 35
Todd, D.K. 68, 287, 298
Toliyat, H. A. 249
Tollner, E. W. 69
Tomlinson, M. J. 138
Tomsic, J. L. 52
Tool and Manufacturing Engineers Handbook 408
Top 100 Magazines: Computer and Software WWW
    Magazines and Journals 224

Top 400 Contractors Sourcebook 349
Top 500 Design Firms Sourcebook 350
Tosheva, T. 68
Totemeier, T.C. 380
Totten, G. E. 407, 415
Tower and the Bridge: The New Art of Structural
    Engineering 170
Townsend, D. P. 411
Toxicology Abstracts 283
TOXLINE 105
Traffic and Highway Engineering 512
Traffic Control Devices Handbook 507
Traffic Engineering and Control 158, 515
Traffic Engineering Handbook 156, 507
Traffic Injury Prevention 515
Traffic Signal Maintenance Handbook 507
Traffic Signal Timing Manual 507
Traffic Signs Manual 508
Traffic Technology International 515
TRAIL (see Technical Report Archive & Image Library)
Transactions of the American Nuclear Society 463
Transactions of the ASABE 76
Transactions of the ASME. Society Records 398
Transactions of the Institution of Mining and Metallurgy
    Section A: Mining Technology 435
Transactions of the Institution of Mining and Metallurgy
    Section B: Applied Earth Science 435
Transactions of the Institution of Mining and Metallurgy
    Section C, Mineral Processing and Extractive
    Metallurgy 435
Transactions of the Newcomen Society. 332
Transforming Undergraduate Education in
    Science, Mathematics Engineering, and
    Technology 271
Transit Capacity and Quality of Service Manual 509
Transmaterial: A Catalog of Materials That Define Our
    Physical Environment 87
Transport 158, 523
Transportation Association of Canada 519, 525
Transport Canada 523
Transportation Energy Data Book 502
Transportation Engineering and Planning 511
Transportation Journal 514
Transportation Noise Reference Book 157
Transportation Research 514
    Part A: Policy and Practice 514
    Part B: Methodological 514
    Part C: Emerging Technologies 514
    Part D: Transport and Environment 514
    Part E: Logistics and Transportation Review 514
    Part F: Traffic Psychology and Behavior 514
Transportation Research Board 5, 157, 162, 519, 525
Transportation Research Record: Journal of the
    Transportation Research Board 514
Transportation Research Thesaurus 499, 500
Transportation Statistics Annual Reports 501
Transport in Porous Media 124
Transport Phenomena 120
Transport Research Laboratory 162, 525
Transport Statistics 503
TRB Publications Index 500
Trajtenberg, M. 26
Transport in Porous Media 124
Treatise on Materials Science and Technology 384

Treatment Wetlands 298
Trek of the oil finders: A history of exploration for petroleum 495
Trenching Practice CIRIA Report 97 152
Trenchless Technology: Pipeline and Utility Design, Construction, and Renewal 513
Trends in Food Science and Technology 76
Trends in Manufacturing 367
Triangle Fraternity 274
Tribology Data Handbook 415
Tribology Handbook 415
Tribology International 422
Tribology Letters 422
Tribology of Mechanical Systems: A Guide to Present and Future Technologies 416
TRID 155, 156, 500
Trimble, S. W. 68
Trinder, B. 327
Trinity College Dublin 204, 240
Trip Generation 156, 507
Trip Generation Handbook: An ITE Recommended Practice 507
TRIS (see also *TRID*)
TRL 525
Troeh, F. R. 68
Troise, F. L. 68, 287
Tropea, C. 407
Trugo, L. 67
Trussed Roof 176
TSO 157
Tschoke, T. 405
Tubular Exchange Manufacturers Association 492
Tucker, A. B. 217
Tuma, J. J. 14
Tunneling and Underground Space Technology 515
Tunnels and Tunnelling International 516
Tunnel Engineering Handbook 436
Turing and the Universal Machine: the Making of the Modern Computer 209
Turner, R. 149, 325, 331
Turner, W. C. 407
Turner–Fairbank Highway Research Center 522
Turovskiy, I. S. 298
Twentieth Century Petroleum Statistics 494
Twitter 1, 5
Twort, A. C. 189
Tyagi, A. 290
Tyler, N. 160

U

UBM Computer Full-text 228
Uff, J. 153
Ugly's Electrical Desk Reference 248
Ugly's Electrical Reference 248
Ulgen, O. 357
Uhlig, H.H. 382
Uhlig's Corrosion Handbook 382
Ullman, D. G. 418
Ullmann's Encyclopedia of Industrial Chemistry 116, 377
Ulrich's Periodicals Directory 14, 15, 486
Ultimate Civil Engineering Directory 84, 90
UMI Dissertation Publishing 19

Underground Storage Tank Management: A Practical Guide 294
Understanding Baccalaureate Completion Rate Increases of Underrepresented Minority Students in Science and Engineering: Three Case Studies 37
Understanding Digital Signal Processing 217
Understanding Metadata 21, 23
Understanding Nanotechnology 167
Understanding Quality Assurance in Construction 151
Understanding Radioactive Waste 461
Understanding Thermodynamics 418
Underwater Investigations: Standard Practice Manual 144
Underwriters Laboratories 254
USBE and Information Technology 39
USGBC Resource Catalog 83
UNESCO 192
United by a Common Vision 368
United Kingdom Environment Agency 189
    Patent Office 27
United Nations Department of Political and Security Council Affairs. Atomic Energy Commission Group 456
    Economic Commission for Europe 503
    Environment Programme 285
United Nations Annual Bulletin of Transport Statistics for Europe and North America 158
United Nations 64, 158, 191, 196, 285, 464, 456, 503
United Negro College Fund 42
U.S. Army Corps of Engineers 28, 183, 308
    Engineer Research and Development Center 135
U.S. Atomic Energy Commission 459
U.S. Black Engineer and Information Technology Magazine 39
U.S. Bureau of Mines 22
    Statistical Compendium 449
U.S. Bureau of Reclamation 138, 189, 308
U.S. Bureau of Standards 24
U.S. Department of Defense 22
U.S. Department of Defense Quicksearch 25
U.S. Department of Energy 308, 388, 464
    Energy Information Administration 494, 501, 502
U.S. Department of Homeland Security 458
U.S. Department of Labor 32
    Occupational Health and Safety Administration 304
    U.S. Department of the Interior 308
U.S. Department of Transportation 160, 162
    Federal Highway Administration 502, 508, 520, 521
    Research and Innovative Technology Administration 501
U.S. Environmental Protection Agency 160, 189, 278, 310–11
U.S. Federal Aviation Administration 162
U.S. Fish and Wildlife Service 308
U.S. Geological Survey 142, 308
    Minerals Information 443
    Mineral Resources Program 449
    Publications Warehouse 283
    SPARROW 190
    Science Center for Coastal and Marine Geology 146
    Water Resources of the U.S. 192
U.S. National Mining Association 446

U. S. National Oceanic and Atmospheric Administration 278, 311
U.S. National Research Council 30
U.S. Nuclear Regulatory Commission 464, 465, 467
U.S. NWIS-W Data Retrieval 192
U.S. River Systems and Meteorology Group 192
U.S. Patent and Trademark Office 4, 26, 27, 90
Unit Operations in Food Engineering 68
Universal History of Computing: From the Abacus to the Quantum Computer 209
University of Alberta Library Resource Guide for Chemical Engineering 115
University of California San Diego Science and Engineering Library Materials Science Guide 374
University of Colorado at Boulder 80
University of Delaware, Internet Resources for Chemical Engineering 126
University of Kansas 80
University of Maryland's Virtual Technical Reports Collection 219
University of Miami 80
University of Michigan 46
    Library Materials Science and Engineering Research 375
University of New Mexico, Centennial Science and Engineering Library 465
University of Pennsylvania 94
University of Texas at Austin 80
    Department of Chemical Engineering
University of Nebraska–Lincoln 80
University of Oklahoma 80
University of Santa Barbara, InfoSurf: Information Resources for Chemical Engineering 126
University of Texas 485
University of Wisconsin Libraries Materials Science and Engineering Research Guide 375
University of Wyoming 80
usa.gov 25
Unsworth, J. F. 510
Upstream 489
Urban Flood Management 189
Urban Flood Protection Benefits 188
Urban Mobility Information 523
Urban Road Pricing: Public and Political Acceptability 160
Urban Transportation Systems: Choices for Utilities 156
Urban Transport Systems 504
Urban Travel Demand Modeling 159
USENIX 233
Use of Engineering Literature 50, 54, 59, 495
User's Guide to Federal Architect Engineer Contracts 151
User's Guide to Vacuum Technology 418
Using the Agricultural, Environmental, and Food Literature 66
Using the Biological Literature: A Practical Guide 97
Utracki, L. A. 383

V

Valentas, K. J. 69
Value of Urban Design 159
Valve Handbook 41
Valve Selection Handbook 411

Valves Manual International: Handbook of Valves and Actuators 410
Valves, Piping and Pipelines Handbook 412
Van Aken, D. C. 33
Van Basshuysen, R. 405
Vance, G.F. 298
Van der Leeden, F. 68, 286, 287
Van der Tuin, J. D. 137, 185
van Doenhoff, A.E. 52
Van Leeuwen, J. 217
Van Ness, H.C. 121, 418
Van Recum, A. 383
Vanderburg, W.H. 347
Vaughn, N. 407
Vault Guide to Engineering Diversity Programs 36
Vector Mechanics for Engineers: Statics and Dynamics 416
Vehicle System Dynamics: International Journal of Vehicle Mechanics and Mobility 517
Veilleux, D. 399
Verger, F. 52
Vertiflite 56
Vesilind, P. A. 31
Vibration Analysis Handbook 412
Vibration and Shock Handbook 411
Vibrationdata.com 423
Vibration Institute 425
ViFaTec (Virtuelle Fachbibliothek Technik) 423
Villars, P. 383
Virtual Library: Engineering 90
Virtual Team Leadership and Collaborative Engineering Advancements: Contemporary issues and Implications 356
Virtual Technical Reports Center 28
Visible Human Project 102
Vižintin, J. 416
VLDB Journal 225
VLSI Design 225
Volpe National Transportation Center 522
Volti, R. 329, 330
von Humboldt, W. 261
von Recum, A. F. 102
Voprosy Istorii Estestvoznaniya i Tekhniki 332
Voskoboinikov, B. S. 401
Vyhnanek, L. 324

W

Wah, B. 216
Wahburn, S.S. 512
Walford, A.J. 456
Walker 343
Walker, J. M. 102, 369
Walker, P. M. B. 52
Walker, R. D. 11, 26
Walker, S. 440
Wallace, I. N. D. 153
Wallace, S. 298
Walsh, R.A. 14, 409, 411
Waltar, A. E. 462
Wand, H.P. 368
Wang, L. K. 290, 296
Wankat, P. C. 268
Ward, J. L. 351

Ward's Communications 503
Warne, D. F. 249
Warren, E.S. 120
Warring, R. H. 407
Wasserman, S. 399, 430
Waste Management 302
Waste Management: Proceedings of the Symposium on Waste Management 463
Waste Management Research Abstracts 455
Wastewater Engineering: Treatment Disposal and Reuse 188
Wastewater Engineering: Treatment and Reuse 298
Wastewater Sludge Processing 298
Water 21 187
Wastewater Reclamation and Reuse 297
Water, Air, and Soil Pollution 303
Water and Environment Journal 187
Water and Environment Magazine 187
Water and Wastewater Calculations Manual 293
Water and Wastewater Control Engineering: Glossary 185
Water Bioengineering Techniques for Watercourse Bank and Shoreline Protection 188
Water Cooled Reactor Technology: Safety Research Abstracts 456
Water, Engineering and Development Centre 192
Water Environment Research: A Research Publication of the Water Environment Federation 303
Water Environment and Technology 187, 303
Water Encyclopedia 68, 286, 287
Water Environment Federation 192, 315
Water Environment Research 187
WateReuse Association 193
Wateright 190
Water Librarian's Homepage 190
Waterman, N. A. 384
Water Management 187
Water Measurement Manual 189
Water Quality and Treatment 187, 293
Water Quality Control Handbook 185, 292
Water-Quality Engineering in Natural Systems 297
Water Research 187, 303
Water Research Commission 192
Water Resources 187
Water Resources Abstracts 184, 283
Water Resources Research 187, 303
Water Science and Technology 187, 303
Water Strategist Community 190
Water Supply 189
Water Treatment Plant Design 297
Water UK 193
Waterways Experiment Station 193
WaterWiser—The Water Efficiency Clearinghouse 308
Watson, D. 30
Watters, G.Z. 513
Wear 422
Wearne, F. 175
Weber, R.D. 36
Web of Knowledge 20, 98
Web of Science (see also Science Citation Index) 10, 19, 33, 98, 123, 230, 246
Webopedia 214
Webster, J. G. 100, 249
Webster, J. K. 11

Webster, L. 85, 133
Webster, L. F. 12, 288
Webster, T.S. 296
Webster's New World Computer Dictionary 213, 240
Wech, L. 510
Weichert, D. 268
Weinberg, G. 207
Weiss, Alan
West Point 260
Welding Institute 183
Weldy, E. A. 37
Well Production Practical Handbook 481
Well Productivity Handbook: Vertical, Fractured, Horizontal, Multilateral, and Intelligent Wells 481
Wells, A. T. 160
Wentling, R. M. 40
Wessel, J. K. 384
West, H. H. 88
West Point Bridge Designer 180
Wetlands 298
Wetlands Ecology and Management 303
Wexelblat, R. L. 210
What Do ChemEs Do? 127
What Every Chemist Know Should about Patents 123
What Every Engineer Should Know about Engineering Information 11
What Is What in the Nanoworld: A Handbook on Nanoscience and Nanotechnology 164
Where to Find Material Safety Data Sheets Online 122
Wherry, T. L. 26
Whitaker, J. 250
Whitaker, J. C. 250
Whitaker Foundation 109
White, F. M. 418
White, G. C. 294
White, J. 33
Whitehouse, D. J. 384
White's Handbook of Chlorination and Alternative Disinfectants 294
Whiting, N.E. 290
Whitelegg, J. 160
Whittles, M. J. 160
Whole Brain Atlas 103
Who's Who in Engineering 11, 400
Who's Who in Environmental Engineering 286
Who's Who in Science and Engineering 11, 84
Why Buildings Fall Down: How Structures Fail 87
Why Buildings Stand Up: The Strength of Architecture 88
Wick, C. 408
Wick, O. J. 460
Wickens, A. H. 513
Wiederrecht, G. 378, 384
Wiersma, G. B. 295
Wikander, O. 183, 328
Wildlife and Roads 160
Wiley Computing 231
Wiley Database of Polymer Properties 103
Wiley Dictionary of Civil Engineering and Construction
Wiley Electrical and Electronics Engineering Dictionary 249
Wiley Encyclopedia of Biomedical Engineering 100
Wiley Encyclopedia of Computer Science and Engineering 216

Wiley Encyclopedia of Electrical and Electronics Engineering 249
Wiley Encyclopedia of Packaging Technology 351
Wiley Engineer's Desk Reference: A Concise Guide for the Professional Engineer 12
Wilkes, J. A. 85
Wilkins, M. V. 209
William, E. 357
Williams, D. F. 99
Williams, F. M. 36
Williams, J. G. 249
Williams, M. R. 210
Williams Dictionary of Biomaterials 99
Williams Natural Gas Company, Engineering Group 414
Willumsen, L. G. 158
Williamson, M. 52
Williamson, T.G. 175
Wilson, E. B. 36
Wilson, D. G. 357
Wilson, J. 250
Wilt, D. E. 430
Wind Energy 422
Wind Energy: Handbook 405
Wind Loading: A Practiced Guide to BS 6399
Wind Loading of Structures 173
Wind Power For Dummies 256
Winds of Change 39
Winter, J. 298
Wireless Solutions for Manufacturing Automation 368
Wirth, N. 207
Wise, J. A. 54
Withington, S. 330, 340
Wnek, G. E. 100
Woldman, N.E. 23
Woldman's Engineering Alloys 23, 380
Wolpert, A. 220
Woman Engineer 39
Women and Minorities in Engineering: A Review of the Literature 35
Women and Minorities in Science, Technology, Engineering, and Mathematics: Upping the Numbers 38
Women and Technology: An Annotated Bibliography 324
Women in Engineering ProActive Network 42
Women in Engineering Programs and Advocates Network 29
Woo, L. 85
Wood, D.F. 511
Wood, D. N. 430
Wood Engineering and Construction Handbook 176
Woodhead, R.W. 152
Woodrow, W. 355
Woods, D.R. 263
Woodward, J. F. 153
WordIQ.com 208
Working in Japan: An Insider's Guide for Engineers and Scientists 8, 32
Work Zone Safety Information Clearinghouse 153
World Aerospace Database 51
World Association for Sustainable Development 198
World Bank 155, 523
World Business Council for Sustainable Development 198

Worldcat 15, 54, 248
World Conference on Transport Research 519
World Congress on Intelligent Transport Systems 519
World Congress on Engineering Education 272
World Commission on Dams 190
World Directory of Nuclear Utility Management 456
World Dredging Congress 148
World Energy and Nuclear Directory: Organizations and Research Activities in Atomic and Non-atomic 457
World Health Organization 193
World Intellectual Property Organization 4, 27
World Journal of Microbiology & Biotechnology 76
World Lecture Hall 234
World of Learning 266
World Nuclear Association 466
World Nuclear Industry Handbook 457
World of Computer Science 213
World Oil 489
World Oil Outlook 494
World Organization of Dredging Associations 142
World Railways 504
World Road Association/Association Mondiale de la Route 162, 505
World Standards Services Network 22, 25
World Road Statistics 502
World Trade Organization 21
World Transport Policy and Practice 160
Worldwide Agricultural Machinery and Equipment Directory 67
Worldwide Automotive Supplier Directory 400
World Wide Web Virtual Library: Industrial Engineering 362
Worterbuch Maschinenbau und Tribologie: Deutsch–Englisch, Englisch–Deutsch 401
WRc 186
WRC information 193
Wright, A. 158
Wright, E. 211
Wright, Orville 45
Wright, P. H. 512
Wright, Wilbur 45
Wright, W.W. 384
Wright State University Libraries, Special Collections and Archives 334
Writing Better Requirements 32
WSEAS International Conference on Automatic Control, Modelling & Simulation 368
Wu, B. 355
Wunsch, F. 401
WWW Computer Architecture Page 239
Wysk, R.A. 368

Y

Yahoo! Chemical Engineering 126
Yahoo! Computers and Internet, Standards 222
Yahoo! Directory: Engineering: Mechanical Engineering 423
Yahoo! Electrical Engineering 254
Yahoo Index of Space Sciences 57
Yam, K. L. 351
Yang, W. J. 54
Yankee Book Peddler 250

Yarin, A. L. 407
Yaws, C. 120
Yeang, K. 85, 87
Yeaple, F. D. 407
Yeomans, D. 176
Yoder, L.W. 412
Young, D.F 418
Young, W.C. 174
YouTube 1
Yu, W.-W. 175

Z

Zaazaa, K.E. 513
Zaknic, I. 174
*Zahlenwerte und Funktionen aus Naturwissenschaften
    und Technik* 120

*Zahlenwerte und Funktionen aus Physik, Chemie,
    Astronomie, Geophysik und Technik* 120
Zandin, K.B. 353
Zappe, R.W. 411
ZDNet 237, 240
Zeman, J. 164
*Zentrum Mensch-Maschine-Systeme–Center of
    Human–Machine–Systems* 362
Zepf, P.J. 351
Zeumer, M. 86
Zevenbergen, C 189
Zhang, Y. 297
Zhongguo Ke ji Shi Za Zhi 332
Ziegler, E.N. 287
Zouridakis, G. 102
Zvelebil, M.J. 99
Zweben, D. 376